Biological Degradation
and
Bioremediation of Toxic Chemicals

Biological Degradation and Bioremediation of Toxic Chemicals

Edited by

G. Rasul Chaudhry

Department of Biological Sciences
Oakland University

DIOSCORIDES PRESS
Portland, Oregon

ISBN 0-931146-27-5
Printed in Hong Kong

DIOSCORIDES PRESS
The Haseltine Building
133 S.W. Second Ave., Suite 450
Portland, Oregon 97204-3527, U.S.A.

Library of Congress Cataloging-in-Publication Data

Biological degradation and bioremediation of toxic chemicals / edited
   by G. Rasul Chaudhry.
       p.   cm.
   Includes bibliographical references and index.
   ISBN 0-931146-27-5
   1. Bioremediation. 2. Hazardous substances. I. Chaudhry, Ghulam
Rasul, 1931-
TD192.5.B55   1994
628.5'2--dc20   93-25453
                                                           CIP

Contents

Preface

Although *biodegradation* or *biological degradation* of natural and synthetic materials and hazardous wastes are quite familiar terms, *bioremediation* is a relatively new word. It is not listed in the 1979 edition of *Webster's New Universal Unabridged Dictionary*. Coining of new words may signal something new, trendy, or recreated. *Bioremediation,* as I understand it, means to cure, to restore, to correct, or to repair by biological processes what we have done to hurt our environment and ecosystems. The concept of bioremediation is not something new—the physician has practiced it for a long time; the farmer is practicing it in restoring farmland and the forester by replanting the forest. It is practiced to treat sewage and waste and to purify water. We have taken bioremediation for granted, believing that natural processes will take care of the poisons, chemicals, and trash we throw away or so hazardously dispose of. New in bioremediation is the realization that the biological processes sustaining nature are overburdened and no longer are able to cope with what we are doing to Mother Earth. This has been understood by a few people for many years but, more recently, also by the general public, some politicians, and some industrialists. We have found that clean air, land, and water as well as productve farms, forests, lakes, and oceans are part of our living standard and keys to survival. We must care for the Earth and prevent destruction of its live-sustaining ability. We must stop polluting and we must clean up. Through bioremediation research we may learn how to do it.

Bioremediation, also called environmental biotechnology, involves microbial processes almost exclusively. Its emergence has caused a renascence within the realm of microbial physiology. New microorganisms are discovered at rates never seen before and with properties not previously imagined. Examples are the very extreme thermophiles growing and reproducing at temperatures well above the boiling point of water, and acidophiles, alkaliphiles, halophiles, and obligate anaerobic fungi. Old microorganisms previously believed to perform only a limited number of fermentations are now found to use much broader ranges of substrates. They are found to generate energy by multiple mechanisms. Bacteria previously believed to be strict heterotrophs have now been shown to grow autotrophically, generating energy under anaerobic conditions by chemosmotic mechanisms with a wide range of electron acceptors and ATPase systems. It may seem odd, but bacteria actually gain energy by dechlorinating oganic halogens and other toxic compounds.

In this book, *Biological Degradation and Bioremediation of Toxic Chemicals,* Rasul Chaudhry presents a wide range of topics authored by researchers on the "cutting edge" of biodegradation and bioremediation. We will read about recent research describing how microorganisms clean soil and water, and how they remove or destroy toxic substances, including thiocarbamates, herbicides, insecticides, organophosphorus compounds, chlorinated aliphatic and aromatic chemicals, aromatice amines, sulfonates, and heavy metals. We will find that genetics and molecular biological methods are used to improve the ability of microorganisms to degrade a variety of substances and are used in attempts to construct microorganims with hybrid catabolic pathways. We will note that lignin and hemicelluloses can be converted to fuel, solvents, single-cell protein, and other useful products. More importantly, the book provides future directions of biotechnologic research and applications for mitigation of pollution and bioremediation of polluted environments.

This is an important book. Rasul Chaudhry and the contributing authors should be congratulated for a job well done. They, together with many others with concerns for the environment, are at a new frontier. In 1962, Rachel Carson published *Silent Spring* and aroused us. Rasul Chaudhry's book seems to be a fitting example that Rachel Carson's challenge to save the environment was not in vain.

Lars Ljundahl
Editor in Chief
Applied and Environmental Microbiology

Introduction

Expanding commercial applications of chemicals in industrial production, agricultural practices, and human health care generate large quantities of hazardous chemical waste that often result in environmental pollution. The efficient treatment of sewage, the recycling of cellulose wastes, and the disposal of toxic chemicals have become of immediate importance in a world faced with increasing pollution and decreasing energy resources. Currently available approaches to the pollution problem such as incineration or excavation and storage are expensive, inefficient, and lead to additional problems. The most promising approach is to biotreat and bioremediate the waste and the polluted environment by exploiting the catabolic capability of microorganisms. Microorganisms exhibit a fascinating metabolic diversity and the ability to evolve new catabolic properties rapidly for the degradation of xenobiotics, and they can be routinely isolated from contaminated sites to metabolize synthetic substances. In order for a microbial cell to utilize xenobiotic substances as sources of carbon, nitrogen, sulfur, or phosphorus, it is often necessary to convert the compounds to normal intermediates of central metabolism, chiefly organic acids that can enter the Krebs cycle.

A major limiting factor in the application of laboratory-adapted microorganisms is their inability to degrade mixtures of pollutants. Mixed microbial cultures (consortia) have shown greater potential than individual isolates to mineralize chemical mixtures. Alternatively, recombinant DNA techniques can be applied to construct bacterial strains with novel metabolic pathways to degrade several synthetic chemicals. Both microbial consortia and genetically engineered microorganisms are increasingly being considered for decontamination of toxic waste and bioremediation of polluted environments.

This book attempts to gather the recent advances in biological degradation of selected pollutants, strategies used for enhancement and expansion of catabolic properties of microorganisms, and potential applications to alleviate pollution problems. It comprises twenty-three chapters covering a wide range of interests, and should prove useful as a source of information for both the scientific community and the educated public. Topics covered include catabolism of well-investigated systems, toulene, napthalene, and the dissimilation diversity and potential of *Pseudomonas*; aerobic and anaerobic degradation of organic compounds, polychlorinated biphenyls (PCBs), pentachlorophenol (PCP), polycyclic aromatic hydrocarbons (PAHs), sulphonated aromatics, morpholines, carbamate and organophosphate pesticides, and low-molecular-weight toxic chemicals; fungal enzyme systems for degradation of toxic chemicals and disposal of lignocellulosic waste; biotransformation of toxic metals, particularly chromium; genetics of hazardous-chemical-degrading bacteria, strategies for selecting and constructing bacterial strains capable of degrading several compounds; and potential applications of microbial isolates in constructing bioreactors (particularly those with encapsulated bacteria) and developing bioremediation strategies and biotreatment of coal to improve quality while reducing pollution.

I am deeply indebted to all contributing authors for their patience and understanding. The efforts of Richard Abel, Suzane Copenhagen, and Karen Kirtley of the Dioscorides Press and Satya Chapalamdugu, Melissa Hixon, and Amatul Mateen of my laboratory made this book a practical undertaking and are greatly appreciated.

G. Rasul Chaudhry

CHAPTER 1

Catabolic Potential of Pseudomonads: A Regulatory Perspective

JOHN E. HOUGHTON and MARK S. SHANLEY

Abstract. The immense potential for microorganisms such as the pseudomonads to degrade xenobiotic compounds does not depend solely on the wealth of catabolic enzymes that they possess, but also upon their capacity for adaptive change. Such a capacity is promoted by their inherent patterns of regulation, which allow for the coincidental induction of different catabolic pathways, and this in turn allows for a positive selection to introduce novel patterns of biodegradation. Other chapters in this book have concentrated upon discrete reactions and specific degradative pathways, clearly defining their possible role in one or more specific process. This chapter takes a somewhat broader perspective on biodegradation, by highlighting three well-characterized dissimilatory pathways of pseudomonads and then reinterpreting these attributes as a network of discrete regulatory units, thereby linking the constituent catalytic processes in ways other than their defined similarity in biochemical function. In so doing we hope to reveal some of the inherent characteristics of Pseudomonas that have allowed them to command such a major role in the dissimilation of various aromatic compounds.

INTRODUCTION

It is of no little consequence that the benefits and ideas derived from any given question are directly related to the perspectives taken in its conception. The importance of perspective is magnified, somewhat, when we try to review a topic as complex as the microbial degradation of aromatic compounds. These processes can be seen from a variety of approaches, each giving some insight into how different bacteria degrade the numerous compounds that populate the microbial landscape. One perspective, which has considerably expanded our understanding of microbial metabolism, has been that of the biochemist: appreciating the overall degradative process as a series of stepwise catalytic rearrangements of complex molecules into less complex, readily usable *central metabolites*. Such an approach has allowed for a description of catalytic mechanisms and a concise definition of some of the more central pathways that comprise the network of microbial degradative activity. In so doing, it has also provided the intuitive framework upon which a biochemical analysis of more exotic aromatic compounds has been built.

Dr. Houghton is in the Department of Biology, Georgia State University, Atlanta, GA 30303, U.S.A. Dr. Shanley is in the Department of Biological Sciences, University of North Texas, Denton, TX 76203, U.S.A.

In this biochemical interpretation of aromatic catabolism, few terms have more definitive meaning than "*ortho-*" and "*meta-*" cleavage of the dioxygenated aromatic ring structure, (Figure 1; 10, 14). Indeed, such definition apparently reflects some significance at the genetic level, since genes that encode enzymes of the *ortho-*cleavage pathways almost invariably reside on the chromosome, whereas those of the *meta-*cleavage pathways are almost always borne on plasmids. Dividing the principal catabolic pathways into these two alternate approaches to ring cleavage has yielded a comprehensive understanding of the catalytic mechanisms preferentially used by different organisms (10, 11, 14). Further discrimination into the eukaryotic and eubacterial metabolic preferences has resulted in some additional definition of catabolic processes in either group, thus giving some appreciation as to why different constituent members of each "urkingdom" can or cannot fully metabolize different aromatic compounds (10). An example of this can be seen in the contrasting *ortho-*catabolism of protocatechuate by soil bacteria (such as pseudomonads) and fungi (Figure 2), in which lactonization of the β-carboxy-*cis,cis*-muconate into the γ-carboxy-muconolactone or the β-carboxymuconolactone, respectively, results in slightly different routes to the common metabolic intermediate, β-ketoadipate (55).

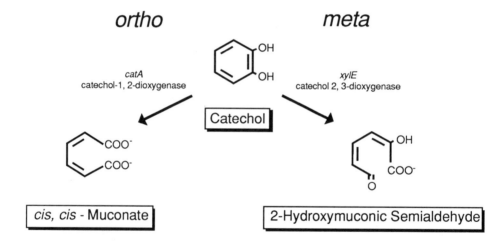

Figure 1. Introduction of molecular oxygen into catechol by two alternative mechanisms, *ortho* cleavage and *meta* cleavage of the ring structure.

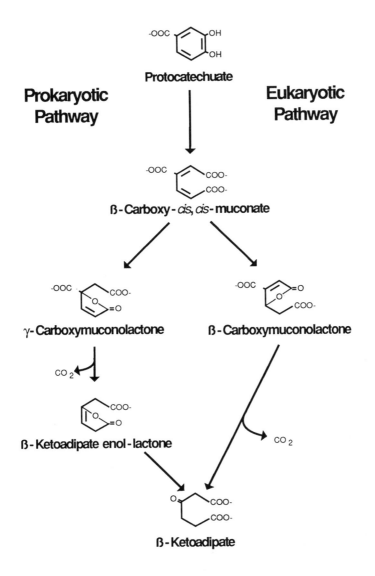

Figure 2. Variations on a metabolic theme. The protocatechuate branch of the β-ketoadipate pathway in bacteria and fungi converts the β-carboxy *cis,cis*-muconate into β-ketoadipate through two distinct catabolic pathways. Redrawn from Ornston and Yeh, 1982 (55).

CENTRAL METABOLIC PATHWAYS

Although such a biochemical outlook upon these catabolic processes is essential to any understanding of biodegradation, it is inherently limited in its evaluation of the substrate management within each individual cell. For example, catechol and protocatechuate are considered to be principal aromatic catabolites. As such, each should define a "node" or focus of enzymatic function and thus a point of reference for a comparison of biochemical mechanisms (10, 13) (Figure 3). To the biochemist, any individual catechol or protocatechuate molecule would be identical to any other catechol or protocatechuate molecule, no matter from where it came. To a prokaryotic cell, however (and indeed to the bacterial physiologist), the source of any particular compound is of considerable importance, as it provides for a metabolic balance (or imbalance) and a definition of the ultimate destination of that compound. Accordingly, the truly central metabolic processes in any organism are not directional pathways but circuits. They are cycles of variable enzymatic activities, designed to shuffle metabolic intermediates where needed and to supplement or replenish nutrients when necessary. In so doing, they maintain a metabolic *metastasis* and provide a buffer against variable environmental conditions. The cyclical nature of these central processes also provides

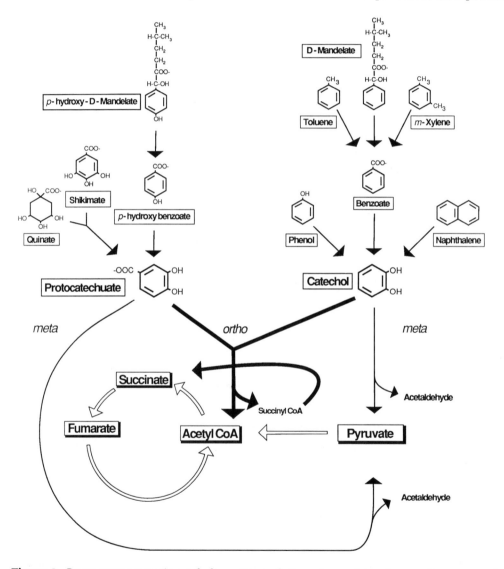

Figure 3. Convergent aromatic catabolism. Dissimilation of a variety of aromatic compounds into the Krebs cycle (*unfilled arrows*) channelled through a few "key" biochemical intermediates, two of which are protocatechuate and catechol. Further catabolism of both these compounds can be achieved by either *ortho*- (*heavy arrows*) or *meta*-pathway enzymes (light arrows). Although these two alternatives for catechol degradation are widely used in a variety of pseudomonads, *ortho*-cleavage is preferentially undertaken in the "fluorescent" pseudomonads such as *P. putida,* whereas the *meta* pathway is more common to the nonfluorescent pseudomonads (20).

for multiple access points into which (and from which) a considerable number of directionally organized pathways feed. With regards to aromatic hydrocarbon metabolism, the ultimate destination of the peripheral degradative pathways is one of the more central of these central metabolic processes, the Krebs' cycle. Not too surprisingly, the amphibolic nature of this cycle requires a major expenditure of cellular resources in regulating the activity of its composite enzymes. This requirement for regulation has been well characterized in a number of reviews and texts and shall receive no further reference here (6, 7, 12, 47).

GLOBAL REGULATORY SCHEMES

The importance of regulation is not distinct to central metabolic processes (especially in soil bacteria that have little preference for their source of carbon). Indeed, in these bacteria regulation may be of greater concern in the peripheral metabolic pathways that feed into these cycles, not only to prevent them from becoming futile but also, apparently, to coordinate the synthesis of functionally related enzymes and to conserve the wasteful expenditure of energy involved in unnecessary protein synthesis. In 1947, Stanier first conceptualized the principle of enzyme regulation, defining the basis for such regulation as a substrate induction of related enzyme synthesis (68). Subsequently, this concept was seen to be an oversimplification, and has been further refined for aromatic catabolism into the more complex induction of catabolic enzymes by discrete intermediates in the degradative process (6, 8, 52, 53). The role of regulation has proven to be critical to our understanding of these catabolic pathways, and has been used to great effect in the selection of a number of catabolic mutations created for either purely scientific interest or industrial use (13).

As the precise definition of the genetic arrangement and organization for a number of the predominant catabolic pathways has progressed, however, regulation has seemed to lose some of its preeminence, and its role has become less clearly defined. Gene duplication, and indeed duplication of almost entire dissimilatory pathways in soil bacteria, has resulted in a duplication of enzyme synthesis and function that seems to belie one of the fundamental regulatory premises: that of conservation of cellular materials against waste. Such an appreciation, however, has been shown to be metabolically naive (6, 52). An additional and presumably important principle of regulatory control is that of *metabolic isolation* of catabolic intermediates that are common to one or more degradative pathways. Whenever an alternative mechanism for the dissimilation of any compound becomes available (*ortho-* versus *meta*-cleavage of ring structures, for example), control of each outcome must be imposed. Although this has been recognized for some time (52), further discussion of its relevance to an understanding of aromatic degradation is desirable in view of the current burgeoning of biochemical and molecular information in this field.

The metabolic separation of compounds casts a very different light upon the biodegradative processes in soil bacteria and is presented here to augment, rather than to reiterate, points raised in some excellent reviews on aromatic degradation that have recently appeared in the literature (4, 20, 46). In mimicking the organellar compartmentalization of higher organisms, the *metabolic isolation* of compounds into distinct aromatic degradative pathways eliminates the creation of "bottlenecks" that might otherwise occur at the various nodes of biochemically defined catalytic focus (Figure 3). Such a metabolic fiasco is avoided by regulatory schemes in which the concentration of any given catabolic intermediate in a peripheral pathway is predetermined by the specific and *coincidental* induction of the enzymes that provided its synthesis, as well as those that provide for its further degradation. But what of the common catabolic intermediates? Is it merely by chance that so many diverse metabolic processes seem to converge upon a few compounds, such as catechol, proto-catechuate, or benzoate? We believe not, and would suggest that their importance lies not only in the variety of ways in which they are formed, but also in the number of pathways that dissimilate them. That is, these principal catabolic intermediates provide alternative routes, or *metabolic options,* the availability of which is dependent upon the concentration of each principal catabolite and the level of enzymatic activity in pathways that might permit their further degradation (Figure 4). The *metabolic option* for these catabolic intermediates (as previously suggested) heavily favors the pathway involved in their formation. However, should alternate sets of catabolic enzymes be present (that have been induced by other related catabolic intermediates or precursors), then alternative routes for their dissimilation are available. Thus, an understanding of the ways in which both the synthesis and degradation of these principal catabolites are controlled might give some indication as to how the "ubiquitous pseudomonads" (8) have thrived in their respective niches, and why they possess such an immense poten-tial for adaptative change and future experiments in bioremediation. To this end, we have chosen to concentrate our efforts on three well-characterized dissimilatory path-ways, each one having a distinct regulatory pattern, and each one having a unique bearing upon the importance of regulation to the catabolic potential of the pseudo-monads.

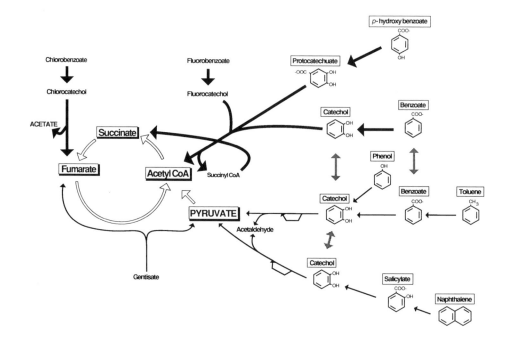

Figure 4. Concentric metabolic processes. Microorganisms that utilize aromatic compounds [degrading them into central metabolites within the Krebs cycle (unfilled arrows)] do so through a variety of pathways. This figure depicts these pathways as a portion of a weblike configura-tion. The pathways are further defined as being *ortho-*, or *meta-* cleavage pathways by *heavy* and *light arrows,* respectively. The peripheral intermediates are represented by the diols, catechol and protocatechuate, in concentric arcs around the Krebs cycle to demonstrate the position of the important catabolically intermediates discussed in this paper. *Hashed arrows* define a "biochemical similarity" between the metabolically isolated catechols and benzoates.

THE β-KETOADIPATE PATHWAY

The branchlike arrangement of the pathway in *Pseudomonas putida* (Figure 5) mimics that of most bacterial species and defines the organization of its constituent enzymes into two distinct but related limbs, the *pca* branch and the *cat* branch. The *pca*-branch breaks down protocatechuate (the dissimilatory product of aromatic compounds such as quinate, shikimate, and p-hydroxybenzoate; Figure 3) into succinate (succinyl CoA) and acetyl CoA. The *cat*-branch, beginning at catechol (which is also an intermediate that is biochemically common to the breakdown of a number of compounds; Figure 3) feeds into the pathway at the level of the enol-lactone, Figure 5 (10, 49, 50, 56).

The ubiquity of the β-ketoadipate pathway within the fluorescent bacteria defines its fundamental metabolic role. It contains an abundant sampling of principal biochemical reactions catalyzed by enzymes such as oxygenases, isomerases, hydrolases, decarboxylases, lactonizing enzymes, thio-transferases, and thiolases (Figure 5A). Consequently this pathway has received a great deal of scientific scrutiny since it acquired its name in 1948 (39), the scope of which has encompassed a number of scientific disciplines, including (A) biochemical analyses of the numerous enzymes and related transport systems from an assortment of bacteria (24, 55, 56); (B) genetic analyses of the attendant genes (2, 3, 21, 30, 48, 67); (C) crystallographic analyses of some of the constitutive enzymes (16, 37, 38, 59); and (D) evolutionary analyses (both within and among different genera) (21, 48, 57). In spite of this wealth of scientific research, until recently relatively little was known concerning the mechanism of regulation within the pathway beyond a physiological determination of the various reporter molecules involved and a definition that the genes from the pathway (in the majority of the bacteria investigated) are apparently positively inducible (5, 56, 70, 71).

REGULATORY ASPECTS OF THE β-KETOADIPATE PATHWAY IN *PSEUDOMONAS PUTIDA*

The two branches of the pathway are independently regulated at the level of transcription by the appropriately named regulatory proteins, PcaR and CatR, in response to the presence of the pathway intermediates, β-ketoadipate and *cis,cis*-muconate, respectively (Figure 5) (3, 30, 54). The PcaR protein is encoded by the *pcaR* gene, which has been localized by Tn5 insertional mutagenesis, cloned, and found to induce the otherwise low-level constitutive expression of all the *pca* genes from *pcaB* on down (30). In addition, protocatechuate, the first substrate distinct to the *pca*-branch of the pathway has also been shown to play a regulatory role by specifically inducing the *pcaH,G* genes (51, 54), although it is not known exactly how this effect is mediated. Unlike the plasmid-borne genes of the *meta*-cleavage pathways, the genes of the β-ketoadipate pathway in pseudomonads are not always neatly packaged within operons, but can be spread throughout the chromosome. There is, however, a more than occasional operonic and superoperonic clustering of two or three genes: *catA-catBC* in *P. aeruginosa* (27, 40); *pcaBDC* (26, 30) and *catR, BCA* in *P. putida* (57, J. E., Houghton, T. M. Brown, E. J. Hughes, and L. N. Ornston, unpublished observations). The *catR* gene (responsible for inducing the three genes that take catechol to β-ketoadipate enol-lactone) has been isolated and further characterized. It is transcribed

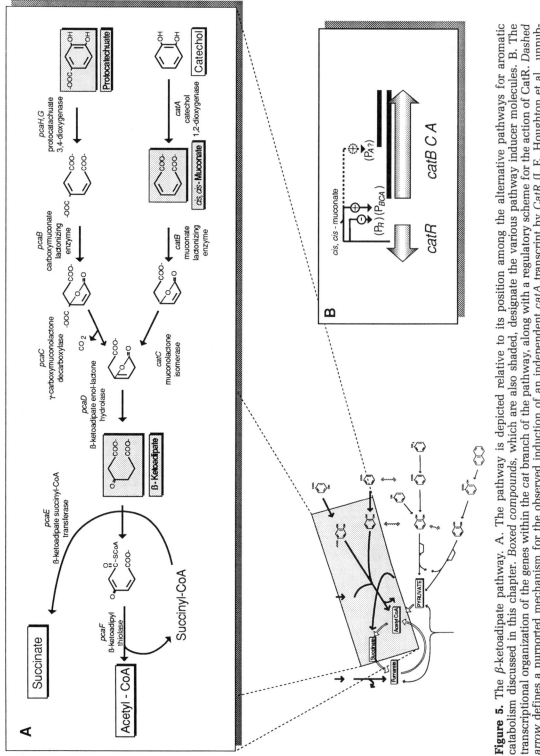

Figure 5. The β-ketoadipate pathway. A. The pathway is depicted relative to its position among the alternative pathways for aromatic catabolism discussed in this chapter. *Boxed compounds*, which are also shaded, designate the various pathway inducer molecules. B. The transcriptional organization of the genes within the *cat* branch of the pathway, along with a regulatory scheme for the action of CatR. *Dashed arrow* defines a purported mechanism for the observed induction of an independent *catA* transcript by *CatR* [J. E. Houghton et al., unpublished observations].

divergently to that of the neighboring *catBCA* operon and encodes a protein of 31.9 kDa (64). Furthermore, CatR apparently autoregulates its own expression while specifically binding to the *catR-B* promoter region (64) and, in the presence of *cis,cis*-muconate, positively induces the divergent *catBCA* transcript (J. E. Houghton, T. M. Brown, E. J. Hughes, and L. N. Ornston, unpublished observations). The *catR* gene from *P. putida* has been sequenced and, from it, the deduced amino acid sequence has been shown to be 289 amino acids in length, forming a protein that exhibits typical structural features of a DNA binding protein, which includes an "α-helix-turn-α-helix" motif at its amino terminus (64). Consistent with its divergent transcriptional expression and regulatory function, the N-terminus of CatR shares considerable amino acid sequence and apparent functional homology with a group of *trans*-acting positive regulatory proteins, collectively known as the the LysR family of gene activators (25, 64). Moreover, recent results suggest that *catR* has the additional capacity to induce *catA* expression independently (Figure 5B). Although it is not yet known how this induction is mediated, Tn5 mutagenic evidence strongly suggests that there is a CatR-sensitive promoter (P_A) residing within the *catC* structural gene, which gives rise to a *catA*-specific transcript, Figure 5B (J. E. Houghton, T. M. Brown, E. J. Hughes, and L. N. Ornston, unpublished observations).

THE TOL PATHWAY

As mentioned above, in *Pseudomonas* the major metabolic route for degradation of toluene and alkyl-substituted toluenes contrasts with that of the β-ketoadipate pathway, in that it involves the *meta*-fission of catechol by introducing oxygen into the ring structure *via* a catechol 2, 3-dioxygenase rather than by a catechol 1, 2-dioxygenase (Figure 1). Like the β-ketoadipate pathway, however, the degradation of toluene has been subject to a great deal of scientific experimentation (4, 20, 74), a large proportion of which has been concerned with regulation of the pathway, especially in *Pseudomonas* sp. (20, 46, 62). In accordance with the *meta*-cleavage bias, almost all the *xyl* genes are borne on plasmids, called, not too surprisingly, TOL plasmids. These TOL plasmids [and there have been a number of different, independent isolates (29, 41, 73)] are large (>110 kbp) and all belong to the same incompatibility group of P9 plasmids. Moreover, the majority of those identified have been shown to bear strong DNA sequence homology to the archetypical TOL plasmid, pWWO, first isolated from *P. arvilla* in 1974 (73). On the occasion when *xyl* genes have been found to be located on the chromosome, evidence still suggests that they have a plasmid origin and have arrived at their present location by some subsequent transpositional event. Indeed, the plasmid-borne *xyl* genes themselves are thought to reside within a transposable element (36, 69). Although the TOL pathway shares a couple of its catabolic intermediates (benzoate and catechol) with its *ortho*-fission counterparts (consistent with the demand for metabolic independence), all the genes necessary for the complete metabolism of toluene and m-xylene to central metabolites are generally carried on the TOL plasmid. These genes are also typically arranged into two functional units on the TOL plasmid within two physically distinct operons, the "upper" and the "lower" pathway operons (Figure 6B).

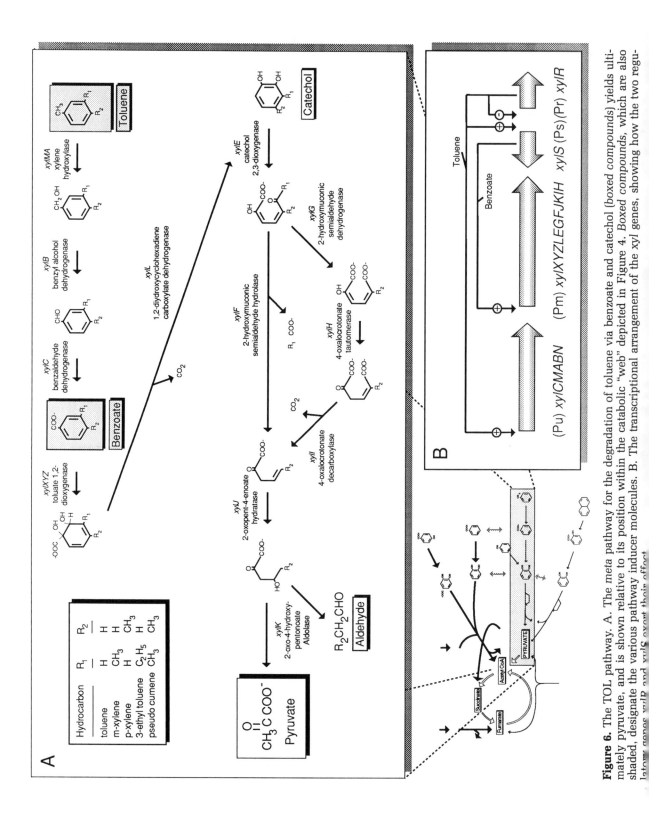

Figure 6. The TOL pathway. A. The meta pathway for the degradation of toluene via benzoate and catechol (boxed compounds) yields ultimately pyruvate, and is shown relative to its position within the catabolic "web" depicted in Figure 4. Boxed compounds, which are also shaded, designate the various pathway inducer molecules. B. The transcriptional arrangement of the *xyl* genes, showing how the two regulatory genes *xylR* and *xylS* exert their effect.

EXPRESSION OF GENES FROM THE "UPPER" PATHWAY

The "upper" operon for toluene degradation encodes the enzymes that oxidize toluene to benzoate. This is accomplished in three enzymatic steps catalyzed by proteins encoded by four genes, xylM, A, xylB, and xylC, which are transcribed in the order xylCMAB (Figure 6A and 6B). The gene designated xylN is an additional open reading frame located 3' proximal to the xylB gene, to which no function has yet been ascribed (18). The enzymes of the "upper" pathway have (relatively) broad substrate specificities and, in addition to catalyzing the oxidation of toluene to benzoate, xylenes can also be oxidized to their corresponding toluates (15). The promoter for the genes of the upper pathway, Pu, is positively induced by one of the two regulatory genes of the pathway, xylR (28, 31). This regulatory gene is located close to both of the pathway operons but is transcribed independently (Figure 6B). It encodes a protein of some 566 amino acids, which, upon binding the inducer molecules (toluene or m-xylene), enhances the transcription from Pu (34). Gene fusions of lacZ with Pu have confirmed the inducible nature of XylR (1). In addition Pu shows distinctive homology to promoters that are recognized by RNA polymerases only when bound to an alternative sigma factor [σ^{54}, ntrA (rpoN) gene product] presenting a comprehensive mechanism for induction. The σ^{54} factor is required to transcribe a diverse set of genes in different bacteria (42). The gene for the analogous sigma factor from Pseudomonas putida (rpoN) has been sequenced and its attendant amino acid sequence shown to share ∿80% protein sequence homology with the σ^{54} from Azotobacter vinelandii (35). When the protein sequence homology between xylR and ntrC from Klebsiella pneumoniae was determined (34), a mechanism for xylR induction was proposed.

EXPRESSION AND CHARACTERIZATION OF A SECOND REGULATORY GENE, xylS

In addition to controlling xylCMABN gene expression from Pu, the XylR protein also controls expression of a second trans-acting positive regulatory element, xylS. The xylS gene product is responsible, as shown below, for coordinating the expression of the lower pathway with that of the upper pathway (32, 44, 61). The xylS gene has been sequenced, and the deduced amino acid sequence has been shown to be a relatively small (331 amino acids) protein with typical structural features of a DNA-binding protein (33, 45). The XylS protein is a basic protein and contains an "σ-helix-turn-σ-helix" motif at its carboxy terminus. Like XylR, it also shares significant homology with other trans- acting positive regulatory proteins from different microorganisms, including AraC from E. coli and other similar positive regulatory proteins (34, 62), but not LysR. The xylS gene is expressed from promoter (Ps), which bears remarkable resemblance to the RpoN-dependent promoter, Pu. It is thus not surprising that expression of xylS is induced in a manner similar to that of the upper pathway from Pu, by an "activated" XylR and σ^{54}/RNA polymerase complex. The physical proximity and divergent nature of the xylR transcript to the xylS promoter, Ps (Figure 6B), also allows for the simultaneous negative autoregulation of the xylR transcript by physically blocking Pr while inducing Ps.

EXPRESSION OF GENES FROM THE "LOWER" PATHWAY

The benzoate produced from toluene (Figure 6) is converted to catechol and then subjected to meta-fission by the enzymes of the lower pathway. These enzymes are encoded by the genes of the lower or "meta- operon" of the TOL plasmid (17). No fewer than nine enzymes are coordinately expressed from a single promoter, Pm, in this polycistronic operon (xylXYZLTEGFJQKIH). The meta-pathway operon lies near the upper operon, separating this operon from the two regulatory genes, xylS and xylR (Figure 6B). Transcription of this meta-pathway operon from Pm (unlike transcription from Pu in the upper pathway) is independent of the rpoN gene product, but, as suggested above, does require the gene product of the xylS gene for efficient high-level expression. Such high levels are mediated by substrates of the lower pathway (benzoate, m-toluate, and p-toluate), which can act as inducers by binding to XylS. Thus it is possible for the cell to conserve the energy necessary to synthesize the upper pathway enzymes when they are not needed by separately inducing the lower pathway when only the intermediate substrates are present within the cell.

Curiously, the independent overexpression of a cloned xylS gene resulted in high-level constitutive expression of the meta-pathway genes (from Pm) in the absence of any inducer (45). Such a phenomenon in the host pseudomonad would in effect provide direct induction of Pm by the xylR gene product when bound to the upper pathway inducer (e.g. toluene). Such high-level constitutive expression of the lower pathway genes has also been observed for recombinant strains harboring mutant XylS proteins (62, 78). Single amino acid substitutions in the "σ-helix-turn-σ-helix" DNA binding motif within the carboxy-terminus of the protein have resulted in constitutive expression of the lower pathway genes from Pm. Moreover, additional XylS mutants exhibited induction by metabolites other than the normal inducers, demonstrating the potential for altering the regulatory properties of meta-pathway induction (60, 62). Careful mutational analysis of these mutant XylS proteins with altered effector binding further showed that substitutions in the amino terminus, the central portion, and the carboxy terminus of the protein all influenced the protein's ability to bind novel inducers with significant degrees of substitution. All the effectors examined had the carboxylate anion at carbon 1. Ring positions 2, 3, and/or 4 allow varying degrees and combinations of substitution for effector binding in the wild type protein (62). The effector binding site of xylS appears to preclude significant substitution of the benzene ring at positions 5 and 6. The effect of substitutions at different ring positions has been correlated to changes in specific portions of the protein. Thus, the effector binding pocket of the functional protein is defined by noncontiguous sequences within the fully assembled polypeptide chains to form the inducer binding site.

With a similar focus, a number of quite elegant genetic manipulations have been used to define the relative roles played by xylS and xylR in the overall regulation of either or both of the two pathways (61, 62, 78). These experiments tend to confirm that, although there is a somewhat coordinated expression of all the xyl genes (through the specific interactions of XylS and XylR), such coordination can be usurped or uncoupled, as shown above. Intriguingly, plasmids devoid of both regulatory genes (xylR and xylS) have been shown to be still subject to induction by the pathway intermediate, benzoate, through interaction with some, as yet unknown, trans-acting factor encoded by the host chromosome in P. putida (9). The story is further complicated by the determination that Pu and Pm promoters exhibit some DNA sequence homology with the specific binding sequence for IHF (integration host factor) from E. coli (28).

IHF has been found to facilitate the activation of the RNA polymerase/σ^{54} complexes in *E. coli*, and a similar role for an IHF homologue in *Pseudomonas* could relate to the expression of these two promoters.

NAPHTHALENE DEGRADATION

In an arrangement reminiscent of the *xyl* genes, in *P. putida* (19) the genes that encode the enzymes responsible for the degradation of naphthalene (the *nah* genes) reside on an 83 kbp plasmid (appropriately named the NAH7 plasmid) and are organized into two separate operons, which encode genes of the "upper" and the "lower" pathway (75). The *nah* operon (upper pathway) expresses genes (*nahABCFDE*) for those enzymatic activities required for the conversion of naphthalene to salicylate (Figure 7A). The *sal* operon (lower pathway) includes the genes (*nahGHINLJK*) responsible for the further dissimilation of salicylate to catechol, and thence, by *meta*-cleavage, to 2-oxo–4-hydroxypentonoate, which can be further degraded into pyruvate and acetaldehyde by a specific aldolase not encoded by any of the *nah* genes (as depicted in Figure 7A). The pathway is regulated by a single regulatory protein, *nahR*, which is positioned "upstream" of the lower pathway and is expressed divergently from the same, Figure 7B; 76). Salicylate fulfills the role of inducer molecule for both pathways and, together with NahR, stimulates transcription from the two respective promoters. The NahR-binding sequence for both operonic promoters has been identified. These two promoters share extensive regions of DNA sequence homology and are functionally identical, as verified by protein/DNA-binding studies (66). In a manner similar to CatR, the in vitro binding of NahR to these promoters is largely independent of the presence of the pathway inducer molecule, salicylate, even though salicylate is required for actual induction of both pathways. The *nahR* gene product is synthesized constitutively and, in a manner similar to both CatR and XylR, its expression is autoregulated, i.e. it represses its own expression while simultaneously inducing transcription of the structural genes from the upper pathway promoter. Although the similarity to XylR ends here, NahR shares extensive amino acid sequence homology with CatR at its amino-terminus (authors' unpublished observations), which mirrors their mutual sequence homology to the LysR family of gene activators (66, 77).

REGULATORY THEMES

The β-ketoadipate pathway from *P. putida* contains two distinct but functionally similar regulatory genes that serve to distinguish between the induction of two related degradative processes, the *pca* branch and the *cat* branch (30; (E. J. Hughes, L. N. Ornston, and authors' unpublished observations). The TOL pathway incorporates the complex interactions of two similar but functionally distinct regulatory genes, *xylR* and *xylS*, which regulate the sequential "upper" and "lower" pathways in toluene degradation (46). Degradation of naphthalene, on the other hand, uses only one regulatory gene to regulate two similarly sequential pathways, and does so in a way entirely different way from that of the TOL pathway (65). Thus, all three pathways collectively provide a synopsis of the variables involved in regulating catabolic pathways in pseudomonads, and thus allow some definitive contrasts and comparisons to be made.

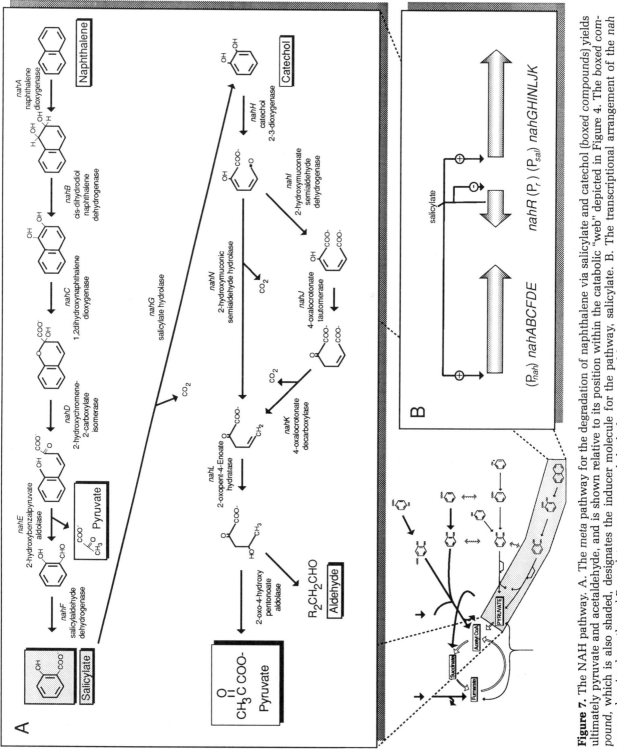

Figure 7. The NAH pathway. A. The meta pathway for the degradation of naphthalene via salicylate and catechol (boxed compounds) yields ultimately pyruvate and acetaldehyde, and is shown relative to its position within the catabolic "web" depicted in Figure 4. The boxed compound, which is also shaded, designates the inducer molecule for the pathway, salicylate. B. The transcriptional arrangement of the *nah* genes, showing how the *nahR* regulatory gene controls both the upper and lower operons.

The most fundamental similarity among all three regulatory units is that they all involve positive induction of their catalytic genes. This is in marked contrast to the *lac* paradigm of *E. coli*, which uses negative inducible regulation dominated by catabolite repression by the "preferred" carbon source, glucose. In the microbial world of the soil, scavenging may be considered a prerequisite for survival, and in this environment such preferences for carbon sources and catabolite repression by alternative carbon sources would appear to be selectively deleterious. That is, of course, unless the preferred source of carbon and catabolite repression is from within a central metabolic pathway (43), which would effectively bypass some of the needs for the peripheral catabolic pathways. The extraordinary nutritional versatility of *Pseudomonas* (notably, *P. cepacia*, which has been shown to utilize 100 out of 135 substrates that were tested) has been well documented (58) and bears testament to the importance of this scavenging ability, thus suggesting that such a selective advantage for positive induction may exist. From a regulatory perspective, the positive inducible nature of the pathways for aromatic catabolism in pseudomonads does seem to impart some selective advantages. It provides for a low-level constitutive expression of all the catabolic enzymes, which allows the dissimilatory products of any "usable" metabolite to "trickle" down through the various catabolic reactions to increase the concentration of one or any number of inducer molecules, which would then induce expression of related catabolic genes. Positive inducible regulation also facilitates the evolution and diversification of the pseudomonads' catabolic repertoire, allowing the development of more numerous and more sophisticated controls of its established pathways. This development could be provided by a simple duplication and modification of a regulatory gene to enable novel compounds to serve as inducer molecules for previously defined pathways (20, 60). Subsequent duplication of some or all of the genes within the defined pathway, together with the newly modified regulatory protein, would then allow for selective pressure to refine the new pathway according to the needs of the cell. Such a feat would not be possible for negatively regulated systems, without at first removing the existing regulatory constraints.

Within the empirically defined regulation of the three pathways reviewed in this chapter lies a number of apparent inconsistencies with some of the arguments put forward thus far. If there are principal catabolites (Figure 5), one might expect that these catabolites would play more of a leading regulatory role and function as the inducer molecules for all the pathways. This is apparently true for the TOL pathway and (by similarity in the structure and position of salicylate) for the NAH pathway. It is not, however, evident in the β-ketoadipate pathway, where the major inducers of each branch are intermediates in the catabolic process (β-ketoadipate and *cis,cis*-muconate), both of which are formed later in the catabolic process, one much later (Figure 5). With regard to the *cat* branch, it may be that the enzymes involved in the *ortho*-degradative pathway for catechol, unlike their *meta*-cleavage counterparts (Figure 6), are highly specific in their substrate requirements (6, 15, 74); hence the pathway is only induced upon the introduction of molecular oxygen into catechol by the specific 1, 2-dioxygenase (Figure 1). Curiously, this reaction also results in a feedback inhibition of the dioxygenase activity by its product, *cis,cis*-muconate (51), thus demonstrating tight metabolic control over this particular fate of catechol. For the *pca* branch of the pathway the explanation may be a little less rigorous. As catabolic pathways evolved away from the central metabolic pathways, due to the increased catabolic availability of what were then "exotic" carbon sources, chromosomally encoded pathways may be considered to have coevolved with their acquired regulatory constraints. In the initial expansion away from the Krebs' cycle intermediates, β-ketoadipate may have been a principal metabolic intermediate (6, 55). Subsequent branching of the pathway might then have resulted in the retention of β-ketoadipate as a usable regulatory "fossil," eventually to be usurped by protocatechuate [which is the preferred reporter molecule for induction of the *pcaH,G* genes in *P. putida* (52), and the inducer molecule of choice for the whole *pca*-branch in a related species of fluorescent bacteria, *Acinetobacter calcoaceticus* (5)]. The question remains, however, how and why are the inducible characteristics of the *meta*-pathways different? An answer to this may be found in the genetic location of these pathways. The propensity for the *meta*-pathways to be plasmid-borne would facilitate their diversification of function and duplication of catabolic roles, permitting their exchange within bacterial populations without too much concern for initial inter pathway competition for cellular resources or regulation. The selective pressure for complete or almost complete pathways to be located on any one plasmid [hence the invariant operonic clustering (72)] would supersede any regulatory constraints imposed by previously established chromosomally borne regulatory schemes. Thus, as these plasmids were shuffled from cell to cell, it is possible that they may have transiently borrowed the regulatory mechanisms from the chromosomes of their temporary host. As these plasmid-borne genes subsequently developed their own independent regulatory schemes, their "regulatory options" need not have been constrained by co-evolving regulatory mechanisms. Consequently, the regulatory mechanisms of these catabolic pathways have been subject to entirely different selective pressures than have their chromosomal counterparts.

What is interesting, in this regard, is the diversity of regulatory elements employed by the three catabolic pathways. Even though all three retain a few of the overall characteristics of positively inducible systems, such as the divergent expression of the primary regulatory genes from related transcripts, they each possess distinctive regulatory characteristics. Contrary to the expected similarity between the regulatory systems of the biochemically similar *meta*-cleavage pathways, the defined functional and sequence similarities are between *catR* and *nahR*. The TOL pathway has apparently acquired its regulatory genes, *xylS* and *xylR*, from alternative sources, through the appropriation of sigma factors and related gene sequences. Further support for the independent development of control for the plasmid borne *meta*-cleavage pathways can be derived from the seemingly modular organization of their genes. Not only are they arranged within operons, but each operon defines the catabolic fate of one principal catabolite to another, e.g. toluene to benzoate, benzoate to pyruvate. Each operonic cluster of catalytic genes is also subject to its own independent regulation. Such a definition of the genes into these *regulatory modules* would suggest that the complete TOL and NAH pathways did not develop as complete systems, but rather as different composites of these independent regulatory modules.

The possible development of the *meta*-cleavage pathways can therefore be further divided into a series of developmental stages: (A) the initial operonic clustering of related genes on plasmids, subject to regulatory constraints from chromosomally bound regulatory systems; (B) the acquisition of regulation by principal catabolites, through a variety of available regulatory schemes; (C) the compilation of sequential pathways by a clustering of related operons and a subsequent redefinition of the independent regulatory schemes into a unitary, sometimes sequential control. Evolution and regulation of the plasmid-borne *meta*-cleavage pathways would therefore tend to reflect more the importance of the "current" principal catabolites, rather than the coevolution of catalytic and regulatory mechanisms, which appears to have been the logic for the development of chromosomally encoded *ortho*-cleavage pathways.

REGULATORY RIDDLES

It is intriguing that the regulatory mechanism of choice for the "upper" operon of the TOL pathway has been transcriptional induction that is normally associated with systems responsive to environmental stimuli (63). Herein lies a curiosity. In nature (and in the laboratory), *P. putida* is chemotactically attracted to aromatic compounds such as m-toluate, p-toluate, and benzoate. *Pseudomonas* cells that carry the TOL plasmid (pWWO) preferentially metabolize aromatic compounds through the *meta*-pathway and maleylacetate, not through the chromosomally encoded *ortho*-pathway, via β-ketoadipate. Harwood has shown that β-ketoadipate is required as the inducer of the taxis genes in *P. putida* (23). In *P. putida* carrying the TOL plasmid, however, no chemotactic response is observed, as in these cells the *ortho*-pathway genes are not expressed, little β-ketoadipate is therefore produced and the taxis genes are not induced (22). Thus, although *P. putida* responds to the environmental presence of benzoate, the regulatory logic for its catabolism through β-ketoadipate remains a perplexing quandary.

Although some of the answers to this quandary may lie in an understanding as to how the different regulatory schemes developed, this curiosity also effectively demonstrates the physiological interaction of the different catabolic pathways. In essence, the regulatory network in pseudomonads does not rely upon the induction of a single set of genes from one single pathway in isolation, but upon a comprehensive regulatory scheme that involves a number of catabolic pathways (6). Though this is not too outrageous a contention, today, when coupled with the positive inducible nature of the aromatic dissimilatory pathways in pseudomonads, it does provide for an inherent potential for catabolic diversity and development. The laws of evolution "have had to comply with chemical boundaries" (6), and therefore there are only a limited number of reactions that any compound can undergo (in spite of the wealth of available metabolites). Thus, whereas pseudomonads encode enzymes that "tinker" specifically or nonspecifically with a variety of diverse compounds (their survival being subject to selective pressure), the majority of these enzymes are simple variations on a theme. The virtuosity of the pseudomonads (that is, their ability to orchestrate these variations for better utilization of the available carbon sources) suggests that *Pseudomonas* have not only mastered the chemistry involved, but are also perfecting their capacity to scavenge. Almost paradoxically, the *metabolic isolation* of catabolites within highly utilized dissimilatory paths channels the peripheral metabolites into the Krebs' cycle through an assortment of preferred pathways. When coupled with the inducible nature of these pathways, the potential for dissimilation of novel or unusual compounds becomes enormous. Initially random, nonspecific catalytic degradation of novel compounds can produce a "trickle" of metabolites, which can feed into any number of the preferred catabolic channels, and in so doing they can assure their further (complete) degradation into the Krebs' cycle by reinforcing the induction of these "preferred" pathways. It is at this level of catabolism that the concentration of key intermediates is no longer defined by the source of the compound, and thus, for example, an increase in concentration of benzoate can serve to induce any one or all of the preferred pathways (Figure 5). This increases the likelihood that a novel compound can be degraded in some manner, thereby initiating the directed utilization of that compound as a future carbon source. Accordingly, the importance of Regulation to aromatic catabolism lies not only in the organization and control of established pathways, but also in the determination of future catabolic routes. It should not be ignored or underestimated while pseudomonads and their relatives are used to degrade an increasing number and variety of xenobiotics in the environment.

Acknowledgments

We would like to thank Gerard O'Donovan, Ahmed Abdelal, Gerard Houghton, and Ron Ferguson for their critical reading of the manuscript and insightful discussions. We would also like to thank J. E. B. Stewart for his considerable time and effort in helping to compose the figures for this review. This work was supported in part by the BRSG Award RRO7171-08 from the National Institutes of Health to J. E. H. and by the University of North Texas Office of Research Administration Grant RIG34211 to M. S. S.

LITERATURE CITED

1. Abril, M.-A., C. Michan, K. N. Timmis, and J. L. Ramos. 1989. Regulator and enzyme specificities of the Tol plasmid-encoded upper pathway for degradation of aromatic hydrocarbons and expansion of the substrate range of the pathway. *J. Bacteriol.* 171:6782–6790.

2. Aldrich, T. L., and A. M. Chakrabarty. 1988. Transcriptional regulation, nucleotide sequence and localization of the promoter of the *catBC* operon in *Pseudomonas putida*. *J. Bacteriol.* 170:1297–1304.

3. Aldrich, T. A., R. K. Rothmel, and A. M. Chakrabarty. 1989. Identification of nucleotides critical for activity of the *Pseudomonas putida catBC* promoter. *Molec. Gen. Genet.* 218:266–271.

4. Burlage, R. S., S. W. Hooper, and G. S. Sayler. 1989. The TOL (pWWO) catabolic plasmid. *Appl. Environ. Microbiol.* 55:1323–1328.

5. Canovas, J. L., and R. Y. Stanier. 1967. Regulation of the enzymes of the β-ketoadipate pathway in *Moraxella calcoaceticus*. *Eur. J. Biochem.* 1:289–300.

6. Clarke, P. H., and L. N. Ornston. 1975a. Metabolic pathways and regulation: Part I. *Richmond Genetics and Biochemistry of* Pseudomonas. Ed. P. H. Clarke, M. H. Richmond. New York: John Wiley and Sons. 191–261.

7. _____ . 1975b. Metabolic pathways and regulation. Part II. In *Genetics and Biochemistry of* Pseudomonas. Ed. P. H. Clarke, M. H. New York: Richmond. John Wiley and Sons. 263–340.

8. Clarke, P. H., and J. H. Slater. 1986. Evolution of enzyme structure and function. In *The Bacteria*, X. Ed. J. R. Sokatch, L. N. New York: Ornston Academic Press. 71–144.

9. Cuskey, S. M., and A. B. Sprenkle. 1988. Benzoate dependent induction from the OP2 operator-promoter region of the TOL plasmid pWWO in the absence of known plasmid regulatory genes. *J. Bacteriol.* 170:3742–3746.

10. Dagley, S. 1986. Biochemistry of aromatic hydrocarbon degradation in *Pseudomonads*. In *The Bacteria*, X. Ed. J. R. Sokatch, L. N. Ornston. New York: Academic Press. 527–555.

11. Dagley, S. 1987. Lessons from biodegradation. *Annu. Rev. Microbiol.* 41:1–23.

12. Gest, H. 1981. Evolution of the citric acid cycle and respiratory conversion in prokaryotes. *FEMS Microb. Letts.* 12:209–215.

13. Gibson, D. T., and V. Subramanian. 1984. Microbial degradation of aromatic hydrocarbons. In *Microbial Degradation of Organic Compounds*. Ed. D. T. Gibson. New York: Dekker. 181–252.

14. Gibson, D. T. 1988. Microbial metabolism of aromatic hydrocarbons and the carbon cycle. In *Microbial Metabolism and the Carbon Cycle*. Ed. S. R. Hagedorn, R. S. Hanson, D. A. Kunz. Chur, Switzerland. Harwood Academic Publishers. 33–58.

15. Gibson, D. T., G. J. Zylstra, and S. Chauhan. 1990. Biotransformations catalysed by toluene dioxygenase from *Pseudomonas putida* F1. In Pseudomonas. *Biotransformations, Pathogenesis, and Evolving Biotechnology.* Ed. R. Silver, A. M. Chakrabarty. Washington, D.C.: ASM Publications. 121–132.

16. Goldman, A., D. L. Ollis, K. Ngai, and T. A. Steitz. 1985. Crystal structure of muconate lactonizing enzyme at 6.5 Å resolution. *J. Mol. Biol.* 182:353–355.

17. Harayama, S., P. R. Lehrbach, and K. N. Timmis. 1984. Transposon mutagenesis analysis of *meta*-cleavage pathway operon genes of the TOL plasmid of *Pseudomonas putida* mt-2. *J. Bacteriol.* 160:251–255.

18. Harayama, S., R. A. Leppick, M. Rekik, N. Mermod, P. R. Lehrbach, and K. N. Timmis. 1986. Gene order of the TOL catabolic plasmid upper pathway operon and oxygenation of both toluene and benzyl alcohol by the *xylA* product. *J. Bacteriol.* 167:455–461.

19. Harayama, S., M. Rekik, A. Wasserfallen, and A. Bairoch. 1987. Evolutionary relationships between catabolic pathways for aromatics: conservation of gene order and nucleotide sequences of catabolic oxidation genes from pWWO and NAH7 plasmids. *Mol. Gen. Genet.* 210:241–247.

20. Harayama, S., and K. N. Timmis. 1989. Catabolism of aromatic hydrocarbons by *Pseudomonas*. In *Genetics of Bacterial Diversity.* Ed. D. A. Hopwood, K. F. Chater. London: Academic Press. 151–174.

21. Hartnett, C., E. L. Neidle, K. Ngai,, and L. N. Ornston. 1990. DNA sequence of genes encoding *Acinetobacter calcoaceticus* protocatechuate 3, 4 dioxygenase: evidence indicating shuffling of genes and of DNA sequences within genes during their evolutionary divergence. *J. Bacteriol.* 172:956–966.

22. Harwood, C. S., and L. N. Ornston. 1984. TOL plasmid can prevent induction of chemotactic responses to aromatic acids. *J. Bacteriol.* 160:797–800.

23. Harwood, C. S., M. Rivelli, and L. N. Ornston. 1984. Aromatic acids are chemoattractants for *Pseudomonas putida*. *J. Bacteriol.* 160:622–628.

24. Harwood, C. S., and L. N. Ornston. 1988. Futile high-level transport activity impairs starvation-survival of *Pseudomonas putida*. *J. Gen. Microbiol.* 134:2421–2427.

25. Henikoff, S. G., G. W. Haughn, J. M. Calco, and J. C. Wallace. 1988. A large family of activator proteins. *Proc. Natl. Acad. Sci. USA.* 85:6602–6606.

26. Holloway, B. W., and A. F. Morgan. 1986. Genome organization in *Pseudomonas*. *Annu. Rev. Microbiol.* 40:79–105.

27. Holloway, B. W., S. Darmsthiti, C. Johnson, A. Kearney, V. Krishnapillai, A. F. Morgan, E. Ratnaningsih, R. Saffery, M. Sinclair, D. Strom, and C. Zhang. 1990. In Pseudomonas. *Biotransformations, Pathogenesis, and Evolving Biotechnology.* Ed. R. Silver, A. M. Chakrabarty. Washington, D.C.: ASM Publications. 269–291.

28. Holtel, A., M.-A. Abril, S. Marques, K. N. Timmis, and J. L. Ramos. 1990. Promoter-upstream activator sequences are required for expression of the *xylS* gene and upper-pathway operon on the *Pseudomonas* TOL plasmid. *Mol. Microbiol.* 4:1551–1556.

29. Hughes, E. J. L., R. C. Batly, and R. A. Skurray. 1984. Characterization of a TOL-like plasmid from *Alcaligenes eutrophus* that controls expression of a chromosomally encoded *p*-cresol pathway. *J. Bacteriol.* 158:73–78.

30. Hughes, E. J., M. J. Shapiro, J. E. Houghton, and L. N. Ornston. 1988. Cloning and expression of *pca* genes from *Pseudomonas putida* in *Escherichia coli*. *J. Gen. Microbiol.* 134:2877–2887.

31. Inouye, S., Y. Ebina, A. Nakazawa, and T. Nakazawa. 1984. Nucleotide sequence surrounding transcription initiation site of *xylABC* operon on TOL Plasmid of *Pseudomonas putida*. *Proc. Natl. Acad. Sci. USA.* 81:1688–1691.

32. Inouye, S., A. Nakazawa, and T. Nakazawa. 1984. Nucleotide sequence of the promoter region of the *xylDEGF* operon on TOL plasmid of *Pseudomonas putida*. *Gene* 29:323–330.

33. Inouye, S., A. Nakazawa, and T. Nakazawa. 1986. Nucleotide sequence of the regulatory

gene *xylS* on the *Pseudomonas putida* TOL plasmid and identification of the protein product. *Gene* 44:235–242.

34. Inouye, S., A. Nakazawa, and T. Nakazawa. 1988. Nucleotide sequence of the regulatory gene *xylR* of the TOL plasmid from *Pseudomonas putida*. *Gene* 66:301–306.

35. Inouye, S., M. Yamada, A. Nakazawa, and T. Nakazawa. 1989. Cloning and sequence analysis of the *ntrA (rpoN)* gene of *Pseudomonas putida*. *Gene* 85:145–152.

36. Jeenes, D. J., and P. A. Williams. 1982. Excision and integration of degradative pathway genes from TOL plasmid pWWO. *J. Bacteriol.* 150:188–194.

37. Katti, S. K., B. Katz, and H. W. Wyckoff. 1989. Crystal structure of muconolactone isomerase at 3.3 Å resolution. *J. Mol. Biol.* 205:557–571.

38. Katz, B., D. L. Ollis, and H. W. Wyckoff. 1985. Low resolution crystal structure of muconolactone isomerase. *J. Mol. Biol.* 184:311–318.

39. Kilby, B. A. 1948. The bacterial oxidation of phenol to β-ketoadipic acid. *Biochem. J.* 43:v–vii.

40. Kukor, J. J., R. H. Olsen, and D. P. Ballou. 1988. Cloning and expression of the *catA* and *catBC* gene clusters from *P. aeruginosa* PAO. *J. Bacteriol.* 170:4458–4465.

41. Kunz, D. A., and P. J. Chapman. 1981. Isolation and characterization of spontaneously occurring TOL plasmid mutants of *Pseudomonas putida* HS1. *J. Bacteriol.* 146:952–964.

42. Kustu, S., E. Santero, J. Keener, D. Popham, and D. Weiss. 1989. Expression of $\sigma54$ (*ntrA*)-dependent genes is probably united by a common mechanism. *Microbiol. Rev.* 53:367–376.

43. Lessie, T. G., and J. Phibbs P. V. 1984. Alternative pathways of carbohydrate utilization in pseudomonads. *Annu. Rev. Microbiol.* 38:359–387.

44. Mermod, N., P. R. Lerhbach, W. Reineke, and K. N. Timmis. 1984. Transcription of the TOL plasmid toluate catabolic pathway operon of *Pseudomonas putida* is deteremined by a pair of co-ordinately and positively overlapping promoters. *EMBO J.* 3:2461–2466.

45. Mermod, N., J. L. Ramos, A. Bairoch, and K. N. Timmis. 1987. The *xylS* gene positive regulator of TOL plasmid pWWO: Identification, sequence analysis and overproduction leading to constitutive expression of *meta* cleavage operon. *Mol. Gen. Genet.* 207:349–354.

46. Nakazawa, T., S. Inouye, and A. Nakazawa. 1990. Regulatory systems for expression of *xyl* genes on the TOL plasmid. In Pseudomonas. *Biotransformations, Pathogenesis, and Evolving Biotechnology.* Ed. R. Silver, A. M. Chakrabarty. Washington, D.C.: ASM Publications. 133–150.

47. Neidhardt, F. C., J. L. Ingraham, and M. Schaechter. 1990. *Physiology of the Bacterial Cell.* New York: Sinauer.

48. Neidle, E. L., C. Hartnett, and L. N. Ornston. 1989. Characterization of *Acinetobacter calcoaceticus catM*, a repressor gene homomologous in seqeunce to transcriptional activator genes. *J. Bacteriol.* 171:5410–5421.

49. Ornston, L. N. 1966a. The conversion of catechol and protocatechuate to β-ketoadipate by *Pseudomonas putida*. II. Enzymes of the protocatechuate pathway. *J. Biol. Chem.* 241:3787–3794.

50. Ornston, L. N. 1966b. The conversion of catechol and protocatechuate to β-ketoadipate by *Pseudomonas putida*. III. Enzymes of the catechol pathway. *J. Biol. Chem.* 241:3795–3799.

51. Ornston, L. N. 1966c. The conversion of catechol and protocatechuate to β-ketoadipate by *Pseudomonas putida*. IV. Regulation. *J. Biol. Chem.* 241:3800–3810.

52. Ornston, L. N. 1971. Regulation of catabolic pathways in Pseudomonas. *Bacteriol. Rev.* 35:87–116.

53. Ornston, M. K., and L. N. Ornston. 1972. The regulation of the β-ketoadipate pathway in *Pseudomonas acidovorans*. *J. Gen. Microbiol.* 73:455–464.

54. Ornston, L. N., and D. Parke. 1976. Evolution of catabolic pathways in *Pseudomonas putida*. *Biochem. Soc. Trans.* 4:468–473.

55. Ornston, L. N., and W. K. Yeh. 1982a. Recurring themes and repeated sequences in metabolic evolution., In A. M. Chakrabarty (ed.), Biodegradation and Detoxification of environmental pollutants, CRC Press, Miami.

56. Ornston, L. N., J. E. Houghton, E. L. Neidle, and, L. A. Gregg. 1990. Subtle selection and novel mutation during divergence of the β-ketoadipate pathway. In Pseudomonas. *Biotransformations, Pathogenesis, and Evolving Biotechnology.* Ed. R. Silver, A. M. Chakrabarty. Ed. Washington, D.C.: ASM Publications. 207–225.

57. Ornston, L. N., E. L. Neidle, and J. E. Houghton. 1990. Gene rearrangements, a force for evolutionary change; DNA sequence rearrangements, a source of genetic constancy. In *The Bacterial Chromosome.* Ed. K. D. M. Riley. Washington, D.C.: ASM Publications. 325–334.

58. Palleroni, N. J. 1986. *Taxonomy of the Pseudomonads.* New York: Academic Press.

59. Pathak, D., K. L. Ngai, and D. Ollis. 1988. X-ray crystallographic structure of dienelactone hydrolase at 2.8 Å. *J. Mol. Biol.* 204:435–445.

60. Ramos, J. L., A. Stolz, W. Reineke, and K. N. Timmis. 1986. Altered effector response of gene expression: TOL plasmid XylS mutants and their use to engineer expansion of the range of aromatics degraded by bacteria. *Proc. Natl. Acad. Sci. USA.* 83:8467–8471.

61. Ramos, J. L., N. Mermod, and K. N. Timmis. 1987. Regulatory circuits controlling transcription of TOL plasmid operon encoding *meta*-cleavage pathway for degradation of alkylbenzoates by *Pseudomonas. Mol. Microbiol.* 1:293–300.

62. Ramos, J. R., F. Rojo, L. Zhou, and K. N. Timmis. 1990. A family of positive regulators related to the *Pseudomonas putida* TOL plasmid *XylS* and the *Escherichia coli* AraC activators. *Nucl. Acids. Res.* 18:2149–2152.

63. Ronson, C. W., B. T. Nixon, L. M. Albright, and F. M. Ausubel. 1987. Conserved domains in bacterial regulatory genes that respond to environmental stimuli. *J. Bacteriol.* 49:571–581.

64. Rothmel, R. K., T. L. Aldrich, J. E. Houghton, W. M. Coco, L. N. Ornston, and A. M. Chakrabarty. 1990. Nucleotide sequencing and characterization of *Pseudomonas putida catR:* A positive regulator of the *catBC* operon is a member of the LysR family. *J. Bacteriol.* 172:922–931.

65. Schell, M. A. 1990. Regulation of the naphthalene degradation genes of plasmid NAH7: Example of a generalized positive control system in *Pseudomonas* and related bacteria. In Pseudomonas. *Biotransformations, Pathogenesis, and Evolving Biotechnology.* Ed. R. Silver, A. M. Chakrabarty. Washington, D.C.: ASM Publications. 1965–1976.

66. Schell, M. A., and M. Sukordhaman. 1989. Evidence that the transcription activator encoded by the *Pseudomonas putida nahR* gene is evolutionarily related to the transcriptional activators encoded by the *Rhizobium nodD* genes. *J. Bacteriol.* 171:1952–1959.

67. Shanley, M. S., E. L. Neidle, R. E. Parales, and L. N. Ornston. 1986. Cloning and expression of *Acinetobacter calcoaceticus catBCDE* genes in *Pseudomonas putida* and *Escherichia coli. J. Bacteriol.* 165:557–653.

68. Stanier, R. Y. 1947. Simultaneous adaptation: A new technique for the study of metabolic pathways. *J. Bacteriol.* 54:339–348.

69. Tsuda, M., and T. Iino. 1987. Genetic analysis of a transposon carrying toluene degrading genes on a TOL plasmid pWWφ. *Mol. Gen . Genet.* 210:270–276.

70. Wheelis, M. L., and R. Y. Stanier. 1970. The genetic control of dissimilatory pathways in *Pseudomonas putida. Genetics.* 66:245–266.

71. Wheelis, M. L., and L. N. Ornston. 1972. Genetic control of enzyme induction in the β-ketoadipate pathway of *Pseudomonas putida:* deletion mapping of cat mutations. *J. Bacteriol.* 108:790–795.

72. Wheelis, M. L. 1975. The genetics of dissimilatory pathways in *Pseudomonas. Annu. Rev. Microbiol.* 29:505–524.

73. Williams, P. A., and K. Murray. 1974. Metabolism of the benzoate and methylbenzoates by *Pseudomonas putida* (arvilla) mt-2: evidence for the existence of a TOL plasmid. *J. Bacteriol.* 120:146–423.

74. Williams, P. A., L. E. Gibb, H. Keil, and D. J. Osborne. 1988. Organisation of and relationships between catabolic genes of TOL plasmids. In *Microbial Metabolism and the Carbon Cycle*. Ed. S. R. Hagedorn, R. S. Hanson, D. A. Kunz. Chur, Switzerland: Harwood Academic Publishers. 339–358.

75. Yen, K.-M., and I. C. Gunsalus. 1982. Plasmid gene organization: napthalene/salicylate oxidation. *Proc. Natl. Acad. Sci. USA.* 79:874–878.

76. Yen, K.-M., and I. C. Gunsalus. 1985. Regulation of naphthalene catabolic genes of plasmid NAH7. *J. bacteriol.* 162:1008–1013.

77. You, I.-S., D. Ghosal, and I. C. Gunsalus. 1988. Nucleotide sequence of plasmid NAH7 gene *nahR* and DNA binding of the *nahR* product. *J. Bacteriol.* 170:5409–5415.

78. Zhou, L., K. N. Timmis, and J. L. Ramos. 1990. Mutations leading to constitutive expression from the TOL plasmid *meta*-cleavage pathway operon are located at the C-terminal end of the positive regulator protein XylS. *J. Bacteriol.* 172:3707–3710.

CHAPTER 2

Genetic Systems in Soil Bacteria for the Degradation of Polychlorinated Biphenyls

KENSUKE FURUKAWA

Abstract. Since their identification in the environment in 1966, polychlorinated biphenyls (PCBs) have become of global concern as serious environmental pollutants. PCB-degrading bacteria are ubiquitous in the environment. They utilize biphenyl as their sole source of carbon and energy and cometabolize PCBs to chlorobenzoic acids through an oxidative route. The genes bphABCD coding for PCB degradation to chlorobenzoates have been cloned from several Pseudomonas strains. These genes are clustered on the chromosome to form a polycistronic operon. DNA-DNA blot hybridization, nucleotide sequence analyses, and immunological studies reveal that nearly identical bph operons are distributed in many different strains, a finding that indicates that these operons have a transposition mechanism of transfer from one strain to another. On the other hand, some bacteria possess bph genes whose homologies are quite low among each other, even though the enzyme system and properties of the enzymes involved in PCB degradation are very similar. Gene-specific transposon mutagenesis allowed us to construct strains that accumulated catabolic intermediates from various biphenyl compounds, including PCBs. In addition, the cloned bphABCD were introduced into benzoate-utilizing bacteria. The recombinant strains utilized biphenyl as a sole source of carbon and degraded PCBs.

INTRODUCTION

Polychlorinated biphenyls (PCBs) were manufactured commercially from 1929 to the early 1970s in many countries. These compounds were synthesized by direct chlorination of biphenyl. Theoretically, 210 compounds containing 0 to 10 chlorines per biphenyl molecule could be produced. Because of their desirable physical and chemical properties, PCBs gained widespread use in a variety of applications, such as transformer oils, capacitator dielectrics, and heat transfer fluids. High contamination of PCBs in wild birds was first reported in Sweden in 1966. Similar to 2,2-bis(p-chlorophenyl)1,1,1-trichloroethane (DDT) and its metabolites, PCBs are now recognized to be present in great abundance in the ecosystem (30).

Dr. Furukawa is at the Department of Agricultural Chemistry, Kyushu University, Hakozaki, Fukuoka 812, Japan.

Metabolic breakdown by microorganisms is considered one of the major mechanisms of degradation for these environmentally widespread pollutants (9). The metabolic pathway and enzyme system involved in their biodegradation have been studied in isolated microorganisms and naturally occurring microbial populations for pure isomers and commercial mixtures of PCBs (9,10). This chapter describes the biochemical and molecular basis of degradation of PCBs in soil bacteria. The genetic manipulation studies described here may serve as a first step to enhance PCB degradation in bacteria.

PCB-degrading Bacteria

A number of PCB-degrading bacteria have been isolated from environmental samples. They are mostly aerobic, gram-negative soil bacteria, and include species of *Pseudomonas*, *Acinetobacter*, *Achromobacter*, *Alcaligenes*, *Moraxella*, and *Acetobacter* (2,6,15,17,18,28,31,34). Gram-positive bacteria such as *Arthrobacter* sp. (13) and *Corynebacterim* sp. (3) have also been described. These strains utilized biphenyl as a sole source of carbon and energy and cometabolized a number of PCB components to chlorobenzoic acids via ring-dioxygenation and meta-cleavage, as illustrated in Figure 1. The biodegradability and catabolic fate of PCBs by bacteria are greatly influenced by chlorine substitution on the biphenyl molecule (21,23). Biphenyl-utilizing strains exhibit many patterns of PCB-congener selectivity for biodegradation and catabolic fate (4,6,21). The degradability of 33 pure congeners of PCBs by two bacterial strains of *Alcaligenes* and *Acinetobacter* allowed us to draw the following general conclusions on the relationship between chlorine substitution and microbial breakdown of PCBs:

1. The degradation rate of PCBs decreases as chlorine substitution increases.

2. PCBs containing two chlorines in the *ortho*-position of a single ring (i.e. 2,6-) and each ring (i.e. 2,2'-) show a striking resistance to degradation.

3. PCBs containing all chlorines on a single ring are generally degraded faster than those containing the same number on both rings.

4. PCBs having two chlorines at the 2,3 position of one ring, such as 2,3,2',3'-, 2,3,2',5'-, 2,4,5,2',3'-chlorobiphenyls, are susceptible to microbial attack, compared with other tetra- and pentachlorobiphenyls, although this series of PCBs is metabolized through an alternative pathway.

5. Initial dioxygenation followed by ring-cleavage of the biphenyl molecules occurs with a nonchlorinated or less chlorinated ring.

Two bacteria, an *Acinetobacter* sp. P6 and an *Alcaligenes* sp. Y42, transformed 33 (chlorine numbers 1 to 5) and 20 PCB congeners among 36 tested, respectively. Analysis of metabolic intermediates showed that a number of PCB components were converted to the corresponding chlorobenzoic acids via an oxidative route, as illustrated in Figure 1 (9). Molecular oxygen is introduced at the 2,3 position of the nonchlorinated or lesser chlorinated ring to produce a dihydrodiol compound (2,3-dihydroxy-4-phenylhexa-4,6-diene) (Fig. 1, compound II) by the action of a biphenyl dioxygenase (product of the *bphA* gene). The dihydrodiol is dehydrogenated to a 2,3-dihydroxybiphenyl (Fig. 1, compound III) by a dihydrodiol dehydrogenase (product of the *bphB* gene). The 2,3-dihydroxybiphenyl is cleaved at the 1,2 position by a 2,3-dihydroxybiphenyl dioxygenase (product of the *bphC* gene) to produce the meta-cleavage compound, 2-hydroxy-6-oxo-6-phenylhexa-2,4-dienoic acid (Fig. 1, compound IV). The meta-cleavage compound is hydrolized to the corresponding chlorobenzoic acid (Fig. 1, compound V) by a hydrolase (product of the *bphD* gene). Thus, four enzymes are involved in the oxidative degradation of PCBs to chlorobenzoic acids. Most of the biphenyl-utilizing strains cannot degrade chlorobenzoic acids any further, and therefore the chlorobenzoates accumulate during PCB catabolism. This oxidative pathway is also common in other pure cultures such as many *Pseudomonas* spp., *Alcaligenes* sp., *Achromobacter* sp., and *Moraxella* sp., in gram-positive *Arthrobacter* sp. and *Corynebacterium* sp., and in naturally occurring mixed microbial cultures for both isomeric chlorobiphenyls and commercial PCB mixtures (6, 10, 22). Some intermediates other than chlorobenzoates accumulated during the degradation of certain PCB congeners, however. For example, 2,4'-, 2,4,4'-, and 2,5,4'-chlorobiphenyls, which all have chlorines at the 2,4' position, were readily converted to the meta-cleavage yellow compounds through the major pathway illustrated in Figure 1. These meta-cleavage compounds accumulated in the reaction mixture during prolonged incubation (24). Dihydroxy compounds accumulated from 2,6-, 2,3,6-, 2,4,2',5'-, 2,5,2',5'-, and 2,4,5,2',5'-chlorobiphenyls. 2,4,6-Trichlorobiphenyl was quickly metabolized to trihydroxy-compound via a dihydroxy-compound by *Acinetobacter* sp. P6, whereas the same compound was converted slowly to 2,4,6-trichlorobenzoic acid through the major oxidative route by *Alcaligenes* sp. Y42 (24). Tetra- and pentachlorobiphenyls which possess chlorines at the 2,3 position, were metabolized by an alternative route, and a large amount of unidentified products of dichloro-compounds from 2,3,2',3'- and 2,3,2',5'-chlorobiphenyls, and trichloro-compound from 2,4,5,2',3'-pentachlorobiphenyl, accumulated in the culture media (21).

The biodegradability and metabolic products of commercial PCB mixtures (Kaneclors®) were investigated with *Acinetobacter* sp. P6 (20). Mono- and dichlorobenzoic acids, dichloro- and trichlorodihydroxy-compounds, and meta-cleavage products with two and three chlorines were observed from KC200 (primarily dichlorobiphenyls), KC300 (primarily trichlorobiphenyls), and KC400 (primarily tetrachlorobiphenyls). The unknown compounds derived from 2,3,2',3'- and/or 2,3,2'5'-tetrachlorobiphenyl were detected from KC400, but KC500 (primarily pentachlorobiphenyls) was resistant to microbial degradation. Only dihydroxy compounds of certain pentachlorobiphenyls were detected. The 25 PCB degrading strains isolated from a variety of sites in the United States included several genera such as *Pseudomonas*, *Alcaligenes*, and *Corynebacterium* (5,7). The strains exhibited many different patterns

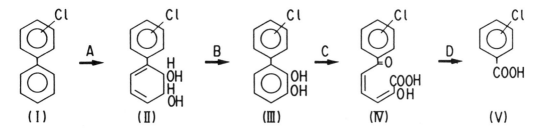

Figure 1. Catabolic pathway for degradation of biphenyl and chlorobiphenyls in *Pseudomonas pseudoalcaligenes* KF707. I, biphenyl (BP); II, 2,3-dihydroxy-4-phenylhexa-2,4-diene (dihydrodiol compound DHDO); III, 2,3-dihydroxybiphenyl (23OHBP); IV, 2-hydroxy-6-oxo-6-phenylhexa-2,4-dienoic acid (meta-cleavage compound, HPDA); and V, benzoic acid. Enzyme activities: A, biphenyl dioxygenase; B, DHDO dehydrogenase; C, 23OHBP dioxygenase; D, HPDA hydrolase.

of PCB congener selectivity. Several bacterial strains were capable of degrading even penta- and hexachlorobiphenyls (5,7). Although the principal route of PCB degradation in aerobic bacteria appeared to involve a 2,3-dioxygenase attack at an unsubstituted 2,3 (or 5,6) position, *Alcaligenes eutrophus* H850 rapidly degraded 2,5,2′,5′-tetrachlorobiphenyl and 2,4,5,2′,5′-pentachlorobiphenyl, which have no unchlorinated 2,3 site. Strain H850 attacked the 2,3-chlorophenyl ring of 2,3,2′5′-tetrachlorobiphenyl to yield 2,5-dichlorobenzoic acid and 2′,3′-dichloroacetophenone. Furthermore, H850 oxidized 2,4,5,2′,4′,5′-hexachlorobiphenyl to 2′,4′,5′-trichloroacetophenone, thus suggesting that a significant route of PCB degradation in H850 involves a 3,4-dioxygenase.

CLONING, GENETIC ORGANIZATION, AND FUNCTION OF THE *BPH* OPERON CODING FOR PCB DEGRADATION

The bphABCXD Operon in Pseudomonas pseudoalcaligenes KF707

A gene cluster coding for biphenyl/PCB-degrading enzymes was first cloned during 1986 from *Pseudomonas pseudoalcaligenes* strain KF707, which was isolated from soil in a biphenyl-manufacturing factory in Kitakyushu, Japan (19). The chromosomal DNA from KF707 was digested with the restriction endonuclease *Xho* I, and ligated into the unique *Xho* I site of a broad-host-range plasmid pKT230, and transformed into *Pseudomonas aeruginosa* PAO1161. Only 1 of 8000 transformants tested contained the chimeric plasmid pMFB1. This transformant was able to convert biphenyl and chlorobiphenyls to the ring-meta-cleavage yellow compounds (compound IV in Fig. 1) via dihydrodiols and dihydroxy compounds. The 6.8-kb *Xho* I fragment contained the genes *bphAB* and *C* (coding for biphenyl dioxygenase, dihydrodiol dehydrogenase, and

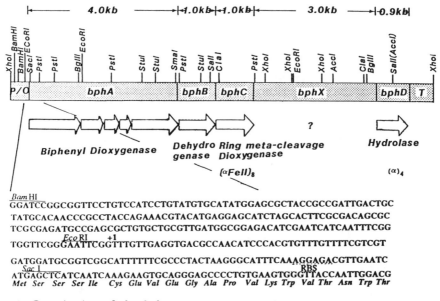

Figure 2. Organization of the *bph* operon in *P. pseudoalcaligenes* KF707 and nucleotide sequence of the upstream region of the gene. +1, Transcriptional start site; RBS, putative ribosome-binding site.

2,3-dihydroxybiphenyl dioxygenase, respectively). Subsequent to these findings, the bphD gene (coding for the meta-cleavage compound hydrolase) was cloned from P. pseudoalcaligenes KF707 (25). Escherichia coli JM109, carrying recombinant plasmid pFYD177 (pHSG396 containing Xho I 3.0-kb DNA, including bphD), readily converted the ring meta-cleavage yellow compound to benzoic acid. Subcloning, deletion analysis, and transposon mutagenesis of the cloned bph genes revealed that the bph operon of P. pseudoalcaligenes KF707 is organized as presented in Figure 2 (14). The first gene bphA starts just downstream of the Eco RI site and extends ca. 4 kb in length, followed by the bphB (ca. 1 kb) and the bphC gene (ca. 1 kb), respectively. An unknown 3,5-kb DNA segment (X) exists between bphC and bphD. The bph operon in P. pseudoalcaligenes KF707 is thus organized as bphABCXD.

Transposon Mutagenesis of the Cloned bphABC of Pseudomonas pseudoalcaligenes KF707 and Replacement of the Chromosomal bph Operon with Tn5-B21 Mutagenized bphABC Genes

Transposons are potentialy useful tools to generate various mutants, particulary for gram-negative bacteria, including Pseudomonas (8,32). The Tn5-B21 is a Tn5 derivative in which the neomycin-resistant (NmR) determinant is replaced with a RP4-tetracycline-resistant (TcR) determinant and the lacZ gene, which lacks the transcriptional start signals (33). Tn5-B21 carried by phage λ467 was used for mutagenesis of the bph genes of P. pseudoalcaligenes KF707. λ467 is replication- and integration-defective in a suppression-deficient Escherichia coli host such as strain CSH52. The CSH52 cells carrying recombinant plasmid pSUP102 bphABC were infected with λ467::Tn5-B21. A number of bphABC::Tn5-B21 mutant plasmids generated by this method were analyzed, and the results are presented in Figure 3 (16). Because Tn5-B21 was randomly transposed into the cloned KF707 bphABC, the bphA, bphB, and bphC loci could be aligned. The results showed that the KF707 bphA gene started within 500 bp of the left Xho I site and extended as far as 4 kb.

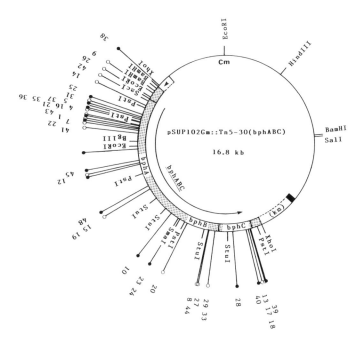

Figure 3. Transposon mutagenesis of the cloned bphABC(KF707). The replication and integration defective λ::Tn5-B21 was infected to E. coli carrying pSUP102Gm::Tn5-B30 (bphABC). A total of 44 Tn5-B21 mutant plasmids with respective transposon integration sites are shown.

Escherichia coli S17–1 cells (chromosomally integrated RP4–2-Tc::Mu-Km::Tn7), carrying Tn5-*B21*-mutagenized *bphABC* on a pSUP vector, were filter-mated with the parent strain *P. pseudoalcaligenes* KF707 cells. Since the pSUP vector cannot replicate in KF707, the Tc^R and Lac$^+$ colonies were screened for chromosomal recombination of the *bph* operon with simultaneous replacement with the Tn5-*B21* mutagenized *bph* DNA carried on the plasmid. The DNA region of sequence homology on both sides of the Tn5-*B21* insertion leads to a high frequency of recombination, which results in the insertion of only Tn5-*B21* into the *bph* operon (Fig. 4). These Tn5-*B21* mutants always show BP$^-$ phenotype.

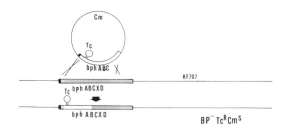

Figure 4. Replacement of chromosomal *bph* genes with the Tn5-*B21*-mutagenized *bphABC* carried on pSUP vector in *P. pseudoalcaligenes* KF707, resulting in the gene-specific transposon insertion.

To confirm the site-specific insertion of Tn5-*B21*, the genomic DNA from BP$^-$ KF707 mutants were analyzed by Southern blot hybridization, with the *Xho* I 6.8-kb *bphABC* DNA as a probe. Two hybridized bands were observed in these strains, since two *Xho* I sites exist in Tn5-*B21*, whereas only the 6.8-kb DNA band was hybridized in the parent KF707. The KF707 mutant strains KF730 (KF707 *bphA*::Tn5-*B21*#30), KF745 (KF707 *bphA*::Tn5-B21#15), KF748 (KF707 *bphB*::Tn5-*B21*#27), and KF744 (KF707 *bphC*::Tn5-*B21*#13) were used for the production of catabolic intermediates from 4-chlorobiphenyl (Fig. 5). KF730 and KF745 did not atttack 4-chlorobiphenyl, since *bphA* is disrupted by transposon insertion. The KF748 produced dihydrodiol. This compound is spontaneously converted to 2-hydroxy- and 3-hydroxycompounds by ethylacetate extraction in acidic pH. The KF744 converted the same compound to a dihydroxy compound.The KF707 mutant strains that accumulate catabolic intermediates from various PCB congeners were constructed with this procedure (16).

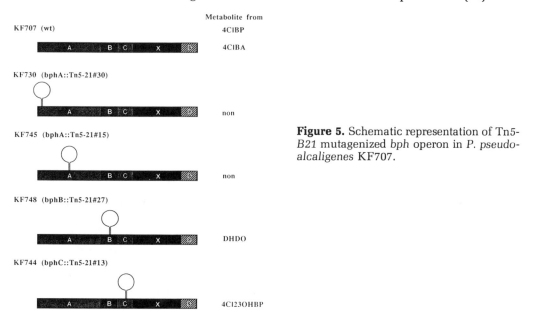

Figure 5. Schematic representation of Tn5-*B21* mutagenized *bph* operon in *P. pseudoalcaligenes* KF707.

The *bphABCD* Operon in *Pseudomonas putida* KF715

Another PCB-degrading strain, *Pseudomonas putida* KF715, grew well on biphenyl and degraded various PCBs to chlorobenzoates. By using *bphABC* and *bphD* from KF707 as probes, *bphABCD* genes were cloned from chromosomal DNA of *P. putida* KF715. Since *bphABC* (KF707) and *bphD* (KF707) probes were hybridized to a 9.4-kb fragment of the *Xho* I-digested chromosomal DNA of KF715, the DNA fragments of 8 to 10 kb in size were electroeluted and ligated into the unique *Xho* I site of pHSG396 and transformed to *Escherichia coli* JM109. The transformants that converted 2,3-dihydroxybiphenyl to a ring meta-cleavage yellow compound were screened. One such transformant carrying pYH715 (11.6 kb) contained the *Xho* I 9.4 kb DNA from *P. putida* KF715 and possessed the capability to convert 4-chlorobiphenyls to 4-chloro-benzoate, thus indicating that the cloned fragment carried the *bphABCD* genes. The cloned 9.4-kb DNA fragment was deleted by exonuclease III in the direction from *bphD* to *bphA*. Eight deletion mutants (Fig. 6) were selected, and products formed from 4-chlorobiphenyl were detected by thin layer chromatography. *E. coli* JM109, carrying either pYH11 or pYH12, converted 4-chlorobiphenyl to 4-chlorobenzoate. Therefore, all the *bphABCD* genes were coded with the 7.8-kb insert DNA of pYH12. *E. coli* JM109, carrying pYH13 or pYH14, yielded a meta-cleavage compound, thus indicating that *bphABC* genes were coded in a 6.5 kb insert DNA of pYH14. Similarly, *bphAB* genes were identified on the 6.0-kb insert of pYH15, yielding 2,3-dihydroxy compound, and the *bphA* gene was coded in the 5.0-kb insert DNA of pYH17, allowing the conversion of 4-chlorobiphenyl to dihydrodiol compound.

The order of *bph* genes in KF715 is thus *bphA-bphB-bphC-bphD*. Sequence analysis reveals that the extra DNA segment X that is observed in the *bph* operon of *Pseudomonas pseudoalcaligenes* KF707 did not exist in *P. putida* KF715. The plasmid pNHF715 was constructed by inserting a 9.4-kb *Xho* I fragment containing *bphABCD* into the unique *Xho* I site of a broad-host-range plasmid pKT230. The plasmid pNHF715 was introduced into various benzoate-utilizing bacteria by the triparental mating method; pRK2013 was used as a helper plasmid. Acquisition of this plasmid by

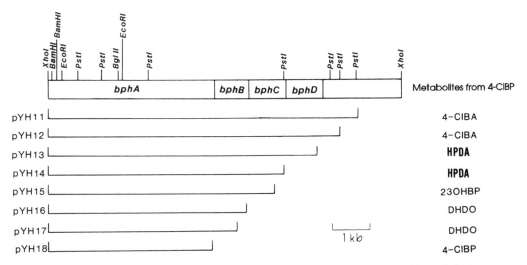

Figure 6. Deletion analysis of *bph* operon in *P. putida* KF715. Deletion plasmid designations are shown in the bottom left column. The bottom right column contains products from 4-chlorobiphenyl by *E. coli* JM109 carrying various deletion plasmids. Abbreviations of metabolites are the same as in Fig. 1.

three strains, *P. aeruginosa* PAO1161, *P. putida* AC30, and *Achromobacter xerosis* IFO12668, allowed the hosts to assimilate biphenyl as a sole carbon source. The recombinant strains degraded many PCB congeners to chlorobenzoates similar to *P. putida* KF715.

The *bphC* Gene from *Pseudomonas paucimobilis* Q1

Pseudomonas paucimobilis Q1 is the only American isolate in our collection of PCB-degrading strains (20). The Q1 grew on toluene as well as biphenyl. The same enzyme systems, via 2,3-dioxygenase, might be involved in toluene catabolism. The *bphC* gene, located on *Xho* I 2.2 kb DNA, was cloned from Q1 chromosomal DNA into *Pseudomonas aeruginosa* PAO1161 by using pKT230 (35).

COMPARATIVE STUDIES OF *bphC* GENES AND 2,3-DIHYDROXYBIPHENYL DIOXYGENASES IN SOME PCB-DEGRADING STRAINS

The *bphC* gene product, 2,3-dihydroxybiphenyl dioxygenase (23OHBPO), catalyzes the oxidation of 2,3-dihydroxybiphenyl to 2-hydroxy-6-oxo-6-phenylhexa-2,4-dienoic acid (Fig. 1., compound IV). This is the key reaction in the degradation of PCBs. The 23OHBPOs were first purified from the two PCB-degrading strains; *Pseudomonas pseudoalcalegenes* KF707 (11) and *P. paucimobilis* Q1 (35). Both enzymes were colorless and contained the ferrous form of iron as the sole cofactor, which is the typical iron state for the extradiol-cleaving enzymes. These enzymes possessed many features in common, such as molecular weight (ca., 260,000), subunit structure ($[\alpha\mathrm{FeII}]_8$), and substrate specificity (Table 1). Both enzymes were specific for 2,3-dihydroxybiphenyl, and they did not oxidize a positional isomer such as 3,4-dihydroxybiphenyl. Despite the close similarities between the two enzymes, some discrepancy existed in terms of substrate specificity; that is, although the 23OHBPO from strain Q1 oxidized catechol with a catalytic efficiency of almost one-quarter ($\frac{1}{4}$ Vmax) that of the natural substrate, 2,3-dihydroxybiphenyl, the catechol was nearly inert to the 23OHBPO from strain KF707. The 23OHBPO purified from the Q1 cells was immunologically different from the KF707 enzyme. Pertinent to this finding was the observation that a *bphC* DNA probe of the KF707 failed to hybridize with the corresponding *bphC* from the Q1 gene. To find the exact molecular differences between these immunologically antagonistic two

Table 1. Enzymatic properties of *bphC* gene products in *P. pseudoalcaligenes* KF707 and *P. paucimobilis* Q1.

	23OHBPO(KF707)	23OHBPO(Q1)
Mol wt	260,000	260,000
Subunit mol wt	33,000	33,000
Cofactor	Fe[II]	Fe[II]
Structure	$(\alpha\mathrm{Fe[II]})8$	$(\alpha\mathrm{Fe[II]})8$
Vmax(s^{-1})		
Substrate		
23OHBP	330	150
34OHBP	1	1
Catechol	3	33

bphC genes, their nucleotide sequences were determined (12,35). The results showed an overall protein homology of 38% between 23OHBPO of KF707 (297 amino acids) and 23OHBPO of Q1 (298 amino acids) (Fig. 7). The *bphC* gene from *Peudomonas putida* KF715 was also sequenced. The DNA-derived protein sequence indicated a primary structure of 292 amino acids and an overall homology betwen 23OHBPOs from *P. pseudoalcaligenes* KF707 and *P. putida* as high as 91.4% (25). The 23OHBPO purified from KF715 showed the same octamer structure of $(\alpha Fe[II])_8$, and cross reacted with the antiserum raised against 23OHBPO(KF707) but formed a spur with the KF707 enzyme, thus indicating that these two enzymes are partially homologous immunologically. The *bphC* gene of *Pseudomonas* sp. KKS102 sequenced by Kimbara et al. (27) shows 65.1 % homology with that of *P. pseudoalcaligenes* KF707 at the amino acid level. This enzyme cross-reacted weakly with the antiserum of the KF707 enzyme.

```
                    25                      50                          75
KF707    --SIRSLGYMGFAVSDVAAWRSFLTQKLGLMEAGTTDNGD---L-FRIDSRAWRIAVQQGEV-DDLAFAGYEVADAAGLAQM
         :*:  *      * ::::     *  ::     ::   *    *   ::::  *  *:*       :
Q1       -VAVTELGYLGLTVTNLDAWRSYAAEVAG-MEIVDEGEGDRLYL--RMDQWHHRI-VLHASDSDDLAYLGWRVADPVEFDAM
         :    *    :  *        :     :       : ::*       *:   :         *   * ***
Catechol NKGVMRPGHVQLRVLDMSKALEHYVELLG-LIEMDRDDQGRVYLKAWTEVDKFSL-VLREADEPGMDFMGFKVVDEDALRQL

                    100                     125                         150
KF707    ADKLKQAG--IAVTTGDASLARRRGVTGLITFADPFGLPLEIYYGASEVFEK----PFLPGAAVSG-FLTGEQGLGHFVRCV
         :*  :*  :  *       ::  *  *    ::* *::  *:         *    :: *      :  : :   :  *
Q1       VAKLTAAG--ISLTVASEAEARERRVLGLAKLADPGGNPTEIFYGPQVDTHK----PFHPGRPMYGKFVTGSEGIGHCILRQ
         *  :   :      ::   :  :*:*       * *  *  : *       *   :         *  :*   *
Catechol ERDLMAYGCAVEQLPAGELNSCGRRVRFQA----PSGHHFE-LYADKEYTGKWGLNDVNPEAWPRDLKGMAAVRFDHALMYG

                    175                     200                         225
KF707    PDSDKAL--AFYTDVLGFQLSDVIDMKMGPDVTV-PVYFLHCNERHHTLAIAAFPLPKRIHHFMLEVASLDDVGFAFDRVDA
         *   :    :*  :   *   :*:   : :::: *    :*  : :* ***   *    ::::   *  * :**:   *
Q1       DDVPAAA--AFY-GLLGLRGS-VEYHLQLPNGMVAQPYFMHCNERQHSVAFGLGPMEKRINHLMFEYTDLDDLGLAHDIVRA
         *   *    :  *    **    :  :      *::  * *   * :*:   *   :  *   :   *:  * *  *  * *
Catechol -DELPATYDLFT-KVLG-FYL-AEQVLDENGTRVAQ--FLSLSTKAHDVAFIHHPEKGRLHHVSFHLETWEDLLRAADLISM

                    250                     275                         300
KF707    -DGLITSTLGRHTNDHMVSFYASTPSGVEVEYGWSARTV-DRSWVVV-RHDSPS--MWGHKSVRDKAAARNKA
         :: *  ::   :*  ***  *  :: ::          *         :*       :
Q1       RKIDVALQLGKHANDQALTFYCANPSGWLWEFGWGARKA-PSQQEYY-TRD-----IFGH--GNEAAGYGMDIPLG
         *   *    ***  *     *  *  : :   : :      *   *  :  ::  *   *
Catechol TDTSIDIGPTRHGLTHGKTIYFFDPSGNRNEVFCGGDYNYPDHKPVTWTTDQLGKAIFYH--DRILNERFM-TVLT
```

Figure 7. Comparison of amino acid sequences of 2,3-dihydroxybiphenyl dioxygenase (KF707), 2,3-dihydroxybiphenyl dioxygenase (Q1), and catechol 2,3-dioxygenase. Identical amino acids are indicated by *colons,* and the *double asterisks* represent conseved amino acid residues among the three dioxygenases.

COMPARISON OF *bph* OPERONS IN VARIOUS PCB-DEGRADING BACTERIA

As described above, the 2,3-dihydroxybiphenyl dioxygenases (*bphC* gene products) from four *Pseudomonas* strains show various degrees of homology at the level of amino acid sequence despite the close similarity of enzymatic properties. In this regard, a number of biphenyl/PCB-degrading bacteria isolated from soils at different locations were investigated to determine how chromosomal *bph* genes are distributed and conserved in various bacteria (15). Biphenyl-utilizing strains studied are listed in Table 2.

Table 2. PCB-degrading strains and their growth characteristics on various biphenyl derivatives.

Strain	Substrates[a]						
	BP	4ClBP	4MeBP	2BrBP	2NO$_2$BP	OPP	DM
Achromobacter xylosoxidans KF70	+++	−	+	−	−	++	−
Pseudomonas sp. KF702	+++	−	+	−	−	−	−
Pseudomonas fluorescens KF703	+++	−	−	−	−	−	+
Moraxella sp. KF704	+++	−	−	−	−	−	+
Pseudomonas paucimobilis KF706	+++	−	+	+	+	−	++
Pseudomonas pseudoalcaligenes KF707	+++	−	+	−	−	−	−
Alcaligenes sp. KF708	+++	+	++	++	++	−	++
Unidentified KF709	+++	−	−	−	−	−	+
Pseudomonas sp. KF710	+++	−	−	−	−	−	−
Alcaligenes sp. KF711	+++	−	−	−	−	−	−
Pseudomonas sp. KF712	+++	−	−	−	−	−	−
Pseudomonas stutzeri KF713	+++	−	−	−	−	−	−
Pseudomonas sp. KF714	+++	+	+	−	−	−	+
Pseudomonas putida KF715	+++	+	+	−	−	−	+
Pseudomonas paucimobilis Q1	+++	+	+	−	+	++	+
Arthrobacter sp. M5	+++	+	++	+	+	−	−

[a]Growth was checked 1 week incubation at 30°C. +++, substantial growth; ++, moderate growth; +, poor growth; −, no growth or very poor growth. BP, biphenyl; 4ClBP, 4-chlorobiphenyl; 4MeBP, 4-methylbiphenyl; 2BrBP, 2-bromobiphenyl; 2NO$_2$BP, 2-nitrobiphenyl; OPP, o-phenylphenol; DM, diphenylmethane.

Pseudomonas pseudoalcaligenes KF707 and *P. putida* KF715 were isolated from soil in Kitakyushu, Japan, and *P. paucimobilis* Q1 was isolated from soil in Chicago, Illinois. *Arthrobacter* sp. M5 was isolated in Chiba, Japan, and was the only gram-positive strain used in this study. The other 12 strains were isolated from various locations in Japan. The newly isolated strains were all gram-negative, including eight species of *Pseudomonas,* two of *Alcaligenes,* and one each of *Achromobacter* and *Moraxella.* The strains all utilized biphenyl as the sole source of carbon and energy. They converted 4-chlorobiphenyl to 4-chlorobenzoic acid and produced 2,3-dihydroxy-biphenyl dioxygenases inducibly, so their catabolism of biphenyl and PCBs proceeded through the major oxidative route as illustrated in Figure 1.

The growth characteristics of the various aromatic compounds of all PCB-degraders used in this study are presented in Table 2. Some strains grew on biphenyl derivatives, such as 4-chloro-, 4-methyl-, 2-bromo-, 2-nitrobiphenyl, and/or diphenylmethane. All the cellular DNA isolated from each biphenyl-utilizing strain was digested with various restriction endonucleases. The genetic homology was examined by Southern blot hybridization, with bphA, bphC, bphABC, and bphD of *P. pseudoalcaligenes* KF707 used as probes. Homologous DNA segments were observed for 10 out of 16 strains, including KF707 itself (Fig. 8). Six strains—KF702, KF703, KF710, KF711, KF713, and KF714 (Table 2)—showed bph operons very similar, if not identical, to bphABCXD of KF707. The DNAs flanking the bph genes were examined by digesting with *Eco* RI, *Sal* I or *Sma* I. These results revealed that the outside bph DNA regions in these PCB strains were occupied by unrelated DNA. The other two strains, KF701 and KF715, carried homologous bph DNA on a 9.4 kb DNA fragment that included bphABCD genes, as described above. Differences in the flanking regions of bph genes of KF701 and KF715 were also observed. KF706 also possesses DNA homologous to KF707 bph genes (Fig. 8). No significant genetic homology could be observed for six strains of KF704, KF708, KF709, KF712, M5, and Q1. All 16 strains harbored one to three

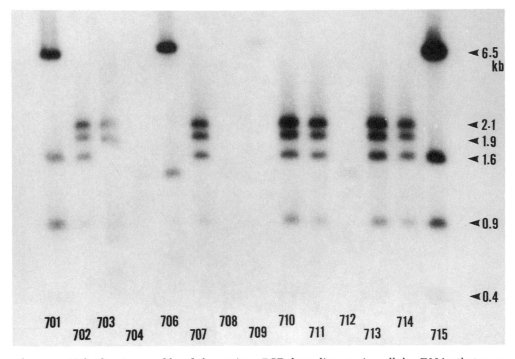

Figure 8. Hybridization profile of the various PCB-degrading strain cellular DNAs that were double-digested with *Xho* I and *Pst* I. The 6.8 kb *Xho* I fragment including *bphABC* genes of *P. pseudoalcaligenes* was used as a probe.

plasmids up to ca. 150 kb in size, but *bph* probes of KF707 did not hybridize with any plasmid DNAs.

The antiserum raised against 2,3-dihydroxybiphenyl dioxygenase (*bphC* gene product) of *P. pseudoalcaligenes* KF707 was used to examine the immunological cross reactivity of the enzymes from various biphenyl-utilizing strains. The immunological reactivity corresponded well to DNA homology (Fig. 9). The enzymes prepared from six strains (KF702, KF703, KF710, KF711, KF713, and KF714) that possessed the homologous DNA on the same *Xho* I 6.8-kb fragment as that of the KF707 showed clear fused precipitin bands against the KF707 antiserum, without forming a spur with the KF707 enzyme. The 2,3-dihydroxybiphenyl dioxygenases from KF701, KF715, and KF706 that possessed the homologous *bph* DNA on different fragments showed distinct precipitin bands forming a spur, thus indicating that these enzymes are partially homologous immunologically with the KF707 enzyme. The enzymes from KF704 and KF712 showed a weak precipitin band and formed a spur. No immunological cross-reactivity was found for the enzymes from KF708, KF709, M5, and Q1. Western blot analysis demonstrated that the 23OHBPO subunits from strains KF701, KF702, KF703, KF710, KF711, KF713, KF714, and KF715 all showed a molecular mass (ca. 33 kDa) similar to that of KF707. In contrast to KF707 *bph* genes, the *bphC* gene of *P. paucimobilis* Q1 did not hybridize with any other biphenyl-utilizing strains tested, and the antiserum raised against 23OHBPO(Q1) did not cross react with any enzymes from the above strains.

Biphenyl/PCB-degrading bacteria are ubiquitously distributed in the environment. Some biphenyl-catabolic genes are conserved as seen in the chromosomal *bphABCXD* genes of *Pseudomonas pseudoalcaligenes* KF707 among soil bacteria. What is the

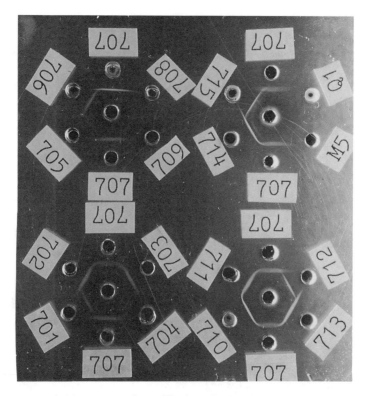

Figure 9. Immunoprecipitin pattern of 2,3-dihydroxybiphenyl dioxygenases from various PCB-degrading strains with antiserum prepared against the purified 2,3-dihydroxybiphenyl dioxygenase of *P. pseudoalcaligenes* KF707. The KF707 2,3-dihydroxybiphenyl dioxygenase antiserum was placed in the central wells. The crude cell extracts obtained from biphenyl-grown cells were placed in the wells surrounding the antiserum.

primary function of *bph* genes? At present, we presume that *bph* genes are involved in the degradation of plant lignin, which is widely and massively distributed in the environment. Wood-rotting microorganisms such as white rot fungi first decompose the complicated lignin molecules into smaller molecules, such as biphenyl and its related compounds. Biphenyl/PCB-degrading soil bacteria then attack the various biphenyl derivatives. If the *bph* genes are thus involved in the latter stage of lignin degradation, the *bph* genes could have a common ancestry, and could have evolved over a long period. The divergence of *bph* genes in various biphenyl/PCB-degrading strains might reflect this hypothesis. On the other hand, Certain *bph* operons may have a mechanism of transfer into other soil bacteria, as can be seen in the *bphABCXD* operon of *P. pseudoalcaligenes* KF707. *bph* genes responsible for the conversion of PCBs to chlorobenzoic acids have recently been cloned in several laboratories (1,26,27,29). Investigating these *bph* genes from the viewpoint of molecular ecology and molecular evolution should be interesting.

Acknowledgments

I thank my colleagues, particularly, T. Miyazaki, N. Arimura, S. Yamashita, T. Okamoto, N. Hayase, H. Kato, and K. Taira for their great contributions to this research.

LITERATURE CITED

1. Ahmad, D., R. Masse, and M. Sylvestre. 1990. Cloning and expression of genes involved in 4-chlorobiphenyl transformation by *Pseudomonas testosteroni*: homology to polychlorinated biphenyl-degrading genes in other bacteria. *Gene* 86:53–61.

2. Ahmed, M, and D. D. Focht. 1973. Degradation of polychlorinated biphenyls by two species of *Achromobacter*. *Can. J. Microbiol.* 19:47–52.

3. Bedard, D. L., M. J. Brennan, and R. Unterman. 1984. Bacterial degradation of PCBs: evidence of distinct pathways in *Corynebacterium* sp. MB1 and *Alcaligenes eutrophus* H850. In *Proceedings of the 1983 PCB Seminar*. Ed. G. Addis, R. Komai. Palo Alto, California: Electrical Power Reserch Institute. 4-101–118.

4. Bedard, D. L., M. L. Haberl, R. J. May, and M. J. Brennan. 1987. Evidence for novel mechanisms of polychlorinated biphenyl metabolism in *Alcaligenes eutrophus* H850. *Appl. Environ. Microbiol.* 53:1103–1112.

5. Bedard, D. L., R. Unterman, L. H. Bopp, M. J. Brennan, M. L. Haberl, and C. Johnson. 1986. Rapid assay for screening and characterizing microorganisms for the ability to degrade polychlorinated biphenyls. *Appl. Environ. Microbiol.* 51:761–768.

6. Bedard, D. L., R. E. Wagner, M. J. Brennan, M. L. Haberl, and J. F. Brown, Jr. 1987. Extensive degradation of Aroclors and environmentally transformed polychlorinated biphenyls by *Alcaligenes euthrophus* H850. *Appl. Environ. Microbiol.* 53:1094–1102.

7. Bopp, L. H. 1986. Degradation of highly chlorinated PCBs by *Pseudomonas* strain LB400. *J. Ind. Microbiol.* 51:761–768.

8. De Bruijin, F. J., and J. R. Lupski. 1984. The use of transposon Tn5 mutagenesis in the rapid generation of correlated physical and genetic maps of DNA segments cloned into multicopy plasmids—a review. *Gene* 27:131–149.

9. Furukawa, K. 1982. Microbial degradation of polychlorinated biphenyls. In *Biodegradation and Detoxification of environmental pollutants*. Ed. A. M. Chakrabarty. Boca Raton, Florida: CRC Press, Inc. 33–57.

10. Furukawa, K. 1986. Modification of PCBs by bacteria and other microorganisms. In *PCBs and the Environment*. Ed. J. Waid. Boca Raton, Fla.: CRC Press Inc. 89–99.

11. Furukawa, K., and N. Arimura. 1987. Purification and properties of 2,3-dihydroxybiphenyl dioxygenase from polychlorinated biphenyl-degrading *Pseudomonas pseudoalcaligenes* and *Pseudomonas aeruginosa* carrying the cloned *bphC* gene. *J. Bacteriol.* 169:924–927.

12. Furukawa, K., N. Arimura, and T. Miyazaki. 1987. Nucleotide sequence of the 2,3-dihydroxybiphenyl dioxygenase gene of *Pseudomonas pseudoalcaligenes*. *J. Bacteriol.* 169:427–429.

13. Furukawa, K., and A. M. Chakrabarty. 1982. Involvement of plasmids in total degradaton of chlorinated biphenyls. *Appl. Environ. Microbiol.* 44:619–626.

14. Furukawa, K., N. Hayase, and K. Taira. 1990. Biphenyl/polychlorinated biphenyl catabolic gene (*bph* operon): organization, function, and molecular relationship in various pseudomonads. In *Pseudomonas*. Ed. S. Simon, A. M. Chakrabarty, B. Iglewski, S. Kaplan. Washington, D.C.: American Society for Microbiology. 111–120.

15. Furukawa, K., N. Hayase, K. Taira, and N. Tomizuka. 1989. Molecular relationship of chromosomal genes encoding biphenyl/polychlorinated biphenyl catabolism: some soil bacteria possess a highly conserved *bph* operon. *J. Bacteriol.* 171:5467–5472.

16. Furukawa, K., S. Hayashida, and K. Taira. 1991. Gene-specific transposon mutagenesis of the biphenyl/polychlorinated biphenyl-degradation-controlling *bph* operon in soil bacteria. *Gene* 98:21–28.

17. Furukawa, K., and F. Matsumura. 1976. Microbial metabolism of polychlorinated biphenyls. Studies on the relative degradability of polychlorinated biphenyl components by *Alcaligenes* sp. *J. Agric. Food Chem.* 42:543–548.

18. Furukawa, K., F. Matsumura, and K. Tonomura. 1978. *Alcaligenes* and *Acinetobacter* strains capable of degrading polychlorinated biphenyls. *Agric. Biol. Chem.* 42:543–548.

19. Furukawa, K., and T. Miyazaki. 1986. Cloning of a gene cluster encoding biphenyl and chlorobiphenyl degradation in *Pseudomonas pseudoalcaligenes*. *J. Bacteriol.* 166:392–398.

20. Furukawa, K., J. R. Simon, and A. M. Chakrabarty. 1983. Common induction and regulation of biphenyl, xylene/toluene, and salicylate catabolism in *Pseudomonas paucimobilis*. *J. Bacteriol.* 154:1356–1362.

21. Furukawa, K., N. Tomizuka, and A. Kamibayashi. 1979. Effect of chlorine substitution on the bacterial metabolism of various polychlorinated biphenyls. *Appl. Environ. Microbiol.* 38:301–310.

22. Furukawa, K., N. Tomizuka, and A. Kamibayashi. 1983. Metabolic breakdown of Kaneclors (polychlorinated biphenyls) and their products by *Acinetobacter* sp. *Appl. Environ. Microbiol.* 46:140–145.

23. Furukawa, K., K. Tonomura, and A. Kamibayashi. 1978. Effect of chlorine substitution on the biodegradability of polychlorinated biphenyls. *Appl. Environ Microbiol.* 35:223–227.

24. Furukawa, K., K. Tonomura, and A. Kamibayashi. 1979. Metabolism of 2,4,4'-trichloro-biphenyl by *Acinetobacter* sp. P6. *Agric. Biol. Chem.* 43:1577–1583.

25. Hayase, N., K. Taira, and K. Furukawa. 1990. *Pseudomonas putida* KF715 *bphABCD* operon encoding biphenyl and polychlorinated biphenyl degradation: cloning, analysis, and expression in soil bacteria. *J. Bacteriol.* 172:1160–1164.

26. Khan, A., and S. Walia. 1989. Cloning of bacterial genes specifying biphenyl and chloro-biphenyl degradation in *Pseudomonas putida* OU83. *Appl. Environ. Microbiol.* 55:798–805.

27. Kimbara, K., T. Hashimoto, M. Fukuda, T. Koana, M. Takagi, M. Oishi, and K. Yano. 1989. Cloning and sequencing of two tandem genes involved in degradation of 2,3-dihydroxybiphenyl to benzoic acid in the polychlorinated biphenyl-degrading bacterium *Pseudomonas* sp. strain KKS102. *J. Bacteriol.* 171:2240–2747.

28. Masse, R., F. Messier, L. Peloquin, C. Ayotte, and M. Sylvestre. 1984. Microbial biodegradation of 4-chlorobiphenyl, a model compound of chlorinated biphenyls. *Appl. Environ. Microbiol.* 47:947–951.

29. Mondello, F. J. 1989. Cloning and expression in *Escherichia coli* of *Pseudomonas* strain LB400 genes encoding polychlorianted biphenyl degradation. *J. Bacteriol.* 171:1725–1732.

30. Riseborough, R. W., P. Reiche, D. B. Peakall, S. G. Hersman, and M. N. Kirven. 1968. Polychlorinated biphenls in the global ecosystem. *Nature* 220:1098–1102.

31. Sayler, G. S., R. Thomas, and R. R. Colwell. 1978. Polychlorinated biphenyl (PCB) degrading bacteria and PCB in esturiane and marine environment. *Estuarine Coastal Mar. Sci.* 6:553–567.

32. Simon, R., M. O'Connell, M. Laves, and A. Pühler. 1986. Plasmid vecters for the genetic analysis and manipulation of rhizobia and other gram-negative bacteria. *Methods Enzymol.* 118:640–659.

33. Simon, R., J. Quandt, and W. Klipp. 1989. New derivatives of transposon Tn5 suitable for mobilization of replicons, generation of operon fusions and induction of genes in gram-negative bacteria. *Gene* 80:161–169.

34. Sylvestre, M. 1985. Total degradation of 4-chlorobiphenyl (PCB) by a two-membered bacterial culture. *Appl. Microbiol. Biotechnol.* 21:192–195.

35. Taira, K., N. Hayase, N. Arimura, S. Yamashita, T. Miyazaki, and K. Furukawa. 1988. Cloning and nucleotide sequence of the 2,3-dihydroxybiphenyl dioxygenase gene from PCB-degrading strain of *Pseudomonas paucimobilis* Q1. *Biochemistry* 27:3990–3996.

CHAPTER 3

Selection of Enhanced Polychlorinated Biphenyl–degrading Bacterial Strains for Bioremediation: Consideration of Branching Pathways

MICHEL SYLVESTRE and MOHAMMAD SONDOSSI

Abstract. One of the prerequisites of applying bioremediation technology to eliminate polychlorinated biphenyls (PCBs) in contaminated soils is the development of microbial strains with enhanced biodegradation capacities. A great deal of information has been obtained since the late 1970s about PCB degradation pathways. Empirical methods such as enrichment were used to develop bacteria with enhanced capacity to degrade PCBs; however, the PCB congeners carrying more than six chlorines, particularly the *ortho*-substituted congeners, remain resistant to biodegradation under aerobic conditions. The same four-step pathway is used in many bacteria for the conversion of biphenyl (BP) and PCBs into corresponding chlorobenzoates (CBAs). The selectivity pattern toward substrates of the (BP) 2,3-oxygenase, which is the first enzyme of the pathway, is among the most important factors limiting PCB degradation. Accumulation of some metabolites, however, such as *meta*-cleavage compounds and chlorobenzoates, and transformation of these metabolites through nonspecific branching pathways can also affect the flux of the BP/PCB degradation pathway. Therefore, in future programs designed to select bacteria with enhanced ability to degrade PCBs, and considering that multiple pathways exist to convert PCBs and CBAs, we believe that only strategies that introduce gradual modifications in the overall flux of the pathway may successfully improve the ability of strains to degrade these substrates.

INTRODUCTION

There has been a growing interest in the use of microorganisms as a means for *in situ* cleanup of contaminated sites. Two approaches can be envisioned for the application of this method: either the contaminated sites can be inoculated with laboratory-trained bacterial populations, or microbial activities toward the substrate to be degraded can be enhanced *in situ* by addition of appropriate nutrients and inducers.

Drs. Sylvestre and Sondossi are at the Université du Québec, Institut National de la Recherche Scientifique, INRS-Santé, Pointe-Claire, Québec, Canada. Dr. Sondossi may be reached at the Department of Biological Sciences, Weber State University, Ogden, UT 84408-2506, U.S.A.

Polychlorinated biphenyls are very persistent in the environment (69, 126). No biological process has yet been successfully applied for the restoration of sites contaminated with high chlorine content PCBs (33, 86, 154), but the progress in research on biodegradation of PCBs has been sufficiently encouraging to lead investigators to believe that bacterial strains with enhanced capacity to degrade these pollutants can be developed. Once such strains are developed, it is essential that they can survive in the soil to be treated. Some of the factors affecting survival of implanted bacteria include predation, competition, resistance to starvation, motility, nutrient (substrate and cofactors) concentration and availability, presence of growth inhibitors, and physical factors such as O_2 concentration, pH, and temperature (15). These factors act either individually or synergistically to exert their effects.

Focht and Brunner (42) and Brunner et. al. (22) have had some success in enhancing endogenous populations to degrade lightly chlorinated PCBs such as Aroclor 1242. When applied to highly chlorinated PCB mixtures, however, the chances of success of the alternate approach of enhancing the indigenous population to degrade PCBs *in situ* are not as good, because the evolution of new catabolic capacities requires acquisition of new or modified enzymatic activities from preexisting enzymes through repeated mutation and selection (145). This evolutionary process can be extremely slow, especially when acquisition of multiple enzymatic steps is required to constitute new pathways.

Although the technology for PCB bioremediation does not yet exist, many features of PCB degradation pathways in bacteria are now known. This knowledge can help in designing novel strategies for the selection of better-performing strains and their application for the bioremediation of PCB-contaminated sites. In this review, we summarize some of the known features of PCB degradation pathways and discuss some of the strategies that can be used for the selection of organisms with enhanced abilities to degrade PCBs.

GENERAL FACTS ABOUT PCBs

PCBs were used extensively in industrial applications, particularly as lubricants, coolants, and electric insulants (68). Between 1930 and 1979, over 600 million kg of PCBs were used in North America alone (86). The general structure of PCBs is shown in Figure 1. Based on the number and position of chlorine atoms on the biphenyl ring, 209 congeners theoretically can be produced. Commercially, PCBs were prepared by catalytic chlorination of biphenyl. This procedure resulted in complex mixtures containing congeners with various chlorine contents (69). PCBs are known by a variety of trade names, such as Aroclor (USA), Kanechlor (Japan), Phenochlor (France), Fenchlor (Italy), Clophen (FRG), and Sovol (the former USSR). The Aroclor products have generally been used in biological studies. Aroclors are characterized by four-digit codes (68). The last two digits of the code give the weight percentage of chlorine in the mixture. Aroclor 1254 and Aroclor 1260 (containing 54% and 60% chlorine by weight, respectively) were used in large amounts in North America. The low chemical reactivity of PCBs and their resistance to microbial metabolic activities make them persistent environmental pollutants (24, 119, 139). It is important to remember that attempts to transform PCBs, whether through chemical or biological means, should address the individual components of the mixtures, and that higher chlorinated components are the most stable.

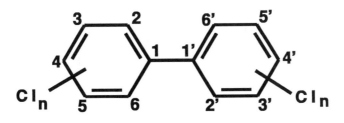

Figure 1. General structure of PCBs.

PCB TOXICITY

Low-level exposure to PCBs has not been shown to cause serious acute toxicity in humans (38), and the relation of low-level exposure to cancer is unclear (19). In all reported cases of human PCB intoxication, direct involvement of PCBs has never really been shown, because the commercial PCB preparations responsible for these intoxications were also contaminated by dioxins and furans (76). The most serious problem from PCB toxicity is related to its general effect on the ecosystem. Because the responses of various species vary greatly, PCBs can cause unbalances of the food chain, since some of the most nutritionally valuable species may count among the most sensitive.

Studies done with individual PCB congeners indicate that PCB toxicity is structure-related (138). The few congeners (priority group 1a, Table 1) sterically resembling 2,3,7,8-tetrachlorodibenzo-p-dioxin, with no *ortho* substituents, are directly toxic (75, 91, 97, 107). Oxygenase attack on these congeners in a 2,3-position produces stable epoxide derivatives that are not further detoxified through conjugation with endogenous substrates in Phase II metabolism (97). These few congeners with a complete co-planar configuration, as well as other partially co-planar forms (priority group 1b, Table 1), may also be involved in toxicity indirectly by a 3-methylcholanthrene-type induction of the aryl hydrocarbon mixed-function oxidases (MFOs) (32, 97). These toxic congeners carry between 5 and 7 chlorine atoms, mostly in *para* and *meta* positions. When the molecule looses its possible co-planar configuration, the oxygenase attack is on the 3,4-position (124). Although other factors such as polarization may also influence the interaction of PCBs with biological material (7), the three-dimensional structure of the molecule can certainly play an important role. Based on the likelihood of toxicity to animals and the prevalence of individual PCB congeners in the environment, it is now generally believed that only 15 to 20 congeners are of real concern (36, 71, 97). Table 1 lists these priority congeners.

Sensitivities toward PCB mixtures such as Aroclor vary greatly among species of microorganisms (16, 18, 78, 90, 150). The effects of individual congeners on microbial species, however, particularly the priority groups of PCB congeners, have yet to be reported. In the light of toxicity studies done with animals, one might expect that the effect of a PCB mixture on microbial growth depends upon the composition of the mixture in terms of amounts and types of congeners present; however, some of the metabolites that are produced from PCBs by microbial conversion of these compounds also are likely to affect microbial growth. These biological interactions should thus be considered among the factors that may affect biological processes involved in PCB degradation.

Table 1. List of the most potent PCB congeners.[a]

Group 1a[b]		Group 1b[c]		Group 2[d]	
IUPAC No.	Structure	IUPAC No.	Structure	IUPAC No.	Structure
77	3,3',4,4'	105	2,3,3',4,4'	87	2,2',3,4,5'
126	3,3',4,4',5	118	2,3',4,4',5	99	2,2',4,4',5
169	3,3',4,4',5,5'	128	2,2',3,3',4,4'	101	2,2',4,5,5'
		138	2,2',3,4,4',5'	153	2,2',4,5,5'
		156	2,3,3',4,4',5'	180	2,2',3,4,4',5,5'
		170	2,2',3,3',4,4',5	183	2,2',3,4,4',5',6
				194	2,2',3,3',4,4',5,5'

[a]The PCB congeners are classified following McFarland and Clarke (97) priority group.
[b]Group 1a includes the three most potent congeners (pure 3-methylcholanthrene-type inducers).
[c]Group 1b comprises mixed type inducers that have been reported frequently in environmental samples.
[d]Group 2 comprises phenobarbital-type inducers that have high abondances in avian and mammalian samples.

MICROBIAL BIODEGRADATION PATHWAYS FOR PCBs

Several studies on the microbial degradation of commercial PCB mixtures have been done (12, 31, 55, 73, 128, 146, 153). Because of the complexity of these mixtures, differentiating between biodegradation and bioaccumulation in bacteria was sometimes difficult (74, 127), and some of the alkane metabolites found in growth media that have been used as proof for PCB biodegradation (73) might have been produced instead from contaminants in the preparation (123). These studies clearly showed, however, that certain patterns of chlorine substitution seriously hinder PCB degradation. Only lightly chlorinated PCB congeners of commercial mixtures are truly degraded by axenic or mixed bacterial populations under aerobic conditions (6, 10, 12, 14, 47, 49, 50, 55, 79, 108, 140, 146).

The sequential enzymatic steps involved in the microbial transformation of PCBs were deduced from studies mainly involving lightly chlorinated biphenyls (4–6, 45, 94, 155). A few systematic studies involving more highly chlorinated PCB congeners, however, have helped to classify the various congeners on the basis of their resistance to biodegradation and allowed grouping of bacterial strains on the basis of their substrate selectivity patterns (11, 14, 46, 52).

The major PCB catabolic pathway found in bacteria is shown in Figure 2. The four structural genes required for transformation of chlorobiphenyls into corresponding chlorobenzoic acids have been cloned from several bacteria (3, 47, 49, 79, 80, 100). The molecular biology of PCB degradation is discussed in other chapters of this book. The features we would like to point out in this review are (a) some of the biphenyl (BP) catabolic genes appear to have been conserved in bacteria, but other bacteria carry distinctly different genes (3, 49, 59, 81, 144, 156); and (b) PCB catabolic genes appear to be organized in an operon (3, 49, 80, 100).

Because BP/PCB genes are conserved in many strains and are organized in an operon, Furukawa et al. (49) and Ahmad et al. (3) have suggested that PCB-degrading

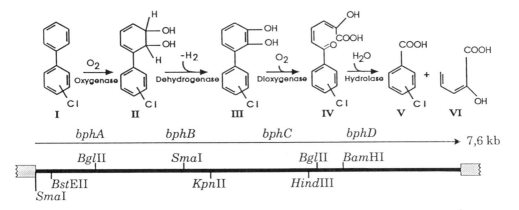

Figure 2. Major steps in the conversion of PCBs into chlorobenzoates and genes controlling their conversion in *P. testosteroni* B-356.

pathways in soil bacteria have evolved for the degradation of BP analogs produced in nature from naturally occurring compounds. The observation that hydroxylated biphenyls, which are generally very toxic, are also degraded by this pathway (136) provided more support for this hypothesis. Moreover, high levels of homology between the genes encoding the 3-methylcatehol 2,3-dioxygenase from *Pseudomonas putida* F1 and genes encoding 2,3-dihydroxybiphenyl 1,2-dioxygenase from *Pseudomonas pseudoalcaligenes* KF707 suggest that some of the bacterial biphenyl and toluene pathways may have evolved from a common ancestor (160). A recent statistical study comparing the nucleotide sequence of *nahC* (1,2-dihydroxynaphtalene dioxygenase) and *nahH* (catechol 2,3-dioxygenase) genes with *bphC* genes from *Pseudomonas alcaligenes* and from *Pseudomonas paucimobilis* clearly showed an evolutionary relationship between these oxygenases (56). It is thus likely that BP/PCB degradation pathways evolved in bacteria from degradation pathways involved in the degradation of aromatic compounds having some homology to BP, perhaps compounds derived from lignin or from other aryl hydrocarbons found in petroleum. These pathways would generally have had a capacity to transform a large array of chemical analogs. The transformation of PCB congeners through these pathways would be a fortuitous event resulting from the relaxed substrate specificity of the enzymes in the pathways.

Although the genes encoding the BP *meta*-cleavage enzyme appear to have evolved from a common ancestor (56, 160), the phylogenetic origins of the first oxygenase have not been established. Based on the substrate selectivity pattern of PCB-degrading strains, it is likely that different lines of BP oxygenase have emerged through evolution. Some observations have yet to be explained, however. For example, the PCB-degrading operon from *Pseudomonas testosteroni* strain B-356 hybridized with *Pseudomonas putida* strain LB-400 DNA, showing about 40% homology (3), even though metabolic studies, as shown below, suggested that these two strains degrade *ortho*-substituted PCB congeners very differently and their chlorine substitution pattern selectivity differs greatly. More precise molecular biology studies and comparisons of the DNA sequences of all genes involved in PCB degradation from various strains may answer these questions.

Whatever its phylogenetic origin, BP oxygenase encoded by the *bphA* gene is certainly a key component determining the biodegradability of the various PCB congeners. The position and number of chlorines on the molecule can influence the rate of the first oxygenase reaction in several ways. The chlorines may prevent access to the enzyme's active site because of steric hindrance (variation in the three-dimensional conformation of the molecule or bulkiness of the chlorine atoms). Chlorine atoms can also prevent oxygenation if they occupy the carbon positions that are the most susceptible to the oxygenase attack (14). Polarization of the molecule has also been shown to affect the biological interaction of PCBs (7), therefore this factor might also affect the rate of the first oxygenase reaction.

The ortho-substituted PCB congeners, which are the least toxic, are also the most resistant to microbial attack under aerobic conditions (11, 52) and anaerobic conditions (20, 21, 110, 111).

Some aerobic strains have acquired the capacity to degrade highly chlorinated PCBs, particularly those having a chlorine substituent in an ortho position, by using a different initial oxygenation reaction involving a 3,4-hydroxylation instead of a 2,3-hydroxylation (Fig. 3). This is apparently the case for P. putida LB-400 (13, 102). In some other strains such as P. testosteroni B-356, the initial oxygenase attack always occurs on ortho and meta carbon, even in the case of ortho-substituted PCBs (Fig. 3). By using P. putida subclones carrying bphA and bphB genes from strain B-356, Ahmad et al. (5) found that oxygenase attack on PCB congeners carrying chlorine atoms on both rings was usually accompanied by a para or ortho dehalogenation of the molecule. In this same study, however, they also found that although there is a preferred ring of attack, the unpreferred ring can also be hydroxylated thus resulting, for some of the congeners, in a mixture of several metabolites (Fig. 4). This was confirmed by the observation that strain B-356 grown in the presence of 2,4'-dichlorobiphenyl usually generated both 2CBA and 4CBA (5). In a recent work based on the accumulation of chlorobenzoate from various PCB congeners that carry chlorine atoms on both rings, Bedard and Haberl (11) suggested that bacterial strains have a preferred ring of attack. The authors used these criteria to classify PCB degraders into four groups.

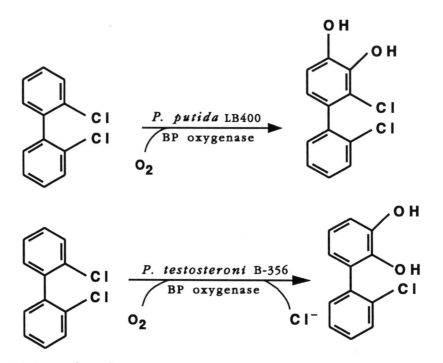

Figure 3. Proposed initial oxygenase step for highly chlorinated PCB congeners carrying ortho substitution on both rings by P. putida strain LB 400 and P. testosteroni strain B-356.

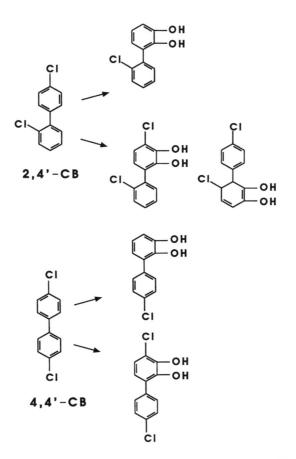

Figure 4. Conversion patterns of 4,4′-dichlorobiphenyl and 2,4′-dichlorobiphenyl by *P. testosteroni* B-356 BP oxygenase. Modified from Ahmad et al. (5).

Strain B-206 provided another example of multiple initial attack on a single substrate; 3,4-dihydro-3,4-dihydroxy-4′-chlorobiphenyl was detected in the growth medium when 4-chlorobiphenyl (4CB) was used as substrate, though the preferred initial oxygenase attack on this substrate is in the 2,3 position (95). It was also shown in strain B-206 that depending on the nitrate concentration of the medium, nitration of the molecule could occur during the oxygenase attack (141).

The rate of transformation of the hydroxylated product generated during the initial oxygenase attack is also very important in maintaining metabolic flux and preventing accumulation of toxic hydroxychlorobiphenyls. The dihydro-dihydroxybiphenyl produced from the initial oxygenase attack appears to be unstable, since it was detected in trace amounts and in only a few instances (95). It probably polymerized quickly in the medium to produce colored metabolites, or was spontaneously dehydrated to produce monohydroxybiphenyls (94). Therefore, the half life of this metabolite is probably very short in the environment. We know nothing about its toxicity. It has recently been suggested that either dihydrodiols or dihydroxylated metabolites generated by microbial conversion of PCBs could be incorporated into the humic acid material through spontaneous or light-catalyzed reactions (9). Although this is a likely route for transformation of PCBs in the environment, more supporting studies are needed to evaluate a precise role of this route and how much it really accounts for PCB detoxification in the environment.

Bacterial transformation of dihydrodiols into dihydroxy compounds is catalyzed by the biphenyl 2,3-dehydrogenase (bphB gene). This enzyme has not yet been isolated from PCB degrading bacterial strains. With reference to benzene glycol and toluene dihydrodiol dehydrogenase, however, this enzyme probably is coded by only one gene (57, 134, 160).

The gene encoding the meta pyrocatechase that catalyzes the meta fission of the hydroxybiphenyl was cloned and sequenced from various bacteria, and it seems to have been highly conserved in the different PCB-degrading strains (48, 56). One important feature about the metabolite produced from this reaction is that the meta-cleavage compound (MCP) is highly unstable (25). There have been no reports on the toxicity of the highly reactive meta-cleavage intermediate, and no study has been done to determine whether this metabolite plays a role in the regulation of the BP/PCB degradation pathway. It was recently shown that MCP can be transformed through several routes in PCB degraders (4, 105), however, and it was suggested that these pathways might be used to remove some of the excess MCP produced (4).

Several metabolites not belonging to the major pathways shown in Figure 2 have been detected in culture media of cells grown in the presence of PCBs (8, 11, 94, 95). Although the origin of these metabolites was not clear (were they part of another minor or major pathway, dead-end products, or products generated during extraction of the less stable metabolites of the major pathway?), some of these metabolites were used as evidence for new catabolic pathways (95, 102). It was recently shown that chromosomal genes from Pseudomonas putida strain KT2440 carrying none of the bph genes were able to transform the MCP (2-hydroxy-6-oxo-6-(chlorophenyl)hexa-2,4-dienoic acid) into many of the PCB metabolites that are unexpected from the pathway encoded by the bph operon (4). A pathway showing the bioconversion pathways of 4CB meta-cleavage product is illustrated in Figure 5. Although the MCP transformation pathways appear to function similarly in both PCB-degrading and nondegrading strains, the rate of transformation of MCPs is much faster in the former organisms (4). The transformation of MCP into corresponding picolinic acid (metabolite 7) is interesting. Although the amount produced varied, it was found in the culture media of most strains analyzed and was produced from a large number of biphenyl analogs. Nonenzymatic conversion of compounds with a 2-hydroxy-6-oxo-hexa-2,4-dienoic moiety in 6N ammonia solution has been reported (53), but because the picolinic acid derivatives obtained during MCP conversion are sometimes produced in large amounts under neutral conditions, in the culture medium, their toxicity should be considered.

The end product of the four-step BP/PCB degrading pathway is the corresponding chlorobenzoic acid. Chloroacetophenone was also suggested to be the end product of a novel BP/PCB degradation pathway in some strains (8, 11). Indeed, production of chlorocinnamic acid and chloroacetophenone from MCP through the pathway illustrated in Figure 5 could provide an alternate pathway to drain some of the metabolites produced from early transformation steps in PCB degradation (5). Since acetophenone is generated by a spontaneous decarboxylation of 3-keto-3-phenyl-propionic acid, which is an unstable normal intermediate in a cinnamic acid degradation pathway (65), it was postulated that chloroacetophenone could be produced through a similar route from chlorocinnamic acid in some PCB-degrading strains (5) (Fig. 5).

Finally, following the major BP/PCB degradation route, the meta-cleavage compound should normally be transformed directly into chlorobenzoic acid (CBA) by a reaction involving a hydrolase enzyme (bphD gene). This reaction should be fast enough to prevent accumulation of MCP, but this is not the case for all PCB congeners, since it was reported (11, 52) that besides CBA, several PCB congeners such as 4,4'-dichlorobiphenyl are transformed into a yellow-colored metabolite. The MCP-

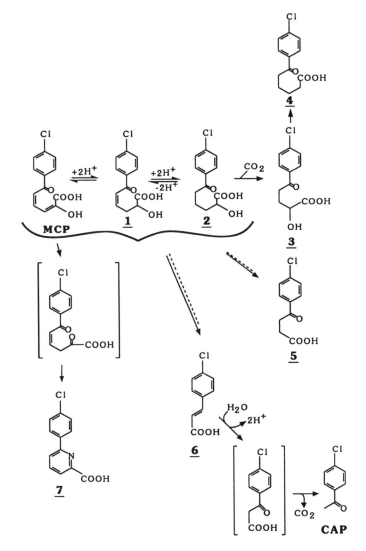

Figure 5. Bioconversion pathways of the *meta*-cleavage product. Proposed pathways for production of chloroacetophenone (CAP). *1*, mono-saturated MCP; *2*, di-saturated MCP; *3*, 2-hydroxy-5-oxo-5 (chlorophenyl)-pentanoic acid; *4*, 5-oxo-5-(chlorophenyl) pentanoic acid, *5*, butanoic acid, *6*, cinnamic acid, *7*, picollinic acid. Modified from Ahmad et al. (4).

hydrolyzing enzyme from the BP-degrading *Pseudomonas cruciviae* 593B1 has been purified (106), but its affinity for chlorinated substrates has not been determined.

CBA appears to play an important role in the regulation of BP/PCB degradation pathways. CBA itself is degraded through pathways involving other sets of genes. There is no single pathway for the degradation of all CBA congeners. Several pathways have been described (for recent reviews, see 29, 103, 116). Some of these pathways involve initial dechlorination, whereas others involve formation of chlorocatechols (Fig. 6).

Degradation of 3CBA through chlorocatechol has been reported for several organisms (28, 34, 35, 61, 67, 112, 159). In *Pseudomonas* sp. B13 (113, 115) and *Pseudomonas putida* AC858 (26, 27), a benzoate oxygenase mediates the transformation of 3CBA into chlorocatechol, and the further degradation of chlorocatechol is

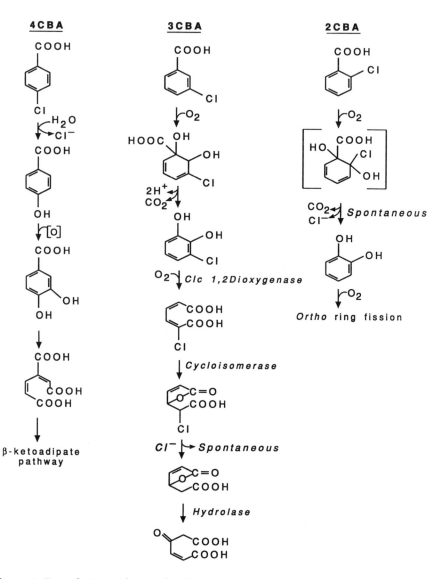

Figure 6. Degradation pathways for the three monochlorobenzoic acids.

directed by plasmid-encoded genes. Chlorocatechol and catechol degradation genes in *Pseudomonas* appear to have evolved from a common ancestor (26, 43, 44, 104, 121) but are substrate-specific.

Benzoate oxygenase from *Pseudomonas* sp. B13 is unable to attack 4CBA (113), but laboratory-constructed recombinants carrying the chlorocatechol degradation plasmid plus the nonspecific *meta*-toluate oxygenase from the TOL plasmid were able to degrade this compound (117).

Most of the 4CBA-degrading isolates described in the literature proceed through an initial dehalogenation of the molecule (1, 2, 77, 82, 92, 122, 133, 148, 157) to generate the corresponding 4-hydroxybenzoate, which is easily degraded by many bacterial strains; only one strain has been described as able to convert 3-chlorobenzoate into 3-hydroxybenzoate (70). The 4CBA dehalogenase in most strains studied was found to function as a hydrolase (93, 101). Some of the components of the *Pseudomonas* sp.

CBS3 dehalogenase system were recently isolated and characterized (37), and the roles of individual components were determined from studies done with *Escherichia coli* clones carrying the 4CBA dehalogenase genes from *Pseudomonas* sp. CBS3 (130).

Degradation of 2CBA is initiated by an oxygenolytic dehalogenation of the molecule which occurs spontaneously during attack by a 1,2-dioxygenase (39, 58, 61, 98, 143, 158); the carboxylic group is lost during this reaction. A nonenzymatic mechanism for this oxygenolytic dehalogenation was initially proposed for the degradation of 2-fluorobenzoate by *Alcaligenes eutrophus* (129). Recently, the same type of reaction was suggested for the degradation of 2-bromobenzoate (63). In *Pseudomonas* strain B-300, besides catechol, 3-chlorocatechol is also produced from 2CBA, resulting from a 1,6- rather than 1,2-dioxygenase attack (143). Recent work showed that 1,6-dioxygenation is not the usual mode of attack, however, and it only occurs under conditions of catabolic repression (unpublished data from our laboratory). Similarly, degradation of 2-fluorobenzoate through 1,6-dioxygenation of the molecule to produce 5-fluoro-3,5-cyclohexadiene-1,2-diol-1-carboxylic acid and 3-fluorocathecol was also shown to be an unproductive pathway in *Pseudomonas* strain B-13, *Alcaligenes* sp. strain A7-2 (129), and strain FLB300, which belongs to a new genus of the *Agrobacterium-Rhizobium* branch (40).

Bedard and Haberl (11) have shown that *P. putida* LB 400 can transform CBAs produced from PCBs, but they did not report any of the metabolites produced from this transformation. Recently, Sondossi et al. (137) showed that *P. testosteroni* B-356 carries biphenyl-induced 1,2-benzoate/chlorobenzoate and 2,3-catechol/chlorocatechol oxygenase activities. The chlorocatechols produced from initial oxygenation of CBA were further degraded through an extradiol cleavage. In the case of 3-chlorocatechol, the resulting product of this reaction, 5-chloroformyl-2-hydroxy-penta-2,4-dienoic acid, inactivates the hydroxybiphenyl *meta*-cleavage enzyme.

The BP-induced chlorobenzoic acid transformation pathway differs from the normal benzoic acid catabolic pathway in strain B-356. Whether the BP-induced oxygenases are part of an organized BP-linked BA-degradation pathway or not, the amounts of unproductive metabolites generated from chlorobiphenyl through these reactions are sufficient to interfere with the BP/CBP catabolic pathway (137). A similar incompatibility between the TOL plasmid *meta*-toluate 2,3-dioxygenase and the chlorobenzoate degradation pathway has been reported (117). Although to a lesser extent than 3-chlorocatechol, 4-chlorocatechol produced from CBAs can also inhibit the 2,3-dihydroxybiphenyl 1,2-dioxygenase. Therefore, when CBAs produced from PCBs accumulate in the growth medium, they can be converted into unproductive metabolites that reduce the flux of BP/PCB degradation pathways. Among the chlorobenzoates tested, 3CBA was the most effective inhibitor of the BP pathway in strain B-356.

Because chlorobenzoates and some of their metabolites can affect PCB degradation pathways, it was suggested that the effective elimination of CBA produced from a PCB-degradation pathway should result in enhanced degradation abilities. For example, a co-culture made of strain B-206 plus *Pseudomonas* sp. CBS3 completely mineralized 33 mg of 4CB within eight days, whereas an axenic culture of strain B-206 could only transform 15 mg of the substrate into a variety of acidic metabolites (142). Only a few bacteria carrying all the genes for total mineralization of chlorobiphenyls have been isolated, however. For example, bacterial isolates able to mineralize 4CB in pure culture have been described (132). The genes specifying the transformation of 4CB in those isolates were carried by degradative plasmids of the pSS series. Upon subculturing of strains, however, these plasmids lost a 15-kilobase DNA fragment associated with the 4CB degradation phenotype (109). Further studies on a recent laboratory-constructed hybrid strain able to mineralize 3-chlorobiphenyl brought further support to the hypothesis that CBA or its metabolites can prevent efficient degradation of chlorinated biphenyls (99). Hence, *Pseudomonas putida* strain BN10 carrying the BP/PCB-degradation pathway was able to degrade 4CB and 2CB with no loss of viability, but a drastic drop in viable cells was noticed in the presence of 3CB. From a mating between strain BN10 and the 3CBA-degrading *Pseudomonas* sp. strain B13, recombinant strains were isolated that were able to grow on 3CB and efficiently degrade that substrate.

Although the presence of CBA in the medium seems to affect PCB biodegradation, it is certainly not the only factor influencing PCB degradation. As mentioned above, Bedard and Haberl (11) have recently classified PCB degraders into four groups based on primary congener-selectivity patterns. They suggest that the reactivity preferences of bacterial strains are primarily dependent upon the specificity of the initial oxygenase of the pathway. Other factors, however, including the permeability of cells to substrate and the availability of substrate to the initial enzyme, also can influence the effectiveness of degradation of the various PCB congeners (88, 151). These factors could also affect the congener-selectivity pattern of strains.

One other important factor that might affect the performance of PCB degradation pathways, and which has so far received little attention, is the flux of the pathway. For any given pathway, flux could be affected by the accumulation of metabolites in the medium (72, 135). The flux of the pathway and the stringent control of all metabolite concentrations are factors that could be of particular importance for the efficient functioning of pathways such as the BP/PCB-degradation pathway that depend upon a patchwork assemblage of various complementary sets of genes and on the acquisition of novel biodegradation capacities by preexisting operons. Therefore, any enrichment program designed to isolate bacteria with enhanced ability to degrade PCB should also take into account that any modification that improves the performance of a single step involved in the PCB degradation pathway should not disturb the overall flux of the pathway. For example, there is no need to improve the rate of conversion of PCBs into the dihydroxylated derivatives if the *bphC* gene product becomes the limiting step. This problem can be avoided by assembling a complete pathway from a pool of well-characterized genes, operons, and gene cassettes. It will be a while, however, before we obtain enough information on the molecular biology of all the genes required to construct a genetically engineered strain able to mineralize highly chlorinated commercial PCB mixtures.

Because of a lack of knowledge of the molecular regulation of the BP/PCB pathway and of the substrate selectivity of each individual enzyme, strategies to enhance PCB biodegradation by direct modifications of specific aspects of the pathway have not yet been possible. We have had to use empirical methods. In the next few sections, we review and evaluate some of the strategies that were described in the literature or tried in our laboratory for the selection of better-performing strains.

SELECTION OF STRAINS WITH ENHANCED ABILITY
TO DEGRADE PCBs

Biphenyl Enrichment

Examples of organisms with abilities to degrade new substrates by using preexisting pathways have been reported. The new catabolic activities apparently were acquired through changes that contributed to broadening the substrate specificities of preexisting enzymes and through mutations of the regulatory genes that increased cytoplasmic enzyme levels. In this way, constitutive mutants for β-galactosidase in *E. coli* (60), for ribitol dehydrogenase in *Klebsiella aerogenes* (118), for mandelate pathway in *P. putida* (60), and for mandelamidase in *P. putida* (87) were obtained by continuous culture of the organisms on low levels of the substrates.

Early studies showed that the BP/PCB pathway is inducible by BP (6, 50); therefore, it is generally recognized that BP increases the enzymatic activities that are required to co-metabolize PCBs. Since BP-degrading enzymes have relaxed specificities and can be used for the co-metabolic transformation of many BP analogs, it is not surprising that enrichment on BP plus a low level of PCB can select for strains with enhanced capacity to degrade these compounds. Several authors have reported that bacteria growing on BP can co-metabolize PCBs more efficiently than corresponding resting cell suspensions (22, 64, 83). Bopp (17) has reported that when used in conjunction with Aroclor 1221, BP was found to favor selection of highly efficient PCB degraders from soil enrichments. *P. putida* LB400 isolated from such an enrichment (17) appears to have an initial oxygenase that differs considerably from many other PCB degraders, based on its congener-selectivity pattern (11).

In our laboratory, we have used this type of enrichment to select for PCB-degrading bacterial populations. Twenty soils with a history of contamination with PCBs were used as inocula to start enrichment cultures on BP plus Aroclor 1242 medium. When tested as soil suspensions in a buffered medium containing BP (0.1%, w/v) plus Aroclor 1242 (10 to 100 ppm), a few of the components (including International Union of Physical and Analytical Chemistry (IUPAC) numbers 5, 8, 18, 15, 17, 31, 33, 53, 42, 48, 74, 66, 91, 95, 90, and 101) in seven of these 20 samples were slightly degraded (it is estimated that between 5 and 20% of these individual congeners were degraded) within 60 days incubation. The percentage of degradation was calculated from changes in the ratio of peak surface with reference to a standard peak (containing congeners 110 and 77, which were not degraded).

These seven cultures were successively transferred on BP (0.1%,w/v) plus Aroclor 1242 (100 ppm) medium 20 times. The cultures were incubated two weeks, then a small aliquot was transferred in fresh medium. The enriched cultures were periodically evaluated for their ability to degrade Aroclor 1242. The first four columns of Table 2 give the percentage of degradation of Aroclor 1242 components as recorded for 15-day-old cultures inoculated with cells from the tenth transfer for four of the seven soils. The data show a gradual increase in the performance of the population toward degradation of the higher chlorinated components with the number of transfers. At the tenth transfer, we reached the maximal capacity of all the populations. As it can be seen from Table 2, all populations performed similarly, independently of the origin of the soil. After the tenth transfer, all the populations were able to degrade most congeners, except congeners 110 (2,3,3′,4′,6), 77 (3,3′,4,4′), 82 (2,2′3,3′,4), 105 (2,3,3′4,4′), 132 (2,2′,3,3′,4,6′), 149 (2,2′,3,4′,5′6), and 118 (2,3′,4,4′,5). These few congeners correspond exactly to the Aroclor 1242 components that were reported most resistant to biodegradation by *Alcaligenes eutrophus* H850 (12). Increasing the number of transfers to 20 did not significantly improve the capacity of these populations to degrade the resistant components.

The efficiency of degradation increased with the number of transfers, and the presence of Aroclor 1242 favored the selection of better-performing populations, since lines of transfers on BP medium without Aroclor did not perform as well (results not shown). Although the role of Aroclor in the enrichment procedure is not clear, its importance as a selecting agent for better-performing strains has already been discussed by Bopp (17). Because there was no apparent improvement in the capacity of

Table 2. Percentage degradation of the various components of Aroclor 1242 by enriched populations.

Peak	Congeners	Populations[a]					
		1	2	3	4	5	6
		% Degradation of peak[b]					
1	4,10	100	100	100	100	61	100
2	7,9	100	100	100	100	0	38
3	6	100	100	100	100	0	32
4	5,8	61	59	76	52	0	72
5	19	0	0	0	0	0	0
6	18	83	78	93	62	7	68
7	17,15	77	81	65	100	0	72
8	24,27	0	39	100	100	0	21
9	16,32	64	70	70	71	0	31
10	34	52	58	24	25	0	37
11	25	45	45	100	100	0	100
12	31,28	78	57	93	85	2	ND
13	33,53	69	78	88	88	0	80
14	22,51	70	71	82	83	0	ND
15	45	ND	ND	ND	ND	ND	ND
16	46	48	49	44	52	0	46
17	52,73	56	60	74	70	0	47
18	42	63	66	77	74	0	44
19	47	ND	ND	ND	ND	ND	ND
20	48	44	61	71	68	0	46
21	44	52	49	61	55	0	34
22	37,41	43	50	70	65	0	37
23	64,71	53	52	67	61	0	35
24	40	37	48	36	24	0	24
25	74	50	57	68	57	10	30
26	70	54	58	67	61	1	24
27	65–95	50	61	75	57	0	27
28	91	7	0	1	5	0	11
29	56,22	43	55	65	49	0	38
30	84,92	ND	ND	ND	ND	ND	ND
31	90,101	30	32	41	39	0	24
32	99	0	0	12	0	0	0
33	97,152	0	6	12	2	0	0
34	87	24	27	32	29	0	0
35	85	4	15	13	9	0	0
36	110,77	0	0	0	0	0	0
37	82	0	0	0	0	0	0
38	149–118	0	15	4	5	0	5
39	105–132	0	10	0	0	0	0

[a]Columns 1 to 4 are enrichment cultures from four different soil populations that were obtained by ten successive transfers on a medium containing BP (0.1% v/v) + Aroclor 1242 (100 ppm). After the tenth transfer, cells were harvested, washed, and transfered in fresh medium containing BP (0.1% v/v) + Aroclor 1242 (100 ppm). The culture was incubated 15 days, then extracted three times with hexane. Controls of cultures with inactivated cells were run in parallel and they showed no degradation. Columns 5 and 6 are cultures of mixed population adapted for 3 months and 11 months, respectively, in a continuous culture containing BP (0.1%, v/v), Aroclor 1242 (100 ppm), and trace amounts of 2-hydroxy-, 3-hydroxy-, and 4-hydroxybiphenyl. The cells were harvested, washed, and transfered in a liquid mineral medium containing Aroclor 1242 (100 ppm) and BP (0.1%, v/v). Cultures were incubated for 15 days, then extracted three times with hexane.
[b]The percentage of degradation of each peak was estimated from the area under the peak with reference to peak #36 which was not degraded. Once the ratio of each peak to peak #36 was established, the percentage degradation was calculated with reference to the value obtained for each peak in a control culture with cells inactivated with sodium azide. N.D.—not determined.

the population to degrade the most resistant components of the Aroclor 1242 mixture after the tenth transfer, other factors, such as bioavailability of congeners and stringent metabolic flux of PCB pathways, may limit the selection of strains with enhanced ability to degrade PCBs.

The enrichment method on BP plus Aroclor 1242 not only can select for better-degrading strains from environmental samples but can also be used to improve the capacity of strains maintained in pure culture. Hence, we were able to isolate a variant strain from the modestly performing strain *P. testosteroni* B-356 that was able to degrade most components of Aroclor 1242 (unpublished). Ongoing work with other pure cultures will help us to compare the capacity of individual strains to improve their ability to degrade selected groups of PCB congeners and, from these data, to determine the limiting factors in the adaptation process.

Bioavailability

That the first oxygenase attack of PCB congeners is affected by their chlorine substitution pattern is not surprising, because the position of chlorine imposes constraints on the three-dimensional structure of the molecules. There is, however, another important factor affecting biodegradation and the rate of the first oxygenase attack: the bioavailability of the substrate. Although the water solubility of the various PCBs congeners cannot be accurately determined (89), the differences in solubility of various PCB congeners are striking. The highly chlorinated PCBs are very insoluble in water (30, 41, 89). This could account for the resistance of highly chlorinated PCB congeners to biodegradation. We found that 4,4'-dichlorobiphenyl is degraded much faster in the presence of Tween 80 than with no surfactants (unpublished data). Some authors have shown that PCB degradation was faster when surfactant material was present in the culture (88, 151); however, to our knowledge, no specific work trying to relate PCB degradation ability to genes specifying surfactant production has been published. There is no reason to believe from published data that the BP/PCB degradation operon in bacteria includes such genes. Therefore, bioavailability through solubilization of substrates appears to be an external factor that one should control to obtain a higher rate of biodegradation.

Enrichment on Toxic Analogs

Because most PCB congeners are poorly soluble in water and by themselves are apparently weakly toxic for bacteria, one might expect these compounds to be rather poor selective agents to favor the expression of improved strains with enhanced degradation capacities. Since the enzymes of the BP/PCB degradation pathway have already been shown to have a very broad substrate specificity with the capacity to transform nonchlorinated analogs of BP such as hydroxybiphenyl (HBP) (136), we chose to use these compounds as selective agents in a continuous culture. We hoped that the toxic HBP would favor the selection of derepressed BP degradation enzymes and of mutated enzymes showing increased affinity to substrates.

We could have used 2,3-dihydroxy-chlorobiphenyls (2,3-diOHCBP) as substrates. These are the preferred substrates for the *bphC*-encoded oxygenase catalyzing the 1,2 *meta* cleavage of the BP ring. In the microtox assay (23), we found that 2,3-diOHCBP showed a toxicity level that was ten times less than phenol (unpublished data). We could also have used hydroxybiphenyls that have at least one unsubstituted ring as substrates, because these compounds are transformed by the BP oxygenase encoded by the *bphA* gene of strain B-356 (136).

Resting cells of strain B-356, or of the clone *Pseudomonas putida* DA1 (3), which carries all BP/PCB degradation genes from strain B-356, accumulated 4CBA in the first 20 minutes after feeding with 4CB with no detectable amounts of *meta*-cleavage product. Under the same conditions, when 2,3-diOH-4'-CBP was used as substrate, the *meta*-cleavage product was the major metabolite, with very small amounts of 4CBA (5). We did not employ 2,3-diOHCBP as selective substrate, because from the above experiment it appears that the *meta*-cleavage reaction is not the limiting step, and because we did not want to increase the *meta*-cleavage activity, which generates a very reactive metabolite that can be converted into a large number of products, as shown in Figure 5.

We started a continuous culture with the PCB-degrading bacterial strains B-206, *P. testosteroni* B356, *Pseudomonas alcaligenes* B-355, and *Pseudomonas* sp. B-300, which can degrade 2CBA; *Pseudomonas putida* B-301, which can degrade 3CBA; and *Pseudomonas putida* B-384 carrying cloned 4CBA dehalogenation genes from *Pseudomonas* sp. CBS3 (125). The cells were immobilized on charcoal to increase the chances of recombination events (85). The culture was fed with fresh medium containing BP (0.1% w/v) plus Aroclor 1242 (100 ppm) and trace amounts of a mixture of 2, 3, and 4-hydroxybiphenyl for a period of 18 months. The capacity of the enriched population to degrade Aroclor 1242 was evaluated by using small aliquots of this culture to inoculate batch cultures containing 0.1% (w/v) BP plus 100 ppm Aroclor 1242. These batch cultures were incubated for 15 days before evaluating the amounts of substrate remaining in the culture. Results showed (Table 2, columns 5 and 6) a gradual improvement of the capacity of the population to degrade Aroclor 1242, but the population grown for 11 months in the presence of hydroxybiphenyls did not degrade Aroclor better than the cells grown in the presence of BP plus Aroclor. Moreover, the capacity of this population to degrade the highly chlorinated components of Aroclor 1242 was not better when compared to the performance of strains developed by enrichment in batch cultures on BP plus Aroclor 1242 with no added hydroxybiphenyls (comparing column 5 with any of the first three columns of Table 2).

There are very few reports on bacterial metabolism of hydroxylated biphenyls (HBP) (51, 62, 84, 136). Depending on the mode of selection of HBP-degrading organisms, at least two types of catabolic pathways have been characterized. Bacterial growth on some selected HBP by *Pseudomonas* sp. HBP1 (84) was initiated by a NADH-dependent monooxygenase that required the presence of a substrate with 2-hydroxyphenyl-R structure. Higson and Focht (62) subsequently reported metabolism of some hydroxybiphenyls by *Pseudomonas* sp. FH12, also initiated by a monooxygenase that introduced a hydroxyl group at the *ortho* position of 3 and 3,3'-hydroxybiphenyls. These strains degraded HBP rather rapidly, but could not degrade PCBs. On the other hand, other organisms such as *P. testosteroni* strain B-356 degraded small amounts of HBP very slowly through the BP/PCB degradation pathway (136).

In natural environments, it is likely that bacteria that evolve to degrade HBP rapidly do so by using a substrate-specific monooxygenase to initiate the degradation; however, with cultures started with PCB-degrading strains such as ours, we expected that the first BP oxygenase of some variants of the population would be mutated to degrade HBP at a faster rate. Nonetheless, data in Table 2 show that enrichment on hydroxybiphenyl did not select for bacteria carrying BP oxygenase with more affinity toward highly chlorinated PCBs.

Anaerobic-aerobic Process

An interesting transformation process that has been suggested for the degradation of PCBs is the use of a two-step, anaerobic-aerobic degradation (11, 110). This process would take advantage of the phenomenon of reductive dehalogenation of halogenated aromatic compounds that occurs in some anaerobic bacterial cultures. The consequence of this reaction is a replacement of the chlorine substituents by hydrogen atoms, resulting in PCB molecules with a lower number of chlorine atoms. Shelton and Tiedje (131) have isolated an anaerobic microbial population that can degrade 3CBA through aryl dehalogenation. The biologically mediated process of reductive dehalogenation of PCBs was first proposed by Brown et al., (20) in upper Hudson river sediments and later confirmed by Quensen et al., (110). More recently, Quensen *et al.,* (111) have presented evidence that there are different PCB-dechlorinating microorganisms with characteristic patterns for PCB dechlorination. Most anaerobic PCB-dechlorinating organisms reported in the literature thus far only dehalogenate the *meta* and *para* positions of PCB molecules (96, 111). Only one anaerobic consortium with the ability to dehalogenate PCB congeners carrying chlorines in the *ortho* position have been reported (149). Studies reported by Quensen et al., (110, 111) and by Van Dort and Bedard (149) were performed with sediments collected from anaerobic sites, and they were treated anaerobically. Studies in our laboratory showed that activated sludge samples can be used to initiate cultures able to dehalogenate 4,4′-dichlorobiphenyl (96). In practical terms, this means that populations that reductively dehalogenate PCBs are not necessarily strictly anaerobic, a quality that may facilitate the initiation of mixed processes using anaerobes plus aerobes for the degradation of chlorobiphenyls. Before this technology can be considered for cleanup of contaminated sites, however, the parameters that will increase the rates of PCB dechlorination, which are rather low (111), must be determined and the engineering system designed that will allow the application of this technology to soils or sediments (66).

Use of Genetically Engineered Strains

Cloning of genes has been suggested as a method to develop bacterial strains endowed with new catabolic abilities toward persistent compounds (54, 152). Two strategies are generally used for the laboratory evolution of new catabolic pathways (145). One strategy is to expand the substrate profile of the first enzyme of the pathway either by mutation or by recruiting a gene from another pathway that encodes for an enzyme with a broader substrate specificity. Transfer of the *meta*-toluate oxygenase from the TOL plasmid into the 3CBA-degrading strain *Pseudomonas* sp. B13 generated a recombinant with the capacity to degrade 4CBA and 3,5CBA (114). In this example, the *meta*-toluate oxygenase from the TOL plasmid was able to convert 3, 4- and 3,5CBA into the corresponding chlorocatechols that were further degraded by the plasmid-encoded chlorocatecol degradation pathway (117).

Another strategy for developing strains of bacteria with new catabolic pathways involves cloning of structural and regulatory genes from various pathways and combining these genes in patchwork fashion into a functional pathway in a host strain (145). Few examples of such genetically engineered pathways were reported (120, 152). Because the PCB-degrading strains can be regrouped on the basis of a congener specificity pattern, it was suggested that the initial BP oxygenase might differ from one strain to another, favoring one pattern of degradation over the other (13). An additive effect of Aroclor 1242 degradation was reported for strains with complementary specificity patterns. For example, a co-culture of *Corynebacterium* sp. MB1 (in which the BP oxygenase attack is in a 2,3 position) plus *P. putida* LB400 (in which the BP oxygenase attack is in 3,4 position) degraded Aroclor 1242 more efficiently than each strain individually (147). Therefore, recruiting in one single strain the genes for initial oxygenase of strains having different patterns of degradation should broaden the substrate specificity. In this respect, further study of the molecular biology of the BP/PCB degradation pathway should help in developing new strategies for the construction of improved genetically engineered strains.

Finally, because of the effect of chlorobenzoates and their derivatives on the BP pathway, strains genetically engineered to degrade PCBs efficiently should probably carry efficient chlorobenzoate degradation pathways.

On the other hand, Mondello suggested for reasons of safety that PCB degradation genes, as any other genetically engineered strain that is to be spread in the environment, be cloned in the inoffensive *E. coli* host strain (100). Successful expression of the *bphA* gene in *E. coli* was reported in PCB clones from *P. putida* LB 400 (100) and in *P. putida* KF715 (59). By analyzing the specific activity of each of the four enzymes of the cloned BP pathway in clones and deleting subclones carrying genes from the BP/PCB pathway of *P. testosteroni* B-356, we showed that *bphC* and *bphD* genes, although they are part of the same operon, are well expressed in *E. coli*, whereas the *bphA* gene from *P. testosteroni* is expressed very poorly in *E. coli* (5). The low water solubility of the substrate may prevent access of the *bphA* gene into *E. coli* cells. Another explanation, such as association of the enzyme with some activating membrane bound protein, might also be considered. In this respect, the expression of the *bphA* gene may also be a constraint to such constructions.

GENERAL CONCLUSIONS

Many factors can improve PCB degradation. The selectivity pattern of the first oxygenase of the pathway is among the most important factors. The bioavailability of substrate is also a constraining factor, particularly in the case of highly chlorinated PCB congeners, which are poorly soluble in water. Moreover, all the enzymes of a pathway must work in harmony to prevent accumulation of metabolites, some of which may be toxic for the cell or inhibitory to selected enzymes of the pathway.

Variant strains carrying an initial oxygenase with enhanced activity toward its substrate should be able to degrade PCB faster, but only if other enzyme rates are not limiting for the conversion of metabolites. Ideally, any mutation affecting the specificity of a given enzyme of the BP/PCB pathway should not affect the flux of the pathway. This implies that unless the mutation affects the regulation of the whole pathway, very few single mutations may significantly improve the efficiency of biodegradation. Also, since double mutations are very rare events, the likelihood that bacteria can adapt to degrade highly chlorinated PCB congeners rapidly through mutations affecting specificity of the enzymes of the pathway is very unlikely. Regulation of pathways, prevention of accumulation of toxic metabolites, and bioavailability of substrates are, therefore, the most important factors to be considered in attempts to improve biodegradation capacity.

Finally, because all components of commercial PCB mixtures are unlikely to be degraded in a single treatment, future research on the microbial destruction of PCBs ought to select for bacterial strains with enhanced capacity to degrade the recognized toxic congeners.

LITERATURE CITED

1. Adriens, P., and D. D. Focht. 1991. Cometabolism of 3,4-dichlorobenzoate by *Acinetobacter* sp strain 4-CB1. *Appl. Environ. Microbiol.* 57:173–179.

2. Adriens, P., H. P. Kohler, D. Kohler-Staub, and D. D. Focht. 1989. Bacterial dehalogenation of chlorobenzoates and co-culture biodegradation of 4-4'-dichlorobiphenyl. *Appl. Environ. Microbiol.* 55:887–892.

3. Ahmad, D., R. Massé, and M. Sylvestre. 1990. Cloning and expression of genes involved in 4-chlorobiphenyl transformation by *Pseudomonas testosteroni*: homology to polychlorobiphenyl degrading genes in other bacteria. *Gene* 86:53–61.

4. Ahmad, D., M. Sylvestre, M. Sondossi, and R. Massé. 1991. Bioconversion of 2-hydroxy-6-oxo-6-(4'-chlorophenyl)hexa-2,4-dienoïc acid, the *meta*-cleavage product of 4-chlorobiphenyl. *J. Gen. Microbiol.* 137:1375–1385.

5. Ahmad, D., M. Sylvestre, and M. Sondossi. 1991. Sub-cloning of *bph* genes from *Pseudomonas testosteroni* B-356 in *Pseudomonas putida* and *Escherichia coli*: evidence for dehalogenation during initial attack of chlorobiphenyls. *Appl. Environ. Microbiol.* 57:2880–2887.

6. Ahmed, M., and D. D. Focht. 1972. Degradation of polychlorinated biphenyls by two species of *Achromobacter*. *Can. J. Microbiol.* 19:47–52.

7. Albro, P. W., and J. D. McKinney. 198l. The relationship between polarizability of polychlorinated biphenyls and their induction of mixed function oxidase activity. *Chem. Biol. Interact.* 34:373–378.

8. Barton, M. R., and R. L. Crawford. 1988. Novel biotransformations of 4-chloro-biphenyl by a *Pseudomonas* sp. *Appl. Environ. Microbiol.* 54:594–595.

9. Baxter, R. A. 1986. Bacterial formation of humus-like materials from polychlorinated biphenyls. *Water Pollut. Res. J. Can.* 21:1–7.

10. Baxter, R. A., P. E. Gilbert, R. A. Lidgett, J. H. Mainprize, and H. A. Vodden. 1975. The degradation of PCBs by microorganisms. *Sci. Total Environ.* 4:53–61.

11. Bedard, D. L., and M. L. Haberl. 1990. Influence of chlorine substitution pattern on the degradation of polychlorinated biphenyl by eight bacterial strains. *Microb. Ecol.* 20:87–102.

12. Bedard, D. L., R. E. Wagner, M. J. Brennan, M. L. Haberl, and J. F. Brown, Jr. 1987. Extensive degradation of Aroclors and environmentally transformed PCBs by *Alcaligenes eutrophus* H850. *Appl. Environ. Microbiol.* 53:1094–1102.

13. Bedard, D. L., M. L. Haberl, R. J. May, and M. J. Brennan. 1987. Evidence for novel mechanisms of PCB metabolism in *Alcaligenes eutrophus*. H850. *Appl. Environ. Microbiol.* 53:1103–1112.

14. Bedard, D. L., R. Unterman, L. H. Bopp, M. J. Brennan, M. L. Haberl, and C. Johnson. 1986. Rapid assay for screening and characterizing microorganisms for the ability to degrade polychlorinated biphenyls. *Appl. Environ. Microbiol.* 51:761–765.

15. Beringer, J. E., and M. J. Bale. 1988. *The Survival and Persistence of Genetically Engineered Microorganisms.* London: Academic Press, Inc.

16. Blakemore, R. P. 1978. Effects of polychlorinated biphenyls on molecular synthesis by heterotrophic marine bacterium. *Appl. Environ. Microbiol.* 35:329–336.

17. Bopp, L. H. 1989. *Pseudomonas putida* capable of degrading PCBs. US patent No. 4,843,009. Filed date: May 23, 1986.

18. Bourquin, A. W., and S. Cassidy. 1975. Effect of polychlorinated biphenyl formulations on the growth of estuarine bacteria. *Appl. Microbiol.* 29:125–127.

19. Brown, D. P. 1987. Mortality of workers exposed to polychlorinated biphenyl—An update. *Arch. Environ. Health* 42:333–339.

20. Brown, J. F., Jr., R. E. Wagner, D. L. Bedard, M. J. Brennan, J. C. Carnahan, and R. J.

May. 1984. PCB transformation in upper Hudson sediments. *Northeast. Environ. Sci.* 3:166–178.

21. Brown, J. F., Jr., R. E. Wagner, H. Feng, D. L. Bedard, M. J. Brennan, J. G. Carnahan, and R. J. May. 1987. Environmental dechlorination of PCBs. *Environ. Toxicol. Chem.* 6:579–593.

22. Brunner, W., F. H. Sutherland, and D. D. Focht. 1985. Enhanced biodegradation of polychlorinated biphenyls in soil by analog enrichment and bacterial inoculation. *J. Environ. Quality* 14:324–328.

23. Bulich, A. A. 1979. Use of luminescent bacteria for determining toxicity in aquatic environment. In *Aquatic Toxicology.* Ed. L. L. Harling, R. A. Kimerlu. *American Society for Testing and Materials* 667:98–106. Philadelphia, PA: ASTM.

24. Bush, B., R. W. Streeter, and R. J. Sloan. 1989. Polychlorobiphenyl (PCB) congeners in striped bass *(Morone saxatiles)* from marine and estuarine waters of New York State determined by capillary gas chromatography. *Arch. Environ. Contam. Toxicol.* 19:49–61.

25. Catelani, D., A. Colombi, C. Sorlini, and V. Treccani. 1973. Metabolism of biphenyl, 2-hydroxy-6-oxo-6-phenylhexa-2,4-dienoate: the *meta*-cleavage product from 2,3-dihydroxybiphenyl by *Pseudomonas putida. Biochem. J.* 134: 1063–1066.

26. Chatterjee, D. K., and A. M. Chakrabarty. 1983. Genetic homology between independently isolated chlorobenzoate degradative plasmids. *J. Bacteriol.* 153:532–534.

27. Chatterjee, D. K., and A. M. Chakrabarty. 1984. Restriction mapping of a chlorobenzoate degradative plasmid and molecular cloning of the degradative genes. *Gene* 27:173–181.

28. Chatterjee, D. K., S. T. Kellogg, S. Hamada, and A. M. Chakrabarty. 1981. Plasmid specifying total degradation of 3-chlorobenzoate by a modified *ortho* pathway. *J. Bacteriol.* 146:639–646.

29. Chaudhry, G. R., and S. Chapalamadugu. 1991. Biodegradation of halogenated organic compounds. *Microbiol. Rev.* 55:59–79.

30. Chu, S. F. J., and R. A. Griffin. 1986. Solubility and soil mobility of polychlorinated biphenyls. In *PCBs and the Environment,* Vol. 1. Ed. J. S. Waid. Boca Raton, FL: CRC Press. 102–118.

31. Clarke, R. R., E. S. K. Chian, and R. A. Griffin. 1979. Degradation of polychlorinated biphenyls by mixed microbial cultures. *Appl. Environ. Microbiol.* 37:680–685.

32. Davis, D., and S. Safe. 1990. Immunosupressive activities of polychlorinated biphenyls in C57BL6N mice: structure-activity relationships as Ah receptor agonists and partial antagonists. *Toxicology* 63:97–111.

33. Donaldson, T. L., G. W. Strandberg, C. P. McGinnes, A. V. Palumbo, D. C. White, D. L. Hill, T. J. Phelps, C. T. Hadden, N. W. Revis, E. D. C. Holdsworth, and T. Osborne. 1988. *Bioremediation of PCB-contaminated Soil at the Y-12 Plant.* Publication ORNL/TM10750 DE89001335. Oak Ridge, TN: Oak Ridge National Laboratory.

34. Dorn, E. M., and H. J. Knackmuss. 1978. Chemical structure and biodegradability of halogenated aromatic compounds. Two catechol 1, 2-dioxygenases from a 3-chlorobenzoate grown pseudomonad. *J. Biochem.* 174:73–84.

35. Dorn, E., M. Hellwig, W. Reineke, and H. J. Knackmuss. 1974. Isolation and characterization of a 3-chlorobenzoate grown pseudomonad. *Arch. Microbiol.* 99:61–70.

36. Duinker, J. C., D. E. Schultz, and G. Petrick. 1988. Selection of chlorinated biphenyl congeners for analysis in environmental samples. *Marine Pollut. Bull.* 19:19–25.

37. Elsner, A., F. Loffler, K. Myashita, R. Müller, and F. Lingens. 1991. Resolution of 4-chlorobenzoate dehalogenase from *Pseudomonas* sp. CBS3 into three components. *Appl. Environ. Microbiol.* 57:324–326.

38. Emmett, E. A., M. Maroni, J. M. Schmith, B. K. Levin, and J. Jefferys. 1988. Studies of transformer repair workers exposed to PCBs: I. study design, PCB concentrations, questionnaire, and clinical examination results. *Am. J. Indust. Med.* 13:415–427.

39. Engesser, K. H., and P. Schulte. 1989. Degradation of 2-bromo, 2-chloro-and 2-fluorobenzoate by *Pseudomonas putida* CLB 250. *FEMS Microbiol. Lett.* 60:143–148.

40. Engesser, K. H., G. Auling, J. Beesse, and H. J. Knakmuss. 1990. 3-Fluorobenzoate enriched bacterial strain FLB300 degrades benzoate and all three isomeric monofluorobenzoates. *Arch. Microbiol.* 153: 193–199.

41. Evans, H. E. 1988. The binding of three PCB congeners to dissolved organic carbon in freshwaters. *Chemosphere* 17:2325–2338.

42. Focht, D. D., and W. Brunner. 1985. Kinetics of biphenyl and polychlorinated biphenyl metabolism in soil. *Appl. Environ. Microbiol.* 50:1053–1058.

43. Frantz, B., and A. M. Chakrabarty. 1987. Organization and nucleotide sequence determination of a gene cluster involved in 3-chlorocatechol degradation. *Proc. Nat. Acad. Sci. USA* 84:4460–4464.

44. Frantz, B., K. L. Ngai, D. K. Chatterjee, N. L. Ornston, and A. M. Chakrabarty. 1987. Nucleotide sequence and expression of *clcD*, a plasmid-borne dienelactone hydrolase gene from *Pseudomonas* sp. strain B-13. *J. Bacteriol.* 169:704–709.

45. Furukawa, K. 1982. Microbial degradation of PCBs. In *Biodegradation and Detoxification of Environmental Pollutants.* Ed. A. M. Chakrabarty. Boca Raton, FL: CRC Press, 33–57.

46. Furukawa, K., and F. Matsumura. 1976. Microbial metabolism of PCBs. Studies on the relative degradability of PCB components by *Alcaligenes* sp. *Agric. Food Chem.* 24:251–2.

47. Furukawa, K., and T. Miyazaki. 1986. Cloning of a gene cluster encoding biphenyl and chlorobiphenyl degradation in *Pseudomonas pseudoalcaligenes.* *J. Bacteriol.* 166:392–398.

48. Furukawa, K., N. Arimura, and T. Miyazaki. 1987. Nucleotide sequence of the 2,3-dihydroxybiphenyl dioxygenase gene of *Pseudomonas pseudoalcaligenes.* *J. Bacteriol.* 169:427–429.

49. Furukawa, K., N. K. Hayase, K. Taira, and N. Tomizuka. 1989. Molecular relationship of chromosomal genes encoding biphenyl/polychlorinated biphenyl catabolism: some soil bacteria possess a highly conserved *bph* operon. *J. Bacteriol.* 171:5467–5472.

50. Furukawa, K., F. Matsumura, and K. Tonomura. 1978. *Alcaligenes* and *Acinetobacter* strains capable of degrading polychlorinated biphenyls. *Agric. Biol. Chem.* 42:543–548.

51. Furukawa, K., J. R. Simon, and A. M. Chakrabarty. 1983. Common induction and regulation of biphenyl, xylene/toluene, and salicylate catabolism in *Pseudomonas paucimobilis.* *J. Bacteriol.* 154:1356–1362.

52. Furukawa, K., K. Tonomura, and A. Kamibayashi. 1978. Effect of chlorine substitution on the biodegradability of polychlorinated biphenyls. *Appl. Environ. Microbiol.* 35:223–227.

53. Gibson, D. T., K. C. Wong, C. J. Sih, and H. Whitock. 1966. Mechanism of steroid oxidation by microorganisms. IX. Mechanism of ring cleavage in the degradation of 9, 10-seco steroids. *J. Biol. Chem.* 241:541–559.

54. Haas, D. 1983. Genetic aspects of biodegradation by Pseudomonads. *Experientia* 39:1199–1213.

55. Hankin, L., and B. L. Sawhney. 1984. Microbial degradation of polychlorinated biphenyls in soil. *Soil Sci.* 137:401–407.

56. Harayama, S., and M. Rekik. 1989. Bacterial aromatic ring-cleavage enzymes are classified into two different gene families. *J. Biol. Chem.* 264:15328–15333.

57. Harayama, S., M. Rekik, and K. N. Timmis. 1986. Genetic analysis of a relaxed substrate specificity aromatic ring dioxygenase, toluate 1,2-dioxygenase encoded by TOL plasmid pWWO of *Pseudomonas putida.* *Mol. Gen.Genet.* 202:226–234.

58. Hartmann, J., K. Engelberts, B. Nordhaus, E. Schmidt, and W. Reineke. 1989. Degradation of 2-chlorobenzoate by *in vitro* constructed hybrid pseudomonad. *FEMS Microbiol. Lett.* 61:17–22.

59. Hayase, N., K. Taira, and K. Furukawa. 1990. *Pseudomonas putida* KF715 *bph ABCD*

operon encoding biphenyl and chlorobiphenyl degradation. Cloning analysis and expression in soil bacteria. *J. Bacteriol.* 172:1160–1164.

60. Hegeman, G. D. 1966. Synthesis of the enzymes of the mandelate pathway by *Pseudomonas putida*. 3. Isolation and properties of constitutive mutants. *J. Bacteriol.* 91:1161–1167.

61. Hickey, W. J., and D. D. Focht. 1990. Degradation of mono-, di-, and trihalogenated benzoic acids by *Pseudomonas aeruginosa* JB2. *Appl. Environ. Microbiol.* 56:3842–3850.

62. Higson, F. K., and D. D. Focht. 1989. Bacterial metabolism of hydroxylated biphenyls. *Appl. Environ. Microbiol.* 55:946–952.

63. Higson, F. K., and D. D. Focht. 1990. Degradation of 2-bromobenzoic acid by a strain of *Pseudomonas aeruginosa*. *Appl. Environ. Microbiol.* 56:1615–1619.

64. Hill, D. L., T. J. Phelps, A. V. Palumbo, D. C. White, G. W. Strandberg, and T. L. Donaldson. 1989. Bioremediation of polychlorinated biphenyls. Degradation capabilities in field lysimeters. *Appl. Biochem. Biotechnol.* 20/21:233–243.

65. Hilton, D. M., and W. J. Cain. 1990. Bioconversion of cinnamic acid to acetophenone by a pseudomonad: microbial production of a natural flavor compound. *Appl. Environ. Microbiol.* 56:623–627.

66. Hooper, S. W., C. A. Pettigrew, and G. S. Sayler. 1990. Ecological fate, effects and prospects for the elimination of environmental polychlorinated biphenyls (PCBs). *Environ. Toxicol. Chem.* 9:655–667.

67. Horvath, R. S., and M. Alexander. 1970. Cometabolism of m-chlorobenzoate by an *Arthrobacter*. *Appl. Environ. Microbiol.* 20:254–258.

68. Hutzinger, O., S. Safe, and V. Zitko. 1972. Polychlorinated biphenyl. *Analabs Res. Notes* 12:1–16.

69. Hutzinger, O., S. Safe, and V. Zitko. 1974. *The Chemistry of PCBs.* Cleveland, Ohio: CRC Press.

70. Johnston, H. W., G. G. Briggs, and M. Alexander. 1972. Metabolism of 3-chlorobenzoic acid by a pseudomonad. *Soil Biol. Biochem.* 4:187–190.

71. Jones, K. C. 1988. Determination of polychlorinated biphenyls in human foodstuffs and tissues: suggestions for a selective congener analytical approach. *Sci. Total Environ.* 68:141–159.

72. Kacser, H., and J. A. Burns. 1973. The control of flux. *Symp. Soc. Exp. Biol.* 27:65–104.

73. Kaiser, K. L. E., and P. T. S. Wong. 1974. Bacterial degradation of polychlorinated biphenyls. I: identification of some metabolic products from Aroclor 1242. *Bull. Environ. Contam. Toxicol.* 11:291–296.

74. Kaneko, M., K. Marimoto, and S. Nambu. 1976. The response of activated sludge to a polychlorinated biphenyl (KC–500). *Water Res.* 10:157–163.

75. Kannan, N., S. Tanabe, and R. Tatsukawa. 1988. Potentially hazardous residues of non-orthochlorine substituted coplanar PCBs in human adipose tissue. *Arch. Environ. Health* 43:11–14.

76. Kashimoto, T., H. Miyata, S. Kunita, T. Tury, S. Hu, K. Chang, S. Tang, G. Ohi, J. Nakagawa, and S. Yamamoto. 1981. Role of polychlorinated dibenzofurans in Yusho (PCB poisoning). *Arch. Environ. Health* 36:321–326.

77. Keil, H., V. Klages, and F. Lingens. 1981. Degradation of 4-chlorobenzoate by *Pseudomonas* sp. CBS3: induction of catabolic enzymes. *FEMS Microbiol. Lett.* 10:213–215.

78. Keil, J. E., S. H. Sandifer, C. D. Graber, and L. E. Priester. 1972. DDT and polychlorinated biphenyl (Aroclor 1242). Effects of uptake on *E. coli* growth. *Water Res.* 6:837–841.

79. Khan, A., and S. Walia. 1989. Cloning of bacterial genes specifying degradation of 4-chlorobiphenyl from *Pseudomonas putida* OU83. *Appl. Environ. Microbiol.* 55:798–805.

80. Khan, A., R. Tewari, and S. Walia. 1988. Molecular cloning of 3-phenylcatechol dioxygenase involved in the catabolic pathway of chlorinated biphenyl from *Pseudomonas putida* and its expression in *Escherichia coli*. *Appl. Environ. Microbiol.* 54:2664–2671.

81. Kimbara, K., T. Hashimoto, M. Fukuda, T. Koana, M. Takagi, M. Oishi, and K. Yano. 1989. Cloning and sequencing of two tandem genes involved in the degradation of 2, 3-dihydroxybiphenyl to benzoic acid in the polychlorinated biphenyl-degrading soil bacterium *Pseudomonas* sp. KKS102. *J. Bacteriol.* 171:2740–2747.

82. Klages, V., and F. Lingens. 1979. Degradation of 4-chlorobenzoic acid by a *Nocardia* species. *FEMS Microbiol. Lett.* 6:201–203.

83. Kohler, H. P. E., D. Kohler-Staub, and D. D. Focht. 1988. Cometabolism of polychlorinated biphenyls: enhanced transformation of Aroclor 1254 by growing bacterial cells. *Appl. Environ. Microbiol.* 54:1940–1945.

84. Kohler, H. P. E., D. Kohler-Staub, and D. D. Focht. 1988. Degradation of 2-hydroxybiphenyl and 2,2'-dihydroxybiphenyl by *Pseudomonas* sp. strain HPB1. *Appl. Environ. Microbiol.* 54:2683–2688.

85. Krockel, L., and D. D. Focht. 1987. Construction of chlorobenzene-utilizing recombinants by progenitive manifestation of a rare event. *Appl. Environ. Microbiol.* 53:2470–2475.

86. Lauber, J. D. 1986. Disposal and destruction of waste PCBs. In *PCBs and the Environment*, Vol. III. Ed. J. S. Waid. Boca Raton, FL: CRC Press. 84–149.

87. Laverack, P. D., and P. H. Clarke. 1979. Selection of mandelamidase constitutive mutants in continuous culture of *Pseudomonas putida*. *Biotechnol. Lett.* 1:353–358.

88. Liu, D. 1980. Enhancement of PCBs biodegradation by sodium ligninsulfonate. *Water Res.* 14:1467–1475.

89. Mackay, D., R. Mascarenhas, and W. Y. Shiu. 1980. Aqueous solubility of polychlorinated biphenyls. *Chemosphere* 9:257–264.

90. Mahanty, H. K., and G. Evans. 1980. Intra-generic variations of sensitivity to polychlorinated biphenyls (PCBs) and DDT in soil bacilli. *Soil Biol. Biochem.* 12:521–522.

91. Marks, T., G. L. Kimmel, and R. E. Staples. 1981. Influence of symmetrical polychlorinated biphenyl isomers on embryo and fetal development in mice. I: teratogenicity of 3,3',4,4',5,5'-hexachloro-biphenyl. *Toxicol. Appl. Pharmacol.* 61:269–276.

92. Marks, T. S., A. R. W. Smith, and A. V. Quirk. 1984. Degradation of 4-chlorobenzoic acid by *Arthrobacter* sp. *Appl. Environ. Microbiol.* 48:1020–1025.

93. Marks, T. S., R. Wait, A. R. W. Smith, and A. V. Quirk. 1984. The origin of the oxygen incorporated during the dehalogenation/hydroxylation of 4-chlorobenzoate by an *Arthrobacter* sp. *Biochem. Biophys. Res. Commun.* 124:669–674.

94. Massé, R., F. Messier, L. Péloquin, C. Ayotte, and M. Sylvestre. 1984. Microbial degradation of 4-chlorobiphenyl, a model compound of chlorinated biphenyls. *Appl. Environ. Microbiol.* 47:947–951.

95. Massé, R., C. Ayotte, M. F. Lévesque, and M. Sylvestre. 1989. A comprehensive gas chromatographic mass spectrometric analysis of 4-chlorobiphenyl bacterial degradation products. *Biomed. Environ. Mass Spectrom.* 18:27–47.

96. Mavoungou, R., R. Massé, and M. Sylvestre. 1991. Microbial dehalogenation of 4,4'-dichlorobiphenyl under anaerobic conditions. *Sci. Total Environ.* 101:263–268.

97. McFarland, V. A., and J. V. Clarke. 1989. Environmental occurrence, abundance, and potential toxicity of polychlorinated biphenyl congeners: considerations for a congener specific analysis. *Environ. Health Perspect.* 81:225–239.

98. Miguez, C. B., C. W. Greer, and J. M. Ingram. 1990. Degradation of mono- and dichlorobenzoic acid isomers by two natural isolates of *Alcaligenes denitrificans*. *Arch. Microbiol.* 154:139–143.

99. Mokross, H., E. Schmidt, and W. Reineke. 1990. Degradation of 3-chlorobiphenyl by *in vivo* constructed hybrid pseudomonads. *FEMS Microbiol. Lett.* 71:179–186.

100. Mondello, F. J. 1989. Cloning and expression in *Escherichia coli* of *Pseudomonas* strain LB400 genes encoding polychlorinated biphenyl degradation. *J. Bacteriol.* 171:1725–1732.

101. Müller, R., J. Thiele, V. Klages, and F. Lingens. 1984. Incorporation of [^{18}0] water into 4-chlorobenzoic acid in the reaction of 4-chlorobenzoate dehalogenase from *Pseudomonas* sp. CBS3. *Biochem. Biophys. Res. Commun.* 24:178–182.

102. Nadim, L. M., M. J. Schocken, F. K. Higson, D. T. Gibson, D. L. Bedard, L. H. J. Bopp, and J. F. Mondello. 1987. Bacterial oxidation of polychlorinated biphenyls. In *Proceedings of the USA EPA 13th Annual Research Symposium on Land Disposal, Remedial Action, Incineration and Treatment of Hazardous Waste*. Publication EPA/600/9–87015. Cincinnati, Ohio: U.S. Environmental Protection Agency.

103. Neilson, A. H. 1990. The biodegradation of halogenated organic compounds. *J. Appl. Bacteriol.* 69:445–470.

104. Ngai, K. L., M. Schlomann, H. J. Knackmuss, and N. L.Ornston. 1987. Dienelactone hydrolase from *Pseudomonas* sp. strain B-13. *J. Bacteriol.* 169:699–703.

105. Omori, T., H. Ishigooka, and Y. Micoder. 1986. Purification and some properties of 2-hydroxy-6-oxo-6-phenylhexa-2,4-dienoic acid (HOPDA) reducing enzyme from *Pseudomonas cruciviae* 593B1 involved in the degradation of biphenyl. *Agric. Biol. Chem.* 50:1513–1518.

106. Omori, T., K. Sugimura, H. Ishigooka, and Y. Minoda. 1986. Purification and some properties of 2-hydroxy-6-oxo-phenylhexa-2,4, dienoic acid hydrolyzing enzyme from *Pseudomonas cruciviae* 593B1 involved in the degradation of biphenyl. *Agric. Biol. Chem.* 50:931–937.

107. Parkinson, A., R. Cockerline, and S. Safe. 1980. Induction of both 3-methylcholanthrene and phenobarbitane-type microsomal enzyme activity by single polychlorinated biphenyl isomers. *Biochem. Pharmacol.* 29:259–262.

108. Parsons, J. R., and D. T. H. M. Sijm. 1988. Biodegradation kinetics of polychlorinated biphenyls in continuous cultures of *Pseudomonas* strain. *Chemosphere* 17:1755–1766.

109. Pettigrew, C. A., A. Breen, C. Corcoran, and G. S. Sayler. 1990. Chlorinated biphenyl mineralization by individual populations and consortia of freshwater bacteria. *Appl. Environ. Microbiol.* 56:2036–2045.

110. Quensen, J. F., J. M. Tiedje, and S. A. Boyd. 1988. Reductive dechlorination of polychlorinated biphenyls by anaerobic microorganisms from sediments. *Science* 242:752–755.

111. Quensen, J. F. III., S. A. Boyd, and J. M. Tiedje. 1990. Dechlorination of four commercial polychlorinated biphenyl mixtures (Aroclors) by anaerobic microorganisms from sediments. *Appl. Environ. Microbiol.* 56:2360–2369.

112. Reber, H. H., and G. Thierbach. 1980. Physiological studies in the oxidation of 3-chlorobenzoate by *Acinetobacter calcoaceticus* strain Bs5. *Eur. J. Appl. Microbiol. Biotechnol.* 10:223–233.

113. Reineke, W., and H. J. Knackmuss. 1978. Chemical structure and biodegradability of halogenated aromatic compounds. Substituents effects on 1,2-dioxygenation of benzoic acid. *Biochim. Biophys. Acta* 542:412–423.

114. Reineke, W., and H. J. Knackmuss. 1979. Construction of haloaromatics utilising bacteria. *Nature* 277:385–386.

115. Reineke, W., and H. J. Knackmuss. 1980. Hybrid pathway for chlorobenzoate metabolism in *Pseudomonas* sp. B-13 derivatives. *J. Bacteriol.* 142:467–473.

116. Reineke, W., and H. J. Knackmuss. 1988. Microbial degradation of haloaromatics. *Annu. Rev. Microbiol.* 42: 263–287.

117. Reineke, W., S. W. Wessels, M. A. Rubio, J. Latorre, U. Schwien, E. Schmidt, M. Schlomann, and H. J. Knackmuss. 1982. Degradation of monochlorinated aromatics following transfer of genes encoding chlorocatechol catabolism. *FEMS Microbiol. Lett.* 14:291–294.

118. Rigby, P. W. J., B. D. Burleigh, Jr., and B. S. Hartley. 1974. Gene duplication in experimental enzyme evolution. *Nature* 231:200–204.

119. Risebrough, R. W., S. G. Herman, D. B. Peakall, and M. N. Kirven. 1968. Polychlorinated biphenyls in global ecosystem. *Nature* 220:1098–1100.

120. Rojo, F., D. H. Pieper, K. H. Engesser, H. J. Knackmuss, and K. W. Timmis. 1987. Assemblage of *ortho*-cleavage route for simultaneous degradation of chloro- and methylaromatics. *Science* 238:1395–1398.

121. Rothmel, R. K., R. A. Haugland, U. M. X. Sangodkar, W. M. Coco, and A. M. Chakrabarty. 1990. Natural versus directed evolution: microbial degradation of chloroaromatic compounds. In *Proceedings of the 6th International Symposium on the Genetics of Industrial Microorganisms.* Strasbourg, France. 1035–1045.

122. Ruisinger, S., V. Klages, and F. Lingens. 1976. Abbau der 4-chloro-benzoesäure durch eine *Arthrobacter* species. *Arch. Microbiol.* 110:253–256.

123. Safe, S. 1984. Microbial degradation of polychlorinated biphenyls. In *Microbial Degradation of Organic Compounds,* Vol. 13. Ed. D. T. Gibson. New York, NY: Marcel Dekker, Inc. 361–369.

124. Safe, S., S. Bandiera, T. Sawyer, L. Robertson, L. Safe, A. Parkinson, P. E. Thomas, D. E. Ryan, L. M. Reik, W. Levin, M. A. Denomme, and T. Fujita. 1985. PCBs: structure function relationships and mechanism of action. *Environ. Health Perspect.* 60:47–56.

125. Savard, P., L. Péloquin, and M. Sylvestre. 1986. Cloning of *Pseudomonas* sp. strain CBS3 genes specifying dehalogenation of 4-chlorobenzoate. *J. Bacteriol.* 168:81–85.

126. Sawhney, B. L. 1986. Chemistry and properties of PCB's in relation to environmental effects. In *PCBs and the Environment,* Vol. 1 Ed. J. S. Waid. Boca Raton, FL: CRC Press. 47–67.

127. Sayler, G. S., M. Shon, and R. Colwell. 1977. Growth of an estuarine *Pseudomonas* sp. on PCB. *Microbiol. Ecol.* 3:241–255.

128. Sayler, G. S., R. Thomas, and R. R. Colwell. 1978. Polychlorinated biphenyl (PCB)—degrading bacteria and PCB in estuarine marine environment. *Estuarine Coastal Marine Sci.* 6:553–567.

129. Schmidt, E., and H. J. Knackmuss. 1984. Production of cis, cis-muconate from benzoate and 2-fluoro-cis,cis- muconate from 3-fluorobenzoate by 3-chlorobenzoate degrading bacteria. *Appl. Microbiol. Biotechnol.* 20:351–355.

130. Scholten, J. D., K. H. Chang, P. C. Babbitt, H. Charest, M. Sylvestre, and D. D. Mariano. 1991. Novel enzymic hydrolytic dehalogenation of a chlorinated aromatic. *Science* 253:182–185.

131. Shelton, D. R., and J. M. Tiedje. 1984. Isolation and partial characterization of bacteria in an anaerobic consortium that mineralize 3-chlorobenzoic acid. *Appl. Environ. Microbiol.* 48:840–848.

132. Shields, M. S., S. W. Hooper, and G. S. Sayler. 1985. Plasmid-mediated mineralization of 4-chlorobiphenyl. *J. Bacteriol.* 163:882–889.

133. Shimao, M., S. Onishi, S. Mizumori, N. Kato, and C. Sarazawa. 1989. Degradation of 4-chlorobenzoic acid by facultative alkalophilic *Arthrobacter* sp. strain SB8. *Appl. Environ. Microbiol.* 55:478–482.

134. Shinji, I., D. Seiji, T. Yorifugi, M. Takagi, and K. Yano. 1987. Nucleotide sequence and characterization of the genes encoding benzene oxidation enzymes of *Pseudomonas putida.* *J. Bacteriol.* 169:5174–5179.

135. Small, R. J., and D. A. Fell. 1989. The matrix method of metabolic control analysis: its validity for complex pathway structures. *J. Theoretical Biology* 136:181–197.

136. Sondossi, M., M. Sylvestre, D. Ahmad, and R. Massé. 1991. Metabolism of hydroxybiphenyl and chloro-hydroxybiphenyl by biphenyl/chlorobiphenyl degrading *Pseudomonas testosteroni,* strain B-356. *J. Indust. Microbiol.* 7:77–88.

137. Sondossi, M., M. Sylvestre, and D. Ahmad. 1992. Effects of chlorobenzoate transformation on the *Pseudomonas testosteroni* biphenyl and chlorobiphenyl degradation pathway. *Appl. Environ. Microbiol.* 58:485–495.

138. Sparling, J., and S. Safe. 1980. The effects of *ortho* chloro substituents on the retention of PCB isomers in rat, rabbit, Japanese quail, guinea pig and trout. *Toxicol. Lett.* 7:23–28.

139. Stratton, C. L., and J. B. Sosebee. 1976. PCB and PCT contamination of the environment near sites of manufacture and use. *Environ. Sci. Technol.* 10:1229–1233.

140. Sylvestre, M., and J. Fauteux. 1982. A new facultative anaerobe capable of growth on chlorobiphenyls. *J. Gen. Appl. Microbiol.* 28:61–72.

141. Sylvestre, M., R. Massé, F. Messier, J. Fauteux, J. G. Bisaillon, and R. Beaudet. 1982. Bacterial nitration of 4-chlorobiphenyl. *Appl. Environ. Microbiol.* 44:871–877.

142. Sylvestre, M., R. Massé, C. Ayotte, F. Messier, and J. Fauteux. 1985. Total degradation of 4-chlorobiphenyl by a bacterial mixed culture. *J. Appl. Microbiol. Biotechnol.* 21:193–197.

143. Sylvestre, M., K. Mailhiot, and D. Ahmad. 1989. Isolation and preliminary characterization of a 2-chlorobenzoate degrading *Pseudomonas*. *Can. J. Microbiol.* 35:439–443.

144. Taira, K., N. Hayase, N. Arimura, S. Yamashito, T. Miyazaki, and K. Furukawa. 1988. Cloning and nucleotide sequence of 2,3-dihydroxy-biphenyl dioxygenase from the PCB degrading strain of *Pseudomonas paucimobilis* Q1. *Biochemistry* 27:3990–3996.

145. Timmis, K. N. 1990. Rational structuring of new catabolic pathways for environmental pollutants. In *Proceedings of the 6th International Symposium on Genetics of Industrial Microorganisms*. Strasbourg, France. 239–298.

146. Tucker, E. S., V. W. Saeger, and O. Hicks. 1975. Activated sludge primary biodegradation of polychlorinated biphenyls. *Bull. Environ. Contam. Toxicol.* 14:705–713.

147. Unterman, K., D. L. Bedard, M. J. Brennan, L. H. Bopp, F. J. Mondello, R. E. Brooks, D. P. Mobley, J. B. McDermott, C. C. Schartz, and D. K. Dietrich. 1988. Biological approaches for polychlorinated biphenyl degradation. *Basic Life Sci.* 45:253–269.

148. Van den Tweel, W. J. J., N. ter Burg, J. B. Kok, and J. A. M. de Bont. 1986. Biotransformation of 4-hydroxybenzoate from 4-chlorobenzoate by *Alcaligenes denitrificans* NTB-1. *Appl. Microbiol. Biotechnol.* 25:289–294.

149. Van Dort, H. M., and D. L. Bedard. 1991. Reductive *ortho* and *meta* dechlorination of polychlorinated biphenyl congener by anaerobic microorganisms. *Appl. Environ. Microbiol.* 57:1576–1578.

150. Véber, K., J. Zahradnik, and I. Breyl. 1980. Efficiency and rate of elimination of polychlorinated biphenyls from waste waters by means of algae. *Bull. Environ. Contam. Toxicol.* 25:841–845.

151. Viney, I., and R. J. F. Bewley. 1990. Preliminary studies on the development of a microbiological treatment for polychlorinated biphenyls. *Arch. Environ. Contam. Toxicol.* 19:789–796.

152. Weightman, A. J., R. H. Don, R. P. Lehrbach, and K. N. Timmis. 1984. The identification and cloning of genes encoding haloaromatic catabolic enzymes and the construction of hybrid pathways for substrate mineralization. In *Genetic Control of Environmental Pollutants,* Vol. 28. Ed. G. S. Omen, A. Hollaender, and M. Alexander. New York, NY: Plenum Press. 47–80.

153. Wong, P. T. S., and K. L. E. Kaiser. 1975. Bacterial degradation of PCBs. Rate studies. *Bull. Environ. Contamin. Toxicol.* 13:424–432.

154. Woodyard, J. P. 1990. PCB detoxification technologies: a critical assessment. *Environmental Progress* 9:131–135.

155. Yagi, D., and R. Sudo. 1980. Degradation of polychlorinated biphenyls by microorganisms. *Water Pollut. Control Fed.* 52:1035–1043.

156. Yates, J. R., and F. J. Mondello. 1989. Sequence similarities in the genes encoding polychlorinated biphenyl degradation by *Pseudomonas* strain LB400 and *Alcaligenes eutrophus* H850. *J. Bacteriol.* 171:1733–1735.

157. Zaïtsev, G. M., and Y. N. Karasevich. 1981. Utilization of 4-chlorobenzoic acid by *Arthrobacter globiformis*. *Mikrobiologyia* 50:35–40.

158. Zaïtsev, G. M., and Y. N. Karasevich. 1984. Utilization of 2-chlorobenzoic acid by *Pseudomonas cepacia*. *Mikrobiologyia* 53:75–80.

159. Zaïtsev, G. M., and Y. N. Karasevich. 1985. Utilization of 3-chlorobenzoic acid by *Arthrobacter calcoaceticus*. *Mikrobiologyia* 54:203–208.

160. Zylstra, G. J., and D. T. Gibson. 1989. Toluene degradation by *Pseudomonas putida* F1. Nucleotide sequence of the *TOD C1C2BADE* genes and their expression in *Escherichia coli*. *J. Biol. Chem.* 264:14940–14946.

CHAPTER 4

Chlorophenol Degradation

D.D. HALE, W. REINEKE, and J. WIEGEL

Abstract. Chlorophenols are a group of toxic compounds that have been widely used as biocides. Although chlorophenols have been detected in air, water, and soil samples, a potential for their degradation by microorganisms exists in both aerobic and anaerobic ecosystems. Several aerobic microorganisms that degrade chlorophenols have been isolated, and in some cases, the mechanisms by which degradation occurs have been elucidated. In contrast, only a few microorganisms that chlorinate chlorophenols under strict anaerobic conditions have been isolated in pure culture. Work with anaerobic mixed cultures indicates that anaerobic and aerobic microorganisms utilize different mechanisms for chlorophenol degradation.

INTRODUCTION

Chlorinated phenols constitute a series of 19 compounds consisting of mono-, di-, tri-, and tetrachloro- isomers and one pentachlorophenol (PCP). Chlorinated phenols with less than three chlorines seem to be of limited use today. Pentachlorophenol and the lower chlorinated phenols, tetra-, tri-, and dichlorophenol, have been in use as biocides to control bacteria, fungi, algae, mollusks, insects, slime, and other biota. They have also been used as precursors in the synthesis of other pesticides, such as the chlorinated phenoxyacetic acids, since the early 1930s. The annual world-wide production of chlorophenols has been estimated to be about 200,000 tons (73), of which PCP represents about 90,000 tons (22). The annual use of PCP in the US has been 23,000 tons (19, 28, 49, 63). The use of polychlorinated phenols has been banned or restricted in several countries since the late 1980s (34b), but because of past practices, chlorinated phenols are widespread in the environment today.

Potential environmental sources of chlorinated phenols include (1) direct soil applications as biocides; (2) leaching or vaporizing from treated wood items; (3) synthesis during routine chlorination processes of drinking water and wastewater at treatment plants, since both water sources can contain aromatic compounds of natural origin; (4) synthesis during production of bleached pulp in which chlorine is used; (5)

Drs. Hale and Wiegel are at the Department of Microbiology, University of Georgia, Athens, GA, U.S.A. Dr. Hale is also at Technology Applications, Inc., Athens, GA, U.S.A. Dr. Reineke is at Chemische Mikrobiologie, Bergische Universität Gesamthochschule Wuppertal, Wuppertal, Germany.

releases from factories into air and water; and (6) incineration of waste materials and burning of fresh lignocellulosic biomass, e.g. forest fires. Variations in the background concentrations of PCP have been reported; e.g. in air (0.25–0.93 ng/m³, Bolivian Andes) (16), water (8.2×10^3 ng/m³) (11), and soil (5.0×10^7 ng/m³) (70). Contaminated areas have been found to contain high concentrations of chlorinated phenols: e.g. air at Antwerp, Belgium (7 ng PCP/m³) (16), air in a wood-preservation factory (up to 1.7 µg PCP/m³) (104), and river sediments from Finland (5–11 ppb), Japan (80–360 ppb), and the US (1518 ppb PCP) (34b). The soil around preserving facilities in Finland have been shown to contain up to several grams of chlorophenols per kilogram of dry soil to a depth of several meters (55, 56, 99).

TOXICITY OF CHLOROPHENOLS

The bactericidal properties of chlorophenols have been known for many years (9, 57, 94). The potential of antibacterial effectiveness generally increases with degree of chlorine substitution, up to trichloro derivatives, tested with strains like *Enterobacter* (syn. Aerobacter) *aerogenes*, *Bacillus mycoides* (29, 100), *Salmonella* 'typhosa', and *Staphylococcus aureus* (9, 94). The growth of *Bacillus mycoides* is completely inhibited by 15 mg/l 2,4,5-trichlorophenol while that of *Pseudomonas putida* is inhibited by 6 mg/l 2,4-dichlorophenol or 64 mg/l of phenol (14, 100). The tetrachloro isomers are generally considerably less active than any of the trichloro isomers against aerobic bacteria. Pentachlorophenol is less effective than the tetrachloro isomers and is about as effective as phenol itself. The sensitivity of anaerobes against the chlorophenols varies (75, 102, 107). Some of the higher chlorophenols, particularly pentachlorophenol, have enhanced fungicidal activity. The bactericidal action of a disinfectant is usually expressed in terms of the phenol coefficient. Thus, disinfectants with high phenol coefficients are more potent compared to phenol. The phenol coefficients for selected chlorophenols are listed in Table 1, along with minimum inhibitory concentrations (MICs) for the bacterium, *Clostridium perfringens* (75) and lethal concentrations (LC_{50}-values) for trout (*Salmo trutta*) (44).

Table 1. Phenol coefficients, minimum inhibitory concentrations (MIC) to the bacterium, *Clostridium perfringens,* and lethal concentrations (LC_{50}-value) to trout (*Salmo trutta*) of selected chlorophenols.

Chlorophenol	Phenol coefficient	MIC (µmol/ml)	LC_{50}-value(ppm)
2-	3.7	>4	—
4-	4.6	>4	—
2,4-di-	13.3	2	1.7
2,6-di-	—	4	4.0
2,3,5-tri-	—	0.25	0.8
2,4,5-tri-	—	0.25	0.9
2,4,6-tri-	25.0	0.25	1.1
2,3,4,6-tetra-	—	0.031	0.5
penta-	—	0.062	0.2

DEGRADATION RATES OF CHLOROPHENOLS

A considerable amount of information is available regarding the stability of chlorinated phenols in the environment. The following generalizations can be made:

1. Chlorophenols are much more environmentally stable than the parent unsubstituted phenol (48).

2. As the number of chlorine substituents increases, the rate of aerobic decomposition decreases (95), whereas the opposite is generally true for the anaerobic degradation.

3. Compounds containing a *meta*-chlorine (i.e. 3-chloro- or 2,4,5-trichlorophenol) are more persistent under aerobic conditions than compounds lacking a chlorine substituent in positions *meta* to the hydroxyl group (1, 2).

Chlorinated phenols may be removed from a water body via: (1) volatilization, (2) photodegradation, (3) adsorption onto suspended (and thus not detachable) or bottom sediments, and (4) microbial degradation.

The time for complete disappearance in aerobic soil measured by Alexander and Aleem (1) was: 2 days for phenol, 14 days for 2-chlorophenol, >72 days for 3-chlorophenol, 9 days for 4-chlorophenol, 9 days for 2,4-dichlorophenol, >72 days for 2,4,5-trichlorophenol, 5 days for 2,4,6-trichlorophenol, >72 for days 2,3,4,6-tetrachlorophenol, and >72 days for pentachlorophenol. PCP can persist in water from 2 hours to 120 days (63), depending on the microbial population present and/or light, since PCP is also photodegraded by sunlight, with a half-life of 3.5 days in water (103).

The primary removal mechanism of PCP in soil is microbial degradation. PCP can persist in soil for 14 days to 5 years, depending on the microbial population present (19) and the environmental parameters, such as pH, temperature, aeration rate, available nutrients, the absence or presence of inhibitory copollutants, and the absence or presence of substances changing the electron flow in the system. Cole and Metcalf (20) and the U.S. EPA (98) reported a half-life for PCP of about 20 days.

MICROBIAL DEGRADATION OF CHLOROPHENOLS

Bacteria use different strategies to degrade chlorophenols:

1. Mono- and dichlorophenols are usually degraded aerobically by hydroxylation to chlorocatechols and by a spontaneous dechlorination after *ortho*-cleavage of the chlorocatechols.

2. Trichloro- and polychlorophenols are degraded aerobically via *para*-hydroquinones, which are subsequently dechlorinated before ring cleavage.

3. All chlorophenols (mono- to pentachlorophenol) are degraded under anaerobic conditions by a variety of microbial communities; degradation is initiated by reductive dechlorination followed by ring cleavage.

Aerobic Degradation of Mono- and Dichlorophenols

Several aerobic bacteria have been shown to grow with mono- and dichlorophenols (12, 13, 58, 82, 97). Chlorophenols, however, have never been used as the enrichment substrate because of their high toxicity. Instead, strains able to grow with mono- and dichlorophenols have been isolated with either 2,4-D or 3-chlorobenzoate. Chlorophenol degradation is part of the pathways involved in the degradation of chlorinated phenoxyacetates or phenoxy propionates (12, 13, 24, 32, 33, 47, 87). Alternatively, the modified *ortho*-pathway used by *Pseudomonas* sp. strain B13 (25) for degradation of 3-chlorobenzoate also functions for the degradation of monochlorophenols (58, 83). Most of the data on the pathways for 2,4-dichloro- and 4-chlorophenol were obtained in the laboratories of M. Alexander and W. C. Evans with the 2,4-D-degrading *Arthrobacter* sp. and *Pseudomonas* sp. (12, 13, 32, 33, 87, 96). Results obtained with *Pseudomonas* sp. strain B13, which grows with 3-chlorobenzoate and 4-chlorophenol and degrades 3-chloro-, 4-chloro-, and 3,5-dichlorocatechol, confirm and complete the pathways (25–27, 80–82). Figure 1 summarizes the data on the aerobic degradation pathways for mono- and dichlorophenols obtained with the three strains. Degradation of mono- and dichlorophenols proceeds by NADH- or NADPH-dependent hydroxylation as the initial step resulting in the formation of chlorocatechols: 3-chlorocatechol results from 2-chloro- and 3-chlorophenol, and 4-chlorocatechol results from 3-chloro- and 4-chlorophenol, whereas 3,5-dichlorocatechol results from 2,4-dichlorophenol (12, 32, 33, 58, 83). Molecular oxygen is required for the reaction. 2,4-Dichlorophenol hydroxylases have been purified and characterized (10, 65). Chlorocatechols are subject to *ortho*-cleavage by catechol 1,2-dioxygenases, which differ from the regular enzymes

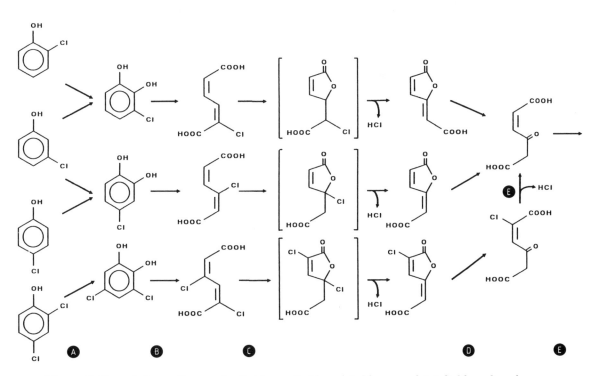

Figure 1. Degradation pathways for 2-chloro-, 3-chloro-, 4-chloro-, and 2,4-dichlorophenol Enzymes involved are phenol hydroxylase (A), catechol 1,2-dioxygenase (B), chloromuconate cycloisomerase (C), 4-carboxymethylenebut-2-en-4-olide hydrolase (D), and maleylacetate reductase (E).

involved in the degradation of aromatic compounds in that they are adapted to the conversion of chlorocatechols. Further degradation of chloro *cis,cis*-muconates by chloromuconate cycloisomerases yields 4-carboxymethylenebut-2-en-4-olides. Spontaneous dechlorinations occur after the lactonization by the cyloisomerases. Thus, no aryl dehalogenase is involved in this degradation. The 4-carboxymethylenebut-2-en-4-olides are converted into maleylacetates by 4-carboxy-methylenebut-2-en-4-olide hydrolases. 2-Chloromaleylacetate, the product of the degradation of 3,5-dichlorocatechol, is dechlorinated by a maleylacetate reductase, as has been shown for *Pseudomonas* sp. strain B13 (51). The same enzyme facilitates the conversion of maleylacetate to 3-oxoadipate (β-ketoadipate), a metabolite in the degradation of aromatic compounds in the so-called β-ketoadipate pathway.

Aerobic Degradation of Tri- and Pentachlorophenol

The present understanding of the degradative pathway for 2,4,5-trichlorophenol is derived from studies with the 2,4,5-T-degrading *Pseudomonas cepacia* strain AC1100 (50, 52, 53). 2,4,5-Trichlorophenol was degraded through 2,5-dichloro-*p*-hydroquinone and 5-chloro-1,2,4-trihydroxybenzene (77) (Figure 2).

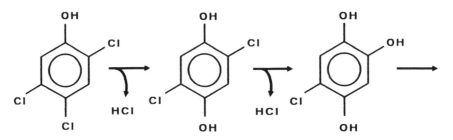

Figure 2. Proposed degradation pathway for 2,4,5-trichlorophenol in *Pseudomonas cepacia* strain AC1100.

A large variety of microorganisms of different genera have been isolated that are capable of degrading pentachlorophenol under aerobic conditions: *Arthrobacter* sp. (30, 79), coryneform-like bacteria (17, 18, 74), *Flavobacterium* sp. (76, 89, 90), *Pseudomonas* sp. (53, 91, 101), *Rhodococcus* sp. (4–8, 38, 39), *Mycobacterium* sp. (39), and some strains with unresolved taxonomic position (88). The pathway involved in the degradation of pentachlorophenol uses hydrolytic as well as reductive dechlorination mechanisms. The data were obtained with strain KC-3 (74), *Rhodococcus chlorophenolicus* strain PCP-I (4, 7, 8, 38), a *Flavobacterium* sp. (89), and *Arthrobacter* sp. strain ATCC 33790 (78) by isolation of metabolites and labeling experiments with $^{18}O_2$ and $H_2^{18}O$. Experiments with *Rhodococcus chlorophenolicus* demonstrated that the initial dechlorination of PCP proceeds by a hydrolytic displacement of chlorine in the *para* position, rather than by an oxygenase-catalyzed mechanism, since the ^{18}O labeled product of the dechlorination, tetrachloro-*p*-hydroquinone, was only found when cell extracts converted PCP in the presence of $H_2^{18}O$ (7). Conversion was only observed in the presence of molecular oxygen, however. Recently, Schenk et al. (78) demonstrated that unlabeled tetra-chloro-*p*-hydroquinone became labeled after incubation with the PCP-dehalogenating enzyme isolated from *Arthrobacter* sp. strain ATCC 33790 in $H_2^{18}O$. Therefore, distinction between an oxygenolytic or a hydrolytic dechlorination mechanism for the initial reaction with PCP is not possible. Two reductive dechlorinations of tetrachloro-*p*-hydroquinone followed, yielding first trichlorohydroquinone and then 2,6-dichlorohydroquinone (74, 89). Reactions whereby complete dechlorination occurred in strains KC-3 and *Flavobacterium* sp. were not elucidated. In contrast, *Rhodococcus chlorophenolicus* degrades tetrachloro-p-hydroquinone through a hydrolytic dechlorination and three reductive dechlorinations, thus producing 1,2,4-trihydroxybenzene (4, 8, 38). The pathways deduced from the metabolites isolated are given in Figure 3.

Synthesis of the enzyme(s) in *Rhodococcus chlorophenolicus* and other *Rhodococcus* sp. for the degradation of chlorinated phenols and hydroquinones, as well as guaiacols and syringols, is inducible (36, 37, 40). The inducer for the degradation pathway of chlorophenols is more likely the substrate than some later metabolite, because it was shown that the first intermediate, chlorohydroquinone, does not induce the degradation of its parent chlorophenol (7).

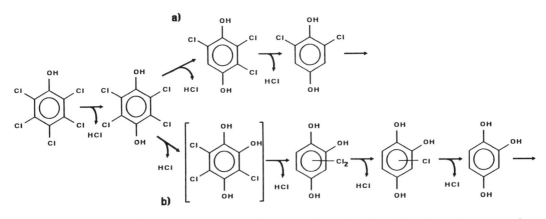

Figure 3. Proposed degradation pathways for pentachlorophenol. (a) *Flavobacterium* sp. and coryneform-like strain KC-3; (b) *Rhodococcus chlorophenolicus* strain PCP-I.

O-Methylation of Chlorophenols

Several bacteria belonging to the genera *Acinetobacter, Mycobacterium, Pseudomonas,* and *Rhodococcus* do not only degrade chlorinated phenols but are also able to O-methylate the compounds and their metabolites (3, 36, 38–40, 71, 72, 92, 93). Chloroanisoles, chloroveratroles, and chloro-1,4-dimethoxybenzenes were formed from chlorinated phenols, catechols, and hydroquinones by the O-methylating enzymes, thus indicating low substrate specificity of the enzymes. In *Rhodococcus chlorophenolicus,* a PCP-degrading organism, O-methylation requires two hydroxyl groups in *para* position to each other, even though only one of them was methylated to give 4-methoxyphenol (38). The enzyme prefers a substrate with two chlorine substituents flanking the hydroxyl group for O-methylation.

Compared to biodegradation, O-methylation is a slow process and is constitutively expressed, as has been shown by Häggblom et al. (38, 39) with chlorophenol-degrading strains of *Rhodococcus* and *Mycobacterium.* O-Methylation has significant environmental relevance (72), however, since the methoxy compounds may be more toxic than their precursors to other organisms of the ecosystem. In addition, O-methylation increases the lipophilicity and thus the bioaccumulation potential of the compound.

Anaerobic Degradation of Mono-, Di-, and Trichlorophenols

In contrast to the strategies discussed above that are used by aerobic bacteria to degrade chlorophenols, anaerobic bacteria initiate degradation of chlorophenols by reductively removing chlorine from the aromatic ring (for a general discussion of reductive dechlorination, see Hale, Jones, and Rogers (42b), this volume). In this transformation, chlorine atoms are replaced with hydrogen atoms. The phenol produced by reductive dechlorination may be subsequently transformed either through cyclohexanol, cyclohexanone, and adipate to succinate, propionate, and acetate (31, 105) or, as presently assumed, predominantly through benzoate and acetate to the end products, methane and carbon dioxide (59, 60, 84–86, 106–108). In contrast to chlorophenol degradation under aerobic conditions, chlorophenol degradation under anaerobic conditions usually requires a sequence of transformations involving more than one organism (Figure 4) (102, 107, 108). To date only a few strict anaerobic and facultative microaerophilic bacteria able to reductively dechlorinate chlorophenols have been isolated. A great deal of information, however, has been collected on anaerobic chlorophenol degradation in natural samples (15, 35, 41–43, 54, 61, 62, 64, 68, 69a) and enrichment cultures (21, 23, 67a).

Hale, Jones, and Rogers (42b) have discussed the anaerobic degradation of chlorophenols in comparison with other halogenated aromatic compounds. Therefore, we focus here mainly on our own research and some additional recent publications.

Dechlorination under Sulfate-reducing Conditions. Data from early studies suggested that degradation of chlorophenols occurs under methanogenic but not under sulfate-reducing conditions (35). King (54) demonstrated dichlorophenol transformation in a bromophenol-degrading marine sediment. Kohring et al. (62) first reported the dechlorination of 2,4-dichlorophenol and 4-chlorophenol in methanogenic pond sediments after initiation of sulfate reduction following the addition of sulfate (25 mM; the naturally occurring sulfate concentrations were between 0.05 and 0.15 mM). The temperature range and the rate of dechlorination in the presence of sulfate were reduced from about 5–55°C to 18–40°C and to approximately 10%, respectively, of that under methanogenic conditions. Subsequently, experiments performed with a highly enriched and heat-treated culture from these sediments indicated that competition for hydrogen between the dechlorinating organisms and the sulfate reducers may explain the inhibitory effect of added sulfate, i.e. creating sulfate-reducing conditions (21). This conclusion was drawn from the observations that (1) uninhibited reductive dechlorination and sulfate reduction occurred simultaneously in the presence of a sulfate reducer when a hydrogen atmosphere (5—25 psi) was supplied in addition to the small amounts of H_2 produced by the yeast extract–utilizing organisms present in the enrichment; (2) uninhibited reductive dechlorination occurred in the absence of externally added H_2 in the presence of the sulfate reducers when the growth of the sulfate reducer was inhibited by the addition of 10 mM molybdate. However, the addition of sulfate plus a sulfate reducer (*Desulfovibrio vulgaris*) caused a strong inhibition. Thus, whether dechlorination ocurs under sulfate-reducing conditions may depend to a great extent on the ratio of the affinities for H_2 of the reductive dechlorinating and the sulfate reducing organisms (see also the effect of sulfuroxy ions on pentachlorophenol degradation in the section, Anaerobic Degradation of Pentachlorophenol, below).

Recently, Haggblom and Young (41) demonstrated that chlorophenol degradation in samples from predominantly sulfate-reducing environments can be coupled to sulfate reduction, i.e. they observed a stoichiometric consumption of sulfate for the complete oxidation of chlorophenol to CO_2. 2,4-Dichlorophenol was dechlorinated to 4-chloro-

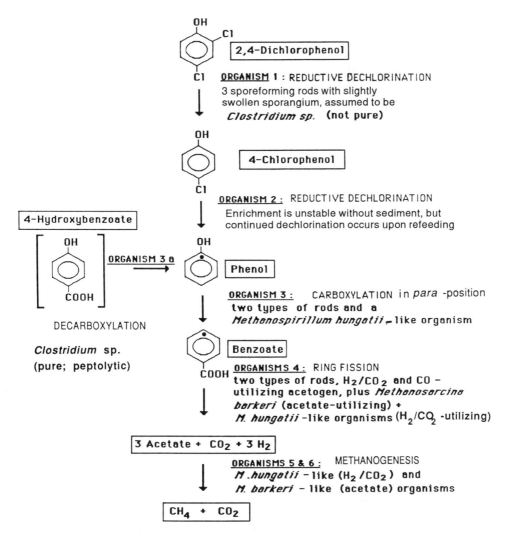

Figure 4. Sequential degradation of 2,4-dichlorophenol under anaerobic (methanogenic) conditions in a lake sediment (Sandy Creek Nature Park, Athens, Georgia) as proposed from identified intermediates, isolated strains, and enrichment cultures.

phenol as a transient intermediate product by estuarine-derived sediment cultures. In contrast to the freshwater enrichment mentioned above, the addition of molybdate to inhibit sulfate reducers also completely inhibited the dechlorination. The observed dechlorination rates by sulfate-reducing estuarine sediment cultures were up to 37 μmol per liter per day. Added monochlorophenols disappeared, but no dechlorination products were observed. Sulfidogenic cultures from an anaerobic bioreactor also transformed 2,4-dichlorophenol to 4-chlorophenol, but the 4-chlorophenol was not further metabolized. In contrast, the 2-chlorophenol formed from 2,6-dichlorophenol was dechlorinated to phenol by these cultures.

Dechlorination in Freshwater Sediments. Extended studies of the microbially mediated dechlorination patterns of chlorophenols in pond sediments revealed (1) a positional preference for chlorine removal from the aromatic ring, and (2) environmental factors that were correlated to the time required for dechlorination of some isomers (43,

61). In sediments collected every other month for a year from five sites in Cherokee Pond (Athens, Georgia), chlorine was preferentially removed in the order *ortho* > *para* > *meta* from 2,4-, 2,5-, and 3,4-dichlorophenol (43). Stepwise regression analysis of data on the physical, chemical, and microbiological characteristics of the sediments indicated that pH, redox potential, and concentration of sulfate and nitrate correlated to the time required to dechlorinate 50% of a dichlorophenol to a monochlorophenol (T_{50}). Studies on the temperature dependence of 2,4-dichlorophenol and 4-chlorophenol dechlorination in sediments from another pond showed a linear relationship in an Arrhenius graph only for the range from about 15–30°C, although degradation was observed between 4 and 55°C (61). In the Cherokee Pond sediments collected over a year (43), the number of microorganisms dechlorinating 2,4-dichlorophenol varied from 1.8×10^2 to 4.8×10^4 cells/gram dry weight sediment (most probable number assay). No statistical correlation was observed between the determined number of dechlorinating organisms and the T_{50} value.

Data on the substrate specificity of microorganisms in methanogenic pond sediments enriched on a single dichlorophenol indicate the presence of different dechlorinating organisms. The activities obtained depended on the enrichment method used for a given pond sediment and the source of the pond sediments. For example, Hale et al. (42a) reported that microorganisms in sediments from Bolton's Pond (Athens, Georgia) enriched on 2,3-, 2,4-, or 2,6-dichlorophenol exhibited different dechlorinating activity toward 2,3-dichlorophenol than similarly enriched Cherokee Trailer Park pond (Athens, Georgia) sediment microorganisms. The enriched Bolton's sediment microorganisms, for example, dechlorinated 2,3-dichlorophenol to 3-chlorophenol, whereas the enriched Cherokee Pond sediment microorganisms dechlorinated this isomer to both 2- and 3-chlorophenol. Bryant et al. (15) also noted that Cherokee sediment microorganisms enriched on 3,4-dichlorophenol dechlorinated 2,4-dichlorophenol to 2-and 4-chlorophenol, whereas sediment microorganisms from this pond enriched on 2,4-dichlorophenol only dechlorinated this isomer to 4-chlorophenol.

Chlorophenol Degradation in Enrichment and Pure Cultures.

A few highly enriched cultures able to dehalogenate dichlorophenols have been reported. Dietrich and Winter (23) enriched an anaerobic mixed culture from sewage sludge that mineralized 2,6-di- and 2-chlorophenol to methane and carbon dioxide and transformed 2,4-dichlorophenol to 4-chlorophenol. The dechlorination required concomitant oxidation of short-chain (n = 4–10) fatty acids. A spirochaete-like organism was assumed to be the dechlorinating bacterium because loss of dechlorinating activity accompanied loss of this organism and vice versa.

A different culture enriched on 2,4,5-, 2,4,6-, and 3,4,5-trichlorophenol was obtained from sewage sludge by Madsen and Aamand (67a). The culture, containing spore-forming organisms, utilized yeast extract, peptone, casamino acids, glucose, galactose, and lactose as carbon sources; 2,4,5- and 3,4,5,-trichlorophenol were dechlorinated to 3,4-dichlorophenol and 3-chlorophenol, respectively, whereas 2,4,6-trichlorophenol was dechlorinated to 2,4-dichlorophenol and 4-chlorophenol. Whether the culture transformed the monochlorophenols, however, was not clear. During the degradation of 2,4,6-trichlorophenol, the most probable number of dechlorinating organisms increased four orders of magnitude from 10^3 to 10^7 cells/ml enrichment culture.

In comparison, Dalton (21) determined 10^8–10^9 2,4-dichlorophenol-dechlorinating microorganisms per milliliter primary enrichment culture from freshwater pond sediment, whereas the sediment-free enrichment culture (using 0.3% yeast extract) contained only about 10^7 cells per milliliter culture. In contrast to the spirochaete-containing enrichment from sludge obtained by Dietrich and Winter (23), the highly enriched culture, containing only sporeformers, obtained by Dalton dechlorinated 2,4-dichlorophenol and 2,4,6-trichlorophenol but not mono- or other dichlorophenols. Optimal dechlorination occurred at pH 8.5 under an atmosphere of 10–25 psi H_2 with yeast extract as the main carbon and energy source. The dechlorinating organisms did not appear to be the dominant organism among the three observed sporulating morphotypes. A heat treatment of the enrichment obtained by Dalton resulted in a significant shift in the fermentation products: from 40–50 mM acetate before the treatment to trace amounts of acetate and other volatile fatty acids after the heat treatment. In contrast to the culture of Dietrich and Winter (23), no requirements for fatty acids were observed.

The addition of common supplements (such as lake water, humic acid, yeast extract, rumen fluid, and some rare compounds such as phenylacetic acid) in purification schemes of anaerobic organisms resisting an easy isolation did not lead to the isolation or even a further enrichment of the dechlorinating organism beyond 10^8 dechlorinating cells per ml compared to a total cell count of about 10^{10} cells per ml culture. Preparing the medium with the culture supernatant from a mixed culture derived from serial dilutions revealed only a minor stimulation of the dechlorination activity. However, by using different concentrations and combinations of antibiotics, we were able to identify conditions that did not inhibit the dechlorination but reduced growth significantly. Consequently, the screening of supernatants from cultures isolated under these conditions led to the isolation of two cultures in which supernatants strongly stimulated the dechlorination activity. The highest stimulation effect was observed with use of a supernatant in which the two organisms had been grown together in a yeast-extract-containing medium (I. Utkin and J. Wiegel, unpublished results). The two clostridial bacteria did not show any dechlorinating activity either alone or in a mixed culture. A pure culture of a dechlorinating organism JW/UI-DCI was finally obtained using 3 chloro-4-hydroxyphenylacetic acid as substrate.

A survey of compounds that were dechlorinated by the further purified culture revealed a simple rule: in the dechlorination of the phenolic *ortho*-chlorine substituent, the phenolic *meta* position must be unsubstituted and the phenolic *para* position must be substituted. Surprisingly, the substituent in the phenolic *para* position can be a halogen-, hydroxyl-, carboxyl-, acyl-, methyl-, or nitro-group; thus, quite different classes of compounds can be dechlorinated. Use of 3-chloro-4-hydroxyphenylacetic acid as an electron acceptor led to substantial growth, but only in medium made with the supernatant of the culture from the combined yeast-extract utilizing organism described above. The improved dechlorinating culture grew with CO_2 as the main carbon source and formate and/or 3-chloro-4-hydroxyphenylacetic acid as electron acceptors.

The predominant substrate specificity of various bacteria for the reductive transformation of chlorophenols remains unclear and requires further study. Because characteristics may vary from mixed microbial communities to pure cultures, as indicated by Dalton (21) with the 2,4-dichlorophenol dechlorinating enrichment, the isolation and characterization of the organisms mediating each of the sequential steps are necessary to prove their distinct identities.

Foxworthy et al. (34a) reported the enrichment and isolation of an organism that reductively dechlorinates 2-chlorophenol under anaerobic conditions. The organism, which is a facultative microaerophile enriched from ditch sediment, grows on a reduced anaerobic mineral medium that contains acetate, formate, vitamins, bromoethanesulfonic acid, and 2-chlorophenol. No growth occurs when either acetate or 2-chlorophenol are omitted from the medium. The thin, nonmotile rods grow singly or in short chains and form small red colonies on solid media. 16S rRNA analysis indicates that this organism belongs to the delta subdivision of the proteobacteria and shares sequence similarity with *Myxococcus xanthus*. This characteristic, in addition to its preference for removing *ortho*-chlorines, differentiates it from the 3-chlorobenzoate-dechlorinating *Desulfomonile tiedjei*, which preferentially removes *meta*-chlorines from highly chlorinated but not monochlorinated phenols (69b), and the chlorophenol-dechlorinating anaerobe recently isolated by Madsen and Licht (67b). [For a description of *D. tiedje*, see the discussion by Hale, Jones, and Rogers, this volume (42b).]

The obligately anaerobic, slightly curved, motile rod obtained by Madsen and Licht (67b) was isolated from a stable trichlorophenol-dechlorinating consortium enriched from municipal sludge. The bacterium utilizes pyruvate as a sole carbon source, producing acetate as the primary fermentation product. The highest growth rate and cell yield are obtained when the organism is grown in media containing yeast extract in addition to pyruvate. The organism stains gram-positive, forms endospores, reduces nitrate to nitrite and ammonium, and does not reduce sulfate. Thus, it is proposed to be related to the genus *Clostridum*. Interestingly, this organism dechlorinates 2,4-dichlorophenol, 2,4,5- and 2,4,6-trichlorophenol, and pentachlorophenol at the *ortho* position but does not remove *para*-chlorines. The bacterium only dechlorinates at the *meta* position of 3,5-dichlorophenol when the substrate concentration is relatively low (i.e. 50 μmol).

The enrichment from Sandy Creek Nature Center pond has been successfully used to initiate 2,4-dichlorophenol dechlorination in a sediment that previously did not exhibit this activity (102), thus suggesting that it should be possible to accelerate dechlorination in contaminated sediments by supplying the proper organisms. About 180 μM 2,4-dichlorophenol was dechlorinated within 6 days when as little as 0.1% (v/v) of the culture was inoculated into a seagrass bed sediment from a mangrove swamp in the Bahamas. In this special case, however, a 1:1 dilution of the sediment with 10 mM phosphate buffer was required to reduce the inhibitory effect of the approximately 3.6% (w/v) NaCl concentration of the marine sample, since salt concentrations above 2% were strongly inhibitory to the enrichment.

Anaerobic Degradation of Pentachlorophenol

A summary of initial research on the anaerobic degradation of PCP has been provided by Hale, Jones, and Rogers (42b); thus, again, we focus mainly on some recent work. Similar to the anaerobic biodegradation of mono- and dichlorophenols, pentachlorophenol may be degraded in a number of anoxic environments by reductive dechlorination and subsequent transformations of the ring to yield carbon dioxide and methane. Methoxylation of pentachlorophenol, as well as its dechlorination products, has also been reported as a minor pathway for anaerobic PCP biotransformation.

Complete dechlorination of PCP to phenol occurred by different pathways in a mixture of sewage sludges adapted to dechlorinate each of the monochlorophenols individually (68, 69a) and in a mixture of sediments adapted to dechlorinate 2,4- and 3,4-dichlorophenol (15). In the sewage sludge mixture, PCP was dechlorinated to 2,3,4,5-tetrachlorophenol, 3,4,5-trichlorophenol, 3,4- and 3,5-dichlorophenol, and 3-chlorophenol. In a mixture of Cherokee Pond sediments that had previously dechlorinated either 2,4-dichlorophenol (*ortho*-dechlorination) or 3,4-dichlorophenol (*para*-dechlorination), PCP was initially dechlorinated to 2,3,5,6-tetrachlorophenol. This chlorophenol was subsequently dechlorinated to 2,3,5-trichlorophenol and 3,5-dichlorophenol. Phenol was the final product of removal of the two *meta*-chlorines from 3,5-dichlorophenol.

These two PCP dechlorination pathways (consisting of two *ortho*-cleavages followed by a *para*-cleavage and of a *para*-cleavage followed by two *ortho*-cleavages), also occurred in samples from a number of environments that were incubated at elevated temperatures (50°C) (64). Although one of the pathways usually predominated in samples from each of the environments incubated at 50°C, both pathways occurred in samples from some environments, including those from an alder swamp and a freshwater stream. Of samples from three freshwater sediments, four anaerobic sewage sludge digesters exposed to industrial inputs, and digested manure from a reactor, those from freshwater environments most readily dechlorinated PCP. For example, in freshwater sediments, PCP was completely transformed to lesser chlorinated compounds after 8 months of incubation, whereas only 3 to 18% of the PCP was transformed in the various sewage sludge samples.

Work by Hendriksen et al. (45, 46) on the dechlorination of PCP in two anaerobic reactors indicated the predominance of one dechlorination pathway, depending on the source of the inoculum. In fixed-film reactors inoculated with anaerobic digested sewage sludge from a municipal treatment plant, the PCP dechlorination pathway was that of two *ortho*-dechlorinations followed by a *para*-dechlorination. In contrast, in upflow anaerobic sludge blanket (UASB) reactors inoculated with granular sludge grown on sugar-containing wastewater, the major pathway was that of a *para*-cleavage followed by two *ortho*-cleavages. Although the addition of glucose to both types of reactors did not alter the predominant dechlorination pathway, the dechlorination of PCP increased in fixed-film and UASB reactors with this carbon source. This enhancement was a result of both higher biomass concentrations and dechlorination rates.

The PCP dechlorination pathway utilized by the mixed culture enriched by Madsen and Aamand (66) from anaerobic sewage sludge was that of two *ortho*-cleavages followed by a *para*-cleavage. This pathway was inhibited by sulfate, thiosulfate, and sulfite and was retarded at dissolved H_2 concentrations less than 0.11 μM under sulfate-reducing conditions. The addition of molybdate to cultures containing sulfate relieved the inhibitory effect of this anion, whereas the addition of H_2 reduced the inhibitory effect of each of the sulfuroxy anions. A competition for H_2 by the sulfate-reducing and dechlorinating microorganisms [similar to that proposed by Dalton (21) for the 2,4-DCP dechlorinating enrichment] was suggested to be the mechanism by which sulfuroxy anions inhibit dechlorination.

Acknowledgments

Research done in the laboratory of J. W. was supported by grants from the U.S. Department of Energy DE-FG09-89-14059, the U.S. Environmental Protection Agency under the University of Georgia Cooperative Agreement, and the U.S.-Naval Research Office. J. W. thankfully acknowledges the skillful work of G. W. Kohring, D. Dalton, and X. Zhang and the helpful discussions with J. E. Rogers. This chapter was written in spring 1992.

LITERATURE CITED

1. Alexander, M., and M. I. H. Aleem. 1961. Effect of chemical structure on microbial decomposition of aromatic herbicides. *J. Agric. Food Chem.* 9:44–47.

2. Alexander, M., and B. K. Lustigman. 1966. Effect of chemical structure on microbial degradation of substituted benzenes. *J. Agric. Food Chem.* 14:410–413.

3. Allard, A.-S., M. Remberger, and A. H. Neilson. 1987. Bacterial O-methylation of halogen-substituted phenols. *Appl. Environ. Microbiol.* 53:839–845.

4. Apajalahti, J. H. A. 1987. Ph.D. Thesis, University of Helsinki, Finland.

5. Apajalahti, J. H. A., P. Kärpänoja, and M. S. Salkinoja-Salonen. 1986. *Rhodococcus chlorophenolicus* sp. nov., a chlorophenol-mineralizing actinomycete. *Int. J. Syst. Bacteriol.* 36:246–251.

6. Apajalahti, J. H. A., and M. S. Salkinoja-Salonen. 1986. Degradation of chlorinated phenols by *Rhodococcus chlorophenolicus*. *Appl. Microbiol. Biotechnol.* 25:62–67.

7. Apajalahti, J. H. A., and M. S. Salkinoja-Salonen. 1987. Dechlorination and para-hydroxylation of polychlorinated phenols by *Rhodococcus chlorophenolicus*. *J. Bacteriol.* 169:675–681.

8. Apajalahti, J. H. A., and M. S. Salkinoja-Salonen. 1987. Complete dechlorination of tetrachlorohydroquinone by cell extracts of pentachlorophenol-induced *Rhodococcus chlorophenolicus*. *J. Bacteriol.* 169:5125–5130.

9. Baker, J. W., I. Schumacher, and D. P. Roman. 1970. Antiseptics and disinfectants. In *Medical Chemistry*, Part I, 4th ed. Ed. A. Burger. New York: John Wiley and Sons, Interscience Publishers. 627–661.

10. Beadle, C. A., and A. R. W. Smith. 1982. The purification and properties of 2,4-dichlorophenol hydroxylase from a strain of *Acinetobacter* species. *Eur. J. Biochem.* 123:323–332.

11. Bevenue, A., J. N. Ogata, and J. W. Hylin. 1972. Organochlorine pesticides in rainwater, Oahu, Hawaii, 1971–1972. *Bull. Environ. Contam. Toxicol.* 8:238–241.

12. Bollag, J.-M., C. S. Helling, and M. Alexander. 1968. 2,4-D-metabolism. Enzymatic hydroxylation of chlorinated phenols. *J. Agric. Food Chem.* 16:826–828.

13. Bollag, J.-M., G. G. Briggs, J. E. Dawson, and M. Alexander. 1968. 2,4-D metabolism: Enzymatic degradation of chlorocatechols. *J. Agric. Food Chem.* 16:829–833.

14. Bringmann, G., and R. Kühn. 1977. Grenzwerte der Schadwirkung wassergefährdender Stoffe gegen Bakterien (*Pseudomonas putida*) and Grünalgen (*Scenedesmus quadricauda*) im Zellvermehrungshemmtest. *Z. Wasser Abwasser Forsch.* 10:87–98.

15. Bryant, F. O., D. D. Hale, and J. E. Rogers. 1991. Regiospecific dechlorination of pentachlorophenol by dichlorophenol-adapted microorganisms in freshwater, anaerobic sediment slurries. *Appl. Environ. Microbiol.* 57:2293–2301.

16. Cautreels, W., K. van Cauwenberghe, and L. A. Guzmann. 1977. Comparison between the organic fraction of suspended matter at a background and an urban station. *Sci. Total Environ.* 8:79–88.

17. Chu, J. P., and E. J. Kirsch. 1972. Metabolism of pentachlorophenol by an axenic bacterial culture. *Appl. Microbiol.* 23:1033–1035.

18. Chu, J. P., and E. J. Kirsch. 1973. Utilization of halophenols by a pentachlorophenol metabolizing bacterium. *Dev. Ind. Microbiol.* 14:264–273.

19. Cirelli, D. P. 1978. Pentachlorophenol position document 1. *Fed. Reg.* 43:48446–48477.

20. Cole, L. K., and R. L. Metcalf. 1980. Environmental destinies of insecticides, herbicides, and fungicides in the plants, animals, soil, air, and water of homologous microcosms. In *Microcosms in Ecological Research*. Ed. J. Giesy. Washington, D.C.: Technical Information Center, US DOE. 971.

21. Dalton, D. D. 1990. M.S. thesis. University of Georgia, Athens, Georgia.

22. Detrick, R. S. 1977. Pentachlorophenol. Possible sources of human exposure. *Forest Product J.* 27:13–16.

23. Dietrich G., and J. Winter 1990. Anaerobic degradation of chlorophenol by an enrichment culture. *Appl. Microbiol. Biotechnol.* 34:253–258.

24. Ditzelmüller, G., M. Loidl, and F. Streichsbier. 1989. Isolation and characterization of a

2,4-dichlorophenoxyacetic acid-degrading soil bacterium. *Appl. Microbiol. Biotechnol.* 31:93–96.

25. Dorn, E., M. Hellwig, W. Reineke, and H.-J. Knackmuss. 1974. Isolation and characterization of a 3-chlorobenzoate degrading pseudomonad. *Arch. Microbiol.* 99:61–70.

26. Dorn, E., and H.-J. Knackmuss. 1978. Chemical structure and biodegradability of halogenated aromatic compounds. Two catechol 1,2-dioxygenases from a 3-chlorobenzoate-grown pseudomonad. *Biochem. J.* 174:73–84.

27. Dorn, E., and H.-J. Knackmuss. 1978. Chemical structure and biodegradability of halogenated aromatic compounds. Substituent effects on 1,2-dioxygenation of catechol. *Biochem. J.* 174:85–94.

28. Dougherty, R. C. 1978. Human exposure to pentachlorophenol. In *Pentachlorophenol: Chemistry, Pharmacology, and Environmental Toxicology.* Ed. K. R. Rao. New York: Plenum Press. 351–361.

29. Dow Chemical Company. 1969. Hazards due to toxicity and precautions for safe handling and use. In *Antimicrobial Agents,* Section I-10, Dowicide B Antimicrobial. Midland, MI: Dow Chemical Company. 3.

30. Edgehill, R. U., and R. K. Finn. 1982. Isolation, characterization and growth kinetics of bacteria metabolizing pentachlorophenol. *Eur. J. Appl. Microbiol. Biotechnol.* 16:179–184.

31. Evans, W. C. 1977. Biochemistry of the bacterial catabolism of aromatic compounds in anaerobic environments. *Nature* 270:17–22.

32. Evans, W. C., B. S. W. Smith, H. N. Fernley, and J. I. Davies. 1971. Bacterial metabolism of 2,4-dichlorophenoxyacetate. *Biochem. J.* 122:543–551.

33. Evans, W. C., B. S. W. Smith, P. Moss, and H. N. Fernley. 1971. Bacterial metabolism of 4-chlorophenoxyacetate. *Biochem. J.* 122:509–517.

34a. Foxworthy, A. L., W. W. Mohn, and J. R. Cole. 1992. Enrichment and isolation of a novel bacterium growing by anaerobic reductive dehalogenation of 2-chlorophenol. Asbtr. Q-197. *Abstr. 92nd Gen. Meet.* American Society for Microbiology. 368.

34b. Gesellschaft Deutscher Chemiker (GDCh). 1985. *Stoffbericht: Pentachlorophenol, GDCh-Beratergremium für umweltrelevante Altstoffe (BUA).* Weinheim: VCH Verlagsgesellschaft.

35. Gibson, S. A., and J. M. Suflita. 1986. Extrapolation of biodegradation results to groundwater aquifers: reductive dehalogenation of aromatic compounds. *Appl. Environ. Microbiol.* 52:681–688.

36. Häggblom, M. M. 1988. Ph.D. Thesis, University of Helsinki, Finland.

37. Häggblom, M. M., J. H. A. Apajalahti, and M. S. Salkinoja-Salonen. 1988. Hydroxylation and dechlorination of chlorinated guaiacols and syringols by *Rhodococcus chlorophenolicus. Appl. Environ. Microbiol.* 54:683–687.

38. Häggblom, M. M., J. H. A. Apajalahti, and M. S. Salkinoja-Salonen. 1988. O-methylation of chlorinated *para* hydroquinones by *Rhodococcus chlorophenolicus. Appl. Environ. Microbiol.* 54:1818–1824.

39. Häggblom, M. M., L. J. Nohynek, and M. S. Salkinoja-Salonen. 1988. Degradation and O-methylation of polychlorinated phenolic compounds by strains of *Rhodococcus* and *Mycobacterium. Appl. Environ. Microbiol.* 54:3043–3052.

40. Häggblom, M. M., D. Janke, P. J. M. Middledorp, and M. S. Salkinoja-Salonen. 1989. O-methylation of chlorinated phenols in the genus *Rhodococcus. Arch. Microbiol.* 152:6–9.

41. Haggblom, M. M., and L. Y. Young. 1990. Chlorophenol degradation coupled to sulfate reduction. *Appl. Environ. Microbiol.* 56:3255–3260.

42a. Hale, D. D., J. E. Rogers, and J. Wiegel. 1990. Reductive dechlorination of dichlorophenols by nonadapted and adapted microbial communities in pond sediments. *Microb. Ecol.* 20:185–196.

42b. Hale, D. D., W. J. Jones, and J. E. Rogers. 1994. Biodegradation of chlorinated homocyclic and heterocyclic compounds in anaerobic environments. In *Biological Degradation and*

Bioremediation of Toxic Chemicals. Ed. G. R. Chaudry. Portland, OR: Dioscorides Press.

43. Hale, D. D., J. E. Rogers, and J. Wiegel. 1991. Environmental factors correlated to dichlorophenol dechlorination in anoxic freshwater sediments. *Environ. Toxicol. Chem.* 10:1255–1265.

44. Hattula, M. L., V. M. Wasenius, H. Reunanen, and A. U. Arstila. 1981. Acute toxicity of some chlorinated phenols, catechols and cresols to trout. *Bull. Environ. Contam. Toxicol.* 26:295–298.

45. Hendriksen, H. V., S. Larsen, and B. K. Ahring. 1991. Anaerobic degradation of PCP and phenol in fixed-film reactors: the influence of an additional substrate. *Water Sci. Technol.* 24:431–436.

46. Hendriksen, H. V., S., Larsen, and B. K. Ahring. 1992. Influence of a supplemental carbon source on anaerobic dechlorination of pentachlorophenol in granular sludge. *Appl. Environ. Microbiol.* 58:365–370.

47. Horvath, M., G. Ditzelmüller, M. Loidl, and F. Streichsbier. 1990. Isolation and characterization of a 2-(2,4-dichlorophenoxy)propionic acid-degrading soil bacterium. *Appl. Microbiol. Biotechnol.* 33:213–216.

48. Ingols, R. S., P. E. Gaffney, and P. C. Stevenson. 1966. Biological activity of halophenols. *J. Water. Pollut. Control Fed.* 38:629–635.

49. International Agency for Research on Cancer (IARC). 1979. *Some Halogenated Hydrocarbons. IARC monographs on the Evaluation of Carcinogen Risk of Chemicals to Humans.* 20:303–325.

50. Karns, J. S., J. J. Kilbane, S. Duttagupta, and A. M. Chakrabarty. 1983. Metabolism of halophenols by 2,4,5-trichlorophenoxyacetic acid-degrading *Pseudomonas cepacia. Appl. Environ. Microbiol.* 46:1182–1186.

51. Kaschabek, S. R. 1990. Diploma Thesis, Universität Wuppertal.

52. Kellogg, S. T., D. K. Chatterjee, and A. M. Chakrabarty. 1981. Plasmid-assisted molecular breeding: new technique for enhanced biodegradation of persistent toxic chemicals. *Science* 214:1133–1135.

53. Kilbane, J. J., D. K. Chatterjee, J. S. Karns, S. T. Kellog, and A. M. Chakrabarty. 1982. Biodegradation of 2,4,5-trichlorophenoxyacetic acid by a pure culture of *Pseudomonas cepacia. Appl. Environ. Microbiol.* 44:72–78.

54. King, G. M. 1988. Dehalogenation in marine sediments containing natural sources of halophenols. *Appl. Environ. Microbiol.* 54:3079–3085.

55. Kitunen, V., R. Valo, and M. S. Salkinoja-Salonen. 1985. Analysis of chlorinated phenols, phenoxyphenols and dibenzofurans around wood preserving facilities. *Int. J. Environ. Anal. Chem.* 20:13–28.

56. Kitunen, V., R. Valo, and M. S. Salkinoja-Salonen. 1987. Contamination of soil around wood-preserving facilities by polychlorinated aromatic compounds. *Environ. Sci. Technol.* 21:96–101.

57. Klarmann, E. G. 1963. Antiseptics and disinfectants. In *Kirk-Othmer Encyclopedia of Chemical Technology,* 2nd ed. Vol. 2. New York: John Wiley and Sons, Interscience Publishers. 623–630.

58. Knackmuss, H.-J., and M. Hellwig. 1978. Utilization and cooxidation of chlorinated phenols by *Pseudomonas* sp. B13. *Arch. Microbiol.* 117:1–7.

59. Knoll, G., and J. Winter. 1987. Anaerobic degradation of phenol in sewage sludge. Benzoate formation from phenol and carbon dioxide in the presence of hydrogen. *Appl. Microbiol. Biotechnol.* 25:384–391.

60. Knoll, G., and J. Winter. 1989. Degradation of phenol via carboxylation to benzoate by a defined, obligate syntrophic consortium of anaerobic bacteria. *Appl. Microbiol. Biotechnol.* 30:318–324.

61. Kohring, G. W., J. E. Rogers, and J. Wiegel. 1989. Anaerobic biodegradation of 2,4-di-

chlorophenol in freshwater lake sediments at different temperatures. *Appl. Environ. Microbiol.* 55:348–353.

62. Kohring, G. W., X. Zhang, and J. Wiegel. 1989. Anaerobic dechlorination of 2,4-dichlorophenol in freshwater sediments in the presence of sulfate. *Appl. Environ. Microbiol.* 55:2735–2737.

63. Kozak, V. P., G. V. Sissiman, G. Chesters, D. Stensby, and J. Harkin. 1979. *Reviews of the Environmental Effects of Pollutants: XI. Chlorophenols.* Cincinnati, Ohio: US EPA Health Effects Research Laboratories, EPA-600/1-79-012.

64. Larsen, S., H. V. Hendriksen, and B. K. Ahring. 1991. Potential for thermophilic (50°C) anaerobic dechlorination of pentachlorophenol in different ecosystems. *Appl. Environ. Microbiol.* 57:2085–2090.

65. Liu, T., and P. J. Chapman. 1984. Purification and properties of a plasmid-encoded 2,4-dichlorophenol hydroxylase. *FEBS Lett.* 173:314–318.

66. Madsen, T., and J. Aamand. 1991. Effects of sulfuroxy anions on degradation of pentachlorophenol by a methanogenic enrichment culture. *Appl. Environ. Microbiol.* 57:2453–2458.

67a. Madsen, T., and J. Aamand. 1992. Anaerobic transformation and toxicity of trichlorophenols in a stable enrichment culture. *Appl. Environ. Microbiol.* 58:557–561.

67b. Madsen, T., and D. Licht. 1992. Isolation and characterization of an anaerobic chlorophenol-transforming bacterium. *Appl. Environ. Microbiol.* 58:2874–2878.

68. Mikesell, M. D., and S. A. Boyd. 1986. Complete reductive dechlorination and mineralization of pentachlorophenol by anaerobic microorganisms. *Appl. Environ. Microbiol.* 52:861–865.

69a. Mikesell, M. D., and S. A. Boyd. 1988. Enhancement of pentachlorophenol degradation in soil through induced anaerobiosis and bioaugmentation with anaerobic sewage sludge. *Environ. Sci. Technol.* 22:1411–1414.

69b. Mohn, W. W., and K. J. Kennedy. 1992. Reductive dehalogenation of chlorophenols by *Desulfomonile tiedjei* DCB-1. *Appl. Environ. Microbiol.* 58:1367–1370.

70. Murray, H. E., L. E. Ray, and C. S. Giam. 1981. Analysis of marine sediment, water and biota for selected organic pollutants. *Chemosphere* 10:1327–1334.

71. Neilson, A. H., A.-S. Allard, P. A. Hynning, M. Remberger, and L. Landner. 1983. Bacterial methylation of chlorinated phenols and guaiacols: formation of veratroles from guaiacols and high-molecular-weight chlorinated lignins. *Appl. Environ. Microbiol.* 45:774–783.

72. Neilson, A., A.-S. Allard, S. Reiland, M. Remberger, A. Tärnholm, T. Viktor, and L. Landner. 1984. Tri- and tetrachloroveratrole, metabolites produced by bacterial O-methylation of tri- and tetrachloroguaiacol: an assessment of their bioconcentration potential and their effects on fish reproduction. *Can. J. Fish. Aquat. Sci.* 41:1502–1512.

73. Paasivirta, J. 1978. Chlorophenols—poisons, possible environmental poisons. *Kemia Kemi* (Helsinki) 9:367–370.

74. Reiner, E. A., J. Chu, and E. J. Kirsch. 1978. Microbial metabolism of pentachlorophenol. In *Pentachlorophenol: Chemistry, Pharmacology and Environmental Toxicology.* Ed. K. R. Rao. New York: Plenum Press. 67–81.

75. Ruckdeschel, G., G. Renner, and K. Schwarz. 1987. Effects of pentachlorophenol and some of its known and possible metabolites on different species of bacteria. *Appl. Environ. Microbiol.* 53:2689–2692.

76. Saber, D. L., and R. L. Crawford. 1985. Isolation and characterization of *Flavobacterium* strains that degrade pentachlorophenol. *Appl. Environ. Microbiol.* 50:1512–1518.

77. Sangodkar, U. M. X., P. J., Chapman, and A. M. Chakrabarty. 1988. Cloning, physical mapping and expression of chromosomal genes specifying degradation of the herbicide 2,4,5-T by *Pseudomonas cepacia* AC1100. *Gene* 71:267–277.

78. Schenk, T., R. Müller, and F. Lingens. 1990. Mechanism of enzymatic dehalogenation of pentachlorophenol by *Arthrobacter* sp. strain ATCC 33790. *J. Bacteriol.* 172:7272–7274.

79. Schenk, T., R. Müller, F. Mörsberger, M. K. Otto, and F. Lingens. 1989. Enzymatic dehalogenation of pentachlorophenol by extracts from *Arthrobacter* sp. strain ATCC 33790. *J. Bacteriol.* 171:5487–5491.

80. Schmidt, E., and H.-J. Knackmuss. 1980. Chemical structure and biodegradability of halogenated aromatic compounds. Conversion of chlorinated muconic acids into maleylacetic acid. *Biochem. J.* 192:339–347.

81. Schmidt, E., G. Remberg, and H.-J. Knackmuss. 1980. Chemical structure and biodegradability of halogenated aromatic compounds. Halogenated muconic acids as intermediates. *Biochem. J.* 192:331–337.

82. Schwien, U., E. Schmidt, E. Remberg, and H.-J. Knackmuss, and W. Reineke. 1988. Degradation of chlorosubstituted aromatic compounds by *Pseudomonas* sp. strain B13: fate of 3,5-dichlorocatechol. *Arch. Microbiol.* 150:78–84.

83. Schwien, U., and E. Schmidt. 1982. Improved degradation of monochlorophenols by a constructed strain. *Appl. Environ. Microbiol.* 44:33–39.

84. Sharak Genthner, B. R., W. A. Price, II, and P. H. Pritchard. 1989. Anaerobic degradation of chloroaromatic compounds in aquatic sediments under a variety of enrichment conditions. *Appl. Environ. Microbiol.* 55:1466–1471.

85. Sharak Genthner, B. R. W. A. Price II, and P. H. Pritchard. 1989. Characterization of anaerobic dechlorinating consortia derived from aquatic sediments. *Appl. Environ. Microbiol.* 55:1472–1476.

86. Sharak Genthner, B. R., G. T. Townsend, and P. J. Chapman. 1989. Anaerobic transformation of phenol to benzoate via *para*-carboxylation: use of fluorinated analogues to elucidate the mechanism of transformation. *Biochem. Biophys. Res. Comm.* 162:945–951.

87. Sharpee, K. W., J. M. Duxbury, and M. Alexander. 1973. 2,4-Dichlorophenoxyacetate metabolism by *Arthrobacter* sp.: Accumulation of a chlorobutenolide. *Appl. Microbiol.* 26:445–447.

88. Stanlake, G. J., and R. K. Finn. 1982. Isolation and characterization of a pentachlorophenol-degrading bacterium. *Appl. Environ. Microbiol.* 44:1421–1427.

89. Steiert, J. G., and R. L. Crawford. 1986. Catabolism of pentachlorophenol by a *Flavobacterium* sp. *Biochem. Biophys. Res. Comm.* 141:825–830.

90. Steiert, J. G., J. J. Pignatello, and R. L. Crawford. 1987. Degradation of chlorinated phenols by a pentachlorophenol-degrading bacterium. *Appl. Environ. Microbiol.* 53:907–910.

91. Suzuki, T. 1977. Metabolism of pentachlorophenol by a soil microbe. *J. Environ. Sci. Health* B 12:113–127.

92. Suzuki, T. 1978. Enzymatic methylation of pentachlorophenol and its related compounds by cell-free extracts of *Mycobacterium* sp. isolated from soil. *J. Pest. Sci.* 3:441–443.

93. Suzuki, T. 1983. Methylation and hydroxylation of pentachlorophenol by *Mycobacterium* sp. isolated from soil. *J. Pest. Sci.* 8:419–428.

94. Sykes, G. 1965. Phenols, soaps, alcohols and related compounds. In *Disinfection and Sterilization*, 2nd ed. London: E. and F. N. Spon Ltd. 311–349.

95. Tabak, H. H., C. W. Chambers, and D. W. Kabler. 1964. Microbial metabolism of aromatic compounds. I. Decomposition of phenolic compounds and aromatic hydrocarbons by phenol-adapted bacteria. *J. Bacteriol.* 87:910–919.

96. Tiedje, J. M., J. M. Duxbury, M. Alexander, and J. E. Dawson. 1969. 2,4-D metabolism: Pathway of degradation of chlorocatechols by *Arthrobacter* sp. *J. Agric. Food Chem.* 17:1021–1026.

97. Tyler, J. E., and R. K. Finn. 1974. Growth rates of a pseudomonad on 2,4-dichlorophenoxyacetic acid and 2,4-dichlorophenol. *Appl. Microbiol.* 28:181–184.

98. U.S. Environmental Protection Agency (U.S. EPA). 1985. *Environmental Profiles and Hazard Indices for Constituents of Municipal Sludge: Pentachlorophenol.* Washington, D.C.: Office of Water Regulation and Standards.

99. Valo, R., V., Kitunen, M. Salkinoja-Salonen, and S. Räisänen. 1984. Chlorinated phenols as contaminants of soil and water in the vicinity of two Finnish sawmills. *Chemosphere* 13:835–844.

100. Walko, J. F. 1972. Controlling biological fouling in cooling systems—Part II. *Chem. Eng.* 79:104.

101. Watanabe, I. 1973. Isolation of pentachlorophenol-decomposing bacteria from soil. *Soil Sci. Plant Nutr.* 19:109–116.

102. Wiegel, J., X. Zhang, D. Dalton, and G. W. Kohring. 1990. Degradation of 2,4-dichlorophenol in anaerobic freshwater lake sediments. In *Emerging Technologies for Hazardous Waste Treatment.* Ed. W. Tedder, F. G. Pohland. Washington, D.C.: American Chemical Society. 119–141.

103. Wong, A. S., and D. G. Crosby. 1978. Photolysis of pentachlorophenol in water. In *Pentachlorophenol: Chemistry, Pharmacology and Environmental Toxicity.* Ed. K. R. Rao. New York: Plenum Press. 19–25.

104. Wyllie, J. A., J. Gabica, W. W. Benson, and J. Yoder. 1975. Exposure and contamination of the air and employees of a pentachlorophenol plant, Idaho-1972. *Pest. Mon. J.* 9:150–153.

105. Young, L. Y. 1984. Anaerobic degradation of aromatic compounds. In *Microbial Degradation of Organic Compounds.* Ed. D. T. Gibson. New York: Marcel Dekker, Inc. 487–523.

106. Zhang, X., T. V. Morgan, and J. Wiegel. 1990. Conversion of ^{13}C-1 phenol to ^{13}C-4 benzoate, an intermediate step in the anaerobic degradation of chlorophenols. *FEMS Microbiol. Lett.* 67:63–66.

107. Zhang, X., and J. Wiegel. 1990. Sequential anaerobic degradation of 2,4-dichlorophenol in freshwater sediments. *Appl. Environ. Microbiol.* 56:1119–1127.

108. Zhang, X., and J. Wiegel. 1990. Isolation and partial characterization of a *Clostridium* species transforming para-hydroxybenzoate and 3,4-dihydroxybenzoate and producing phenols as the final transformation products. *Microb. Ecol.* 20:103–121.

CHAPTER 5

Microbial Metabolism of Polycyclic Aromatic Hydrocarbons

JAIRAJ V. POTHULURI and CARL E. CERNIGLIA

Abstract. Polycyclic aromatic hydrocarbons (PAHs) are ubiquitous environmental pollutants. PAHs contain carbon and hydrogen. The carbon atoms are arranged in a series of adjoining, six-membered benzene rings. They are formed during the incomplete combustion of a wide variety of organic materials such as fossil fuels, cigarette smoke, diesel engine exhaust, oil pollution and industrial processes, and urban air pollution. Extensive research over the past decade has shown that PAHs undergo metabolic activation that induces toxicity, mutagenicity, or carcinogenicity in mammalian systems. Microorganisms have been found to be very useful in the degradation and detoxification of PAHs in terrestrial and aquatic ecosystems. Both prokaryotic and eucaryotic microorganisms have the ability to metabolize PAHs via dioxygenase, monooxygenase, or peroxidase-catylazed reactions to ring fission products. In this chapter, we describe biodegradation and detoxification processess of selected PAHs.

INTRODUCTION

Polycyclic aromatic hydrocarbons (PAHs) occur as common constituents of petroleum, coal tar, and shale oil but are mainly formed as products from the combustion of fossil fuels. Due to increases in anthropogenic sources and atmospheric deposition from natural sources, pollution levels of PAHs in the environment have increased steadily during the last century (123). Some PAHs are carcinogens and bacterial mutagens (2, 158). PAHs migrate to the sediments in aquatic ecosystems due to their hydrophobic nature and low water solubility, and readily adsorb to particulate matter (34). A high input source of PAHs in urban terrestrial and marine sediments is industrial wastes, especially those derived from coal gasification and liquification processes, coke, carbon black, waste incineration, and other petroleum-derived products (34, 115).

Due to their mutagenic and carcinogenic properties, 16 PAHs have been listed as priority pollutants by the United States Environmental Protection Agency. PAHs are typically monitored in industrial effluents via capillary column gas chromatography (123). Because they pose a human health risk, significant interest has been generated regarding the development of technologies leading to the containment and detoxification of PAH-contaminated wastes.

Drs. Pothuluri and Cerniglia are at the Division of Microbiology, National Center for Toxicological Research, Food and Drug Administration, Jefferson, AR 72079, U.S.A.

Several comprehensive reviews have been written on the metabolism, biochemistry, genetics, and regulation of PAH degradation (23, 24, 34, 60, 78, 79, 82). In this chapter, we discuss in detail the microbial metabolism of PAHs in aquatic and terrestrial environments and the potential for microbial detoxification of some selected PAHs such as naphthalene, acenaphthene, anthracene, phenanthrene, benz[a]anthracene, fluoranthene, pyrene, benzo[a]pyrene, 3-methylcholanthrene, and 1-nitropyrene.

TOXICOLOGY OF PAHs

The mechanisms of metabolic activation of chemical carcinogens have been reviewed extensively (98, 146). Metabolism of PAHs by mammalian systems demonstrates that metabolic activation is required for the initiation of carcinogenesis (146). The mammalian metabolism of certain PAH compounds results in organic-soluble metabolites, typically, epoxides, phenols and their sulfate ester conjugates, dihydrodiols, quinones, dihydrodiol epoxides, tetraols, and water-soluble metabolites, which have been determined to be glucuronide, sulfate, and glutathione conjugates (21). The chemical structures, genotoxicity, and carcinogenicity of selected PAHs are given in Table 1. The most extensively studied PAH is benzo[a]pyrene, which is also

Table 1. Chemical structures and toxicological characteristics of selected PAHs.[a]

Chemical Structure	Genotoxicity	Carcinogenicity	Chemical Structure	Genotoxicity	Carcinogenicity
Naphthalene	−	−	Benz[a]anthracene	+ Ames + SCE	+ CA + Carcinogen
Acenaphthene	−	−	3-Methylcholanthrene	+ Ames	+ Carcinogen
Anthracene	−	−	Fluoranthene	+ Ames	Weak Carcinogen
Phenanthrene	−	−	Pyrene	+/? Ames	− + UDS + SCE
Benzo[a]pyrene	+ Ames	+ Carcinogen + UDS + SCE + CA + DA	1-Nitropyrene	+ Ames	+ Carcinogen

Ames = Salmonella typhimurium reversion assay
CA = chromosomal abberations
DA = DNA adducts
UDS = unscheduled DNA synthesis
SCE = sister chromatid exchange
? = inadequate/inconclusive

[a]Ames, *Salmonella typhimurium* reversion assay; CA, chromosomal aberrations; DA, DNA adducts; UDS, unscheduled DNA synthesis; SCE, sister chromatid exchange; ?, inadequate/inconclusive.

the most carcinogenic PAH known. In rat liver microsomes and rat liver nuclear enzymes, metabolism of benzo[a]pyrene is highly stereospecific to form (−)-trans-7,8-dihydroxy-7,8-dihydrobenzo[a]pyrene (110, 186, 203). This (−)-enantiomer of benzo[a]pyrene 7,8-dihydrodiol is more potent as a tumor initiator than (+)-benzo[a]pyrene 7,8-dihydrodiol (118, 136, 137, 142, 175). The ultimate carcinogen formed from benzo[a]pyrene is benzo[a]pyrene 7,8-dihydrodiol–9,10-epoxide-2, which is highly mutagenic toward bacterial and mammalian cells (Figure 1) (21).

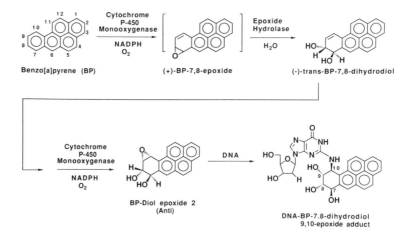

Figure 1. The metabolic activation of benzo[a]pyrene in mammalian systems. Adapted from Cerniglia (23).

Mammalian systems have been shown to oxidize PAHs to reactive electrophilic intermediates (187). Jerina et al. (113) suggest that the ultimate carcinogens formed from some PAHs are compounds derived from the preferential enzymatic oxidation at the bay region (10, 11-position) to form bay-region diol epoxides (Figure 1). Even though four optically active bay-region diol epoxides can be formed from a single bay-region PAH, the most tumorigenic enantiomer for a PAH like benz[a]anthracene is almost always the (R,S)-diol-(S,R)-oxide (112). Studies on mouse skin, DNA binding, and mutagenicity (55, 184, 201) have shown that for benzo[a]pyrene, the metabolic activation leading to biologically active compounds occurs at the 7,8,9,10-positions of the molecule (Figure 1).

GENERAL CONCEPTS IN THE MICROBIAL DEGRADATION OF PAHs

The degradation of PAHs containing up to four or five fused rings by bacteria and fungi has been well documented. The processes involving biodegradation are inversely proportional to the ring size of the PAH molecule. The lower weight PAHs are degraded more rapidly than the higher weight PAHs containing three or more fused rings. Normally, the higher molecular weight PAHs do not serve as amenable substrates for microbial metabolism.

During the microbial catabolism of PAHs, two important groups of enzymes, mono- and dioxygenases, have been found essential (82, 199). Therefore, microorganisms require molecular oxygen to catalyze initial hydroxylation of unsubstituted PAHs. Dioxygenases incorporate both atoms of the oxygen molecule into the PAH (Figure 2);

this initial oxidative attack on PAH by bacteria leads to the formation of *cis*-dihydrodiols (34, 59, 78, 79, 82). Another highly stereoselective reaction during bacterial oxidation is the rearomatization of the *cis*-dihydrodiol by dehydrogenases to form a dihydroxylated intermediate (Figure 2) (157). Dihydroxylation of the benzene nucleus has been found to be essential for cleavage of the aromatic ring (61). In contrast, the enzymatic fission of the aromatic ring is catalyzed by dioxygenases, and for the cleavage to occur, the hyroxyl groups must be located in positions either *ortho* or *para* to one another. For example, if the hydroxyl substituents are located *ortho* to one another, oxygenolytic ring cleavage can occur either between the two hydroxyl groups or adjacent to the two hydroxyl groups by *ortho* or intradiol-cleaving dioxygenases or by *meta* or extradiol-cleaving dioxygenases, respectively (34, 59).

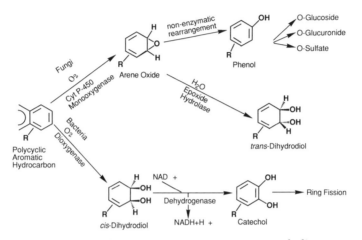

Figure 2. Microbial oxidation of PAHs via monooxygenase and dioxygenase pathways. Adapted from Cerniglia and Heitkamp (34).

Methane-monooxygenase may also be important in the catabolism of PAHs. Methanotrophs oxidize methane to carbon dioxide by a pathway that involves two-electron oxidation steps via methanol, formaldehyde, and formate (104). The methane mono-oxygenase isolated from *Methylococcus capsulatus* (Bath) was shown to catalyze oxidations of a variety of substrates, including alkanes, alkenes, and arenes. Methane mono-oxygenase also hydroxylates aromatic compounds (50, 51). Because methane mono-oxygenase of *M. capsulatus* is a very nonspecific enzyme system resistant to inhibition by carbon monoxide, Colby and Dalton (49) suggested that it does not contain cytochrome P-450. Subsequent studies with cyclopropane and arenes indicated that *M. capsulatus* operated via a mechanism in which dioxygen was converted presumably into a metal-bound oxygen species capable of insertion into a C–H bond and addition to a C=C bond. The monosubstituted benzenes were oxidized to *para*-substituted phenols via a reaction that has been termed the National Institute of Health (NIH) shift mechanism but required either an arene 3,4-oxide or a *p*-substituted cationic δ-complex as intermediate (62).

Brusseau et al. (13) demonstrated naphthalene oxidation by methanotrophs. *Methylosinus trichosporium* OB3b oxidized naphthalene to 1- and 2-naphthol. Methanotrophs expressing soluble methane mono-oxygenase (sMMO) also produced 1- and 2-naphthol. A rapid colorimetric assay was developed in this study that appears to be useful for monitoring the biodegradative potential of methanotrophs that produce sMMO.

Mammalian cytochrome P-450, which contains the important group of heme-enzymes (155), is also found in yeasts and filamentous fungi (105, 119, 127, 130, 149). Fungi are able to oxidize PAHs via a cytochrome P-450 monooxygenase. One atom of the oxygen molecule is incorporated into the PAH, while the other oxygen atom is reduced to water (21, 24). The arene oxide formed then becomes a substrate for further metabolism. The enzymatic hydration of the arene oxide leads to the formation of a dihydrodiol with a *trans*-configuration (Figure 2). Another pathway involves isomerization of arene oxide to form phenols that can be conjugated with sulfate, glucuronic acid, glucose, and glutathione. The initial oxidative reactions involved in the fungal metabolism of the PAHs are shown in Figure 2.

Many wood-decaying microorganisms have been found to produce lignin peroxidase, which is a heme-containing enzyme with protoporphyrin IX that oxidizes lignin in wood, so the wood carbohydrates may be used as nutrients. Many researchers have elucidated the role of the lignin peroxidases in the degradation of lignin by *Phanerochaete chrysosporium* and other white-rot fungi (84, 120, 140, 167). Also, lignin peroxidase has shown the ability to oxidize a variety of xenobiotiotics in addition to lignin (15, 91). Purified lignin peroxidase from *P. chrysosporium* has been shown to oxidize PAHs such as benzo[a]pyrene, benz[a]anthracene, pyrene, anthracene, and perylene (90, 92, 165). The oxidation products reported for benzo[a]pyrene by *P. chrysosporium* are the 1,6-, 3,6-, and 6,12-quinones (90).

The hydroxylation of PAHs by fungi is an important reaction, since it is a prelude to detoxification via phase II enzymatic reactions. However, bacteria oxidize PAHs to dihydroxylated compounds as a prelude to ring fission and assimilation (34, 61). Interestingly, algal incubations with various PAHs have produced both *cis*- and *trans*-dihydrodiols as products of transformation (48, 139, 150).

Denitrification involves processes that are biologically mediated wherein oxidized forms of nitrogen may be reduced while organic carbon is oxidized. The source of organic carbon may be either from natural material such as soil organic fraction or from an organic contaminant such as PAH compounds. Therefore, depending upon the characteristics of the organic carbon content of the soil and the nature and quantity of the PAH compounds present, PAH compounds may or may not become the major source of organic carbon for utilization during microbial denitrification reactions. Many PAHs persist in the groundwater and subsurface soil. Microbial degradation of PAHs in these environments is largely dependent on the availability of suitable nutrients and soil microorganisms capable of degrading such compounds via utilization of electron acceptors other than molecular oxygen.

PAHs may degrade slowly in anoxic soil-water systems, however, especially in the absence of nitrate or a suitable substituent group on the aromatic ring (144, 145). Therefore, the role of soil organic carbon as an electron donor in the process of denitrification is important (163).

Unsubstituted PAHs are resistant to microbial attack under strictly anaerobic conditions. A methanogenic consortium, however, was able to metabolize aromatic hydrocarbons by incorporating oxygen derived from water into the aromatic ring (85, 191). Several studies have shown that monoaromatic hydrocarbons are biodegradable under anaerobic conditions. This degradation may occur via anaerobic respiration, fermentation, and photometabolism (71, 97, 98, 117, 191). For example, Vogel and Grbić-Galić reported the anaerobic degradation of benzene and toluene to CO_2 and nonvolatile intermediates by adapted methanogenic cultures (191). They observed that the oxygen incorporation into cresol and phenol, formed from the anaerobic transformation of toluene and benzene, respectively, was derived from water and not molecular oxygen. This phenomenon appears to be important, since many unsubstituted PAHs are not degraded in the water column and remain persistent in a concentration higher than normally found in the sediments. Anaerobic catabolism of PAHs could be an important pathway in the ultimate removal of PAHs, although this has not yet been demostrated.

BIODEGRADATIVE PATHWAYS OF SELECTED PAHs

Naphthalene

Bacterial metabolism of naphthalene in soil was first reported in 1927 (182, 183). Since then, many examples of the ability of bacteria to utilize naphthalene as the sole source of carbon and energy have been reported (21). Numerous studies have used *Pseudomonas* sp. successfully to degrade naphthalene (7, 64, 109, 179, 189, 192).

The naphthalene dioxygenase from *Pseudomonas* sp. strain NCIB 9816 was characterized by Ensley and Gibson (69) and Ensley et al. (70). The metabolic pathway involves the incorporation of both atoms of molecular oxygen to form cis-naphthalene dihydrodiol (20, 109). Naphthalene dioxygenase consists of three protein components that are essential for the formation of cis-naphthalene dihydrodiol. This multicomponent enzyme system has properties similar to those reported for benzene dioxygenase isolated from a strain of *P. putida* (3, 58). Since naphthalene dioxygenase is very unstable, rapid purification in the presence of dithiothreitol, ethanol (10% v/v), and glycerol in Tris hydrochloride buffer (10% v/v) is necessary to maintain the enzymatic activity (70). The terminal naphthalene dioxygenase component is an iron-sulfur protein consisting of iron (6 g-atoms) and acid-labile sulfur (4 g-atoms) per mole of the purified enzyme. In the presence of oxygen, NADH, and two other components of the naphthalene dioxygenase system, the terminal oxygenase forms cis-naphthalene dihydrodiol. Subsequently, during the oxidation of naphthalene by *P. putida*, cis-naphthalene dihydrodiol is converted to 1,2-dihydroxynaphthalene (157). This reaction is catalyzed by cis-naphthalene dihydrodiol dehydrogenase reductant. Interestingly, cis-naphthalene dihydrodiol dehydrogenase is highly stereoselective for the (+)-isomer of the cis-naphthalene dihydrodiol and therefore cannot metabolize trans-naphthalene dihydrodiol (157).

Earlier studies have reported that 1,2-dihydroxynaphthalene was cleaved enzymatically by a dioxygenase to yield cis-2'-hydroxybenzalpyruvate by *Pseudomonas* sp. (6, 64). Further cleavage of cis-2'-hydroxybenzalpyruvate to pyruvate and salicylaldehyde by an aldolase and subsequent oxidation by a dehydrogenase to salicylate was postulated. The oxidation of salicylate by salicylate hydroxylase enzyme yields catechol, which then undergoes either *ortho* or *meta* fission depending on the bacterial species involved (5, 64). Another study showed a similar degradative sequence for naphthalene metabolism by *Aeromonas* sp. (131). Figure 3 illustrates the sequence of reactions involved in the mineralization of naphthalene by bacteria. Even though the two metabolic pathways shown in Figure 3 are completely separated, both can occur in the same microorganism (68). Tagger at al. (179) recently isolated two bacterial strains *Pseudomonas* Lav. 4 and *Moraxella* Lav. 7 from a bacterial community of a marine sediment on a seawater medium with naphthalene as the sole carbon source. These bacteria oxidized naphthalene into catechol, which was degraded only via the *meta* pathway. Naphthalene oxygenase and salicylate hydroxylase were inducible in *Pseudomonas* Lav. 4, whereas catechol 2,3-dioxygenase was constitutive. For

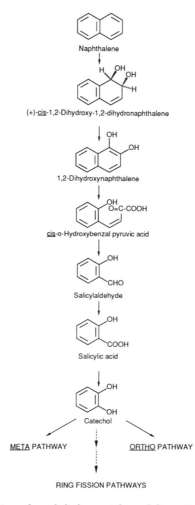

Figure 3. Bacterial degradation of naphthalene. Adapted from Cerniglia and Heitkamp (34).

Moraxella Lav. 7, however, naphthalene oxygenase was constitutive, but salicylate hydroxylase and catechol 2,3-oxygenase were inducible. Garcia-Valdes et al. (75) also found that *Pseudomonas* strains isolated from samples of marine sediment polluted with naphthalene had no catechol 1,2-dioxygenase; however, catechol 2,3-dioxygenase was responsible for the aromatic ring cleavage. In a recent study, Kelley et al. (125) found that *Mycobacterium* sp. converted naphthalene to *cis*- and *trans*-naphthalene dihydrodiol by dioxygenase- and monooxygenase-catalyzed reactions, respectively. The formation of both *cis*- and *trans*-naphthalene dihydrodiols by a *Mycobacterium* sp. is unique and was reported for the first time in this study (125).

Many investigators have reported that the genes that code for naphthalene oxidation in pseudomonads are found in plasmids (9, 19, 53, 204). Also, the genes encoding the enzymes that are responsible for the degradation of naphthalene have been cloned from *Pseudomonas putida* and have been expressed in *Escherichia coli* (70, 166). Devereux and Sizemore (65) examined a plasmid-containing bacteria *Pseudomonas* sp. isolated from chronically polluted sites and cured successfully one marine strain of a naphthalene-degradative plasmid. Serdar and Gibson (171) have reported the genes encoding the upper pathway (Nah→Sal) for naphthalene degradation located on a 15 kb *EcoRI* fragment that was cloned into pKT230. The resulting pDTG113 recombinant was nick-translated and was used as a radioactive probe to study nucleotide sequence homology of the cloned fragment with DNA that was isolated from other naphthalene-utilizing organisms, *P. putida* G7, *P. putida* NP, and strain PL6.

Phylogenetic relationships among isofunctional extradiol dioxygenases of catechol 2,3-dioxygenases from different *Pseudomonas* sp. was demonstrated by Harayama et al. (94) and Taira et al. (180). Later, Harayama and Rakik (93) studied the complete nucleotide sequence of the *nahC* gene, the structural gene for the extradiol enzyme 1,2-dihydroxynaphthalene dioxygenase encoded in the NAH7 plasmid of *Pseudomonas putida*. This enzyme is an extradiol ring-cleaving enzyme and cleaved the first ring of the 1,2-dihydroxynaphthalene. A study by King et al. (129) demonstrates the role of *lux* transcriptional fusions with catabolic genes and offers a useful molecular tool for direct analysis of microbial degradative activity in complex environmental matrices. Interestingly, a bioluminescent reporter plasmid for naphthalene catabolism (pUTK21) was developed by transposon (Tn4431) insertion of the *lux* gene cassette from *Vibrio fischeri* into a naphthalene catabolic plasmid in *Pseudomonas fluorescens*. The insertion site of the *lux* transposon was the *nahG* gene encoding for salicylate hydroxylase (129).

A variety of fungi have also been found to transform naphthalene to metabolites that are similar to those produced by mammalian enzymes and laboratory animals (38, 42, 73, 153, 176).

In contrast to procaryotes, fungi do not utilize naphthalene as a sole source of carbon and energy but utilize cytochrome P-450 monooxygenase to form naphthalene 1,2-oxide. Epoxide hydrolase catalyzes the conversion of this arene oxide to form *trans*-naphthalene dihydrodiol. Naphthalene 1,2-oxide can also isomerize spontaneously to 1-naphthol (Figure 4) (21).

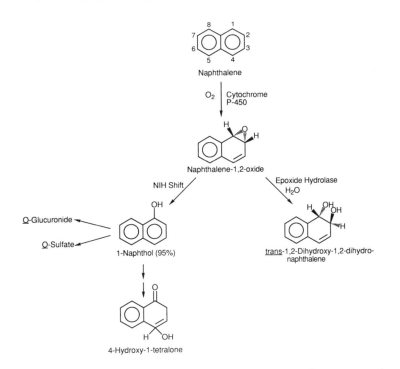

Figure 4. Fungal transformation of naphthalene. The enzymatic and nonenzymatic pathways shown are adapted from Cerniglia and Heitkamp (34).

Cunninghamella elegans and many other fungal strains metabolize naphthalene via a branched pathway to naphthalene *trans*-1,2-dihydrodiol, 1-naphthol, 2-naphthol, 4-hydroxy-1-tetralone, 1,4-naphthoquinone, and 1,2-naphthoquinone (21, 42). Additionally, two conjugates, 1-naphthyl glucuronide and 1-napthyl sulfate, have also been reported (37). Isolation of *trans*-naphthalene dihydrodiol by Ferris and co-workers (73) provided the initial evidence for an arene oxide formation. The rearrangement of arene oxide to form 1-naphthol via the NIH shift mechanism indicates that naphthalene 1,2-oxide is an intermediate in naphthalene metabolism (25). *Cunninghamella elegans* metabolizes naphthalene by mechanisms similar to *in vivo* and *in vitro* mammalian systems. Experiments with $^{18}O_2$ and $H_2^{18}O$ demonstrated that only one atom of molecular oxygen was incorporated into the C-1 hydroxyl group during the formation of *trans*-naphthalene 1,2-dihydrodiol. The other oxygen atom at the C-2 position catalyzed by epoxide hydrolase was derived from water (25). This is consistent with the mechanism of *trans*-dihydrodiol formation via cytochrome P-450 monooxygenase and epoxide hydrolase activities.

Enantiomeric resolution of *trans*-naphthalene 1,2-dihydrodiol from *C. elegans* indicates that the major enantiomer has an *S,S* absolute stereochemistry when compared to the predominant *R,R* form produced by rat liver microsomes (25). Studies conducted with other PAHs have also shown such stereospecificity for *trans*-dihydrodiols formed during fungal oxidation (23, 27, 44).

Cyanobacteria grown photoautotrophically in the presence of naphthalene oxidized naphthalene to 1-naphthol (35–37). Cyanobacterial oxidation of naphthalene also produced *cis*-naphthalene dihydrodiol as a minor metabolite (34).

Acenaphthene

Acenaphthene and its metabolites are not carcinogenic but have been shown to produce nuclear and cytological changes in plants and microbial species (18). Only a few studies have been conducted to determine the microbial metabolism of acenaphthene.

Chapman (45) found that a naphthalene-grown *Pseudomonas* sp. metabolized acenaphthene to 1-acenaphthenol and 1-acenaphthenone. Schocken and Gibson (168) also found that *Pseudomonas putida* metabolized acenaphthene to 1-acenaphthenol, 1-acenaphthenone, and acenaphthenequinone. Also, their study with a *Beijerinckia* sp. showed cooxidation of the polycyclic aromatic hydrocarbon acenaphthene to the same spectrum of metabolites mentioned above. The proposed pathway for the oxidation of acenaphthene by the *Beijerinckia* sp. is shown in Figure 5. The oxidation of acenaphthene occurred via the aliphatic ring to form 1-acenaphthenone; however, a second hydroxylation step formed 1,2-dihydroxyacenaphthylene, and the subsequent dehydrogenation resulted in the formation of acenaphthenequinone as the end product. Acenaphthenequinone could not undergo further oxidation, however. Thus, *Beijerinckia* was unable to cleave the fused ring or utilize acenaphthene as a sole source of carbon and energy for growth. In contrast, the ring cleavage to form naphthalic anhydride was observed in mammalian systems (67, 106, 107, 173).

Mihelcic and Luthy (145) studied the microbial degradation of acenaphthene under denitrification conditions in soil-water systems. Acenaphthene decreased from initial aqueous-phase concentrations of 1 mg/liter to nondetectable levels within 9 weeks, under nitrate-excess conditions. They attributed the microbial degradation of the PAH compound to the dependence on the interrelationships among (a) the desorption kinetics as well as the reversibility of the desorption of the sorbed compound from the soil, (b) the concentration of the microorganisms responsible for the PAH degradation,

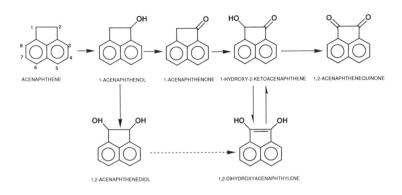

Figure 5. Proposed oxidative pathway of acenaphthene by *Beijerinckia* sp. Adapted from Schocken and Gibson (168).

and (c) the competing reaction for nitrate utilization through mineralization of the labile fraction of soil organic carbon (144).

There is a paucity of information on the fungal metabolism of acenaphthene. Recently, Pothuluri et al. (161) studied the transformation of acenaphthene by the fungus *Cunninghamella elegans*. *C. elegans* metabolized acenaphthene to 1,5-dihydroxyacenaphthene, *trans*-1,2-dihydroxyacenaphthene, 1,2-acenaphthenedione, *cis*-1,2-dihydroxyacenaphthene, 6-hydroxyacenaphthene, and 1-acenaphthenone. The results indicated that the metabolism of acenaphthene by *C. elegans* proceeded via hydroxylation of both the aromatic and aliphatic ring to phenolic derivatives of acenaphthene. This is similar to mammalian metabolism, since the primary site of enzymatic attack is on the aliphatic ring. The oxidation of acenaphthene by *Beijerinckia* sp. was shown to occur via two successive monooxygenations and two dehydrogenation steps (168). Pothuluri et al. (161) found initial oxidative attack at the C-1 and C-2 positions of acenaphthene, with subsequent hydroxylation to form *cis*- and *trans*-dihydroxyacenaphthene and oxidation at the C-5 and C-6 positions to form 1,5-dihydroxyacenaphthene and 6-hydroxyacenaphthenone by *C. elegans*.

Anthracene and Phenanthrene

Anthracene and phenanthrene degradative pathways have been widely studied. Even though these two PAHs and their metabolites are not carcinogenic, their emission into the environment in trace quantities during coal gasification and liquification processes have aroused considerable interest. Additionally, both anthracene and phenanthrene are used as model substrates in studies pertaining to environmental degradation of PAHs, since the structures of both of these PAHs are also found in carcinogenic PAHs such as benzo[a]pyrene, benz[a]anthracene, and 3-methylcholanthrene (Table 1).

Bacterial oxidation of anthracene and phenanthrene and their degradative sequences are similar to those described previously for naphthalene. Some microorganisms isolated from soil have the ability to utilize anthracene and phenanthrene as the sole source of carbon and energy (21).

The degradative pathway during bacterial mineralization of anthracene is shown in Figure 6. Several species of *Pseudomonas* and a *Beijerinckia* strain are able to oxidize anthracene initially in the 1,2-position to form (+)-*cis*-1R,2S-dihydroxy-1,2-

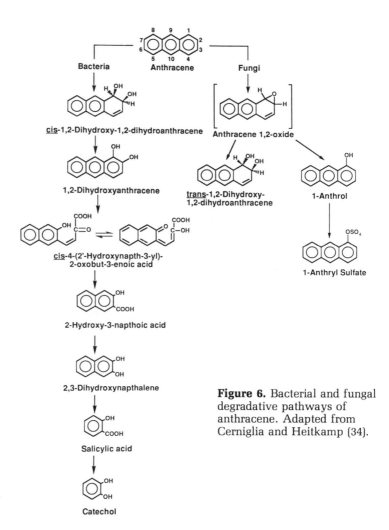

Figure 6. Bacterial and fungal degradative pathways of anthracene. Adapted from Cerniglia and Heitkamp (34).

dihydroanthracene (1, 111). The second step in the bacterial oxidation of anthracene is the conversion of *cis*-1,2-dihydroxy-1,2-dihydroanthracene to 1,2-dihydroxy-anthracene (Figure 6). A *Pseudomonas* sp. was found to cleave 1,2-dihydroxy-anthracene enzymatically by a dioxygenase to yield *cis*-4-(2-hydroxynaphth-3-yl)-2-oxo-but-3-enoic acid. Further metabolism of this ring fission product leads to the formation of 2-hydroxy-3-naphthoic acid (72). This compound is further metabolized through salicylate and catechol by the enzymes of the naphthalene pathway shown in Figure 3.

Fungi also metabolize anthracene (Figure 6). *Cunninghamella elegans* initially oxidizes anthracene in the 1,2-position to form *trans-1S,2S*-dihydroxy-1,2-dihydro-anthracene (22, 44). Anthracene 1,2-oxide is rapidly rearranged to form 1-anthrol, which is subsequently conjugated with sulfate (22). Recently, Sutherland et al. (178) studied the metabolism of anthracene by *Rhizoctonia solani*. One of the principal metabolites was identified as *trans*-1,2-dihydroxy-1,2-dihydroanthracene. The other metabolites were the novel conjugates of 1-O-anthracene *trans*-1,2-dihydrodiol β-D-xylopyranoside, 2-O-anthracene *trans*-1,2-dihydrodiol β-D-xylopyranoside, and 1-O-anthryl-β-D-xylopyranoside.

Phenanthrene metabolism by bacteria has been reported by several investigators (7, 52, 72, 111, 131–133, 164). Phenanthrene can be degraded by two pathways (Figure 7). The initial enzymatic attack is in the 1,2- and 3,4-positions to form (+)-cis-1R,2S-dihydroxy-1,2-dihydrophenanthrene   and   (+)-cis-3S,4R-dihydroxy-3,4-dihydrophenanthrene (Figure 7) (111, 134). The major isomer formed is (+)-cis-3,4-dihydroxy-3,4-dihydrophenanthrene. Pseudomonads and a *Nocardia* strain oxidize cis-3,4-dihydroxy-3,4-dihydrophenanthrene to 3,4-dihydroxyphenanthrene, which is subsequently cleaved and converted to 1-hydroxy-2-naphthoic acid. This product of ring cleavage is decarboxylated oxidatively to give 1,2-dihydroxynaphthalene, which can then enter the naphthalene pathway (Figure 3) (72). Some microorganisms like *Aeromonas, Alcaligenes, Micrococcus,* and *Vibrio* strains have an alternate pathway for 1-hydroxy-2-naphthoic acid catabolism (76, 77, 131, 133). The oxidation of 1-hydroxy-2-naphthoic acid is through *ortho*-phthalic acid to form protocatechuic acid (Figure 7). Guerin and Jones (88) have reported that a *Mycobacterium* sp. strain BG1 isolated from estuarine sediment enriched with PAH was able to utilize phenanthrene as the sole carbon and

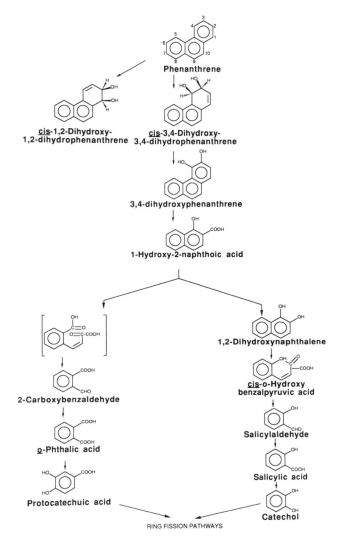

Figure 7. Bacterial degradation pathways of phenanthrene. Adapted from Cerniglia and Heitkamp (34).

energy source. This bacterium was able to degrade phenanthrene via 1-hydroxy-2-naphthoic acid and *meta* cleavage of protocatechuic acid without accumulating 1-hydroxy-2-naphthoic acid or other aromatic intermediates. Guerrin and Jones (88) postulated that plasmids are involved in the rapid mineralization of phenanthrene.

Recently, Bumpus (14) reported that the ligninolytic fungus *Phanerochaete chrysosporium* oxidizes phenanthrene to CO_2 and several unidentified polar metabolites. Since phenanthrene was found to be the most abundant PAH present in anthracene oil (a distillation product obtained from coal tar), experiments with [^{14}C]phenanthrene after 27 days of incubation with *P. chrysosporium* showed that 7.7% of the recovered radiolabeled carbon was metabolized to $^{14}CO_2$. The recovery from the aqueous fraction was 25.2%, whereas those from the methylene chloride and particulate fractions were 56.1 and 11.0%, respectively. Of the methylene chloride fraction, 91.9% of the material was composed of polar metabolites of [^{14}C]phenanthrene. These results suggested that *P. chrysosporium* can be useful for decontaminating PAH polluted sites (14). Sutherland et al. (177) investigated the metabolism of phenanthrene by *P. chrysosporium* grown under nonligninolytic conditions and identified the metabolites as *trans*-9,10-dihydroxy-9,10-dihydrophenanthrene, *trans*-3,4-dihydroxy-3,4-dihydrophenanthrene, 9-phenanthrol (9-hydroxyphenanthrene), 3-phenanthrol, 4-phenanthrol, and 9-O-phenanthryl β-D-glycopyranoside. Five of the metabolites produced from phenanthrene by *P. chrysosporium* were similar to those reported from mammalian metabolism.

The fungus *Cunninghamella elegans* initially oxidizes phenanthrene predominantly at the 1,2-positions (Figure 8) to form phenanthrene *trans*-1,2-dihydrodiol and a glucoside conjugate of 1-phenanthrol (26, 44). The oxidation may also occur at the 3,4- and 9,10-positions to form dihydrodiols with a *trans*-configuration (Figure 8). The major enantiomers formed by *C. elegans* for each of the dihydrodiols have an S,S-absolute configuration (44). The cyanobacterial oxidation of phenanthrene indicates, however, that *Agmenellum quadruplicatum* strain PR-6 and *Oscillatoria* sp. strain JCM oxidized phenanthrene predominately to *trans*-9,10-dihydroxy-9,10-dihydrophenanthrene (150), a result that is quite similar to that obtained from mammalian cytochrome P-450 monooxygenase-mediated reactions (11, 46, 151).

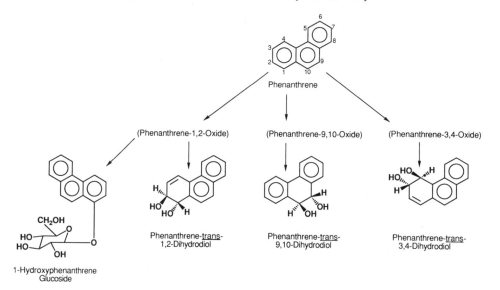

Figure 8. Fungal transformation pathway of phenanthrene. Adapted from Cerniglia and Heitkamp (34).

Benzo[a]pyrene

The most potent carcinogen of the PAHs known, benzo[a]pyrene, was first isolated from coal tar by Cook et al. (54). Since then, mammalian metabolism, biological activities of metabolites, and the interaction of the metabolites of benzo[a]pyrene with nucleic acids have been studied extensively (42, 56). In contrast to numerous reports on mammalian studies, little is known about the microbial degradation of PAHs that contain more than three aromatic rings (21). Although bacteria capable of utilizing tetra- and pentacyclic aromatic hydrocarbons as the sole carbon and energy source are limited, microorganisms can oxidize these insoluble PAHs when they are grown on an alternative carbon source. Using biotransformation techniques, Gibson and co-workers (81) were the first to elucidate the structural information on the metabolites of PAHs formed by bacteria. A mutant strain of *Beijerinckia* sp., grown on succinate in the presence of biphenyl, oxidized benzo[a]pyrene to *cis*-9,10-dihydroxy-9,10-dihydrobenzo[a]pyrene and *cis*-7,8-dihydroxy-7,8-dihydrobenzo[a]pyrene. However, the ring cleavage products of benzo[a]pyrene and its reaction sequence for bacterial mineralization have not yet been identified.

Many fungi have also been shown to oxidize benzo[a]pyrene by mechanisms similar to those observed in mammals (28, 31, 40, 41, 77, 138, 200). Some fungi that can metabolize benzo[a]pyrene by the inducible cytochrome P-450 monooxygenases are *Aspergillus ochraceus, Chrysosporium pannorum, Cunninghamella bainieri, C. elegans, Mortierella verrucosa, Neurospora crassa, Penicillium* sp., *Saccharomyces cerevisiae,* and *Trichoderma viride* (42).

A frequently studied fungus, *C. elegans,* oxidizes benzo[a]pyrene to arene oxides, epoxides, phenols, *trans*-dihydrodiols, dihydrodiol epoxides, tetraols, and quinones (31–33). These primary metabolites (Figure 9) are further transformed to sulfate and glucuronide conjugates (31). Although benzo[a]pyrene is postulated to be transformed

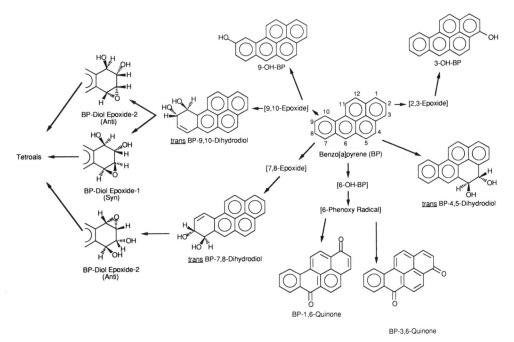

Figure 9. Fungal transformation of benzo[a]pyrene. Pathway adapted from Cerniglia and Heitkamp (34).

ultimately into carcinogenic metabolites in higher organisms (32, 33), the overall effect of incubation of benzo[a]pyrene with *C. elegans* was to transform this genotoxic compound into detoxified derivatives, such as sulfate conjugates (23, 43).

Similar to *C. elegans, Aspergillus ochraceus* also oxidize benzo[a]pyrene, but the predominant dihydrodiol produced is the *trans*-4,5-dihydrodiol (63). The lignin-degrading fungus *Phanerochaete chrysosporium* oxidized benzo[a]pyrene to CO_2 (16, 165). In the presence of glucose, however, benzo[a]pyrene rapidly oxidized to CO_2 and several unidentified metabolites (165). Haemmerli et al. (90) reported that crude and purified lignin peroxidase preparations from *P. chrysosporium* oxidized benzo[a]pyrene to 1,6-quinone, 3,6-quinone, and 6,12-quinone. The addition of veratryl alchohol enhanced the oxidation of benzo[a]pyrene.

Studies with freshwater green alga *Selenastrum capricornutum,* grown photo-autotrophically, showed that benzo[a]pyrene was oxidized to *cis*-4,5-dihydroxy-4,5-dihydrobenzo[a]pyrene, *cis*-7,8-dihydroxy-7,8-dihydrobenzo[a]pyrene, *cis*-9,10-dihydroxy-9,10-dihydrobenzo[a]pyrene, and *cis*-11,12-dihydroxy-11,12-dihydroben-zo[a]pyrene (48, 139). The predominant isomer was the *cis*-11,12-dihydrodiol. The formation of *cis*-dihydrodiols suggests that a dioxygenase similar to that found in prokaryotes is the pathway rather than the monooxygenase pathway found in other eukaryotes (Figure 1) (196). Further study by Warshawsky et al. (195) showed that the freshwater green alga metabolized the PAH benzo[a]pyrene via a dioxygenase pathway with subsequent conjugation and excretion. Eighty-six percent of the conjugates were acid labile. Previously, Keenan et al. (121) had reported for the first time that freshwater green algae conjugated xenobiotics. Approximately, 75% of the metabolites of benzo[a]pyrene formed by *S. capricornutum* were conjugated.

Benz[a]anthracene

The bay-region 1,2-position is the preferred site of attack for the bacterial oxidation of benz[a]anthracene (82); however, oxidation at the 8,9- and 10,11-positions can also occur (81). The absolute stereochemistry of the *cis*-1,2-, *cis*-8,9-, and *cis*-10,11-dihydrodiols formed from benz[a]anthracene by a strain of *Beijerinckia* has been reported by Jerina et al. (113).

More recently, Mahaffey et al. (141) studied the bacterial oxidation of benz[a]an-thracene after induction with biphenyl, m-xylene, and salicylate. Biotransformation studies with benz[a]anthracene showed that after 14 hr, 56% was converted to an isomeric mixture of three o-hydroxypolyaromatic acids by *Beijerinckia* strain B1. The major metabolite was identified as 1-hydroxy-2-anthranoic acid; the two minor metabolites were identified as 2-hydroxy-3-phenanthroic acid and 3-hydroxy-2-phenanthroic acid (Figure 10). The formation of the two minor metabolites is presumed to have occurred by oxidative cleavage of catechol formed at the 10,11- and 8,9-positions of benz[a]anthracene, respectively (80). Both of these acids appear to have formed by a series of reactions analogous to those proposed for the oxidation of anthracene to 2-hydroxy-3-naphthoic acid (72). Mahaffey et al. (141) suggest that since mineralization experiments with [12-^{14}C]benz[a]anthracene showed formation of $^{14}CO_2$, the hydroxy acids can be oxidized whereby at least two rings of the benz[a]anthracene molecule can be further degraded.

Other studies with *Pseudomonas putida* NCIB 9816 have shown that salicylate enhances the degradation of benz[a]anthracene (4), probably because of the increased expression of naphthalene-degrading enzymes by salicylate (5). Furukawa et al. (74) explains that the catabolism of biphenyl, xylene-toluene, and salicylate is regulated by

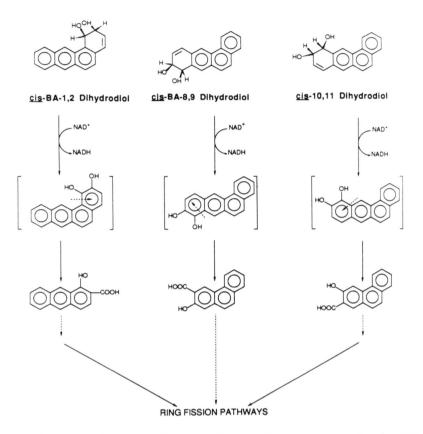

Figure 10. The proposed major pathway for benz[a]anthracene degradation by *Beijerinckia* strain B1. Adapted from Mahaffey et al. (141).

a common unit in *P. paucimobilis* (Q1). Metabolism of these compounds was postulated to be interrelated, since benzoate and toluate were common intermediates of biphenyl and xylene-toluene metabolism and salicylate was produced from *o*-phenylphenol (74). Similar coordinated regulation of biphenyl, *m*-xylene, and salicylate metabolism by *Beijerinckia strain* B1 is postulated (141). The major proposed pathway for benz[a]anthracene degradation by *Beijerinckia* strain B1 is shown in Figure 10 (141).

Biotransformation studies conducted on the fungal oxidation of benz[a]anthracene have shown that cells of *Cunninghamella elegans* grown in the presence of this compound formed *trans*-8,9-dihydroxy-8,9-dihydrobenz[a]anthracene, *trans*-10,11-dihydroxy-10,11-dihydrobenz[a]anthracene, and a trace amount of *trans*-3,4-dihydroxy-3,4-dihydrobenz[a]anthracene (66). Figure 11 shows the structures of these metabolites. The major dihydrodiol, however, *trans*-5,6-dihydroxy-5,6-dihydrobenz[a]anthracene, produced by rat liver microsomes (185), was not found in experiments on fungal metabolism. Similarly, a study of the metabolism of benzo[a]pyrene by *C. elegans* showed no "K-region" dihydrodiol (80); thus, a basic difference in the specificity of the fungal monooxygenase enzyme system is implied. Analogous to the fungal transformation of benzo[a]pyrene, the metabolite of benz[a]anthracene, *trans*-3,4-dihydroxy-3,4-dihydrobenz[a]anthracene formed from *C. elegans,* was also highly mutagenic (21). Additionally, benz[a]anthracene 3,4-dihydrodiol–1,2-epoxides are thought to be the most reactive forms of this PAH (135, 201). The proposed pathway of benz[a]anthracene metabolism by fungi is shown in Figure 11.

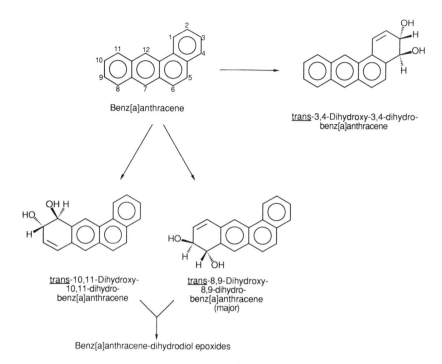

Figure 11. Fungal oxidative pathways of benz[a]anthracene. Adapted from Cerniglia (23).

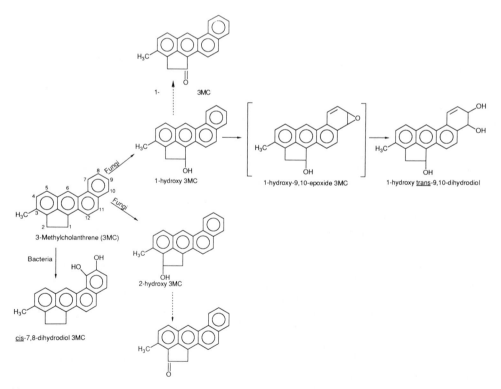

Figure 12. Bacterial and fungal oxidation of 3-methylcholanthrene. Pathways adapted from Cerniglia (24).

3-Methylcholanthrene

3-Methylcholanthrene is an alkyl-substituted derivative of benz[a]anthracene. It is a highly potent carcinogen. The microbial metabolism of 3-methylcholanthrene has not been studied extensively. A *Beijerinckia* sp. strain B-836 transformed 3-methylcholanthrene to *cis*-7,8-dihydroxy-7, 8-dihydro-3-methylcholanthrene (Figure 12) (128), whereas the filamentous fungus, *Cunninghamella elegans,* oxidized 3-methylcholanthrene primarily on the methylene bridge to form 1-hydroxy- and 2-hydroxy-3-methylcholanthrene (Figure 12). Also, the formation of two diastereomeric 1-hydroxy-*trans*-9,10-dihydrodiols of 3-methylcholanthrene was reported (29). In higher organisms, the above-mentioned compounds have been postulated to be proximate carcinogenic metabolites of 3-methylcholanthrene.

Fluoranthene

Fluoranthene is a PAH containing four fused rings. It occurs as a natural constituent of unaltered fossil fuels. Fluoranthene is consistently the most abundant of the PAHs measured in environmental samples (114, 115, 143) and has been reported to be cytotoxic, weakly mutagenic, and potentially carcinogenic (10, 17, 116, 174, 188, 190).

Mueller and coworkers (147) have isolated *Pseudomonas paucimobilis,* strain EPA505, shown to utilize fluoranthene as the sole source of carbon and energy from a bacterial consortium enriched from creosote-contaminated soil (148). Other researchers (198) reported the degradation of fluoranthene by a mixed culture of bacteria. A pure strain of *Alcaligenes denitrificans* was isolated and identified as the fluoranthene-degrading bacterium from the mixed culture. A *Mycobacterium* sp. has been found to mineralize fluoranthene significantly, following enzymatic induction by pyrene (99, 102). Recently, Kelley and Cerniglia (124) found the *Mycobacterium* sp. to degrade fluoranthene in excess of 95% within a 24-hour period after an initial 6- to 12-hour lag phase. The major fluoranthene metabolite formed by the *Mycobacterium* sp. was identified as 9-fluorenone-1-carboxylic acid. The initial attack of fluoranthene occurs presumably via a dioxygenase in the 1,2- or 2,3-positions of fluoranthene (126).

In contrast, the initial oxidative attack on fluoranthene by the fungus *Cunninghamella elegans* occurred predominantly at the 2,3-position, resulting in the formation of fluoranthene *trans*-2,3-dihydrodiol. Since fungi have been found to oxidize PAHs via cytochrome P-450 and to catalyze monoxygenase and epoxide hydrolase reactions to form *trans*-dihydrodiols, further oxidation of fluoranthene *trans*-2,3-dihydrodiol resulted in the formation of two stereoisomers. The two isomers, 9-hydroxyfluoranthene and 8-hydroxyfluoranthene *trans*-2,3-dihydrodiols, accounted for 24% of the total radioactive organic-soluble metabolites (159). The proposed pathway is shown in Figure 13.

The formation of 3-hydroxyfluoranthene with subsequent glycosylation is the major pathway found in fluoranthene metabolism by *C. elegans.* The two glucoside conjugates, identified as 3-fluoranthene-β-glucopyranoside and 3-(8-hydroxyfluoranthene)-β-glucopyranoside, together accounted for 52% of the total ethyl acetate-extractable metabolites (Figure 13) (159). Furthermore, these two glucoside conjugates were not mutagenic toward *Salmonella typhimurium* strain TA100 and TA104, with or without S9 activation at 2.5 to 10.0 μg fluoranthene per plate (160). These studies indicated that the conjugation reactions are common metabolic pathways in *C. elegans* and that the pathway involving glycosylation of the hydroxylated metabolites may be a significant detoxification pathway in the fungal metabolism of fluoranthene.

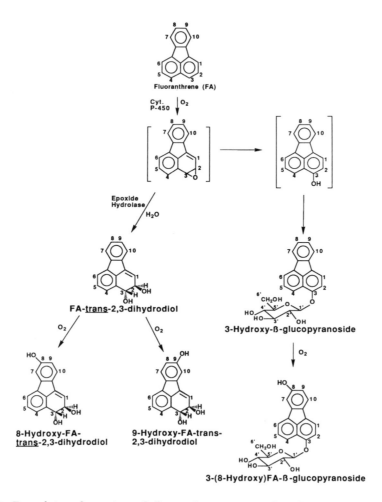

Figure 13. Fungal transformation of fluoranthene. Proposed pathways were adapted from Pothuluri et al. (159).

Pyrene

Even though pyrene is not a genotoxic PAH, it has been detected in environmental samples and has been used as an indicator compound in the monitoring of the PAH-contaminated wastes (83, 87, 152). Because of its high symmetry, pyrene has been used in metabolism studies to determine the induction of various cytochrome P-450 isozymes (108). It also serves as the model PAH for examining the photochemical and biological degradation of other compounds of this class of PAHs (181). Additionally, pyrene structure is found in carcinogenic PAHs, and therefore pyrene is used as a model compound for measuring binding to DNA (47).

Despite the several studies on the mammalian metabolism of pyrene (12, 39, 86, 95, 96, 108, 122, 154, 172), only a few have described the microbial metabolism of pyrene. A mineralization study of PAHs by a bacterium isolated from sediment below an oil field showed 63% pyrene mineralization in two weeks in pure culture with organic nutrients (99). Heitkamp et al. (101) isolated a pyrene-mineralizing bacterium from sediments collected near a point source for petrogenic chemicals. This pyrene-induced *Mycobacterium* sp. mineralized 5% of the pyrene after 6 hr and attained a maximum of 48% mineralization within 72 hr. This first report showed that pyrene-degrading enzymes were inducible in the *Mycobacterium* sp. Since Crawford and Bates (57) were able to isolate plasmid DNA from several strains of *Mycobacterium avium–M. intracellulare*, Heitkamp et al. (101) proposed that the pyrene-degrading plasmid in *Mycobacterium* sp. may have practical application in cloning studies to construct genetically engineered bacteria capable of degrading higher molecular weight PAHs.

The chemical pathway for the degradation of pyrene by *Mycobacterium* sp. was first reported by Heitkamp et al. (101). In this study, over 60% of [^{14}C]pyrene was mineralized to CO_2 after 96 hr of incubation. The metabolites, *cis*- and *trans*-4,5-dihydrodiols and pyrenol, were identified as initial microbial ring-oxidation products of pyrene (Figure 14). The principal metabolite, 4-phenanthroic acid, and 4-hydroxy-perinaphthenone and cinnamic and phthalic acids were identified as ring fission products. The pyrene *trans*-4,5-dihydrodiol reported in this study was the first report on the monooxygenase-catalyzed *trans*-dihydrodiol formed from a PAH by a bacterium.

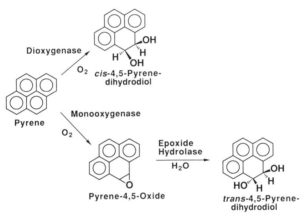

Figure 14. Mono- and dioxygenase pathways of pyrene. Adapted from Cerniglia et al. (42).

Studies were also conducted in microcosms to evaluate the survival and performance of *Mycobacterium* sp. added to sediment and water from a pristine ecosystem. The *Mycobacterium* survived in the microcosms for 6 weeks both with and without preexposure to PAH and mineralized multiple doses of pyrene; however, inorganic nutrient enrichment had little effect on pyrene mineralization (100).

Recently, Walter et al. (193) reported the degradation of pyrene by *Rhodococcus* sp. UW1. This *Rhodococcus* sp. was isolated from a contaminated soil and was able to mineralize 72% of pyrene to CO_2 within a period of 2 weeks. Walter et al. (193) proposed two possible pathways for the initial oxidation and ring fission of pyrene by *Rhodococcus* sp. UW1. They postulated that ring dihydroxylation occurs in either the 1,2- or 4,5-position of pyrene, since pyrene is a highly symmetrical and compact PAH.

Microbial metabolism of pyrene by the fungus *Cunninghamella elegans* was first reported by Cerniglia et al. (39), who suggested that the fungal metabolism of pyrene proceeds via hydroxylation of the aromatic ring to phenolic derivatives of glucoside conjugates (Figure 15). The glucoside conjugate of 1-hydroxypyrene accounted for 61% of the ethyl acetate–soluble pyrene metabolites. Cerniglia et al. (39) suggested that the glucoside conjugates of the mono- and dihydroxylated pyrenes (Figure 15) are the novel metabolites and may be important in the detoxification of pyrene.

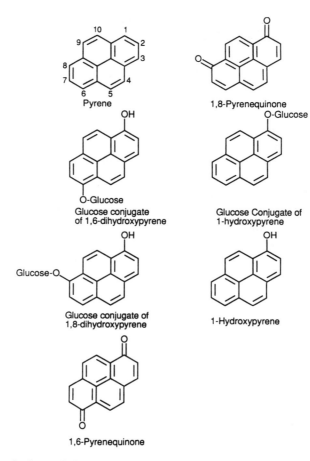

Figure 15. Identified metabolites from the fungal transformation of pyrene. Adapted from Cerniglia et al. (39).

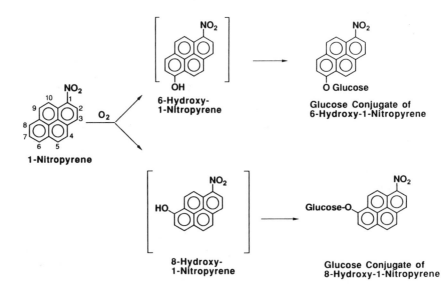

Figure 16. Fungal transformation of 1-nitropyrene.

1-Nitropyrene

Nitrated PAHs have been detected in urban areas in air (162, 194), coal-fly ash (197), and diesel exhaust particulates (156, 169, 170, 202). The most abundant of the nitro-PAHs was 1-nitropyrene. 1-Nitropyrene is mutagenic to bacteria (8), mutagenic or genotoxic to mammalian cells (8), and potentially carcinogenic to humans.

Few studies have reported the microbial metabolism of nitro-PAHs. Heitkamp et al. (103) found that 1-nitropyrene degraded very slowly in sediments compared to other PAHs. Previous studies with pure cultures of *Mycobacterium* sp. indicated that pyrene was rapidly mineralized by the *Mycobacterium* sp. (102); however, this microorganism degraded 1-nitropyrene to a limited extent (103). The major ethyl acetate–extractable metabolites identified were 1-nitropyrene *cis*-9,10- and 4,5-dihydrodiols. Heitkamp et al. (103) suggest that the slow mineralization of 1-nitropyrene could be due to the nitro substituent in the 1-position of pyrene, which significantly alters the metabolism.

Cerniglia et al. (30) investigated the fungal metabolism of 1-nitropyrene. *C. elegans* metabolized 1-nitropyrene to form glucoside conjugates of 6-hydroxy-1-nitropyrene and 8-hydroxy-1-nitropyrene (Figure 16). These glucoside conjugates of 6- and 8-hydroxy-1-nitropyrene were nonmutagenic in the *Salmonella* reversion assay and indicated that *C. elegans* metabolizes 1-nitropyrene to detoxified products.

CONCLUSIONS

Microorganisms, including bacteria, fungi, yeasts, and algae, have the ability to metabolize both lower and higher molecular weight polycyclic aromatic hydrocarbons (PAHs) found in the natural environment. Molecular oxygen is necessary for degradation of unsubstituted PAHs. Most bacteria have been found to oxygenate the PAH initially to form dihydrodiol with a *cis*-configuration. Both the oxygen atoms are incorporated into the aromatic substrate via a dioxygenase. This leads to the formation of *cis*-dihydrodiols, which can be further oxidized to catechols. The dihydroxylated intermediates undergo intra- or extradiol ring cleavage to ring fission products. Interestingly, a *Mycobacterium* recently isolated contains both a mono- and dioxygenase to metabolize PAHs.

Most fungi oxidize PAHs via a cytochrome P-450–catalyzed monooxygenase reaction. One atom of molecular oxygen is incorporated into the aromatic substrate via a cytochrome P-450 monooxygenase to form reactive arene oxides that can isomerize to phenols. The arene oxide can undergo enzymatic hydration to yield *trans*-dihydrodiols, which are optically active predominately in S,S configurations. These reactions are similar to those reported for PAH oxidation by mammalian enzyme systems except that the absolute stereochemistry of the dihydrodiols are different. The fungus *Cunninghamella elegans* has shown promise, since it can oxidize potentially carcinogenic PAHs predominately to detoxified products rather than the proximate or ultimate carcinogens typically produced in animal metabolism. White-rot fungi oxidize PAHs via ligninases to form highly reactive quinones; however, preliminary studies have suggested some evidence of P-450 activity.

Information pertaining to the ability of microorganisms to degrade aromatic hydrocarbons under anaerobic conditions is limited. The anaerobic pathways defined to date, however, may be helpful in evaluating future studies on anaerobic biodegradation of PAHs under *in situ* conditions.

Many aspects of environmental bioremediation of PAHs remain to be understood thoroughly, because only a few PAHs have been studied and the mechanisms of enzymatic oxygenation of PAHs in natural microbial habitats have not been explored significantly. In addition, little is known about the biodegradation of mixtures of PAHs present in growth-supporting concentrations.

LITERATURE CITED

1. Akhtar, M. N., D. R. Boyd, N. J. Thomas, M. Koreeda, D. T. Gibson, V. Mahadevan, and D. M. Jerina. 1975. Absolute stereochemistry of the dihydroanthracene-cis- and trans-1,2-diols produced from anthracene by mammals and bacteria. *J. Chem. Soc. Perkins Trans.* 1:2506–2511.

2. Autrup, H. 1990. Carcinogen metabolism in cultured human tissues and cells. *Carcinogenesis* 11:707–712.

3. Axcell, B. C., and P. J. Geary. 1975. Purification and some properties of a soluble benzene-oxidizing system from a strain of *Pseudomonas. Biochem. J.* 146:173–183.

4. Barnsley, E. A. 1975. Bacterial degradation of fluoranthene and benz[a]pyrene. *Can. J. Microbiol.* 21:1004–1008.

5. Barnsley, E. A. 1976. Role and regulation of the *ortho* and *meta* pathways of catechol metabolism in pseudomonads metabolizing naphthalene and salicylate. *J. Bacteriol.* 125:404–408.

6. Barnsley, E. A. 1976. Naphthalene metabolism by *Pseudomonas*: the oxidation of 1,2-dihydroxynaphthalene to 2-hydroxychromene-2-carboxylic acid and the formation of 2-hydroxybenzalpyruvate. *Biochem. Biophys. Res. Commun.* 72:1116–1121.

7. Barnsley, E. A. 1983. Bacterial oxidation of naphthalene and phenanthrene. *J. Bacteriol.* 153:1069–1071.

8. Beland, F. A., R. H. Heflich, P. C. Howard, and P. P. Fu. 1985. The in vitro metabolic activation of nitro polycyclic aromatic hydrocarbons. In *Polycyclic Hydrocarbons and Carcinogenesis.* Ed. R. G. Harvey. *ACS Symp. Ser.,* No. 283. Washington, D.C.: American Chemical Society. 371–396.

9. Boronin, A. M., V. V. Kochetkov, and G. K. Skryabin. 1980. Incompatibility groups of naphthalene degradative plasmids in *Pseudomonas. FEMS Microbiol. Lett.* 7:249–252.

10. Bos, R. P., W. J. C. Prinsen, J. G. M. van Rooy, F. J. Jongeneelen, J. L. G. Theuws, and P. Th. Henderson. 1987. Fluoranthene, a volatile mutagenic compound, present in creosote and coal tar. *Mutat. Res.* 187:119–125.

11. Boyland, E., and P. Sims. 1962. The metabolism of phenanthrene in rabbits and rats: dihydroxy compounds and related glucosiduric acids. *Biochem. J.* 84:571–582.

12. Boyland, E., and P. Sims. 1964. Metabolism of polycyclic compounds 23. The metabolism of pyrene in rats and rabbits. *Biochem. J.* 90:391–398.

13. Brusseau, G. A., Hsien-Chyang Tsien, R. S. Hanson, and L. P. Wackett. 1990. Optimization of trichloroethylene oxidation by methanotrophs and the use of a colorimetric assay to detect soluble methane monooxygenase activity. *Biodegradation* 1:19–29.

14. Bumpus, J. A. 1989. Biodegradation of polycyclic aromatic hydrocarbons by *Phanerochaete chrysosporium. Appl. Environ. Microbiol.* 55:154–158.

15. Bumpus, J. A., and S. D. Aust. 1987. Biodegradation of chlorinated organic compounds by *Phanerochaete chrysosprium,* a wood-rotting fungus. In *Solving Hazardous Waste Problems: Learning from Dioxins.* Ed. J. H. Exner. Washington, D.C.: American Chemical Society. 340–349.

16. Bumpus, J. A., M. Tien, D. Wright, and S. D. Aust. 1985. Oxidation of persistent environmental pollutants by a white rot fungus. Science 228:1434–1436.

17. Busby, W. F. Jr., M. E. Goldman, P. M. Newberne, and G. N. Wogan. 1984. Tumorigenicity of fluoranthene in a newborn mouse lung adenoma bioassay. Carcinogenesis 5:1311–1316.

18. Byrne, M., S. Coons, M. Goyer, J. Harris, and J. Perwak. 1985. Exposure and risk assessment for benzo[a]pyrene and other polycyclic aromatic hydrocarbons. Govt. Reports Announcements Index 3:21.

19. Cane, P. A., and P. A. Williams. 1982. The plasmid-coded metabolism of naphthalene and 2-methylnaphthalene in Pseudomonas strains. Phenotypic changes correlated with structural modification of the plasmid PWW60–1. J. Gen. Microbiol. 128:2281–2290.

20. Catterall, F. A., K. Murray, and P. A. Williams. 1971. The configuration of the 1,2-dihydroxy-1,2-dihydronaphthalene formed by the bacterial metabolism of naphthalene. Biochem. Biophys. Acta 237:361–364.

21. Cerniglia, C. E. 1981. Aromatic hydrocarbons: metabolism by bacteria, fungi and algae. Biochem. Toxicol. 3:321–361.

22. Cerniglia, C. E. 1982. Initial reactions in the oxidation of anthracene by Cunninghamella elegans. J. Gen. Microbiol. 128:2055–2061.

23. Cerniglia, C. E. 1984. Microbial metabolism of polycyclic aromatic hydrocarbons. Adv. Appl. Microbiol. 30:31–71.

24. Cerniglia, C. E. 1984. Microbial transformation of aromatic hydrocarbons. In Petroleum Microbiology. Ed. R. M. Atlas. New York: Macmillan Publishing Company. 99–128.

25. Cerniglia, C. E., J. R. Althaus, F. E. Evans, J. P. Freeman, R. K. Mitchum, and S. K. Yang. 1983. Stereochemistry and evidence for an arene oxide-NIH shift pathway in the fungal metabolism of naphthalene. Chem. Biol. Interact. 44:119–132.

26. Cerniglia, C. E., W. L. Campbell, J. P. Freeman, and F. E. Evans. 1989. Identification of a novel metabolite in phenanthrene metabolism by the fungus Cunninghamella elegans. Appl. Environ. Microbiol. 55:2275–2279.

27. Cerniglia, C. E., W. L. Campbell, P. P. Fu, J. P. Freeman, and F. E. Evans. 1990. Stereoselective fungal metabolism of methylated anthracenes. Appl. Environ. Microbiol. 56:661–668.

28. Cerniglia, C. E., and S. A. Crow. 1981. Metabolism of aromatic hydrocarbons by yeasts. Arch. Microbiol. 129:9–13.

29. Cerniglia, C. E., R. H. Dodge, and D. T. Gibson. 1982. Fungal oxidation of 3-methylcholanthrene: formation of proximate carcinogenic metabolites of 3-methylcholanthrene. Chem. Biol. Interact. 38:161–173.

30. Cerniglia, C. E., J. P. Freeman, G. L. White, R. H. Heflich, and D. W. Miller. 1985. Fungal metabolism and detoxification of the nitropolycyclic aromatic hydrocarbon 1-nitropyrene. Appl. Environ. Microbiol. 50:649–655.

31. Cerniglia, C. E., and D. T. Gibson. 1979. Oxidation of benzo[a]pyrene by the filamentous fungus Cunninghamella elegans. J. Biol. Chem. 254:12174–12180.

32. Cerniglia, C. E., and D. T. Gibson. 1980. Fungal oxidation of benzo[a]pyrene and (+)-trans,7,8-dihydroxy-7,8-dihydrobenzo[a]pyrene: evidence for the formation of benzo[a]pyrene 7,8-diol-9,10-epoxide. J. Biol. Chem. 255:5159–5163.

33. Cerniglia, C. E., and D. T. Gibson. 1980. Fungal oxidation of (+)-9,10-dihydroxy-910-dihydrobenzo[a]pyrene: formation of diastereomeric benzo[a]pyrene 9′10-diol-7,8-epoxides. Proc. Natl. Acad. Sci. USA. 77:4554–4558.

34. Cerniglia, C. E., and M. A. Heitkamp. 1989. Microbial degradation of polycyclic aromatic hydrocarbons in aquatic environment. In Metabolism of Polycyclic Aromatic Hydrocarbons in the Aquatic Environment. Ed. U. Varanasi. Boca Raton, FL: CRC Press, Inc. 41–68.

35. Cerniglia, C. E., D. T. Gibson, and C. Van Baalen. 1979. Algal oxidation of aromatic hydrocarbons: formation of 1-naphthol from naphthalene by *Agmenellum quadruplicatum* strain PR-6. *Biochem. Biophys. Res. Commun.* 88:50–58.

36. Cerniglia, C. E., D. T. Gibson, and C. Van Baalen. 1980. Oxidation of naphthalene by cyanobacteria and microalgae. *J. Gen. Microbiol.* 116:495–500.

37. Cerniglia, C. E., D. T. Gibson, and C. Van Baalen. 1982. Naphthalene metabolism by diatoms isolated from the Kachemak Bay region of Alaska. *J. Gen. Microbiol.* 128:987–990.

38. Cerniglia, C. E., R. L. Herbert, P. J. Szaniszlo, and D. T. Gibson. 1978. Fungal transformation of naphthalene. *Arch. Microbiol.* 117:135–143.

39. Cerniglia, C. E., D. W. Kelly, J. P. Freeman, and D. W. Miller. 1986. Microbial metabolism of pyrene. *Chem. Biol. Inter.* 57:203–216.

40. Cerniglia, C. E., C. Van Baalen, and D. T. Gibson. 1980. Metabolism of naphthalene by the cyanobacterium *Oscillatoria* sp. strain JCM. *J. Gen. Microbiol.* 116:485–495.

41. Cerniglia, C. E., W. Mahaffey, D. T. Gibson. 1980. Fungal oxidation of benzo[a]pyrene: formation of (−)−trans-7,8,dihydroxy-7,8-dihydrobenzo[a]pyrene by *Cunninghamella elegans*. *Biochem. Biophys. Res. Commun.* 94:226–232.

42. Cerniglia, C. E., J. B. Sutherland, and S. A. Crow. 1991. Fungal metabolism of aromatic hydrocarbons. In *Microbial Degradation of Natural Products*. Ed. G. Winkelmann. VCH Verlagsgesellschaft. Weinheim: In press.

43. Cerniglia, C. E., G. L. White, and R. H. Heflich. 1985. Fungal metabolism and detoxification of polycyclic aromatic hydrocarbons. *Arch. Microbiol.* 50:649–655.

44. Cerniglia, C. E., and S. K. Yang. 1984. Stereoselective metabolism of anthracene and phenanthrene by the fungus *Cunninghamella elegans*. *Appl. Environ. Microbiol.* 47:119–124.

45. Chapman, P. J. 1979. Degradation mechanisms. In *Proceedings of the Workshop: Microbial Degradation of Pollutants in Marine Environments*. Ed. A. W. Bourquin, P. H. Pritchard. Gulf Breeze, FL.: U.S. Environmental Protection Agency. 28–66.

46. Chaturapit, S., and C. M. Holder. 1978. Studies on the hepatic microsomal metabolism of ^{14}C-phenanthrene. *Biochem. Pharmacol.* 27:1865–1871.

47. Chen, F. M. 1983. Binding of pyrene to DNA, base sequence specificity and its implications. *Nucl. Acid Res.* 11:7232–7250.

48. Cody, T. E., M. J. Radike, and D. Warshawsky. 1984. The phytotoxicity of benzo[a]pyrene in the green alga, *Selenastrum capricornutum*. *Environ. Res.* 35:122–131.

49. Colby, J., and H. Dalton. 1976. Some properties of a soluble methane monooxygenase from *Methylococcus capsulatus* strain Bath. *Biochem. J.* 157:495–497.

50. Colby, J., and H. Dalton. 1978. Resolution of the methane monooxygenase of *Methylococcus capsulatus* (Bath) into three components. Purification and properties of component C, a flavoprotein. *Biochem. J.* 171:461–468.

51. Colby, J., D. I. Stirling, and H. Dalton. 1977. The soluble methane monooxygenase of *Methylcoccus capsulatus* (Bath): Its ability to oxygenate n-alkanes, n-alkenes, ethers and alicyclic, aromatic and heterocyclic compounds. *Biochem. J.* 165:395–402.

52. Colla, A., A. Fiecchi, and V. Treccani. 1959. Ricerche sul metabolismo ossidativo microbico dell'antracene e del fenantrene. *Ann. Microbiol.* 9:87–91.

53. Connors, M. A., and E. A. Barnsley. 1982. Naphthalene plasmids in Pseudomonads. *J. Bacteriol.* 149:1096–1101.

54. Cook, J. W., C. L. Hewett, and I. Hieger. 1933. The isolation of a cancer-producing hydrocarbon from coal tar. Parts I, II, and III. *J. Chem. Soc.* 394–405.

55. Cooper, C. S., P. Vigny, M. Kindts, P. L. Grover, and P. Sims. 1980. Metabolic activation of 3-methylcholanthrene in mouse skin: Fluorescence spectral evidence indicates the involvement of diol-epoxides formed in the 7,8,9–10 ring. *Carcinogenesis* 1:855–860.

56. Cooper, C. S., P. L. Grover, and P. Sims. 1983. The metabolism and activation of benzo[a]pyrene. *Prog. Drug Metab.* 7:295–396.

57. Crawford, J. T., and J. H. Bates. 1979. Isolation of plasmids from *Mycobacteria. Infect. Immun.* 24:979–981.

58. Crutcher, S. E., and P. J. Geary. 1979. Properties of the iron-sulfur proteins of the benzene dioxygenase system from *Pseudomonas putida. Biochem. J.* 177:393–400.

59. Dagley, S. 1971. Catabolism of aromatic compounds by microorganisms. *Adv. Microbiol. Physiol.* 6:1–46.

60. Dagley, S. 1975. A biochemical approach to some problems of environmental pollution. *Essays Biochem.* 11:81–138.

61. Dagley, S. 1981. New perspectives in aromatic catabolism. In *Microbial Degradation of Xenobiotics and Recalcitrant Compounds.* Ed. T. Leisinger, R. Hutter, A. M. Cook, J. Nuesch. New York: Academic Press. 181–186.

62. Dalton, H., B. T. Golding, B. W. Waters, R. Higgins, and J. A. Taylor. 1981. Oxidations of cyclopropane, methylcyclopropane, and arenes with the mono-oxygenase system from *Methylococcus capsulatus. J. Chem. Soc. Chem. Comm.* 482–483.

63. Datta, D., and T. B. Samantha. 1988. Effects of inducers on metabolism of benzo[a]pyrene in vivo and in vitro analysis by high-pressure liquid chromatography. *Biochem. Biophys. Res. Commun.* 155:493–502.

64. Davies, J. I., and W. C. Evans. 1964. Oxidative metabolism of naphthalene by soil pseudomonads. *Biochem. J.* 91:251–261.

65. Devereux, R., and R. K. Sizemore. 1981. Incidence of degradative plasmids in hydrocarbon-utilizing bacteria isolated from the gulf of Mexico. *Dev. Indust. Microbiol.* 22:409–414.

66. Dodge, R. H., and D. T. Gibson. 1980. Fungal metabolism of benz[a]anthracene. *Abstr. Annu. Am. Soc. Microbiol. Meet.* 138.

67. Drummond, E. C., P. Callaghan, and R. P. Hopkins. 1972. Metabolic dehydrogenation of *cis-* and *trans-*acenaphthene-1,2-diol. *Xenobiotica* 2:529–538.

68. Dunn, N. W., and I. C. Gunsalus. 1973. Transmissible plasmid coding early enzymes of naphthalene oxidation in *Pseudomonas putida. J. Bacteriol.* 114:974–979.

69. Ensley, B. D., and D. T. Gibson. 1983. Oxidation of naphthalene by a multicomponent enzyme system from *Pseudomonas* sp. strain NCIB 9816. *J. Bacteriol.* 155:505–511.

70. Ensley, B. D., D. T. Gibson, and L. A. LaBorde. 1982. Naphthalene dioxygenase: purification and properties of a terminal oxygen component. *J. Bacteriol.* 149:948–954.

71. Evans, W. C. 1977. Biochemistry of the bacterial catabolism of aromatic compounds in anaerobic environments. *Nature* 270:17–22.

72. Evans, W. C., H. N. Fernley, and E. Griffiths. 1965. Oxidative metabolism of phenanthrene and anthracene by soil pseudomonads; the ring fission mechanism. *Biochem. J.* 95:819–821.

73. Ferris, J. P., M. J. Fasco, F. L. Stylianopoulou, D. M. Jerina, J. W. Daly, and A. M. Jeffrey. 1973. Mono-oxygenase activity in *Cunninghamella bainieri*: evidence for a fungal system similar to liver microsomes. *Arch. Biochem. Biophys.* 156:97–103.

74. Furukawa, K., J. Simon, and A. M. Chakrabarty. 1983. Common induction and regulation of biphenyl, xylene/toluene, and salicylate metabolism in *Pseudomonas paucimobilis. J. Bacteriol.* 154:1356–1362.

75. Garcia-Valdes, E., E. Cozar, R. Rotger, J. Lalucat, and J. Ursing. 1988. New naphthalene-degrading marine *Pseudomonas* strains. *Appl. Environ. Microbiol.* 54:2478–2485.

76. Ghosh, D. K., and A. K. Mishra. 1983. Oxidation of phenanthrene by a strain of *Micrococcus*: evidence of protocatechuate pathway. *Curr. Microbiol.* 9:219–224.

77. Ghosh, D. K., D. Dutta, T. B. Samanta, and A. K. Mishra. 1983. Microsomal benzo[a]pyrene hydroxylase in *Aspergillus ochraceus* TS: assay and characterization of the

enzyme system. *Biochem. Biophys. Res. Commun.* 113:497–505.

78. Gibson, D. T. 1977. Biodegradation of aromatic petroleum hydrocarbons. In *Fate and Effect of Petroleum Hydrocarbons in Marine Ecosystems and Organisms.* Ed. D. A. Wolfe. New York: Pergamon Press.36–46.

79. Gibson, D. T. 1982. Microbial degradation of hydrocarbons. *Toxicol. Environ. Chem.* 5:237–250.

80. Gibson, D. T., J. R. Koch, and R. E. Kallio. 1968. Oxidative degradation of aromatic hydrocarbons by microorganisms. I. Enzymatic formation of catechol from benzene. *Biochemistry* 7:2653–2661.

81. Gibson, D. T., V. Mahadevan, D. M. Jerina, H. Yagi, and H. J. C. Yeh. 1975. Oxidation of the carcinogens benzo[a]pyrene and benz[a]anthracene to dihydrodiols by a bacterium. *Science* 189:295–297.

82. Gibson, D. T., and V. Subramanian. 1984. Microbial degradation of aromatic hydrocarbons. In *Microbial Degradation of Organic Compounds.* Ed. D. A. Wolfe. New York: Marcel Dekker. 181–252.

83. Giger, W., and M. Blumer. 1974. Polycyclic aromatic hydrocarbons in the environment: isolation and characterization by chromatography, visible, ultraviolet and mass spectrometry. *Anal. Chem.* 46:1663–1671.

84. Gold, M. H., H. Wariishi, and K. Valli. 1989. Extracellular peroxidases involved in lignin degradation by the white rot basidiomycete *Phanerchaete chrysosporium.* In *Biocatalysis in Agricultural Biotechnology.* Ed. J. R. Whitaker, P. E. Sonnet. Washington, D.C.: American Chemical Society.127–140.

85. Grbić-Galić, D., and T. M. Vogel. 1987. Transformation of toluene and benzene by mixed methanogenic cultures. *Appl. Environ. Microbiol.* 53:254–260.

86. Grover, P. L., A. Hewer, and P. Sims. 1972. Formation of K-region epoxides as microsomal metabolites of pyrene and benzo[a]pyrene. *Biochem. Pharmacol.* 21:3316.

87. Gschwend, P. M., and R. A. Hites. 1981. Fluxes of polycyclic aromatic hydrocarbons to marine and lacustrine sediments in the northeastern United States. *Geochim. Cosmochim. Acta* 45:2359–2367.

88. Guerin, F. W., and G. E. Jones. 1988. Mineralization of phenanthrene by a *Mycobacterium* sp. *Appl. Environ. Microbiol.* 54:937–944.

89. Guyer, M., and G. Hegeman. 1969. Evidence for a reductive pathway for the anaerobic metabolism of benzoate. *J. Bacteriol.* 99:906–907.

90. Haemmerli, S. D., M. S. A. Leisola, D. Sanglard, and A. Fiechter. 1986. Oxidation of benzo[a]pyrene by extracellular ligninases of *Phanerochaete chrysosporium*: veratryl alchohol and stability of ligninase. *J. Biol. Chem.* 261:6900–6903.

91. Hammel, K. E. 1989. Organopollutant degradation by ligninolytic fungi. *Enzyme Microb. Technol.* 11:776–777.

92. Hammel, K. E., B. Kalyanaraman, and T. K. Kirk. 1986. Oxidation of polycyclic aromatic hydrocarbons and dibenzo[p]dioxins by *Phanerochaete chrysosporium. J. Biol. Chem.* 261:16948–16952.

93. Harayama, S., and M. Rakik. 1989. Bacterial aromatic ring-cleavage enzymes are classified into two different gene families. *J. Biol. Chem.* 264:15328–15333.

94. Harayama, S., M. Rakik, A. Wasserfallen, and A. Bairoch. 1987. Relationships between catabolic pathways for aromatics: conservation of gene order and nucleotide-sequences of catechol oxidation genes of PWWO and NAH7 plasmids. *Mol. Gen. Genet.* 210:241–247.

95. Harper, K. H. 1957. The metabolism of pyrene. *Br. J. Cancer* 11:499–507.

96. Harper, K. H. 1958. The intermediary metabolism of pyrene. *Br. J. Cancer* 12:116–120.

97. Healy, J. B., Jr., and L. Y. Young. 1979. Anaerobic biodegradation of eleven aromatic compounds to methane. *Appl. Environ. Microbiol.* 38:84–89.

98. Heidelberger, C. 1975. Chemical carcinogenesis. *Annu. Rev. Biochem.* 44:79–121.

99. Heitkamp, M. A., and C. E. Cerniglia. 1988. Mineralization of polycyclic aromatic hydrocarbons by a bacterium isolated from sediment below an oil field. *Appl. Environ. Microbiol.* 54:1612–1614.

100. Heitkamp, M. A., and C. E. Cerniglia. 1989. Polycyclic aromatic hydrocarbon degradation by a *Mycobacterium* sp. in microcosms containing sediment and water from pristine ecosystem. *Appl. Environ. Microbiol.* 55:1968–1973.

101. Heitkamp, M. A., W. Franklin, and C. E. Cerniglia. 1988. Microbial metabolism of polycyclic aromatic hydrocarbons: Isolation and characterization of a pyrene-degrading bacterium. *Appl. Environ. Microbiol.* 54:2549–2555.

102. Heitkamp, M. A., J. P. Freeman, D. W. Miller, and C. E. Cerniglia. 1988. Pyrene degradation by a *Mycobacterium* sp.: Identification of ring oxidation and ring fission products. *Appl. Environ. Microbiol.* 54:2556–2565.

103. Heitkamp, M. A., J. P. Freeman, D. W. Miller, and C. E. Cerniglia. 1991. Biodegradation of 1-nitropyrene. *Arch. Microbiol.* In press.

104. Higgins, I. J., D. J. Best, and R. C. Hammond. 1980. New findings in methane-utilizing bacteria highlight their importance in the biosphere and their commercial potential. *Nature* 286:561–564.

105. Honeck, H., W. H. Schunuck, and H.-G. Muller. 1985. The function of cytochrome P-450 in fungi and prospects of application. *Pharmazie* 40:221–227.

106. Hopkins, R. P. 1968. Microsomal ring-fission of *cis*- and *trans*-acenaphthene-1,2-diol. *Biochem. J.* 108:577–582.

107. Hopkins, R. P., and L. Young. 1966. Biocehmical studies of toxic agents:metabolic ring-fission of *cis*- and *trans*-acenaphthene-1,2-diol. *Biochem. J.* 98:19–24.

108. Jacob, J., G. Grimmer, G. Raab, and A. Schmoldt. 1982. The metabolism of pyrene by rat liver microsomes and the influence of various mono-oxygenase inducers. *Xenobiotica* 12:45–53.

109. Jeffrey, A. M., H. J. C. Yeh, D. M. Jerina, T. R. Patel, J. F. Davey, and D. T. Gibson. 1975. Initial reactions in the oxidation of naphthalene by *Pseudomonas putida*. *Biochemistry* 14:575–584.

110. Jennette, K. W., W. Bornstein, A. H. L. Chuang, and E. Bresnick. 1979. Stereospecificity of metabolism of benzo[a]pyrene (BaP) to (+)–7,8-dihydroxy-7,8-dihydro-BaP by rat liver nuclear enzymes. *Biochem. Pharmacol.* 28:338–339.

111. Jerina, D. M., H. Selander, H. Yagi, M. C. Wells, J. F. Davey, V. Mahadevan, and D. T. Gibson. 1976. Dihydrodiols from anthracene and phenanthrene. *J. Am. Chem. Soc.* 98:5988–5996.

112. Jerina, D. M., R. Lehr, M. Schaefer-Ridder, H. Yagi, J. M. Karle, D. R. Thakker, A. W. Wood, A.-Y. H. Lu, D. Ryan, S. West, W. Levin, and A. H. Conney. 1977. Bay region epoxides of dihydrodiols: a concept which explains the mutagenic and carcinogenic activity of benzo[a]pyrene and benzo[a]anthracene. In *Origins of Human Cancer*. Ed. H. H. Hiatt, J. D. Watson, J. A. Winstein. Cold Spring Harbor, NY: Cold Spring Harbor Laboratory. 639–658.

113. Jerina, D. M., P. J. van Bladeren, H. Yagi, D. T. Gibson, V. Mahadevan, A. S. Neese, M. Koreeda, N. D. Sharma, and D. Boyd. 1984. Synthesis and absolute configuration of *cis*-1,2-, 8,9- and 10,11-dihydrodiol metabolites of benz[a]anthracene formed by a strain of *Beijerinckia*. *J. Org. Chem.* 49:1075–1082.

114. Jerina, D. M., H. Yagi, D. R. Thakker, J. M. Sayer, P. J. van Bladeren, R. E. Lehr, D. L. Whalen, W. Levin, R. L. Chang, A. W. Wood, and H. Conney. 1984. Identification of the ultimate carcinogenic metabolites of polycyclic aromatic hydrocarbons: bay region (R,S)-diol-(S,R)-epoxides. In *Foreign Compound Metabolism*. Ed. J. Caldwell, G. D. Paulsen. London: Taylor and Francis Ltd. 257–280.

115. Jones, K. C., A. Stratford, K. S. Waterhouse, and N. B. Vogt. 1989. Organic contaminants in Welsh soils: Polynuclear aromatic hydrocarbons. *Environ. Sci. Technol.* 13:540–550.

116. Kaden, D. A., R. A. Hites, and W. G. Thilly. 1979. Mutagenicity of soot and associated

polycyclic aromatic hydrocarbons to *Salmonella typhimurium. Cancer Res.* 39:4152–4159.

117. Kaiser, J. P., and K. W. Hanselmann. 1982. Fermentative metabolism of substituted monoaromatic compounds by a bacterial community from anaerobic sediments. *Arch. Microbiol.* 133:185–194.

118. Kapitulnik, J., P. G. Wilslocki, W. Levin, H. Yagi, D. M. Jerina, and A. H. Conney. 1977. Tumerigenicity studies with diol epoxides of benzo[a]pyrene which indicate that (+)-*trans*-7β,8-dihydroxy-9, 10-epoxy-7,8,9,10-tetrahydrobenzo[a]pyrene is an ultimate carcinogen in newborn mice. *Cancer Res.* 38:354–358.

119. Kappeli, O. 1986. Cytochrome P-450 of yeasts. *Microbiol. Rev.* 50:244–258.

120. Kaushik, V., and V. S. Bisaria. 1989. Ligninases: biosynthesis and applications. *J. Sci. Ind. Res.* 48:276–290.

121. Keenan, T. H., T. Cody, M. Radike, and D. Warshawsky. 1986. Conjugation of benzo[a]pyrene by green algae. In *Polynuclear Aromatic Hydrocarbons: Chemistry, Characterization and Carcinogenesis.* Ed. M. Cooke, A. Dennis. Columbus, OH: Battelle Press. 427–435.

122. Keimig, S. D., K. W. Kirby, and D. P. Morgan. 1983. Identification of 1-hydroxypyrene as a major metabolite of pyrene in pig urine. *Xenobiotica* 13:415–420.

123. Keith, L. H., and W. A. Telliard. 1979. Priority pollutants. I. A perspective view. *Environ. Sci. Technol.* 13:416–423.

124. Kelley, I., and C. E. Cerniglia. 1991. The metabolism of fluoranthene by a species of *Mycobacterium. J. Indust. Microbiol.* 7:19–26.

125. Kelley, I., J. P. Freeman, and C. E. Cerniglia. 1991. Identification of metabolites from degradation of naphthalene by a *Mycobacterium* sp. *Biodegradation* 1:283–290.

126. Kelley, I., J. P. Freeman, F. E. Evans, and C. E. Cerniglia. 1991. Identification of a carboxylic acid metabolite from the catabolism of fluoranthene by a *Mycobacterium* sp. *Appl. Environ. Microbiol.* 57:636–641.

127. Kelly, S. L., and D. E. Kelly. 1988. Analysis and exploitation of cytochrome P-450 in yeast. *Biochem. Soc. Trans.* 16:1086–1088.

128. Kilbourn, R. 1980. M.S. thesis, University of Texas at Austin, Austin, Texas.

129. King, J. M. H., P. M. DiGrazia, B. Applegate, R. Burlage, J. Sanseverino, P. Dunbar, F. Larimer, and G. S. Sayler. 1990. Rapid, sensitive bioluminescent reporter technology for naphthalene exposure and biodegradation. *Science* 249:778–781.

130. King, D. J., and A. Wiseman. 1987. Yeast cytochrome P-488 enzymes and the activation of mutagens, including carcinogens. In *Enzyme Induction, Mutagen Activation and Carcinogen Testing in Yeast.* Ed. A. Wiseman. Chichester, Ellis Horwood. 115–167.

131. Kiyohara, H., and K. Nagao. 1978. The catabolism of phenanthrene and naphthalene by bacteria. *J. Gen. Microbiol.* 105:69–75.

132. Kiyohara, H., K. Nagao, K. Kuono, and K. Yano. 1982. Phenanthrene degrading phenotype of *Alcaligenes faecalis* AFK2. *Appl. Environ. Microbiol.* 43:458–461.

133. Kiyohara, H., K. Nagao, and R. Nomi. 1976. Degradation of phenanthrene through o-phthalate by an *Aeromonas* sp. *Agric. Biol. Chem.* 40:1075–1082.

134. Koreeda, M., M. N. Akhtar, D. R. Boyd, J. D. Neill, D. T. Gibson, and D. M. Jerina. 1978. Absolute stereochemistry of *cis*-1,2-, *trans*-1,2 and *cis*-3,4-dihydrodiol metabolites of phenanthrene. *J. Org. Chem.* 43:1023–1027.

135. Levin, W., D. R. Thakker, A. W. Wood, R. L. Chang, R. E. Lehr, D. M. Jerina, and A. H. Conney. 1978. Evidence that benz[a]anthracene 3,4-diol-1,2-epoxide is an ultimate carcinogen on mouse skin. *Cancer Res.* 38:1705–1710.

136. Levin, W., A. W. Wood, R. L. Chang, T. J. Slaga, H. Yagi, D. M. Jerina, and A. H. Conney. 1977. Marked differences in the tumor-initiating activity of optically pure (+)- and (−)-*trans*-7,8-dihydroxy-7,8-dihydrobenzo[a]pyrene on mouse skin. *Cancer Res.* 37:2721–2725.

137. Levin, W., A. W. Wood, G. Yagi, D. M. Jerina, and A. H. Conney. 1976. Benzo[a]pyrene

7,8-dihydrodiol: a potent skin carcinogen when applied topically to mice. *Proc. Natl. Acad. Sci. USA* 73:3867–3871.

138. Lin, W. S., and M. Kapoor. 1979. Induction of aryl hydrocarbon hydrolase in *Neurospora crassa* by benzo[a]pyrene. *Curr. Microbiol.* 3:177–181.

139. Lindquist, B., and D. Warshawsky. 1985. Identification of the 11,12-dihydroxyben-zo[a]pyrene as a major metabolite produced by the green alga, *Selenastrum capricornutum*. *Biochem. Biophys. Res. Commun.* 130:71–75.

140. Lobarzewski, J. 1990. The characteristics and functions of the peroxidases from *Trametes-versicolor* in lignin biotransformation. *J. Biotechnol.* 13:111–117.

141. Mahaffey, R. W., D. T. Gibson, and C. E. Cerniglia. 1988. Bacterial oxidation of chemical carcinogens: formation of polycyclic aromatic acids from benz[a]anthracene. *Appl. Environ. Microbiol.* 54:2415–2423.

142. Malaveille, G., and H. Bartsch. 1975. Mutagenicity of non-K-region diols and diol-epoxides of benz[a]anthracene and benzo[a]pyrene. *Biochem. Biophys. Res. Commun.* 6:693–700.

143. McElroy, A. E., J. W. Farrington, and J. M. Teal. 1989. Metabolism of polynuclear aromatic hydrocarbons in the aquatic environment. In *Metabolism of Polynuclear Aromatic Hydrocarbons in the Aquatic Environment.* Ed. U. Varanasi. Boca Raton, FL: CRC Press. 119–125.

144. Mihelcic, J. R., and R. G. Luthy. 1987. Degradation of polycyclic aromatic hydrocarbon compounds under various redox conditions in soil-water systems. *Appl. Environ. Microbiol.* 53:1182–1187.

145. Mihelcic, J. R., and R. G. Luthy. 1988. Microbial degradation of acenaphthene and naph-thalene under denitrification conditions in soil-water systems. *Appl. Environ. Microbiol.* 54:1188–1198.

146. Miller, E. C., and J. A. Miller. 1974. Biochemical mechanisms of chemical carcinogenesis. In *The Molecular Biology of Cancer.* Ed. H. Busch. New York: Academic Press. 377–403.

147. Mueller, J. G., P. J. Chapman, B. O. Blattmann, and P. H. Pritchard. 1990. Isolation and characterization of a fluoranthene-utilizing strain of *Pseudomonas paucimobilis*. *Appl. Environ. Microbiol.* 56:1079–1086.

148. Mueller, J. G., P. J. Chapman, and P. H. Pritchard. 1989. Action of fluoranthene-utilizing bacterial community on polycyclic aromatic hydrocarbon components of creosote. *Appl. Environ. Microbiol.* 55:3085–3090.

149. Muller, H.-G., W.-H. Schunck, P. Riege, and H. Hoenck. 1982. *Cytochrome P-450.* Ed. K. Ruckpaul, H. Rein. Berlin: Akademie-Verlag. 337–369.

150. Narro, M. L., C. E. Cerniglia, D. T. Gibson, and C. Van Baalen. 1985. The oxidation of aromatic compounds by microalgae. In *Environmental Regulation of Microbial Metabolism.* Ed. I. S. Kulaev, E. A. Dawes, D. W. Tempest. New York: Academic Press. 249–254.

151. Nordquist, M., D. R. Thakker, K. P. Vyas, H. Yagi, W. Levin, D. E. Ryan, P. E. Thomas, A. H. Conney, and D. M. Jerina. 1981. Metabolism of chrysenes and phenanthrenes to bay-region diol epoxides by rat liver enzymes. *Mol. Pharmacol.* 19:168–178.

152. Obana, H., S. Hori, and T. Kashimoto. 1981. Determination of polycyclic aromatic hydrocarbons in marine samples by high-performance liquid chromatography. *Bull. Environ. Contam. Toxicol.* 26:613–620.

153. Oesch, F., D. M. Jerina, J. W. Daly, A. W. H. Lu, R. Kuntzman, and A. J. Cooney. 1972. A reconstituted microsomal enzyme system that converts naphthalene to *trans*-1,2-dihydroxy-1,2-dihydronaphthalene via naphthalene-1,2-oxide: presence of epoxide hydrolase in cytochrome P-450 and P-448 fractions. *Arch. Biochem. Biophys.* 153:62–67.

154. Okamoto H., and D. Yoshida. 1981. Metabolic formation of pyrenequinones as enhancing agents of mutagenicity in *Salmonella*. *Cancer Lett.* 11:215–220.

155. Omura, T., and R. Sato. 1964. The carbon monoxide-binding pigment of liver microsomes. *J. Biol. Chem.* 239:2370–2378.

156. Paputa-Peck, M. C., R. S. Marano, D. Schuetzel, T. L. Riley, C. V. Hampton, T. J. Prater, L. M. Skewes, T. E. Jensen, P. H. Ruehle, L. C. Bosch, and W. P. Duncan. 1983. Determination of nitrated polynuclear aromatic hydrocarbons in particulate extracts by capillary column gas chromatography with nitrogen selected detection. *Anal. Chem.* 55:1947–1954.

157. Patel, T. R., and D. T. Gibson. 1974. Purification and properties of (+)-*cis*-naphthalene dihydrodiol dehydrogenase of *Pseudomonas putida*. *J. Bacteriol.* 119:879–888.

158. Pelkonen, O., and D. W. Nebert. 1982. Metabolism of polycyclic aromatic hydrocarbons: etiologic role in carcinogenesis. *Pharm. Rev.* 34:189–222.

159. Pothuluri, J. V., J. P. Freeman, F. E. Evans, and C. E. Cerniglia. 1990. Fungal transformation of fluoranthene. *Appl. Environ. Microbiol.* 56:2974–2983.

160. Pothuluri, J. V., R. H. Heflich, and C. E. Cerniglia. 1991. Fungal metabolism and detoxification of fluoranthene. *Appl. Environ. Microbiol.* 58:937–41.

161. Pothuluri, J. V., J. P. Freeman, F. E. Evans, and C. E. Cerniglia. 1992. Fungal metabolism of acenaphthene by *Cunninghamella elegans*. *Appl. Environ. Microbiol.* 58:3654–3659.

162. Ramdahl, T., G. Becher, and A. Bjorseth. 1982. Nitrated polycyclic aromatic hydrocarbons in urban air particles. *Environ. Sci. Technol.* 16:861–865.

163. Reddy, K. R., W. H. Patrick, Jr., and R. E. Phillips. 1978. The role of nitrate diffusion in determining the order and rate of denitrification in flooded soil. I. Experimental results. *Soil Sci. Soc. Am. J.* 42:268–272.

164. Rogoff, M. H., and I. Wender. 1957. Microbiology of coal. I. Bacterial oxidation of phenanthrene. *J. Bacteriol.* 73:264–268.

165. Sanglard, D., M. S. A. Leisola, and A. Fliechter. 1986. Role of extracellular ligninases in biodegradation of benzo[a]pyrene by *Phanerochaete chrysosporium*. *Enzyme Microbiol. Technol.* 8:209–212.

166. Schell, M. A. 1983. Cloning and expression in *Escherichia coli* of the naphthalene degradative genes from plasmid NAH7. *J. Bacteriol.* 153:822–829.

167. Schoemaker, H. E., and M. S. A. Leisola. 1990. Degradation of lignin by *Phanerochaete chrysosporium*. *J. Biotechnol.* 13:101–109.

168. Schocken, M. J., and D. T. Gibson. 1984. Bacterial oxidation of the polycyclic aromatic hydrocarbons acenaphthene and acenaphthylene. *Appl. Environ. Microbiol.* 48:10–16.

169. Schuetzle, D. 1983. Sampling of vehicle emissions for chemical analysis and biological testing. *Environ. Health Perspect.* 47:65–80.

170. Schuetzle, D., T. L. Riley, T. J. Prater, T. M. Harvey, and D. F. Hunt. 1982. Analysis of nitrated polycyclic aromatic hydrocarbons in diesel particulates. *Anal. Chem.* 54:265–271.

171. Serdar, C. M., and D. T. Gibson. 1989. Studies of nucleotide sequence homology between naphthalene-utilizing strains of bacteria. *Biochem. Biophys. Res. Commun.* 164:772–779.

172. Simmons, C. J., R. P. Hopkins, and P. Callaghan. 1973. Metabolic oxidation of acenaphthene-1-ol. *Xenobiotica* 3:633–642.

173. Sims, P. 1970. Qualitative and quantitative studies on the metabolism of a series of aromatic hydrocarbons by rat liver preparations. *Biochem. Pharmacol.* 19:795–818.

174. Sims, P., and P. L. Grover. 1981. Involvement of dihydrodiols and diolepoxides in the metabolic activation of polycyclic hydrocarbons other than benzo[a]pyrene. In *Polycyclic Hydrocarbons and Cancer*. Ed. H. V. Gelboin, P. O. P. Tso. New York: Academic Press. 117–181.

175. Slaga, T. J., A. Viaje, D. L. Barry, W. Bracken, S. G. Buty, and J. D. Scribner. 1976. Skin tumor initiating ability of benzo[a]pyrene 4,5-, 7,8-, and 7,8-diol-9,10-epoxides, and 7,8-diol. *Cancer Lett.* 2:115–120.

176. Smith, R. V., and J. P. Rosazza. 1974. Microbial models of mammalian metabolism. *Arch. Biochem. Biophys.* 161:551–558.

177. Sutherland, J. B., A. L. Selby, J. P. Freeman, F. E. Evans, and C. E. Cerniglia. 1991.

Metabolism of phenanthrene by *Phanerochaete chrysosporium*. *Appl. Environ. Microbiol.* 57:3310–3316.

178. Sutherland, J. B., A. L. Selby, J. P. Freeman, P. P. Fu, D. W. Miller, and C. E. Cerniglia. 1992. Identification of xyloside conjugates formed from anthracene by *Rhizoctonia solani*. *Mycol. Res.* 96:509–517.

179. Tagger, S., N. Truffaut, and J. Le Petit. 1990. Preliminary study on relationships among strains forming a bacterial community selected on naphthalene from marine sediment. *Can. J. Microbiol.* 36:676–681.

180. Taira, K. N. Hayase, N. Arimura, S. Yashamita, T. Miyazaki, and K. Furukawa. 1988. Cloning and nucleotide-sequence of the 2,3-dihydroxybiphenyl dioxygenase gene from the PCB-degrading strain of *Pseudomonas paucimobilis* Q1. *Biochemistry* 27:3990–3996.

181. Takeda, N., K. Teranishi, and K. Hamada. 1984. Mutagenicity of the sunlight exposed sample of pyrene in *Salmonella typhimurium* TA98. *Bull. Environ. Contam. Toxicol.* 33:410–417.

182. Tattersfield, F. 1927. The decomposition of naphthalene in the soil and the effect upon its insecticidal action. *Ann. Appl. Biol.* 15:57–67.

183. Tausson, W. O. 1927. Napthalin als Kohlenstoffqelle für Bakterine. *Planta* 4:214–256.

184. Thakker, D. R., W. Levin, A. W. Wood, A. H. Conney, T. A. Stoming, and D. M. Jerina. 1978. Metabolic formation of 1,9,10-trihydroxy-9,10-dihydro-3-methylcholanthrene: a potential proximate carcinogen from 3-methylcholanthrene. *J. Amer. Chem. Soc.* 100:645–647.

185. Thakker, D. R., W. Levin, H. Yagi, D. Ryan, P. E. Thomas, J. M. Karle, R. E. Lehr, D. M. Jerina, and A. H. Conney. 1979. Metabolism of benz[a]anthracene to its tumorigenic 3,4-dihydrodiol. *Mol. Pharmacol.* 15:138–153.

186. Thakker, D. R., H. Yagi, H. Agaki, M. Koreeda, A. Y. H. Lu, W. Levin, A. W. Wood, A. H. Conney, and D. M. Jerina. 1977. Metabolism of benzo[a]pyrene. VI. Stereoselective metabolism of benzo[a]pyrene 7,8-dihydrodiol to diol epoxides. *Chem. Biol. Interact.* 16:281–300.

187. Thakker, D. R., H. Yagi, W. Levin, A. W. Wood, A. H. Conney, and D. M. Jerina. 1985. Polycyclic aromatic hydrocarbons: metabolic activation to ultimate carcinogens. In *Bioactivation of Foreign Compounds*. Ed. M. W. Anders. New York: Academic Press. 177–242.

188. Thilly, W. G., J. G. DeLuca, E. E. Furth, H. Hoppe, IV, D. A. Kaden, J. J. Krolewski, H. L. Liber, T. R. Slopek, S. A. Slepikoff, R. J. Tizard, and B. W. Perman. 1980. Gene-locus mutation assays in diploid human lymphoblast lines. In *Chemical Mutagenicity*. Ed. F. J. de Serres, A. Hollaender. New York: Plenum Publishing. 331–364.

189. Treccani, V., N. Walker, and G. H. Wiltshire. 1954. The metabolism of naphthalene by soil bacteria. *J. Gen. Microbiol.* 11:341–348.

190. Van Duuren, B. L., and B. M. Goldschmidt. 1976. Carcinogenic and tumor-promoting agents in tobacco carcinogenesis. *J. Natl. Cancer Inst.* 56:1237–1242.

191. Vogel, T. M., and D. Grbić-Galić. 1986. Incorporation of oxygen from water into toluene and benzene during anaerobic fermentative transformation. *Appl. Environ. Microbiol.* 52:200–202.

192. Walker, N., and G. H. Wiltshire. 1953. The breakdown of naphthalene by a soil bacterium. *J. Gen. Microbiol.* 8:273–276.

193. Walter, U., M. Beyer, J. Klein, and H.-J. Rehm. 1991. Degradation of pyrene by *Rhodococcus* sp. UW1. *Appl. Microbiol. Biotechnol.* 34:671–676.

194. Wang, C. Y., M.-S. Lee, C. M. King, and P. O. Warner. 1980. Evidence for nitroaromatics as direct-acting mutagens of airborne particulates. *Chemosphere* 9:83–87.

195. Warshawsky, D., T. M. Keenan, R. Reilman, T. E. Cody, and M. J. Radike. 1990. Conjugation of benzo[a]pyrene metabolites by freshwater green alga *Selenastrum capricornutum*. *Chem. Biol. Interact.* 74:93–105.

196. Warshawsky, D., M. Radike, K. Jayasimhulu, and T. Cody. 1988. Metabolism of benzo[a]pyrene by a dioxygenase enzyme system of the freshwater green alga *Selenastrum capricornutum. Biochem. Biophys. Res. Commun.* 152:540–544.

197. Wei, C.-I., O. G. Raabe, and L. S. Rosenblatt. 1982. Microbial detection of mutagenic nitroorganic compounds in filtrates of coal fly ash. *Environ. Mutagen.* 4:249–258.

198. Weissenfels, W. D., M. Beyer, and J. Klein. 1990. Degradation of phenanthrene, fluorene and fluoranthene by pure bacterial cultures. *Appl. Microbiol. Biotechnol.* 32:479–484.

199. Wiseman, A., and D. J. King. 1982. Microbial oxygenases and their potential application. In *Topics in Enzyme and Fermentation Biotechnology.* Ed. A. Wiseman. New York: Halstead Press. 151–178.

200. Wiseman, A., and L. F. J. Woods. 1979. Benzo[a]pyrene metabolites formed by the action of yeast cytochrome P-450/P-448. *J. Chem. Biotechnol.* 29:320–324.

201. Wood, A. W., R. L. Chang, W. Levin, R. E. Lehr, M. Schaefer-Ridder, J. M. Karle, D. M. Jerina, and A. H. Conney. 1977. Mutagenicity and cytotoxicity of benz[a]anthracene diol-epoxides and tetrahydro-epoxides: Exceptional activity of the bay region 1,2-epoxides. *Proc. Natl. Acad. Sci. USA.* 74:2746–2750.

202. Xu, X. B., J. P. Nachtman, Z. L. Jin, E. T. Wei, and S. M. Rappaport. 1982. Isolation and identification of mutagenic nitro-PAH in diesel-exhaust particulates. *Anal. Chim. Acta* 136:163–174.

203. Yang, S. K., D. W. McCourt, P. P. Roller, and H. V. Gelboin. 1976. Enzymatic conversion of benzo[a]pyrene leading predominantly to the diol-epoxide r-7,t-8-dihydroxy-t-9,10-oxy-7,8,9,10-tetrahydrobenzo[a]pyrene. through a single enantiomer of r-7,t-8-dihydroxy-7,8-dihydrobenzo-[a]pyrene. *Proc. Natl. Acad. Sci. USA* 73:2594–2598.

204. Yen, K.-M., and I. C. Gunsalus. 1982. Plasmid gene organization: naphthalene/salicylate oxidation. *Proc. Natl. Acad. Sci. USA* 79:874–879.

CHAPTER 6

Microbial Ecology of Polycyclic Aromatic Hydrocarbon (PAH) Degradation in Coastal Sediments

A. RONALD MacGILLIVRAY and MICHAEL P. SHIARIS

Abstract. Contamination of coastal ecosystems with polycyclic aromatic hydrocarbons (PAHs) is an acknowledged environmental problem. To address this problem, a two-pronged approach is ideal. First, source-reduction is needed to prevent the contamination of healthy coastal environments and, second, remediation measures are needed to ameliorate contaminated areas. Remediation based on microbial biotechnology has considerable promise as a relatively inexpensive and effective approach.

PAHs, as products of combustion and diagenesis, have been present since the origin of life. Thus, it is not surprising that microorganisms have evolved biochemical pathways to use PAHs as a source of carbon or to oxidize the parent compound to a more water-soluble form. In coastal sediments, microorganisms are most likely the principal agents of PAH removal and detoxification. To understand and exploit this remediation potential of microorganisms, a thorough understanding of the microbial ecology of PAH degraders is essential. This endeavor challenges our knowledge of marine microbial ecology on several fronts. First, the microbial fate of PAHs depends on their physico-chemical properties and the characteristics of the affected environment. Second, the ability of microorganisms to degrade PAHs is affected by numerous ecological factors: microbial interactions with other living components of the ecosystem, competition for nutrients, bioavailability of the PAHs, and other environmental factors. Current understanding of microbial PAH degradation in natural environments is still fragmentary, but ample information has been gathered to initiate rational bioremediation approaches.

This chapter reviews microbial degradation of PAHs in coastal ecosystems. We include a review of the environmental factors affecting the fate and microbial transformation of PAHs, and we stress the importance of ecological considerations in developing bioremediation options.

INTRODUCTION

Polycyclic aromatic hydrocarbons (PAHs) are ubiquitous contaminants of aquatic and terrestrial ecosystems. The singular feature that all PAHs share is a structure based on three or more fused benzene rings. The chemical structures of individual PAHs discussed in this chapter are shown in Figure 1. PAHs originate in part from

Dr. MacGillivray is at Roy F. Weston, Inc., Lionville, PA 19341, and Dr. Shiaris is at the Department of Biology, University of Massachusetts at Boston, Boston, MA 02125, U.S.A.

natural processes, but their concentrations are elevated in many coastal ecosystems as a direct result of human activities. High PAH levels are harmful to the well-being of marine biota and to human health because many PAHs have toxic, mutagenic, or carcinogenic properties (7, 118). Individual PAH compounds are not very soluble in water and their water solubility decreases with increasing molecular weight (77). Consequently, they display a high affinity for organic matter and organic-laden particles that are common in aquatic ecosystems (82). Because particles settle, PAHs often reach elevated levels in aquatic sediments, particularly near PAH-generating activities. Urban and industrial areas, in particular, manifest PAH levels that are several thousand-fold higher than pristine marine environments. Thus, there has been an increased interest in the past 15 years to describe fully the fate and toxicity of PAHs in coastal ecosystems, to set standards for water and sediment quality, and to develop approaches for reducing inputs and increasing their rates of degradation (91).

The focus of this chapter is on the microbial degradation of PAHs in coastal ecosystems. PAHs may be transformed to metabolic intermediates or mineralized (i.e. completely degraded) to carbon dioxide and water by the microbial community. The rate and extent of degradation are dependent on the physico-chemical properties of the individual PAH compound and its interaction with the living and nonliving components of the coastal ecosystem. In this chapter, we review the chief attributes of PAHs

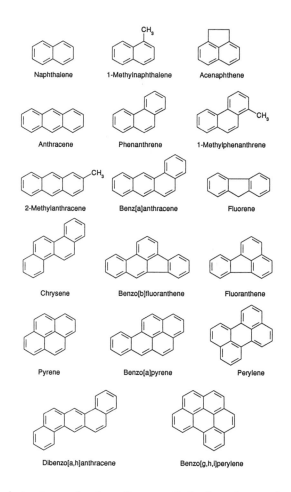

Figure 1. Chemical structures of polycyclic aromatic hydrocarbons referred to in the text.

that affect their concentration, fate, and transport in coastal areas. We review PAH-degraders that have been identified in marine environments and compare rates of PAH biodegradation estimated in coastal sediments and waters. Naphthalene, a hydrocarbon composed of only two fused benzene rings, is included in our treatment of PAHs because it shares many of their chemical properties. For further information on the biochemistry and genetics of aromatic hydrocarbon degradation, we refer readers to reviews by Cerniglia (24), Gibson and Subramanian (41), and Smith (107). A recent review by Cerniglia and Heitkamp (1989) covers general aspects of PAH degraders in aquatic ecosystems. The chapter in this book by Pothuluri and Cerniglia (95a), may also be consulted for further information on microbial degradation of PAHs.

PHYSICAL-CHEMICAL PROPERTIES OF PAHs

PAHs are relatively insoluble in water. Banerjee (8) calculated the solubility of a wide range of PAHs and found it to vary greatly. For example, solubility in 25 to 27°C distilled water varies from 31,690 ng/g (parts per billion) for naphthalene to 1.8 ng/g for the four-ring chrysene (77). This hydrophobic property of PAHs is a major factor in their environmental fate and rates of degradation. Low water solubility also causes methodological difficulty in attempts to determine experimentally their fate and rates of degradation.

As mentioned above, hydrophobic PAHs strongly associate with particles in the water column, thus resulting in their subsequent transport with the particles to the sediments. Low solubility also reduces bioavailability and degradation rates by limiting access and uptake by marine biota (65, 79, 111). In general, PAH solubility decreases with increasing molecular weight or number of rings (77). Temperature and salinity typical of marine environments can affect the solubility of PAHs (116). Even a 4°C rise in temperature can increase the solubility of phenanthrene, anthracene, 2-methylanthracene, 2-ethylanthracene, benz[a]anthracene, and benzo[a]pyrene in seawater (116). Salinity has a compound-specific effect on solubility. Phenanthrene, anthracene, 2-methylanthracene, 2-ethylanthracene, and benzo[a]pyrene display decreasing solubility with increasing salt concentration (salting-out) but are insensitive to small changes in salinity (3 o/oo change). In contrast, benz[a]anthracene has increased solubility upon the addition of salt (salting-in), and its solubility varies with small changes in salinity, especially at lower temperatures. Whitehouse (116) suggested that the salinity effect is not substantial in most marine environments but may be of concern at interfaces introduced by terrestrial run-off, atmospheric fallout, oil spills, sewage discharge, and offshore drilling.

PAHs are sensitive to light. Photooxidation of PAHs has been observed in aqueous systems (38, 95). As with solubility, PAH photodegradation is influenced by temperature, salinity, and type of PAH. For example, the rate of photolysis of alkylated naphthalenes increases in artificial seawater compared to distilled water (38). Nitrogen-substituted PAHs photooxidize more slowly than unsubstituted PAHs, and photooxidation rates are faster at 27°C than 10°C (95). Although PAH photooxidation should not be substantial in coastal sediments (except possibly in light-exposed intertidal sediments), experiments have shown that photooxidation influences the fate and effect of PAHs in both water column and sediment (53). Photolysis may prime biodegradation of larger PAHs (87). Photooxidation products may also sorb to particles and persist in sediment. Further investigation of PAH photooxidation products in coastal ecosystems is needed to clarify their role in PAH fate and rates of degradation.

Photodegradation also can lead to erroneously high estimates of PAH degradation during experimental manipulation (119). Sample extraction and processing are typically conducted under filtered light to prevent photooxidation. Internal standard controls are incorporated to account for photooxidation.

Lower molecular weight PAHs, especially naphthalene, are volatile. PAHs in the molecular weight range encompassing naphthalene to dibenz[a]anthracene volatilize from aqueous solutions at a rate inversely proportional to molecular weight (67). Volatilization occurs in surface water, may occur in bioturbated or resuspended sediments, and should be carefully monitored during experimental manipulation.

PAHs IN THE COASTAL ENVIRONMENT

Inputs and Concentrations

Sediments are sinks for naturally derived PAHs, but anthropogenic sources contribute the predominant share of PAHs to urban waters. The primary natural routes of PAHs into the environment are through oil seeps and forest fires (122). Major anthropogenic sources of PAH include combustion and transportation of fossil fuels (91). PAHs enter estuarine environments through atmospheric fall-out, non-point source runoff, wastewater discharge, commercial shipping, recreational boating, and major oil spills (43, 92). The depositional history of industrial and energy-generated PAH input can be readily chronicled in undisturbed sediment adjacent to urban areas (55). Urban estuaries, such as Boston Harbor, attain PAH concentrations exceeding 100 mg/g sediment (54, 59, 105, 122). Once trapped in aquatic sediments, PAHs persist for long durations. Recalcitrance of PAHs depends on environmental conditions and structure of the compound.

Distribution and Partitioning

The transport, coastal distribution, and fate of PAHs has been linked to their molecular weight, volatility, and the concentration of organic matter in a given ecosystem. In a detailed study of the distribution and degradation of PAHs in the Tamar Estuary, England (98), PAHs could be divided into two groups based on their fate: (1) low molecular weight PAHs (naphthalene, phenanthrene, and anthracene); and (2) larger molecular weight PAHs (fluoranthrene, pyrene, chrysene, benz[a]anthracene, benzo[b]fluoranthrene, benzo[k]fluoranthrene, and benzo[a]pyrene). The distribution of the low molecular weight PAHs exhibited no correlation with particle concentration in the water column. Volatilization from the water column appeared to be the most important factor that determined their fate. In contrast, the distribution of higher molecular weight PAHs correlated with the concentration of suspended particulates. Sediment burial was predicted as the primary fate of these PAHs. In other studies, the partitioning of PAHs between water and sediment has been observed to be related to the organic content of both the sediment and water (82). Organic contaminant concentrations also correlate with the clay content of sediment (61, 94). In a study of riverine sediment, both organic matter and PAHs had bimodal distributions (92). The smallest and largest particle size classes of sediment had the highest concentrations of organic matter and PAHs. This was explained by the presence of two classes of organic matter in the sediment. The ratio of organic matter to PAHs and composition of the PAH mixtures did not change in the sediments with time, not even in sediments of varying PAH concentration. As a consequence, sediment particle size and organic content may act to focus PAH contamination in particular sites of an estuary or coastal area. For instance, depositional sites with small particle size sediment often contain high organic content and elevated PAH concentrations.

Other factors affecting the partitioning of PAHs are dissolved organic matter (DOM) and colloidal matter. Since DOM is typically separated from particulate organic matter by filtration, the operationally defined DOM includes the colloidal fraction. Boehm and Quinn (16) suggested that DOM increased the solubility of PAHs in aquatic environments. Specific interactions of hydrocarbon-dissolved organic matter have been observed with respect to compound and origin of DOM. The most hydrophobic PAHs show increased association with DOM, and the DOM-PAH "complex" can enhance the transport of the PAH (76). Interaction is also greater with DOM from terrestrial rather than marine origin (117). This is perhaps due to humic compounds associated with terrestrial runoff. PAHs were also observed to bind reversibly to the humic component of DOM during 2- to 7-day contact periods. Binding rates of PAHs to DOM are comparable with binding to organic coated particles, and the process is more rapid than PAH binding to sediment (78). Over longer duration (4, 7, and 70 days), PAHs become more tightly bound to aquatic humic materials (58). Natural colloids appear to be important to the transport and fate of hydrophobic compounds like PAHs in estuarine environments (17, 83). Thus, PAHs tend to be carried by small particles coated with organics that settle in depositional areas of coastal ecosystems; however, some portion of PAHs may remain in the water column or efflux out of sediments when associated with DOM and colloidal material.

Bioavailability

Recent advances in the knowledge of PAH partitioning among water, particulate, colloid, and DOM fractions have clarified the underlying mechanisms of PAH distribution and PAH flux in sediment and coastal ecosystems. Similar processes also affect the availability of PAHs for microbial uptake and metabolism. Irreversible binding to particulates reduces PAH bioavailability to microorganisms and benthic organisms. Thus, chemical measures of the PAH content in coastal sediments and waters do not necessarily reflect the bioavailability of PAH to bacteria.

The processes responsible for the transfer of PAHs from the environment to organisms and the environmental factors that govern these processes are not well understood. Several recent reviews (33, 66, 79) summarize bioavailability of PAH in aquatic ecosystems, but the information is almost exclusively limited to benthic animals, which may take up PAH by different mechanisms than benthic microorganisms. PAH bioavailability is dependent on the physical and chemical characteristics of the PAH. The most useful parameter is the PAH octanol/water partition coefficient (K_{ow}) which is related to the aqueous solubility (65). As the aqueous solubility of a particular PAH decreases, the K_{ow} increases, and the compound is more slowly desorbed from sediment particles, rendering it less available for biological uptake. Some organic compounds display two-phase (fast and slow) desorption kinetics (99) and PAH uptake may be rate-limited by desorption. Farrington (33) warns that the relationship between K_{ow} and bioavailability is often site specific because it is affected by the composition and characteristics of the sediments. These characteristics include organic carbon content, particle size distribution, clay type and content, cation exchange capacity, and pH. Experimental determination of PAH bioavailability is further complicated by the behavior and physiology of the benthic organism under study (79).

Microbial degradation of organic compounds may be rate-limited by bioavailability and the rate of desorption from soils (99), clays (75), and humic materials (58). Microbial degradation rates of low molecular weight PAHs, such as naphthalene, may be governed by slow desorption rates as demonstrated in anaerobic soils (3, 85). It is likely that the desorption rate becomes increasingly slow with higher molecular weight PAHs in sediments, but this remains to be demonstrated.

The question of PAH bioavailability is crucial to understanding PAH degradation, yet quantitative data are lacking. There is a need to develop methods for determining bioavailability of PAHs to microorganisms. A recent and novel approach by King et al. (63) in which a bioluminescent reporter plasmid for naphthalene catabolism is used may prove useful in assessing the bioavailability of naphthalene in a variety of environments.

ABIOTIC ENVIRONMENTAL FACTORS AFFECTING PAH DEGRADATION

The degradative fate of PAH in sediments, like its distribution, is under the control of complex biological and environmental factors (5). The role of some variables, oxygen and temperature for example, is better understood than others such as infaunal activity or protozoan grazing. We are still a long way from incorporating the effects of these factors into any predictive model of PAH degradation. PAH biodegradation is affected by oxygen, temperature, nutrients, salinity, pH, exposure duration, and the concentration and composition of the PAH mixture.

Oxygen

Oxygen is a master environmental variable that controls degradation rates (31, 46). Well-characterized eukaryotic and prokaryotic PAH degradation pathways incorporate bimolecular oxygen into the PAH ring during the initial enzymatic attack (24), yet oxygen is often limited to the top few millimeters in PAH-contaminated sediments. Enhanced oxygen penetration into deeper sediments, in part, probably explains the apparent stimulation of PAH degradation by benthic worms that mechanically rework the sediment (12).

Although the preponderance of evidence from the field indicates negligible rates or lack of PAH biodegradation under anaerobic conditions (46, 86), evidence for anaerobic biodegradation of hydrocarbons (13, 85) indicate the possibility of PAH transformation linked to anaerobic microbial metabolism in nature. Mihelcic and Luthy (85) observed biodegradation of naphthol, naphthalene, and acenaphthene at mg/l concentrations in soil-water systems under denitrifying conditions. Single-ring aromatic hydrocarbons can be degraded anaerobically, even under methanogenic conditions (74, 112). The aerobic-anaerobic interface in coastal sediments may also be a site of PAH biotransformation. Oxidation of PAHs initiated in the aerobic zone may lead to continued transformations under anaerobic conditions. Given the limited data available on the subject and recent advances in the field of anaerobic biodegradation, the fate of PAHs in the anaerobic zone of coastal sediments and the role of the aerobic-anaerobic interface in PAH biodegradation warrant further study.

Temperature

PAH degradation can probably occur over a wide range of environmental temperatures, although few studies have examined the effect of temperatures on PAH degradation rates. Bauer and Capone (9) found significant increases (up to four-fold) in the degradation rates of naphthalene, phenanthrene, and anthracene when the incubation temperature of estuarine sediment slurries was increased from 10 to 30°C. Haines and Atlas (45) attributed the slow disappearance rates of the PAH fraction of petroleum in Arctic sediments to low temperatures (−1.8 to 4°C). Seasonal studies also support a significant role for temperature. Lee and Ryan (70) reported 10- to 15-fold higher rates of naphthalene mineralization in a coastal river as temperature increased. Shiaris (102) found a significant correlation ($p < 0.01$) between temperature and the rate of degradation of naphthalene and phenanthrene, but not of benzo[a]pyrene, in three Boston Harbor sediments.

Nutrients

There is confusion about the effects of nutrients, in particular nitrogen and phosphorus, on the rates of degradation of petroleum hydrocarbons in the environment. As discussed by Atlas (5) in his comprehensive review of microbial degradation of petroleum hydrocarbons in nature, hydrocarbons in bulk concentrations, such as oil slicks, show limited degradation due to nutrient limitation whereas the degradation of hydrocarbons in the soluble state is probably not nutrient-limited. This appears to be the case for PAHs as well, although few investigations of nutrient limitation on the degradation of PAH in the environment appear in the scientific literature.

It is likely that the effect of nutrient limitation may vary from one locale to another as nutrient regimes change with habitat. In addition, the effect may vary from one PAH compound to another in the same environment. For example, Roubal and Atlas (101) found that nitrogen and phosphorus additions stimulated naphthalene but not benz[a]anthracene degradation in sediment slurries. Fedorak and Westlake (35) also reported a stimulation of naphthalene degradation in seawater amended with nutrients. On the other hand, for estuarine sediments, the addition of nitrogen, phosphorus, or glucose had only minimal stimulatory effects on the degradation of naphthalene, phenanthrene, anthracene, and benzo[a]pyrene (9, 102). Thus, the nutrient regime of a bioremediation site should be determined before a nutrient addition scheme is decided.

Salinity and pH

In marine environments, salinity also appears to have a significant effect on microbial degradation of PAHs. Kerr and Capone (62) reported that naphthalene and anthracene degraders in the Hudson River estuary were highly adapted to their salinity regime. Degradation of naphthalene and phenanthrene in three urban estuarine sediments was significantly related to seasonal salinity changes over the course of a 15-month period (102). The same trend was observed for benzo[a]pyrene degradation for one of the sites. It is likely that salinity may affect the species composition of the PAH-degrading consortia (103), but the answer awaits population studies of PAH degraders.

PAH degradation appears to be influenced by pH, but few studies have been conducted. The pH levels recorded at three estuarine sites (pH 7.2 to 8.2) did not affect transformation rates of naphthalene, phenanthrene, or benzo[a]pyrene (102). In two other estuarine systems, naphthalene and benzo[a]pyrene were most rapidly mineralized at the *in situ* pH (8.0). Degradation rates were reduced when pH was experimentally adjusted to 5.0, 6.5, and 9.0 (31, 46). The evidence indicates that PAH-degrading populations are well-adapted to their environments.

Concentration and Composition of PAHs

In general, the rate of PAH degradation increases with increasing concentration of PAH added to the test system (9). Lee and Ryan (70) observed that the turnover time of benzene and toluene but not naphthalene followed Michaelis-Menten kinetics as described by Wright and Hobbie (120). This general conclusion is also supported by rates of PAH degradation observed in clean versus PAH-contaminated freshwater (52) and estuarine (102) sediments. It is important to note, however, that though the degradation rate may increase with increasing PAH concentration, the turnover rate also may increase with increasing PAH concentration. In other words, the greater the PAH concentration in a sediment, the more slowly it is removed as a percentage of the total PAH burden.

PAHs exist in nature as complex mixtures of substituted and unsubstituted fused benzene rings. The relative ratios of the different components reflect the nature of the PAH sources, the physico-chemical properties of the PAHs, "weathering" effects, and the susceptibility of the PAHs to microbial degradation (15). Several general conclusions can be derived from the various studies on PAH degradation in waters (35, 68, 70, 98), freshwater sediments (47, 52), estuarine sediments (9, 40, 45, 97), and soils (18, 35). First, the PAH fraction of petroleum is typically more recalcitrant than the alkane fraction. Second, PAHs are increasingly resistant to biological attack as the number of rings increases. Third, the PAH in a mixture is more resistant to degradation with increasing alkyl substitution. The position of the alkyl substitution on the aromatic ring also plays a role.

A review of the recent literature suggests a hierarchy of PAH susceptibility to biodegradation. The majority of scientific reports, whether performed with pure cultures or with environmental inocula, show agreement. Table 1 lists PAHs from the most readily degradable components to the most resistant. These relationships should be the same in marine ecosystems as in soils and freshwater sediments.

One aspect of PAH degradation that is poorly understood is the effect of one PAH compound on the degradation of another. In a study of the effects of co-occurring aromatic hydrocarbons on degradation of individual PAHs, Bauer and Capone (11) found that pre-exposure of marine sediment to a single PAH or benzene enhanced the ability of the microbial community to degrade the pre-exposed PAH and only selected other PAHs. Interestingly, the PAHs that were affected varied depending on the PAH to which they were exposed. The authors suggest that enhanced degradation of a second PAH compound after exposure to a first PAH may be due to a shift to a microbial population with either broad specificity for PAH degradation, common pathways of PAH degradation, or both. Mixtures of PAHs and other compounds in the environment may enhance or inhibit degradation. The potential PAH-degrading bac-

Table 1. Recalcitrance of selected PAH compounds to biodegradation from the most readily degradable to the most recalcitrant.

PAH compound	Number of rings
Naphthalene	2
1-Methyl naphthalene	2
2-Methyl naphthalene	2
C_2 Naphthalenes	2
Phenanthrene	3
C_3 Naphthalenes, 1-methylphenanthrene	2, 3
Fluorene	3
C_2 Phenanthrenes	3
Anthracene	3
Benzanthracene, fluoranthene, chrysene	4
Pyrene	4
Benzopyrenes, benzofluoranthenes, perylene	5
Benzo[g,h,i]perylene	6

terial flora of contaminated estuarine sediments includes PAH degraders, PAH cooxidizers (104), and perhaps closely associated bacteria that do not degrade the parent compounds (109). Synergistic effects may occur if cooxidation occurs to increase the microbial attack of more recalcitrant PAHs. On the other hand, some compounds may repress PAH degradation by a phenomenon known as diauxie. For example, the addition of phenol retarded the degradation of naphthalene by bacteria in a settling pond, most likely by repression of the naphthalene-degrading system (84). Toxicity to the microbial population by a single compound or an additive effect of a combination of compounds may inhibit degradation of a PAH mixture (10, 22, 57). Known pathways of aromatic hydrocarbon catabolism are inducible at the level of DNA transcription (4) and may, therefore, be repressed by the presence of a more readily utilizable carbon source for individual bacterial species. In this regard, the concentration and nature of the organic matter in the environment may affect rates of biodegradation. This is an area important to PAH degradation in the environment and deserves more investigation.

BIOTIC ENVIRONMENTAL FACTORS AFFECTING PAH DEGRADATION

Benthic Fauna

While chemical oxidation, photooxidation, and volatilization may be important routes of PAH loss in the water column, biological transformation is probably the prevailing mechanism of PAH loss in sediments. Benthic fauna take up and metabolize PAHs and affect the activities of PAH-degrading microorganisms. Bioavailability and metabolism of PAHs by these benthic organisms is species-specific, related to the source of PAHs (pyrogenic vs. oil-spill), and dependent on the type of contaminated sediment (6, 79). With this in mind, a look at how benthic organisms may influence the microbial ecology of sediments is justified.

Infauna, animals living within the sediment, play an important role in the fate of PAHs in sediment. Perturbation of sediments by burrowing animals such as worms and clams may distribute PAHs in sediment and increase net flux out of the sediment (80). The disturbance or "bioturbation" of sediments may enhance microbial activity and PAH degradation rates in several ways: by increasing aeration of the sediment, by refining detrital material, thus making it more accessible to microorganisms; and by stimulating microbial growth rates and increasing microbial biomass as a result of grazing coupled with the excretion of nutrients (12, 40). The latter process is called "microbial gardening." Water-soluble metabolites formed by the transformation of PAHs by benthic organisms also may enhance the release of PAHs from sediments, thus increasing bioavailability (80). Furthermore, the degree of bioturbation is influenced by the number, types, and seasonal succession of benthic animals in sediments (71). The interaction of benthic organisms and microorganisms may be particularly important for larger molecular weight PAHs, but these relationships are usually inferred. There is little experimental evidence.

Protozoa

Protozoa also may stimulate bacterial degradation of aromatic compounds (56) and crude oil (100). The participation of protozoa in PAH degradation has not been evaluated, however. An examination of the role of protozoa should be a rewarding area of investigation for understanding the microbial ecology of PAH degradation and for developing bioremediation approaches.

The stimulation of bacterial activity or growth by protozoans is usually equated with the effects of grazing (36). Grazing of bacteria by protozoans increases their growth rates and, thus, their metabolic activities. Other bacterial-protozoan interactions of importance to PAH degradation may occur, however. For example, Huang et al. (56) observed that p-aminobenzoate degradation by an *Alcaligenes* spp. was stimulated by bacterial growth factors excreted by *Tetrahymena pyriformis*. Although experiments that separate the relative contribution of component members in the sediment community are difficult to design, further studies on interactions within the microbial community and the impact of marine fauna as it relates to microbial degradation of PAHs in coastal sediments are needed.

MICROORGANISMS AND MICROBIAL COMMUNITIES
DEGRADING PAHs

The ability to oxidize PAHs at environmentally significant rates does not appear to be a common attribute of microorganisms but appears to be sparsely dispersed throughout a diverse group of bacteria (23), fungi (23, 25, 28, 81, 96), and perhaps algae (23, 26, 72). PAH mineralization, the complete catabolism of PAHs to carbon dioxide and water, is generally attributed to bacteria and not fungi or algae; however, the white rot fungus, *Phanerochaete chrysoporium,* has been reported to partially mineralize a wide range of environmental contaminants, including PAHs (benzo[a]pyrene and phenanthrene), when nitrogen is limiting (20, 21). Two other white-rot fungi (*Chrysosporium lignorum* and *Trametes versicolor*) have also been reported to partially mineralize phenanthrene (88).

Information on PAHs as sole carbon and energy sources for PAH-degrading bacterial cultures is also expanding. Smaller 2-and 3- ring PAHs are known sole carbon and energy sources for PAH-degrading bacterial cultures (24). A few isolates of bacteria can also oxidize the 4- and 5-ring PAHs but they cannot use them as sole carbon and energy sources (27, 50, 76a). Recently, strains of *Pseudomonas paucimobilis* and *Alcaligenes denitrificans* were demonstrated to be capable of utilizing fluoranthene as a sole source of carbon and energy (89, 114).

The abundant information on the biochemistry and genetics of PAH catabolism is in large part derived from bacteria isolated from oil-contaminated soils. The use of PAH enrichment culture with soil inocula frequently yields *Pseudomonas*, a bacterial genus that grows rapidly under laboratory conditions. This genus has thus become the best-studied model for PAH degradation. Systematic analysis of aquatic and terrestrial ecosystems for PAH-degrading microorganisms is still lacking, so the list of PAH degraders in the environment is probably far from complete. We review here the microorganisms that have been described in marine ecosystems.

Yeast and Fungi

Filamentous fungi and yeast are common in marine environments (1, 44, 64, 73, 93) and the majority of yeasts examined possess cytochrome P-450 (60). This suggests a potential role for eukaryotic microorganisms in PAH degradation, but only a few attempts to isolate PAH-degrading marine fungi have been reported. Ahearn et al. (2) examined a variety of yeasts isolated from oil-contaminated marine environments. Most could grow on alkane as a sole carbon source but none displayed increased O_2 uptake in the presence of aromatics. Similarly, Fedorak et al. (4) isolated 298 fungi and yeasts from uncontaminated coastal waters and sediments, but none of them could mineralize radiolabeled naphthalene or phenanthrene as sole carbon sources. Kirk and Gordon (64) identified 4 genera of beach-adapted fungi that could utilize hexadecane as a sole carbon source, but none of the 54 strains of filamentous marine fungi that they examined could grow on naphthalene. Yet, yeast that can grow on simple aromatic compounds were readily isolated from a polluted estuary (29).

The limited evidence suggests that yeasts and fungi do not play a predominant role in most marine ecosystems. Because of their common occurrence and observed activity, however, the potential for fungal PAH degradation in marine ecosystems deserves further attention. The role of fungi versus bacteria in PAH degradation may be distinct. Cerniglia et al. (28) postulated that fungal oxidation of PAHs is a prelude to detoxification, whereas bacteria oxidize PAHs as an ultimate source of carbon and energy.

Bacteria

Several bacterial genera that degrade PAHs have been isolated and characterized from marine environments although efforts to isolate degraders from marine environments are considerably fewer than from terrestrial and freshwater environments. Sisler and Zobell (106) were the first to isolate bacteria with the potential for PAH degradation from marine waters, but the bacteria were not identified. Other early reports provided only indirect evidence for the existence of PAH-degrading marine bacteria (90, 108).

Several laboratory groups have isolated PAH-degrading pseudomonads from marine environments. Pseudomonads that grow on naphthalene as a sole carbon source and oxidized phenanthrene and anthracene have been isolated from oil-polluted estuarine waters (30) and marine sediments (109). A naphthalene-degrading *Pseudomonas putida*, a species commonly isolated from soils, was also isolated from marine sediments (37). García-Valdéz and coworkers (39) described two species of naphthalene-degraders, *Pseudomonas stutzeri* and *Pseudomonas testosteroni*, isolated from coastal sediments. Thus, pseudomonads, frequently isolated PAH degraders in terrestrial ecosystems, are also frequently isolated as marine degraders.

As in terrestrial systems, the variety of PAH-degrading marine bacteria is probably large. Two true marine bacteria, *Vibrio parahaemolyticus* and *Vibrio fluvialis*, as well as a variety of unidentified members of the Enterobacteriaceae were identified by numerical taxonomy in a study of phenanthrene-degrading bacteria isolated from estuarine sediments and waters (115). A naphthalene-degrading *Moraxella* sp. has been isolated from marine sediments (109). Guerin and Jones (42) described a phenanthrene-degrading *Mycobacterium* sp. BG1 isolated from estuarine sediments. Another *Mycobacterium* sp. isolated from coastal sediments (48) was capable of oxidizing an extended range of PAHs from naphthalene to pyrene. This organism showed promise as a seed culture in the removal of PAHs from contaminated sediments (49).

Many bacterial species capable of PAH biodegradation have been isolated from marine waters and sediments, although whether they were autochthonous marine bacteria or whether they had entered coastal waters from land runoff and stormwater/sewage outlets was not possible to discern. Evidence from PAH biodegradation studies (10, 103), suggests that active PAH-biodegrading microbial communities are well-adapted to the marine regime. Thus, the presence of indigenous populations of PAH-degraders in marine sediments is a likelihood.

The isolation of specific microorganisms, especially by enrichment methods, does not necessarily indicate the importance of their role in *in situ* PAH degradation. An emerging paradigm in microbial ecology is that many of the key bacteria involved in marine carbon-cycling are not readily culturable on common bacteriological media (19). The identity of key PAH-degraders in marine ecosystems has yet to be confirmed.

Basic ecological data on PAH-degraders in marine ecosystems is sparse. Information on their distribution, densities, population dynamics, population genetics, and interaction with other microorganisms awaits future investigation. Evidence suggests, for example, that two or more naphthalene degraders may interact with non-PAH degraders as a naphthalene-degrading consortium in estuarine sediments (109). Evolving methodology such as gene probes (14) should facilitate these studies.

In this laboratory, Christopher Smith (M.S. thesis, University of Massachusetts, Boston, 1988) employed a DNA probe consisting of the *nahABCD* genes from the NAH7 plasmid (121) to characterize 130 naphthalene-degrading isolates from PAH-contaminated soils and from freshwater and marine sediments. Isolates were obtained

Table 2. Distribution of *nahABCD* among naphthalene-degrading bacteria isolated from soils and sediments

| Genotype | All isolates | Habitat[a] | | |
		Estuarine	Freshwater	Soil
nahABCD+	63 (48.5)[b]	15 (28.8)	7 (22.6)	41 (87.2)
nahABCD−	67 (51.5)	37 (71.2)	24 (77.4)	6 (12.8)

[a]$\chi^2 = 44.6$ (2 d.f.), $p < 0.00001$, for H_o = all habitats have same distribution.
$\chi^2 = 0.14$ (1 d.f.), $p < 0.7$, for H_o = estuarine distribution is the same as freshwater distribution.
$\chi^2 = 31.9$ (1 d.f.), $p < 0.00001$, for H_o = estuarine distribution is the same as soil distribution.
$\chi^2 = 30.3$ (1 d.f.), $p < 0.00001$, for H_o = freshwater distribution is the same as soil distribution.
[b]Percentage of isolates in habitat with corresponding genotype.

from primary spread plates on dilute-nutrient medium of soil and sediment dilutions. Plates were sprayed with a naphthalene solution (104) and colonies that displayed zones of naphthalene clearing were randomly picked and isolated. Results of DNA-DNA colony hybridization of the naphthalene-degrading isolates (Table 2) indicate that a NAH7-like pathway is not evenly distributed among degraders from different habitats. The results also suggest that as yet undescribed catabolic systems may be important in marine and freshwater environments.

RATES OF PAH DEGRADATION

Information on the biochemical pathways of PAH catabolism and the kinetics of PAH degradation by pure and mixed cultures abound, but accurate rates of biodegradation and the fate of PAHs in natural sediments have been difficult to ascertain. PAHs are transformed in nature by a variety of mechanisms. Photooxidation, chemical oxidation, and biological transformation are among the prevailing mechanisms. This section summarizes the scientific literature that provides estimates of microbial PAH degradation rates in sediments.

Determination of degradation rates in nature is fraught with difficulty. Several obstacles impede the collection of accurate PAH degradation rates. First, the types of environments in which PAHs are found vary greatly, yet studies to assess rates in nature are few. Second, the environmental parameters that affect rates vary daily and seasonally, but environmental parameters are rarely taken into account. Third, individual PAH compounds often occur in trace concentrations, so rate determination becomes analytically difficult. Most researchers, therefore, use high doses of PAH in their experimental systems, and it is not possible to extrapolate accurate rates to the environment. Fourth, the methodology is arduous and not amenable to large-scale experimentation. Finally, the rates are often so slow that long incubation times are required to detect degradation. Long incubation can lead to "bottle effects," which include changes in microbial abundance, microbial community structure, and enrichment-selection for PAH-degraders. For these reasons, a brief review of the approaches used to determine PAH degradation rates is necessary to understand the shortcomings of environmental rate estimates and to evaluate the rates reported in the literature.

Methodology

The approaches used to estimate PAH degradation rates in nature fall into two main categories: (1) degradation based on the disappearance of the parent compound; and (2) the use of radiolabeled tracers to detect the products of PAH degradation. The former approach has been favored by investigators, but it is replete with problems. Since only the disappearance of parent material is followed, the fate of the PAH is unknown and it is incorrect to assume that PAH disappearance is caused by biodegradation. Recovery of the parent material is typically low depending on volatilization, irreversible binding to particulates, and losses in handling during extraction. In field studies, the input of parent PAHs from external sources cannot be controlled, nor is input normally determined. This approach has been useful, however, in cases of massive input of PAH to an ecosystem; for example, oil spills in the field or oil spill simulations in the laboratory.

The use of radiolabeled PAH in degradation studies is the preferred approach because a mass balance of the parent compound and the resulting catabolic products can be calculated. Serious flaws in experimental design may go undetected unless an account of the total starting material is performed. Some early studies and reports suffer from unaccounted losses of PAH. For example, PAH can partition to flask caps and tubing of experimental vessels. Investigators incorrectly assumed that the reduced recovery of PAH was degradation; mass balances were not included, nor were appropriate controls conducted. The fate of the PAH is uncertain unless a mass balance is performed. Herbes and Schwall (52) also demonstrated that analysis of only the carbon dioxide produced by mineralization can severely underestimate the degradation rate of PAH. A substantial or even predominant portion of the degraded PAH may be in the form of polar metabolites or converted to cell-bound residue. The fate of the parent compound can be traced using radioisotopic techniques.

In spite of the advantages of employing radiolabeled PAH, caution is still necessary in the interpretation of results produced from this approach. Most PAH compounds are degraded only very slowly, i.e. turnover times are on the order of weeks or years. Thus, the ability to detect trace levels of the PAH degradation products becomes a major consideration. One solution to this problem that has been employed is to dose environmental samples with levels of radiolabeled PAH that are often orders of magnitude higher than *in situ* concentrations. Since degradation rates are dependent on the PAH concentration, calculated rates may be severely overestimated.

Another approach for increasing the detection levels of catabolic products is to incubate samples for long periods, up to weeks and months, in the laboratory; however, extended incubation in small enclosed experimental systems no doubt leads to significant changes in the abundance and composition of the microbial communities. Effects such as "wall growth" on the surfaces of experimental flasks leads to significantly increased microbial activity in laboratory experiments; more so with prolonged incubation periods. The use of large-scale experimental systems such as the Marine Ecosystem Research Laboratory (MERL) tanks (53) or the CEPEX bag enclosures (68, 69) allow prolonged incubation under natural environmental conditions. Unfortunately, only a handful of studies have been conducted in these "mesocosms." The cost of mesocosm construction and maintenance is prohibitive for routine experimentation.

Another factor in the experimental design that may influence the rates of PAH degradation is the choice between sediment slurries or intact sediment cores. Sediment-slurries are popular because variation is reduced in replicate experiments in contrast to the variation observed in sediment cores. Slurry experiments are also easier to set up and conduct than experiments with cores. Slurries lack the steep redox gradients, however, and the resulting spatial heterogeneity of microbial communities that are characteristic of intact sediments. This aspect is particularly important in the degradation of PAHs, a process dependent on oxygen concentration and redox conditions. Therefore, rate estimates based on slurry experiments may overestimate *in situ* PAH degradation rates by a factor of ten or more (110).

A final shortcoming of reported PAH degradation rates is lack of seasonal sampling. Only a few studies (53, 68, 70) have taken into account seasonal changes in degradation rates. Estimated turnover times based on a single sampling time are not likely to yield a realistic annual mean turnover rate.

Estimates of PAH Turnover Times in Water, Sediments, and Soils

With the limitations of reported PAH degradation rates in mind, we calculated turnover times (T_t) of some PAH compounds from reports in the literature (Table 3). Turnover time is calculated by dividing the length of incubation by the fraction of parent compound remaining in the experimental system. This calculation assumes that

Table 3. A survey of estimated turnover times of selected PAH compounds in waters, sediments, and soils.

SPAH	Environment	Temperature (°C)	T_t (days)	Ref.
Naphthalene	estuarine water	13	500	12
	estuarine water	24	30–79	12
	estuarine water	10	1–30	98
	seawater	24	330	12
	seawater	12	15–800	12
	estuarine sediment	25	21	9
	estuarine sediment		287	97
	estuarine sediment[a]	22	34	51
	estuarine sediment[a]	30	15–20	46
	estuarine sediment[a]	2–22	14–18	102
	stream sediment	12	>42	52
	stream sediment[a]	12	0.3	52
	reservoir sediment	22	62	51
	reservoir sediment	22	45	51
Methylnaphthalene	estuarine water	10	130–2100	68
	estuarine water	24	50	12
	estuarine water	13	500	12
	estuarine sediment	22	196	51
	reservoir sediment	22	280	51
	reservoir sediment	22	224	51
Phenanthrene	estuarine sediment[a]	22	56	51
	estuarine sediment[a]	2–22	7–15	102
	reservoir sediment	22	252	51
	reservoir sediment	22	112	51
	sludge-treated soil	20	282	18
Anthracene	estuarine sediment	25	85	9
	stream sediment	12	157	52
	stream sediment[a]	12	17	52
	sludge-treated soil	20	658	18
Benzanthracene	estuarine sediment	4–25	730	53
	stream sediment	12	10417	52
	stream sediment[a]	12	417	52
	sludge-treated soil	20	658	18
Pyrene	estuarine sediment[a]	22	476	51
	reservoir sediment	22	1260	51
	sludge-treated soil	20	8900	18
Benzo(a)pyrene	estuarine water	10	2000–9000	98
	estuarine sediment[a]	22	>2800	51
	estuarine sediment[a]	2–22	52–3700	102
	stream sediment	12	>20800	52
	stream sediment[a]	12	>1250	52
	reservoir sediment	22	>4200	51
	sludge-treated soil	20	>2900	18

[a] PAH-contaminated environments.

the input of parent PAH is equal to the loss by biodegradation. The calculations are based on PAH concentrations given in the reports, usually based on PAH added to water and sediment samples by the investigators. Thus, the turnover times are actually potential turnover times for the given sites.

Several conclusions can be inferred from the available information. First, the turnover times for individual PAHs may vary greatly. For example, naphthalene turnover ranges from 0.3 to 800 days. Variations in the environmental parameters of the sites and differences in the methods that were employed may explain, in large part, the variation in turnover times. Yet some trends emerge from the collective data. As expected, T_t tends to be shorter in sediments than in the water column. This is probably a consequence of two factors: (1) higher sediment concentrations of PAHs, and (2) greatly elevated numbers of PAH-degrading bacteria in the sediments (104). Also, the turnover times tend to be shorter in contaminated sediments than they are in sediments with low PAH levels. Again, this is most likely a function of PAH concentration and the elevated numbers of bacteria capable of PAH degradation.

A second observation from Table 3 is a general increase in T_t for compounds with increasing molecular weight and alkyl substitution. This supports the results observed in pure culture experiments and experiments that employ mixed PAH enrichments of environmental samples as summarized in Table 1. The higher molecular weight PAHs such as the five-ring benzo[a]pyrene have very long turnover times, on the order of years, even in surficial sediments. In nature, these compounds may be buried in anoxic sediments before they are significantly degraded, so actual turnover times may be much longer than has been estimated in these studies.

It follows from the discussion of the methodology used to assess degradation rates in nature, that the calculated turnover times in Table 3 probably underestimate actual biological turnover times in the environment. That is, these turnover times are probably a best-case scenario for the removal of PAHs in nature. Most laboratory studies cited used optimal conditions for PAH degradation (e.g. slurries, high temperatures, and high oxygen concentrations). Clearly, there is a need to obtain more realistic turnover values by establishing experimental conditions that better simulate the environment.

From the information in Table 3, it is obvious that only a few of the environmentally prevalent PAH compounds have been assessed. Most of the data are for the naphthalenes, the three-ring PAHs, and benzo[a]pyrene. Slow degradation rates, expensive and arduous analytical techniques, and the unavailability of suitable radiolabeled PAH compounds have restricted research in this area to these few model PAHs.

SUMMARY

A diverse group of microorganisms capable of PAH degradation can be isolated from coastal waters and sediments. Most PAH-degraders isolated from coastal sediments are bacteria, a finding that indicates a predominant role in PAH catabolism by the prokaryotes. The majority of bacterial PAH-degraders that have been examined are capable of mineralizing two- to three-ring PAHs, although a few isolates can oxidize four- and five-ring PAHs. Eukaryotic microorganisms may play a role, but they have received insufficient attention. The basic ecology of PAH degraders in marine ecosystems has also received little attention. Almost nothing is known on the distribution, densities, population dynamics, and microbial interactions of PAH-degraders in aquatic ecosystems.

PAH degradation is controlled by the physico-chemical properties of the PAHs, the duration of exposure, the concentration of PAHs, and numerous environmental and seasonal factors. The relative insolubility of PAHs in water and their binding capacity to natural materials results in their eventual transport to the sediment in aquatic ecosystems. Once in the sediment, the rate of PAH degradation may be limited by desorption from particles and organic matter, i.e. PAH bioavailability. The duration of PAH exposure together with the concentration and composition of PAH mixtures strongly contribute to establish a PAH-degrading community in the environment. Oxygen, temperature, nutrient availability, salinity, and benthic faunal activity are environmental factors that have been shown to affect the rate of PAH biodegradation in sediments.

Although studies have been conducted in microcosms and mesocosms, measurements of *in situ* degradation rates are few. For coastal sediments in general, many fundamental ecological questions remain. For example, what is the fate of PAH photooxidation products? How does DOM affect PAH degradation? How do PAH mixtures and other organic contaminants affect the degradation of individual PAH compounds? Does PAH degradation occur at the aerobic/anaerobic interface? In the anaerobic zone? To what extent do marine fauna affect microbial degradation of PAHs?

A solid foundation of the microbial ecology of PAH degradation is essential for developing effective bioremediation protocols. Efforts to remove PAHs from coastal sediments by microbial activity will require accurate measures of contamination, degradation rates, and the fate of transformation products. The exposure history and physical-chemical characteristics of a site must be determined. Environmental factors such as seasonal changes in temperature, nutrients, and benthic organisms may enhance or deter bioremediation activities. Undoubtedly, advances in PAH bioremediation technologies and a more complete understanding of the microbial ecology of PAH-degraders are reciprocal goals.

LITERATURE CITED

1. Ahearn, A. N., and S. P. Meyers. 1976. Fungal degradation of oil in the marine environment. In *Recent Advances in Aquatic Mycology*. Ed. E. B. Ganth Jones. New York, NY: John Wiley and Sons. 125–133.

2. Ahearn, D. G., S. P. Meyers, and P. G. Standard. 1971. The role of yeasts in the decomposition of oils in marine environments. *Dev. Indust. Microbiol.* 12:126–134.

3. Al-Bashir, B., T. Cseh, R. Leduc, and R. Samson. 1990. Effect of soil/contaminant interactions on the biodegradation of naphthalene in flooded soil under denitrifying conditions. *Appl. Microbiol. Biotechnol.* 34:414–419.

4. Assinder, S. J., and P. A. Williams. 1990. The TOL plasmids: Determinants of the catabolism of toluene and the xylenes. *Adv. Microbial Physiol.* 31:1–69.

5. Atlas, R. M. 1981. Microbial degradation of petroleum hydrocarbons: an environmental perspective. *Microbiol. Rev.* 45:180–209.

6. Augenfeld, J. M., J. W. Anderson. 1982. The fate of polycyclic aromatic hydrocarbons in an intertidal sediment exposure system: bioavailability to *Macoma inquinata* (Mollusca:Pelecypoda) and *Abarenicola pacifica* (Annelida:Polychaeta). *Mar. Environ. Res.* 7:31–50.

7. Baker, J. M., Ed. 1976. *Marine Ecology and Oil Pollution*. New York: John Wiley and Sons.

8. Banerjee, S. 1985. Calculation of water solubility of organic compounds with UNIFAC-derived parameters. *Environ. Sci. Technol.* 19:369–370.

9. Bauer, J. E., and D. G. Capone. 1985. Degradation and mineralization of the polycyclic aromatic hydrocarbons anthracene and naphthalene in intertidal marine sediments. *Appl. Environ. Microbiol.* 50:81–90.

10. Bauer, J. E., and D. G. Capone. 1985. Effects of four aromatic organic pollutants on microbial glucose metabolism and thymidine incorporation in marine sediments. *Appl. Environ. Microbiol.* 49:828–835.

11. Bauer J. E., and D. G. Capone. 1988. Effects of co-occurring aromatic hydrocarbons on degradation of individual polycyclic aromatic hydrocarbons in marine sediment slurries. *Appl. Environ. Microbiol.* 54:1649–1655.

12. Bauer, J. E., R. P. Kerr, M. F. Bautista, C. J. Decker, and D. G. Capone. 1988. Stimulation of microbial activities and polycyclic aromatic hydrocarbon degradation in marine sediments inhabited by *Capitella capitata*. *Mar. Environ. Res.* 25:63–84.

13. Bertrand, J. C., P. Caumette, G. Mille, M. Gilewicz, and M. Denis. 1989. Anaerobic biodegradation of hydrocarbons. *Sci. Prog. Oxf.* 73:333–350.

14. Blackburn, J. W., R. K. Jain, and G. S. Sayler. 1987. Molecular microbial ecology of a naphthalene-degrading genotype in activated sludge. *Environ. Sci. Technol.* 21:884–890.

15. Blumer, M. 1976. Polycyclic aromatic compounds in nature. *Sci. Am.* 234:34–44.

16. Boehm, P. D., and J. G. Quinn. 1973. Solubilization of hydrocarbons by dissovled organic matter in seawater. *Geochim. Cosmochim. Acta* 37:2459–2466.

17. Brownawell, B. J., and J. W. Farrington. 1986. Biogeochemistry of PCBs in interstitial waters of a coastal marine sediment. *Geochim. Cosmochim. Acta* 50:157–169.

18. Bossert, I., W. M. Kachel, and R. Bartha. 1984. Fate of hydrocarbons during oily sludge disposal in soil. *Appl. Environ. Microbiol.* 47:763–767.

19. Britschgi, T., and S. Giovannoni. 1991. Phylogenetic analysis of a natural marine bacterioplankton population by rRNA gene cloning and sequencing. *Appl. Environ. Microbiol.* 57:1707–1713.

20. Bumpus, J. A., M. Tien, D. Wright, and S. Aust. 1985. Oxidation of persistent environmental pollutants by a white rot fungus. *Science* 228:1434–1436.

21. Bumpus, J. A. 1989. Biodegradation of polycyclic aromatic 3-hydrocarbons by *Phanerochaete chrysosporium*. *Appl. Environ. Microbiol.* 55:154–158.

22. Calder, J. A., and J. H. Lader. 1976. Effect of dissolved aromatic hydrocarbons on the growth of marine bacteria in batch culture. *Appl. Environ. Microbiol.* 32:95–101.

23. Cerniglia, C. E. 1981. Aromatic hydrocarbons: metabolism by bacteria, fungi, and algae. In *Reviews in Biochemical Toxicology*. Ed. E. Hodgson, J. R. Bend, and R. M. Philpot. New York: Elsevier/North Holland. 321–361.

24. Cerniglia, C. E. 1984. Microbial transformation of aromatic hydrocarbons. In *Petroleum Microbiology*. Ed. R. M. Atlas. New York: MacMillan Publ. Co. 99–128.

25. Cerniglia, C. E., and S. A. Crow. 1981. Metabolism of aromatic hydrocarbons by yeast. *Arch. Microbiol.* 129:9–13.

26. Cerniglia, C. E., D. T. Gibson, and C. Van Baalen. 1980. Oxidation of naphthalene by cyanobacteria and microalgae. *J. Gen. Microbiol.* 116:495–500.

27. Cerniglia, C. E., and M. A. Heitkamp. 1989. Microbial degradation of polycyclic aromatic hydrocarbons (PAH) in the aquatic environment. In *Metabolism of Polycyclic Aromatic Hydrocarbons (PAH) in the Aquatic Environment*. Ed. U. Varanasi. Boca Raton, FL: CRC Press. 41–68.

28. Cerniglia, C. E., G. L. White, and R. H. Heflich. 1985. Fungal metabolism and detoxification of polycyclic aromatic hydrocarbons. *Arch. Microbiol.* 143:105–110.

29. Da Silveria Pinto, A., M. E. P. Bomfim, A. N. Hagler, and J. Angluster. 1979. Metabolism

of aromatic compounds by yeasts isolated form a polluted esturary in Rio de Janeiro, Brazil. *Rev. Bras. Pesquisas Méd. Biol.* 12:339–346.

30. Dean-Raymond, D., and R. Bartha. 1975. Biodegradation of polynuclear aromatic petroleum components by marine bacteria. *Dev. Indust. Microbiol.* 16:97–110.

31. DeLaune, R. D., W. H. Patrick, and M. E. Casselman. 1981. Effect of sediment pH and redox conditions on degradation of benzo(a)pyrene. *Mar. Pollut. Bull.* 12:251–253.

32. Evans, K. M., R. A. Gill, and P. W. J. Robotham. 1990. The PAH and organic content of sediment particle size fractions. *Water Air Soil Pollut.* 51:13–31.

33. Farrington, J. W. 1991. Biogeochemical processes governing exposure and uptake of organic pollutant compounds in aquatic organisms. *Environ. Health Perspect.* 90:75–84.

34. Fedorak, P. M., K. M. Semple, and D. W. S. Westlake. 1984. Oil-degrading capabilities of yeast and fungi isolated from coastal marine environments. *Can. J. Microbiol.* 30:565–571.

35. Fedorak, P. M., and D. W. S. Westlake. 1981. Microbial degradation of aromatics and saturates in Prudhoe Bay crude oil as determined by glass capillary gas chromatography. *Can. J. Microbiol.* 27:432–443.

36. Fenchel, T. 1987. *Ecology of Protozoa.* Madison, WI: Science Tech Publ./Springer-Verlag.

37. Ferrer, C., E. Cózar, E. García-Valdés, and R. Rotger. 1986. IncP-7 naphthalene-degradative plasmids from *Pseudomonas putida. FEMS Microbiol. Lett.* 36:21–25.

38. Fukuda, K., Y. Inagaki, T. Maruyama, H. I. Kojima, and T. Yoshida. 1988. On the photolysis of alkylated naphthalenes in aquatic systems. *Chemosphere* 17:651–659.

39. García-Valdés, E., E. Cozar, R. Rotger, J. LaLucat, and J. Ursing. 1988. New naphthalene-degrading marine *Pseudomonas* strains. *Appl. Environ. Microbiol.* 54:2478–2485.

40. Gardner, W. S., R. F. Lee, K. R. Tenore, and L. W. Smith. 1979. Degradation of selected polycyclic aromatic hydrocarbons in coastal sediments: importance of microbes and polychaete worms. *Water Air Soil Pollut.* 11:339–347.

41. Gibson, D. T., and V. Subramanian. 1984. Microbial degradation of aromatic hydrocarbons. In *Microbial Degradation of Organic Compounds.* Ed. D. T. Gibson. New York: Marcel Dekker. 361–369.

42. Guerin, W. F., and G. E. Jones. 1988. Mineralization of phenanthrene by a *Mycobacterium* sp. *Appl. Environ. Microbiol.* 54:937–944.

43. Guerin, W. F., and G. E. Jones. 1989. Estuarine ecology of phenanthrene-degrading bacteria. *Estuar. Coast. Shelf Sci.* 29:115–130.

44. Hagler, A. N., S. S. Santos, and L. C. Mendonca-Hagler. 1979. *Rev. Microbiol.* 10:36–41.

45. Haines, J. R., and R. M. Atlas. 1982. *In situ* microbial degradation of Prudhoe Bay crude oil in Beaufort Sea sediments. *Mar. Environ. Res.* 7:91–102.

46. Hambrick, G. A., R. D. DeLaune, and W. H. Patrick. 1980. Effect of estuarine sediment pH and oxidation-reduction potential on microbial hydrocarbon degradation. *Appl. Environ. Microbiol.* 40:365–369.

47. Heitkamp, M. A., and C. E. Cerniglia. 1987. Effects of chemical structure and exposure on the microbial degradation of polycyclic aromatic hydrocarbons in freshwater and estuarine ecosystems. *Environ. Toxicol. Chem.* 6:535–546.

48. Heitkamp, M. A., and C. E. Cerniglia. 1988. Mineralization of polycyclic aromatic hydrocarbons by a bacterium isolated from sediment below an oil field. *Appl. Environ. Microbiol.* 54:1612–1614.

49. Heitkamp, M. A., and C. E. Cerniglia. 1989. Polycyclic aromatic hydrocarbon degradation by a *Mycobacterium* sp. in microcosms containing sediment and water from a pristine ecosystem. *Appl. Environ. Microbiol.* 55:1968–1973.

50. Heitkamp, M. A., W. Franklin, and C. E. Cerniglia. 1988. Microbial metabolism of polycyclic aromatic hydrocarbons: isolation and characterization of a pyrene-degrading bacterium. *Appl. Environ. Microbiol.* 54:2549–2555.

51. Heitkamp, M. A., J. P. Freeman and C. E. Cerniglia. 1987. Naphthalene biodegradation in environmental microcosms: estimates of degradation rates and characterization of metabolites. *Appl. Environ. Microbiol.* 53:129–136.

52. Herbes, S. E., and L. R. Schwall. 1978. Microbial transformation of polycyclic aromatic hydrocarbons in pristine and petroleum-contaminated sediments. *Appl. Environ. Microbiol.* 35:306–316.

53. Hinga, K. R., M. E. Q. Pilson, R. F. Lee, J. W. Farrington, K. Tjessem, and A. C. Davis. 1980. Biogeochemistry of benzanthracene in an enclosed marine ecosystem. *Environ. Sci. Technol.* 14:1136–1143.

54. Hites, R. A., and K. Biemann. 1972. Water pollution: organic compounds in the Charles River, Boston. *Science* 178:158–160.

55. Hites, R. A., R. E. LaFlamme, and J. W. Farrington. 1977. Polycyclic aromatic hydrocarbons in recent sediments: the historical record. *Science* 198:829–831.

56. Huang, T. C., M. C. Chang, and M. Alexander. 1981. Effect of protozoa on bacterial degradation of an aromatic compound. *Appl. Environ. Microbiol.* 41:229–232.

57. Hudak, J. P., and J. A. Fuhrman. 1988. Effects of four organic pollutants on the growth of natural marine bacterioplankton populations. *Mar Ecol. Prog. Ser.* 47:185–194.

58. Johnsen, S. 1987. Interactions between polycyclic aromatic hydrocarbons and natural aquatic humic substances. contact time relationship. *Sci. Total Environ.* 67:269–278.

59. Johnson, A. C., P. F. Larson, D. F. Gadbois, and A. W. Humason. 1985. The distribution of polycyclic aromatic hydrocarbons in the surficial sediments of Penobscot Bay (Maine, USA) in relation to possible sources and to other sites worldwide. *Mar. Environ. Res.* 15:1–16.

60. Kärenlampi, S. O., E. Marin, and O. O. P. Hänninen. 1980. Occurence of cytochrome P-450 in yeast. *J. Gen. Microbiol.* 120:529–533.

61. Karickhoff, S. W., D. S. Brown, and T. A. Scott. 1979. Sorption of hydrophobic pollutants on natural sediments. *Water Res.* 13:241–248.

62. Kerr, R. P., and D. G. Capone. 1988. The effect of salinity on the microbial mineralization of two polycyclic aromatic hydrocarbons in estuarine sediments. *Mar. Environ. Res.* 26:181–198.

63. King, J. M. H., P. M. DiGrazia, B. Applegate, R. Burlage, J. Sanseverino, P. Dunbar, F. Larimer, and G. S. Sayler. 1990. Rapid, sensitive bioluminescent reporter technology for naphthalene exposure and biodegradation. *Science* 249:778–781.

64. Kirk, P. W., and A. S. Gordon. 1988. Hydrocarbon degradation by filamentous marine higher fungi. *Mycologia* 80:776–782.

65. Landrum, P. F. 1989. Bioavailability and toxicokinetics of polycyclic aromatic hydrocarbons sorbed to sediments for the amphipod *Pontoporeia hoy. Environ. Sci. Technol.* 23:588–595.

66. Landrum, P. F., and J. A. Robbins. 1990. Bioavailability of sediment-associated contaminants to benthic invertebrates. In *Sediments: Chemistry and Toxicity of In-Place Pollutants.* Ed. R. Baudo, J. Giesy, H. Muntau. Chelsea, MI: Lewis Publishers, Inc. 237–263.

67. Laughlin, R. B., O. Linden, and J. M. Neff. 1979. A study on the effects of salinity and temperature on the disappearance of armatic hydrocarbons from the water-soluble fraction of No. 2 fuel oil. *Chemosphere* 10:741–749.

68. Lee, R. F., and J. W. Anderson. 1977. Fate and effect of naphthalenes: controlled ecosystem pollution experiment. *Bull. Mar. Sci.* 27:127–134.

69. Lee, R. F., W. S. Gardner, J. W. Anderson, J. W. Blaylock and J. Bardwell-Clarke. 1978. Fate of polycyclic aromatic hydrocarbons in controlled ecosystem enclosures. *Environ. Sci. Technol.* 12:832–838.

70. Lee, R. F., and C. Ryan. 1976. Biodegradation of petroleum hydrocarbons by marine

microbes. In *Proceedings of the Third International Biodegradation Symposium.* Ed. J. M. Sharpley, A. M. Kaplan. London: Applied Science Publishers. 119–125.

71. Levinton, J. S. 1982. *Marine Ecology.* Englewood Cliffs, NJ: Prentice-Hall Inc.

72. Lindquist, B., and D. Warshawsky. 1985. Identification of the 11,12-dihydro–11,12-dihydroxybenzo[a]pyrene as a major metabolite produced by the green algae, *Selenastrum capricornutum. Biochem. Biophys. Res. Commun.* 130:7–12.

73. Litchfield, C., and G. Floodgate. 1975. Biochemistry and microbiology of some Irish Sea sediments. II. Bacteriological analyses. *Mar. Biol.* 30:97–103.

74. Loveley, D., and D. Lonergan. 1990. Anaerobic oxidation of toluene, phenol, and p-cresol by the dissimilatory iron-reducing organism GS-15. *Appl. Environ. Microbiol.* 56:1858–1864.

75. MacIntyre, W. G., and P. O. deFur. 1985. The effect of hydrocarbon mixtures on desorption of substituted naphthalenes by clay and sediment from water. *Chemosphere* 14:103–111.

76. Magee, B. R., L. W. Lion, and A. T. Lemley. 1991. Transport of dissolved organic macro-molecules and their effect on the transport of phenanthrene in porous media. *Environ. Sci. Technol.* 25:323–331.

76a. Mahaffey, W. R., D. T. Gibson, and C. E. Cerniglia. 1988. Bacterial oxidation of chemical carcinogens: formation of polycyclic aromatic acids frm benz[a]anthracene. *Appl. Environ. Microbiol.* 54:2415–2423.

77. May, W. E. 1980. The solubility behavior of polycyclic aromatic hydrocarbons in aqueous systems. In *Petroleum in the Marine Environment.* Ed. L. Petrakis, F. T. Weiss. Washington, D.C.: American Chemical Society. 143–192.

78. McCarthy, J. F., and B. D. Jimenez. 1985. Interactions between polycyclic aromatic hydrocarbons and dissolved humic materials: binding and dissociation. *Environ. Sci. Technol.* 19:1072–1076.

79. McElroy, A. E., J. W. Farrington, and J. M. Teal. 1989. Bioavailability of polycyclic aromatic hydrocarbons in the aquatic environment. In *Metabolism of Polycyclic Aromatic Hydrocarbons (PAHs) in the Aquatic Environment.* Ed. U. Varanasi. Boca Raton, FL: CRC Press. 1–40.

80. McElroy, A. E., J. W. Farrington, and J. M. Teal. 1990. Influence of mode of exposure and the presence of a tubiculous polychaete on the fate of benz[a]anthracene in the benthos. *Environ. Sci. Technol.* 24:1648–1655.

81. McMillan, D. C., P. P. Fu, J. P. Freeman, D. W. Miller, and C. E. Cerniglia. 1988. Microbial metabolism and detoxification of 7,12-dimethylbenz[a]anthracene. *J. Ind. Microbiol.* 3:211–225.

82. Means, J. C., J. J. Hasett, S. G. Wood, and W. L. Banwart. 1980. Sorption properties of polynuclear aromatic hydrocarbons by sediments and soils. *Environ. Sci. Technol.* 14:1524–1528.

83. Means, J. C., and R. Wijayaratne. 1982. Role of natural colloids in the transport of hydrophobic pollutants. *Science* 215:968–970.

84. Meyer, J. S., M. D. Marcus, and H. L. Bergman. 1984. Inhibitory interactions of aromatic organics during microbial degradation. *Environ. Toxicol. Chem.* 3:583–587.

85. Mihelcic, J. R., and R. G. Luthy. 1988. Degradation of polycyclic aromatic hydrocarbon compounds under various redox conditions in soil-water systems. *Appl. Environ. Microbiol.* 54:1182–1187.

86. Mille, G., M. Mulyono, T. El Jammel, and J. C. Bertrand. 1988. Effects of oxygen on hydrocarbon degradation studies *in vitro* in surficial sediments. *Est. Coastal Shelf Sci.* 27:283–295.

87. Miller, R. M., G. M. Singer, J. D. Rosen, and R. Bartha. 1988. Photolysis primes biodegradation of benzo[a]pyrene. *Appl. Environ. Microbiol.* 54:1724–1730.

88. Morgan, P., S. T. Lewis, and R. J. Watkinson. 1991. Comparison of abilities of white-rot fungi to mineralize selected xenobiotic compounds. *Appl. Microbiol. Biotechnol.* 34:693–696.

89. Mueller, J. G., P. J. Chapman, B. O. Blattmann, and P. H. Pritchard. 1990. Isolation and characterization of a fluoranthrene-utilizing strain of *Pseudomonas paucimobilis*. *Appl. Environ. Microbiol.* 56:1079–1086.

90. Nagata, S. 1982. Degradation of aliphatic and aromatic hydrocarbons by marine bacteria. *Bull. Japan. Soc. Sci. Fish.* 45:781–786.

91. National Academy of Sciences. 1983. *Polycyclic Aromatic Hydrocarbons: Evaluation of Sources and Effects*. Washington, D.C.: National Academy Press.

92. Neff, J. M., Ed. 1979. *Polycyclic aromatic hydrocarbons in the Environment. Sources, Fates and Biological Effects*. London: Applied Science Publishers, Ltd.

93. Norkrans, B. 1966. On the occurrence of yeasts in an esturary off the Swedish Westcoast. *Svensk Bot. Tdiskr.* 60:462–482.

94. O'Connor, D. J., and J. P. Connolly. 1980. The effect of concentration of adsorbing solids on the partition coefficient. *Water Res.* 14:1517–1523.

95. Paalme, L., N. Irha, E. Urbas, A. Tsyban, and U. Kirso. 1990. Model studies of photochemical oxidation of carcinogenic polyaromatic hydrocarbons. *Mar. Chem.* 30:105–111.

95a. Pothuluri, J. V., and C. E., Cerniglia. 1993. Microbial metabolism of polycyclic aromatic hydrocarbons. In *Biological Degradation and Bioremediation of Toxic Chemicals*. Ed. G. R. Chaudhry. Portland, OR: Dioscorides Press.

96. Pothuluri, J. V., J. P. Freeman, F. E. Evans, and C. E. Cerrniglia. 1990. Fungal transformation of fluoranthrene. *Appl. Environ. Microbiol.* 56:2974–2983.

97. Pruell, R. J., and J. G. Quinn. 1985. Polycyclic aromatic hydrocarbons in surface sediments held in experimental mesocosms. *Toxicol. Environ. Sci.* 10:183–200.

98. Readman, J. W., R. F. C. Mantoura, M. M. Rhead, and L. Brown. 1982. Aquatic distribution and heterotrophic degradation of polycyclic aromatic hydrocarbons (PAH) in the Tamar Estuary. *Estuar. Coast. Shelf Sci.* 14:369–389.

99. Robinson, K. G., W. S. Farmer, and J. T. Novak. 1990. Availability of sorbed toluene in soils for biodegradation by acclimated bacteria. *Wat. Res.* 24:345–350.

100. Rogerson, A. W., and J. Berger. 1983. Enhancement of the microbial degradation of crude oil by the ciliate *Colpidium colpoda*. *J. Gen. Appl. Microbiol.* 29:41–50.

101. Roubal, G., and R. M. Atlas. 1978. Distribution of hydrocarbon-utilizing microorganisms and hydrocarbon biodegradation potentials in Alaskan continental shelf areas. *Appl. Environ. Microbiol.* 35:897–905.

102. Shiaris, M. P. 1989. Seasonal biotransformation of naphthalene, phenanthrene, and benzo[a]pyrene in surficial estuarine sediments. *Appl. Environ. Microbiol.* 55:1391–1399.

103. Shiaris, M. P. 1989. Phenanthrene mineralization along a natural salinity gradient in an urban estuary, Boston Harbor, Massachusetts. *Microb. Ecol.* 18:135–146.

104. Shiaris, M. P., and J. J. Cooney. 1983. Replica plating method for estimating phenanthrene-utilizing and phenanthrene-cometabolizing microorganisms. *Appl. Environ. Microbiol.* 45:706–710.

105. Shiaris, M. P., and D. Jambard-Sweet. 1986. Distribution of polycyclic aromatic hydrocarbons in surficial sediments of Boston Harbor, Massachusetts, USA. *Mar. Pollut. Bull.* 17:469–472.

106. Sisler, F. D., and C. E. Zobell. 1947. Microbial utilization of carcinogenic hydrocarbons. *Science* 521–522.

107. Smith, M. R. 1990. The biodegradation of aromatic hydrocarbons by bacteria. *Biodegradation* 1:191–206.

108. Soli, G., and E. M. Bens. 1973. Selective substrate utilization by marine hydrocarbonoclastic bacteria. *Biotechnol. Bioengin.* 15:285–297.

109. Tagger, S., N. Truffaut, and J. LePetit. 1990. Preliminary study on relationships among strains forming a bacterial community selected on naphthalene from a marine sediment. *Can. J. Microbiol.* 36:676–681.

110. Van Es, F. B., and L. A. Meyer-Reil. 1982. Biomass and metabolic activity of heterotrophic marine bacteria. *Adv. Microb. Ecol.* 4:111–170.

111. Varanasi, U., W. L. Reichert, J. E. Stein, D. W. Brown, and H. R. Sanborn. 1985. Bioavailability and biotransformation of aromatic hydrocarbons in benthic organisms exposed to sediment from an urban estuary. *Environ. Sci. Technol.* 19:836–841.

112. Vogel, T. M., and D. Grbić-Galić. 1986. Incorporation of oxygen into toluene and benzene during anaerobic fermentative transformation. *Appl. Environ. Microbiol.* 52:200–206.

113. Walker, J. D., R. R. Colwell, and L. Petrakis. 1975. Microbial petroleum degradation: application of computerized mass spectrometry. *Can. J. Microbiol.* 21:1760–1767.

114. Weissenfels, W. D., M. Beyer, and J. Klein. 1990. Degradation of phenanthrene, fluorene and fluoranthrene by pure bacterial cultures. *Appl. Microbiol. Biotechnol.* 32:479–484.

115. West, P. A., G. C. Okpokwasili, P. R. Brayton, D. J. Grimes, and R. R. Colwell. 1984. Numerical taxonomy of phenanthrene-degrading bacteria isolated from the Chesapeake Bay. *Appl. Environ. Microbiol.* 48:988–993.

116. Whitehouse, B. G. 1984. The effects of temperature and salinity on the aqueous solubility of polynuclear aromatic hydrocarbons. *Mar. Chem.* 14:319–332.

117. Whitehouse, B. G. 1985. The effects of dissolved organic matter on the aqueous partitioning of polynuclear aromatic hydrocarbons. *Estuar. Coast. Shelf Sci.* 20:393–402.

118. Wolfe, D. E., Ed. 1977. Fates and effects of petroleum hydrocarbons in marine organisms and ecosystems. New York: Pergamon Press.

119. Wood, J. L., C. L. Barker, and C. J. Grubbs. 1979. Photooxidation products of 7,12-dimethylbenz[a]anthracene. *Chem. Biol. Interactions* 26:339–347.

120. Wright, R. T., and J. E. Hobbie. 1966. Use of glucose and acetate by bacteria and algae in aquatic ecosystems. *Ecology* 47:447–464.

121. Yen, K.-M., and C. M. Serdar. 1988. Genetics of naphthalene catabolism in pseudomonads. *CRC Crit. Rev. Microbiol.* 15:247–268.

122. Youngblood, M. W., and M. Blumer. 1975. Polycyclic aromatic hydrocarbons in the environment: homologous series in soils and recent marine sediments. *Geochim. Cosmochim. Acta* 39:1303–1314.

CHAPTER 7

Biodegradation of Chlorinated Homocyclic and Heterocyclic Compounds in Anaerobic Environments

DOROTHY D. HALE, W. JACK JONES, and JOHN E. ROGERS

Abstract. A variety of mechanisms have been identified for the aerobic transformation of chlorinated aromatic compounds, including hydrolytic, oxidative, and reductive dechlorination reactions; however, reductive dechlorination is the only reported mechanism for the removal of chlorines from aromatic compounds under anaerobic conditions.

Many chlorinated aromatic compounds are amenable to reductive dechlorination and are biotransformed in anaerobic soils, sediments, and sewage sludge. These include chlorinated homocyclic and heterocyclic aromatic compounds, such as chloroanilines, chloroanisoles, chlorobenzenes, chlorobenzoates, chlorobiphenyls, chlorinated pesticides, chlorophenols, and chloropyridines. The initial step in the anaerobic biodegradation of these compounds is often the reductive removal of a chlorine atom from the aromatic ring. The dechlorination reaction occurs preferentially at the *ortho* and *para* positions of chloroanilines and chlorophenols and at the *meta* position of chlorobenzoates. Chlorobenzenes are predominantly transformed by *ortho*-directed dechlorination reactions, whereas chlorobiphenyls are usually not dechlorinated at the *ortho* position. In contrast, no preference for removal of chlorines from chloropyridines is apparent. Chloroanisoles and some chloropesticides may be transformed (e.g. *o*-demethylated) prior to dechlorination. Reducing conditions that permit methanogenesis typically favor reductive dechlorination reactions, although recent evidence suggests reductive dechlorination of some compounds under conditions of sulfate or nitrate reduction. Knowledge of the preference for chlorine removal and nutritional/physiological conditions that favor reductive dechlorination reactions may be useful for the development of bioremediation strategies for contaminated environments such as sediments and soils.

INTRODUCTION

With few exceptions, chlorinated aromatic compounds are completely or partially degraded by metabolic pathways that have evolved in microorganisms to degrade naturally occurring, nonchlorinated aromatic compounds (78). Under aerobic conditions, the pathways are drawn either from catabolic processes that completely

Dr. Hale is at Technology Applications, Inc., 960 College Station Road, Athens, GA 30605, U.S.A. Drs. Jones and Rogers are at the U.S. Environmental Protection Agency, Environmental Research Laboratory, Athens, GA 30605, U.S.A.

mineralize the compound to produce energy or from reactions that transform the compound into precursors for biosynthetic pathways (6, 7, 24, 25). The initial steps in this process are generally oxidative or hydrolytic [Reineke (74a), this volume]. Under anaerobic conditions, however, the initial degradative steps are reductive and generally involve the loss of a chlorine atom from the compound (84). In some cases, the chlorinated aromatic compound may act as a terminal electron acceptor in anaerobic respiration (22).

A variety of mechanisms have been identified for the aerobic degradation of chlorine-substituted aromatic compounds. To some degree, these mechanisms can be differentiated based on the number and position of chlorine substituents on the compound. For one- and two-chlorine-substituted compounds, chlorines are lost following ring cleavage, whereas chlorines can be removed from more highly chlorinated compounds by either hydrolytic, oxygenolytic, or reductive dechlorination reactions (74a).

Beginning with the seminal report by Suflita et al. (84), reductive dechlorination is the only biological mechanism that has been reported for the loss of chlorine atoms from aromatic compounds in methanogenic environments. Since then, many chlorinated aromatic compounds have been demonstrated to be reductively dechlorinated in anaerobic environments. These compounds include chloroanilines, chloroanisoles, chlorobenzenes, chlorobenzoates, chlorobiphenyls, chlorophenoxyacetates, and chlorophenols (Figure 1). In this chapter, we summarize the available information on the anaerobic degradation of these compounds.

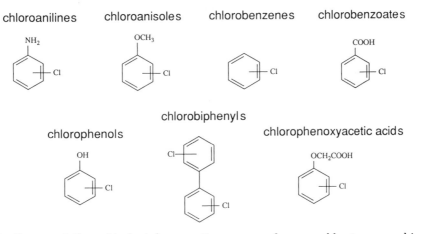

Figure 1. Representative chlorinated aromatic compounds amenable to anaerobic dechlorination.

REDUCTIVE DECHLORINATION UNDER
METHANOGENIC CONDITIONS

Chloroanilines

The reductive dechlorination of chloroanilines was initially reported in 1989 in studies by Struijs and Rogers (83) and by Kuhn and Suflita (54). Struijs and Rogers (83) observed the dechlorination of 3,4-dichloroaniline (to 3-chloroaniline), but not 2,4-dichloroaniline, in anoxic pond sediment slurries. Kuhn and Suflita (54) and Kuhn et al.

(55) also observed the dechlorination of dichloroanilines as well as tri-and tetrachloro-anilines (but not monochloroanilines) in methanogenic aquifer slurries. Although both research teams demonstrated that 3,4-dichloroaniline was converted to 3-chloro-aniline, Kuhn and Suflita (54) observed that the 3,5-dichloroaniline produced by sequential removal of the *para* and *ortho* chlorines of 2,3,4,5-tetrachloroaniline was persistent. A second pathway for dechlorination of 2,3,4,5-tetrachloroaniline by removal of a *meta* chlorine to yield 2,4,5-dichloroaniline was observed in slurries with added butyrate or sulfate (55). Kuhn et al. (55) also observed two pathways for the dechlorination of 2,3,4-trichloroaniline in aquifer slurries: (a) *ortho* dechlorination to 3,4-dichloroaniline followed by *para* dechlorination to 3-chloroaniline and (b) *meta* dechlorination to 2,4-dichloroaniline, which was not further dechlorinated.

Chloroanisoles

Under anaerobic conditions, chloroanisoles can be the products of chlorophenol methoxylation or substrates for reductive dechlorination and o-demethylation. Although considered to be a minor pathway for pentachlorophenol (PCP) transformation under anaerobic conditions, methoxylation of PCP to pentachloroanisole has been reported (45, 56, 66a, 90a). Weiss et al. (90a) identified several tri- and tetrachloro-anisoles in flooded rice field soil treated with PCP but were unable to determine whether the chloroanisoles resulted from the dechlorination of pentachloroanisole or from the methoxylation of tri- and tetrachlorophenols. Murthy and coworkers (66a), however, reported that methoxylation of PCP predominates under aerobic conditions, whereas o-demethylation of pentachloroanisole to PCP predominates under anaerobic conditions.

Studies in our laboratory have shown that reductive dechlorination and o-demethylation reactions can occur in the same sediment sample (Figure 2). Following a lag period, 2,4-dichloroanisole added to fresh sediment slurries was converted to 2,4-di-chlorophenol; after continued incubation, small amounts of 4-chlorophenol were detected, thus indicating reductive dechlorination of 2,4-dichlorophenol. When a second addition of 2,4-dichloroanisole was added, the principal product observed was 4-chloroanisole.

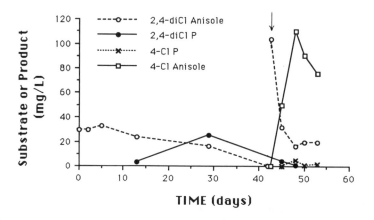

Figure 2. Anaerobic biotransformation of 2,4-dichloroanisole in freshwater sediment. Both o-demethylating and dechlorinating reactions are evident. *Arrow* indicates time of 2,4-dichloro-anisole addition.

Chlorobenzenes

Fathepure et al. (26) were the first to demonstrate in laboratory investigations that hexachlorobenzene (HCB) could be degraded under anaerobic conditions. Before this report, anaerobic degradation had been inferred from field data collected from the Great Lakes (3, 67). The ratios of dichlorobenzene and trichlorobenzene to pentachlorobenzene and HCB increased dramatically from younger to older sediment layers, thereby indicating reductive dechlorination of the higher chlorinated compounds in the deeper anaerobic sediments. In recent studies at our laboratory, Van Hoof and Jafvert (89) observed the reductive dechlorination of HCB in anaerobic pond sediments. Preferential dechlorination occurred at positions that were vicinal to another chlorine substituent. Intermediates of HCB dechlorination are presented in Figure 3. Relative to the anaerobic sludges studied by Fathepure et al. (26), more complete dechlorination occurred in anoxic pond sediments; 1,3,5-trichlorobenzene was degraded to 1,3-dichlorobenzene and the dichlorobenzenes were degraded to monochlorobenzene and benzene.

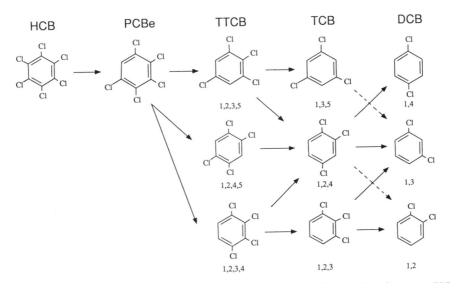

Figure 3. Proposed pathways for the reductive dechlorination of hexachlorobenzene (HCB) in pond sediments (89). *Dashed lines* represent slower reactions. Only aromatic reactants are shown. PCBe, pentachlorobenzene; TTCB, tetrachlorobenzene; TCB, trichlorobenzene; DCB, dichlorobenzene.

Chlorobenzoates

In 1982, Suflita and coworkers (84) reported the reductive dehalogenation of chloro-, bromo-, iodo-, and fluorobenzoates in anaerobic lake sediment and sewage sludge. Although microorganisms from both environments dechlorinated mono-, di-, and trichlorobenzoates, dehalogenation of bromo- and iodobenzoates proceeded more rapidly (and after a shorter lag period) than dehalogenation of the corresponding chloro- and fluorobenzoates. In addition, dehalogenation of chlorobenzoates occurred preferentially at the *meta* position (Figure 4), whereas dehalogenation of bromo- and iodobenzoates occurred at the *ortho, meta,* and *para* positions. Further characterization of the reductive dehalogenation of halobenzoates by sediment and sludge microorganisms was provided by Horowitz et al. (42) and Suflita et al. (85).

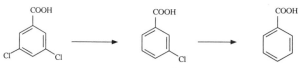

Figure 4. Proposed reaction scheme for the reductive dechlorination of 3,5-dichlorobenzoate (30).

Other researchers also have reported the reductive dechlorination of chlorobenzoates in a variety of anaerobic environments. Gibson and Suflita (30) studied the dechlorination of 3- and 4-chlorobenzoate and 3,4- and 3,5-dichlorobenzoate in sewage sludge, pond sediment, and methanogenic aquifer material. Following three months of incubation, less than 20% of each of the substrates was lost from sewage sludge. In contrast, extensive dechlorination occurred in methanogenic aquifer material and pond sediment. In these samples, dechlorination of 3,4-dichlorobenzoate occurred at both the *meta* and *para* positions to yield 4- and 3-chlorobenzoate, respectively. Although 4-chlorobenzoate was not dechlorinated in any of the samples, work by Genthner et al. (28) provided evidence for the dechlorination of both 4- and 2-chlorobenzoate in anaerobic sediments.

The chlorobenzoates are the class of compounds for which an anaerobic dechlorinating microorganism was initially isolated. *Desulfomonile tiedjei*, formerly named DCB-1 for "dechlorinating bacterium number 1," was isolated from a methanogenic, sewage sludge enrichment culture (79). This gram-negative, nonmotile, nonspore-forming rod has a unique collar-like structure involved in cell division (65b). Although *D. tiedjei* grows at temperatures between 30 and 38°C, its optimum growth temperature is 37°C. 16S rRNA sequence analysis has resulted in the placement of this sulfate-reducing bacterium, which contains desulfoviridin and cytochrome c_3, into a new genus (20). Unlike its closest relatives *Desulfuromonas acetoxidans* and *Desulfobacter postgatei*, *D. tiedjei* is unable to grow with acetate as an electron donor and sulfate or sulfur as the electron acceptor. Instead, *D. tiedjei* metabolizes acetate slowly with thiosulfate and grows on H_2/CO_2 with sulfate or thiosulfate. *D. tiedjei* can also use sulfite as an electron acceptor, and growth substrates include pyruvate, 3- and 4-anisate, vanillate, isovanillate, and 3-methoxyvanillate. Additional physiological characterization of this microorganism was reported by Stevens et al. (81) and Stevens and Tiedje (82).

Deweerd and Suflita (21) initially reported that the inducible, membrane-associated dehalogenating activity of *D. tiedjei* was restricted to the *meta* position of halogenated benzoates. Thus, 3-bromo-, 3-iodo- and 3-chlorobenzoate (but not 3-fluorobenzoate) were dehalogenated by this organism, whereas the *ortho* and *para* isomers were not (19, 21). Monochlorophenols were also not dehalogenated by *D. tiedjei*, although more highly chlorinated phenols were dechlorinated (65a). No relationship appears to exist between the reactions for dehalogenation and aromatic substituent removal, since *D. tiedjei* is unable to remove carboxyl, hydroxyl, and methyl functional groups from *meta* substituted benzoates (19). Although this organism can o-demethylate *meta*-methoxybenzoate, different enzymes appear to mediate the dechlorinating and o-demethylating activities.

Conservation of energy from the reductive dechlorination of 3-chlorobenzoate may occur by chemiosmotic coupling (23). In this tentative scheme, hydrogen is the electron donor and 3-chlorobenzoate is the electron acceptor. Transport of electrons produced by a membrane-associated hydrogenase to a membrane-associated dehalogenase results in dechlorination of the cytoplasmic chlorobenzoate to benzoate. Concomitant extrusion of protons generated by the hydrogenase forms a proton gradient across the membrane, which subsequently drives the formation of ATP by a proton driven ATPase.

Chlorobiphenyls

The reductive dechlorination of polychlorinated biphenyls (PCBs) was first inferred from the analysis of field data collected from the Hudson River, which was originally contaminated with Aroclor 1242 (11–13). Analysis of surface sediments led to capillary gas chromatograms that were very similar to Aroclor 1242. Analysis of three sub-surface sediment samples indicated a marked reduction in the concentrations of tri-, tetra-, and pentachlorobiphenyls. Quantitation of individual peaks indicated a 2- to 6-fold increase in dichlorobiphenyls and a 7 to 70-fold increase in 2-chlorobiphenyl. The observed transformations were congener-specific and suggested that chlorines were selectively removed from *meta* and *para* positions.

The conclusions reached in the field studies have since been confirmed in laboratory studies through an investigation of the reductive dechlorination of various Aroclors and selected PCB congeners (66b, 73a, 73b, 88). For example, an 85% conversion of tri- and tetrachlorobiphenyls to di- and monochlorinated products was observed in Aroclor-amended mixtures (73a). Chlorine atoms were selectively removed from the *meta* and *para* positions that are known to contribute to PCB toxicity (76). The specific congener 2,3,4,3',4'-pentachlorobiphenyl was converted to the products shown in Figure 5 when added to Hudson River sediments (1). The observed pathway consisted of the sequential loss of specific chlorine atoms from the *meta* and *para* positions to yield ultimately 2-chlorobiphenyl. Van Dort and Bedard (88) provided evidence for the biologically mediated *ortho* and *meta* dechlorination of 2,3,6-trichlorobiphenyl to produce 2,5-dichlorobiphenyl and 2,6-dichlorobiphenyl. *Ortho* dechlorination, however, was not the preferred pathway. Removal of the *meta* chlorine atom to produce 2,6-dichlorobiphenyl was the principal dechlorination reaction. Little is known about the microorganisms responsible for the reductive dechlorination of PCBs. Ye et al. (90b), using pasteurized Hudson River microorganisms, suggested that anaerobic spore-formers may be an important physiological group for PCB dechlorination.

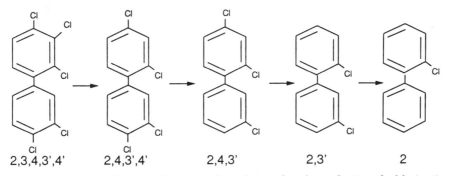

Figure 5. Major intermediates and proposed pathway for the reductive dechlorination of 2,3,4,3',4'-pentachlorobiphenyl in Hudson River sediments (1).

Chlorinated Pesticides

The chlorinated herbicides 2,4-dichlorophenoxyacetic acid (2,4-D) and 2,4,5-trichlorophenoxyacetic acid (2,4,5-T) and the insecticide 1,1,1-trichloro-2,2-bis(p-chlorophenyl)ethane (DDT) are known contaminants of various natural ecosystems, including soils, sediments, and groundwaters. There has long been considerable interest and concern regarding the environmental fate of these and related compounds in anaerobic environments because of their widespread use, toxicity, and possibility of bioaccumulation. In this section, we examine in detail the anaerobic transformations of 2,4-D, 2,4,5-T, and DDT.

Anaerobic degradation of 2,4-D has been investigated in both environmental and sewage sludge samples (Table 1). The reductive dechlorination of 2,4-D in anaerobic sludges from three wastewater treatment plants receiving differing mixtures of industrial and residential waste was examined by Mikesell and Boyd (61). 2,4-D was rapidly degraded in all sludge samples. The major intermediate product, 4-chlorophenol, also was gradually transformed over an incubation period of 56 days. The transformation of 4-chlorophenol was contrary to previous studies of Boyd and Shelton (10) that suggested 4-chlorophenol persisted in anaerobic sludge incubations. 2,4-Dichlorophenol was not detected as a product of 2,4-D transformation, thus indicating that cleavage of the ether bond and removal of the *ortho* chlorine were rapid events.

The biotransformation of 2,4-D in pond sediment, anaerobic sludge, and from a shallow, leachate-contaminated, anoxic aquifer was examined by Gibson and Suflita (30). 2,4-Dichlorophenol was the common initial product, thereby indicating that deacylation occurred before reductive dechlorination in these samples. In our laboratory, the anaerobic biotransformation of 2,4-D in sediment (Wacissa Springs, Florida) slurries occurred after a lag period of approximately two weeks. Subsequent additions of 2,4-D resulted in immediate biotransformation with production of both 2,4-dichlorophenol and 4-chlorophenoxyacetate (Table 1). Apparently, both reductive dechlorination and deacylation were occurring in these samples. Continued incubation of the Wacissa Springs samples led to the production of 4-chlorophenol and phenol, concomitant with the disappearance of the initial intermediate products 2,4-dichlorophenol and 4-chlorophenoxyacetate (14). The source of microbial inocula may be an important determinant in the development of consortia for the transformation of these environmental pollutants.

The anaerobic degradation of 2,4,5-T also has been investigated in environmental and sewage sludge samples (Table 1). Mikesell and Boyd (61) experimented with anaerobic sludges from three wastewater treatment plants receiving differing mixtures of industrial and residential waste. Although differences in the rates of 2,4,5-T

Table 1. Biotransformations of 2,4-D and 2,4,5-T in anaerobic environments.[a]

Inoculum source	Substrates	Products	Lag time	References
Sewage sludge	2,4-D	2,4-diCl-P, 4-Cl-P, P	none	30, 61
	2,4,5-T	2,4,5-triCl-P, 3,4-diCl-P, 4-Cl-P, P	none	30, 61
Aquifer, pond sediment	2,4-D	2,4-diCl-P, 4-Cl-P, P	3–4 weeks	30, unpublished data
	2,4,5-T	2,4-D, 2,5-D, mono and di-Cl-P, P	3–4 weeks	40, 31, unpublished data
Freshwater sediment	2,4-D	(4-Cl-phenoxy)acetate, 2,4-diCl-P, 4-Cl, P	2 weeks	unpublished data
Freshwater sediment (2,4-diCl-P-adapted)	2,4-D	(4-Cl-phenoxy)acetate	none	unpublished data

[a]Abbreviations: P, phenol; D, (dichlorophenoxy)acetate; T, (trichlorophenoxy)acetate.

degradation were evident among the sludges tested, results indicated that the transformation reaction schemes were consistent. In sludge samples from a site receiving significant waste input from an industrial source, the phenoxy-ether linkages of 2,4-D and 2,4,5-T were initially and rapidly cleaved (seven days' incubation) to produce 2,4,5 trichlorophenol, 3,4-dichlorophenol, and 4-chlorophenol, thus suggesting that the sequence of chlorine removal following ether cleavage was *ortho* > *meta* > *para*. Continued incubation of samples for a total of 21 days resulted in the complete loss of 2,4,5-trichlorophenol; 3,4- and 4-chlorophenol were detected as major products. Although some 3,4-dichlorophenol persisted after 70 days of incubation, the eventual production of 4-chlorophenol suggested dehalogenation of the *meta* chlorine of 3,4-dichlorophenol.

These results are in contrast to previous studies of Boyd and Shelton (10) and Suflita et al. (86), who observed only *ortho* chlorine removal from various chlorophenols and 2,4,5-T, respectively. The biotransformation of 2,4,5-T in environmental samples such as pond sediments, anaerobic sludges, and from a shallow, leachate-contaminated, anoxic aquifer were examined by Gibson and Suflita (30). Their results from sewage sludge samples were consistent with those of Mikesell and Boyd (61), in that 2,4,5-trichlorophenol was the major intermediate product as indicated previously. Pond sediment and methanogenic-aquifer samples, however, dehalogenated 2,4,5-T at either the *para* or *meta* position to produce 2,4- or 2,5-D. Continued incubation of these samples resulted in the production of mono- and dichlorophenols as well as phenol. These results suggest the possibility of the ultimate mineralization of 2,4,5-T by anaerobic methanogenic consortia, since transformation of phenol to CH_4 and CO_2 has been documented. In studies from our lab with freshwater sediment, an acclimation period of at least 14 days was observed in most experiments before the onset of 2,4,5-T biotransformation. Intermediates detected included 3,4-D, 2,5-D, 3,4-dichlorophenol, 3-chlorophenoxyacetic acid, 3-chlorophenol, and phenol. None of the intermediates accumulated to an appreciable extent, thus indicating further degradation of the reaction products.

Numerous studies have implicated the role of microorganisms in the transformation of DDT under anoxic conditions. The reductive removal of a single chlorine atom of DDT to DDD [1,1-dichloro-2,2-bis(p-chlorophenyl)-ethane] is the most commonly reported biotransformation reaction in anaerobic incubations. A number of studies, however, have demonstrated the chemical dechlorination of DDT by reduced porphyrins or by electrochemical and additional catalytic reactions (4, 5, 17, 32, 64, 75, 92).

Chemical and/or biochemical dechlorination reactions were reported by Castro (17) and Miskus et al. (64). These researchers provided preliminary evidence for the involvement of reduced iron porphyrins in reductive dechlorination reactions and demonstrated that anaerobic solutions of dithionite, reduced hematin, and hemoglobin catalyzed the transformation of DDT to DDD. The involvement of these biomolecules established a biological link for DDT dechlorination. The dehalogenating mechanism was further examined in a variety of biological and chemical systems by Zoro et al. (92) who postulated that the dehalogenation of DDT to DDD in natural environments was a consequence of cellular decay and was mediated by the release of iron two reacporphyrins in a reducing environment. The authors concluded that dehalogenation of DDT was not an essential part of cellular metabolism. Based on findings similar to those of Zoro et al. (92), Glass (32) proposed a mechanism for the reductive dehalogenation of DDT to DDD in soils (in the absence of oxygen) mediated by an iron redox system. The mechanism involved the formation of reduced iron (Fe^{2+}) from the oxidation of organic substrates and the subsequent transfer of an electron to DDT, thereby generating a chloride ion and a free radical that was subsequently converted to DDD. These authors suggested that microorganisms in the soil played a major role in DDT transformation via the consumption of oxygen, the generation of reduced iron, and the maintenance of low redox conditions.

A number of biologically active systems have been examined with regard to anaerobic DDT transformation, including anoxic soils (16, 18, 33, 35, 50, 68, 69), sewage sludge (2, 40, 46, 92), bovine rumen fluid (64), and a variety of isolated and mixed microorganisms (27, 47, 48, 57, 71–72). DDD (TDE) and DDE [1,1- dichloro-2,2-bis-(p-chlorophenyl)ethylene] are most often reported as the predominant metabolites of DDT biotransformation; DDE is the usual product of aerobic DDT transformation. Based on the profile of dehalogenation products, it has been generally accepted that two reaction mechanisms are prominent: the reductive dehalogenation reaction (Reaction 1) and the dehydrodechlorination reaction (Reaction 2), shown below.

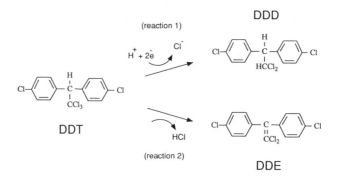

Additional products that have been detected from the previously described biological and chemical tranformations include DDMS [1-chloro-2,2-bis(p-chlorophenyl) ethane], DDMU [1-chloro-2,2-bis(p-chlorophenyl)ethylene], and DDNU [2,2-bis(p-chloro-phenyl)ethylene], likely formed from the reductive dehalogenation of DDD, the dehydrodechlorination of DDD, and the dehydrodechlorination of DDMS, respectively. A potential reaction sequence is presented in Figure 6. Other evidence from

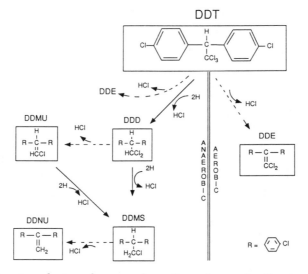

Figure 6. Significant products and proposed reaction schemes for the chemical and biological transformations of DDT. Refer to text for chemical abbreviations. *Solid lines,* dehalogenation reactions; *dashed lines,* dehydrodechlorination reactions.

pure microbial cultures suggests that different sequences of dehydrodechlorination and reductive dehalogenation reactions are possible for generating the variety of dechlorinated products listed above. The apparent complete dehalogenation of the aliphatic carbon of DDT was observed in studies by Matsumura et al. (60), Pfaender and Alexander (71), and Patil et al. (70). Further, Guenzi and Beard (34) detected p-chlorobenzoate as a metabolite of DDT biotransformation in anaerobic soil incubations. These results suggest that DDT may be transformed extensively to nontoxic products under conditions of anaerobiosis.

Chlorophenols

Reports by a number of investigators (9, 10, 28–30, 44, 51, 52, 62, 63, 87) have provided evidence for reductive dechlorination of chlorophenols in various anaerobic environments, including sewage sludge, aquifer materials, soils, and freshwater, estuarine, and marine sediments. That reductive dechlorination may be rate-limiting in the mineralization of chlorophenols, as reported by Zhang and Wiegel (91) for 2,4-dichlorophenol, provides additional impetus for investigations of this transformation.

In 1983, Boyd and coworkers (9) reported the degradation of all three monochlorophenols in 10% anaerobic sewage sludge in a mineral salts medium. The relative degradability of the monochlorophenols followed the order ortho > meta > para and required 3, 7, and 16 weeks for complete dechlorination, respectively. Detection of phenol from 2-chlorophenol by HPLC and mass spectrometry confirmed that reductive dechlorination was the initial transformation process. In a subsequent study, Boyd and Shelton (10) also noted the same relative order of biodegradation of the monochlorophenols (50 ppm) in fresh sewage sludge. They observed that dichlorophenols were more readily dechlorinated at the ortho position than at the meta or para positions. The ortho-substituted dichlorophenols were dechlorinated without a lag, whereas the 3,4- and 3,5-dichlorophenol isomers persisted during a six-week incubation. Dechlorination of the 2,6- and 2,3-dichlorophenol isomers yielded 2- and 3-chlorophenol, respectively, as transient intermediates; however, the 3- and 4-chlorophenol products of 2,5- and 2,4-dichlorophenol dechlorination were persistent.

In contrast to the work of Boyd and coworkers (9, 10), Hrudey et al. (44) found an ortho > para > meta preference for degradation of the chlorophenols in anaerobic sewage sludge. Complete degradation of 2-chlorophenol at 3 and 10 ppm occurred within one week, whereas complete degradation of this isomer at 285 ppm required 33 weeks. The onset of dechlorination for all concentrations of 2-chlorophenol tested was no greater than one week. Although the phenol that initially accumulated was subsequently degraded, phenol supplementation at the time of 2-chlorophenol addition significantly reduced the dechlorination rate. The transformation of 3- and 4-chlorophenol at 3 to 250 ppm required 16 to 20 weeks and occurred in only some of the replicate cultures. Although subsequent additions of these isomers resulted in more rapid dechlorination, concurrent phenol degradation was slower.

In addition to research on the degradation of chlorophenols in sewage sludge, several investigations have focused on chlorophenol degradation in sediments. Genthner et al. (28, 29), for example, studied the anaerobic degradation of monochlorophenols in freshwater and estuarine sediments, some of which had been exposed to industrial effluents. The relative preference for the degradation of the monochlorophenols (ortho > meta > para) was similar to that reported by Boyd et al. (9), with acclimation periods ranging from 2 to 12 months.

Work in our laboratory on the reductive dechlorination of dichlorophenols in methanogenic sediments from two ponds indicated that distinctly different dechlorinating activities were present (39). In sediment from the Cherokee Trailer Park pond, for example, 2,3-, 2,4-, and 2,6-dichlorophenol were dechlorinated at a faster rate and after a shorter lag phase than in Bolton's Pond sediment, which was characterized by a low organic carbon content. Although no lag phase was observed in either sediment after adaptation to one of these dichlorophenols, adapted Cherokee sediments exhibited faster rates of dechlorination and a broader substrate specificity (tested against all six dichlorophenols) than adapted sediment from Bolton's Pond. Similar to the findings of Boyd and Shelton (10) for sewage sludge, dechlorination of the dichlorophenols in both sediments occurred more readily at the *ortho* than the *meta* or *para* positions.

The degradation of chlorophenols in anaerobic aquifer material and saturated soil has been confirmed by several studies. Suflita and Miller (87) observed the degradation of 2-, 3-, and 4-chlorophenol in addition to 2,4- and 2,5-dichlorophenol in samples from a methanogenic aquifer. Dechlorination at the *ortho* position of chlorophenols occurred most readily and without an identifiable lag period, whereas dechlorination of 3- and 4-chlorophenol followed a lag period of one month. Gibson and Suflita (30) also reported the degradation of the three monochlorophenols and 2,4- and 2,5-dichlorophenol during a three-month incubation in methanogenic aquifer material. 3,4-Dichlorophenol, however, persisted under these conditions. The observed preference for *ortho* chlorine removal from the aromatic ring was in agreement with the results of Suflita and Miller (87). Smith and Novak (80) also observed a significant decrease in 2-chlorophenol concentration (from 50 ppm to 20 ppm) within 45 days after addition to saturated subsurface soil (31 m) from a site in Virginia; because no products were identified, the transformation pathway was not clear.

Until recently, no microbial isolates had been described that were capable of reductive dechlorination of chlorophenols. Madsen and Licht (58b) isolated an anaerobic spore-forming organism related to the genus *Clostridium* that preferentially dechlorinated *ortho*-substituted di- and trichlorinated phenols. The isolate is also capable of removal of the *meta*-chlorine of 3,5-dichlorophenol.

Similar to the degradation of the mono- and dichlorophenols, PCP degradation occurs in a number of anaerobic environments. Ide and coworkers (45) were the first to demonstrate PCP biodegradation under anaerobic conditions. Other studies with flooded soil and sewage sludge confirmed anaerobic PCP biodegradation (36, 56, 66a, 90a). Recently, Guthrie et al. (36) and Mikesell and Boyd (62) observed the mineralization of ^{14}C-labeled PCP to ^{14}CO$_2$ and ^{14}CH$_4$ in sewage sludge. High concentrations of PCP, however, have been shown to inhibit methane production in anaerobic digestors and in pure and mixed cultures of microorganisms (36, 41, 74b).

Dechlorination of PCP occurs at the *ortho, meta,* or *para* positions, as indicated by the identification of products lacking chlorine atoms in each of these positions. Indeed, all three tetrachlorophenols and all six trichlorophenols have collectively been identified as PCP degradation products in a number of different studies (45, 56, 66a, 90a). In several investigations, these tri- and tetrachlorophenols have been shown to accumulate from PCP degradation (56, 90a). In contrast, only two dichlorophenols (the 3,4- and 3,5-isomers) and one monochlorophenol (the 3-isomer) have been identified as intermediates in the anaerobic biodegradation of PCP (45, 61, 62). Mikesell and Boyd (61) identified 3,5-dichlorophenol as the terminal product of PCP degradation in anaerobic sewage sludge.

Chloropyridines

Chlorinated heterocyclic compounds, such as chloropyridines, can be reductively dechlorinated in a manner similar to that of homocyclic compounds. Liu and Rogers (58a) reported that 2,3- and 3,5-dichloropyridine were dechlorinated in samples of methanogenic pond sediment. Both 2- and 3-chloropyridine were produced from 2,3-dichloropyridine, whereas only 3-chloropyridine was detected as a dechlorination product of 3,5-dichloropyridine. Subsequent dechlorination of the monochloropyridines to pyridine was not evident. In contrast, the loss of 2,5- and 2,6-dichloropyridine from pond sediments was slow during six months of incubation and paralleled that of autoclaved sediments.

REDUCTIVE DECHLORINATION IN THE PRESENCE OF ALTERNATIVE ELECTRON ACCEPTORS

Only in recent years has reductive dechlorination been demonstrated in other than methanogenic conditions. King (49) was the first to report the reductive dechlorination of 2,4-dichlorophenol in marine sediments. Approximately 78% of the 2,4-dichlorophenol (1 mM) added to these sediments was dechlorinated within a 200-hour period. Kohring et al. (52) observed dechlorination of 4-chlorophenol and 2,4-dichlorophenol in sulfate-reducing freshwater sediments; sulfate was concomitantly reduced from an initial concentration of 25 mM to approximately 6 mM. Haggblom and Young (37) also observed dechlorination of 2,4-dichlorophenol under sulfate-reducing conditions with an estuarine sediment culture. The intermediate product, 4-chlorophenol, was also transformed, and no additional metabolites were detected. Likewise, both 2- and 3-chlorophenol were degraded by these cultures, but no metabolites were detected. Thus, it is not clear whether the monochlorophenols were dechlorinated to phenol or are degraded through an alternate pathway by these estuarine-derived cultures. Haggblom and Young (37), however, reported that in sulfidogenic cultures derived from an anaerobic bioreactor, the product (2-chlorophenol) of 2,6-dichlorophenol dechlorination was degraded through phenol.

Although reductive dechlorination has been observed under sulfate-reducing conditions, the addition of sulfate to actively dechlorinating methanogenic environments inhibits the reductive dechlorination process. Gibson and Suflita (30) observed an apparent inhibitory effect of sulfate on 2,4,5-T dehalogenation. Aquifer samples from a sulfate-reducing site did not transform 2,4-D or 2,4,5-T, and addition of sulfate to methanogenic aquifer samples resulted in significant inhibition of 2,4,5-T dehalogenation. Further, combinations of solids and groundwater from both the methanogenic and sulfate-reducing sites were tested; the addition of sulfate, or conditions that stimulated sulfate reduction, inhibited 2,4,5-T transformation. In a second study at this site (31), aquifer samples collected from a sulfate-reducing zone again exhibited no significant dehalogenation of 2,4,5-T with or without organic amendments. Addition of molybdate, an inhibitor of biological sulfate reduction, alone or in combination with acetate and/or sulfate, enhanced dehalogenation of 2,4,5-T.

Kuhn et al. (55) also reported that addition of sulfate to aquifer slurries slowed the initial dechlorination of tetrachloroaniline to trichloroaniline and blocked further dechlorination. Studies with methanogenic pond sediments also have demonstrated that the dechlorination of 2,4-dichlorophenol is affected by the presence of alternative electron acceptors (W. N. Howard, personal communication). In fact, the addition of sulfate (20 mM $(NH_4)_2SO_4$) or nitrate (20 mM $NaNO_3$) dramatically inhibited dechlorination, but the addition of inhibitors of methanogenesis [20 mM bromoethane-sulfonic acid (BESA)] or sulfate reduction (20 mM $NaMoO_4$) had little effect (Figure 7). Although the mechanism of sulfate inhibition of reductive dechlorination has not yet been determined, it is likely that sulfate is a preferred electron acceptor and that sulfate-reducing bacteria more effectively compete for electron donors, thus reducing the potential for reductive dehalogenation.

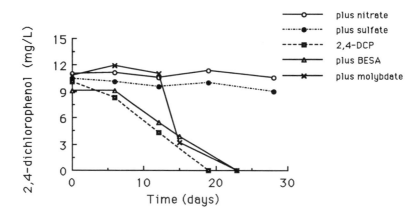

Figure 7. Effects of alternative electron acceptors (sulfate, nitrate) and inhibitors of sulfate reduction (molybdate) and methanogenesis (bromoethanesulfonic acid, BESA) on the reductive dechlorination of 2,4-dichlorophenol in methanogenic pond sediments (W. N. Howard, unpublished data).

BIOREMEDIATION APPLICATIONS

Anaerobic upflow bioreactors alone or in combination with aerobic trickling filters have been used to harness the degradative capacity for chlorophenols found in some sewage sludges (38, 53, 77). Krumme and Boyd (53) utilized anaerobic upflow bioreactors to degrade three monochlorophenols individually or in a mixture. Although 3,4,5-trichlorophenol was also degraded in these reactors, minimal degradation of 2,4,6-trichlorophenol and PCP was reported. Finnish researchers (38, 77), however, reported the mineralization of PCP to carbon dioxide in a two-stage reactor consisting of an anaerobic upflow bioreactor and an aerobic trickling filter. Sulfate-reducing cultures enriched from similar bioreactors used to treat pulp-bleaching effluents also dechlorinated 2,4- and 2,6-dichlorophenol to 4- and 2-chlorophenol, respectively (37). Only 2-chlorophenol was further dechlorinated to phenol during the 84-day incubation. Collectively, the results of these studies suggest that bioreactor treatment systems may be useful in removing mixtures of chlorophenols from contaminated water.

Several laboratories have had mixed success with the addition of organic substrates as a source of reducing equivalents for the reductive dechlorination process. Kuhn et al. (55) reported the effects of organic carbon supplements on chloroaniline dechlorination. Tetrachloroaniline dechlorination was inhibited by addition of methanol to subcultures from an aquifer slurry in which 3,4-dichloroaniline was initially dechlorinated to 3-chloroaniline. The dechlorination of 3,4-dichloroaniline also was inhibited by supplementation with either butyrate, a mixture of volatile fatty acids, a mixture of glucose, benzoate, succinate, and citrate, or groundwater from a methanogenic site. In contrast, subcultures amended with rumen fluid, yeast extract, pyruvate, or trypticase dechlorinated 3,4-dichloroaniline to 3-chloroaniline in one to two weeks.

The addition of typical fermentation intermediates to aquifer samples from a methanogenic site significantly reduced the acclimation period before the onset of 2,4,5-T dehalogenation (31). Butyrate, propionate, ethanol, and formate amendments had the greatest stimulatory effect and reduced the acclimation period from four months to less than one month in most samples. Addition of the supplemental carbon sources also resulted in an increase in the extent of 2,4,5-T transformation, as evidenced by the detection of lesser chlorinated aromatic compounds as major products. For example, addition of butyrate to samples collected from a methanogenic aquifer site resulted in the production of only monochlorinated products during a 4-month incubation period.

Studies with pond sediments demonstrated that the dechlorination of 2,4-dichlorophenol is affected by the presence of specific organic compounds (43). Although the addition of 0.1% yeast extract stimulated dechlorination in fresh and adapted sediment, acetate (0.3%) stimulated dechlorination only in fresh sediment. In contrast, propionate inhibited dechlorination in both sediments, and butyrate had no effect on the rate of dechlorination in either sediment. Also, the addition of clarified rumen fluid (5, 10, or 20%), mixed vitamins, or 1,4-naphthoquinone did not affect dechlorination.

Another approach for enhancing reductive dechlorination processes has been the use of acclimated cultures. The success of this approach arises from the use of structure-activity relationships to match acclimated cultures and a number of structurally similar chlorinated compounds.

The potential for using acclimated and cross-acclimated microbial populations in bioremediation applications is supported by the observations of Struijs and Rogers (83) for dechlorination of dichloroanilines and dichlorophenols. Sediments acclimated to 2,4- or 3,4-dichlorophenol rapidly dechlorinated 2,4- and 3,4-dichloroaniline without a lag. The reductive dechlorination of 3,4-dichloroaniline in non-acclimated or 3,4-dichlorophenol acclimated sediments is presented in Figure 8. In nonacclimated sediment slurries, 3,4-dichloroaniline was dechlorinated to 3-chloroaniline only after a three-week lag period, and 2,4-dichloroaniline was not dechlorinated during the eight-week study. The chemical characteristics of the two compound classes are consistent with cross-acclimation. Amino and hydroxyl groups are *ortho-para* directing (59), are capable of forming hydrogen bonds (59), and have similar Van der Waals radii (8).

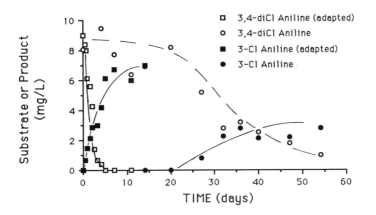

Figure 8. Dechlorination of 3,4-dichloroaniline to 3-chloroaniline in non-acclimated or 3,4-dichlorophenol acclimated sediment (83). Results are the average of duplicate experiments.

Anaerobic transformation of the chlorophenoxy herbicide 2,4,5-T was demonstrated by Suflita et al. (86), who used a microbial consortium enriched from anoxic sewage sludge with 3-chlorobenzoate as the primary substrate. The primary product of 2,4,5-T biotransformation was identified by mass spectrometry and nuclear magnetic resonance spectrometry as 2,5-dichlorophenoxyacetic acid (2,5-D), indicating dechlorination only at the *para* position. No further transformation of 2,5-D was detected. In addition, the microbial consortium was unable to transform related tri- and dichlorinated aromatic compounds, including 2,4-D, 2,3-D, 2,4-dichlorophenol, and 2,4,5-trichlorophenol. Diluted anaerobic sludge that had not been previously acclimated to either 2,4,5-T or 3-chlorobenzoate did not degrade 2,4,5-T within five to six weeks, nor was 2,4,5-T transformed in autoclaved control samples. Addition of fresh or autoclaved anaerobic sludge to the adapted dehalogenating consortium, however, stimulated the rate of 2,4,5-T transformation compared to unamended controls, thus indicating that a heat-stable chemical factor (nutrient) was probably responsible for the enhanced transformation activity.

We recently have investigated the biotransformation of 2,4-D and 2,4,5-T with nonadapted as well as dichlorophenol-adapted freshwater sediment slurries (14). 2,4-D was dechlorinated at the *ortho* position, without a lag, in 2,4-dichlorophenol-adapted sediments to form 4-chlorophenoxyacetic acid; however, an acclimation time equivalent to that required for nonadapted sediments was necessary in 3,4-dichlorophenol-adapted sediment before the onset of 2,4-D dehalogenation. The results with mixtures of 2,4- and 3,4-dichlorophenol-adapted sediments paralleled the results with 2,4-dichlorophenol-adapted sediments with regard to 2,4-D biotransformation. In sediment samples collected from the same site at a different time of the year (March rather than November), results were similar except that the intermediate product, 4-chlorophenoxyacetate, was further degraded in 2,4-dichlorophenol-acclimated samples.

In contrast to 2,4-D biotransformation, the anaerobic transformation of 2,4,5-T was not enhanced by dichlorophenol-acclimation of sediments. A lag period of at least 14 days was observed with both acclimated and non acclimated sediment microorganisms before the onset of 2,4,5-T biotransformation. Intermediates detected included 3,4-D, 2,5-D, 3,4-dichlorophenol, 3-chlorophenoxyacetate, 3-chlorophenol, and phenol. None of the intermediates accumulated to an appreciable extent, thus indicating further degradation of the intermediate products.

Complete dechlorination of PCP has been achieved by using sewage sludge microorganisms adapted to dechlorinate monochlorophenols. Mikesell and Boyd (62, 63) demonstrated that sewage sludge microorganisms that were adapted to any of the monochlorophenols dechlorinated PCP rapidly at the same ring position(s) as the monochlorophenol to which they were previously exposed. Thus, in sludge with microorganisms adapted to 2- or 4-chlorophenol, PCP was primarily dechlorinated at the *ortho* or *para* positions, respectively. Sludge microorganisms adapted to 3-chlorophenol only dechlorinated PCP at the *meta* positions. PCP was completely dechlorinated in a mixture of sludges with microorganisms adapted to dechlorinate each of the monochlorophenols individually. This dechlorinating activity was subsequently transferred to soil contaminated with PCP, resulting in enhanced PCP degradation rates. The dechlorination products were 2,3,4,5-tetrachlorophenol, 3,4,5-trichlorophenol, 3,5-dichlorophenol, 3,4-dichlorophenol, and 3-chlorophenol, the same as those observed in sludge.

Complete dechlorination of PCP also has been achieved by using freshwater sediments with microorganisms adapted to dechlorinate dichlorophenols (15). In sediments with microorganisms adapted to dechlorinate 2,4- or 3,4-dichlorophenol at the *ortho* and *para* positions, respectively, PCP was initially dechlorinated at the same relative positions on the aromatic ring. A combination of the adapted sediments dechlorinated PCP to phenol in the order *para* > *ortho* > *meta*. The chlorinated intermediates identified included 2,3,5,6-tetrachlorophenol, 2,3,5-trichlorophenol, 3,5-dichlorophenol, and 3-chlorophenol (Figure 9). These intermediates of PCP dechlorination are less mutagenic than the intermediates detected in sewage sludge by Mikesell and Boyd (62, 63).

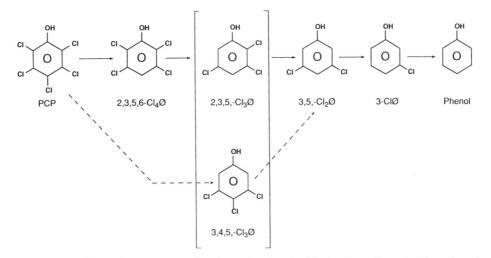

Figure 9. Proposed reaction sequence for the reductive dechlorination of pentachlorophenol (PCP) in a mix of 2,4- and 3,4-dichlorophenol adapted sediments (15). *Dashed lines* depict the reaction sequence in anaerobic sewage sludge as reported by Mikesell and Boyd (62). Abbreviation: Ø, phenol.

Acknowledgments

The authors thank Dr. Chad Jafvert and Dr. Juergen Wiegel for helpful comments and discussions as well as Wren Howard for unpublished data.

LITERATURE CITED

1. Abramowicz, D. A. 1990. Aerobic and anaerobic biodegradation of PCBs: a review. *Crit. Rev. Biotech.* 10:241–251.

2. Albone, E. S., G. Eglinton, N. C. Evans, and M. M. Rhead. 1972. Formation of bis(p-chlorophenyl)-acetonitrile (p,p'-DDCN) from p,p'-DDT in anaerobic sewage sludge. *Nature* 240:420–421.

3. Bailey, R. E. 1983. Comment on chlorobenzenes in sediments, water, and fish from Lakes Superior, Huron, Erie, and Ontario. *Environ. Sci. Technol.* 17:504.

4. Beland, F. A., S. O. Farwell, and R. D. Geer. 1974. Anaerobic degradation of 1,1,1,2-tetra-chloro-2,2-bis(p-chlorophenyl)ethane (DTE). *J. Agric. Food Chem.* 22:1148–1149.

5. Berry, J. D., and D. A. Stotter. 1977. Dechlorination of DDT by vitamin B_{12} under mild reducing conditions. *Chemosphere* 6:783–787.

6. Bollag, J.-M., G. G. Briggs, J. E. Dawson, and M. Alexander. 1968. 2,4-D metabolism: Enzymatic degradation of chlorocatechols. *J. Agric. Food Chem.* 16:829–833.

7. Bollag, J.-M., C. S. Helling, and M. Alexander. 1968. 2,4-D metabolism: Enzymatic hydroxylation of chlorinated phenols. *J. Agric Food Chem.* 16:826–828.

8. Bondi, A. 1964. Van der Waals volumes and radii. *J. Chem. Phys.* 68:441–451.

9. Boyd, S. A., D. R. Shelton, D. Berry, and J. M. Tiedje. 1983. Anaerobic biodegradation of phenolic compounds in digested sludge. *Appl. Environ. Microbiol.* 46:50–54.

10. Boyd, S. A., and D. R. Shelton. 1984. Anaerobic biodegradation of chlorophenols in fresh and acclimated sludge. *Appl. Environ. Microbiol.* 47:272–277.

11. Brown, J. F. Jr., R. E. Wagner, D. L. Bedard, M. J. Brennan, J. C. Carnahan, and R. J. May. 1984. PCB transformations in upper Hudson sediments. *Northwest Environ. Sci.* 3:176–179.

12. Brown, J. F. Jr., D. L. Bedard, M. J. Brennan, J. C. Carnahan, H. Feng, and R. E. Wagner. 1987. Polychlorinated biphenyl dechlorination in aquatic sediments. *Science* 236:709–712.

13. Brown, J. F. Jr., R. E. Wagner, H. Feng, D. L. Bedard, M.J. Brennan, J. C. Carnahan, and R. May. 1987. Environmental dechlorination of PCBs. *Environ. Toxicol. Chem.* 5:579–593.

14. Bryant, F. O., and J. E. Rogers. 1989. Dechlorination of pentachlorophenol, 2,4-dichlorophenoxyacetic acid and 2,4,5-trichlorophenoxyacetic acid in anaerobic freshwater sediments. *Abstr. 198th Annu. Meet. Am. Chem. Soc.* 1989.

15. Bryant, F. O., D. D. Hale, and J. E. Rogers. 1991. Regiospecific dechlorination of pentachlorophenol by dichlorophenol-adapted microorganisms in freshwater, anaerobic sediment slurries. *Appl. Environ. Microbiol.* 57:2293–2301.

16. Burge, W. D. 1971. Anaerobic decomposition of DDT in soil: acceleration by volatile components of alfalfa. *J. Agric. Food Chem.* 19:375–378.

17. Castro, C. E. 1964. The rapid oxidation of iron (II) porphyrins by alkyl halides. A possible mode of intoxication of organisms by alkyl halides. *J. Am. Chem. Soc.* 86:2310–2311.

18. Castro, T. F., and T. Yoshida. 1971. Degradation of organochlorine insecticides in flooded soils in the Phillipines. *J. Agric. Food Chem.* 19:1168–1170.

19. DeWeerd, K. A., J. M. Suflita, T. Linkfield, J. M. Tiedje, and P. H. Pritchard. 1986. The relationship between reductive dehalogenation and other aryl substituent removal reactions catalyzed by anaerobes. *FEMS Microbiol. Ecol.* 38:331–339.

20. DeWeerd, K. A., L. Mandelco, R. S. Tanner, C. R. Woese, and J. M. Suflita. 1990. *Desulfomonile tiedjei* gen. nov. and sp. nov., a novel anaerobic, dehalogenating, sulfate-reducing bacterium. *Arch. Microbiol.* 154:23–30.

21. DeWeerd, K. A., and J. M. Suflita. 1990. Anaerobic aryl reductive dehalogenation of halobenzoates by cell extracts of "*Desulfomonile tiedjei.*" *Appl. Environ. Microbiol.* 56:2999–3005.

22. Dolfing, J., and J. M. Tiedje. 1986. Hydrogen cycling in a three-tiered food web growing on the methanogenic conversion of 3-chlorobenzoate. *FEMS Microbiol. Ecol.* 38:293–298.

23. Dolfing, J. 1991. *Desulfomonile tiedjei* and other anaerobic bacteria with a taste for halogenated aromatic compounds. In *Proc. COST 641 Workshop, Organic Micropollutants in the Aquatic Environment,* November 1990, Copenhagen, Denmark.

24. Evans, W. C., B. S. W. Smith, and H. N. Ferney. 1971. Bacterial metabolism of 4-chlorophenoxyacetate. *Biochem. J.* 122:509–517.

25. Evans, W. C., B. S. W. Smith, H. N. Ferney, and J. I. Davies. 1971. Bacterial metabolism of 2,4-dichlorophenoxyacetate. *Biochem. J.* 122:543–551.

26. Fathepure, B. Z., J. M. Tiedje, and S. A. Boyd. 1988. Reductive dechlorination of hexachlorobenzene to tri- and dichlorobenzenes in anaerobic sewage sludge. *Appl. Environ. Microbiol.* 54:327–330.

27. Fries, G. R., G. S. Marrow, and C. H. Gordon. 1969. Metabolism of *o,p'*- and *p,p'*-DDT by rumen microorganisms. *J. Agric. Food Chem.* 17:860–862.

28. Genthner, B. R. Sharak, W. A. Price II, and P. H. Pritchard. 1989. Anaerobic degradation of chloroaromatic compounds in aquatic sediments under a variety of enrichment conditions. *Appl. Environ. Microbiol.* 55:1466–1471.

29. Genthner, B. R. Sharak, W. A. Price II, and P. H. Pritchard. 1989. Characterization of anaerobic dechlorinating consortia derived from aquatic sediments. *Appl. Environ. Microbiol.* 55:1472–1476.

30. Gibson, S. A., and J. M. Suflita. 1986. Extrapolation of biodegradation results to groundwater aquifers: reductive dehalogenation of aromatic compounds. *Appl. Environ. Microbiol.* 52:681–688.

31. Gibson, S. A., and J. M. Suflita. 1990. Anaerobic biodegradation of 2,4,5-trichlorophenoxyacetic acid in samples from a methanogenic aquifer: stimulation by short-chain organic acids and alcohols. *Appl. Environ. Microbiol.* 56:1825–1832.

32. Glass, B. L. 1972. Relation between the degradation of DDT and the iron redox system in soils. *J. Agric. Food Chem.* 20:324–327.

33. Guenzi, W. D., and W. E. Beard. 1967. Anaerobic biodegradation of DDT to DDD in soil. *Science* 156:116–119.

34. Guenzi, W. D., and W. E. Beard. 1968. Anaerobic conversion of DDT to DDD and aerobic stability of DDT in soil. *Soil Sci. Am. Proc.* 32:522–527.

35. Guenzi, W. D., and W. E. Beard. 1976. DDT degradation in flooded soil as related to temperature. *J. Environ. Qual.* 5:391–394.

36. Guthrie, M. A., E. J. Kirsch, R. F. Wukasch, and C. P. L. Grady Jr. 1984. Pentachlorophenol biodegradation—II. *Water Res.* 18:451–461.

37. Haggblom, M. M., and L. Y. Young. 1990. Chlorophenol degradation coupled to sulfate reduction. *Appl. Environ. Microbiol.* 56:3255–3260.

38. Hakulinen, R., and M. Salkinoja-Salonen. 1982. Treatment of pulp and paper industry wastewaters in an anaerobic fluidized bed reactor. *Process Biochem.* 17:18–22.

39. Hale, D. D., J. E. Rogers, and J. Wiegel. 1990. Reductive dechlorination of dichlorophenols by non-adapted and adapted microbial communities in pond sediments. *Microb. Ecol.* 20:185–196.

40. Hill, D. W., and P. L. McCarty. 1967. Anaerobic degradation of selected chlorinated hydrocarbon pesticides. *J. Water Poll. Cont. Fed.* 39:1259–1277.

41. Horowitz, A., D. R. Shelton, C. P. Cornell, and J. M. Tiedje. 1982. Anaerobic degradation of aromatic compounds in sediments and digested sludge. *Dev. Indust. Microbiol.* 23:435–444.

42. Horowitz, A., J. M. Suflita, and J. M. Tiedje. 1983. Reductive dehalogenations of halobenzoates by anaerobic lake sediment microorganisms. *Appl. Environ. Microbiol.* 45:1459–1465.

43. Howard, W. N., and J. E. Rogers. 1990. Effect of fatty acids and yeast extract on the dechlorination of 2,4-dichlorophenol in anaerobic sediments, Q-106. *Abstr. 90th Annu. Meet. Am. Soc. Microbiol.* 1990.

44. Hrudey, S. E., E. Knettig, S. A. Daignault, and P. M. Fedorak. 1987. Anaerobic biodegradation of monochlorophenols. *Environ. Tech. Lett.* 8:65–75.

45. Ide, A., Y. Niki, F. Sakamoto, I. Watanabe, and H. Watanabe. 1972. Decomposition of pentachlorophenol in paddy soil. *Agric. Biol. Chem.* 36:1937–1944.

46. Jensen, S., R. Gothe, and M. O. Kindstedt. 1972. Bis-(*p*-chlorophenyl)-acetonitrile (DDN), a new DDT derivative formed in anaerobic digested sewage sludge and lake sediment. *Nature* 240:421–423.

47. Johnson, B. T., R. N. Goodman, and H. S. Goldberg. 1967. Conversion of DDT to DDD by pathogenic and saprophytic bacteria associated with plants. *Science* 157:560–561.

48. Kallman, B. J., and A. K. Andrews. 1963. Reductive dechlorination of DDT to DDD by yeast. *Science* 141:1050–1051.

49. King, G. M. 1988. Dehalogenation in marine sediments containing natural sources of halophenols. *Appl. Environ. Microbiol.* 54:3079–3085.

50. Ko, W. H., and J. L. Lockwood. 1968. Conversion of DDT to DDD in soil and the effect of these compounds on soil microorganisms. *Can. J. Microbiol.* 14:1069–1073.

51. Kohring, G. W., J. E. Rogers, and J. Wiegel. 1989. Anaerobic biodegradation of 2,4-dichlorophenol in freshwater lake sediments at different temperatures. *Appl. Environ. Microbiol.* 55:348–353.

52. Kohring, G. W., X. Zhang, and J. Wiegel. 1989. Anaerobic dechlorination of 2,4-dichlorophenol in freshwater sediments in the presence of sulfate. *Appl. Environ. Microbiol.* 55:2735–2737.

53. Krumme, M. L., and S. A. Boyd. 1988. Reductive dechlorination of chlorinated phenols in anaerobic upflow bioreactors. *Water Res.* 22:171–177.

54. Kuhn, E. P., and J. M. Suflita. 1989. Sequential reductive dehalogenation of chloroanilines by microorganisms from a methanogenic aquifer. *Environ. Sci. Technol.* 23:848–852.

55. Kuhn, E. P., G. T. Townsend, and J. M. Suflita. 1990. Effect of sulfate and organic carbon supplements on reductive dehalogenation of chloroanilines in anaerobic aquifer slurries. *Appl. Environ. Microbiol.* 56:2630–2637.

56. Kuwatsuka, S., and M. Igarashi. 1975. Degradation of PCP in soils. II. The relationship between the degradation of PCP and the properties of soils, and the identification of the degradation products of PCP. *Soil Sci. Plant Nutr.* 21:405–414.

57. Langlois, B. E., J. A. Collins, and K. G. Sides. 1970. Some factors affecting degradation of organochlorine pesticides by bacteria. *J. Dairy Sci.* 53:1671–1675.

58a. Liu, S. M., and J. E. Rogers. 1991. Anaerobic biodegradation of chlorinated pyridines, abstr. K-3, p. 215. *Abstr. 91st Gen. Meet. Amer. Soc. Microbiol. 1991.*

58b. Madsen, T., and D. Licht. 1992. Isolation and characterization of an anaerobic chlorophenol-transforming bacterium. *Appl. Environ. Microbiol.* 58:2874–2878.

59. March, J. 1977. *Advanced Organic Chemistry: Reactions, Mechanisms and Structure.* New York: McGraw-Hill Book Co.

60. Matsumura, F., K. C. Patil, and G. M. Boush. 1971. DDT metabolized by microorganisms from Lake Michigan. *Nature* 230:325.

61. Mikesell, M. D., and S. A. Boyd. 1985. Reductive dechlorination of the pesticides 2,4-D, 2,4,5,-T, and pentachlorophenol in anaerobic sludges. *J. Environ. Qual.* 14:337–340.

62. Mikesell, M. D., and S. A. Boyd. 1986. Complete reductive dechlorination and mineralization of pentachlorophenol by anaerobic microorganisms. *Appl. Environ. Microbiol.* 52:861–865.

63. Mikesell, M. D., and S. A. Boyd. 1988. Enhancement of pentachlorophenol degradation in soil through induced anaerobiosis and bioaugmentation with anaerobic sewage sludge. *Environ. Sci. Technol.* 22:1411–1414.

64. Miskus, R. P., D. P. Blair, and J. E. Casida. 1965. Conversion of DDT to DDD by bovine rumen fluid, lake water, and reduced porphyrins. *J. Agric. Food Chem.* 13:481–483.

65a. Mohn, W. S., and K. J. Kennedy. 1992. Reductive dehalogenation of chlorophenols by *Desulfomonile tiedjei* DCB-1. *Appl. Environ. Microbiol.* 58:1367–1370.

65b. Mohn, W. W., T. G. Linkfield, H. S. Pankratz, and J. M. Tiedje. 1990. Involvement of a collar structure in polar growth and cell division of strain DCB-1. *Appl. Environ. Microbiol.* 56:1206–1211.

66a. Murthy, N. B. K., D. D. Kaufman, and G. F. Fries. 1979. Degradation of pentachlorophenol (PCP) in aerobic and anaerobic soil. *J. Environ. Sci. Health* B14:1–14.

66b. Nies, L., and T. M. Vogel. 1990. Effects of organic substrates on dechlorination of Aroclor 1242 in anaerobic sediments. *Appl. Environ. Microbiol.* 56:2612–2617.

67. Oliver, B. G., and N. D. Nicol. 1982. Chlorobenzenes in sediments, water, and selected fish from Lakes Superior, Huron, Erie, and Ontario. *Environ. Sci. Technol.* 16:532–536.

68. Parr, J. F., and S. Smith. 1974. Degradation of DDT in an Everglades muck as affected by lime, ferrous iron, and anaerobiosis. *Soil Sci.* 118:45–52.

69. Patil, K. C., F. Matsumura, and G. M. Boush. 1970. Degradation of endrin, aldrin, and DDT by soil microorganisms. *Appl. Microbiol.* 19:879–881.

70. Patil, K. C., F. Matsumura, and G. M. Boush. 1972. Metabolic transformation of DDT, dieldrin, aldrin, and endrin by marine microorganisms. *Environ. Sci. Technol.* 6:629–632.

71. Pfaender, F. K., and M. Alexander. 1972. Extensive microbial degradation of DDT *in vitro* and DDT metabolism by natural communities. *J. Agric. Food Chem.* 20:842–846.

72. Plimmer, J. R., P. C. Kearney, and D. W. Von Endt. 1968. Mechanism of conversion of DDT to DDD by *Aerobacter aerogenes*. *J. Agric. Food Chem.* 16:594–597.

73a. Quensen, J. F. III, J. M. Tiedje, and S. A. Boyd. 1988. Reductive dechlorination of polychlorinated biphenyls by anaerobic microorganisms from sediment. *Science* 242:752–754.

73b. Quensen, J. F. III, S. A. Boyd, and J. M. Tiedje. 1990. Dechlorination of four commercial polychlorinated biphenyl mixtures (Aroclors) by anaerobic microorganisms from sediments. *Appl. Environ. Microbiol.* 56:2360–2369.

74a. Reineke, W. 1994. Degradation of chlorinated aromatic compounds by bacteria: strain development. In *Biological Degradation and Bioremediation of Toxic Chemicals*. Ed. G. R. Chaudry. Portland, OR: Dioscorides Press.

74b. Roberton, A. M., and R. S. Wolfe. 1970. Adenosine triphosphate pools in Methanobacterium. *J. Bacteriol.* 102:43–51.

75. Rosenthal, I., and R. J. Lacoste. 1959. A systematic polarographic study of the aromatic chloroethanes. *J. Am. Chem. Soc.* 81:3268–3270.

76. Safe, S., S. Bandiera, T. Sawyer, L. Robertson, L. Safe, A. Parkinson, P. E. Thomas, D. E. Ryan, L. M. Reik, W. Levin, M. A. Denomme, and T. Fujita. 1985. PCBs: structure-function relationships and mechanisms of action. *Environ. Health Perspect.* 60:47–56.

77. Salkinoja-Salonen, M. S., R. Hakulinen., R. Valo, and J. Apajalahti. 1983. Biodegradation of recalcitrant organochlorine compounds in fixed film reactors. *Wat. Sci. Technol.* 14:309–319.

78. Sangodkar, U. M. X., T. L. Aldrich, R. A. Haugland, J. Johnson, R. K. Rothmel, P. J. Chapman, and A. M. Chakrabarty. 1989. Molecular basis of biodegradation of chloroaromatic compounds. *Acta Biotechnol.* 4:301–316.

79. Shelton, D. R., and J. M. Tiedje. 1984. Isolation and partial characterization of bacteria in an anaerobic consortium that mineralizes 3-chlorobenzoic acid. *Appl. Environ. Microbiol.* 48:840–848.

80. Smith, J. A., and J. T. Novak. 1987. Biodegradation of chlorinated phenols in subsurface soils. *Wat. Air Soil Pollut.* 33:29–42.

81. Stevens, T. O., T. G. Linkfield, and J. M. Tiedje. 1988. Physiological characterization of strain DCB-1, a unique dehalogenating sulfidogenic bacterium. *Appl. Environ. Microbiol.* 54:2938–2943.

82. Stevens, T. O., and J. M. Tiedje. 1988. Carbon dioxide fixation and mixotrophic metabolism by strain DCB-1, a dehalogenating anaerobic bacterium. *Appl. Environ. Microbiol.* 54:2944–2948.

83. Struijs, J., and J. E. Rogers. 1989. Reductive dehalogenation of dichloroanilines by anaerobic microorganisms in fresh and dichlorophenol-acclimated pond sediment. *Appl. Environ. Microbiol.* 55:2527–2531.

84. Suflita, J. M., A. Horowitz, D. R. Shelton, and J. M. Tiedje. 1982. Dehalogenation: a novel pathway for the anaerobic biodegradation of halogenated compounds. *Science* 218:1115–1117.

85. Suflita, J. M., J. A. Robinson, and J. M. Tiedje. 1983. Kinetics of microbial dehalogenation of haloaromatic substrates in methanogenic environments. *Appl. Environ. Microbiol.* 45:1466–1473.

86. Suflita, J. M., J. Stout, and J. M. Tiedje. 1984. Dechlorination of (2,4,5-trichlorophenoxy)acetic acid by anaerobic microorganisms. *J. Agric. Food Chem.* 32:218–221

87. Suflita, J. M., and G. D. Miller. 1985. Microbial metabolism of chlorophenolic compounds in ground water aquifers. *Environ. Toxicol. Chem.* 4:751–758.

88. Van Dort, H. M., and D. L. Bedard. 1991. Reductive *ortho* and *meta* dechlorination of a polychlorinated biphenyl congener by anaerobic microorganisms. *Appl. Environ. Microbiol.* 57:1576–1578.

89. Van Hoof, P., and C. T. Jafvert. 1991. Influence of nonionic surfactants on the anaerobic dechlorination of hexachlorobenzenes. *Symposium on Bioremediation of Hazardous Wastes: EPA's Biosystems Technology Development Program,* 16–18 April, 1991, Falls Church, Va.

90a. Weiss, U. M., I. Scheunert, W. Klein, and F. Korte. 1982. Fate of pentachlorophenol-^{14}C in soil under controlled conditions. *J. Agric. Food Chem.* 30:1191–1194.

90b. Ye, D, J. F. Quensen III, J. M. Tiedje, and S. A. Boyd. 1992. Anaerobic dechlorination of polychlorobiphenyls (Aroclor 1242). *Appl. Environ. Microbiol.* 58:1110–1114.

91. Zhang, X., and J. Wiegel. 1990. Sequential anaerobic degradation of 2,4-dichlorophenol in freshwater sediments. *Appl. Environ. Microbiol.* 56:1119–1127.

92. Zoro, J. A., J. M. Hunter, G. Eglinton, and G. C. Ware. 1974. Degradation of *p,p'*-DDT in reducing environments. *Nature* 247:235–237.

CHAPTER 8

Biodegradation of Sulfonated Aromatics

SCOTT W. HOOPER

INTRODUCTION

Sulfonated aromatics constitute a class of compounds in which a sulfone function ($-SO_3H$) is covalently attached to benzene or a benzene-like ring by a carbon–sulfur bond. This compound class is considered to be of anthropogenic origin, as only a single case of natural synthesis [that of a phenazine sulfonic acid (Aeruginosin B, 2-amino-6-carboxy-10-methyl-8-sulfophenaziniumbetain) produced by *Pseudomonas aeruginosa*] is known (20).

Commonly appearing as a byproduct of many industrial and domestic processes, sulfonated aromatics present a nearly ubiquitous pollution problem. As precursors for, and breakdown products of, dyes (e.g. Orange II), optical brighteners, pharmaceuticals (e.g. sulfanilamide), softening agents (e.g. phenol sulfonates), and preservatives (e.g. sulfanilic acid), sulfoaromatics enter the environment from a variety of industrial processes. As the linear alkylbenzene sulfonates (LAS) of detergents, large amounts of aromatic sulfonates are released through domestic use. Additionally, these compounds are generated as a byproduct of pulp-paper production and processing. Although the exact quantity of sulfoaromatics entering the environment from anthropogenic sources is unknown, a general idea of the scope of the problem emanating solely from dyestuffs is available. Of the 6.4×10^5 metric tons of dyestuffs produced in 1975, approximately 6×10^4 of those tons ended up in the environment (36). For LAS, between 8×10^5 and 1×10^6 metric tons are produced by the United States, Western Europe, and Japan each year (51). The fraction of this material that is eventually released to the environment is unknown, but is suspected to be substantial.

Although massive production of a particular class of compounds does not necessarily make those chemicals of environmental concern, sulfoaromatics have recalcitrance and toxicity factors that compound their potential for environmental insult. Early studies of sulfoaromatic degradation suggested that these compounds were recalcitrant to normal degradative processes (12), and they have been labeled as non-biodegradable as recently as 1981 (4). It is now clear, however, that most sulfoaromatics are biodegradable. The dual problems of high production and low to

---

Dr. Hooper is at the Department of Biology, University of Mississippi, University, MS 38677, U.S.A.

moderate rates of degradation have apparently resulted in the persistence of sulfoaromatics in the environment. By one coarse estimate, sulfoaromatics may represent 7 to 15% of the organic pollution load of the Rhine River (34).

SULFONATED AZO-DYES

Sulfonated azo-dyes are produced primarily for the dyeing of textiles. These dyes consist of an aromatic moiety bonded by an azo-bridge to a second aromatic function. Sulfoaromatic rings are rare but not unknown in nature [witness the above-mentioned natural phenazine sulfonate, aeruginosin B (20)], and naturally produced azo-bridges are also very rare to absent in nature. A similar structure, the azoxy-bridge, has been noted, however, in a product excreted by the fungal insect pathogen *Entomophthora virulenta* [4,4'-dicarboxyazooxybenzene (11)]. Anaerobic cleavage of the azo bond is known to occur in a variety of biological systems (1, 13, 40, 42, 44, 50). This section deals with bacterial degradation of sulfonated azo-dyes.

Pure Culture Degradation of Sulfo-azo-dyes

The xenobiotic nature of sulfonated azo-dyes is reflected by the lack of sulfo-azo-dye degraders directly isolable from nature. Because of this, adaptation of natural strains to growth on sulfo-azo-dyes is usually necessary. Kulla et al. (26, 28) have reported the development of azo-dye degrading cultures by long-term adaptation of chemostat cultures. Nonsterile culturing of a 4,4'-dicarboxyazobenzene-(DCAB)-degrading *Flavobacterium* strain on limiting amounts of DCAB, coupled with increasing input of the Orange II carboxy dye, resulted in the *Flavobacterium* strain eventually being out-competed by a *Pseudomonas*-like organism. This pseudomonad (KF46) was able to utilize Orange II carboxy as the sole carbon, energy, and nitrogen sources. A parallel experiment resulted in an organism (KF24) able to degrade a similar dye, Orange I carboxy. Although both organisms were able to use the carboxylated azo-dyes as sole carbon, energy, and nitrogen sources, they were unable to use the sulfonated forms of the dyes for growth (Figure 1). The strains did, however, cleave the azo-bridge of the sulfo-azo-dyes, thus resulting in the production of a mono-cyclic sulfo-aromatic (sulfanilic acid) and either 1-amino-4-naphthol or 1-amino-2-naphthol (from Orange I and Orange II, respectively). Neither azo-reductase induced by one Orange dye had activity against the other.

The azo-reductases of both strains have been purified and biochemically examined. The first characterized reductase, Orange II azo-reductase [NAD(P)H: 1-(4'-sulfophenylazo)-2-naphthol oxidoreductase], is an oxygen-insensitive monomeric protein with a molecular weight of 30,000. It has an absolute requirement for a hydroxyl group in the 2-position of the naphthol ring (54). Similarly, the Orange I azo-reductase is also an oxygen-insensitive monomeric protein, but it is smaller (21,000 Daltons) and has a requirement for the hydroxyl function to be in the 4-position on the naphthol ring (55). Immunologically, the two azo-reductases do not antigenically cross react with each other, but do cross react with precursor strains that were isolated from the chemostat cultures as the KF24 and KF46 strains were being selected. This cross reaction of the antibodies with the precursor strains, but not the final strains, may indicate that the enzymes have a common ancestry that has become antigenically obscured due to selective pressures (55).

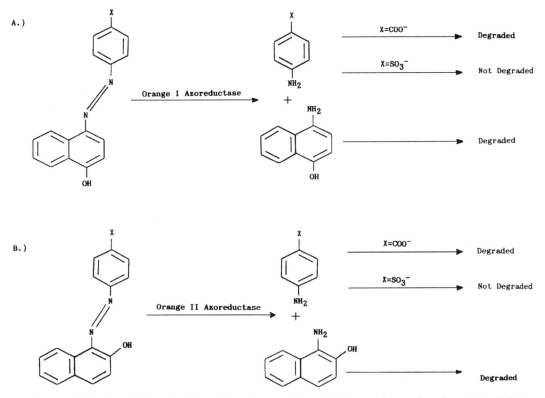

Figure 1. Biodegradation of sulfonated and non-sulfonated Orange dye analogs by strains KF24 (part A) and KF46 (part B) through the action of azoreductases (28). In both cases, the strains were able to grow on the respective carboxylated analogs, but were only able to transform the sulfonated dye to sulfanilic acid and an aminonaphthol.

The result of the cleavage of the azo-bridge of either Orange I or II is the release of sulfanilic acid (sulfanilate) and the production of an amino-naphthol. Sulfanilate has antibacterial activity. Indeed, azo-dyes have been noted to have therapeutic value due to the production of sulfanilic acid from ingested azo-dyes (49) via the action of detoxifying azo-reductase systems of the liver (37, 38).

Even if the degradative bacteria are able to withstand the effects of the sulfanilate, additional metabolic problems may arise. As an example, selective permeability of azo-dyes into the cell have been noted (26, 35). A lack of entry of the xenobiotic into the cell precludes its metabolism except by excreted enzymes or reactive substances (neither has been directly implicated as a widespread mechanism for azo-dye degradation). Additional potential metabolic problems are the creation of deadend metabolites of sulfanilic acid (27) and the inhibition of further metabolism due to a high intracellular sulfo-aromatic concentration (27). These results were not noted when the nonsulfonated carboxy-analogues were metabolized.

Mixed Culture Degradation of Sulfo-azo-dyes

The first report of the complete biological mineralization of a sulfonated azo-dye was published in 1991 (19). In this study, a consortium able to grow on 6-aminonaphthalene-2-sulfonate (see the discussion on sulfonate naphthalene degradation below) was found to be able to mineralize the sulfonated dye Mordant Yellow 3 when the culture was first held under anaerobic conditions, and then shifted to aerobic conditions. The ability of this culture to degrade Mordant Yellow 3 was not entirely unexpected, as the expected metabolites from azo-bridge reduction (6-aminonaphthalene-2-sulfonate and 5-aminosalicylate) were previously shown to be mineralizable by two members of the consortium (39). The 6-aminonaphthalene-2-sulfonate degrader, BN6, was shown in the Mordant Yellow 3 study to have significant azo-reductase activity. Therefore, the initial reduction of the azo-bridge by BN6 under anaerobic conditions produced metabolites that could be mineralized when the culture was shifted to aerobic culture conditions. This resulted in the complete mineralization of the sulfonated dye. Similar sequential anaerobic/aerobic processing of sulfonated dyes may well prove to be the method of choice for the disposal of waste sulfonated azo-dyes.

SULFONATED NAPHTHALENES

Naphthylsulfonates are released to the environment from degradative, synthetic, and industrial processes. Since they are precursors and breakdown products for many of the azo-dyes, naphthylsulfonates are released from both ends of the dye processing industry. Additionally, naphthylsulfonates are widely used as wetting and emulsifying agents because of their surface-active properties.

Naphthalene sulfonate-degrading bacteria have been isolated by continuous enrichment of an activated sewage sludge–derived naphthalene-degrading consortium (5). From this consortium, a 2-naphthalenesulfonate (2NS) degrading strain (*Pseudomonas testosteroni* sp. A3) was enriched through stepwise increases of the concentration of 2NS. The ability of the strain to utilize 2NS could be lost by nonselective culturing of the A3 strain (50% loss after 10 generations). Addition of mitomycin C increased the rate of 2NS utilization loss to 80%, thus indicating possible plasmid mediation of the pathway. The utilization of 2NS could be restored at a frequency of 0.6×10^{-6} to 1.0×10^{-6} by filter matings with a 2NS$^+$ strain, thereby suggesting that a 2NS to salicylate pathway is plasmid encoded in this strain (5).

Replacement of 2NS in the 2NS-adapted culture with 1-naphthalenesulfonate (1NS) enriched the culture for organisms able to use 1NS as a carbon source but not as a sole carbon source. One of these strains, C22, only used 1NS after the complete exhaustion of 2NS from 1NS-2NS mixtures (5). Examination of the characteristics of the C22 and A3 strains showed that they were closely related and that both strains utilized the naphthalenesulfonates via identical enzymes with salicylate and gentisate as intermediates (6). The organisms were unusual (with respect to other known naphthalene degraders) because they had high gentisate 1,2-dioxygenase and low catechol dioxygenase activities when grown on the naphthalenic substrates. Both the sulfonated naphthalenes and naphthalene efficiently induced the pathway enzymes but, when simultaneously present, the compounds were sequentially used in the order naphthalene, 2NS, 1NS (6). Presumably, this sequential utilization of naphthalenes is due to differences in uptake specificity, oxygenation rate, or elimination of the sulfone function from the unstable naphthalenesulfonate-dihydrodiol, since following sulfone elimination, all three parent compounds would yield the same intermediate. Of the three listed options, differences in uptake specificities would seem the most probable (see the section on monocyclic sulfoaromatics for further discussion of sulfo-aromatic transport selectivity).

Zürrer and co-workers (56) were able to demonstrate that sulfonated naphthalenes and benzenesulfonic acids could serve as sole sources of sulfur for growth. Of the 18 prepared enrichments on substituted naphthalene sulfonates, 15 were positive for utilization of these substances as sulfur sources. This is somewhat surprising, because isolating degradative strains by using these compounds as carbon sources is apparently very difficult, hence a high proportion of positive cultures for sulfur utilization is the opposite of what is expected. Furthermore, four of the isolated strains were reported to utilize a wide range of sulfo-aromatics as sulfur sources. Based on the incorporation of ^{18}O into naphthol intermediates, a broad-substrate-specificity mono-oxygenolytic cleavage of the C–S bond was proposed to be the mechanism by which sulfur was made available to the bacterium. This again suggests that elimination of the sulfone function is the initial step in the degradation of the sulfonaphthalenes.

Combinations of substituents on the bicyclic ring of naphthalene present additional degradative challenges to microorganisms. Enrichments of water samples from the Elbe River (Germany) have yielded a two-membered consortium (both members of the genus *Pseudomonas*) able to degrade an azo-dye precursor, 6-aminonaphthalene-2-sulfonic acid (6A2NS) (39). The first strain, BN6, could grow on 6A2NS but in pure culture accumulated a polymeric material that consisted of polymerized 5-aminosalicylate. The second member of the consortium, strain B9, mineralized 5-aminosalicylate thereby eliminating the accumulation of the black polymer. Following a long adaptation period, strain BN6 was adapted to conversion of all of the naphthalene-2-sulfonates having $-NH_2$ or $-OH$ functions in the 5, 6, 7, or 8 positions. The intermediates detected from the differing starting compounds apparently arise via the same basic pathway. As with the A3 and C22 strains, the sulfone moiety was eliminated in the initial dioxygenation step of the catabolic pathway (39).

A partial exception to the apparent general mechanism by which sulfonaphthalenes are metabolized via initial elimination of the sulfonate function was noted for disulfonic naphthalene degradation by a *Moraxella* strain (53). This strain was capable of growth on 2,6- and 1,6-naphthalene disulfonic acids (2,6NDS and 1,6NDS, respectively). The organism made the initial attack on the ring at one sulfonate function, eliminating it, as in monosulfonic naphthalenes. Due to the presence of a sulfonate on the opposite ring, however, a resulting metabolite was 5-sulfosalicylic acid (5SS). Wittich et al. (53) proposed that the sulfone group was eliminated from 5SS by an oxygenolytic mechanism, because they noted that a small amount of gentisate (the expected product from an oxygenolytic elimination) accumulated in the medium.

LINEAR ALKYLBENZENE SULFONATES

Several excellent comprehensive reviews of the linear alkylbenzene sulfonates (LAS) are available in the literature (e.g. 8, 18, 45). For this reason, this section provides a brief overview of LAS degradation that focuses primarily on the degradation of the alkyl side chain. The following section (Benzene Sulfonates) deals with degradation of the sulfonated benzene compounds that result from elimination of side chains.

Although linear alkylbenzene sulfonates are now one of the most widely utilized surface-active agents in detergents, they have not always been used so extensively. The rise of LAS owes a great deal to the failure of some closely related compounds, the alkylbenzene sulfonates (ABS), to be generally biodegradable. ABS [or tetra-

propylenebenzene sulfonates [TBS]] are similar to LAS except that the side chain of
ABS is branched, thus rendering them less accessible to degradation. In the early
1950s, ABS were finding widespread use in detergent formulations due to the availability of cheap alkylbenzenes for use as synthesis precursors. By the mid-1950s, it was
recognized that ABS were not degrading in sewage-treatment systems and were exiting
those systems in concentrations sufficiently high to cause foaming of the receiving
waters. Through the mid-1960s, detergent formulations were voluntarily changed from
ABS- to LAS-containing formulations. Although LAS mixtures are more degradable
than the previous ABS compounds, LAS are still not an environmentally transparent
choice for use and release to the environment.

LAS are more easily degraded because the side-chain, which is basically linear and
resembles a simple alkane or carbohydrate, is more accessible to degradation by
general mechanisms of metabolism. Elimination of the branch-points in the molecule
also eliminates the need for specialized enzymes to deal with the branched chains. In
general, initial attack on the linear side chain is by ω-oxidation through mono-oxygenase attack. Subsequent reduction of side chain length generally occurs through β-
oxidation. Substitutions at the β-carbon or dimethyl substitutions anywhere on the
chain prevent β-oxidations. In these cases, oxidations can proceed via an α-oxidation,
by removing a single carbon from the side chain. By performing successive α- and β-
oxidations, LAS degradation can occur (Figure 2); however, certain combinations of
substitutions in LAS side-chains cannot be degraded by α- and β-oxidations alone.
LAS molecules having these configurations are known to biodegrade, so additional
degradative mechanisms (e.g. hydrolases, broad-substrate range mono-oxygenases)
must also be used.

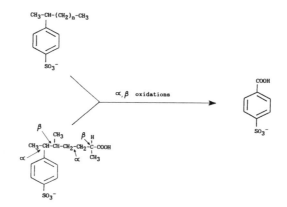

Figure 2. Side chain degradation of simple linear alkylbenzenesulfonates.

In addition to the number and length of the branchings of the alkyl side chain, the
degradation of LAS is dependent on the position of the sulfonate group on the benzene
ring. In systems having a greater size and diversity of biomass, the effects of sulfonate
position appear to be less pronounced. Presumably this is a reflection of the diversity
of enzymatic activities present in the biomass. A recent study of LAS degradation by
the diverse microbiota of pond detritus demonstrated mineralization of LAS (15). The
authors suggested that sorptive and degradative activities associated with detritus may
make it significant for the removal of LAS in detritus-rich environments. Similar
physico-biological mechanisms probably account for the LAS removal from activated
sludges, soil shake flasks, etc. that have been previously noted by other authors (e.g. 8,
18, 45).

BENZENE SULFONATES

In many ways, degradation of benzene sulfonates is an extension of the degradation of azo-dyes, sulfonated naphthalenes, and LAS, as benzene sulfonates are potential metabolites of such compounds (Figure 3). Depending on the extent, type, and configuration of the substitutions on the benzene ring, the produced benzene sulfonate can have altered toxicological significance and can be more persistent in nature than the parent compound.

Toluene and benzene sulfonates are often used as "simple case" models for sulfonoaromatic degradation. In mixed culture, sewage sludge microflora that have not been adapted to sulfonated aromatics appear to be generally incapable of degrading them, but adapted sludges have been shown to degrade the model compounds at least partially (21, 33). This difference in degradation appears to be due to the addition of the sulfonate function, because its presence drastically reduces the rate of degradation (as compared to the rate for the nonsubstituted analog) by soil microflora (2).

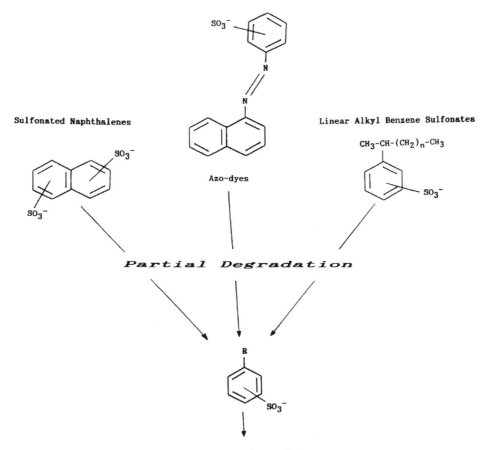

Figure 3. Benzenesulfonates are central to many sulfoaromatic degradation pathways. The degradation of complex sulfonated aromatics usually results in a mono-cyclic sulfoaromatic intermediate. That intermediate is then typically desulfonated by the action of a dioxygenase. R is a substituent group; the actual type is dependent on the starting compound.

Pure Culture Degradation of Benzene Sulfonates

With pure cultures, sulfonated aromatics were initially thought to be nondegradable (e.g. 12); however, single strains degrading a number of forms of sulfonated benzenes have been isolated.

Cain and Farr (9) reported degradation of benzensulfonate by *Pseudomonas aeruginosa* strain A and other *P. aeruginosa* strains able to degrade p-toluenesulfonate by way of 4-methyl catechol. Shortly thereafter, reports documented the degradation of p-toluenesulfonate via 3-methyl catechol by a pseudomonad (17) and benzenesulfonate by a strain of *Pseudomonas testosteroni* (41). The first report of desulfonation by extracts of active cells was presented by Endo et al. in 1977 (14). Because it is the sulfonate function that transforms the sulfonoaromatics into xenobiotic compounds, desulfonation is the critical step in compound detoxification. The sulfone group is released as sulfite, and the sulfite is then oxidized either spontaneously or by an inducible sulfite oxidase to form sulfate (10, 52). Liberation of sulfite and its subsequent conversion to sulfate can be monitored by noting the yellow coloration of Ellman's reagent in the presence of sulfite (25).

Additional substituents that share the aromatic ring have the potential for further complicating the degradative process. In the case of sulfone and amino substitutions in the same ring, however, degradation by pure cultures has been detected for each of the isomers. For 2-aminobenzenesulfonate (orthanilic acid), Thurnheer et al. (47) isolated two strains from enrichments that were able to grow on the compound. Other strains isolated during the enrichment procedure were able to grow on other sulfonated substrates (e.g. two strains grew on benzenesulfonate, two grew on 4-amino-benzenesulfonate, three grew on 4-hydroxybenzenesulfonate, four grew on 4-methyl-benzenesulfonate, two grew on 4-sulfobenzoic acid, and one strain grew on 4-amino-benzenesulfonamide—this strain was indirectly enriched). Interestingly, Thurnheer et al (47) were unable to isolate strains able to grow on 3-aminobenzenesulfonate. Strains able to degrade this compound were later reported by Locher et al. (32). Furthermore, the Thurnheer strains degrading 4-aminobenzenesulfonate were not reported to give a colored product. This is in contrast to the results noted by Feigel and Knackmuss (16) for 4-aminobenzenesulfonate-degrading strain S1. This strain converted 4-amino-benzenesulfonate to catechol-4-sulfonate, which is intensely violet colored. Strain S1 appears to be an exception to the general mechanism of sulfonate release prior to ring cleavage.

Although 4-methylbenzenesulfonate (p-toluenesulfonic acid) had been noted to act as a growth substrate before, the degradation pathway had been described as occurring only via 3-methyl (17) or 4-methyl catechol (9). A third mechanism for degradation, that of side-chain oxidation followed by *meta*-ring cleavage with concomitant loss of the sulfonate group, was described in 1989 (30). This pathway, shown in Figure 4, is initiated by a monooxygenase, yielding an alcohol that is then dehydrogenated to first form the aldehyde, then the carboxylate form of the sulfobenzenoid (p-sulfobenzoic acid). The next step, dioxygenation [carried out by 4-sulfobenzoate 3,4-dioxygenase (29)], is the critical step for detoxification of the molecule, as at this step the sulfone group leaves in the form of sulfite. The product of this dioxygenation is the nonxenobiotic compound protocatechuic acid. Through genetic manipulation of the genes for this pathway, it was determined that the nonsulfonated analog compound, p-toluic acid, was probably metabolized via the identical pathway, with the exception of dioxygenation (22). Subsequent examination of the enzymes' activities confirmed that this is the case (31). This process of side-chain oxidation had not been previously described for the metabolism of p-toluic acid. The relationship between the metabolism

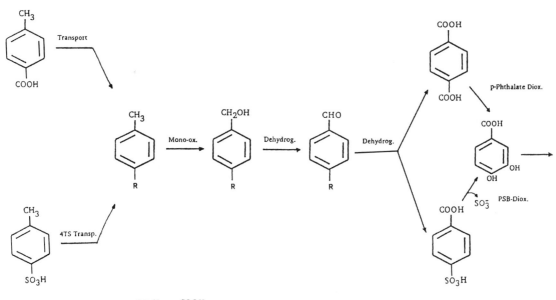

R = -SO₃H or -COOH

Figure 4. Degradation of 4-toluenesulfonate and *p*-toluic acid by *Comamonas testosteroni* T2. The monooxygenase, alcohol dehydrogenase, and aldehyde dehydrogenase are able to modify either compound. At the point of dioxygenation, separate specific enzymes are used to dioxy-genate *p*-phthalate (*p*-phthalate dioxygenase) and *p*-sulfobenzoic acid (PSB dioxygenase). The resulting compound from both reactions is protocatechuic acid. This nonxenobiotic is subsequently mineralized. R = $-SO_3$ or $-COOH$.

of the sulfonated and nonsulfonated forms of 4-methylbenzesulfonate (4TS) by *Pseudomonas testosteroni* T2 is also depicted in Figure 4.

Given that the type strain of this organism does not metabolize either 4-methylbenzesulfonate or *p*-toluic acid, and that at least some of the genes involved in this pathway have been cloned from a plasmid from strain T2 (22), it appears that the 4TS side-chain oxidative pathway of strain T2 is plasmid encoded. Plasmid-associated mineralization of orthanilic acid has previously been reported (24). Given the abundance of plasmid-encoded xenobiotic degradative pathways (43) and the general architecture and behavior of catabolic plasmids (cf. 7), it will not be surprising to discover that additional sulfonate degradation systems are plasmid-associated.

In addition to the amino- and methyl- benzenesulfonate-degrading organisms, strains degrading 3-nitrobenzenesulfonate [N–1, (32)] and 4-chlorobenzenesulfonate [Cl–F, (3)] have been isolated from enrichments. The N–1 strain had a narrow range of substrate utilization, and was unable to use any of the monoaminobenzenesulfonates or the 2- or 4- nitro- isomers of 3-nitrobenzenesulfonate. On 3-nitrobenzenesulfonate, the growth yield for this bacterium was 3.0 grams per mole of carbon. The Cl–F strain was the only 4-chlorobenzenesulfonate-degrading bacterium isolated from the continuous culture reactor enrichments described below. It bears a great deal of similarity to an orthanilic-acid-degrading organism [strain O–1, (24, 47)] originally residing in the reactor (3). Strain O–1 has been shown to have selective permeability for various sulfonated aromatics that is thought to be due to substrate-specific transport (47, 48). The possibility of mutations in specific transport could provide a parsimonious explanation for the apparent adaptation of this strain to 4-chlorobenzenesulfonate.

Mixed Culture Degradation of Benzene Sulfonates

Although the isolation of specific strains for the metabolism of specific compounds does allow one to make simple predictions of compound fate in the environment, real-life systems are not one-compound, one-organism systems. To address this problem, Thurnheer et al. (46) set up a co-culture of five defined bacteria that were able to degrade at least seven substituted sulfonated benzenes. The reactor was run continuously for over two and one half years. After 18 months of operation, it was noted that the degradative capacity of the reactor strains had shifted: many isolates had greater degradative capacity than any of the originally introduced bacteria. Since there had been no sign of contamination of the reactor by external microorganisms, Thurneer et al (46) hypothesized that the newly appearing strains had arisen from those originally present in the reactor. The reactor system also showed an increased ability to degrade compounds in the feed mixture. As an example, even though the co-culture had not been previously exposed to 4-chlorobenzenesulfonate, this compound was added on day 903 of continuous culturing and disappeared within 15 days. The CL–F strain was subsequently isolated from this enrichment. These experiments demonstrated that mixtures of sulfonated benzenes can be efficiently degraded in reactor systems.

SUMMARY AND CONCLUSIONS

The physico-chemical properties of a compound directly affect its bioavailability. With highly hydrophobic compounds, partitioning of the compound away from the aquatic environment removes it from the milieu of microorganisms, thus resulting in recalcitrance due to nonavailability [e.g. PCBs, (23)]. Fortunately, the addition of the sulfonate function to sulfonated aromatics does not impart a highly hydrophobic nature. This results in higher solubility and therefore greater bioavailability. Perhaps primarily for this reason, a number of strains degrading a number of less complex sulfonated aromatics have been described.

Although exceptions have been noted, degradation of sulfonated aromatics appears to follow a general pattern of dioxygenation of the ring with simultaneous release of the sulfone as sulfite. This process converts the sulfonated aromatic to a less unusual compound that theoretically should be degradable by a wider range of microorganisms. Degradation of more complex sulfoaromatics also tends to produce simpler sulfoaromatics. In the example of Orange II dye and disulfononaphthalene degradation, the degradation products include sulfanilic acid and sulfosalicylic acid, respectively. Reduction of the side-chain length of many LAS molecules results in products similar to p-sulfotoluene. Since monocyclic sulfonated aromatics appear to be central to the degradation pathways for many of the more complex sulfoaromatics, monocyclic sulfoaromatic-degrading microorganisms would seem to provide a "best bet" for the development (either by selection or genetic manipulation) of organisms able to degrade more complex sulfoaromatics such as the components of ABS.

The successful degradation of mixtures of sulfonated aromatics by sludges, detritus communities, and reactor systems bodes well for the development and improvement of disposal systems for sulfonated aromatics. Indeed, based on the ability of many investigators to isolate sulfonate degraders from acclimated sewage systems, it would appear that a substantial amount of degradation capacity exists in seasoned systems. A near universal failure of enrichments to turn up sulfonate degraders from non-acclimated systems suggests, however, that sulfonate-degrading organisms are not ubiquitously distributed. Perhaps by using labeled sulfonate degradation genes, tracking systems could be developed that would allow the quantification of the potential of a system for sulfonated aromatic degradation. These gene probes would be particularly useful to system operators managing systems that are occasionally dosed by sulfoaromatics, both for monitoring the remaining capacity and for monitoring the efficacy of inoculations with sulfonate-degrading microorganisms.

Given proper precaution and preparation, simpler sulfonated aromatics can be disposed of in waste treatment systems. Through a better understanding of the biology of the sulfonated aromatic degrading subpopulations of these systems, better treatment schemes will be developed. The result, it is hoped, will lessen the impact of man's activities on the environment.

Acknowledgments

This work was supported in part by the Department of Biology of The University of Mississippi. The author thanks Thomas Leisinger and Alasdair M. Cook for reviewing this manuscript prior to its submission.

LITERATURE CITED

1. Adamson, R. H., R. L. Dixon, F. L. Francis, and D. P. Rall. 1965. Comparative biochemistry of drug metabolism by azo and nitro reductase. *Proc. Natl. Acad. Sci. USA* 54:1386–1391.

2. Alexander, M., and B. K. Lustigman. 1966. Effect of chemical structure on microbial degradation of substituted benzenes. *J. Agric. Food Chem.* 14:410–413.

3. Arnold, P., A. Jenny, T. Lüthi, and K. Walliman. 1989. *Biodegradation Sulfonierter Aromaten.* Zürich, Switzerland: POST Schlussbericht, Mikrobiologisches Institut, Swiss Federal Institute of Technology.

4. Bretcher, H. 1981. Waste disposal in the chemical industry. In *Microbial Degradation of Xenobiotics and Recalcitrant Compounds.* Ed. T. Leisinger, A. M. Cook, R. Hütter, J. Nüesch. London: Academic Press. 65–74.

5. Brilon, C., W. Beckmann, M. Hellwig, and H.-J. Knackmuss. 1981. Enrichment and isolation of naphthalenesulfonic acid utilizing pseudomonads. *Appl. Environ. Microbiol.* 42:39–43.

6. Brilon, C., W. Beckmann, and H.-J. Knackmuss. 1981. Catabolism of naphthalenesulfonic acids by *Pseudomonas* sp. A3 and *Pseudomonas* sp. C22. *Appl. Environ. Microbiol.* 42:44–55.

7. Burlage, R. S., S. W. Hooper, and G. S. Sayler. 1989. The TOL (pWW0) catabolic plasmid. *Appl. Environ. Microbiol.* 55:1323–1328.

8. Cain, R. B. 1981. Microbial degradation of surfactants and builder compounds. In *Microbial Degradation of Xenobiotics and Recalcitrant Compounds.* Ed. T. Leisinger, A. M. Cook, R. Hütter, and J. Nüesch. London: Academic Press. 325–370.

9. Cain, R. B., and D. R. Farr. 1968. Metabolism of arylsulphonates by micro-organisms. *Biochem. J.* 106:859–877.

10. Cain, R. B., A. J. Willetts, and J. A. Bird. 1972. Surfactant biodegradation: metabolism and enzymology. In *Biodeterioration of Materials,* Vol. 2, *Proceedings of the Second International Biodeterioration Symposium.* Ed. A. H. Waters, E. H. Hueck-van der Plas. London: Applied Science Publishers. 136–144.

11. Claydon, N. 1978. Insecticidal secondary metabolites from entomogenous fungi: *Entomoph-thora virulenta. J. Invert. Pathol.* 32:319–324.

12. Czekalowski, J. W., and B. Skarzynsky. 1948. The breakdown of phenols and related compounds by bacteria. *J. Gen. Microbiol.* 2:231–238.

13. Dubin, P., and K. L. Wright. 1975. Reduction of azo food dyes in cultures of *Proteus vulgaris.* Xenobiotica 5:563–571.

14. Endo, K., H. Kondo, and M. Ishimoto. 1977. Degradation of benzenesulfonate to sulfite in bacterial extract. *J. Biochem.* 82:1397–1402.

15. Federle, T. W., and R. M. Ventullo. 1990. Mineralization of surfactants by the microbiota of submerged plant detritus. *Appl. Environ. Microbiol.* 56:333–339.

16. Feigel, B. J., and H.-J. Knackmuss. 1988. Bacterial catabolism of sulfanilic acid via catechol-4-sulfonic acid. *FEMS Microbiol. Lett.* 55:113–118.

17. Focht, D. D., and F. D. Williams. 1970. The degradation of *p*-toluenesulphonate by a *Pseudomonas. Can. J. Microbiol.* 16:309–316.

18. Gledhill, W. E. 1974. Linear alkylbenzene sulfonate: biodegradation and aquatic interaction. *Adv. Appl. Microbiol.* 17:265–293.

19. Haug, W., A. Schmidt, B. Nörtemann, D. C. Hempel, A. Stoltz, and H.-J. Knackmuss. 1991. Mineralization of the sulfonated azo dye Mordant Yellow 3 by a 6-aminonaphthalene-2-sulfonate-degrading bacterial consortium. *Appl. Environ. Microbiol.* 57:3144–3149.

20. Herbert, R. B., and F. G. Holliman. 1964. Aeruginosin B—a naturally occuring phenazinesulphonic acid. *Proc. Chem. Soc.* 1964:19.

21. Heukelekian, H., and M. C. Rand. 1955. Biochemical oxygen demand of pure organic compounds. *Sewage Ind. Wastes* 27:1040–1053.

22. Hooper, S. W., H. H. Locher, A. M. Cook, and T. Leisinger. 1990. Genetic and functional analysis of the 4-toluene sulfonate pathway of *Comamonas (Pseudomonas) testosteroni* T2. *Abstr. 90th Annu. Meet. Am. Soc. Microbiol.,* 1990. Abstr. Q72, p. 300. Washington, D.C.: American Society for Microbiology.

23. Hooper, S. W., C. A. Pettigrew, and G. S. Sayler. 1990. Ecological fate, effects and prospects for the elimination of environmental polychlorinated biphenyls (PCBs). *Environ. Toxicol. Chem.* 9:655–667.

24. Jahnke, M., T. ElBanna, R. Klintworth, and G. Auling. 1990. Mineralization of orthanilic acid is a plasmid-associated trait in *Alcaligenes* sp. O-1. *J. Gen. Microbiol.* 136:2241–2249.

25. Johnston, J. B., K. Murray, and R. B. Cain. 1975. Microbial metabolism of aryl sulphonates. A reassessment of colorimetric methods for the determination of sulphite and their use in measuring desulphonation of aryl and alkylbenzene sulphonates. *Antonie van Leeuwenhoek* 41:493–511.

26. Kulla, H. G. 1981. Aerobic bacterial degradation of azo dyes. In *Microbial Degradation of Xenobiotics and Recalcitrant Compounds.* Ed. T. Leisinger, A. M. Cook, R. Hütter, J. Nüesch. London: Academic Press. 387–399.

27. Kulla, H. G., F. Klausener, U. Meyer, B. Lüdeke, and T. Leisinger. 1983. Interference of aromatic sulfo groups in the microbial degradation of the azo dyes Orange I and Orange II. *Archiv. Microbiol.* 135:1–7.

28. Kulla, H. G., R. Krieg, T. Zimmerman, and T. Leisinger. 1984. Experimental evolution of azo dye-degrading bacteria. In *Current Perspectives in Microbial Ecology.* Ed. M. J. Klug and C. A. Reddy. Washington D.C.: American Society for Microbiology. 663–667.

29. Locher, H. H., T. Leisinger, and A. M. Cook. 1991. 4-Sulphobenzoate 3,4-dioxygenase. Purification and properties of a desulphonative two-component enzyme system from *Comamonas testosteroni* T-2. *Biochem. J.* 274:833–842.

30. Locher, H. H., T. Leisinger, and A. M. Cook. 1989. Degradation of *p*-toluenesulphonic acid via sidechain oxidation, desulphonation and *meta* ring cleavage in *Pseudomonas (Comamonas) testosteroni* T-2. *J. Gen. Microbiol.* 135:1969–1978.

31. Locher, H. H., C. Malli, S. W. Hooper, T. Vorherr, T. Leisinger, and A. M. Cook. 1991. Degradation of p-toluic acid (p-toluenecarboxylic acid) and p-toluenesulphonic acid via oxygenation of the methyl sidechain is initiated by the same set of enzymes in *Comamonas testosteroni* T-2. *J. Gen. Microbiol.* 137:2201–2208.

32. Locher, H. H., T. Thurnheer, T. Leisinger, and A. M. Cook. 1989. 3-Nitrobenzenesufonate, 3-aminobenzenesufonate, and 4-aminobenzenesulfonate as sole carbon sources for bacteria. *Appl. Environ. Microbiol.* 55:492–494.

33. Ludzack, F. J., and M. B. Ettinger. 1960. Chemical structures resistant to aerobic biochemical stabilization. *J. Water Pollut. Control Fed.* 32:1173–1200.

34. Malle, K. G. 1978. Wie schmutzig ist der Rhein? *Chem. Unserer Zeit* 12:111–122.

35. Mechsner, K., and K. Wuhrmann. 1982. Cell permeability as a rate limiting factor in the microbial reduction of sulfonated azo dyes. *Eur. J. Appl. Microbiol. Biotechnol.* 15:123–126.

36. Meyer, U. 1981. Biodegradation of synthetic organic colorants. In *Microbial Degradation of Xenobiotics and Recalcitrant Compounds*. Ed. T. Leisinger, A. M. Cook, R. Hütter, J. Nüesch. London: Academic Press. 371–385.

37. Mueller, G. C., and J. A. Miller. 1948. The metabolism of 4-dimethylaminoazobenzene by rat liver homogenates. *J. Biol. Chem.* 176:535–544.

38. Mueller, G. C., and J. A. Miller. 1949. Reductive cleavage of 4-dimethylaminoazobenzene by rat liver tissue. Intracellular distribution of the enzyme system and its requirements for triphosphopyridine nucleotide. *J. Biol. Chem.* 180:1125–1136.

39. Nörtemann, B., J. Baumgarten, H. G. Rast, and H.-J. Knackmuss. 1986. Bacterial communities degrading amino- and hydroxynaphthalene-2-sulfonates. *Appl. Environ. Microbiol.* 52:1195–1202.

40. Rafii, F., W. Franklin, and C. E. Cerniglia. 1990. Azoreductase activity of anaerobic bacteria isolated from human intestinal microflora. *Appl. Environ. Microbiol.* 56:2146–2151.

41. Ripin, M. J., K. F. Noon, and T. M. Cook. 1971. Bacterial metabolism of aryl sulphonates. 1. Benzenesulfonate as a growth substrate for *Pseudomonas testosteroni* H-8. *Appl. Environ. Microbiol.* 21:495–499.

42. Roxon, J. J., A. J. Ryan, and S. E. Wright. 1967. Enzymatic reduction of tartrazine by *Proteus vulgaris* from rats. *Food Cosmet. Toxicol.* 5:645–656.

43. Sayler, G. S., S. W. Hooper, A. C. Layton, and J. M. H. King. 1990. Catabolic plasmids of environmental and ecological significance. *Microb. Ecol.* 19:1–20.

44. Scheline, R. R., R. T. Nygaard, and B. Longberg. 1970. Enzymatic reduction of the azo dye, Acid Yellow, by extracts of *Streptococcus faecalis* isolated from rat intestine. *Food Cosmet. Toxicol.* 8:55–58.

45. Swisher, R. D. 1987. *Surfactant Biodegradation*. 2nd Edition, New York: Marcel Dekker.

46. Thurnheer, T., A. M. Cook, and T. Leisinger. 1988. Co-culture of defined bacteria to degrade seven sulfonated aromatic compounds: efficiency, rates and phenotypic variations. *Appl. Microbiol. Biotechnol.* 29:605–609.

47. Thurnheer, T., T. Köhler, A. M. Cook, and T. Leisinger. 1986. Orthanilic acid and analogues as carbon sources for bacteria: growth physiology and enzymic desulphonation. *J. Gen. Microbiol.* 132:1215–1220.

48. Thurnheer, T., D. Zürrer, O. Höglinger, T. Leisinger, and A. M. Cook. 1990. Initial steps in the degradation of benzene sulfonic acid, 4-toluene sulfonic acids, and orthanilic acid in *Alcaligenes* sp. strain O-1. *Biodegradation* 1:55–64.

49. Tréfouël, J., J. Tréfouël, F. Nitti, and D. Bovet. 1935. Action of p-aminophenylsulfamide in experimental *Streptococcus* infections of mice and rabbits. *C. R. Soc. Biol.* 120:756–758.

50. Walker, R. 1970. The metabolism of azo compounds: a review of the literature. *Food Cosmet. Toxicol.* 8:659–676.

51. Werdelmann, B. W. 1984. Tenside in unserer Welte—heute und morgen. 3–21. In

Proceedings of the Second World Surfactants Congress, Vol. 1. Paris: Syndicat National des Fabricants d'Agents de Surface et de Produits Auxiliaires Industriels.

52. Willetts, A. J., and R. B. Cain. 1972. Microbial metabolism of alkylbenzene sulphonates. Bacterial metabolism of undecylbenzene-p-sulphonate and dodecylbenzene-p-sulphonate. *Biochem. J.* 129:389–402.

53. Wittich, R. M., H. G. Rast, and H.-J. Knackmuss. 1988. Degradation of naphthalene-2,6- and naphthalene-1,6-disulfonic acid by a *Moraxella* sp. *Appl. Environ. Microbiol.* 54:1842–1847.

54. Zimmerman, T., H. G. Kulla, and T. Leisinger. 1982. Properties of purified Orange II azo-reductase, the enzyme initiating azo dye degradation by *Pseudomonas* KF46. *Eur. J. Biochem.* 129:197–203.

55. Zimmerman, T., F. Gasser, H. G. Kulla, and T. Leisinger. 1984. Comparison of two bacterial azo-reductases acquired during adaptation to growth on azo dyes. *Archiv. Microbiol.* 138:37–43.

56. Zürrer, D., A. M. Cook, and T. Leisinger. 1987. Microbial desulfonation of substituted naphthalenesulfonic acids and benzenesulfonic acids. *Appl. Environ. Microbiol.* 53:1459–1463.

CHAPTER 9

The Biodegradation of Morpholine

W. A. VENABLES and J. S. KNAPP

Abstract. Morpholine is a simple, heterocyclic, xenobiotic compound that is used widely and in large quantities by industry and is a common constituent of industrial effluents. Owing to its potential for reaction with nitrite to form the potent carcinogen N-nitroso-morpholine, it is important that morpholine should be efficiently removed from industrial effluents before they are discharged to the environment. Removal of xenobiotes from effluents can often be achieved by biodegradation in activated sludge plants, but in the case of morpholine this process is slow and can be unreliable. This problem appears to derive from the nature of the organisms involved in morpholine degradation. They are almost invariably mycobacteria that grow slowly and lack the competitive ability to survive in an activated sludge plant unless there is a continual supply of morpholine in the influent. It is likely that morpholine degradation is plasmid-encoded in most mycobacterial isolates, and the main features of the degradative pathway involved have been elucidated in *Mycobacterium chelonei* MorG. The technology for the genetic manipulation of mycobacteria, and our understanding of the biochemistry and genetics of morpholine biodegradation, have now reached a stage at which it would be realistic to contemplate the use of genetic manipulation to construct faster growing, more efficient, and more competitive morpholine-degrading bacteria.

INTRODUCTION

Morpholine (1,4-tetrahydro-oxazine) is a relatively simple heterocyclic molecule, comprising a fully saturated six-membered ring of four carbon atoms, one oxygen, and one nitrogen (Figure 1). It is an important industrial chemical whose major use is in the manufacture of rubber additives, but it is also employed widely as an anticorrosive agent, a versatile solvent (exceeding benzene, pyridine, and dioxane in solvent power),

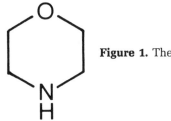

Figure 1. The structure of morpholine.

Dr. Venables is at the School of Pure and Applied Biology, University of Wales, College of Cardiff, Cathays Park, Cardiff CF1 3TL, United Kingdom. Dr. Knapp is at the Department of Microbiology, University of Leeds, Leeds LS2 9JT, United Kingdom.

and an emulsifying agent. Additionally, it is a substituent in a wide range of chemicals, including drugs and herbicides, and is an intermediate in the production of optical brighteners for detergents. Owing to its very varied applications, morpholine is produced and used in large quantities in many countries; in the United States, for example, demand is currently as high as 11,000 metric tons per annum (9, 28). Given such a high level of usage, morpholine is inevitably a constituent of many industrial effluents, and this is presumably the origin of the morpholine residues that have been shown to occur widely in the environment, including a variety of foods (Table 1).

Table 1. Food items shown to contain morpholine.

Rice[a]	Baked ham[b,c]
Canned tuna[b]	Frankfurters[b]
Frozen ocean perch[b]	Evaporated milk[b]
Frozen cod[b]	Coffee[b]
Spotted trout[b]	Canned beer[b]
Small-mouth bass[b]	Bottled beer[b]
Salmon[b]	Wine[b]

Data from: [a] Mohri (29); [b] Singer and Lijinsky (35); [c] Hamano et al. (15).

Although morpholine is not intrinsically toxic in dilute solution, its occurrence in foods or drinking water is of considerable concern owing to its potential for N-nitrosation to give the potent mutagen and carcinogen N-nitroso-morpholine (30). Nonenzymic N-nitrosation of morpholine by nitrite, together with carcinogenesis arising from it, has been reported in a number of laboratory animals fed on a diet including morpholine and nitrite (9, 27). This nitrosation reaction has also been shown to occur in human gastric juice under conditions that approximate those found in the human stomach (45), and indirect experiments have strongly suggested that the reaction also occurs *in situ* (31). In addition, enzymes that catalyze the N-nitrosation of morpholine have been reported from a range of bacteria that inhabit the mammalian gut, and also from bacteria commonly encountered in the wider environment, such as *Pseudomonas aeruginosa* (5, 38). Morpholine is therefore a pollutant of the environment that has potentially serious consequences for health, and its biodegradation, particularly with regard to its effective removal from industrial effluents, is of considerable interest. That morpholine has acquired a reputation of being resistant to biodegradation (see below) enhances this interest.

SYSTEMATICS AND ECOLOGY OF
MORPHOLINE-DEGRADING BACTERIA

Following several earlier failures, the biodegradability of morpholine was first indicated by the observation of Mills and Stack (26) that it exerted a biological oxygen demand when exposed to a seed of microbes from river water. The river water had previously been acclimated to diethanolamine over a period of 116 days. Morpholine biodegradation was not unequivocally demonstrated, however, until Knapp et al. (23) isolated a morpholine-degrading, Gram-positive, acid-fast, rod-shaped bacterium from an activated sludge plant that was receiving a morpholine-containing industrial effluent. This bacterium, designated MorG, grew in pure culture and utilized morpholine as a sole source of carbon, nitrogen, and energy for growth.

Morpholine Biodegradation in the Environment

Subsequent to the initial demonstrations of morpholine biodegradation, bacteria that carry out the process have been found in a variety of environments. For example, they appear to be invariably present in activated sludge plants used to treat either domestic or industrial effluent (1, 24). They have also been found in most samples of water taken from a range of rivers in eastern and northern England; Table 2 shows typical data for the distribution of morpholine-degrading microbes in a river in Yorkshire (England) as it progresses downstream and changes in quality from clean to highly polluted. Furthermore, morpholine degraders have been found in many samples of soil collected from urban and suburban environments in Cardiff (Wales) (K. V. Waterhouse, personal communication). The overall picture is therefore of wide distribution of morpholine-degrading bacteria within the environment.

Table 2. Numbers of morpholine-degrading microbes, and the lag period in die-away tests (DA-lag), for water samples taken from different sites on the River Aire, United Kingdom.[a,b]

Sampling site		Chemical classification of water	MPN ml^{-1}	DA-lag (days)
A	Gargrave	1A/B	<0.4	22/24
B	Bingley	3	0.4	16/20
C	Shipley	2	2.6	18/18
D	Apperly Bridge	4	2.6	16/20
E	Swillington	3	450	8/8
F	Beal	3	1100	6/6

[a] After Knapp and Whytell (24).
[b] MPN is the number of morpholine-degrading microbes determined by the "most probable number" method. In die-away tests, morpholine is mixed with river water, incubated, and time-samples assayed for morpholine; the die-away lag is the time taken before measurable morpholine degradation is noted (values given are duplicate determinations). Sites A to F span the length of the river from near its source to its tidal limit. In chemical classification of river water, 1A/B is clean to very clean; 2 is fairly clean; 3 is badly polluted; 4 is grossly polluted.

As an alternative to their complete assimilative degradation, certain xenobiotic compounds have sometimes been found to be degraded (partially) by co-metabolism, that is, the organism concerned makes no use of the energy or chemical products that are made available by degradation of the xenobiote, and it therefore needs to utilize another compound for its growth requirements. To date, studies by the authors have provided no evidence to indicate that morpholine is ever degraded co-metabolically. On the contrary, studies in model activated sludge plants and in river water have shown strong correlation between the ability of a sample to degrade morpholine and the growth of morpholine-assimilating organisms within that sample, thus suggesting that any contribution from co-metabolism is of minor significance (1, 24). In contrast, Dmitrenko et al. (11) reported the isolation of several bacteria that degraded morpholine only when they were provided with an additional co-substrate.

The Prevalence of Mycobacteria

Whenever morpholine-degrading bacteria have been obtained in pure culture, they have been characterized as Gram-positive, and in almost all cases they have been slow-growing, acid-fast rods with no hyphal development. These characteristics are indicative of the genus *Mycobacterium* (1, 8, 10, 23, 24). Interestingly, this slow-growing genus of bacteria in the last few years has been increasingly implicated in the environmental degradation of recalcitrant organic compounds. In addition to morpholine, the list of xenobiotes recently found to be degraded mainly or exclusively by mycobacteria includes phenanthrene (13), pyrene (16), chlorinated phenolic compounds (14), and chlorinated hydrocarbons (42).

In only two cases have authors attempted to assign isolates of morpholine-degrading mycobacteria to known species: Cech et al. (8) considered their bacterium to be *M. aurum,* whereas Franklin (12), on the basis of mycolic acid analysis, assigned organism MorG to the *M. chelonei* group. Morpholine-degrading mycobacteria have been found to have mean generation times in the range 9–20 hr at 30°C, which is very slow in comparison with bacteria of more classical degradative genera such as *Pseudomonas.* The slow growth of these organisms appears to be an intrinsic characteristic that is equally manifest during growth on central metabolites such as acetate or succinate, or even on rich media such as Nutrient Broth®.

Problems Encountered in Morpholine Biodegradation

The previous sections have demonstrated that there is abundant evidence for the biodegradability of morpholine and for the frequent occurrence of morpholine-degrading bacteria in activated sludge plants. Furthermore, full-scale industrial activated sludge plants exist that successfully remove morpholine from factory effluents. Nonetheless, morpholine is frequently reported as being resistant to biodegradation, and several authors have reported difficulties in removing it from effluents by biological treatment (e.g. 4, 37, 40, 41).

Knapp and Whytell (24) propose that this apparent paradox is a consequence of the slow growth of bacteria that degrade morpholine and their very low numbers in the environment (although they are widespread) unless their proliferation has been selected by the presence of morpholine. Thus, these authors found fewer than 70 morpholine degraders per milliliter of activated sludge at most sewage treatment works. When samples of such sludge were incubated with morpholine, lag periods of up to 30 days were observed before morpholine degradation became detectable. It follows that these treatment works would have been quite incapable of dealing effectively with a sudden load of morpholine in their influent. Owing to the extremely low numbers of morpholine-degraders in the sludge, the morpholine would have passed through the system essentially unchanged in concentration.

Brown and Knapp (1) also studied morpholine degradation in a model activated sludge system. In this system, efficient morpholine biodegradation became well established and a stable, high-density population of morpholine degraders was demonstrated. When morpholine was removed from the influent, however, numbers of morpholine degrading mycobacteria fell by five logs, and this was inversely reflected by the lag period for morpholine degradation in sludge samples, which increased from zero to *ca* 1000 hr (Figure 2).

In both the above cases (the sewage works and the model activated sludge plants following removal of morpholine from the influent), the long lag periods observed for morpholine degradation represented the time taken for an initially small population of slow-growing morpholine degraders to grow large enough to effect a detectable decrease in morpholine concentration.

Brown and Knapp (1) concluded therefore that morpholine-degrading bacteria compete successfully in effluent treatment systems only when morpholine is available as a

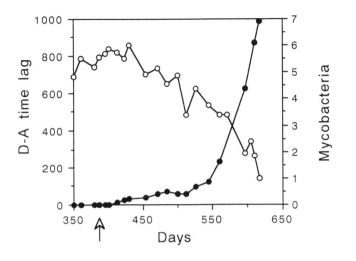

Figure 2. The effect of removal of morpholine from the influent of a model activated sludge plant that had been successfully degrading morpholine. After Brown and Knapp (1). The *arrow* indicates the time of morpholine removal. Log_{10} numbers of morpholine-degrading mycobacteria (o) and die away lag times (hours) (●) are shown. The term *die-away* lag is explained in Table 2.

nutrient source to which they have sole access. In the absence of morpholine, they appear to be out-competed in the utilization of other nutrient sources by faster growing genera of degradative bacteria, such as the *Pseudomonads*. The population of morpholine-degraders then declines to biodegradatively ineffective levels.

Even when morpholine is continuously present in the influent of an activated sludge plant and a high population of morpholine-degraders is maintained, the slow growth of these degraders can still impose limits on the operation of the plant. This applies particularly to the mean solids retention time (MSRT), which has to be of considerable length to achieve effective treatment. MSRT represents the average time that a particle of activated sludge is retained within the system; if its value is low, slow-growing bacteria will be flushed out of the treatment plant, and their substrate will remain either undegraded or incompletely degraded. The importance of MSRT in the control of the degradation of xenobiotic compounds by activated sludge plants has only recently been appreciated, though its role in the control of nitrification (another process catalyzed by slow-growing bacteria) has long been known. The crucial importance of this parameter in the degradation of morpholine is well-illustrated by the observations of Cech and Chudoba (7). They found that in model activated sludge plants, morpholine degradation became detectable only if the MSRT was greater than 3 days, and that degradation was complete only if the MSRT exceeded 8 days. This finding complements those of Brown and Knapp (1) in explaining the problems frequently encountered in morpholine degradation by activated sludge plants.

It seems likely that morpholine biodegradation in activated sludge systems would be made more efficient and reliable by the introduction of morpholine-degrading bacteria that have been genetically manipulated to utilize morpholine at a faster rate and to compete better under growth conditions in which they are not directly selected. A rational approach to the construction of such an organism would require the elucidation of the biodegradative pathway involved (to identify, for instance, the rate-limiting steps) and a detailed understanding of its genetic control. The remainder of this chapter reviews the current state of knowledge in these areas.

BIOCHEMISTRY OF MORPHOLINE BIODEGRADATION

Studies of the pathway(s) by which bacteria degrade morpholine have been limited to *Mycobacterium chelonei* MorG. The main features of the inducible catabolic pathway employed by this organism have now been elucidated, and are shown in Figure 3. The evidence for the early reactions remains to some extent indirect, however, and further investigation is required to complete our understanding of them (39).

As would be predicted, morpholine is cleaved at two sites: first, at its secondary amine linkage to create an open-chain ether and, second, at its ether linkage, which is the more stable of the two. Thus, two nonidentical 2-carbon cleavage products are generated, one of which contains the nitrogen atom. These are further metabolized by separate pathway branches to yield molecules that can enter the pathways of intermediary metabolism via the tricarboxylic acid cycle. Evidence for the proposed catabolic pathway is discussed below under three headings that represent the perceived temporal sequence of the pathway reactions.

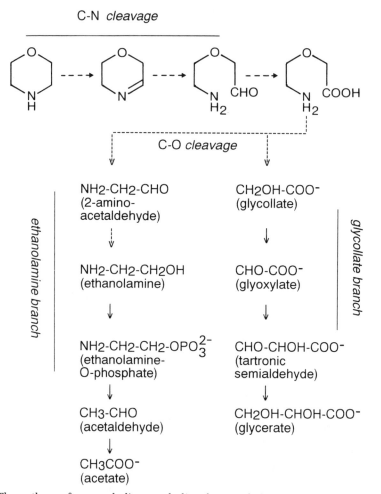

Figure 3. The pathway for morpholine catabolism by *M. chelonei* MorG as proposed by Swain et al. (39). 2(2-Aminoethoxy)acetic acid is cleaved oxidatively in a position anticlockwise to the ether oxygen, yielding 2-amino-acetaldehyde and glycollate. *Solid arrows* indicate proven reaction steps; *dashed arrows* indicate reaction steps for which evidence is indirect.

Cleavage of the C–N Bond

Catabolism of morpholine requires the cleavage of the secondary amine linkage. There are several precedents for the bacterial cleavage of amine linkages in saturated heterocyclics, including the compounds hydroxyproline, piperidine, and pyrrolidine (3). In each case, partial oxidation of the ring has been shown to occur, giving a compound with a single unsaturated link between the nitrogen heteroatom and one of its neighboring carbons. Such compounds (for example, 1-pyrroline), when in aqueous solution, exist in equilibrium with a hydrated open-chain structure, so that no additional enzymic mechanism is required for ring opening. The use of this ring-opening mechanism in the catabolism of pyrrolidine by *Pseudomonas fluorescens* is shown in Figure 4.

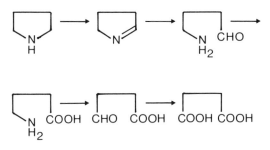

Figure 4. The pathway for pyrrolidine catabolism by *Pseudomonas fluorescens* as proposed by Jacoby and Fredericks (19).

MorG has been shown to catabolize pyrrolidine by the same pathway as utilized by *P. fluorescens* (39), and there is substantial indirect evidence for the use of the same initial reaction steps in its catabolism of morpholine. The first observation to support this view came from Knapp et al. (23), who showed that intact cell suspensions of MorG that had been grown on morpholine were pre-induced for pyrrolidine oxidation, while those grown on pyrrolidine were pre-induced for morpholine oxidation. Such cross-specificity of induction indicates relatedness of the two pathways, either through possession of common regulatory mechanisms, common enzymes, or both. Relatedness of the two pathways is also indicated by the analysis of mutants of MorG defective in morpholine utilization. Most of these Mor⁻ mutants were found to have to have lost the ability to utilize not only morpholine, but also pyrrolidine. This indicates that a single genetic event can frequently affect the catabolic pathways for both these heterocycles (39).

If it is assumed that the early steps of morpholine and pyrrolidine catabolism by MorG are analogous, then the morpholine ring would be opened by partial oxidation at the nitrogen atom to generate the compound 2(2-aminoethoxy)acetaldehyde. If the pathway continued to follow that of pyrrolidine catabolism, this intermediate would subsequently be oxidized to the corresponding carboxylic acid 2(2-aminoethoxy)acetate. Swain et al. (39) synthesized this latter compound and demonstrated that its oxidation by cell suspensions is induced by growth of the cells on morpholine, thus implicating it directly in the morpholine catabolic pathway. Therefore, the partial oxidation of the secondary amine linkage, giving rise, after a further oxidation step, to 2(2-aminoethoxy)acetate, is strongly indicated as the initial stage in morpholine catabolism by MorG, and may be catalyzed by the same enzyme that performs the equivalent reaction in pyrrolidine.

Cleavage of the C–O Bond

Further metabolism of 2(2-aminoethoxy)acetate requires cleavage of one of the C–O bonds of the ether linkage. Bacteria generally cleave such linkages oxidatively by the use of mono-oxygenase enzymes. This mechanism has been found in the degradation of both phenolic and alkyl ethers (6, 17, 33). The observations of J. S. Knapp (personal communication) that growth of MorG on morpholine strongly induces synthesis of cytochrome p450, the usual cofactor of bacterial mono-oxygenases, and of Swain et al. (39) that 2(2-aminoethoxy)acetate is oxidatively metabolized by this bacterium, indicate that MorG very probably utilizes a mono-oxygenase to cleave this intermediate. The involvement of ethanolamine and glycollate (see below) as early post-cleavage intermediates is also consistent with this cleavage mechanism.

Metabolism of the Ether-Cleavage Products

Cleavage of the ether link gives rise to two 2-carbon units that are further catabolized by separate pathway branches. The involvement of ethanolamine in the pathway branch that processes the nitrogen-containing cleavage product, and of glycollate in the other branch, was first indicated some years ago by co-induction experiments with morpholine, ethanolamine, and glycollate (22). These studies showed that MorG utilized all three of these compounds as sole sources of carbon for growth, and that their inducible oxidation by washed-cell suspensions could be demonstrated in a respirometer. Oxidation of ethanolamine, however, was induced by growth on either ethanolamine or morpholine but not by growth on glycollate, whereas oxidation of glycollate was induced by growth on either glycollate or morpholine but not ethanolamine. This indicates that ethanolamine and glycollate are both intermediates of morpholine degradation, but in different pathway branches. For convenience, these two pathway branches are referred to subsequently as the "ethanolamine branch," and the "glycollate branch" respectively (Figure 3).

The work of Swain et al. (39) has shown that induction of ethanolamine oxidation in MorG (by growth on either ethanolamine or morpholine) involves the de novo synthesis of the enzymes of the ethanolamine kinase/phospholyase pathway, which has previously been shown to bring about ethanolamine catabolism in both Pseudomonas and Erwinia (20, 21). Swain et al. (39) have also shown that the induction of glycollate oxidation in MorG (by growth on either glycollate or morpholine) involves the de novo synthesis of the enzymes of the "glycerate pathway" (25). In this pathway, glycollate is converted to glycerate, which then enters the tricarboxylic acid cycle by the reaction sequence of glycerate kinase, phosphoglyceromutase, enolase, and pyruvate kinase. The above results indicate that these two pathways, the kinase/phospholyase pathway and the glycerate pathway, form the ethanolamine and glycollate branches, respectively, of the morpholine degradation pathway (Figure 3).

Further evidence for the operation of the glycollate branch as constituted in Figure 3 comes from the study of Mor⁻ mutants of MorG. Such mutants, almost invariably, fail to utilize glycollate as a growth substrate, owing to their loss of either glycollate oxidase, glyoxylate carboligase, tartronic semialdehyde reductase, or combinations of these activities, thus indicating that all three enzymes are essential in catabolism of both glycollate and morpholine (39). Additional support for this conclusion is provided by the observation that in one such mutant, which has lost tartronic semialdehyde reductase and reverts at detectable frequencies to Mor⁺, reversion is associated with restoration of this enzyme activity (12).

Although the above work has established that both ethanolamine and glycollate are early post-cleavage intermediates in morpholine catabolism, theoretical considerations indicate that only one of them can be a direct product of mono-oxygenase cleavage of 2(2-aminoethoxy)acetate. Figure 5(a) shows that depending on which side of the ether oxygen the cleavage occurs, two alternative pairs of products will be produced. These pairs comprise either glycollate plus aminoacetaldehyde on the one hand, or ethanolamine plus glyoxylate on the other. Swain et al. (39) have pointed out that glyoxylate is unlikely to be a cleavage product, however, as this would render glycollate oxidase redundant when all other relevant evidence indicates that this activity is an essential pathway component. Consequently, they have proposed that glycollate and aminoacetaldehyde are the immediate cleavage products, and that the latter is subsequently reduced to ethanolamine. It is also possible, however, that 2(2-aminoethoxy)acetate is not directly involved in morpholine catabolism but is merely an alternative substrate for a mono-oxygenase that normally cleaves 2(2-aminoethoxy)acetaldehyde. In this case, the cleavage products could be ethanolamine and glyoxal (Figure 5b), and the latter could be converted to glycollate by a dismutase reaction as catalyzed by glyoxalase (25).

Clearly, the details of the early reactions of morpholine catabolism by MorG remain to be elucidated; however, the instability of several of the proposed early intermediates makes progress difficult to achieve by classical approaches. Further investigation would profit from the use of labeled morpholine in ^{13}C nuclear magnetic resonance spectroscopy studies.

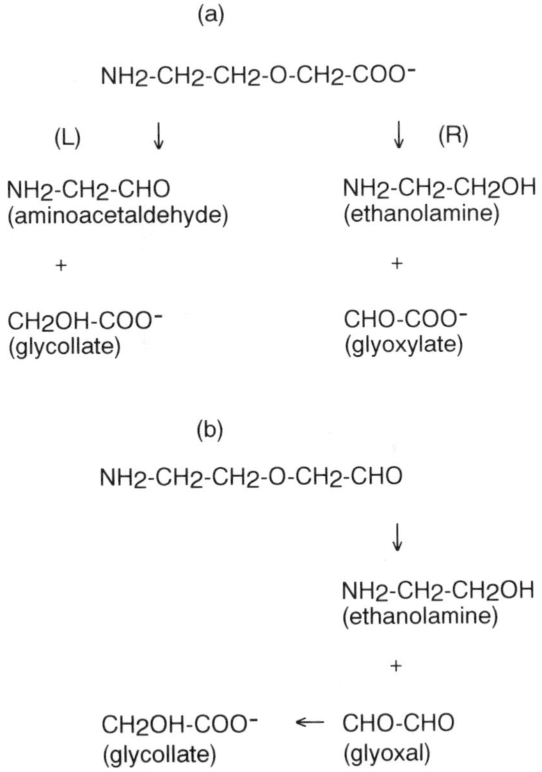

(a)

$$NH_2-CH_2-CH_2-O-CH_2-COO^-$$

(L) ↓   ↓ (R)

NH_2-CH_2-CHO
(aminoacetaldehyde)

$NH_2-CH_2-CH_2OH$
(ethanolamine)

+   +

$CH_2OH-COO^-$
(glycollate)

$CHO-COO^-$
(glyoxylate)

(b)

$$NH_2-CH_2-CH_2-O-CH_2-CHO$$

↓

$NH_2-CH_2-CH_2OH$
(ethanolamine)

+

$CH_2OH-COO^-$ ← $CHO-CHO$
(glycollate)   (glyoxal)

Figure 5. Possible ether cleavage reactions in morpholine catabolism. (a) Products of the oxidative cleavage of 2(2-aminoethoxy)acetate are shown: (L) cleavage to the left of, and (R) cleavage to the right of the ether oxygen. (b) Oxidative cleavage of 2(2-aminoethoxy)acetaldehyde to yield ethanolamine and glyoxal, and conversion of the latter to glycollate by dismutation.

THE GENETICS OF MORPHOLINE BIODEGRADATION

Genetic studies of morpholine biodegradation are still at a very preliminary stage, and have been limited mainly to the isolation of mutants for use in pathway elucidation and the identification of the genetic elements (i.e. a specific plasmid or the chomosome) on which the degradative genes (*mor* genes) reside.

The Involvement of Plasmids

Variants that have lost the ability to grow on morpholine (Mor⁻ variants) have been readily isolated from cultures of all morpholine-degrading mycobacteria after growth on nonselective media (i.e. having a carbon source other than morpholine). Brown et al. (2) studied nine environmental isolates, all of which showed marked instability of the Mor character (e.g. Figure 6). Waterhouse et al. (43) found that *M. chelonei* MorG also showed instability; during Nutrient Broth® culture, the Mor⁺ character was lost at a rate of 0.4% per generation, and a 10-fold increase of this frequency was observed in the presence of acriflavin. This degree of instability in a catabolic phenotype has often (though not always) been found to indicate that the genes concerned reside on a plasmid, and are lost at high frequency owing to the segregational loss of the plasmid. This is seen, for example, in the TOL (toluene) catabolic plasmids of *Pseudomonas putida* (44).

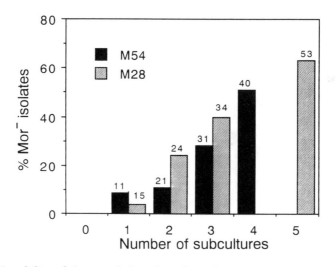

Figure 6. Instability of the morpholine-degrading phenotype: Morpholine-degrading mycobacteria M28 and M54, which had been grown previously on morpholine/mineral salts medium, were subcultured in Nutrient Broth® and the numbers of Mor⁻ variants that arose were monitored. Numbers on the tops of columns indicate the number of generations completed at each subculture.

Proof of the involvement of a plasmid in a catabolic pathway inevitably requires the physical demonstration of the existence of the plasmid, together with a demonstration that expression of the pathway is linked inextricably to plasmid presence. It is of importance, therefore, that at least six of the nine mycobacterial strains studied by Brown et al. (2) were shown to contain plasmids (in some cases three or more different plasmids in the same strain). Mor⁻ variants of these strains appeared to retain all plasmids, but in two cases, large plasmids (approximately 200 kbp in size) appeared to have undergone sizeable deletions. It was concluded that these plasmids probably encoded morpholine catabolism in the parent strains, and that the Mor⁻ variants had arisen through deletion of the *mor* genes from these plasmids. The lack of reversion among Mor⁻ strains, indicating that the *mor* genes had been irreversibly lost, supported this interpretation. Deletion of this sort is another type of instability that is commonly found among catabolic plasmids, the deletion often arising from recombination between homologous base sequences that flank the catabolic genes, as found (for example) in the pTDN1 (toluidine) plasmid of *P. putida* (34).

M. chelonei MorG has also been found to contain plasmids (43). These have been designated pMOR1, 2, 3, and 4, and have been shown to have sizes of 54 (approximately), 27.7, 22.8, and 22.6 kbp, respectively. Mor⁻ variants of MorG retained all four plasmids, and most of the variants that arose spontaneously showed no detectable differences in plasmid profile from wild-type MorG. In all Mor⁻ variants isolated after acriflavin treatment (together with a small fraction of spontaneous isolates), however, plasmid pMOR2 showed an increase in size from 27.7 to 29.5 kbp. In all cases this resulted from an insertion of 1.8 kbp of additional DNA into a specific 5.9 kbp fragment of pMOR2 (Figure 7), thus indicating that acriflavin treatment must trigger the site-specific insertion of a transposable element of DNA into this region of the plasmid. Waterhouse et al. (43) considered this to indicate that mor genes are carried by plasmid pMOR2, and that the transposable element interferes with their expression by inserting at the site where they are located. This interpretation is very plausible, but the possibility that the insertion of the transposable element into pMOR2 and the loss of Mor phenotype are separate (possibly unrelated) events cannot be completely ruled out at this stage.

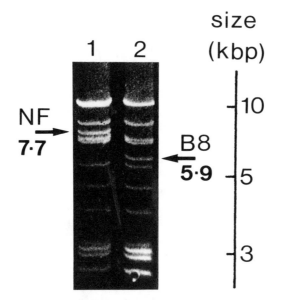

Figure 7. The rearrangement of plasmid DNA in *Mycobacterium chelonei* MorG associated with acriflavin-induced loss of the Mor⁺ character. Electrophoresis of a Bam H1 restriction digest of mixed MorG plasmids is shown. In the Mor⁻ mutant (track 1), a new 7.7 kbp fragment (NF) has replaced the 5.9 kbp fragment B8 of the wild-type (track 2). After Waterhouse et al. (43).

Pathway Mutants

The enzymology of Mor⁻ variants has been investigated only in the case of *Mycobacterium chelonei* MorG (39). Information yielded by these strains was limited by their lack of phenotypic variety. All acriflavin-induced and spontaneous variants failed to grow on morpholine, pyrrolidine, and glycollate, and most (possibly all) were defective in glyoxylate carboligase and tartronic semialdehyde reductase. Enzyme activities of the ethanolamine branch, together with growth on ethanolamine, were always retained. It is assumed that all these Mor⁻ strains have as yet uncharacterized deficiencies in early steps of the degradative pathway that prevent growth on pyrrolidine. Such deficiencies would also explain the failure of Mor⁻ strains to assimilate morpholine carbon atoms (and hence utilize morpholine for growth) via the ethanolamine branch of the pathway, which might be expected to occur if the glycollate branch alone was defective (Figure 3).

MorG is mutable by nitrosoguanidine, but with the exception of a few mutants that retained glyoxylate carboligase and tartronic semialdehyde reductase, Mor⁻ mutants induced by this agent appeared to possess the same phenotypes as those which arise spontaneously or are induced by acriflavin. Such mutants contribute little to the investigation of the genetic organization of the morpholine pathway, and further progress will require the application of more incisive molecular techniques. Detailed knowledge of the organisation and control of *mor* genes will be an essential prerequisite for the successful outcome of the type of genetic manipulation program outlined below.

THE FUTURE

For the reasons outlined at the beginning of this chapter, it is obviously important that morpholine should be removed effectively from effluents before they are discharged into the environment, and it is generally recognized that biodegradation is the most cost effective means for removing biodegradable xenobiotes from effluents.

Morpholine is certainly biodegradable, yet its removal from industrial effluents by biological treatment can be problematical owing to the slow growth and uncompetitive nature of the mycobacterial species that degrade it. Problems are unlikely to occur when a large population of morpholine degraders has been established in a treatment plant that is receiving a regular and fairly continuous input of morpholine. In such circumstances, there will be a strong selection for the maintenance of morpholine-degrading bacteria, and adequate biodegradation levels should be maintained. Intermittent input of morpholine will preclude the initial establishment of a high population of degraders, however, and will cause the decline of a pre-established population, as demonstrated in Figure 2.

The possibility of improving the removal of morpholine from industrial effluents by genetically engineering a bacterium to degrade it more efficiently and reliably has already been mooted in this chapter. A theoretically plausible approach to the construction of such an organism could be to clone the genes of the morpholine degradative pathway from *M. chelonei* MorG, and to introduce them (with appropriate provision for their effective expression) into a fast-growing pseudomonad. This latter organism should be isolated from an activated sludge plant to ensure that it is adequately adapted to survival and growth in such an environment. Once accomplished, genetic transfer and expression could be followed by a program to manipulate the pathway in order to improve or replace rate-limiting enzymes.

In the last few years, considerable progress has been made in the development of technology for the genetic manipulation of mycobacteria. This includes the development of versatile cloning vectors (18, 32, 46) and techniques for genetic transformation by electroporation (36). Understanding of the biochemistry and genetics of morpholine degradation by *M. chelonei* MorG is now at a stage at which the application of these newly developed techniques to produce an improved morpholine degrading bacterium is a realistic objective. This objective is currently being pursued in the laboratories of the authors.

LITERATURE CITED

1. Brown, V. R., and Knapp, J. S. 1990. The effect of withdrawal of morpholine from the influent and its reinstatement on the performance and microbial ecology of a model activated sludge plant treating a morpholine-containing effluent. *J. Appl. Bacteriol.* 69: 43–53.

2. Brown, V. R., Knapp, J. S., and Heritage, J. 1990. Instability of the morpholine degradative phenotype in mycobacteria isolated from activated sludge. *J. Appl. Bacteriol.* 69:54–62.

3. Callely, A. G. 1978. The microbial degradation of heterocyclic compounds. In *Progress in Industrial Microbiology,* vol. 14. Ed. M. J. Bull. Amsterdam/Oxford-New York: Elsevier Scientific Publishing Company. 205–281.

4. Calmari, D., Da Gasso, R., Galassi, S., Provini, A. and Vighi, M. 1980. Biodegradation and toxicity of selected amines on aquatic organisms. *Chemosphere* 9:753–762.

5. Calmels, S., Oshima, H., and Bartsch, H. 1988. Nitrosamine formation by denitrifying and non-denitrifying bacteria: implication of nitrite reductase and nitrate reductase in nitrosation catalysis. *J. Gen. Microbiol.* 134:221–226.

6. Cartwright, N. J., Holdom, K. S., and Broadbent, D. A. 1971. Bacterial attack on phenolic ethers: dealkylation of higher ethers and further observations on O-demethylases. *Microbios* 3:113–130.

7. Cech, J. S., and Chudoba, J. 1988. Effect of solids retention time on the rate of biodegradation of organic compounds. *Acta Hydrochim. Hydrolog.* 16:313–323.

8. Cech, J. S., Hartmann, P., Slosarek, M., and Chudoba, J. 1988. Isolation and identification of a morpholine degrading bacterium. *Appl. Environ. Microbiol.* 54:619–621.

9. CIR Expert Panel. 1989. Final report on the safety assessment of morpholine. *J. Am. Coll. Toxicol.* 8:707–748.

10. Dmitrenko, G. N., and Gvozdyak, P. 1988. Destruction of morpholine by mycobacteria. In *Proceedings of Conference on Microbiological Methods for Protecting the Environment.* Scientific Centre for Biological Research, Puschino, USSR.

11. Dmitrenko, G. N., Gvozdyak, P. I., and Udod, V. M. 1987. Selection of destructor microorganisms for heterocyclic xenobiotics. *Khim. Teknol. Vody* 9:442–445 (Russian version); 77–81 (English language version).

12. Franklin, A. 1985. *Genetic and biochemical studies of morpholine degradation by a Mycobacterium species.* Ph.D. Thesis, University of Wales, U.K.

13. Guerin, W. F., and Jones, G. E. 1988. Mineralization of phenanthrene by a *Mycobacterium* sp. *Appl. Environ. Microbiol.* 54: 937–944.

14. Häggblom, M. M., Nohynek, L. J., and Salkinoja-Salonen, M. S. 1988. Degradation and O-methylation of chorinated phenolic compounds by *Rhodococcus* and *Mycobacterium* strains. *Appl. Environ. Microbiol.* 54:3043–3052.

15. Hamano, T., Mitsuhashi, Y., and Mutsuki, Y. 1981. Glass capilliary chromatography of secondary amines in foods, with flame photometric detection after derivatization with benzene sulphonyl chloride. *Agric. Biol. Chem.* 45:2237–2243.

16. Heitkamp, M. A., Freeman, J. P., Miller, D. W. and Cerniglia, C. E. 1988. Pyrene degradation by a *Mycobacterium* sp.: identification of ring oxidation and ring fission products. *Appl. Environ. Microbiol.* 54:2556–2565.

17. Heydeman, M. T. 1974. Growth of soil bacteria on diethyl ether. *J. Gen. Microbiol.* 81:ix–x.

18. Jacobs, W. R. Jr., Snapper, S. B., Tuckman, M., and Bloom, B. R. 1989. Mycobacteriophage vector systems. *Rev. Infect. Dis.* 11(suppl. 2):S404-S410.

19. Jacoby, W. B., and Fredericks, J. 1959. Pyrrolidine and putrescine metabolism: gamma-aminobutyraldehyde dehydrogenase. *J. Biol. Chem.* 234:2145–2150.

20. Jones, A., Faulkner, A., and Turner, J. M. 1973. Microbial metabolism of amino alcohols: metabolism of ethanolamine and 1-aminopropan-2-ol in species of *Erwinia* and the roles of amino alcohol kinase and amino alcohol O-phosphate phospho-lyase in aldehyde formation. *Biochem. J.* 134:959–968.

21. Jones, A., and Turner, J. M. 1973. Microbial metabolism of amino alcohols: 1-aminopropan-2-ol and ethanolamine metabolism via propionaldehyde and acetaldehyde in a species of *Pseudomonas. Biochem. J.* 134:167–182.

22. Knapp, J. S. 1975. *The microbiology of an effluent system treating heterocyclic compounds.* Ph.D. Thesis, University of Wales, U.K.

23. Knapp, J. S., Callely, A. G., and Mainprize, J. 1982. The microbial degradation of morpholine. *J. Appl. Bacteriol.* 52:5–13.

24. Knapp, J. S., and Whytell, A. J. 1990. The biodegradation of morpholine in river water and activated sludge. *Environ. Pollut.* 68: 67–79.

25. Kornberg, H. L. 1966. Anaplerotic sequences and their role in metabolism. In *Essays in Biochemistry,* vol. 2. Ed. P. N. Campbell, G. D. Greville. London and New York: Academic Press, 1–31.

26. Mills, E. J., and Stack, V. T. 1955. Suggested procedure for evaluation of biological oxidation of organic chemicals. *Sewage Ind. Wastes* 27:1061–1064.

27. Mirvish, S. S., Salmasi, S., Cohen, S. M., Patil, K., and Mahboubi, E. 1983. Liver and forestomach tumors, and other forestomach lesions in rats treated with morpholine and sodium nitrite, with and without sodium ascorbate. *J. Natl. Cancer Inst.* 71:81–85.

28. Mjos, K. 1978. Cyclic amines. *Kirk-Othmer Encyclop. Chem. Technol.* 2:295–308.

29. Mohri, T. 1987. Dietary intakes of nitrosamine precursors. *Kyusha Yakugakki Katho* 41:105–112.

30. National Research Council 1981. Selected aliphatic amines and related compounds: an assessment of the biological and environmental effects. Washington D.C.: National Research Council.

31. Ohshima, H., and Bartsch, H. 1981. Quantitative estimation of endogenous nitrosation in humans by monitoring N-nitrosoproline excreted in urine. *Cancer Res.* 41:3658–3662.

32. Rauzier, J., Moniz-Pereira, J., and Gicquel-Sanzey, B. 1988. Complete nucleotide sequence of pAL5000: a plasmid from *Mycobacterium fortuitum. Gene* 71:315–321.

33. Ribbons, D. W. 1970. Stoichiometry of O-demethylase activity in *Pseudomonas aeruginosa. FEBS Lett.* 8:101–104.

34. Saint, C. P., and Venables, W. A. 1990. Loss of Tdn catabolic genes by deletion from, and curing, of plasmid pTDN1 in *Pseudomonas putida:* rate and mode of loss are substrate and pH dependent. *J. Gen. Microbiol.* 136:627–636.

35. Singer, G. M., and Lijinsky, W. 1976. Naturally occurring nitrosatable compounds. I. secondary amines in foodstuffs. *J. Agric. Food Chem.* 24:550–553.

36. Snapper, S. B., Lugosi, L., Jekkel, A., Melton, R. E., Kieser, T., Bloom, B. R., and Jacobs, W. R. 1988. Lysogeny and transformation in mycobacteria: stable expression of foreign genes. *Proc. Natl. Acad. Sci. (USA)* 85:6987–6991.

37. Subramanyan, P. V. R., Khadakkar, S. N., Chakrabarti, T., and Sundaresan, B. B. 1983. Wastewater treatment of a phthalate plasticiser, ethanolamine and morpholine manufacturing plant: a case study. In *Proceedings, 37th Industrial Waste Conference.* Purdue University, Lafayette, IN. 13–20.

38. Suzuki, S., and Mitsuaka, T. 1984. N-nitrosamine formation by intestinal bacteria. *IARC Sci. Publ.* 57:275–282.

39. Swain, A., Waterhouse, K. V., Venables, W. A., Callely, A. G., and Lowe, S. E. 1991. Biochemical studies of morpholine catabolism by an environmental mycobacterium. *Appl. Microbiol. Biotechnol.* 35:110–114.

40. Tanaka, M., Okada, Z., Katuji, M., and Seiko,Y. 1968. Industrial waste treatment by activated sludge, XV. Treatment of waste from an organic vulcanisation accelerator producing plant. *Kogyo Gijutsuin, Hakko Kenkushko Kenkyo Hokoko* 33:19–29 (in Japanese) cited *Chem. Abstr.* (1970) 73:123333a.

41. Tolgyessy, P., Kollar, M., Vanco, D., and Piatrok, M. 1986 The effect of gamma radiation on biodegradability of morpholine in aqueous solutions. *J. Radioanal. Nucl. Chem. Lett.* 107:291–295.

42. Wackett, L. P., Brusseau, G. A., Householder, S. R. and Hanson, R. S. 1989. Survey of microbial oxygenases: trichloroethylene degradation by propane-oxidising bacteria. *Appl. Environ. Microbiol.* 55:2960–2964.

43. Waterhouse, K. V., Swain, A., and Venables, W. A. 1991. Physical characterisation of plasmids in a morpholine-degrading mycobacterium. *FEMS Microbiol. Lett.* 80:305–310.

44. Williams, P. A., and Murray, K. 1974. Metabolism of benzoates and methylbenzoates by *Pseudomonas putida (arvilla)* mt–2: evidence for the existence of a TOL plasmid. *J. Bacteriol.* 120: 433–437.

45. Zeibarth, D. 1975. Studies on the nitrosation of secondary amines in buffer solutions and in human gastric juice. *Arch. Geschwulfstforsch.* 43:42–51.

46. Zhang, Y., Lathigra, R., Garbe, T., Catty, D., and Young, D. 1991. *Molec. Microbiol.* 5:381–391.

CHAPTER 10

Microbial Metabolism of Carbamate and Organophosphate Pesticides

A. MATEEN, S. CHAPALAMADUGU, B. KASKAR, A. R. BHATTI, and G. R. CHAUDHRY

Abstract. Several carbamate and organophosphate compounds are used to control a wide variety of insect pests, weeds, and disease-transmitting vectors. These chemicals were introduced to replace the recalcitrant and hazardous chlorinated pesticides. Although newly introduced pesticides were considered to be biodegradable, some of them are highly toxic and their residues are found in certain environments. In addition, degradation of some of the carbamates generates metabolites that are also toxic. In general, hydrolytic metabolites are less toxic than oxidative metabolites. Although microorganisms capable of degrading many of these pesticides have been isolated, knowledge about the biochemical pathways and respective genes involved in the degradation is sparse. The mechanisms of biodegradation of carbamate and organophosphate compounds hold great interest for the following reasons: (1) an efficient mineralization of the pesticides used for insect control could mitigate the problems of environmental pollution; (2) a balance between degradability and efficacy of pesticides could result in the dual benefits of safer application and effective insect control; (3) understanding the mechanisms of biodegradation could help in proper bioremediation of polluted environments. In this chapter we discuss the recent advances in the microbial degradation of carbamate and organophosphate pesticides and the work on genetic engineering of microbial strains for complete degradation of these pesticides.

INTRODUCTION

Pesticides are widely used for improving crop productivity and public health by controlling insect pests and disease vectors. Halogenated hydrocarbons were the first such group of compounds, but their intensive use resulted in enormous environmental contamination (34). For example, dichlorodiphenyl trichloroethane (DDT), the best known among the chlorinated pesticides, was extensively applied for pest control from the 1930s until its ban in 1979. Metabolites of this pesticide were found to contaminate soil and groundwater and were even detected in human populations (1, 8, 18, 106, 159). Dieldrin, heptachlor, benzene, hexachloride, chlordane, and a number of other

---

Drs. Mateen, Chapalamadugu, and Chaudhry are at the Department of Biological Sciences, Oakland University, Rochester, MI, U.S.A. Dr. Kaskar is at Ash Stevens, Inc., Detroit, MI, U.S.A. Dr. Bhatti is at the Biomedical Defence Section, Defence Research Establishment, Suffield, Ralston, Alberta, Canada.

halogenated pesticides have also caused widespread contamination of biota (4, 12, 24, 163). The recalcitrance and susceptibility to biomagnification and the toxicity, mutagenicity, and carcinogenicity of halogenated compounds raised public health concerns (5). This concern led to the development of highly potent but biodegradable carbamate and organophosphate pesticides. The newly developed compounds have gradually replaced most of the chlorinated pesticides. Carbamates and organophosphates are the active ingredients of most of the insecticides and some of the herbicides in use. Their uses include crop protection, control of insects and weeds in recreational facilities, and eradication of insect vectors of animal and human diseases.

Despite the many benefits of carbamates and organophosphates, their widespread use has also caused environmental and health problems. First, most of the synethic carbamates and organophosphates are highly toxic and potent inhibitors of acetylcholinesterase, a vital enzyme involved in neurotransmission. Therefore, concerns have been raised regarding even the judicious commercial use of these chemicals. Chemical residues of pesticides and their metabolites that result from biodegradation concentrate in the food chain and cause short- and long-term human health problems (6, 20, 37, 68, 160). A national monitoring survey conducted by the U.S. Environmental Protection Agency found that 0.6% and 0.8% of the rural domestic and community water system wells, respectively, were polluted above the health advisory levels with at least one pesticide (161). Because of groundwater contamination, aldicarb [2-methyl-2-(methylthio)-propionaldehyde-O-(methyl-carbomyol)oxime] and parathion (O,O-diethyl-O,p-nitrophenyl phosphorothioate) are banned or classified as restricted use pesticides in many parts of the United States (55, 176). Since many carbamate and organophosphate compounds with properties similar to aldicarb and parathion are still in use, the risk of environmental contamination remains. Second, repeated application of these pesticides often causes enhanced degradation of the pesticides (49), thus resulting in inefficient pest control, poor crop yields, and economic losses. Third, target insects develop a resistance to pesticides (91, 106, 162) similar to the drug resistance acquired by human pathogens. This resistance in turn leads to application of higher doses of the pesticides to achieve the desired level of insect control, thus further increasing the risk of environmental pollution. Finally, degradation of some pesticides such as aldicarb yield metabolites as toxic as the parent compound (55). Generally, the oxidation products of these compounds are more toxic than the hydrolytic metabolites. Extensive work has been carried out on the persistence and fate of carbamates and organophosphates. We refer the reader to several excellent reviews of the earlier literature (29a, 49, 52, 68, 80, 115). We limit coverage in this review to studies on the biochemical aspects of degradation of carbamates and organophosphates.

CARBAMATES

Biologically active carbamate chemicals have a long history. Physostigmine, a neurotoxin from calabar bean seeds (*Physostigma venenosum*), was reported to have been used as an ordeal poison on West African witchcraft trials during the 17th and 18th centuries (77). Physostigmine is the only known naturally occurring carbamate ester (27). The first group of synthetic carbamate esters to exhibit insecticidal potential were derivatives of dithiocarbamic acids. Carbamic acid, the monoamide of carbon dioxide, is the backbone of all carbamate structures (Table 1). It does not exist in free form; its salts, which are more stable, are referred to as carbamates or carbaminates.

Carbamates possess the broadest specturm of biological activity. They are used as insecticides, herbicides, fungicides, and nematicides. The major classes of carbamate chemicals used as pesticides include N-methylcarbamates, thiocarbamates, and phenylcarbamates.

Methylcarbamates

The methylcarbamates include the most commonly used insecticides, such as carbofuran (2,3-dihydro-2,2-dimethyl-7-benzofuranyl-N-methylcarbamate), carbaryl (1-naphthyl-N-methylcarbamate), and aldicarb (Table 1).

Table 1. Structure and toxicity (rat, oral) of carbamates.

Common name (trade name)	Structure	LD_{50} (mg/kg of body wgt).
Aldicarb (Temik)	$H_3CS\text{-}C(CH_3)_2\text{-}CH=NOCONHCH_3$	1
Aminocarb (Matacil)		30
Barban (Carbyne)		600
Benomyl (Benlate)		9000
Butylate (Sutan)		4000
Carbaryl (Sevin)		307
Carbofuran (Furadan)		8

Table 1. Continued.

Common name (trade name)	Structure	LD_{50} (mg/kg of body wgt).
Chlorpropham (CIPC)	Cl—C6H3—NH-COOCH(CH3)2	5000
Cycloate (Ro-Neet)	H5C2\N-C(=O)-S-C2H5 (phenyl)	3000
Desmedipham (Betanil AM)	OCONH-phenyl ; OCONHC2H5	9600
Diallate (Avadex)	(CH3)2CH\N-C(=O)-S-H2C-ClHC=CHCl / (CH3)2CH	395
EPTC (Eptam)	H7C3\N-C(=O)-S-C2H5 / H7C3	1367
Phenmedipham (Betanal)	OCONH-C6H4-CH3 ; OCONHCH3	8000
Promecarb (Carbamult)	H3C, (H3C)2HC-C6H3-OCONHCH3	74
Propham (IPC)	phenyl—NH-COOCH(CH3)2	5000
Propoxur (Baygon)	OCONHCH3 ; OCH(CH3)2	90
SWEP	Cl, Cl-C6H3-NHCOOCH3	522
Vernolate (Vernam)	H7C3\N-C(=O)-S-C3H7 / H7C3	1170

Carbofuran. Carbofuran was introduced in 1967 under the registered trademark, Furadan®, by the Niagara Chemical Division of FMC Corporation (Table 1). It is a broad-spectrum, residual insecticide and nematicide effective by contact, ingestion, and systemic action. Carbofuran is usually applied to foliage for insect control and is placed in or above seed furrows or broadcast for nematode control. Some of the pests controlled by carbofuran include corn rootworm, rice water weevil, wire-worm, sugar cane borer, alfalfa weevil, snout beetle, armyworms, budworms, corn flea beetle, pea aphid, thrips, and hornworms. This insecticide has also been used in the production of bananas, coffee, and sugar beets (27, 77).

Even though it is effective in controlling many insects, carbofuran has shown poor insect control in certain areas. The loss of efficacy was attributed to rapid degradation of the pesticide by microorganisms when used for controlling phylloxera in a vineyard (171). The rapid degradation of carbofuran was associated with high levels of actinomycetes in the soil (171), prior use of the pesticide (26, 50, 59, 158), and soil conditions (122, 166, 169). Carbofuran-degrading microorganisms have been isolated from various terrestrial and aquatic environments (Table 2). Soil enrichment cultures were used to study the degradation of carbofuran under laboratory conditions (115, 120, 121). The incubation of uniformly ring-labeled ^{14}C[carbofuran] with an enrichment culture developed at 35°C converted 90% of the radioactivy to $^{14}CO_2$ in 5 days (121). Likewise, enrichment cultures obtained from three flooded soils previously exposed to carbofuran or its hydrolytic product, carbofuan phenol, degraded carbofuran rapidly more than enrichment cultures obtained from unexposed soils (116). An *Arthrobacter* sp. isolated from an enrichment culture that was obtained from a pesticide-treated flooded soil, mineralized the ring-labeled ^{14}C[carbofuran] to $^{14}CO_2$ within 72 to 120 hr by utilizing the chemical as a sole source of carbon and nitrogen under aerobic conditions (121). No degradation of carbofuran occurred under anaerobic conditions, and the mineralization was more rapid at 35°C than at 20°C. In similar studies, an *Azospirillum liopferum,* two *Streptomyces* spp. (166), and species of *Achromobacter* and *Pseudomonas* were also implicated in the degradation of carbofuran (50).

Attempts have been made to understand the biochemical mechanisms of inactivation of this pesticide in soil environments. Camper et al. (26) identified the degradation products of carbofuran as 3-ketocarbofuran and 3-hydroxycarbofuran (Fig. 1). Another study found that even though carbofuran phenol and 3-hydroxycarbofuran were identified as metabolites of carbofuran, these compounds never accumulated in large quantities, thus suggesting their further degradation through ring cleavage (121). Alternatively, in flooded soils, carbofuran phenol was found to be the major product of bacterial metabolism of carbofuran; 3-hydroxy carbofuran was the minor product (167). The amount of carbofuran phenol decreased after incubation for 30 days, thereby indicating its slow transformation. Contrary to these studies, carbofuran phenol was found to be persistent in a carbofuran-amended soil (119). An *Arthrobacter* sp. isolated from this soil rapidly degraded 3-hydroxy carbofuran and 3-ketocarbofuran but not carbofuran phenol. These findings suggest that the major pathway involved in the initial breakdown of carbofuran is hydrolysis, yet little is known about the fate of hydrolytic metabolites of carbofuran.

Initial studies on the fate of carbofuran in enrichment cultures led to the isolation of microorganisms that were used for investigating the metabolic pathways and respective enzymes involved in the degradation of the pesticide. Venkateswarlu and Sethunathan (166) isolated *Pseudomonas cepacia* and a *Nocardia* sp. from a flooded alluvial soil. The *Nocardia* sp. metabolized carbofuran to colored, water soluble, nonextractable metabolites. An *Achromobacter* sp. hydrolyzed the pesticide to carbofuran phenol, and the hydrolytic enzyme was regulated and repressed during growth

Table 2. Microbial degradation of carbamates.

Compound	Microorganism	Reference
Carbofuran	*Arthrobacter* sp.	121
	Actinomycetes	171
	Azospirillum lipoferum *Streptomyces* spp.	166
	Achromobacter sp. *Pseudomonas* sp.	50
	Arthrobacter sp. *Micrococcus* sp. *Bacillus* sp.	119
	Pseudomonas cepacia *Nocardia* sp.	169
	Achromobacter sp.	65
	Pseudomonas spp. *Flavobacterium* spp.	32
Carbaryl	*Pseudomonas* spp. *Rhodococcus* sp.	78
	Pseudomonas aeruginosa strain 50581	30
	Bacillus sp. *Micrococcus* sp.	119
	Several fungal and bacterial isolates	126
1-Naphthol	*Pseudomonas aeruginosa* strain 50552	30
Aldicarb	*Acrhomobacter* sp.	65
	Pseudomonas sp.	32
	Pseudomonas sp. *Nocardia* sp. *Arthrobacter* sp.	123
EPTC	*Fusarium oxysporum* *Epicoccum purpurascens* *Paecilomyces lilacinus* *Penicillum* spp. *Diheterospora* spp. *Bacillus* sp. *Alcaligenes* sp. *Micrococcus* sp. *Pseudomonas* sp.	81
	Arthrobacter sp. strain TE1	156
	Flavobacterium sp. strain Vl.15	96
	Rhodococcus sp. strain JE1	43
Butylate, vernolate	*Flavobacterium* sp. strain Vl.15	96
Barban, CIPC, IPC, and SWEP	*Pseudomonas alcaligenes*	89
BIPC, CIPC	*Pseudomonas cepacia*	165
Propoxur	*Achromobacter* sp.	65
Aniline	*Moraxela* sp. strain G	179
	Rhodococcus erythropolis	10
	Pseudomonas multivorans	61
	Alcaligenese faecalis	152
	Pseudomonas sp. strain CIT1	9
	Pseudomonas sp. strain SB3	64
3-Chloroaniline	*Pseudomonas cepacia*	165
	Moraxela sp.	179
3,4-Dichloroaniline	*Pseudomonas putida*	175
Desmedipham, Phenmedipham, Promecarb	*Pseudomonas putida* *Flavobacterium* sp. *Aspergillus versicolor*	76

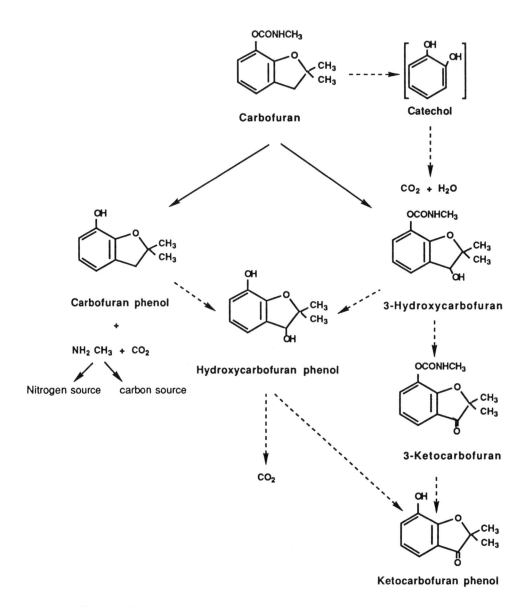

Figure 1. Proposed pathways for the microbial degradation of carbofuran.

on nitrogen-rich medium (65). Whether this hydrolase was an esterase (cleaving between the carbonyl group of N-methyl carbamic acid and the phenol) or an amidase (cleaving between the carbonyl and amine moieties of N-methyl carbamic acid) is not clear. The purified enzyme, with a molecular size of 150 kD (67), also catalyzed the degradation of other N-methyl carbamates, such as carbaryl, aldicarb, and baygon (2-isopropoxyphenyl-N-methylcarbamate) but not chloropropham (isopropyl-N-3-chloro-phenylcarbamate) or EPTC (S-ethyl dipropylthiocarbamate) (42).

Study of soil enrichment cultures and isolated microorganisms has proved useful in defining the fate of carbofuran in soil, but the features that could be common in various soils, enrichment cultures, and their individual members have been poorly investigated. Our studies therefore focused on comparing enrichment cultures and their members obtained from various soils and comparing the mechanisms of pesticide degradation simultaneously in enrichment cultures and isolated bacteria (32). In this investigation, 17 soil samples collected from various geographical areas yielded 12 enrichment cultures that metabolized carbofuran to a variable degree. Some cultures hydrolyzed carbofuran to carbofuran phenol only, whereas no other metabolites could be detected in other cultures. Further study of these enrichment cultures led to the isolation of 15 bacteria belonging to either *Pseudomonas* or *Flavobacterium*. Six isolates, placed in group I, utilized carbufuran as sole source of nitrogen whereas seven isolates, placed in group II, used carbofuran as a sole source of carbon. Two other isolates, placed in group III, also utilized the pesticide as sole source of carbon but degraded carbofuran completely and more rapidly. When ring-labeled [^{14}C]carbofuran was used as carbon source for group II isolates, up to 40% of the pesticide was lost as $^{14}CO_2$ in 1 hr, and no metabolic product was detected in the culture medium. The isolates of this group metabolized carbofuran via an oxidative pathway. The proposed pathways involved in the catabolism of carbofuran are shown in Figure 1. Isolates of groups I and II hydrolyzed carbofuran to carbofuran phenol. Crude cell extracts prepared from selected isolates of these groups exhibited hydrolase activity. The enzyme activity was greater in isolates of group II compared to isolates of group I. The bacteria utilizing carbofuran as a nitrogen source exhibited a mechanism of carbofuran degradation similar to that of the *Achromobacter* sp. strain WM111 reported by Karns et al. (65). The strain WM111 harbored several plasmids, and the carbofuran-degradation function was found to be plasmid-encoded (157). In a manner similar to the hydrolase from *Achromobacter* sp., the hydrolase enzyme from the isolates of group I and II also was able to hydrolyze other methylcarbamates, including carbaryl and aldicarb. The plasmid-encoded gene (*mcd*) for the mehylcarbamate degradation enzyme (hydrolase) has been cloned (157); however, the cloned gene expressed poorly in the other microorganisms tested, including *Achromobacter pestifer*, *Acinetobacter calocaceticus*, *Alcaligenes eutrophus*, and *P. putida*. A probe with the clone *mcd* gene of *Achromobacter* sp. failed to hybridize with plasmid or chromosomal DNA of any of the above 15 bacteria isolated by Chaudhry and Ali (32). These results suggest that the hydrolase gene in these bacteria is different from the hydrolase gene of *Achromobacter* sp.

Pseudomonas sp. 50432 of group III effectively decontaminated carbamate-polluted water samples. This result suggests a potential use of this bacterium for the detoxification of pesticide-polluted environments (35). The isolate harbored several plasmids. A study on the function of these plasmids may help to elucidate the oxidative pathway involved in degradation of carbofuran. Recent studies with the *Pseudomonas* sp. 50432 showed that at least three distinct metabolites of carbofuran can be detected in the culture medium. The first metabolite accumulates transiently during the first 4 to 8 hr of growth of the bacterium on carbofuran. This metabolite is not detected upon further incubation of the cultures, however. The cells grown in minimal medium containing glucose or in LB medium had long lag phases when used as inocula in minimal medium containing carbofuran as a sole source of carbon, but the induced cell suspensions metabolized carbofuran completely without any lag period, and the crude extracts of induced cells rapidly converted carbofuran to the transient metabolite. The crude extracts exhibited an enzyme activity responsible for the transient metabolite. These results suggest that the transient metabolite may be the first intermediate of the carbofuran metabolic pathway in strain 50432. The other two metabolites were consistently detected in stationary cultures. Both metabolites were separated by HPLC based upon their retention time on reverse phase column and color. The yellow-colored metabolite was detected first and could be purified up to approximately 90%. This was followed immediately by a pink compound. The peak containing pink compound appeared not to be homogenous, however. Further identification of these metabolites should help in defining the pathway of carbofuran in *Pseudomonas* sp. 50432.

Carbaryl. Carbaryl is the common name for the active ingredient of the insecticide manufactured by Rhone-Poulenc (formerly Union Carbide Corporation) under the trade name, Sevin® (Table 1). This pesticide is available as wettable powder, sprayable powder, granule, dust, and other formulations. Carbaryl is the least toxic among the *N*-methylcarbamates and is by far the most widely used carbamate insecticide. Some of the important uses of carbaryl include applications to cotton, corn, soybeans various fruit and vegetable crops, bananas, pineapple, olives, cacao, coffee, rice, and sugar cane for the control of various insects. It is also used in forests, range land, livestock, poultry, and buildings. In response to public concern regarding the safety of carbaryl, the New Jersey Department of Environmental Protection temporarily suspended use of this insecticide (86).

Although carbaryl has been effective against several insects, it also has shown reduced efficacy in soils previously treated with the pesticide. Three applications of carbaryl to a submerged soil resulted in the rapid degradation of the pesticide (117). Studies of other soils showed that the residual pesticide was detected to be 28% and 90% after four days of incubation with ring-labeled [^{14}C]carbaryl in samples from the treated and untreated soils, respectively (126). In another study, an untreated soil that failed to degrade carbaryl was induced to metabolize the pesticide rapidly when treated with 1-naphthol, a hydrolysis product of carbaryl (118). These investigations suggest that like other carbamates, carbaryl degradation is also enhanced by repeated applications of this pesticide and its metabolites. Investigation of carbaryl-polluted river samples indicated that both the abiotic and biotic degradation processes play an important role in the degradation of carbaryl (86).

The degradation of carbaryl has been reported to be microbially mediated (30, 78, 97, 118, 119, 126). The observation prompted studies to isolate microorganisms capable of degrading carbaryl from soils with a history of the pesticide treatment (Table 2). A *Pseudomonas* sp. (NCIB 12042) and a *Rhodococcus* sp. (NCIB 12038) isolated from a garden soil utilized carbaryl as the sole source of carbon and nitrogen (78). Although both bacteria hydrolyzed the pesticide, the pathways for further metabolism of 1-naphthol were found to be different in these bacteria. Strain NCIB 12042 metabolized 1-naphthol via salicylic acid only, whereas the strain NCIB 12038 metabolized it via both salicylic and gentisic acids (Fig. 2). In contrast, another *Pseudomonas* sp. (NCIB 12043), isolated from the same soil by perfusion column enrichment, metabolized carbaryl rapidly to 1-naphthol, which was further degraded via gentisic acid alone. The pathway for 1-naphthol degradation in NCIB 12042 exhibited similarities to that of naphthalene catabolism in *Pseudomonas* spp. (36). In a separate study, up to 70% of carbaryl was found to be hydrolyzed by a *Bacillus* sp. isolated by enrichment culture techniques from a flooded soil (119), whereas the second bacterial strain, *Micrococcus* sp. isolated from the same soil, hydrolyzed carbaryl poorly. It degraded only 15% of the added carbaryl under similar conditions. These reports indicate the functional diversity among the bacteria in adapting to catabolism of the same substrate.

A newly isolated soil bacterium, *P. aeruginosa* strain 50581, also hydrolyzed carbaryl to 1-naphthol and utilized the pesticide as a sole source of carbon (30). Unilke the carbofuran-hydrolyzing bacteria (32, 65, 121), this isolate did not degrade other carbamates. This result indicates that the hydrolase in *P. aeruginosa* isolate 50581 is different from the *mcd* gene product which has a wide substrate range. This was confirmed by DNA hybridization studies in which the *mcd* gene was used as the probe. No homology between the total DNA from the isolate 50581 and that *mcd* gene was

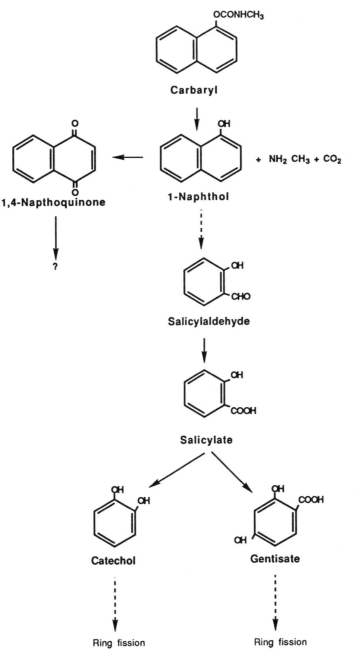

Figure 2. Proposed pathways for the microbial degradation of carbaryl.

observed. Mulbry and Eaton (97) purified a cytosolic enzyme from *Pseudomonas* strain CRL-OK isolated from sewage sludge. The bacterium utilizes carbaryl as its sole source of carbon and energy. The enzyme hydrolyzes carbaryl and other N-methylcarbamates such as carbofuran and aldicarb, but the purified enzyme did not hydrolyze thiocarbamate or phenylcarbamate, such as EPTC and CIPC, respectively. This enzyme consists of two identical subunits of 85 kD, and the size of the native enzyme was determined to be 187 kD (97). It has a pH optimum of 8.5 and temperature optimum of 60°C. The carbaryl-hydrolyzing strain 50581 isolated from soil (40) also was found to have a novel enzyme activity. This activity was specific to carbaryl but not to N-methyl-carbamate and was associated with the membrane fraction. The partially purified membrane-bound hydrolase had activity over a broad pH range of 6.0 to 8.5 and showed the highest activity at pH 8.5. It showed an optimal activity at 45°C. The enzyme activity was precipitated between 30 to 60% saturation with ammonium sulphate. Further, partial purification of the enzyme was carried out via DEAE sepharose and G-200 sephadex column chromatography. The partially purified enzyme had K_m and V_{max} values of 833 μM and 9 μmol/min, respectively.

Strain 50581 harbored a 50 kb plasmid, pCD1. Initial attempts to cure the plasmid DNA were not successful, but conjugation experiments between 50581 and 50552 led to the isolation of derivatives of the latter bacterium that completely metabolized carbaryl (29a, 30). Analysis of these transconjugants should yield information on the gene(s) of 50581 involved in the hydrolysis of carbaryl in isolate 50581.

Aldicarb.   Aldicarb was the first oxime carbamate pesticide, introduced under the trade name of Temik by Union Carbide Corporation in 1967 (Table 1). Aldicarb has been used on tobacco, sugar beets, sugar cane, potatoes, and peanuts for the control of aphids, thrips, mealybugs, white flies, mites, and nematodes. Only the granular formulation of aldicarb is marketed, mostly because of its high mammalian toxicity (Table 1).

Although microbial degradation of aldicarb is poorly understood, it has been shown to be oxidized to its sulfoxide and sulfone metabolites and hydrolyzed to its oxime and nitrile metabolites (84, 94, 108, 143, 146). Both the oxidative and hydrolytic products have been detected in the soil (123). The oxidation products may be as toxic as the parent compound, but the hydrolytic metabolites have low toxicity (55). Therefore, the mechanism of its tranformation under various environmental conditions is important to know.

Attempts have been made to isolate microorganisms responsible for aldicarb degradation in soil. Several isolates of *Fusarium, Penicillium, Arthrobacter, Pseudomonas, Nocardia, Achromabacter,* and *Bacillus* have been isolated from aldicarb-treated soil (123). The fungal isolates, *Fusarium* and Penicillium, slowly metabolized aldicarb. Although the bacterial isolates collectively degraded the pesticide rapidly, none of the single bacterial strains degraded aldicarb or its toxic metabolites. The degradation of aldicarb was concentration-dependant. Concentrations of the pesticide higher than 800 ppm and 5000 ppm inhibited bacterial and fungal growth, respectively. The microorganisms capable of degrading carbofuran were also found to metabolize aldicarb (32, 59, 65, 121). Aldicarb was reported to be hydrolyzed under anaerobic conditions as well (74, 108). Aldicarb stimulated methanogenesis, since methylamine, a metabolite of the pesticide, was utilized as the source of energy by the methanogenic bacteria (73, 74).

Thiocarbamates

Thiocarbamates are used as herbicides and fungicides. These compounds differ structurally from other carbamate insecticides in that they are primarily thio- or dithiocarbamates (93). The most widely used chemicals in this group are butylate (S-ethyl diisobutylthio-carbamate), cycloate (S-ethyl N-ethylthiocyclohexane carbamate), diallate [S-(2,3-dichloro-allyl) diisopropylthiocarbamate], EPTC, and vernolate (S-propyldipropyl-thiocarbamate) (Table 1). EPTC was the first among the thiocarbamates to be developed and registered for its use in alfalfa, corn, cotton, potatoes, fruit crops, ornamentals, and a number of bean crops for control of weeds.

As with methylcarbamates, repeated exposure of soils to thiocarbamates results in a reduction in the efficacy of these compounds (76, 127, 141, 142). The reduced efficacy has been attributed to the rapid microbial degradation of the herbicides (40, 57, 60, 69, 114). Microbial populations evolved traits responsible for the degradation of structurally related thiocarbamates, even though the soils were exposed to only one of the herbicides (172). If, however, the herbicides are structurally different, adaptation of microorganisms to degrade one herbicide may not confer upon them the ability to degrade other herbicides of the same group. Since vernolate and EPTC are similar in structure (Table 1), prolonged exposure of soil to either herbicide resulted in adaptation of microorganisms that degrade both the compounds (172); however, microorganisms adapted to the degradation of butylate, which is somewhat structurally different from EPTC and vernolate, failed to degrade EPTC and vernolate. Similarly, cycloate degradation was not influenced by either EPTC, vernolate, or butylate, because of the presence of a benzene ring (Table 1). Several microorganisms capable of degrading EPTC have been isolated (81, 82, 107, 114), as listed in Table 2.

Lee and his co-workers (82) isolated 29 fungal and 9 bacterial isolates capable of degrading EPTC. Although all the fungal isolates retained the ability to degrade EPTC, the bacterial isolates lost this function after 15 months of storage. Lee et al. speculated that the loss of EPTC degradation may be associated with the loss of the plasmids harbored by these bacteria. The loss of plasmid-associated degradative function has also been reported in other herbicide-degrading bacteria (45). These findings prompted a search for plasmids involved in degradation of EPTC. Tam et al. (156) isolated an EPTC-degrading bacterium, *Arthrobacter* sp. strain TE1, that harbored four plasmids of 65.5, 60.0, 50.5, and 2.5 mega daltons (MD) in size. One of these plasmids (50.5 MD) was cured (either spontaneously or by acridine orange treatment), and the cured derivatives of the strain TE1 lacked the ability to degrade EPTC. The results of the curing studies suggested that the 50.5 MD plasmid may be associated with catabolism of EPTC. The involvement of this plasmid in EPTC degradation was confirmed by conjugal transfer of the 50.5 MD plasmid back into the mutants of strain TE1, since the transconjugants recovered the ability to degrade EPTC. Dick et al. (43) isolated a *Rhodococcus* strain JE1 from a loamy soil that was capable of metabolizing EPTC and exhibited a plasmid profile similar to *Arthrobacter* strain TE1. *Rhodococcus* plasmids could not be cured, however, and no function could be attributed to its plasmids. In another study, a *Flavobacterium* sp. strain Vl.15, isolated from soil with a history of vernolate use, utilized butylate, EPTC, or vernolate as sole sources of carbon (96). This bacterium also harbored two plasmids, pSMB1 and pSMB2, but only one of the plasmids, pSMB2, was thought to have butylate-utilizing ability. The above studies demonstrate that genes for degradation of thiocarbamates are encoded on plasmids in the isolated bacteria.

Recent investigations of the microbial metabolism of EPTC indicated that the *Rhodococcus* strain JE1 metabolized EPTC to propionaldehyde and N-dipropyl EPTC (43). N-dipropyl EPTC was further degraded to mercaptan by the same isolate. Behki and Khan (19) proposed a different pathway for EPTC degradation in the *Arthrobacter* strain TE1. They reported that dipropyl amine was found in culture media as one of the metabolites of EPTC. Mutants of the TE1 strain that were deficient in EPTC degradation metabolized dipropyl amine or propyl amine as efficiently as the parent strain. This finding demonstrated that the 50.5 MD plasmid of TE1 codes for the determinants involved only in the initial cleavage of the thioester linkage in EPTC.

Phenylcarbamates

Phenylcarbamates, or carbanilates, are used as herbicides and insecticides (Table 1). Phenylcarbamates were among the earliest herbicides developed (68). Like other carbamates, phenylcarbamates are rapidly degraded in soils, particularly those with a history of herbicide treatment. Again, the rapid degradation or loss in efficacy of these herbicides has been shown to be associated with the adaptation of microbial populations (Table 2). Vega et al. (165) isolated *Pseudomonas cepacia,* which utilized two phenylcarbamate herbicides, BIPC (1-methyl-prop-2-ynyl-3-chlorophenyl carbamate) and chlorpropham (isopropyl-*N*-3-chlorophenyl carbamate), as the sole source of carbon and energy. Similarly, a *P. alcaligenes* was found to hydrolyze four phenylcarbamate herbicides—BIPC, chlorpropham, propham (isospropyl-*N*-phenylcarbamate), and SWEP (methyl 3,4-dichloro-phenylcarbamate)—to corresponding anilines: aniline from propham, 3-chloroaniline from BIPC and chlorpropham, and 3,4-dichloroaniline from SWEP (89) (Fig. 3). This isolate also metabolized barban (4-chlorobutynyl 3-chlorocarbanilate), but without the production of 3-chloroaniline, even though barban has the same 3-chlorophenyl carbamate group as BIPC and chloropropham (88) (Table 1). In both the above isolates, the herbicide-degrading function was found to be stable and inducible. *P. alcaligenes* was used to purify the enzyme responsible for the hydrolysis of phenylcarbamates (90). The isolated enzyme, an amidase, catalyzed the hydrolysis of all of the structurally related herbicides but failed to use barban and carbetamide as substrates.

Aniline and its ring-substituted derivatives were also found to be the degradative products of other herbicides that accumulate in soil and represent important pollutants. Since spontaneous transformation of anilines in soil is very slow (153), studies have been conducted on the microbial degradation of aniline and substituted anilines (10, 61, 75, 79, 152, 174, 177–179). A *Moraxella* sp. strain G, isolated from a chemostat inoculated with soil, utilized aniline, 2-chloroaniline, 3-chloroaniline, and 4-chloro-, 4-fluoro-, and 4-bromoanilines but not 3,4-dichloraoaniline as sole source of carbon and nitrogen (178, 179). In another study, rapid degradation of 3,4-dichloroaniline was observed in soils treated with aniline (174). 3,4-Dichloroaniline was also rapidly mineralized in soil samples incubated with *Pseudomonas putida* isolated from sewage by enrichment on propionanilide. Degradation of 3,4-dichloroaniline occurred involving the metabolite 4,5-dichlorocatechol, 3,4-dichloromuconate, 3-chlorobutenolide, 3-chloromaleylacetate, and 3-chloro-4-ketoadipate with succinate and acetate (175).

Biochemical characterization of the isolated microorganisms (9, 11, 64) revealed that anilines are generally degraded to catechols, which in turn are subjected to well established *ortho-* or *meta-*cleavage pathways (24, 61), except 4-chloroaniline, which was degraded via a modified *ortho-*cleavage pathway (75, 179) (Fig. 3).

The ability of microorganisms to degrade the phenylcarbamate herbicides depends on the presence of broad-specificity hydrolases and oxygenases. These enzymes are involved in the transformation of herbicides first to their corresponding anilines and then to the catechols. The catechols are further metabolized to tricarboxylic acid (TCA) cycle intermediates. Lack of one or more of these enzymes in the microorganisms may be the cause of incomplete degradation of phenylcarbamate herbicides.

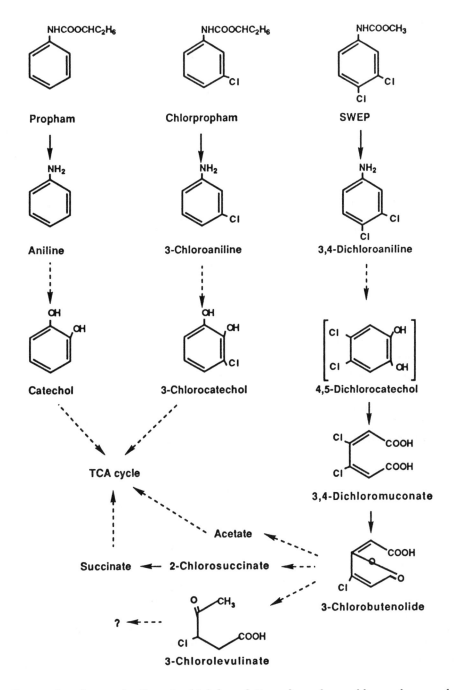

Figure 3. Proposed pathways for the microbial degradation of propham, chlorpropham, and SWEP.

Other Carbamates

Microbial degradation of other carbamates such as desmedipham, phenmedipham, and promecarb (Table 1) was investigated by Knowles and Benezet (76). Both bacterial and fungal species were isolated from soil that degrade these pesticides. Metabolites resulting from the cleavage of the ester linkage of the three carbamates were identified as ethyl-N-(3-hydroxyphenyl)carbamate, methyl-N-(3-hydroxyphenyl)carbamate, and isothymol from desmedipham, phenmedipham, and promecarb, respectively. In three other independent studies, benfuracarb, bufencarb, carbosulfan, cleothocarb, furathiocarb, propoxur, and trimethacarb were found to be ineffective against cabbage and sorghum pests in soils previously treated with carbofuran (59, 151, 170). The loss of efficacy was attributed to the rapid degradation of these pesticides by microorganisms. Yardon et al. (173) investigated the degradation of methyl-benzimidazol-2-ylcarbamate (MBC) in five soils. MBC is the hydrolytic product of benomyl [methyl 1-(butylcarbamoyl) benzimidazole-2-ylcarbamate]. MBC was rapidly degraded in soils previously treated with benomyl; however, degradation was delayed in soils treated with the fungicide, tetramethylthiuram disulfide. Based upon these obsevations, Yardon et al. proposed that fungal species may be involved in the rapid degradation of MBC.

Degradation of another carbamate, aminocarb (4-dimethylamino-3-methylphenyl-N-methylcarbamate), was reported in cultures of human intestinal bacteria with nutrient broth supplemented with 250–1000 μg aminocarb/ml 921). Only traces of aminocarb were detected in aerobic cultures of bacteria after 5 days.

Most of the studies on the degradation of carbamate pesticides are concerned with aerobic microorganisms. Very little is known about the degradation of these compounds by anaerobic microorganisms. Since anaerobic conditions may be present in waterlogged soils, aquatic sediments, and subsurface soil, efforts should be focused on degradation of carbamates under anaerobic conditions.

ORGANOPHOSPHATES

A number of synthetic organophosphates such as lubricants, plasticizers, and pesticides are produced for commercial use. These compounds are relatively less persistent but more effective compared to the chlorinated compounds used for similar purposes (93). They have replaced some of the recalcitrant chlorinated compounds and are one of the widely used classes of pesticides (48). Approximately 140 organophosphate compounds are currently used as pesticide and plant growth regulators world-wide, and over 60,000 tons of these chemicals are produced annually in the United States alone. Microbial degradation of selected organophosphates is discussed below.

Parathions

Parathion (Table 3) was first synthesized in 1944. Parathion and its methyl analog, methyl parathion (O,O-dimethyl O-p-nitrophenyl phosphorothioate) (Table 3), have been widely used for controlling insects of agricultural and public health importance. Although parathions are considered to be less persistent, these pesticides and their metabolites, particularly p-nitrophenol, have caused environmental pollution (16, 72). p-Nitrophenol imparts odor problems in water resources.

Table 3. Structure and toxicity (rat, oral) of organophosphates.

Common name (trade name)	Structure	LD$_{50}$ (mg/kg of body wgt).
Acephate (Orthene)	H$_3$CO, O P, H$_3$CS, NHCOCH$_3$	866
Carbophenothion (Trithion)	H$_3$CO, S P, H$_3$CO, S-CH$_2$-S—⟨ring⟩-Cl	6
Chlorpyriphos (Dursban)	Cl, Cl ⟨pyridine ring⟩ N, O-P(OC$_2$H$_5$)$_2$ (S)	97
Coumaphos (Co-Ral)	(H$_5$C$_2$O)$_2$ P-O—⟨coumarin ring⟩, CI, CH$_3$ (S)	13
Diazinon	(H$_5$C$_2$O)$_2$ P-O—⟨pyrimidine ring⟩-CH(CH$_3$)$_2$, CH$_3$ (S)	66
Dichlorovos (Vapona)	(CH$_3$O)$_2$ P-O-CH=CCl$_2$ (O)	25
Dimethoate (Cygon)	(CH$_3$O)$_2$ P-S-CH$_2$C-NH-CH$_3$ (O)(O)	250
EPN	H$_5$C$_2$O, S P-O—⟨ring⟩-NO$_2$, H$_5$C$_6$	7
Ethoprop (Mocap)	H$_5$C$_2$O-P(S-C$_3$H$_7$)$_2$ (O)	61
Fenitrothion (Sumithion)	(H$_3$CO)$_2$ P-O—⟨ring⟩-NO$_2$, CH$_3$ (S)	250
Fensulfothion (Dasanit)	(H$_5$C$_2$O)$_2$ P-O—⟨ring⟩-S-CH$_3$ (S)(O)	2
Fonofos (Dyfonate)	H$_5$C$_2$O, S P-S—⟨ring⟩, H$_5$C$_2$	8
Isofenphos (Oftanol)	H$_5$C$_2$O, S P-O—⟨ring⟩, (H$_3$C)$_2$CH-NH, COO-CH(CH$_3$)$_2$	38

continued

Table 3. Continued.

Common name (trade name)	Structure	LD$_{50}$ (mg/kg of body wgt).
Malathion (Cythion)	$(H_3CO)_2$ $\overset{S}{\underset{\parallel}{P}}$-S-$\overset{CH_2\text{-}\overset{O}{\overset{\parallel}{C}}\text{-}OC_2H_5}{\underset{\overset{\parallel}{O}}{CH\text{-}C\text{-}OC_2H_5}}$	885
Methidathion (Supracide)	H_3CO-ring with N−N-CH_2-S-$\overset{S}{\overset{\parallel}{P}}(OCH_3)_2$	25
Methylparathion	$(H_3CO)_2$ $\overset{S}{\overset{\parallel}{P}}$-O-⟨ring⟩-$NO_2$	9
Monocrotophos (Azodrin)	$(H_3CO)_2$ $\overset{O}{\overset{\parallel}{P}}$-O-$\overset{CH_3}{\underset{}{C}}$=CH$\overset{O}{\overset{\parallel}{C}}$-NH-$CH_3$	21
Parathion (Niram)	$(H_5C_2O)_2$ $\overset{S}{\overset{\parallel}{P}}$-O-⟨ring⟩-$NO_2$	3
Phorate (Thimet)	$(H_5C_2O)_2$ $\overset{S}{\overset{\parallel}{P}}$-S-$CH_2$-S-$C_2H_5$	2

Parathion can be hydrolyzed chemically at high pH, and the hydrolytic product escapes into the atmosphere by volatilization. The bulk of organophosphate pesticides are removed from the environment by microbial degradation, however (68, 103). Microorganisms (Table 4) capable of utilizing parathion as sole source of carbon were isolated from pesticide-treated soils (56). Nelson et al. (105) observed an increase in the bacterial population of a loamy soil proportional to an increase in the concentration of parathion applied. These observations suggested that microbial growth was stimulated and that parathion was used as a carbon source. In another study, enrichment cultures from flooded soils and water and sediment from river, lake, and pond were found to hydrolyze parathion (16). Flooded soil conditions favored the hydrolysis of parathion. More $^{14}CO_2$ was evolved from ring-labeled [^{14}C]parathion in the rhizosphere of rice seedlings under flooded conditions than under nonflooded conditions, thus indicating that soil planted with rice under flooded conditions permits significant ring cleavage by microorganisms (124). Similar to carbamates, enhanced degradation of parathions was reported in soils previously exposed to the pesticide. Ferris and Lichtenstein (51) used soil samples treated with parathion or p-nitrophenol for 4 days to investigate the fate of ring-labeled [^{14}C]parathion. On the average, 37% and 2% of the spiked [^{14}C]parathion was released as $^{14}CO_2$ in treated and untreated samples, respectively, after 24 hr of incubation.

Table 4. Microbial degradation of organophosphates.

Compound	Microorganism	Reference
Parathion	*Bacillus* strains *Arthrobacter* strains	104
	Flavobacterium sp. ATCC 27551 *Pseudomonas* sp. ATCC 29353	2
	Pseudomonas spp.	128
	Pseudomonas sp.	33
	Pseudomonas strain CTP-01	29
	Pseudomonas diminuta MG	134
	Pseudomonas sp.	102
	Strain SC	98
Methyl parathion	*Pseudomonas* sp.	33
	Pseudomonas sp. isolate 50541	39
	Flavobacterium sp. ATCC 27551	2
	Aufwuchs	85
p-Nitrophenol	*Flavobacterium* sp.	33
	Pseudomonas spp. PNP-1, 2, and 3	62
	Pseudomonas sp. isolate 50445	39
	Moraxella sp.	147
	Pseudomonas sp. strain 24	133
DETP	*Pseudomonas acidovorans*	38
Diazinon	*Flavobacterium* sp. ATCC 27551 *Pseudomonas* sp. ATCC 29353	2
	Pseudomonas spp.	128
	Pseudomonas sp.	14
	Arthrobacter sp. SB3 and SB4	17
Isofenphos	*Arthrobacter* sp.	113
	Pseudomonas sp.	112
Fenetrothion	*Flavobacterium* sp. ATCC 27551	2
Dichlorovos	*Pseudomonas aeruginosa* sp. *Pseudomonas* sp.	83
Coumaphos	Isolates B-1, B-2, and B-3	139
Methidathion	*Bacillus coagulans*	54

Nelson (105) isolated 50 microorganisms from a loamy soil with a history of parathion use. He found that 8 of the isolates hydrolyzed 75% of the added 10 μg parathion/g dry soil in 5 days and appeared to be *Bacillus* strains. Ten of the isolates of the *Arthrobacter* genus hydrolyzed all of the parathion after 5 days. *Arthrobacter* sp. degraded parathion even in rich medium, suggesting that the enzymes are constitutively expressed. The *Bacillus* sp. lost the capability for degradation upon subculturing, whereas *Arthrobacter* sp. did not. This result suggested that genes for hydrolysis of parathion may be plasmid-borne in *Bacillus* sp. In another study, two *Pseudomonas* strains isolated from soil and sewage utilized parathion and malathion [S-(1,2-dicarboxy-ethyl)-O,O-dimethyl dithiophosphate] as sources of carbon (128). These bacteria hydrolyzed parathion to p-nitrophenol. A *Flavobacterium* sp ATCC 27551 isolated from flooded soil hydrolyzed both diethyl (parathion and diazinon) and dimethyl (methyl parathion and fenitrothion) phosphorothioates, while another bacterium, *Pseudonomas* sp. ATCC 29353, isolated from the same soil, hydrolyzed only diethyl (parathion and diazinon) phosphorothioates (2). Similar observations were recorded by Chaudhry et al. (33), who isolated two mixed bacterial cultures that utilized both methyl parathion and parathion as a sole source of carbon. A *Pseudomonas* sp. isolated from these mixed cultures degraded the pesticides to *p*-nitrophenol only, whereas the mixed cultures degraded the pesticides completely. An investigation of methyl parathion degradation by *aufwuchs,* an aquatic microbial growth attached to submerged surfaces and suspended in streamers or mats, demonstrated a rapid transformation of the pesticide (85). These and other (17, 29, 103) studies showed that parathions are generally first hydrolyzed to *p*-nitrophenol and diethylthiophosphate (DETP), which are further metabolized (Fig. 4).

The fate of *p*-nitrophenol in the environment has been extensively investigated. Several microorganisms have been reported to readily degrade *p*-nitrophenol in soil (87, 154), sediment (111, 140, 145, 148, 164), activated sludge (22, 23), water (63, 109, 149, 164), and groundwater (3). The degradation of *p*-nitrophenol has also been examined in the presence of inorganic nutrients (131, 154, 169), by bacteria in granular activated-carbon columns (144), at low concentrations (132, 169), and at high concentrations by immobilized bacteria in an aqueous waste stream (62). A *Moraxella* sp. isolated from activated sludge utilized *p*-nitrophenol as the sole source of carbon, releasing nitrite (147). Similarly, a *Flavobacterium* sp., degrading *p*-nitrophenol, released nitrite, which was assimilated into the bacterial cells (33). Spain and Gibson (145) investigated the enzymology and pathway of *p*-nitrophenol degradation in *Moraxella* sp. They found that membrane-bound *p*-nitrophenol oxygenase mediated the conversion of *p*-nitrophenol to hydroquinone, which was further metabolized to β-ketoadipate via γ-hydroxymuconic semialdehyde and maleyl acetate (Fig. 4).

The second metabolite of parathions, DETP, was metabolized by enrichment cultures obtained from cattle dip solution containing coumaphos (137). These cultures mineralized DETP to sulfate and phosphate while utilizing ethyl moieties as a carbon and energy source. Cook et al. (38) isolated *P. acidovorans* from sewage sludge, which utilized DETP as a sole source of sulfur. These studies suggest that the hydrolysis of parathions and subsequent degradation of *p*-nitrophenol and DETP by soil microorganisms may be responsible for the rapid removal of these chemicals from the environment.

Biochemical studies of microorganisms capable of degrading parathions showed the presence of a parathion hydrolase enzyme (25, 33, 103, 134, 135). *P. diminuta* cells grown for 48 hr exhibited 3400 units of parathion hydrolase activity per liter of culture (134). Cell free extracts of another *Pseudomonas* strain hydrolyzed parathion at a rate of 1×10^4 nmol/min/mg of protein (29). In another study, crude cell extracts prepared from a methyl parathion–hydrolyzing *Pseudomonas* sp. showed an optimum pH range from 7.5 to 9.5 from the enzymatic hydrolysis of methyl parathion (33). Mulbry and Karns (98) conducted extensive studies to characterize the parathion hydrolase from three bacteria isolated from different locations. The hydrolase from a *Flavobacterium* strain (2) was found to be membrane bound, having a single subunit of 35 kD. This enzyme was inhibited by sulfhydryl reagents such as dithiothreitol (DDT) and by metal salts such as cupric (II) chloride ($CuCl_2$). The enzyme from SC strain (98) was also membrane-bound but was composed of four identical subunits of 67 kD. It was inhibited by DDT but stimulated by $CuCl_2$. Unlike the above, parathion hydrolase was found in the cytosol of the B-1 strain (139). It is composed of a single subunit of about 43 kD and was stimulated by DTT but inhibited by $CuCl_2$. The relative affinities of the enzymes to parathion and *O*-ethyl-*O*-4-nitrophenyl phenylphosphonothioate (EPN), another organophosphate insecticide, also differed. The hydrolase from the B-1 strain

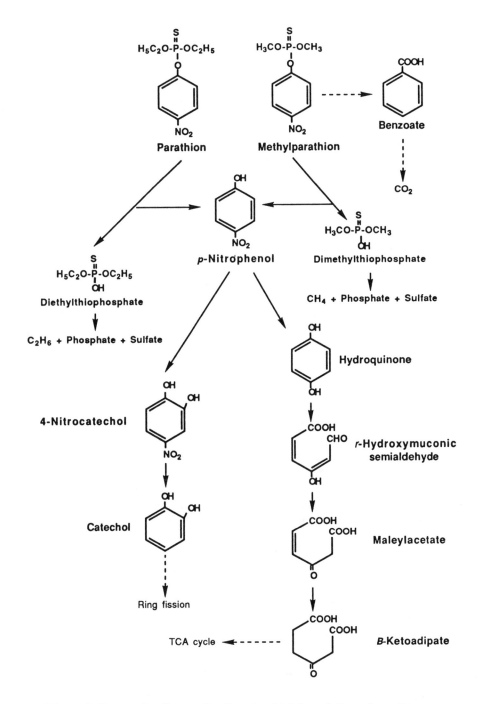

Figure 4. Proposed pathways for the microbial degradation of parathions.

showed equal affinity for both the insecticides, while the *Flavobacterium* enzyme displayed two-fold lower affinity for EPN than for parathion. The hydrolase for strain SC exhibited no afinity for EPN, however. The hydrolase from all of the tested strains was produced constitutively and had similar temperature optima (40°C) (98). In another independent study, the yield of parathion hydrolase was found to be improved 22-fold by growing *Pseudomonas* sp. on a complex medium rather than on parathion in minimal medium (102). Since the enzymes esterase, aryl esterase, or phospho-triesterase, involved in the inactivation of parathions from the isolated bacteria appeared to be isofunctional, their spread among the microorganisms may have a common ancestral origin.

Although methyl parathion has been known to be metabolized by oxidation also, little is known about the biochemistry of the oxidative pathway. A *Pseudomonas* sp. strain 50541 isolated from a pesticide-waste disposal site was found to oxidize methyl parathion rapidly and completely (39). Since mineralization of parathion in this bacterium is complete without accumulation of intermediate metabolites, it may have greater potential for detoxification of pesticide-contaminated sites than the microorganisms that hydrolyze parathions, in which p-nitrophenol accumulates as the hydrolysis product. Investigation of this isolate may be helpful in further defining the pathway for the degradation of parathions.

Parathion Hydrolase Gene (opd). In bacteria, genes encoding unique degradative functions are often plasmid-encoded and range in size from a few kilobases of DNA to several hundred kilobases (28, 110). Initial studies on parathion-hydrolyzing *Pseudomonas diminuta* strain MG showed that expression of parathion hydrolase activity from this strain was lost at a high frequency (9 to 12%) after treatment of the cells with mitomycin C 9134). The hydrolase-negative derivatives were found to be missing a plasmid, pCMS1, which was present in the wild-type organism, thus indicating that parathion hydrolase is encoded by the pCMS1 plasmid. Similarly, a plasmid pPDL2 from *Flavobacterium* sp. ATCC 27551 was found to be involved in the hydrolysis of parathion (100). The parathion hydrolase gene from both the *P. diminuta* and the *Flavobacterium* sp. has been cloned (58, 100, 133, 135) and sequenced (92, 99, 136). The cloned gene was located on a 1.5 kb BamHI (133) and a 7.3 kb EcoRI (100) fragment in *P. diminuta* and *Flavobacterium* sp., respectively. Hybridization studies demonstrated not only that the opd gene from both sources was homologous (58, 100) but also that it showed homology with total DNA from a *Pseuodomonas* sp. that hydrolyzed methylparation (33). Mulbry et al. 9101) characterized the plasmids pCMS1 and pPDL2 to determine the regions of homology other than the opd gene. The sizes of the pCMS1 and pPDL2 were reported to be 70 and 39 kb, respectively. Further, the opd genes of both plasmids were located within a highly conserved region of approximately 5.1 kb. This region of homology extended approximately 2.6 kb upstream and 1.7 kb downstream from the opd genes. No homology between the two plasmids was found outside this region. To characterize the cloned gene in detail, Mulbry and Karns (99) sequenced a 1.6 kb BamHl-Pstl fragment containing the opd gene of plasmid pPDL2. This fragment contained only one open reading frame large enough to encode the 35 kd protein. The amino acid composition of the purified protein corresponded well with that predicted from the nucleotide sequence, and the data suggest that the parathion hydrolase protein is procesed at its amino-terminus in *Flavobacterium* sp. Although a promoter region with perfect match at −35 but a less favorable match at −10 consensus sequences was identified upstream of the opd coding region, it failed to express in *P. putida* (99). The cloned gene, however, fused with lacZ expressed at higher levels in *E. coli* compared with the parent strain. Similarly, the cloned opd gene from *P. dimunita* expressed poorly in *E. coli* and in a *Pseudomonas* sp., even though a nif-type promoter sequence (44) was present upstream to the opd coding region (99). High levels of hydrolase activity were obtained, however, when the opd gene was placed under the control of lambda P_L promoter in *E. coli* (136), thus indicating that the opd gene in *E. coli* and *P. putida* requires an exogenous promoter. DNA sequencing studies showed that the nucleotide sequence of the two opd genes is identical (99, 136).

The *opd* gene has also been cloned in other microorganisms, and its product has been characterized (46, 130, 150). A comparision of the cloned and native hydrolases showed that the hydrolase synthesized in *E. coli* was larger in size than the *Flavobacterium* enzyme (99). In contrast, the cloned gene in gram-positive *Streptomyces lividans* had a gene product similar in size to the wild-type enzyme (150). In addition, the enzyme produced in *S. lividans* was excreted into the culture medium, thus suggesting that the enzyme required processing. In fact, the DNA sequencing results suggest the presence of a signal sequence in the precursor protein (99, 136). Rowland et al. (130) purified parathion hydrolase to homogeneity from the recombinant *S. lividans*. The recombinant hydrolase and the native hydrolase had similar characteristics, including molecular weight, temperature optima, and K_M values, except that the recombinant enzyme had a higher affinity for ethyl parathion (K_M value of 46 μmol/min/mg of protein) than the native enzyme (K_M of 211 μmol). Both proteins appeared to have been processed similarly in the native *Flavobacterium* sp. and the recombinant strain of *S. lividans* with the result that both had the same N-terminal amino acid sequence (130). Analysis of the native membrane-bound parathion hydrolases from *Flavobacterium* sp. and *P. diminuta* and the recombinant hydrolases from *E. coli* and *S. lividans* indicate that the *S. lividans* system can be used effectively to produce high levels of enzyme. Because the enzyme is secreted into the culture medum, it offers more potential for toxic waste treatment, since purifying the enzymes secreted into culture medium is relatively easier than purifying from the whole cell. Further, the excreted enzyme may have greater accessibility to chemicals that are not actively taken up by the bacterium or metabolites that are toxic to the host. Comparison of the properties of this enzyme with the isofunctional enzyme from another gram-positive bacterium, *S. pilosus*, should also be interesting (53).

Diazinon

Diazinon (*O,O*-diethyl *O*-2-isopropyl-4-methyl-6-pyrimidyl phosphorothioate) (Table 3) is used as a soil and foliar insecticide and is effective against a broad range of insect pests of crops and ornamental plants. It provides a good residual treatment for control of flies in barns and is used also in household sprays and dusts for ant and cockroach control (93). It has also been widely used on golf courses and sod farms and nurseries. Two *Pseudomonas* spp. isolated from sewage sludge were found to degrade diazinon, producing diethyl phosphorothioate as the metabolite in the culture medium (128). A *Pseudomonas* sp. that hydrolyzed several organophosphates, including parathion, methyl parathion, dursban, paraxon, aminoparathion, and other methoxy- or ethoxy-substituted organophosphates, was found to hydrolyze diazinon as well (14). The crude cell extracts of this bacterium exhibited an enzyme activity that hydrolyzed diazinon at an optimal pH of 9.0. The specific activity of the enzyme in the crude extract was 0.44 μmol/mg protein/min, whereas the optimum temperature for this activity was determined to be between 35°C and 47°C. In another study, Barik et al. (15) found that the cell free extracts from two strains of *Arthrobacter* spp. SB3 and SB4 hydrolyzed diazinon, and the K_M values for the enzyme activity for the two strains were 1.3 and 2.0 μmol/mg protein/min, respectively. This enzyme activity had a broad pH and temperature optima of 6 to 9 and 25°C to 36°C, respectively.

Because of the potential toxicity of diazinon to birds, its use on golf courses and sod farms has been suspended twice since 1988 (7). In spite of the concerns regarding use of diazinon, little is known about the microbial degradation of this pesticide. Although *Flavobacterium* sp. ATCC 27551 was originally isolated as a diazinon-degrading bacterium, it was characterized with respect to parathion hydrolysis. Clearly, additional studies are needed to elucidate the metabolic fate and biochemical pathways involved in the degradation of diazinon.

Isofenphos

Isofenphos {O-ethyl O-[2-(isopropoxycarbonyl)phenyl] N-isopropyl phosporo-amido-thioate} (Table 3) is a systemic nematicide (52). Isofenphos was degraded more rapidly in soils with a history of the insecticide use than in unexposed soils (113). Soils with enhanced isofenphos degradation contained an adapted population of soil micro-organisms. Two bacterial isolates, an *Arthrobacter* and a *Pseudomonas* species isolated from the adapted cultures, metabolized the pesticide in pure culture (112, 113).

Dichlorovos

Dichlorovos (2,2-dichlorovinyl O,O-dimethyl phosphate) (Table 3) is used extensively in Vapona strips, a preparation in which the insecticide is impregnated in a resin and volatizes at a fairly uniform rate to give control of household pests, especially flies (93). Dichlorovos is effective against ectoparasites and is used in flea collars for dogs and cats and a number of veterinary applications. Information of the microbial degradation of dichlorovos is limited. In a single study, a microbial enrichment was found to convert dichlorovos to dichloroethanol, dichloroacetic acid, and ethyl dichloroacetate (83). This enrichment culture was obtained from sewage. Three members of the mixed culture—*P. aeruginosa*, a *Pseudomonas* sp., and a gram-positive, spore-forming bacterium—individually had less activity than the enrichment culture.

Coumaphos

Coumpahos [O,O-diethyl O-(3-chloro-4-methyl-2-oxo-2H-1-benzo-pyran-7-yl)phos-phoro-thioate] (Table 3) is used as in acaricide for the control of the southern cattle tick (*Boophilus microplus*) and the cattle tick (*Boophilus annulatus*). It is used by the Animal and Plant Health Inspection Service (APHIS), U.S. Department of Agriculture, in its tick eradication program. The APHIS dips several hundred thousands of cattle annually for tick control along the U.S.–Mexican border. Each of the 42 vats used contains ca. 12,000 liters of flowable solution (42% coumaphos, 58% inert ingredients) (139), presenting a considerable waste disposal problem. Since the half-life of coumaphos in soil and water is about 300 days (70), safe and effective methods for disposal of coumaphos waste are required.

Shelton and Karns (138) observed rapid degradation of coumaphos in several cattle dipping vats, which resulted in the loss of its efficacy against ticks. In a separate study, three bacteria were isolated from enrichment cultures developed by using the dip solutions as an inoculum (139). These isolates, B-1, B-2, and B-3, hydrolyzed coumaphos to chlorferon (3-chloro-4-methyl-7-hydroxy-coumarin) and diethylthiophosphate. Chlorferon was further metabolized by B-1 and B-2 strains to α-chlor-β-methyl 2,3,4-trihy-droxy-*trans*-cinnamic acid. Parathion hydrolase enzyme produced by the *Flavobac-terium* sp. ATCC 27551 was also found to hydrolyze coumaphos, yielding chlorferon and DETP (66, 70). Chlorferon, thus generated, was further degraded by a UV-ozonation process whereby pesticide suspensions are pretreated with ultraviolet light in the presence of oxygen prior to soil disposal (71). UV-ozonation of coumaphos resulted in only limited degradation of the pesticide, however (70). When microbial hydrolysis was followed with ozonation on large volumes of coumaphos waste under field conditions, complete hydrolysis of coumaphos was achieved in 48 hr and more than 20% of the chlorferon produced was degraded in 20 hr. Hence, the combined treatment was very effective in eliminating coumaphos waste (66).

Other Organophosphates

Some other organophosphates have also been extensively used for insect and weed control. Glylphosate (N-phosphonomethyl glycine), the active ingredient of Roundup, is a broad-spectrum organophosphate herbicide. This herbicide was used as a sole source of phosphorus by Pseudomonas spp. (95, 155), and an Alcaligenes sp. (155). One of the Pseudomonas spp. (95) completely degraded glyphosate, and equivalent cellular yields were obtained with equimolar amounts of either inorganic phosphate or glyphosate as the phosphorus source. Methidathion {[S-(S-methoxy-2-oxo-1,3,4-thiadiazol-3-(2H)-yl)methyl]O,O-dimethyl phosphorodithioate} (Table 3) is used for the control of insects on alfalfa, cotton, and fruit crops. It is also used in greenhouses, mainly for rose cultures, against thysanopterae and lepidopterae, and in vegetable nurseries. A sewage bacterium, Bacillus coagulans, degraded methidathion (54). Desmethyl methidathion was identified as one of the major metabolites found in the culture medium supplemented with the pesticide. This suggested that methidathion is converted by demethylation.

Use of the pesticides in the production of agricultural commodities has increased with the world-wide demand for food and fiber, thus increasing the potential for contamination of terrestrial and aquatic environments by agricultural chemicals. In recent years, several pesticides have been reported to contaminate groundwater (31, 37, 160). Aldicarb residues ranging from 1 to 50 ppb have been found in groundwater in Arizona, California, Florida, Maine, Missouri, New York, North Carolina, Oregon, Virginia, Washington, and Wisconsin (13, 31, 41, 129, 176). Similarly, carbofuran has also been found in groundwater in New York and Wisconsin at levels of 1 to 50 ppb (41). Pesticide residues have been detected in vegetables and fruits that were sprayed with these chemicals for insect control. Three outbreaks of illnesses associated with aldicarb sulfoxide–contaminated watermelons and cucumbers in California and one in Nebraska have been reported (55). Although the concentrations of these chemicals required to cause acute toxicity in humans are generally higher than those found in the contaminated environment, conditions present in the human gut favor the formation of N-nitrosocarbamates (47, 125). The potential for the formation of such mutagens in the gut cautions that carbamate contamination may pose health problems. The importance of investigating the enzymology and biochemistry of microbial degradation of these pesticides to help design strategies for detoxifying pollutants in the environment is thus clear.

GENETIC ENGINEERING AND BIOREMEDIATION POTENTIAL OF MICROORGANISMS

The bioremediation potential of isolated microorganisms is generally limited by their strict substrate specificity. For example, a bacterial strain, *Pseudomonas* sp. B13, can metabolize 3-chlorobenzoate (3CBA) but not 4-chlorobenzoate (4CBA) because of the stringent specificity of the 3CBA oxygenase. When the TOL plasmid or the gene encoding 1,2-dioxygenase activity was transferred into the *Pseudomonas* sp. B13, the transformants or recombinant strains oxidized both 3CBA and 4CBA (30a, 124a). Similarly, introduction of a broad-specificity salicylate hydroxylase gene (*nah*G) into a *Pseudomonas* sp. capable of degrading chlorocatechols resulted in strains that were able to mineralize 3-chloro, 4-chloro, and 5-chlorosalicylates (168a). More recently, the transfer of *xyl*D, *xyl*L, and *nah*G into *A. eutrophus* harboring pRC10 (encoding genes for 2,4-D degradation) or its derivatives yielded strains capable of degrading several new compounds (34) (Table 5).

The genes encoding catabolic pathways are often clustered and plasmid-borne. These clusters can be cloned; if they are readily expressed in new hosts, microorganisms with a broader substrate range can be developed. Aside from cloning individual plasmids, genes, or gene clusters for expanding the substrate range of microorganisms, portable catabolic cassettes (Fig. 5) may be constructed and placed in indigenous microorganisms for bioremediation and cleaning toxic wastes. Strategies for constructing recombinant strains are as follows:

1. Recruitment of cloned genes encoding broader substrate range enzymes: *bph*C (3-phenylcatechol dioxygenase), *nah*G (salicylate hydroxylase), *tfd*A (2,4-dichlorophenoxyacetic acid monooxygenase), *xyl*D (toluate 1,2-dioxygenase), *xyl*L (dihydroxycyclohexidiene carboxylate dehydrogenase).

2. Recruitment of cloned operons: *bph*ABCD, *clc*ABD, *tfd*CDEF, *xyl*AABC, *cyl*DEFG.

3. Recruitment of constructed portable gene cassettes: *xyl*D *xyl*L, Pm *xyl*D *xyl*L (regulated), Pl *xyl*D *xyl*L (constitutive), *xyl*D *nah*G, *tfd*A *nah*G, *xyl*D *tfd*A *nah*G, *xyl*D *xyl*L *tfd*A *nah*G.

As discussed above, microorganisms may convert pesticides to hazardous metabolites. Although the hazardous intermediates are often further metabolized by members of the microbial community, they may persist under certain environmental conditions. To bioremediate such environments, microbial consortia or recombinant strains capable of degrading hazardous pollutants completely are desired. We have used the carbaryl-hyrdolyzing isolate 50581 (which converts carbaryl to 1-naphthol) and 1-naphthol-degrading isolate 50552 to construct transconjugants that completely metabolize carbaryl (29a, 40). These transconjugants (derivatives of isolate 50552) may be more suitable for bioremeddiation purposes.

The use of microorganisms for decontamination of groundwater and restoration of other contaminated water resources as well as terrestrial environments has elicited much interest. Soil microbial isolates generally do not grow well in groundwater. Therefore, they have had limited application for the treatment of waters. The addition of nutrients has been recommended to enable microorganisms to decontaminate waters (29a), but this may cause the additional problem of removing excess nutrients from the reclaimed waters. *Pseudomonas* sp. 50432 capable of degrading carbamates was found to grow well in a nutrient-poor medium and to grow minimally in LB medium (32); this suggested that it may grow in fresh water containing low nutrient levels and thus may be potentially useful for treatment of contaminated waters and waste waters. Surprisingly, it degraded the carbofuran well in sewage, oligotrophic lake water, and groundwater (33), although the rate of carbofuran degradation was low in groundwater (Fig. 5), but it did not degrade the pesticide in the eutrophic lake water. Apparently the nutrient constituents of the water were sufficient to support bacterial growth and/or activity. All the tested water samples had near-neutral pH and low conductivity except the eutrophic lake water. *Pseudomonas* sp. 50432 may be unable to degrade the pesticide in eutrophic lake water because the nutrient concentration or one of its constituents may be inhibitory for the microbial activity.

Table 5. Growth of *Alcaligenese eutrophus* recombinant strains in the presence of different substrates.

Strains	Genotype	Substrates[a]					
		TFC	PAA	4CB	CPH	SA	CSA
Flavobacterium sp.	Hg^r *tfd*$^+$ (pRC10)	+	+	−	+	−	−
A. eutrophus	Hg^r *tfd*$^+$ (pJP4)	+	+	−	+	−	−
A. eutrophus	(cured)	−	−	−	−	−	−
A. eutrophus	Hg^r *tfd*$^+$ (pRC10)	+	+	−	+	−	−
A. eutrophus	Hg^r *tfd*$^+$ (pRC301)	+	+	−	+	−	−
A. eutrophus	Hg^r *tfd*$^+$ nahG	+	+	−	+	+	+
A. eutrophus	Hg^r *tfd*$^+$ xylD	+	+	+	+	−	−

[a]Abbreviations: TFD, 2,4-D; PAA, phenoxyacetic acid; 4CB, 4-chlorobenzoate, CPH, chlorophenol; SA, salicylate, CSA, chlorosalicylate.

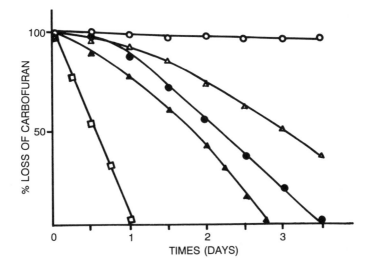

Figure 5. Degradation of carbofuran in fortified waters. Shown are eutrophic lakewater (*open circles*), oligotophic lakewater (*filled circles*), groundwater (*filled triangles*), trickling filter sewage effluent (*open triangles*), and minimal medium (*square*).

CONCLUSIONS

The range of metabolic activities of microorganisms is enormous. Microbial mineralization of xenobiotic compounds, including pesticides, is largely responsible for removal of the bulk of these compounds from soil and aquatic environments. Like other biological processes, biodegradation is governed by physical parameters such as oxygen, pH, temperature, and the nutrients present in a particular environment and also to a large extent by the nature of the microhabitat with respect to microbial community and xenobiotic compound. Prior exposure of microbial populations to pesticides is also important, as it often is responsible for the adaption of microorganisms that evolve new genetic functions under stress, such as in limited nutrient conditions. In general, microorganisms responsible for degradation of xenobiotics belong to a few genera: *Achromobacter, Alcaligenes, Arthrobacter, Bacillus, Flavobacterium, Nocardia,* and *Pseudomonas.* Numerous microorganisms belonging to these genera have been isolated that can eliminate commmercially used hazardous chemicals such as pesticides from the environment. These microorganisms have been used to study the enzymology of biodegradation of a few pesticides only. The pathways and the genes involved in degradation of these chemicals are less understood. Learning more about these aspects of microbial degradation of pesticides should be helpful in developing recombinant derivatives of indigenous microorganisms with broad substrate range. Such microorganisms may degrade pesticide wastes or decontaminate environments polluted with hazardous chemical mixtures.

Acknowledgments

Work in our laboratory is supported by National Science Foundation grant DMB-9020525, Public Service Health grant BSRG 50RR713 from the National Institutes of Health, and the Michigan State Research Excellence Fund.

LITERATURE CITED

1. Adeshima, F., and Todd, E. L. 1990. Organochlorine compounds in human adipose tissue from north Texas. *J. Toxicol. Environ. Heath* 29:147–156.

2. Adhya, T. K., Barik, S., and Sethunathan, N. 1981. Hydrolysis of selected organophosphorous insecticides by two bacterial isolates from flooded soil. *J. Appl. Bacteriol.* 40:167–172.

3. Aelion, C. M., Swindoll, C. M., and Pfander, F. K. 1987. Adaptation and biodegradation of xenobiotic compound by microbial communities from a pristine aquifer. *Appl. Environ. Microbiol.* 55:2212–2217.

4. Ahraf, A. E., Marzouk, M., Willis, W. V., and Saleh, R. 1990. Dieldrin in the food chain; potential health effects of recycled manure. *J. Environ. Health* 53:17–19.

5. Alexander, M. 1981. Biodegradation of chemicals of environmental concern. *Science* 211:132–138.

6. Anonymous. 1922. Fatal poisoning with naphthol salve. *J. Am. Med. Assoc.* 79:51.

7. Anonymous. 1990. Currents. *Environ. Sci. Technol.* 24:1275.

8. Ansari, G. A. S., James, G. P., Hu, A. L., and Reynolds, E. S 1986. Organochlorine residues in adipose tissue of residents of the Texas gulf coast. *Bull. Environ. Contam. Toxicol.* 36:311–316.

9. Anson J. G., and Mackinnon, G. 1984. Novel *Pseudomonas* plasmid involved in aniline degradation. *Appl. Environ. Microbiol.* 48:868–869.

10. Aoki, K., Ohtsuka, K., Shinke, R., and Nishira, H. 1983a. Isolation of aniline assimilation bacteria and physiological characterisatoin of aniline biodegradation in *Rhodococcus erythropolis* AN-13. *Agric. Biol. Chem.* 47:2569–2575.

11. Aoki, K., Shinke, R., and Nishara, H. 1983b. Metabolism of aniline by *Rhodococcus erythropolis* AN-13. *Agric. Biol. Chem.* 47:1611–1616.

12. Atuma, S. S. 1985. Organochlorine pesticides in some Nigerian food materials. *Bull Environ. Contam. Toxicol.* 35:735–738.

13. Baier, J. H., and Robbins, S. F. 1982. Report on the occurrence and movement of agricultural chemicals in ground water: South Fork of Suffolk County. Suffolk County Dept. Health Services, Hauppauge, New York. 11788:50–51.

14. Barik, S., and Munnecke, D. M. 1982. Enzymatic hydrolysis of concentrated diazinon in soil. *Bull. Environ. Contam. Toxicol.* 29:235–239.

15. Barik, S., Munnecke, D. M., and Fletcher, J. S. 1982. Enzymatic hydrolysis of malathion and other dithioate pesticides. *Biotechnol. Let.* 4:795–798.

16. Barik, S., and Sethunathan, N. 1978. Biological hydrolysis of parathion in natural ecosystems. *J. Environ. Qual.* 7:346–348.

17. Barik, S. Wahid, P. A. Ramakrishna, C., and Sethunathan, N. 1979. A change in the degradation pathway of parathion after repeated applications to flooded soil. *J. Agric. Food Chem.* 27:1391–1392.

18. Barquet, A., Morgade, C., and Paffenberger, C. D. 1981. Determination of organochlorine and metabolites in drinking water, human blood serum, and adipose tissue. *J. Toxicol. Environ. Health* 7:469–479.

19. Behki, R. M., and Khan, S. U. 1991. Degradation of [1-^{14}C-propyl] EPTC (s-ethyldipropylthiocarbamate) by a soil bacterial isolate. *Chemosphere* 21:1457–1462.

20. Belluck, D. A., and Benjamin, M. S. 1990. Pesticides and human health. *J. Environ. Health* 53:11–13.

21. Blaszczyk, M., Boileau, S., Fournier, M., Chevalier, G., and Krzystyniak, K. 1987. Growth of facultative anaerobic population of human intestinal bacteria with the carbamate pesticide aminocarb. *Pesticide Biochem. Physiol.* 29:233–243.

22. Boatman, R. J., Cunningham, S. L., and Ziegler, D. A. 1984. A method for measuring the biodegradation of organic chemicals. *Environ. Toxicol. Chem.* 5:233–243.

23. Boyd, S. A., Shelton, D. R., Berry, D., and Tiedje, J. M. 1983. Anaerobic biodegradation of phenolic compounds in digested sludge. *Appl. Environ. Microbiol.* 47:50–54.

24. Boyle, M. 1989. The environmental microbiology of chlorinated aromatic decomposition. *J. Environ. Qual.* 18:395–402.

25. Brown, K. A. 1980. Phosphotriesterases of *Flavobacterium* sp. *Soil Biol. Biochem.* 12:105–112.

26. Camper, N. D., Fleming, M. M., and Skipper, H. D. 1987. Biodegradation of carbofuran in pretreated and nonpretreated soils. *Bull. Environ. Contam. Toxicol.* 39:571–578.

27. Casida, J. E. 1963. Mode of action of carbamates. *Annu. Rev. Entomol.* 8:39–58.

28. Chakrabarty, A. M. 1976. Plasmids in *Pseudomonas. Annu. Rev. Genet.* 10:7–30.

29. Change, Y., Tan, Y., and Sun, M. 1981. Biodegradation mechanism of organophosphate pesticides in the aquatic ecosystem. *Huanjing Kexue xuebao.* 1:115-125.

29a. Chapalamadugu, S., and Chaudhry, G. R. 1992. Microbiological and biotechnological aspects of metabolism of carbamates and organophosphates. *Crit. Rev. Biotech.* 12:357–389.

30. Chapalamadugu, S., and Chaudhry. G. R. 1991. Hydrolysis of carbaryl by a *Pseudomonas* sp. and construction of a microbial consortium that completely metabolizes carbaryl. *Appl. Environ. Microbiol.* 57:744–750.

30a. Chatterjee, D. K., and A. M. Charkrabarty. 1982. Genetic rearrangements in plasmids specifying total degradation of chlorinated benzoic acids. *Mol. Gen. Genet.* 188:279–285.

31. Chaudhry, G. R. 1988. Isolation and characterization of pesticide-degrading microorganisms from soil and ground water. Report of investigation. Florida Department of Environmental Regulation. 1–77.

32. Chaudhry, G. R., and Ali, A. N. 1988. Bacterial metabolism of carbofuran. *Appl. Environ. Microbiol.* 54:1414–1419.

33. Chaudhry, G. R., Ali, A. N., and Wheeler, W. B. 1988. Isolation of a methyl parathion-degrading *Pseudomonas* sp. that possesses DNA homologous to the *opd* gene from a *Flavobacterium* sp. *Appl. Environ. Microbiol.* 54:288–293.

34. Chaudhry, G. R., and Chapalamadugu, S. 1991. Biodegradation of chlorinated compounds. *Microbiol. Rev.* 57:59–79.

35. Chaudhry, G. R. and Wheeler, W. B. 1988. Biodegradation of carbamates. *Wat. Sci. Tech.* 20:89–94.

36. Clarke, P. H., and Richmond, M. H. 1975. *Genetics and Biochemistry of Pseudomonads.* London: Wiley.

37. Cohen, S. Z., Creeger, S. M., Carsel, R. F., and Enfield, C. G. 1984. Potential pesticide contamination of groundwater from agricultural uses. In *Treatment and Disposal of Pesticide Waste.* Ed. R. R. Kreuger, J. N. Seiber. Washington, D.C.: Amer. Chem. Soc. 297–325.

38. Cook, A. M., Doughton, C. G., and Alexander, M. 1980. Desulfuration of fialkyl thiophosphoric acids by a Pseudomonad. *Appl. Environ. Microbiol.* 39:463–465.

39. Cortez, L., Chapalamadugu, S., and Chaudhry, G. R. 1989. The biodegradation of methyl parathion and p-nitrophenol by *Pseudomonas* spp., Abstr. K-178, p. 274. *Abstr. 89th Annu. Meet. Am. Soc. Microbiol.*

40. Cotterill, E. G., and Own, P. G. 1989. Enhanced degradation in soil of triallate and other carbamate pesticides following application of triallate. *Weed Res.* 29:65–68.

41. Data summarized from EPA pesticide registration files, unpublished.

42. Derbyshire, M. K. Karns, J. S. Kearney, P. C., and Nelson, J. O. 1987. Purification and characterization of an N-methylcarbamate pesticide hydrolyzing enzyme. *J. Agric. Food Chem.* 35:871–877.

43. Dick, W. A., Ankumah, R. A., McClung, G., and Abou-Assaf, N. 1990. Enhanced degradation of s-ethyl N,N-dipropyl carbamothioate in soil and by an isolated soil microorganism. In *Enhanced Biodegradation of Pesticides in the Environment.* Ed. K. D. Racke, J. R Coats. Washington, D.C.: 98–112.

44. Dixon, R. 1986. The xylABC promoter from the *pseudomonas putida* TOL plasmid is activated by nitrogen regulatory genes in *Eschrichia coli*. *Mol. Gen. Genet.* 203:129–136.

45. Don, R. J., and Pemberton, J. M. 1981. Properties of six pesticide degradation plasmids isolated from *Alcaligenes paradoxus* and *Alcaligenes eutrophus*. *J. Bacteriol.* 145:681–686.

46. Dumas, D. P., Caldwell, S. R., Wild, J. R., and Raushel, F. M. 1989. Purification and properties of the phosphotriesterase from *pseudomonas diminuta*. *J. Biol. Chem.* 264:19659–19665.

47. Elespuru, R., Lijnsky, W., and Seetlow, J. K. 1974. Nitrosocarbaryl as a potent mutagen of environmental significance. *Nature* 247:386–397.

48. Eto, M. 1974. *Organosphosphorus Pesticides: Organic and Biological Chemistry.* Cleveland, Ohio: CRC Press Inc.

49. Felsot, A. S. 1989. Enhanced biodegradation of insecticides in soil: Implications for agrosystems. *Annu. Rev. Entomol.* 34:453–476.

50. Felsot, A., Maddox, J. V., and Bruce, W. 1981. Enhanced microbial degradation of carbofuran in soils with histories of furadan use. *Bull. Environ. Contamin. Toxicol.* 26:781–788.

51. Ferris, I. G., and Lichtenstein, E. P. 1980. Interactions between agricultural chemicals and soil microflora and their effects on the degradation of [¹⁴C]parathion in a cranberry soil. *J. Agric. Food Chem.* 28:1011–1140.

52. Fest, C., and Schmidt, K. J. 1983. Organophosphorus insecticides In *Chemistry of Pesticides.* Ed. E. H. Buchel. New York: John Wiley & Sons. 48–125.

53. Gauger, W. K., MacDonald, J. M., Adrian, N. R., Matthees, D. P., and Walgenbach, D. D. 1986. Characterization of a streptomycete growing on organophosphate and carbamate insecticide. *Arch. Environ. Contam. Toxicol.* 15;137–141.

54. Gauthier, M. J., Berge, J. B., Cauany, A., Breittmayer, B., and Fournier, D. 1988. Microbial degradation of methidathion in natural environments and metabolization of this pesticide by Bacillus coagulans. *Pestic. Biochem. Physiol.* 31:61–66.

55. Goldman, L. R., Beller, M., and Jackson, R. J. 1990. Aldicarb food poisonings in California, 1985–1988: toxicity estimates for humans. *Arch. Environ. Health* 45:141–147.

56. Gorder, G. W., and Lichtenstein, E. P. 1980. Degradation of parathion in culture by microorganisms found in cranberry bogs. *Can J. Microbiol.* 26:475–481.

57. Gray, R. A., and Joo, G. K. 1985. Reduction in weed control after repeated application of thiocarbamate and other herbicides. *Weed Sci.* 33:698–702.

58. Harper, L. L., McDaniel, C. S., Miller, C. E., and Wild, J. R. 1988. Dissimilar plasmids isolated from *Pseudomonas diminuta* MG and a *Flavobacterium* sp. (ATCC 27552) contain identical *opd* genes. *Appl. Environ. Microbiol.* 54:2586–2589.

59. Harris, C. R., Chapman, R. A., Harris, C., and Tu, C. M. 1984. Biodegradation of pesticides in soil: Rapid induction of carbamate degrading factors after carbofuran treatment. *J. Environ. Sci. Health Part B* 19:1–11.

60. Harvey, R. G., Dekker, J. H., Fawcett, R. H., Roeth, F. W., and Wilson, G. R. 1987. Enhanced biodegradation of herbicides in soil and effects on weed control. *Weed Technol.* 1:349–361.

61. Helm, V., and Reber, H. 1979. Investigation on the regulation of aniline utilization in *Pseudomonas multivorans* strain AN-1. *J. Appl. Microbiol. Biotechnol.* 7:191–199.

62. Heitkamp, M. A. Camel, V., Reuter, T. J., and Adamas, W. J. 1991. Biodegradation of p-nitrophenol in an aqueous waste stream by immobilized bacteria. *Appl. Environ. Microbiol.* 56:2967–2973.

63. Jones, S. H., and Alexander, M. 1986. Kinetics of mineralization of phenols in lake water. *Appl. Environ. Microbiol.* 51:891–897.

64. Karns, J. S., Mulbry, W. W., Nelson, J. O., and Kearney, P. C. 1986. Metabolism of carbofuran by a pure bacterial culture. *Pestic. Biochem. Physiol.* 25:211–217.

65. Karns, J. S., Mulbry, W. W., Nelson, J. O., and Kearney, P. C. 1986. Metabolism of carbofuran by a pure bacterial culture. *Pestic. Biochem. Physiol.* 25:211–217.

66. Karns, J. S., Muldoon, M. T., Mulbry, W. W., Derbyshire, M. K., and Kearney, P. C. 1987. Use of microorganisms and microbial systems in the degradation of pesticides. In *Biotechnology in Agricultural Chemistry.* Ed. H. M. Le Baron, R. O. Mumma, R. C. Honeycutt, J. H. Huesing. *ACS Symp. ser.* 334:156–170. Washington, D.C.: American Chemical Society.

67. Karns, J. S., and Tomasek, P. H. 1991. Carbofuran hydrolase-purification and properties. *J. Agric. Food Chem.* 39:1004–1008.

68. Kaufman, D. D. 1974. Degradation of pesticides by soil microorganisms. In *Pesticides in Soil and Water.* Ed. W. D. Guenzi. Madison, WI: Soil Sci. Soc. Ameri. 133–156.

69. Kaufman, D. D., Katan, Y., Edwards, D. F., and Jordan, E. G. 1985. Microbial adaptation and metabolism of pesticides. In *Agricultural Chemicals of the Future.* Ed. J. L. Hilton. NJ: Rowman and Allenhold. 255–272.

70. Kearney, P. C., Karns, J. S., Muldoon, M. T., and Ruth, J. M. 1986. Coumaphos disposal by combined microbial and uv-ozonation reactions. *J. Agric. Food Chem.* 3:702–706.

71. Kearney, P. C., Plimmer, J. R., and Li, A. M. 1983. UV-ozonation and land disposal of aqueous pesticide waste. In *Pesticide Chemistry–Human Welfare and the Environment.* Ed. J. Miyamoto. Vol. 4. Oxford, U.K.: Pergamon Press.

72. Keith, L. H., and Telliard, W. A. 1979. Priority pollutants. I. A perspective view. *Environ. Sci. Technol.* 13:416–423.

73. Kiene, R. P., and Capone, D. G. 1984. Effects of organic pollutants on methanogenesis, sulfate reduction and carbondioxide evolution in salt marsh sediments. *Marine Environ. Res.* 13:141–160.

74. Kiene, R. P., and Capone, D. G. 1986. Stimulation of methanogenesis by aldicarb and several other N-methyl carbamate pesticides. *Appl. Environ. Microbiol.* 51:1247–1251.

75. Knackmuss, H.-J. 1983. Xenobiotic degradation in industrial sewage: haloaromatics as target substrates. pp. 173–190. In *Biotechnology.* Ed. C. F. Phelps, P. H. Clarke. *Biochem. Soc. Symp.* no. 48. London: Biochemical Society.

76. Knowles, C. O., and Benezet, J. J. 1981. Microbial degradation of the carbamate pesticides desmedipham, phenmedipham, promecarb, and propamocarb. *Bull. Environ. Contam. Toxicol.* 27:529–533.

77. Kuhr, R. J., and Dorough, H. W. 1976. *Carbamate Insecticides: Chemistry, Biochemistry, and Toxicology.* Cleveland Ohio: CRC Press, Inc.

78. Larkin, M. J., and Day, M. J. 1986. The metabolism of carbaryl by three bacterial isolates, *Pseudomonas* spp. (NCIB 12042 & 12043) and *Rhodococcus* sp. (NCIB 12038) from garden soil. *J. Appl. Bacteriol.* 60:233–242.

79. Latorre, J. 1982. Ph.D. thesis, University of Göttingen, Federal Republic of Germany.

80. Laveglia, J., and Dahm, P. A. 1977. Degradation of organo-phosphorous and carbamate insecticides in the soil by the soil microorganisms. *Annu. Rev. Entomol.* 22:483–513.

81. Lee, A. 1984; EPTC degrading microorganisms isolated from a soil previously exposed to EPTC. *Soil Biol. Biochem.* 16:529–531.

82. Lee, A., Rahman, A., and Holland, P. T. 1984. Decomposition of the herbicide EPTC in soils with a history of previous EPTC applications. *N. Z. J. Agric. Res.* 27:201–206.

83. Leiberman, M. T., and Alexander, M. 1983. Microbial and nonenzymatic steps in the decomposition of dichlorvos (2,2-dichlorovinyl O,O-dimethyl phosphate). *J. Agric. Food Chem.* 31:265–267.

84. Lemley, A. T., and Zhong, W. 1984. Hydrolysis of aldicarb, aldicarb sulfoxide and aldicarb sulfone at parts per billion levels in aquous mediums. *J. Agric. Food. Chem.* 4:714–719.

85. Lewis, D. L., and Holm, H. W. 1981. Rates of transformation of methyl parathiion and diethyl phthalate by aufwuchs microorganisms. *Appl. Environ. Microbiol.* 42:698–703.

86. Liu, D., Thomson, K. and Strachan, W. M. J. 1981. Biodegradation of carbaryl in simulated aquatic environments. *Bull. Environ. Contam. Toxicol.* 27:412–417.

87. Lokke, H. 1985. Degradation of 4-nitrophenol in two Danish soils. *Environ. Pollu. Ser. A* 38:171–181.

88. Marty, J. L., and Bastide, J. 1988. Transformation of barban and analogues by *Pseudomonas alcaligenese* isolated from soil. *Pestic. Sci.* 24:221–230.

89. Marty, J. L., Khafif, T., Vega, D., and Bastide, J. 1986. Degradation of phenyl carbamate herbicides by *Pseudomonas alcaligens* isolated from soil. *Soil Biol. Biochem.* 18:649–653.

90. Marty, J. L., and Vogues, J. 1987. Purification and properties of a phenylcarbamate herbicide degrading enzyme of *Pseudomonas alcaligenes* isolated from soil. *Agric. Biol. Chem.* 51:3287–3291.

91. Matsumura, F., and Brown, A. W. A. 1961. Biochemistry of malathion resistance in *Culex tarsalis.* *J. Econ. Entomol.* 54:1176–1179.

92. McDaniel, C. S., Harper, L. L., and Wild, J. R. 1988. Cloning and sequencing of a plasmid-borne gene (*opd*) encoding a phosphotriesterase. *J. Bacteriol.* 170:2306–2311.

93. McEwen, F. L., and Stephenson, G. R. 1979. *The Use and Significance of Pesticides in the Environment.* New York: John Wiley & Sons, Inc.

94. Miles, C. J., and Delfino, J. J. 1985. Fate of aldicarb, aldicarb sulfoxide and aldicarb sulfone in Floridian groundwater. *J. Agric. Food Chem.* 33:455–460.

95. Moore, J. K., Braymer, h. D., and Larson, A. D. 1983. Isolation of a *Pseudomonas* sp. which utilizes the phosphonate herbicide glyphosate. *Appl. Environ. Microbiol.* 46:316–320.

96. Mueller, J. G., Skipper, H. D., and Kline, E. L. 1988. Loss of butylate-utilizing ability by a *Flavobacterium* sp. *Pestic. Biochem. Physiol.* 32:189–196.

97. Mulbry, W. M., and Eaton, R. W. 1991. Purification and characterization of the N-methylcarbamate hydrolase from *Pseudomonas* strain CRL-OK. *Appl. Environ. Microbiol.* 57:3679–3682.

98. Mulbry, W. W., and Karns, J. S. 1989. Purification and characterization of threE parathion hydrolases from gram-negative bacterial strains. *Appl. Environ. Microbiol.* 55:289–293.

99. Mulbry, W. W., and Karns, J. S. 1989. Parathion hydrolase specified by the *Flavobacterium opd* gene: relationship between the gene and protein. *J. Bacteriol.* 171:6740–6746.

100. Mulbry, W. W., Karns, J. C., Kearney, P. C., Nelson, J. O., McDaniel, C. S., and Wild, J. R. 1986. Identification of plasmid-borne parathion hydrolase gene from *Flavobacterium* sp. by southern hybridization with opd from *Pseudomonas diminuta*. *Appl. Environ. Microbiol.* 51:926–930.

101. Mulbry, W. W., Kearney, P. C., Nelson, J. O., and Karns, J. S. 1987. Physical comparison of parathion hydrolase plasmids from *Pseudomonas diminuta* and *Flavobacterium* sp. *Plasmid* 18:173–177.

102. Munnecke, D. M., and Fischer, H. F. 1979. Production of parathion hydrolase activity. *Eur. J. Appl. Microbiol.* 8:1103–112.

103. Munnecke, D. M., Johnson, L. M., Talbot, H. W., and Barik, S. 1982. Microbial metabolism and enzymology of selected pesticides. In *Biodegradation and Detoxification of Environmental Pollutants.* Ed. A. M. Chakrabarty. Boca Raton, FL: CRC Press. 1–32.

104. Nelson, L. M 1982. Biologically induced hydrolysis of parathion in soil: isolation of hydrolyzing bacteria. *Soil Biol. Biochem.* 14:219–222.

105. Nelson, M. L., Yaron, B., and Nye, P. H 1982. Biologically induced hydrolysis of parathion in soil: kinetics and modelling. *Soil Biol. Biochem.* 14:223–227.

106. O'Brien, R. D. 1967. *Insecticides: Action and Metabolism.* New York: Academic Press. 290–306.

107. Obrigawitch, T., Martin, A. R., and Roeth, F. W. 1983. Degradation of thiocarbamate herbicides in soils exhibiting rapid EPTC breakdowns. *Weed Sci.* 31:187–192.

108. Ou L. T., Edvarsson, K. S. V., and Rao, P. S. C. 1985. Aerobic and anaerobic degradation of aldicarb in soils. *J. Agric. Food. Chem.* 33:72–78.

109. paris, D. F., Wolfe, N. L., and Steen, W. C. 1982. Structure-activity relationships in microbial transformation of phenols. *Appl. Environ. Microbiol.* 44;153–158.

110. Pemberton, J. M. 1983. Degradative plasmids. *Internatl. Rev. Cytol.* 84:155–183.

111. Portier, R. J., Chen, H. M., and Meyers, S. P. 1983. Environmental effect and fate of selected phenols in aquatic ecosystems using microcosm approaches. *Dev. Ind. Microbiol.* 24:409–424.

112. Racke, K. D., and Coats, J. R. 1987. Enhanced degradation of isophenphos by soil microorganisms. *J. Agric. Food Chem.* 35:94–99.

113. Racke, K. D., and Coats, J. R. 1988. Comparative degradation of organophosphorus insecticides in soil: specificity of enhanced microbial degradation. *J. Agric. Food Chem.* 36:193–199.

114. Rahman, A., and James, T. K. 1983. Decreased activity of EPTC + R-25788 following repeated use in some New Zealand soils. *Weed Sci.* 31:783–786.

115. Rajagopal, B. S., Brahmaprakash, G. P., Reddy, B. R., Singh, U. D., and Sethunathan, N. 1984. Effect and persistence of selected carbamate pesticides in soil. *Res. Rev.* 93:1–199.

116. Rajagopal, B. S., Brahmaprakash, G. P., and Sethuanathan, N. 1984. Degradation of carbofuran by enrichment cultures and pure cultures of bacteria from flooded soils. *Environ. Pollut.* 36A:61–73.

117. Rajagopal, B. S., Chendrayan, K., Reddy, B. R., and Sethuanathan, N. 1983. Persistence of carbaryl in flooded soils and its degradation by soil enrichment cultures. *Plant Soil* 73:35–45.

118. Rajagopal, B. S., Panda, S., and Sethuanathan, N. 1984. Accelerated degradation of carbaryl and carbofuran in a flooded soil pretreated with hydrolysis products, 1-naphthol and carbofuran phenol. *Bull. Environ. Contam. Toxicol.* 36:827–32.

119. Rajagopal, B. S., Rao, V. R., Nagendrappa, G., and Sethuanathan, N. 1984. Metabolism of carbaryl and carbofuran by soil enrichment cultures. *Can. J. Microbiol.* 30:1458–1456.

120. Ramanand, K., Panda, S., Sharmila, M. Adhya, T. K., and Sethunathan, N. 1988. Development and acclimatization of carbofuran-degrading soil enrichment cultures at different temperatures. *J. Agric. Food Chem.* 36:200–205.

121. Ramanad, K., Sharmila, M., and Sethuanathan, N. 1988. Mineralization of carbofuran by a soil bacterium. *Apple. Environ. Micirobiol.* 54:2129–2133.

122. Read, D. C. 1986. Accelerated microbial breakdown of carbofuran in soil from previously treated fields. *Agric. Ecosyst. Environ.* 15:51–61.

123. Read, D. C. 1987. Greatly accelerated microbial degradation of aldicarb in re-treated field soil, in flooded soil, and in water. *J. Econ. Entomol.* 80:156–163.

124. Reddy, B. R., and Sethunathan, N. 1983. Mineralization of parathion in the rice rhizosphere. *Appl. Environ. Microbiol.* 45:826–829.

124a. Reineke, W., and H.-J. Knackmuss. 1980. Hybrid pathway for chlorobenzoate metabolism in *Pseudomonas* sp. B13 derivatives. *J. Bacteriol.* 142:467–473.

125. Rickard, R. W., and Dorough, W. H. 1984. *In vivo* formation of nitrosocarbamates in the stomach of rats and guinea pigs. *J. Toxicol. Environ. Health* 14:279–290.

126. Rodriguez, D. D., Dorough, H. W. 1977. Degradation of carbaryl by soil microorganisms. *Arch. Environ. Contam. Toxicol.* 6:47–56.

127. Roeth, F. W. 1986. Enhanced herbicide degradation in soil with repeat application. *Rev. Weed Sci.* 2:45–65.

128. Rosenberg, A., and Alexander, M. 1979. Microbial cleavage of various organophosphorus insecticides. *Appl. Environ. Microbiol.* 37:88–891.

129. Rothschild, E. R., Manser, R. J., and Anderson, M. P. 1982. *Ground Water* 20:437–445.

130. Rowland, S. S., Speedie, M. K., and Pogell, B. M. 1991. Purification and characterization of a secreted recombinant phosphotriesterase (parathion hydrolase) from *Streptomyces lividans. Appl. Environ. Microbiol.* 57:440–444.

131. Rubin, H. E., and Alexander, M. 1983. Effect of nutrients on the rates of mineralization of trace concentrations of phenol and p-nitrophenol. *Environ. Sci. Technol.* 17:104–107.

132. Scow, K. M., Simkins, S., and Alexander, M. 1986. Kinetics of mineralization of organic compounds at low concentration soil. *Appl. Environ. Microbiol.* 51:1028–1035.

133. Serdar, C. M., and Gibson, D. T. 1985. Enzymatic hydrolysis of organophosphates: cloning and expression of a parathion hydrolase gene from *Pseudomonas diminuta. Bio/Technology,* 3:567–571.

134. Serdar, C.. M., Gibson, D. T., Munnecke, D. M., and Lancaster, J. H. 1982. Plasmid involvement in parathion hydrolysis by *Pseudomonas diminuta. Appl. Environ. Microbiol.* 44:246–249.

135. Serdar, C. M., and Murdock, D. M. 1989. Plasmid encoded parathion hydrolase activity. In *Enzymes Hydrolyzing Organophosphorus Compounds.* Ed. E. Reiner, W. Aldridge, F. Hoskin. New York: Ellis Harwood Publishers. 143–154.

136. Serdar, C. M., Mudruck, D. C., and Rohde, M. F. 1989. Parathion hydrolase gene from *Pseudomonas diminuta* MG: subcloning, complete nucleotide sequence, and expression of the mature portion of the enzyme in *Escherichia coli. Bio/Technology* 7:1151–1155.

137. Shelton, D. R. 1988. Mineralization of diethylthiophosphoric acid by an enriched consortium from cattle dip. *Appl. Environ. Microbiol.* 54:2572–2573.

138. Shelton, D. R., and Karns, J. S. 1988. Coumpahos degradation in cattle-dipping vats. *J. Agric. Food Chem.* 36:831–834.

139. Shelton, D. R., and Somich, C. J. 1968. Isolation and characterization of coumaphos-metabolizing bacteria from cattle dip. *Appl. Environ. Microbiol.* 54:2566–2571.

140. Siragusa, G. R., and DeLaune, R. D. 1986. Mineralization and sorption of *p-nitrophenol* in estuarine sediment. *Environ. Toxicol. Chem.* 4:175–178.

141. Skipper, H. D. 1990. Enhanced biodegradation of carbamothiote herbicides in South Carolina. In *Enhanced Biodegradation of Pesticides in the Environment.* Ed. K. D. Racke, and R. R. Coats. Washington, D.C.: American Chemical Society.

142. Skipper, H. D., Murdock, E. C., Gooden, D. T., Zublena, J. P., and Amakiri, M. A 1986. Enhanced herbicide biodegradation in South Carolina soils previously treated with butylate. *Weed Sci.* 34:558–563.

143. Smelt, J. H., Crum, S. J. H., Tenunissen, W., and Leistra, M. 1987. Accelerated transformation of aldicarb, oxamyl and ethoprophos after repeated soil treatments. *Crop Prot.* 6:295–303.

144. Speitel, G. E., Lu, C. J., Turakhia, M., and Zhu, X. J. 1989. Biodegradation of trace concentrations of substituted phenols in granular activated carbon columns. *Environ. Sci. Technol.* 23:68–74.

145. Spain, J. C., and Gibson, D. T. 1991. Pathway for biodegradation of *p-nitrophenol* in a *Moraxella* sp. *Appl. Environ. Microbiol.* 57:812–819.

146. Spain, J. C., Pritchard, P. H., and Bourqin, A. W. 1980. Effects of adaptation on biodegradation rates in sediment/water cores from estuarian and freshwater environments. *Appl. Environ. Microbiol.* 40:726–734.

147. Spain, J. C., Wyss, O., and Gibson, D. T. 1979. Enzymatic oxidation of *p-nitrophenol. Appl. Environ. Microbiol.* 48:944–950.

148. Spain, J. C., and Van Veld, P. A. 1983. Adaptation of natural microbial communities to degradation of xenobiotic compounds: effects of concentration, exposure, time, inoculum, and chemical structure. *Appl. Environ. Microbiol.* 45:428–435.

149. Spain, J. C., Van Veld, P. A., Monti, V. A., Pritchard, P. H., and Cripe, C. R. C. 1984. Comparison of *p-nitrophenol* biodegradation in field and laboratory test systems. *Appl. Environ. Microbiol.* 48:944–950.

150. Steiert, J. G., Pogell, B. M., Speedie, M. K., and Laredo, J. 1989. A gene coding for a membrane-bound hydrolase is expressed as a secreted, soluble enzyme in *Streptomyces lividans. Bio/Technology* 7:65–68.

151. Suett, D. L. 1987. Influence of treatment of soil with carbofuran on the subsequent performance of insecticides against cabbage root fly (*Delia radicum*) and carrot fly (*Psila rosae*). *Crop Prot.* 6:371–378.

152. Surovtseva, E. G., and Vol'nova, A. I. 1980. Aniline as the sole source of carbon, nitrogen and energy for *Alcaligenes faecalis. Microbiologiya* 49:49–53.

153. Suss, A., Fuschsbichler, G., and Eben, C., 1978. Degradation of analine, 4-chloroaniline, and 3,4-dichloroaniline in various soils. *Z. Pflanzenernahr. Bodenkd.* 141:57–66.

154. Swindoll, C. M., Aelion, C. M., and Pfaender, F. K. 1988. Influence of inorganic and organic nutrients on aerobic biodegradation and on the adaptation response of subsurface microbial communities. *Appl. Environ. Microbiol.* 54:212–217.

155. Talbot, H. W., Johnson, L. M., and Munnecke, D. M 1984. Glyphosate utilization by a *Pseudomonas* and an *Alcaligenes* spp. isolated from environmental sources. *Curr. Microbiol.* 10:255–259.

156. Tam, A. C., Behki, R. M., and Khan, S. U. 1987. Isolation and characterization of an EPTC-degrading *Arthrobacter* strain and evidence for plasmid-associated EPTC degradation. *Appl. Environ. Microbiol.* 53:1088–1093.

157. Tomasek, P. H., and Karns, J. S. 1989. Cloning of a carbofuran hydrolase gene from *Achromobacter* sp. strain WM111 and its expression in gram-negative bacteria. *J. Bacteriol.* 171:4038–4044.

158. Turco, R. F., and Konopka, A. 1990. Biodegradation of carbofuran in enhanced and non-enhanced soils. *Siol Biol. Biochem.* 22:195–201.

159. U.S. Environmental Protection Agency. 1979. Suspended and cancelled pesticides. EPA 159/9. Pesticides and Toxic Substances Enforcement Division, U.S. EPA, Washington, D.C.

160. U.S. Environmental Protection Agegncy. 1987. National Pesticide Survey, Pilot Study Evaluation Summary Report. Office of Drinking Water, Office of Pesticide Programs, U.S. EPA, Washington, D.C.

161. U.S. Environmental Protection Agency. 1989. EPA Results of National Survey of Pesticides in Drinking-Water Wells. EPA Press Advisory, Office of Public Affairs (A-107). Washington, D.C.

162. Van den Heuvel, M. J., and Cochram, D. G. 1965. Cross resistance to organophosphorus compounds in malathion and diazinon-resistant strains of *Blattella germanica*. *J. Econ. Entomol.* 58:872.

163. Van Raalte, H. G. S. 1977. Human experience with dieldrin in perspective. *Ecotoxicol. Environ. Safety* 1:203–210.

164. Van Veld, P. A., and Spain, J. C. 1983. Degradation of selected xenobiotic compounds in three types of aquatic test systems. *Chemosphere* 12:1291–1305.

165. Vega, D., Bastide, J., and Coste, C. 1985. Isolation from soil and growth characteristics of a CIPC-degrading strain of *Pseudomonas cepacia*. *Soil Biol. Biochem.* 17:541–545.

166. Venkateswarlu, K., and Sethunathan, N. 1984. Degradation of carbofuran by *Azospirillum lipoferum* and *streptomyces* spp. isolated from flooded soil. *Bull. Environ. Contam. Toxicol.* 33:556–560.

167. Venkateswarlu, K., and Sethunathan, N. 1985. Enhanced degradation of carbofuran by *Pseudomonas cepacia* and *Nocardia* sp. in the presence of growth factors. *Plant Soil* 84:445–449.

168. Venkateswarlu, K., Siddarama Gowda, T. K., and Sethunathan, N. 1977. Persistence and biodegradation of carbofuran in flooded soils. *J. Agric. Food Chem.* 25:533–537.

168a. Weightman, A. J., Don, R. H., Lehrbach, P. R., and Timmis, K. N. 1984. The identification and cloning of genes encoding haloaromatic cataboic enzymes and the construction of hybrid pathways for substrate mineralization. In *Genetic Control of Environmental Pollutants*. Ed. G. S. Omenn, A. Hollaender. New York: Plenum Press. 47–80.

169. Wiggins, B. A., and Alexander, M. 1988. Role of chemical concentration and second carbon sources in acclimation of microbial communities for biodegradation. *Appl. Environ. Microbiol.* 54:2803–2807.

170. Wilde, G., and Mize, T. 1984. Enhanced microbial degradation of systemic pesticides in soil and its effect on chinch bug *Blissus leucopterus leucopterus* (Say) (Heteroptera: Lygaedae) and greenbug *Schizaphis graminum rondani* (Homoptera: Aphididae) control in seedling sorghum. *Environ. Entomol.* 13:1079–1082.

171. Williams, I. H., Pepsin, H. S., and Brown, M. J. 1976. Degradation of carbofuran by soil microorganisms. *Bull. Environ. Contam. Toxicol.* 14:244–249.

172. Wilson, R. G. 1984. Accelerated degradation of thiocarbamate herbicides in soil with prior thiocarbamate exposure. *Weed Sci.* 32:264–268.

173. Yardon, O., Aharonson, N., and Katan, J. 1987. Accelerated microbial degradation of MBC (methyl benzimidazol-2-xlcarbamate) in soil and its control. *Soil Biol. Biochem.* 19:735–739.

174. You, I.-S., and Bartha, R. 1982. Stimulation of 3,4-dichloroaniline mineralization by aniline. *Appl. Environ. Microbiol.* 44:679–681.

175. You, I.-S., and Bartha, R. 1982. Metabolism of 3,4-dichloroaniline by *Pseudomonas putida*. *J. Agric. Food Chem.* 30:274–277.

176. Zaki, M. H., Moran, D., and Harris, D. 1982. Pesticides in ground water: the aldicarb story in Suffolk County, New York. *Amer. J. Public Health* 72:1391–1395.

177. Zeyer, J., and Kearney, P. C. 1982. Microbial degradation of para-chloroaniline as sole carbon and nitrogen source. *Pestic. Biochem. Physiol.* 17:215–223.

178. Zeyer, J., and Kearney, P. C. 1982. Microbial metabolism of propanil and 3,4-dichloroaniline. *Pestic. Biochem. Physiol.* 17:224–231.

179. Zeyer, J., Wasserfallen, A., and Timmis, K. N. 1985. Microbial mineralization of ring-substituted anilines through an *ortho*-cleavage pathway. *Appl. Environ. Microbiol.* 50:447–453.

CHAPTER 11

Degradation of Thiocarbamate Herbicides and Organophosphorus Insecticides by *Rhodococcus* Species

RAM M. BEHKI

Abstract. Three *Rhodococcus* strains that degraded thiocarbamate herbicides initiated the degradation of EPTC (S-ethyl dipropylthiocarbamate) by different pathways. They all harbored similar plasmids. A transmissible plasmid in one of the bacteria, *Rhodococcus* TE1, was shown to be associated with the biodegradation of the herbicides. Mutants of *Rhodococcus* TE1 lacking the plasmid could not degrade EPTC or other thiocarbamate herbicides. The extender dietholate, used to prolong the persistence of thiocarbamates in soil, inhibited degradation of the herbicides through its inhibitory effect on the growth of the bacterium. The duration of inhibition of bacterial growth and the biodegradation of thiocarbamate herbicides lasted until dietholate itself was degraded to subinhibitory concentrations by the same bacterium. Similar inhibitory effects were seen with parathion, an organophosphorus insectide and an analog of dietholate. Other organophosphorus insecticides had little effect on the growth or thiocarbamate degradation by *R.* TE1, however, because *R.* TE1 degraded these insecticides rapidly. The plasmid associated with EPTC degradation was not involved in parathion degradation.

INTRODUCTION

Until recently, many bacterial strains now assigned to the *Rhodococcus* genus were described as species of *Nocardia, Mycobacterium,* and *Corynebacterium* under the general name of "nocardioform bacteria." The *Rhodococcus* genus is now well characterized with many type strains available. This has been made possible by detailed examination of nocardioform bacteria and the development of tests to distinguish between the closely related genera (35, 36, 59). For the purpose of this chapter, the bacterial species mentioned under different designations in the cited literature are now considered to belong to the *Rhodococcus* genus.

Dr. Behki is at the Centre for Land and Biological Resources Research, Research Branch, Agriculture Canada, Ottawa K1A 0C6, Canada. This chapter is LRRC Contribution No. 91-63.

GENERAL METABOLIC ACTIVITIES OF RHODOCOCCI

Rhodococci are aerobic, gram-positive actinomycetes that show considerable morphological diversity. They are ubiquitous in natural environments and have been readily isolated from soil and aquatic systems by use of selective conditions. Iizuka and Komagata (47) isolated and characterized hydrocarbon-utilizing bacteria from oil brine and bottom water of oil storage tanks and from soils in the oil-bearing zones in Japan. Initially designated *Corynebacterium hydrocarboclastus*, the isolates were later identified as species of *Rhodococcus* (53; ATCC Catalog 1988). The bacteria utilized complex hydrocarbons as the sole source of carbon for growth. They have since been commercially exploited, under various patents in Japan, for the industrial production of many amino acids (100). Hydrocarbon-utilizing Rhodococci were also isolated from an area of the Baltic sea contaminated by black oil spillage (55). *Rhodococcus* bacteria can survive under conditions of complete starvation and under harsh environments for many months (54, 78).

Rhodococci are recognized as an important group of bacteria notable for their diverse metabolic activities (15, 30, 31). They have been reported to degrade alkanes, halogenated aliphatics and aromatics (15, 24, 62, 98, 99), lignin (19, 57), toxic compounds like acrylamide, phenolic compounds, and other pollutants, including aromatic amines and phthalate esters (3–5, 37, 39, 49, 56, 71, 72, 85, 92). Other *Rhodococcus* strains synthesize emulsifying agents (82, 102) and antibiotics (71, 95), transform steroids, and degrade cholesterol (28). Many *Rhodococcus* bacteria show very high levels of inducible nitrilase and nitrile-hydratase activities that are potentially important for the production of vitamins and other chemicals (12, 44, 64, 65, 70, 94). An important characteristic of the *Rhodococcus* strains is that they are active toward both aromatic and aliphatic nitrile compounds. A *Rhodococcus* sp. M4 used for the efficient and economic production of L-phenylalanine, an important amino acid in the artificial sweetener aspartame, has been reported (45).

Biodegradation or biotransformation reactions catalyzed by Rhodococci normally require a cometabolite (31). The nature of the cometabolite used during growth or during the studies of the processes can determine not only the efficiency but also the nature of the product(s) formed. The concentration of the cometabolite used can also greatly influence the rate of transformations. High concentrations tend to affect the processes adversely.

Although the presence of plasmids in the heterogenous group of nocardioform bacteria including Rhodococci has been reported (21, 69, 101), most of the plasmids are known to be cryptic, and no phenotypic traits have been shown to be associated with them. A plasmid-associated transformation of *cis*-abenol in *R. erythropolis* JTS-131 was reported by Heida et al. (42). The involvement of plasmid-associated cadmium resistance in *R. fascians* has been suggested (21), and a plasmid in a *conynebacterium* sp. (most probably *Rhodococcus*) is associated with the degradation of the herbicide benthiocarb (66).

Rhodococci are involved in the degradation of some pesticides (Table 1). In this chapter I discuss the degradation of thiocarbamate herbicides and organophosphorus insecticides catalyzed by the two strains of *Rhodococcus* used in our laboratory.

Table 1. Pesticides degraded or transformed by Rhodococci.

Pesticide	Species or strain	Reference
Aldrin	UFM 30 and 48	27
Carbaryl	NCIB-12038	58
Deethylsimazine	*corralinus*	16, 17
Glufosinate	DX-35	7, 90
Methoxychlor	*erythropolis*	33
Molinate	*erythropolis*	32
Phenylamide herbicides	b3	60
Organophosphorus	TE1	11
EPTC	TE1, JE1	22, 88
	ATCC 15961	R. M. Behki, unpublished data
Butylate	TE1	89
Vernolate	TE1	89
Cycloate	TE1, ATCC 15961	R. M. Behki, unpublished data
Molinate	TE1, ATCC 15961	R. M. Behki, unpublished data
Diallate	TE1	8
Triallate	TE1	8

BIODEGRADATION OF THIOCARBAMATE HERBICIDES

Microorganisms are mainly responsible for the degradation of thiocarbamate herbicides in soil (25, 50, 63, 76). Until recently, however, a soil microbial isolate capable of degrading EPTC (the first thiocarbamate herbicide introduced by Stauffer Chemicals in 1956 and extensively studied since) had not been identified. The metabolism of EPTC in plants and animals has recently been reviewed (96).

The enhanced biodegradation of thiocarbamate herbicides in soils after repeated application (79, 83, 97) and their consequent reduced efficacy (38, 41, 74, 75) has been attributed to the adaptation of soil microflora to the herbicides (51). The mechanism(s) involved in the development of "adaptation-associated" enhanced microbial degradation of pesticides in soil remains unknown. Nonetheless, soils exhibiting an enhanced rate of pesticide degradation and containing the adapted microflora would be ideally suited as inocula for enrichment cultures.

Lee (61) used soil showing enhanced EPTC degradation for his enrichment cultures and isolated various classes of bacteria (*Bacillus, Alcaligenes, Pseudomonas, Micrococcus*) and fungi capable of degrading EPTC. He determined EPTC degradation in a nutrient-broth-supplemented medium and found that only 4–6 μg mL^{-1} was degraded after 4 days of incubation. He was unable to characterize any EPTC-degrading bacterial system because all his isolates lost their ability to degrade the herbicide after storage for 15 months. He attributed this to the loss of EPTC degradation-associated plasmids from his bacterial isolates similar to the phenomenon reported by Don and Pemberton (23) for 2,4-D-degrading bacteria. He speculated that the enhanced degradation of EPTC in soil following successive applications of the herbicide may result from the rapid transfer of EPTC-degrading information on the plasmids of microorganisms in soil.

Degradation of EPTC by *Rhodococcus* TE1

Tam et al. (88) isolated an EPTC-degrading bacterial strain by enrichment cultures from clay loam soil from a cornfield exposed to four successive annual applications of Eradicane (commercial EPTC formulation). Originally assigned to the genus *Arthrobacter*, the bacterial isolate designated TE1 was subsequently found to belong to the *Rhodococcus* genus based on morphology and cell-wall composition (34). TE1 could utilize EPTC as the sole source of carbon and energy for growth (Fig. 1). The isolate could degrade EPTC effectively in basal salts medium, BMN (9) or in BMN supplemented with glycerol, glucose, or acetate. EPTC was not degraded by the bacterium in nutrient broth, however. This may explain the minimal degradation found by Lee (61) in his assays in nutrient broth medium.

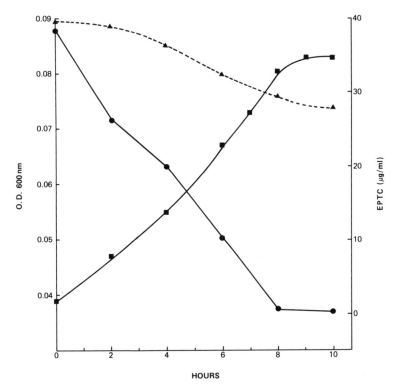

Figure 1. Utilization of EPTC as sole carbon source for growth by *Rhodococcus* TE1 incubated in BMN (9). The *dotted line* shows EPTC loss from an uninoculated control. Line with *circles* indicate loss of pesticide; line with *squares* shows growth of bacteria.

The possible involvement of an indigenous plasmid with EPTC degradation was suspected when mutants unable to grow on EPTC as the sole source of carbon (designated E⁻ mutants) arose at a rather high frequency of about 4% when an overnight nutrient broth culture of TE1 was plated on BMN-agar plates supplemented with EPTC. The frequency of E⁻ mutants increased to about 14% when the nutrient broth culture was treated with the plasmid-curing agent, acridine orange. A further increase to about 20% E⁻ mutant production was obtained when the culture was incubated at 32°C. The E⁻ mutant did not revert to E⁺ phenotype.

Rhodococcus TE1 harbors four plasmids of relative molecular masses 65.5, 60, 50.5, and 2.5 MDa (88). All E⁻ mutants, irrespective of the method used for their isolation, were lacking the 50.5 MDa plasmid (Fig. 2). This plasmid was therefore suspected to be associated with EPTC degradation.

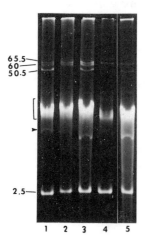

Figure 2. Plasmid profiles: 1. *Rhodococcus* TE1; 2. EPTC degradation-deficient mutant of *R.* TE1; 3. transconjugant from mating of 1 with 2; 4. another EPTC degradation mutant of *R.* TE1 (lacking two plasmids); 5. transconjugant from mating of 1 with 4.

To confirm the association of the 50.5 MDa plasmid with EPTC degradation, the plasmid was transferred from gentamycin-resistant *R.* TE1 to rifampicin- and streptomycin resistant E$^-$ mutants by conjugation. The E$^+$ transconjugants, which were gentamycin-sensitive and rifampicin- and streptomycin-resistant, were obtained at a frequency of about 1×10^{-7} per donor cell. The transconjugants had acquired the 50.5 MDa plasmid missing from the recipient E$^-$ cells. This was the first report of a successful plasmid transfer by bacterial conjugation in Rhodococci. The transconjugants degraded EPTC as efficiently as TE1 (88). The conjugal transfer of plasmid and the restoration of the herbicide-degradation ability confirmed the plasmid-associated degradation of EPTC in *Rhodococcus* TE1. Other E$^-$ mutants missing the 60 MDa plasmid in addition to the 50.5 MDa plasmid have been isolated. Plasmid transfer by conjugation showed the acquisition of only the 50.5 MDa plasmid by the E$^-$ mutant, but complete restoration of EPTC-degradation ability in the transconjugants, however. The 60 MDa plasmid apparently is not self-transmissible and not associated with EPTC degradation. It has not been possible to produce a plasmid-free *Rhodococcus* TE1.

The development of EPTC-degrading capability in soil in response to the flux and/or stress caused by the introduction of the herbicide remains a mystery. A plasmid already present in the cell may acquire additional genes to initiate the first step in the degradation of EPTC while the subsequent catabolic steps are encoded by the chromosomal genes. This possibility appears likely in view of the fact that E$^-$ mutants (deficient in EPTC degradation) are equally effective in the degradation of EPTC metabolites, like dipropylamine or propylamine (R. M. Behki, unpublished data). The functional role of other plasmids present in the EPTC-degrading *Rhodococcus* TE1 is not yet known. The availability of this system, nonetheless, should facilitate investigations into the genetic evolution of events leading to the development of enhanced pesticide degradation in soil.

Degradation of EPTC by *Rhodococcus* JE1

Dick et al. (22) have recently described an EPTC-degrading *Rhodococcus* JE1 isolated from EPTC-adapted Jimtown loam soil by batch culture enrichment technique. This isolate, like TE1, utilized EPTC as the sole source of carbon and energy for growth. *Rhodococcus* JE1 harbors plasmids of approximately the same molecular masses as *R.* TE1, but it does not give rise to E⁻ mutants following continued growth in nutrient broth or by treatment with plasmid-curing agents. Apparently the plasmids in JE1 are maintained stably, and the association of any plasmid with EPTC degradation has not yet been reported. *Rhodococcus* JE1, however, differs from TE1 and the other EPTC-degrading *Rhodococcus* sp. in the metabolic pathway of EPTC degradation (see below).

Degradation of EPTC by a *Rhodococcus* sp. (ATCC 15961)

Iizuka and Komagata (47) isolated hydrocarbon-utilizing bacteria from oil brine and soil samples from oil fields in Japan. Among them were species identified as *Corynebacterium hydrocarboclastus*. They are now listed in the ATTC Catalog as *Rhodococcus* species. One of these (ATCC 15961), when grown and subcultured in basal salts-glycerol medium supplemented with EPTC, was found to degrade EPTC. It degrades EPTC as effectively as *R.* TE1, without the addition of a cometabolite like glucose or glycerol. Unlike TE1, however, this *Rhodococcus* sp. cannot utilize EPTC as the sole source of carbon for growth, possibly because of the inability of this strain to utilize propylamine or dipropylamine, the probable metabolites of EPTC, for growth (R. M. Behki, unpublished data). This *Rhodococcus* sp. also metabolized EPTC by a pathway different from those proposed for *R.* TE1 or *R.* JE1 (discussed below). Although this strain harbors plasmids similar to those of *Rhodococcus* TE1, mutants deficient in EPTC degradation were not obtained under the various experimental procedures used for plasmid curing. None of the resident plasmids in this bacterium was conjugally transferable to other *Rhodococcus* strains, including E⁻ mutants of *R.* TE1.

Metabolism of EPTC

Microorganisms could attack EPTC at several possible sites on the molecule: the thioester linkage, the amide linkage, or the alkyl group (25). According to Kaufman (50), the hydrolytic cleavage at the thioester linkage is the most probable site of initial microbial attack. This would result in the formation of mercaptan, CO_2 and dipropylamine. Kaufman suggested the dealkylation of dipropylamine to propylamine and the formation of a corresponding aldehyde as a possible pathway of EPTC degradation (Fig. 3). More than one metabolic pathway appears to be involved in the microbial degradation of EPTC. Three *Rhodococcus* strains degrade EPTC by different pathways.

EPTC Metabolism by *Rhodococcus* JE1

Dick et al. (22) reported the formation of N-depropyl EPTC and propionaldehyde from EPTC with *Rhodococcus* JE1. They suggested that N-depropyl EPTC was further metabolized as shown in Figure 3, and they provided evidence for the initial progressive increase in the amount of N-depropyl EPTC formed with a corresponding decrease in EPTC concentration. A strong mercaptan-like odor that developed in JE1 cultures that were actively degrading EPTC indicated mercaptan formation. Dick et al. found traces of propylamine in the aqueous fraction of JE1 culture during EPTC degradation but obtained no evidence for dipropylamine formation.

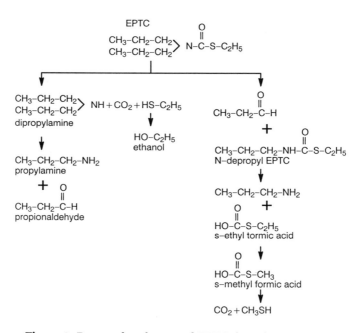

Figure 3. Proposed pathways of EPTC degradation.

EPTC Metabolism by *Rhodococcus* TE1

The metabolism of 1-^{14}C-propyl EPTC by *Rhodococcus* TE1 (Fig. 4) shows that about 70% of the label is evolved as $^{14}CO_2$ and 23% is found in the cell biomass. The E[-] mutant does not metabolize the label at all. Similar results were obtained when the metabolism of 1-^{14}C propionic acid was monitored with TE1 cells; 70% of the label was evolved as $^{14}CO_2$ and 22% was incorporated into the cell biomass (R. M. Behki, unpublished data), thus indicating propionic acid (or propionaldehyde) is one of the metabolites in EPTC degradation.

Formation of propionic acid from EPTC may be preceded by formation of dipropylamine (Fig. 3). Amines are volatile, and they have been detected and quantitated following derivatization with pentafluorpropionic anhydride (PFPA) to form stable derivatives (46). This method was used to show the formation of dipropylamine during EPTC degradation by *Rhodococcus* TE1. The gas chromatographic evidence was further confirmed by gas chromatographic/mass spectrometric analysis (10). Further support for dipropylamine formation during EPTC metabolism was provided by the isotope dilution effect of added dipropylamine (as hydrochloride salt) on the metabolism of 1-^{14}C-propyl EPTC. The evolution of $^{14}CO_2$ and the incorporation of the label into cellular biomass was inhibited in direct proportion to the amount of dipropylamine added. This effect was not due to the inhibition of EPTC degradation by dipropylamine. Similar amounts of EPTC were degraded in both cases. Propylamine formation was not detected in the PFPA-derivatized sample. Addition of propylamine-HCl did not show the expected concentration-dependent isotope dilution effect on $^{14}CO_2$ evolution or the incorporation of the radioactivity into the cell biomass during 1-^{14}C-propyl EPTC degradation. Propylamine utilization by *R.* TE1, moreover, is inducible, and *R.* TE1 cells grown in EPTC are not induced for propylamine utilization (Fig. 5). Subsequent metabolism of dipropylamine by *R.* TE1 has not been determined.

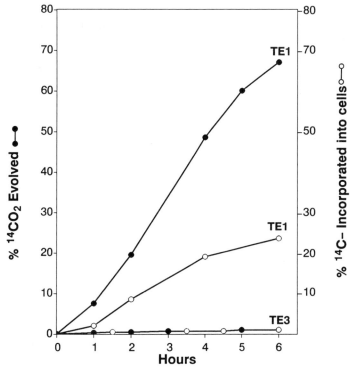

Figure 4. Utilization of (1-^{14}C-propyl) EPTC by R. TE1 and E$^-$ mutant, TE3.

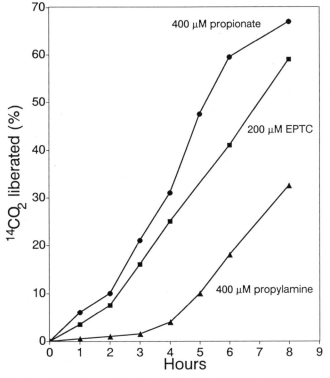

Figure 5. Utilization of 1-^{14}C-propylamine-hydrochloride (400 μM), 1-^{14}C propionate (400 μM), and 1-^{14}C-propyl EPTC (200 μM) by R. TE1 cells grown overnight in EPTC-supplemented medium.

EPTC-degradation-deficient mutants of R. TE1 metabolize dipropylamine, propionate, and propylamine as effectively as the parent strain. The plasmid missing in the mutants and associated with EPTC degradation is obviously involved only in the initial step of thioester cleavage.

The sulfur moiety of EPTC is metabolized rather slowly and contributes to the growth of TE1 cells only after a lag period of approximately 60 hours in a sulfur-free basal salts-glucose medium. This may explain the results reported for the metabolism of ethyl-labeled EPTC (50) when the $^{14}CO_2$ evolution from the label was much slower than the inactivation of EPTC. None of the sulfur metabolites has been positively identified.

The labeled carbonyl-carbon of EPTC has been shown to be totally dissipated as $^{14}CO_2$; this dissipation has been used as an index of EPTC degradation in soil (87). No information is available, however, on the kinetics of CO_2 evolution from the carbonyl-carbon by bacterial species.

EPTC Metabolism by *Rhodococcus*, ATCC 15961

The *Rhodococcus* sp. (ATCC 15961) metabolizes EPTC by yet another undetermined pathway. There is no evolution of $^{14}CO_2$ from 1-^{14}C-propyl EPTC degradation by this bacterium. Unlike *Rhodococcus* TE1 or JE1, the bacterium cannot metabolize propylamine and cannot use dipropylamine or propylamine as the sole carbon source for growth. This *Rhodococcus* sp. can metabolize 1-^{14}C-propionate, however, which is obviously not formed during EPTC degradation by this bacterium. Since EPTC does not serve as the sole carbon source for the growth of this bacterium, it was assumed that the first metabolite was dipropylamine or propylamine and that the cell's inability to metabolize these amines may account for the failure to mineralize 1-^{14}C propyl EPTC. No accumulation of the amines was detected, however. Incubation of EPTC with this *Rhodococcus* sp. resulted in complete recovery of the metabolized radioactivity in the aqueous fraction. This radioactivity was quite stable during evaporation of the cell-free aqueous extract under N_2. Under these conditions, the amines are unstable and not recovered. The radioactive metabolite(s) of EPTC could be recovered from the aqueous fraction at pH 9.5 but not at pH 2 by extraction with organic solvents. The exact nature of the metabolite(s) formed with this bacterium remains to be determined.

Degradation of Other Thiocarbamates

Rhodococcus TE1 degraded the herbicides butylate and vernolate as rapidly as EPTC. These thiocarbamate herbicides, structurally related to each other, could serve as the sole source of carbon for the growth of the bacterium. Like EPTC, mutants lacking the 50.5 MDa plasmid associated with EPTC degradation in R. TE1 were unable to degrade butylate and vernolate, thus demonstrating the relationship between the phenotype associated with thiocarbamate degradation and the 50.5 MDa plasmid. This broad specificity can explain the development of cross-adaptation of enhanced biodegradation in soil treated with EPTC, butylate, and vernolate (41).

Cycloate and molinate are thiocarbamate herbicides with hexyl and azapene rings. Neither of the two herbicides are able to serve as the sole source of carbon for the growth of *Rhodococcus* TE1 (R. M. Behki, unpublished data). The degradation of cycloate, unlike EPTC, but similar to the numerous catabolic functions associated with Rhodococci, required a cometabolite (31) like glycerol, glucose, or acetate.

Rhodococcus TE1 and *Rhodococcus* ATCC 15961 degraded molinate cometabolically. Unlike EPTC degradation, molinate was degraded at a higher rate in basal salts medium supplemented with diluted nutrient broth than in medium containing glucose or glycerol (R. M. Behki, unpublished data). Imai and Kawatsuka (48) reported an enhanced rate of molinate degradation in their bacterial cultures induced by growth in vernolate or molinate. Molinate degradation was the same in similarly induced or uninduced R. TE1. This probably represents the characteristics of the bacterial species. No *Rhodococcus* bacteria were tested by these workers. Mutants of TE1 that are deficient in EPTC degradation were also ineffective in degrading molinate or cycloate.

DEGRADATION OF DIALLATE AND TRIALLATE

Dissipation of diallate and triallate and the effect of other pesticides on their degradation in soil has been documented (2). Cotterill and Owen (18) reported enhanced degradation of triallate in soil exposed to the herbicide for 17 years. The soil was cross-adapted to degrade EPTC and carbofuran; however, no bacteria were isolated that could explain this cross-adaptation. *Rhodococcus* TE1 degraded both diallate and triallate quite effectively but only in the presence of a cometabolite (8). The rate and the extent of degradation was higher in yeast extract or nutrient broth than in glycerol or glucose supplemented-medium (Table 2). *Rhodococcus* TE1 could not utilize these herbicides as the sole source of carbon for growth. The degradation products of diallate were inhibitory to the continued growth of the bacterial culture even in the presence of a readily utilizable energy source (Fig. 6). *Rhodococcus* TE1 metabolizes dipropyl and propylamines but was unable to utilize diisopropyl or isopropylamines, the probable metabolites of diallate degradation. Mutants of TE1 defective in EPTC degradation did not degrade either of the halogenated thiocarbamates. EPTC-adapted soils exhibiting enhanced degradation showed cross-adaptation for diallate degradation (8).

Table 2. Degradation of diallate and triallate by *Rhodococcus* TE1.

Treatments	Degradation (%)	
	Diallate (24 hr)	Triallate (70 hr)
No cells + glycerol only	2.1	2.7
TE1 cells only	7.9	4.9
TE1 + glycerol	35.3	32.3
TE1 + glucose	34.7	30.4
TE1 + yeast extract	63.2	50.1
TE1 + 1/10 diluted nutrient broth	72.2	66.4
TE3 + glycerol	1.7	1.5
TE3+ yeast extract	2.7	2.7

The cultures were adjusted to O. D. (600 nm) of about 0.2 The supplements were added at 500 μg mL^{-1}. TE3 is an EPTC-degradation deficient mutant of R. TE1.

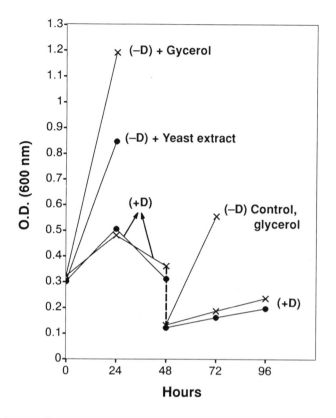

Figure 6. Inhibitory effect of diallate or its metabolites on the growth of *Rhodococcus* TE1. The cultures incubated with diallate (+D) for 48 hr were centrifuged, and the supernatants were freshly inoculated with R. TEI.

INHIBITION OF THIOCARBAMATE DEGRADATION

Successive application of the same thiocarbamate herbicide to the soil results in the development of "problem" soils. These soils develop the capacity for an accelerated rate of degradation of thiocarbamate herbicides (see Biodegradation of Thiocarbamate Herbicides, above). This phenomenon has been attributed to the adaptation of soil microbial species to the herbicides (51). Moorman (67) hypothesized that the adaptation was due not to an increase in the general microbial population but to the selective stimulation of specific microbial groups in the general population. The adaptation may result from the induction and/or derepression of specific enzymatic activities in the soil microflora or from some genetic alterations in the microflora to produce new metabolic capabilities. This may include plasmid exchange with its associated catabolic functions. Plasmid exchange in soil and its broad implications have recently been reviewed (86).

Kaufman et al. (52) suggested the use of microbial inhibitors to control biodegradation and thereby prolong the persistence of pesticides and the duration of their efficacy. To combat the problem of enhanced EPTC degradation, when first reported by Rahman et al. (75), diethiolate (R-33865) was introduced as an extender by Stauffer Chemicals and found to prolong significantly the duration of persistence of EPTC and related thiocarbamates like butylate and vernolate in soil (73, 79, 84). It was suggested (83) that dietholate exerted this effect through the inhibition of soil microbial activity. Dietholate (0,0-diethyl 0-phenyl phosphorothioate) has been reported to exert different effects in soils with history of thiocarbamate use. It improved significantly the efficacy of EPTC in EPTC-adapted soil but did not improve the performance of vernolate in such soil (74). Obrigawitch et al. (73) reported cross-adaptation of EPTC-adapted soil microorganisms for vernolate but not for butylate. Skipper et al. (83) reported cross-adaptation of butylate-adapted microorganisms for EPTC but little crossadaptation for vernolate in South Carolina soils. Wilson (97) also reported enhanced EPTC degradation in butylate-adapted soil but no effect on butylate degradation in EPTC-adapted soil or in soil previously exposed to vernolate. These studies indicated some microbial and enzymatic specificity for EPTC and butylate degradation.

Rhodococcus TE1 degraded EPTC, butylate, and vernolate equally effectively, and dietholate inhibited the degradation of all three herbicides severely (89). Similar inhibition of EPTC, butylate, and vernolate was found in EPTC-adapted soil, thus indicating the extended persistence of the herbicides but no specificity.

Effect of Dietholate on Bacterial Growth

Dietholate inhibited the growth of Rhodococcus TE1 cultures severely for about 48 hr. Little EPTC was degraded. The cultures subsequently resumed growth at a slow rate initially; they were growing at a normal rate of growth by 96 hr. The resumption in cell growth almost paralleled the resumption in the rate of EPTC degradation. This result showed that dietholate exerted its inhibitory effect on thiocarbamate degradation through inhibition of the growth and cellular metabolism of Rhodococcus TE1 (89). Dietholate was not a general inhibitor of cell growth of other soil bacterial species, however (Table 3). Instead, dietholate specifically inhibited Rhodococcus strains irrespective of their ability to degrade the thiocarbamates. Only TE1, out of the Rhodococcus strains tested, was a thiocarbamates degrader. The inhibitory effect of degradation and R. TE1 growth was also dependent upon its concentration. The reasons for the specific inhibitory effects of dietholate on the growth of Rhodococci are not understood.

Table 3. Effect of dietholate (15 μg mL^{-1}) on the growth of various bacterial species.

Bacterial cultures	Ohr	Optical density (600 nm) 24 hr	
		−DE[a]	+DE
E. coli K12	0.12	2.30	2.20
Pseudomonas putida PAWI	0.09	1.95	1.88
Rhizobium meliloti	0.07	1.24	1.14
B. thuringiensis HD-2	0.12	1.70	1.65
B. cereus	0.15	1.85	1.80
A. globiformis	0.07	1.42	1.10
A. oxidans	0.07	1.31	1.04
R. TE1	0.08	1.22	0.11
EPTC$^-$ mutant of TE1	0.07	1.10	0.08
Rhodococcus erythropolis	0.07	2.1	0.083
Rhodococcus rhodochrous (ATCC 13859)	0.07	1.59	0.119

[a] DE is dietholate.

The release of inhibition of growth and EPTC degradation (Fig. 7) also coincided with the degradation of dietholate by R. TE1. The breakdown product of dietholate was identified as dietholate oxon, which was formed in progressively increasing amounts, with a corresponding decrease in dietholate concentration. Dietholate oxon was noninhibitory for the growth of TE1. The results provide an insight into the probable mode of action of dietholate as an extender of thiocarbamate herbicides in soil. Since dietholate itself is also degraded by the bacterial species, the duration of the effect is limited. This is consistent with the results obtained by Obrigawitch et al. (73), who reported that dietholate increased the half life of EPTC by 2.5-fold only.

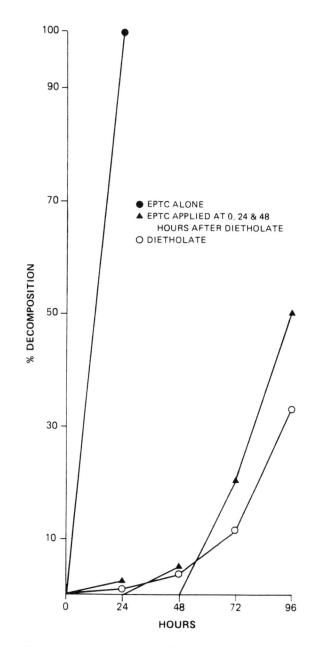

Figure 7. The effect of R. TE1 exposure to dietholate on EPTC and dietholate degradation.

EFFECT OF ORGANOPHOSPHORUS INSECTICIDES ON EPTC DEGRADATION

The use of combinations of pesticides to improve efficacy and increase the activity spectrum of pesticides is now an accepted practice in agriculture. The effects of these combinations have been attributed to the modification of the microbial activity of diverse populations in soil (26, 77, 91).

Several organophosphorus insecticides structurally resemble dietholate (not listed as an insecticide) and are often simultaneously or successively applied with certain herbicides in agricultural practice. The various organophosphorus insecticides were found to affect the degradation of EPTC by *Rhodococcus* TE1 differently (Table 4). Parathion was the most potent inhibitor of EPTC degradation. Butylate degradation was also inhibited by parathion. Similar inhibition of EPTC and butylate degradation by parathion was evident in EPTC-adapted soil (11). Parathion inhibited the growth of *Rhodococcus* TE1 severely. The other organophosphorus insecticides had variable effects similar to their effects on EPTC degradation. The inhibitory effect on bacterial

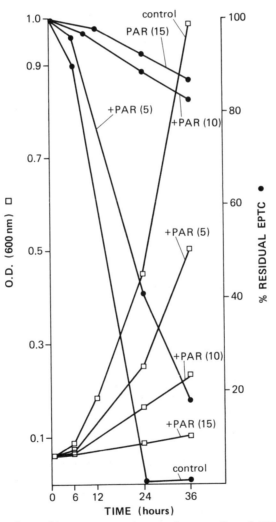

Figure 8. The effect of parathion concentration on the growth and EPTC degradation by *R.* TE1. The numbers in parenthesis are μg mL⁻¹ parathion (PAR).

growth by parathion was species-specific and limited to the bacterial strains of the *Rhodococcus* genus irrespective of their ability to degrade the thiocarbamate herbicides. Although the effect of parathion on EPTC degradation was concentration-dependent and paralleled the effect on the growth of the bacteria (Fig. 8), the inhibitory effects were completely reversible. The cells exhibited normal growth and EPTC-degrading capacity following removal of parathion by centrifugation, similar to results found with the extender dietholate.

Table 4. Effect of various organophosphorus insecticides on EPTC degradation by *Rhodococcus* TE1[a].

Addition	EPTC degraded (%)	
	24 hr	48 hr
Control	100	—
Parathion	9.7	15.8
Fonofos	10.8	59.8
Isofenphos	16.5	83.5
Malathion	100	—
Phorate	100	—

[a]Cell density at 0 time = 0.1 (600 nm). EPTC, 200 μM and insecticides, 40 μM.

DEGRADATION OF ORGANOPHOSPHORUS INSECTICIDES BY *RHODOCOCCUS* TE1

The inhibitory effect of parathion on bacterial growth and EPTC degradation persisted for a long time before the growth resumed and EPTC degradation started (Fig. 8). This indicated that parathion was degraded by R. TE1 slowly. Paraoxon and p-nitrophenol were identified as the degradation products of parathion. Paraoxon also inhibited the growth of EPTC degradation by TE1 and could account for the prolonged inhibitory effect of parathion on this system. This is different from the results with dietholate. The degradation product of dietholate was not inhibitory to the growth or EPTC degradation by TE1. The other insecticides were readily degraded by TE1 cells. The order of degradation of these insecticides by TE1 was as follows: phorate and malathion > fonofos > isofenfos > parathion.

Parathion was also degraded as effectively by the E⁻ mutant of R. TE1. This would rule out the involvment of the EPTC degradation-associated plasmid (missing in the mutant) of TE1 in parathion degradation. The presence of *opd* (organophosphate-degrading) genes on the other plasmids in R. TE1 cannot be ruled out. Identical *opd* genes on the plasmids of *Flavobacterium* and *Pseudomonas* have been reported (40). Although parathion degradation in soil bacteria primarily involves hydrolysis to diethyl thiophophoric acid and p-nitrophenol (6, 63), *Rhodococcus* TE1 degrades parathion by a secondary pathway involving oxidation to paraoxon (11), which is short-lived in the environment and is hydrolyzed to diethyl phosphoric acid and p-nitrophenol (29). Microbial degradation of paraxon and p-nitrophenol has been reported (1, 68). These results suggest the feasibility of employing a suitably formulated combination of thiocarbamate herbicides with organophosphorus insecticides in modulating herbicide degradation in agricultural practice.

BIOREMEDIATION POTENTIAL OF RHODOCOCCI

Rhodococci show great metabolic diversity in their ability to degrade pollutants and xenobiotics. Most strains harbor plasmids (101), and some strains are plant pathogens or animal pathogens. The genetics of this genus are poorly understood, however, and this has limited the exploitation of the bacteria either for bioremediation of toxic chemicals or for understanding pathogenicity. Like the EPTC degradation-associated plasmid in TE1, it is not unlikely that the other indigenous plasmids, so far designated as cryptic, carry genetic determinants for other catabolic functions. The lack of advances in genetic exchange systems and recombinant DNA technology in this genus has hindered the detailed genetic examination of Rhodococci.

The earlier studies were focused on mapping the chromosome of R. erythropolis and the development of a Rhodococcus-actinophage system for its potential as a vector for gene cloning (13, 14). Dabbs (20) described a generalized transducing phage that shows promise for genetic mapping in Rhodococcus species. The development of a Escherichia coli–Rhodococcus shuttle vector and a more efficient transformation system promises a very significant improvement (81).

In earlier studies, the frequency of protoplast regeneration after DNA-transformation or transduction was low. In addition, protoplasting and protoplast regeneration could lead to curing of resident plasmids as shown with many gram-positive bacteria (21). Thus, the lack of an appropriate and efficient system for the introduction of DNA and the exchange of genetic material has been an impediment. Any increase in the frequency of protoplast regeneration can have a marked beneficial effect on transformation. The protoplasts that are most competent for DNA uptake may also be the ones difficult to regenerate. Singer and Finnerty (81) were the first to report heterologous gene expression in three Rhodococcus species following plasmid transformation with an E. coli–Rhodococcus shuttle plasmid. Another recent report (43) described the cloning and expression of Rhodococcus genes (pigment-producing genes) in E. coli to produce pigmented E. coli cells.

There are only two reports (21, 88) of plasmid transfer by mating-type procedures between the Rhodococcus species. Electroporation of Rhodococcus DNA into other Rhodocci is being investigated in our laboratory via the smallest (about 5 kb) plasmid, which is almost universally present in Rhodococcus isolates from soil and is also present in Corynebacterium hydrocarboclastus strains (101). This plasmid may prove to be quite important in developing as a cloning vector or for constructing a hybrid derivative to study the expression of relevant genes. Similar plasmids in a closely related genus, Corynebacterium, have been successfully developed as cloning vectors (80, 101).

Another promising application is the use of a DNA probe (prepared by W. A. Dick, personal communication) that will specifically hybridize with this small plasmid. This application would allow investigations into the development and/or genetic evolution of Rhodococcus bacteria in soil in response to the herbicide stress that leads to enhanced thiocarbamate degradation following repeated application in soil. The processes leading to accelerated degradation of pesticides in soil remain unknown. Additionally, a probe specifically designed to follow the emergence in soil of microbial DNA containing a thiocarbamate-degrading gene sequence(s) (like the one in R. TE1 and other degraders) may lead to the understanding of microbial gene transfer systems in soil.

The earliest available bioremediation technique utilizing Rhodococci is more likely to be the decontamination of toxic waste as demonstrated (93) with Rhodococcus bacteria. Use of thiocarbamate-decomposing enzymes isolated from the *Rhodococcus* species described in this chapter could be used for the controlled remediation of the herbicide or organophosphorus insecticide–contaminated soil and water.

LITERATURE CITED

1. Adhya, T. K., S. Barik, and N. Sethunathan. 1981. Hydrolysis of selected organophosphorous insecticides by two bacterial species. *J. Appl. Bacteriol.* 50:167–172.

2. Anderson, J. P. E., and K. H. Domsch. 1980. Influence of selected pesticides on the microbial degradation of ^{14}C-triallate and ^{14}C-diallate in soil. *Arch. Environ. Contam. Toxicol.* 9:115–123.

3. Apajalahti, J. H. A., and M. S. Salkinoja-Salonen. 1986. Degradation of polychlorinated phenols by *Rhodococcus chlorophenolicus*. *Appl. Microbiol. Biotechnol.* 24:62–67.

4. Appel, M., T. Raabe, and F. Lingens. 1984. Degradation of toluidine by *Rhodococcus rhodochrous*. *FEMS Microbiol. Lett.* 24:123–126.

5. Arai, T., S. Kuroda, and I. Watanabe. 1981. Biodegradation of acrylamide monomers by a *Rhodococcus* strain. In *Actinomycetes.* Ed. K. P. Schaal, G. Pulverer. Stuttgart: Gustav Fischer Verlag. 297–308.

6. Barik, S. 1984. Metabolism of insecticides by microorganisms. In *Insect Microbiology.* Ed. R. Lal. New York: Springer-Verlag. 87–130.

7. Bartsch, K., and C. C. Tebbe. 1989. Initial steps in the degradation of phosphinothricin glufosinate by soil bacteria. *Appl. Environ. Microbiol.* 55:711–716.

8. Behki, R. M. 1991. Diallate degradation by an EPTC-degrading *Rhodococcus,* and in EPTC-treated soil. *Soil Biol. Biochem.* 23–789–793.

9. Behki, R. M., and S. U. Khan. 1986. Degradation of atrazine by *Pseudomonas*: N-dealkylation and dehalogenation of atrazine and its metabolites. *J. Agric. Food Chem.* 34:746–749.

10. Behki, R. M., and S. U. Khan. 1990. Degradation of [1-^{14}C-propyl] EPTC (s-ethyldipropylthiocarbamate) by a soil bacterial isolate. *Chemosphere* 21:1457–1462.

11. Behki, R. M., and S. U. Khan. 1991. Inhibitory effect of parathion on the bacterial degradation of EPTC. *J. Agric. Food Chem.* 39:805–808.

12. Bengis-Garber, C., and A. L. Gutman. 1989. Selective hydrolysis of dinitriles into cyanocarboxylic acids by *Rhodococcus rhodochrous* N.C.I.B. 11216. *Appl. Microbiol. Biotechnol.* 32:11–16.

13. Brownell, G. H., and K. Denniston. 1984. Genetics of nocardioform bacteria. In *The Biology of the Actinomycetes.* Ed. M. Goodfellow, M. Mordarski, S. T. Williams. London: Academic Press. 201–228.

14. Brownell, G. H., J. A. Saba, K. Denniston, and L. W. Enquist. 1982. The development of a *Rhodococcus*-actinophage gene cloning system. *Develop. Industr. Microbiol.* 23:287–298.

15. Cain, R. B. 1981. Regulation of aromatic and hydroxyaromatic catabolic pathways in nocardioform actinomycetes. In *Actinomycetes.* Ed. K. P. Schaal, G. Pulverer. Stuttgart: Gustav Fischer Verlag. 335–354.

16. Cook, A. M., and R. Hütter. 1984. Deethylsimazine: Bacterial dechlorination, deamination and complete degradation. *J. Agric. Food Chem.* 32:581–585.

17. Cook, A. M., and R. Hütter. 1986. Ring dechlorination of deethylsimazine by hydrolases from *Rhodococcus corallinus*. *FEMS Microbiol. Lett.* 34:335–338.

18. Cotterill, E. G., and P. G. Owen. 1989. Enhanced degradation in soil of tri-allate and other carbamate pesticides following application of tri-allate. *Weed Res.* 29:65–68.

19. Crawford, R., L. Robinson, and M. Chet. 1981. ^{14}C-labeled lignin as substrate for the study of lignin biodegradation. In *Lignin Biodegradation* Vol. I. Ed. T. K. Kirk, T. Higuchi, M. Chang. Boca Raton: CRC Press. 61–76.

20. Dabbs, E. R. 1987. A generalized transducing bacteriophage for *Rhodococcus erythropolis*. *Mol. Gen. Genet.* 206: 116–120.

21. Desomer, J., P. Dhaese, and M. V. Montagu. 1988. Conjugal transfer of cadmium resistance in *Rhodococcus fascians* strains. *J. Bacteriol.* 170:2401–2405.

22. Dick, W. A., R. A. Ankumah, G. McClung and N. Abou-Assaf. 1990. Enhanced degradation of s-ethyl N,N-dipropyl carbamothioate in soil and by an isolated soil microorganism. In *Enhanced Biodegradation of Pesticides in the Environment*. Ed. K. D. Racke, and J. R. Coats. Washington, D.C.: American Chemical Society. 98–112.

23. Don, R. J., and J. M. Pemberton. 1981. Properties of six pesticide degradation plasmids isolated from *Alcaligenes paradoxus* and *Alcaligenes eutrophus*. *J. Bacteriol.* 145:681–686.

24. Engesser, K. H., R. B. Cain, and H. J. Knackmuss. 1988. Bacterial metabolism of side chain flourinated aromatics: Cometabolism of 3-trifluromethylbenzoate by *Pseudomonas putida* arvilla Mt. 2 and *Rhodococcus rubropertinctus* N657. *Arch. Microbiol.* 149:188–197.

25. Fang, S. C. 1969. Thiocarbamates. In *Herbicides*. Ed. P. C. Kearney, D. D. Kaufman. New York: Marcel Dekker Inc. 323–348.

26. Felsot, A. F. 1989. Enhanced biodegradation of insecticides in soil: Implications for agroecosystems. *Annu. Rev. Entomol.* 34:453–476.

27. Ferguson, J. A., and F. Korte. 1981. Transformation of aldrin by soil microorganisms. *Appl. Environ. Microbiol.* 34:7–15.

28. Ferreira, N. P., and R. P. Tracey. 1984. Numerical taxonomy of cholesterol degrading soil bacteria. *J. Appl. Bacteriol.* 57:429–446.

29. Fuhremann, T. W., and E. P. Lichtenstein. 1980. A comparative study of the persistence, movement and metabolism of six carbon-14 insecticides in soils and plants. *J. Agric. Food Chem.* 28:446–452.

30. Golovlev, L., L. A. Golovleva, N. V. Eroshina, and G. K. Skryabin. 1978. Microbiological transformation of xenobiotics by *Nocardia*. In *Nocardia and Streptomyces*. Ed. M. Mordarski, W. Kurytowicz, J. Jeljaszewicz. Stuttgart: Gustav Fischer Verlag. 284–301.

31. Golovlev, L. 1980. Biochemical activity of *Rhodococi*, In *Genetics and Physiology of Actinomycetes*. Ed. S. G. Bradley. Washington, D.C.: American Society for Microbiology. 284–301.

32. Golovleva, L. A., Z. I. Finkelshtein, N. A. Popovich, and G. K. Skryabin. 1980. Conversion of ordam by microorganisms. *Izvest. Akadem. Nauk SSSR Ser. Biol.* 3:368–355.

33. Golovleva, L. A., A. B. Polyakova, R. N. Pertsova, and Z. I. Finkelshtein. 1984. The fate of methoxychlor in soils and transformation by soil microorganisms. *J. Environ. Sci. Health B* 19:423–438.

34. Goodfellow, M., and G. Anderson. 1977. The actinomycete genus *Rhodococcus*: A home for the 'rhodochrous' complex. *J. Gen. Microbiol.* 106:99–122.

35. Goodfellow, M., and D. E. Minnikin. 1981. The genera *Norcardia* and *Rhodococcus*. In *The Prokaryotes, Vol. II. A Handbook on Habitats, Isolation and Identification of Bacteria*. Ed. N. P. Starr, H. Stolp, H. G. Trüper, A. Balows, H. G. Schlegal. New York: Springer-Verlag. 2016–2027.

36. Goodfellow, M., and T. Cross. 1984. Classification. In *The Biology of Actinomycetes*. M. Goodfellow, M. Mordarski, S. T. Williams. London: Academic Press. 7–164.

37. Gorlatov, S. N., O. V. Maltseva, V. I. Shevchenko, and L. A. Golovleva. 1989. Degradation of chlorophenols by a culture of *Rhodococcus erythropolis*. *Mikrobiologiya* 58:802–806.

38. Gray, R. A., and G. K. Joo. 1985. Reduction in weed control after repeat application of thiocarbamate and other herbicides. *Weed Sci.* 33:698–702.

39. Haggblom, M. M., J. H. A. Apajalahti, and M. S. Salkinoja-Salonen. 1988. Degradation of chlorinated phenolic compounds occurring in pulp mill effluents. *Water Sci. Technol.* 20:205–208.

40. Harper, L. L., C. S. McDaniel, C. E. Miller, and J. R. Wild. 1988. Dissimilar plasmids isolated from *Pseudomonas diminuta* MG and a *Flavobacterium* sp. (ATCC 27551) contain identical opd genes. *Appl. Environ. Microbiol.* 54:2586–2589.

41. Harvey, R. G., J. H. Dekker, R. S. Fawcett, F. W. Roeth, and R. G. Wilson. 1987. Enhanced biodegradation of herbicides in soil and effects on weed control. *Weed Technol.* 1:349–361.

42. Heida, T., Y. Mikami, and Y. Obi. 1983. Plasmid-determined transformation of cis-abienol and sclareol in *Rhodococcus erythropolis* JTS-131. *Agric. Biol. Chem.* 47:781–786.

43. Hill, R., S. Hart, N. Illing, R. Kirby, and D. R. Woods. 1989. Cloning and expression of *Rhodococcus* genes encodingpigment production in *Escherichia coli*. *J. Gen. Microbiol.* 135:1507–1514.

44. Hjort, C. M., S. E. Godtfresden, and C. Emborg. 1990. Isolation and characterization of a nitrile-hydratase from a *Rhodococcus* sp. *J. Chem. Technol. Biotechnol.* 48:217–226.

45. Hummel, W., H. Schütte, E. Schmidt, C. Wandrey, and M. Kula. 1987. Isolation of L-phenylalanine dehydrogenase from *Rhodococcus* sp. M4 and its application for the production of L-phenylalanine. *Appl. Microbiol. Biotechnol.* 26:409–416.

46. Ibrahim, K. E., M. W. Couch, C. M. Williams, M. B. Budd, R. A. Yost, and J. M. Midgley. 1984. Quantitative measurement of octopamines and synephrines in urine using capillary column gas chromatography negative ion chemical ionization mass spectrometry. *Anal. Chem.* 56:1695–1699.

47. Iizuka, H., and K. Komagata. 1964. Microbiological studies of petroleum and natural gas. I. Determination of hydrocarbon-utilizing bacterium. *J. Gen. Appl. Microbiol.* 10:207–221.

48. Imai, Y., and S. Kuwatsuka. 1986. Substrate specificity and induction of degrading activity for the herbicide molinate in three microbes isolated from soil. *J. Pest. Sci.* 11:563–572.

49. Janke, D., and W. Ihn. 1989. Cometabolic turnover of aniline, phenol and some of their monochlorinated derivatives by the *Rhodococcus* mutant strain AM144. *Arch. Microbiol.* 152:347–352.

50. Kaufman, D. D. 1967. Degradation of carbamate herbicides in soil. *J. Agric. Food Chem.* 15:582–591.

51. Kaufman, D. D., Y. Katan, D. F. Edwards, and E. G. Jordan. 1985. Microbial adaptation and metabolism of pesticides. In *Agricultural Chemicals of the Future*. Ed. J. L. Hilton. New Jersey: Rowman and Allenhold. Totowa. 255–272.

52. Kaufman, D. D., P. C. Kearney, D. W. Von Endt, and D. E. Miller. 1970. Methyl carbamate inhibition of phenylcarbamate metabolism in soil. *J. Agric. Food Chem.* 18:513–519.

53. Komura, I., K. Komagata, and K. Mitsugi. 1973. A comparison of *Corynebacterium hydrocarboclastus* Iizuka and Komagata 1964 and *Nocardia* erythropolis (Gray and Thornton) Walksman and Henrici 1948. *J. Gen. Appl. Microbiol.* 19:161–170.

54. Koronelli, T. V., S. G. Dermicheva, and E. V. Korotaeva. 1988. Survival of hydrocarbon-oxidizing bacteria in conditions of complete starvation. *Mikrobiologiya* 57:298–304.

55. Koronelli, T. V., V. V. Il'inskii, V. A. Yanushka, and T. I. Krasnikova. 1985. Hydrocarbon-oxidizing microflora of areas of the Baltic sea and Kurshkii bay contaminated by black-oil spillage. *Mikobiologiya* 56:472–478.

56. Kurane, R., T. Szuki, and Y. Takahara. 1979. Removal of phthalate ester by activated sludge inoculated with a strain of *Nocardia erythropolis*. *Agric. Biol. Chem.* 43:421–427.

57. Kuwahara, M. 1981. Metabolism of lignin-related compounds by bacteria. In *Lignin Biodegradation*, Vol. II. Ed. K. T. Kirk, T. Higuchi, M. Chang. Boca Raton: CRC Press. 127–146.

58. Larkin, M. J., and M. J. Day. 1986. The metabolism of carbaryl by three bacterial isolates, *Pseudomonas* spp. NCIB 12042–12043 and *Rhodococcus* sp. NCIB-12038 from garden soil. *J. Appl. Bacteriol.* 60:233–242.

59. Lechevalier, H. 1986. Nocardioform. In *Bergey's Manual of Systematic Bacteriology.* Vol. 2. Ed. Ed. P. H. A. Sneath, N. S. Mair, N. E. Sharp, J. G. Holt. Baltimore: Williams and Wilkins Co. 1451–1506.

60. Lechner, U., and G. Straub. 1984. Influence of substrate concentration on the induction of amidases in herbicide degradation. *Z. Allegemie Mikrobiol.* 24:581–584.

61. Lee, A. 1984. EPTC (s-ethyl N,N-dipropylthiocarbamate)-degrading microorganisms isolated from a soil previously exposed to EPTC. *Soil Biol. Biochem.* 16:529–531.

62. Lloyd-Jones, G., and P. W. Trudgill. 1989. The degradation of alicyclic hydrocarbons by a microbial consortium. *Internat. Biodeter.* 25:197–206.

63. MacRae, I. C. 1989. Microbial metabolism of pesticides and structurally related compounds. *Rev. Environ. Contam. Toxicol.* 109:1–87.

64. Mathew, C. D., T. Nagasawa, M. Kobayashi, and M. H. Yamada. 1988. Nitrilase-catalyzed production of nicotinic acid from 3-cyanopryridine in *Rhodococcus rhodochrous* JI. *Appl. Environ. Microbiol.* 54:1030–1032.

65. Mauger, J., T. Nagasawa, and H. Yamada. 1988. Nitrile hydratase-catalyzed production of isonicotinamide, picolinamide and pyrazinamide from 4-cyanopyridine, 2-cyanopyridine and cyanopyrazine in *Rhodoccus rhodochrous* J. I. *J. Biotechnol.* 8:87–96.

66. Miwa, N., Y. Takeda, and S. Kuwatsuka. 1988. Plasmid in the degrader of the herbicide thiobencarb (benthiocarb) isolated from soil. A possible mechanism for enrichment of pesticide degraders in Soil. *J. Pest. Sci.* 13:291–293.

67. Moorman, T. B. 1986. Microbial response to EPTC in a soil showing accelarated degradation of EPTC. *Abstr. Weed Sci. Soc. America* No. 252.

68. Munnecke, D. M., and D. P. H. Hsieh. 1976. Pathways of microbial metabolism of parathion. *Appl. Environ. Microbiol.* 31:63–69.

69. Murai, N. 1981. Cytokinin biosynthesis and its relationship to the presence of plasmids in strains of *Corynebacterium fascians*. In *Metabolism and Molecular Activities of Cytokinins*. Ed. J. Guern, C. Péaud-Lenoël. Berlin: Springer-Verlag KG. 17–26.

70. Nagasawa, T., C. D. Mathew, J. Mauger, and H. Yamada. 1988. Nitrile hydratase-catalyzed production of nicotinamide from 3-cyanppyridine in *Rhodococcus rhodochrous* JI. *Appl. Environ. Microbiol.* 54:1766–1769.

71. Nakano, H., F. Tomita, K. Yamaguchi, M. Nagashima, and T. Suzuki. 1977. Corynecin (chloramphenicol analogs) fermentation studies: selective production of Corynecin I by *corynebacterium hydrocarboclastus* grown on acetate. *Biotechnol. Bioeng.* 19:1009–1018.

72. Ninnekar, H. G., and B. G. Pujjar. 1985. Degradation of dimethylterephthalate by a *Rhodococcus* spp. *Ind. J. Biochem. Biophys.* 22:232–235.

73. Obrigawitch T., A. R. Martin, and F. W. Roeth. 1983. Degradation of thiocarbamate herbicides in soils exhibiting rapid EPTC breakdown. *Weed Sci.* 31:187–192.

74. Rahman, A., and T. K. James. 1983. Decreased activity of EPTC + R-25788 following repeated use in some New Zealand soils. *Weed Sci.* 31:783–789.

75. Rahman, A., G. A. Atkinson, J. A. Douglas, and D. P. Sinclair. 1979. Eradicane causes problems. *New Zealand J. Agric.* 139:47–49.

76. Rajagopal, B. S., G. P. Brahmaprakash, B. R. Reddy, U. D. Singh, and N. Sethunathan. 1984. Effect and persistence of selected carbamate pesticides in soil. *Residue Rev.* 93:62–75.

77. Reddy, B. V. P., P. S. Dhanraj, and V. V. S. N. Rao. 1984. Effect of insecticides on soil microorganisms. In *Insecticide Microbiology*. Ed. R. Lal. New York: Springer-Verlag. 169–201.

78. Robertson, J. G., and R. D. Batt. 1973. Survival of *Nocardia corallina* and degradation of constituents during starvation. *J. Gen. Microbiol.* 78:109–118.

79. Roeth, F. W. 1986. Enhanced herbicide degradation in soil with repeat application. *Rev. Weed Sci.* 2:45–65.

80. Sandoval, H., A. Aguilar, C. Paniagua, and J. F. Martin. 1984. Isolation and physical characterization of plasmid pCC1 from corynebacterium callunae and construction of hybrid derivatives. *Appl. Microbiol. Biotechnol.* 19:409–413.

81. Singer, M. E. V., and W. R. Finnerty. 1988. Construction of an *Escherichia coli-Rhodococcus* shuttle vector and plasmid transformation in *Rhodoccus* spp. *J. Bacteriol.* 170:638–645.

82. Singer, M. E. Vogt., and W. R. Finnerty. 1990. Physiology of biosurfactant synthesis by *Rhodococcus* species H13-A. *Can. J. Microbiol.* 36:741–745.

83. Skipper, H. D., E. C. Murdock, D. T. Gooden, J. P. Zublena, and M. A. Amakiri. 1986. Enhanced herbicide biodegradation in South Carolina soils previously treated with butylate. *Weed Sci.* 34:558–563.

84. Skipper, H. D. 1990. Enhanced biodegradation of carbamothioate herbicides in South Carolina. In *Enhanced Biodegradation of Pesticides in the Environment*. Ed. K. D. Racke, and J. R. Coats. Washington, D.C.: American Chemical Society. 37–52.

85. Slizen, Z. M., T. G. Zimenko, A. S. Samsonova, and G. M. Volkova. 1989. Utilization of dimethylterephthalate by *Rhodococcus erythropolis*. *Microbiologiya* 58:382–386.

86. Stotzky, G., M. A. Devanas, and L. R. Zeph. 1990. Methods for studying bacterial gene transfer in soil by conjugation and transduction. *Adv. Applied Microbiol.* 35:57–169.

87. Subba-Rao, R. V., T. H. Cromartie, and R. A. Gray. 1897. Methodology in accelerated biodegradation of herbicides. *Weed Technol.* 1:333–340.

88. Tam, A. C., R. M. Behki, and S. U. Khan. 1987. Isolation and characterication of an s-ethyl-N,N-dipropylthiocarbamate degrading *Arthrobacter* strain and evidence for plasmid-associated s-ethyl-N,N-dipropylthiocarbamate degradation. *Appl. Environ. Microbiol.* 53:1088–1093.

89. Tam, A. C., R. M., Behki, and S. U. Khan. 1988. Effect of dietholate (R-33865) on the degradation of thiocarbamate herbicides by an EPTC-degrading bacterium. *J. Agric. Food Chem.* 36:654–657.

90. Tebbe, C. C., and H. H. Reber. 1988. Utilization of herbicide phosphinothricin as a nitrogen source by soil bacteria. *Appl. Microbiol. Biotechnol.* 29:103–105.

91. Tu, C. M., and J. R. W. Miles. 1976. Interaction between insecticides and soil microbes. *Residue Rev.* 64:17–65.

92. Valo, R. J., and M. Salkinoja-Salonen. 1986. Microbial transformation of polychlorinated phenoxyphenols. *J. Gen. Appl. Microbiol.* 32:505–518.

93. Valo, R. J., and M. Salkinoja-Salonen. 1986. Bioreclamation of chlorophenol-contaminated soil by composting. *Appl. Microbiol. Biotechnol.* 25:68–75.

94. Vaughan, P. A., P. S. J. Cheetham, and C. J. Knowleds. 1988. The utilization of pyridine carbonitriles and carboxamides by *Nocardia rhodochrous* LL100-21. *J. Gen. Microbiol.* 134:1099–1107.

95. Wakisuki, Y., K. Koizumi, Y. Nishimoto, M. Kobayashi, and N. Tsuji. 1980. Hygromycin and epihygromycin from a bacterium *corynebacterium equi*. *J. Antibiot.* 33: 695–704.

96. Wilkinson, R. E. 1988. Carbamothioates. In *Herbicides, Chemistry, Degradation and Mode of Action*. Vol. 3. Ed. P. C. Kearney, D. D. Kaufman. New York: Marcel Dekker Inc. 245–300.

97. Wilson, R. G. 1984. Accelerated degradation of thiocarbamate herbicides in soil with prior thiocarbamate exposure. *Weed Sci.* 32:264–268.

98. Woods, N. R., and J. C. Murrel. 1989. The metabolism of propane in *Rhodococcus rhodochrous* PNK b1. *J. Gen. Microbiol.* 135:2335–2344.

99. Woods, N. R., and J. C. Murrel. 1990. Epoxidation of gaseous alkenes by *Rhodococcus* sp. *Biotechnol. Lett.* 12:409–414.

100. Yamada, K. 1977. Bioengineering report. Recent advances in industrial fermentation in Japan. *Biotechnol. Bioeng.* 19:1563–1621.

101. Yoshihama, M., K. Higashiro, E. A. Rao, M. Akedo, W. G. Shanebruch, M. T. Follettie, G. C. Walker, and A. J. Sinskey. 1985. Cloning vector system for *Corynebacterium glutamicum. J. Bacteriol.* 162:591–597.

102. Zajic, J. E., H. Guignard, and D. F. Gerson. 1977. Emulsifying and surface active agents from *Corynbacterium hydrocarboclastus. Biotechnol. Bioeng.* 19:1295–1301.

CHAPTER 12

Anaerobic Biodegradation of Chlorinated Organic Compounds

LARRY MONTGOMERY, NADA ASSAF-ANID, LORING NIES,
PAUL J. ANID, and TIMOTHY M. VOGEL

Abstract. Many chlorinated organic compounds are potentially hazardous environmental pollutants. Knowledge regarding their movement and fate in the environment is critical to understanding their impact and to developing remediation technologies. A better understanding is needed of the biological processes by which they are transformed into other compounds, which are usually less toxic but occasionally more so. Because bacteria are important agents in transforming most organic compounds, exploring their specific roles in biotransformations of chlorinated compounds holds great interest. In this chapter, we review research into the anaerobic dechlorination of chlorinated aliphatic compounds and polychlorinated biphenyls (PCBs). The major mechanism for anaerobic transformation of chlorinated organic compounds is reductive dechlorination, in which a chlorine substituent is replaced with a hydrogen atom. This reaction is more exergonic, and usually more rapid, with the more highly chlorinated members of a homologous series. A growing body of evidence indicates that reductive dechlorination of some compounds occurs by nonenzymatic catalysis, which is unusual in biological systems, whereas dechlorination of other compounds requires enzymatic catalysis. Much remains to be learned about the mechanisms of dechlorination, the microbes responsible, and the physiological conditions that encourage this type of detoxification. Nonetheless, knowledge gained in recent years provides a basis for initiating testing of bioremediation schemes. For example, sequential anaerobic/aerobic treatment of various chlorinated compounds involves the stimulation of reductive dechlorination as a preparatory step for subsequent aerobic mineralization.

INTRODUCTION

Chlorinated organic compounds are among the most problematic of the anthropogenic xenobiotic chemicals (55,62,81). They have been produced and released into the environment in large quantities and are relatively persistent.

In spite of the environmental novelty of these compounds, a growing body of research demonstrates that many, if not all, can be degraded biologically, albeit slowly.

Drs. Montgomery, Assaf-Anid, Nies, Anid, and Vogel are in Environmental and Water Resources Engineering, Department of Civil and Environmental Engineering, University of Michigan, Ann Arbor, MI 48109-2125, U.S.A. Drs. Montgomery and Vogel are also at the NSF Center for Microbial Ecology, University of Michigan.

Environmental and public health concerns, reflected in environmental regulations, are creating a demand for safe and practical techniques to clean up polluted sites. Thus, there is a great need for information as to which chlorinated compounds can be biodegraded and under what conditions, and about the microorganisms and biochemical mechanisms involved. An understanding of these factors will not only add to our scientific knowledge, but will help in assessing natural detoxification processes and in designing and improving bioremediation methods.

Two major classification schemes should be considered in discussing the biological dehalogenation reactions: the nature of the catalyst and the type of chemical reaction. Most biological reactions are catalyzed by enzymes that are considered to have evolved to accomplish particular physiological functions. Enzymes generally exhibit remarkable specificity for their physiological substrates, although some enzymes are less specific. Microorganisms have had relatively little time to evolve specific enzymes to degrade anthropogenic halogenated compounds. Marine microbes are exposed to naturally occurring chlorinated compounds, but these compounds have only one or, rarely, two chlorine substituents. Establishing that an enzyme has (or has not) evolved specifically for a particular substrate, other than by teleological arguments, is difficult. In advanced phases of research, however, genetic and biochemical investigations may provide information about the specificity and physiological functions of such enzymes. Genetic aspects of halogenated compound degradation have recently been reviewed by Boyle (20) and Chaudhry and Chapalamadugu (22).

Assuming that some or all microbial dehalogenation reactions are not catalyzed by specific enzymes, one must postulate either nonspecific enzyme activity or non-enzymatic reactions. The latter possibility may receive less consideration because of the near ubiquity of enzyme involvement in biological reactions. Therefore, scientists may have approached this issue with some bias in favor of enzymatic mechanisms, albeit sometimes nonspecific, of dechlorination. As discussed below, recent research suggests that certain classes of halogenated compounds, such as the aliphatic compounds, are likely to be dechlorinated by nonenzymatic catalysts, whereas others, such as the chlorophenols, are probably dechlorinated by enzymes, whether specific or not.

BIOLOGICAL DEHALOGENATION REACTIONS

There are two major types of microbially mediated reactions involved in the removal of halogen substituent groups from organic molecules (reviewed in 55,81). These are oxidation and reduction reactions involving the transfer of electrons from or to the halogenated compound, respectively. The electronegative character of halogen substituent groups makes halogenated compounds more oxidized than the corresponding unhalogenated compounds, and thus somewhat less susceptible to oxidative reactions but more susceptible to reductive reactions.

Oxidative reactions usually do not directly cause loss of a halogen group but often serve as essential first steps in the aerobic biodegradation of halogenated compounds; dehalogenation occurs subsequently (reviewed in 62). Oxidative reactions may involve the transfer of electrons from the substrate (electron donor) to an electron acceptor, or the inclusion of an oxygen molecule into the substrate, as catalyzed by mono- and dioxygenases. Microbial oxidations of halogenated compounds generally are catalyzed by enzymes that presumably have evolved (teleologically speaking) for catabolism of natural substrates. In such cases, the capacity for degrading anthropogenic compounds may depend on a low degree of substrate specificity of the enzymes. Because this chapter focuses on anaerobic processes, dehalogenation reactions by aerobic bacteria are not discussed except for comparison.

Under anaerobic conditions, reductive dehalogenation is the dominant mechanism of halogen removal. The transfer of electrons to the halogenated compound causes the replacement of a halogen substituent with a hydrogen atom:

$$R-Cl + 2e^- + H^+ \rightarrow R-H + Cl^-.$$

Within a family of chlorinated compounds, the ease of reductive dechlorination is generally proportional to the degree of chlorination. For example, tetrachloromethane (carbon tetrachloride) is the most highly substituted of the chloromethanes. Thus, it has the highest oxidation-reduction (redox) potential, and is the most favorable electron acceptor, of the chloromethanes; dechlorination of that compound is the most exergonic, as represented by the most negative value for change in Gibbs free energy ($\Delta G^{0'}$) (81). Conversely, oxidative reactions are more exergonic for substrates (electron donors) with a lower degree of halogenation.

Although reaction rates are not directly controlled by $\Delta G^{0'}$, there is an indirect relationship for reactions in which the first step is a one-electron transfer to form a carbon radical transition state (72). The activation energy for formation of the transition state is proportional to the affinity of the substrate for the electron, as reflected in the reduction potential for the half reaction. The first reduction step generally determines the rate for the overall reaction, because the transition state radical is much more reactive than the original substrate. A linear plot of pseudo-first-order rate constants vs. standard reduction potentials for members of the chloroethylene family, when reductive dechlorination is catalyzed by various enzymatic cofactors, supports this interpretation (44).

The significance of reductive dehalogenation of halogenated xenobiotic compounds by anaerobic bacteria has gained recognition only in the last decade. In most cases, the activities have been observed in mixed cultures (see below) and the responsible microbes have not been identified. Under anaerobic conditions, the fate of halogenated compounds is greatly influenced by the presence or absence of potential electron acceptors (76). Nitrate, sulfate, and carbonate can serve as terminal electron acceptors but generally do not support initial oxidative metabolism of halogenated compounds for several purely chemical reasons. First, molecular oxygen is required in most cases as a substrate for oxygenase attack. Secondly, since these inorganic ions have lower redox potentials than oxygen, and since halogenated compounds have higher redox potentials than the corresponding nonhalogenated compounds, electron transfer is generally unfavorable thermodynamically.

Rather than serving as electron donors, the halogenated compounds can, in many cases, serve as electron acceptors. Therefore, the inorganic electron acceptors compete with the halogenated compounds for electrons removed from other substrates (electron donors). In fact, at equivalent concentrations, the more highly chlorinated aliphatic compounds, such as tetrachloromethane (TTCM), trichloroethane (TCE), and tetrachloroethylene, are potentially better (more thermodynamically favorable) electron acceptors (i.e. they have higher redox potentials) than nitrate, sulfate, or carbon dioxide (81). The presence of specific enzymes for transfer of electrons to the natural electron acceptors confers a kinetic advantage, however, so that most of the electron flow would be to the natural electron acceptors when available. Furthermore, the concentration of halogenated compounds in most environments is far lower than that of one or more natural electron acceptors. Nevertheless, there are conditions under which some of the electron flow is diverted to the halogenated compounds, as evidenced by the occurrence of reductive dehalogenation.

In a highly reduced anaerobic environment, many anaerobic bacteria face a challenge to dispose of electrons in order to allow their energy-yielding metabolism to continue. Since polyhalogenated organic compounds have relatively high redox potentials, the possibility exists for microbes to profit from reducing those compounds. One process is analogous to respiration, in which electrons are transferred from a donor to a terminal electron acceptor such as oxygen, nitrate, sulfate, or Fe(III). In respiration, the organism conserves energy by electron transport–linked phosphorylation of adenosine diphosphate (ADP) to adenosine triphosphate (ATP). The only organism known to gain energy from "respiratory" reductive dehalogenation is *Desulfomonile tiedjei* strain DCB-1, which can use pyruvate, formate, or hydrogen as electron donor in the reductive dehalogenation of chlorobenzoates (28,57,61). Thiosulfate and sulfite also serve as electron acceptors for *D. tiedjei,* which shares some characteristics with the sulfate-reducing bacteria (75).

Even microbes unable to use chlorinated compounds in energy-conserving respiration might benefit from simply disposing of electrons by reducing the chlorinated compounds. For example, butyrate-fermenting syntrophic organisms normally must dispose of electrons by reducing protons to H_2 (30,58). The redox potential for that half-reaction at pH 7 (E_0') is −414 mV (58), whereas it is 580 mV for tetrachloroethylene (perchloroethylene, PCE). Thus, PCE is a much better electron acceptor than protons at almost any concentration; as always, the actual value of ΔG is dependent on PCE and trichlorethylene (the product) concentrations. If the butyrate-fermenting organism was capable of "dumping" electrons on PCE, proton reduction (and, hence, syntrophic H_2 consumption by methanogenic bacteria) would be obviated, thus allowing the butyrate-fermenting organisms to grow in pure culture, a condition that is not normally possible. Considering that PCE might inhibit microbial growth, there may be a very narrow concentration range in which this hypothetical metabolism could occur. This may explain observations with a benzoate-degrading, methanogenic mixed culture (71). Inhibition of methanogenesis with bromoethane-sulfonic acid (BES) halted benzoate degradation, thus suggesting that H_2 consumption was necessary. Benzoate degradation resumed when PCE was added to BES-inhibited cultures, however, and PCE was reductively dechlorinated to dichloroethylene.

Most reports of reductive dehalogenation by microbes involve the activities of mixed methanogenic cultures, although mixed sulfidogenic (sulfate-reducing) cultures may also be involved (Table 1). Much of the research to date has been aimed at characterizing the general environmental conditions conducive to reductive dechlorination and at determining the range of compounds subject to such transformation. Microbial populations are generally tested without extensive adaptation to chlorinated compounds. Inhibitors, such as BES (for methanogens) or molybdate (for sulfate-reducing bacteria), are sometimes used to limit the metabolic activities of certain groups essentially to determine which groups dechlorinate by a process of elimination. Results may be ambiguous because of the interlocking metabolisms of microbial groups, e.g. obligate proton-reducing bacteria are indirectly inhibited by BES, since H_2 consumption slows and the H_2 concentration rises. In addition, BES is debrominated by methanogenic bacteria (15) and often inhibits methanogenesis only partially (18). Inhibitor effectiveness may also be compromised by soil or sediments, which are often present in environmental research. Although the screening of pure cultures of anaerobes has begun only recently, the diversity of microbes known to dechlorinate halocompounds is becoming substantial (Table 2). For most of these cultures, the relationship between metabolism and dechlorination has not been investigated.

Table 1. Substrates reported to be dehalogenated by mixed cultures of anaerobic bacteria.

Overall metabolism	Chlorinated substrates[a,b]	References
Methanogenic	halomethanes	(19,25)
	TTCM, TCM, TTCA, PCE	(17)
	TCA, DCA, DCE, VC	(82)
	PCE, TCE, DCE, VC	(8,27,40,71,83)
	chloroacetic acids	(33)
	chlorobenzoates	(10,73,74,77,78)
	chlorophenols	(45,48,50,51,53,59,68,73,74,84)
	chlorocatechols	(3,63)
	chloroesorcinol	(37)
	2,4-D	(45)
	chloroanilines	(51,54,56)
	(hexa)chlorobenzenes	(16,38)
	halovanillins	(63,64)
	halobenzaldehydes	(64)
	bromacil	(2)
	PCB	(66,69,70,80)
Sufidogenic	PCE, TCE	(9)
	chlorophenols	(46,47)

[a]Some articles discussed additional chlorinated substrates.
[b]BA, bromoethane; DBA, dibromoethane; DBE, dibromoethylene; DCA, dichloroethane; DCE, dichloroethylene; DCM, dichloromethane; PCE, perchloroethylene; TCA, trichloroethane; TCE, trichloroethylene; TCM, trichloromethane; TTCA, tetrachloroethane; TTCM, tetrachloromethane; VC, vinyl chloride (chloroethylene).

Some observations do not fit into the neat pattern described above. One unusual case is the transformation of TTCM by *Pseudomonas* sp. strain KC under denitrifying conditions (23). Criddle et al. (23) postulate that TTCM was reductively dechlorinated to a dichlorocarbene radical by non-specific activity of a trace-metal scavenging system, followed by abiotic hydrolysis of the radical to formate. The formate could then be oxidized to CO_2 with nitrate as the terminal electron acceptor. Another phenomenon is the partial transformation of TTCM to CO_2 by an uncharacterized "substitutive" pathway in the acetogenic bacterium, *Acetobacterium* woodii, and in *Methanobacterium thermoautotrophicum* and *Desulfobacterium autotrophicum* (32,34). That transformation occurs concurrently with reductive dechlorination of TTCM to trichloromethane (TCM) and dichloromethane (DCM). Furthermore, a separate pathway for "oxidative" transformation of TCM to CO_2 may be present in *A. woodii* (32). In that organism, reductive dechlorination of TTCM is abolished by autoclaving the cells, whereas the substitutive conversion is not, thus suggesting that the latter process is not protein-catalyzed. Still another unusual reaction is the hydrolytic dechlorination of chloroacetic acid to glycolate by a methanogenic mixed culture (33). Freedman and Gosset recently discovered a methanogenic enrichment culture able to use DCM as sole carbon and energy source (41). Without another source of electrons, most of the DCM was oxidized to CO_2, whereas $< 20\%$ was reductively dechlorinated. Hydrogen removed from DCM was largely used to produce CH_4. A similar disproportionation of chloromethane by an acetogenic bacterium has recently been discovered (79).

Table 2. Substrates reported to be reductively dehalogenated by pure cultures of bacteria.

Metabolic type, genus	Chlorinated substrate[a]	Reference
Methanogenic		
Methanosarcina	TTCM, TCM, DCM, bromoform	(60)
	PCE	(35)
Methanobacterium	TTCM, TCM, TCA, DCA, PCE	(31)
	DCA, DBA, DCA, DBE	(15)
Methanococcus	DCA, DBA, BA, DBE	(15)
Acetogenic		
Acetobacterium	TTCM, TCM, DCM	(34)
Clostridium	TTCM, TCM, DCM	(34)
Sulfidogenic		
Desulfobacterium	TTCM, TCM	(34)
	TTCM, TCM, TCA	(31)
Desulfomonile	chlorobenzoates	(29)
	PCE	(36)
Heterotrophic		
Clostridium	TTCM, TCM, TCA	(43)
Denitrifying		
Pseudomonas	TTCM	(23)
Facultative		
Escherichia	TTCM	(24)

[a] BA, bromoethane; DBA, dibromoethane; DBE, dibromoethylene; DCA, dichloroethane; DCE, dichloroethylene; DCM, dichloromethane; PCE, perchloroethylene; TCA, trichloroethane; TCE, trichloroethylene; TCM, trichloromethane; TTCA, tetrachloroethane; TTCM, tetrachloromethane; VC, vinyl chloride (chloroethylene).

DECHLORINATION OF PCE BY UNADAPTED
ANAEROBIC CULTURES

One approach to physiological characterization of reductive dechlorination is to determine whether that activity requires adaptation by the microbes. When soil samples obtained from four sites presumed to be unexposed to chlorinated compounds were incubated in anaerobic batch cultures with acetate as the primary substrate, PCE was dechlorinated during, but not after, the period of active methanogenesis. When these cultures were incubated again with PCE and acetate, the amounts of methane produced and PCE dechlorinated were not different from those in parallel cultures not previously exposed to PCE; methane production was also similar in cultures without PCE. These results suggest that PCE dechlorination by these mixed methanogenic cultures does not require a detectable period of acclimation or adaptation. In addition, the requirement for an electron donor (acetate) for reductive dechlorination was confirmed (N. Assaf-Anid and T.M. Vogel, unpublished data).

Similarly, Belay and Daniels (15) found that three species of methanogenic bacteria completely dehalogenated mono- and dihalogenated ethanes and ethylenes. Although the question of adaptation was not specifically addressed, lag periods before the onset of dehalogenation are not evident in the data presented graphically. In both studies, the concentrations of the chlorinated compounds were ca. 1000-fold lower than those of the electron donors.

REDUCTIVE DECHLORINATION OF ALIPHATIC COMPOUNDS BY PORPHYRINS AND CORRINS

As mentioned above, some compounds can be dechlorinated by nonenzymatic catalysts produced by microbes (reviewed in 11). These catalysts are metallo-organic ligands, including porphyrins and cobalamins, which serve as enzymatic cofactors (coenzymes) or their precursor vitamins. The discovery of the ability of coenzymes to dechlorinate compounds in the absence of enzymes *in vitro* has inspired the hypothesis that they may catalyze dechlorinations observed *in vivo*, perhaps while serving as enzyme prosthetic groups.

Recently, hematin, an iron(II) porphyrin, was shown to catalyze the dechlorination of TTCM, TCM, and trichloroethane with sulfide or cysteine as the ultimate electron donor (49). Less-chlorinated analogs were not dechlorinated under the experimental conditions. Tetrachloromethane was also dechlorinated by $Fe(II)SO_4$ with cysteine, although far more slowly than by hematin.

Several corrinoids have been shown to dechlorinate TTCM sequentially to methane with titanium(III) citrate or dithiothreitol as electron donor (52). The rate of dechlorination decreased with each chlorine removed, consistent with the decrease in redox potential (81). Evidence for the formation of alkylcorrinoid intermediates was presented.

Baxter (12) confirmed earlier work showing that γ—1,2,3,4,5,6-hexachloro-cyclohexane (lindane), DDT, and several other xenobiotic compounds were dechlori-nated by hematin, with sodium dithionite as reductant. Chlorobenzoic acids and poly-chlorinated biphenyls were not dechlorinated within two weeks in that system.

Recently, vitamin B_{12} and the methanogenic coenzyme F_{430} were demonstrated to catalyze the complete dechlorination of PCE to ethylene, whereas hematin removed only three chlorines, producing vinyl chloride (44). Titanium(III) citrate was a more effective reductant than dithiothreitol. In the aforementioned coenzymes, the transi-tion metals are coordinated to nonprotein ligands. Three other transition metal coenzymes—azurin (copper), and two-iron and four-iron ferredoxins—in which the metal is coordinated directly to protein ligands, did not catalyze the reductive de-chlorination of PCE. Vitamin B_{12}, hematin, and F_{430} also catalyzed the dechlorination of TTCM, at rates 10- to-25 fold greater than for PCE. Vitamin B_{12} and hematin de-chlorinated hexachlorobenzene to pentachlorobenzene and tetrachlorobenzenes; the rate for the second dechlorination step was 100-fold lower than for the first step.

To gain further understanding of the role of biological molecules in nonenzymatic dehalogenation reactions, Assaf-Anid and Vogel have investigated a variety of porphyrins and corrins and the metals capable of forming complexes with them (6,7). The major goal was to extend previous findings by testing different metals with the organic ligands and alone. With vitamin B_{12} as the catalyst and dithiothreitol as the reductant, TTCM was rapidly dechlorinated, approaching 90% transformation within 160 minutes (Figure 1a). Trichloromethane (TCM) appeared rapidly and remained at approximately 0.2 mM. Dichloromethane (DCM) appeared soon thereafter and

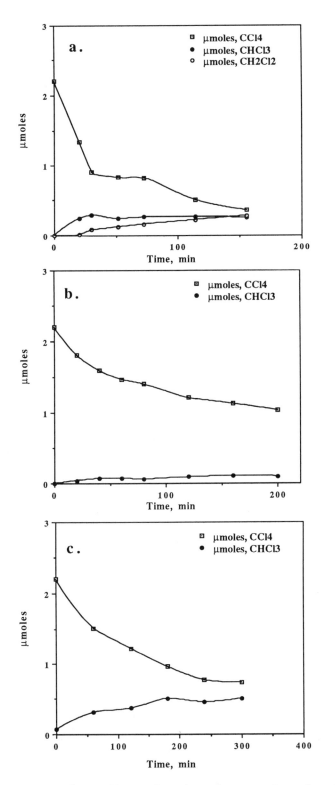

Figure 1. Dechlorination of tetrachloromethane by cofactors and metals and appearance of products with dithiothreitol as reductant. Catalysts were vitamin B_{12} (a); Co(II)—hematoporphyrin (b); and Co(II) without organic ligand (c).

increased gradually. Methane was also produced (data not shown), probably via chloromethane, which could not be detected due to coelution with other substances during gas chromatography.

By comparison, dechlorination of TTCM catalyzed by hematin or hematoporphyrin-Co(II) was ca. one fifth as fast, and DCM was not detected (Figure 1b). Hematoporphyrin-Fe(II) was much less effective than hematoporphyrin-Co(II) or hematin, even though the latter is coordinated with Fe(II). Hematoporphyrin was barely active when coordinated with Mg(II) and was inactive without a coordinating metal.

Substantial activity was also detected with Co(II) or Fe(II) in the absence of organic ligand (Table 3, Figure 1c). This demonstrates that abiotic catalysis may occur under some circumstances, although it is likely that the reductants must be produced by microbes in environments free of geological sulfide. Assuming that biological molecules, such as the cofactors tested here, are accessible to the chlorinated compounds, they might be quantitatively more important as catalysts than the uncomplexed metals. Cobalt was more effective than Fe(II), whether alone or complexed with hematoporphyrin. When the same catalysts were incubated with PCE, only vitamin B_{12} gave detectable activity (not shown).

The demonstration of nonenzymatic dechlorination of TTCM and PCE by various cofactors does not prove that this is the only, or even a significant, mechanism of dechlorination occurring in situ. Nonenzymatic and enzymatic reactions may both be important; the relative importance may differ among chlorinated compounds. For example, the preceding discussion shows that reductive dechlorination of chloroalkanes and chloroalkenes can occur without enzymatic catalysis, whereas the ability of Desulfomonile tiedjei to conserve energy from the dechlorination of chlorobenzoates indicates enzyme involvement (28). More importantly, aryl reductase activity is heat-labile (26) and nonviable cells of D. tiedjei are inactive (57). PCE-dechlorinating activity is apparently induced by 3-chlorobenzoate (76), thus suggesting that this activity is also enzymatic.

Table 3. Pseudo-first-order constants (hr^{-1}) for dehalogenation of tetrachloromethane (TTCM) and tetrachloroethylene (PCE).

Catalyst	TTCM	PCE
Vitamin B12	1.0	0.10
Co(II)-hematoporphyrin	0.22	$\ll 0.1$
Hematin	0.20	$\ll 0.1$
Co(II)[a]	0.16	$\ll 0.1$
Fe(II)	0.12	$\ll 0.1$
Ni(II)	0.03	$\ll 0.1$
Ni(II)-hematoporphyrin	0.03	$\ll 0.1$
Fe(II)-hematoporphyrin	0.01	$\ll 0.1$

[a] Metals precipitated when added in the absence of organic ligands, confounding comparisons of rate constants. The total amounts of the metals added were much higher than when added with organic ligands.

PCB DECHLORINATION BY VITAMIN B_{12}

Since the discovery that polychlorinated biphenyls are reductively dechlorinated by unknown anaerobic bacteria in sediments, the mechanism of dechlorination has been of great interest. Under aerobic conditions, degradation of PCBs is apparently catalyzed by nonspecific activity of oxygenases whose normal physiological function is catabolism of nonchlorinated aromatic compounds (1,13,14).

Considering that some compounds can be reductively dechlorinated by non-enzymatic systems (above), it was of interest to determine whether the far more complex PCBs might be similarly susceptible to nonenzymatic transformation. During a four-week incubation, 2,3,4,5,6-pentachlorobiphenyl was dechlorinated by vitamin B_{12} (with dithiothreitol as reductant) to 2,3,5,6- and 2,3,4,6-tetrachlorobiphenyls (Figure 2) (5). The reaction was carried out in 1:1 water:1,4-dioxane at 30°C. This is the first reported demonstration of PCB dechlorination by an acellular system.

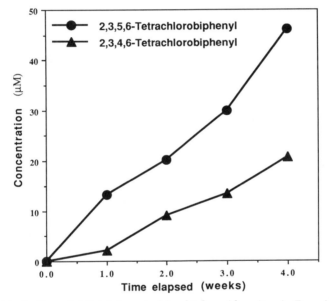

Figure 2. Dechlorination of 2,3,4,5,6-pentachlorobiphenyl by vitamin B_{12} with dithiothreitol as reductant. Line shows time course of product appearance.

These results are consistent with the hypothesis that microbial PCB dechlorination observed in anaerobic environments is due, at least in part, to nonspecific reduction by enzyme cofactors, i.e. a fortuitous side reaction of normal anaerobic metabolism. If so, virtually any microbe could dechlorinate PCBs in the presence of an adequate supply of reducing power, normally from the growth substrate. Coenzyme B_{12} is a wide-spread, if not ubiquitous, component of procaryotic cells. These results do not prove, however, that coenzymes are the only (or major) catalysts of PCB dechlorination. For example, it is not clear whether coenzyme B_{12}, with a deoxyadenosyl prosthetic group in place of the cyano group of vitamin B_{12}, would behave the same; the cobalt has a +1 valence in the coenzyme versus a +3 valence in the vitamin.

PCBs are rather large and may enter bacterial cells only slowly, if at all. The molecules are sufficiently hydrophobic that they may be concentrated in the plasma membrane. Therefore, electron transfer components, imbedded in the plasma membrane, may be important in reductive dechlorination of PCBs *in vivo*.

Coenzyme B_{12} is not primarily involved in electron transfer reactions but rather in rearrangements, typically the intramolecular exchange of an hydrogen atom for an alkyl group. When the reaction was carried out in deuterium oxide, the tetrachloro-biphenyl products were deuterated, thus demonstrating that hydrogen atoms were not directly transferred from dithiothreitol to the PCB molecule. That *meta-* and *para*-chlorine atoms were removed, but not *ortho*-chlorine atoms, may be related to the relative stability of the different substituent locations. This pattern is also found in PCBs dechlorinated in anaerobic sediments, although a low level of reductive *ortho* dechlorination has recently been reported for sediments incubated in batch culture (80).

These results will spark further investigation to determine whether other cofactors, particularly those more commonly associated with redox reactions, catalyze the dechlorination of PCBs. The mild conditions required may encourage the development of detoxification technology utilizing the B_{12}-catalyzed dechlorination of PCBs.

PCB DECHLORINATION BY ANAEROBIC MICROBES

Polychlorinated biphenyls (PCBs) in the environment are of great concern not only because of their toxicity but also because of their persistence. They are recalcitrant to aerobic degradation, although aerobic isolates have now been obtained that can oxidize and dechlorinate some congeners, particularly those with relatively few chlorine substituents (14,42). The more highly chlorinated congeners are less susceptible to aerobic degradation because their high redox potentials make oxidative reactions less thermodynamically favorable, as with halogenated aliphatic compounds (81), and because of increased steric hindrance of oxygenases (a kinetic effect). Free coenzymes, because they are smaller than enzymes, may be less subject to steric hindrance.

Recently, evidence has been uncovered indicating that PCBs have been modified in contaminated anaerobic sediments of the Hudson River and Silver Lake, Massachusetts (21). Compared to the congener distribution in the original commercial mixture of PCBs, the sediments were found to contain relatively less of the highly chlorinated congeners (tri-, tetra-, and higher chlorobiphenyls) and more of the less chlorinated congeners (mono-, di-, and certain trichlorobiphenyls), i.e. the average degree of chlorination had decreased. The decreased chlorination level was consistent with reductive dechlorination, which would not destroy the biphenyl moiety. The ability of anaerobic microbes of the aforementioned sediments to dechlorinate PCBs was subsequently confirmed under laboratory conditions (70). Evidence was obtained that the sediment microbial populations (Hudson River and Silver Lake) had adapted to the particular congener mixtures to which they had been exposed (69).

The mechanism of the conversion is of considerable interest, largely because of the desire to enhance biodegradation of PCBs. Current regulations require that some sediments with high concentrations of PCBs be detoxified. Lacking methods to treat PCBs *in situ,* vast quantities of sediments would probably have to be excavated from rivers and lakes. Since there are no cost-effective methods to destroy the PCBs *ex situ* either, contaminated sediments would have to be incinerated or sequestered in hazardous-waste landfills indefinitely. In addition to the extreme costliness of such procedures, it is likely that the excavation, shipping, and landfilling of the sediments would release some fraction of the PCBs back into circulation, thus temporarily increasing the risk to human and animal populations.

Any process that can detoxify and destroy the PCBs *in situ,* thereby minimizing exposure, would be beneficial. To this end, anaerobic dechlorination is being investigated in the hope that bioremediation of PCB contamination may become feasible. As indicated above, reductive (anaerobic) dechlorination presumably does not decrease the total amount of PCBs present but only decreases the (average) number of chlorines per molecule. Superficially, this might seem to limit the usefulness and importance of the anaerobic process, but two points can be made. First, reductive dechlorination makes PCBs more susceptible to total degradation (mineralization) by aerobic microbes. Anaerobic bacteria generally attack the more highly chlorinated, ergo most oxidized, congeners; they produce less chlorinated, more reduced, congeners that are more susceptible to aerobic oxidation, which destroys the (poly)chlorinated biphenyl moiety and reduces the total amount of PCBs (1). Removal of chlorine groups from PCBs would also decrease steric hindrance of enzymatic attack by oxidases of aerobic bacteria, further increasing susceptibility to oxidative degradation. Thus, any practical scheme for bioremediation would presumably exploit this situation by promoting anaerobic dechlorination followed by aerobic degradation.

The second point is that the aerobic step may not be required in every situation. Current toxicological information indicates that carcinogenicity is roughly correlated with the degree of chlorination, i.e. the most highly chlorinated congeners are the most carcinogenic (see 1). In addition, the meta and para chlorines, which are those generally removed in anaerobic sediments (but see 80), are most associated with toxicity (70). Therefore, anaerobic dechlorination decreases carcinogenicity even though the total amount of PCBs does not decrease. Current regulatory limits do not differentiate between congeners based upon carcinogenicity, and so do not recognize any benefit from dechlorination. Refinement of the regulations, however, might lead to a situation in which limits were based on the actual health risk rather than the total amount of PCBs. Thus, in some situations, benefit/risk analysis might show that anaerobic dechlorination in situ, leaving residual lowly chlorinated PCBs, would be preferable to excavating and landfilling vast amounts of sediments, inevitably releasing some PCBs.

EFFECTS OF ORGANIC SUBSTRATES ON PCB DECHLORINATION

Neither the anaerobic microbes nor the electron donors responsible for reductive dechlorination of PCBs are known. Therefore, some common substrates were tested for their ability to support PCB dechlorination (65,67). Although these simple substrates are not found in high concentrations in situ, testing them aids in understanding how the process operates and provides preliminary information about which physiological groups of microbes might be responsible for PCB dechlorination. Hudson River sediments collected from an area contaminated with a commercial PCB mixture, Arochlor 1242, were incubated in serum bottles with anaerobic medium and with additional Aroclor 1242.

Acetone and methanol, fed at 1700 µg/g sediment/week, each produced 35% dechlorination during 20 weeks of incubation (Figure 3, Table 4). Cultures fed 230 µg/g sediment/week glucose removed 32% of the chlorines, whereas a ten-fold higher level of glucose allowed only 25% dechlorination, possibly due to a rapid pH drop. Butanol (135 µg/g/week) and acetate (140 µg/g/week) allowed only 13 and 11% dechlorination, respectively; with those substrates, ten-fold higher concentrations were also less effective, for reasons unknown.

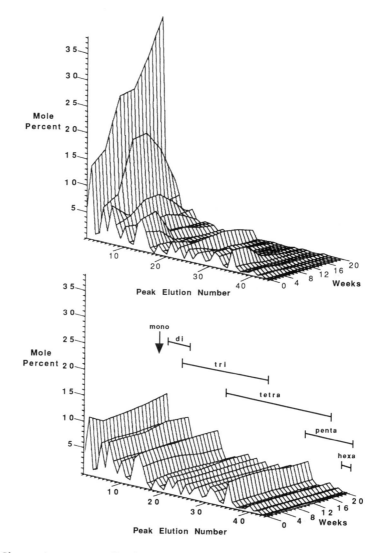

Figure 3. Change in congener distribution of Aroclor 1242 incubated with anaerobic Hudson River sediments. *Top:* methanol was provided as substrate. *Bottom:* no exogenous substrate was added. *Bars* indicate the overlapping range of homologs (mono stands for monochlorobiphenyls, etc.).

Table 4. Homolog distribution for Aroclor 1242 after anaerobic incubation with various substrates for 20 weeks.

Homolog class	Change in mole percent with:				
	Acetone[a]	Methanol	Glucose	Butanol	Acetate
MonoCB[b]	+387%	+383%	+388%	+21%	+25%
DiCB	+106%	+103%	+85%	+67%	+55%
TriCB	−57%	−55%	−52%	+3%	−4%
TetraCB	−86%	−85%	−77%	−58%	−42%
PentaCB	−92%	−89%	−73%	−54%	−40%
HexaCB	−100%	−97%	−85%	−71%	−62%

[a] Values are from incubations with the most effective substrate concentration.
[b] MonoCB, monochlorobiphenyl, etc.

With acetone and methanol in particular, a substantial decrease in the average degree of chlorination was observed, as evidenced by a shift toward less chlorinated congeners (Figure 4). Hexa-, penta-, and tetrachlorobiphenyls decreased by 85% or more, and trichlorobiphenyls by ca. 55% (Table 4); dichlorobiphenyls doubled and monochlorobiphenyl increased nearly four-fold. Dechlorination was entirely accounted for by removal of the *meta* and *para* chlorines, whereas *ortho* chlorines were not removed (within the limits of detection) (Figure 5). Only a minor amount of dechlorination occurred during 54 weeks of incubation in the absence of added substrate, indicating that these sediments were severely limited for substrate (65).

In an attempt to gain more understanding of the mechanism of reductive dechlorination of PCBs, an acetone-fed culture was subjected to two treatments during the dechlorination of 2,3,4,5,6-pentachlorobiphenyl (66). When deuterated acetone was fed, the mass spectrum of the molecular ion of the dechlorination product, 2,3,4,6-tetrachlorobiphenyl, was the same as when normal acetone was fed, demonstrating that the hydrogen atom used in reduction is not derived directly from the electron-donating substrate. When D_2O was present, however, the mass spectrum of the product was shifted toward higher molecular weights. This suggests that reduction of the PCB molecule results in formation of a carbanion intermediate, which abstracts a proton (or deuterium ion) from water.

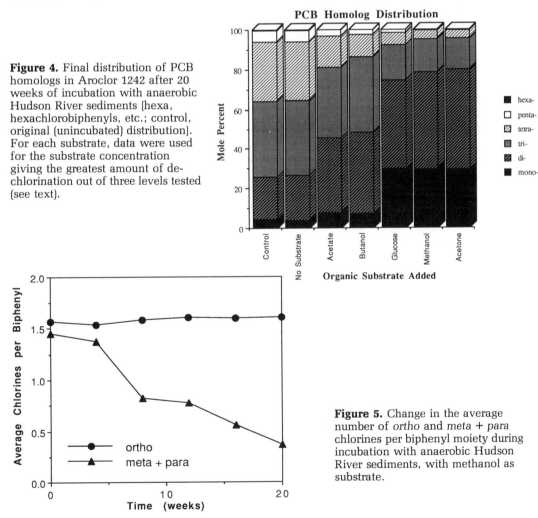

Figure 4. Final distribution of PCB homologs in Aroclor 1242 after 20 weeks of incubation with anaerobic Hudson River sediments [hexa, hexachlorobiphenyls, etc.; control, original (unincubated) distribution]. For each substrate, data were used for the substrate concentration giving the greatest amount of dechlorination out of three levels tested (see text).

Figure 5. Change in the average number of *ortho* and *meta* + *para* chlorines per biphenyl moiety during incubation with anaerobic Hudson River sediments, with methanol as substrate.

SEQUENTIAL ANAEROBIC/AEROBIC DEGRADATION OF POLYCHLORINATED HYDROCARBONS

As discussed above, reductive dechlorination is most effective with the more highly chlorinated members of a structural family, whereas aerobic (oxidative) degradation is most effective with the less chlorinated members. This relationship might be exploited in practical methods to detoxify highly chlorinated compounds by sequentially exposing them to the activities of appropriate anaerobic and aerobic cultures. Such a procedure has been tested on a bench scale with hexachlorobenzene (HCB), PCE, and trichloromethane (TCM; chloroform) (39). The chlorinated compounds were added to the influent of a constant flow bioreactor [(37.5 hour hydraulic retention time (HRT)] containing a mixed methanogenic culture developed with acetate as substrate. More than 90% of each of the chlorinated compounds underwent reductive dechlorination. HCB was converted ($> 70\%$) to 1,2,3-trichlorobenzene and 1,2-dichlorobenzene, consistent with results in batch culture (38). PCE and TCM were converted to cis-1,2-dichloroethylene and dichloromethane, respectively, although low recoveries suggested that further metabolism to nonchlorinated products may have occurred. Dechlorination was less extensive with glucose or methanol as substrate.

To complete the detoxification, the effluent from the anaerobic bioreactor was pumped into an aerobic bioreactor (2.25 hr HRT) along with additional aerobic medium containing 200 ppm hydrogen peroxide. The aerobic bioreactor contained a mixed culture developed with glucose as substrate and adapted to the expected partial-dechlorination products (dichlorobenzenes, dichloroethylene, and dichloromethane). When carbon-14-radiolabeled HCB, trichloroethylene (TCE), and TCM were fed individually to the linked columns, at least 83 to 96% was transformed to CO_2 or to water-soluble, nonvolatile intermediates. Such high degrees of apparent detoxification in this system suggests that sequential anaerobic/aerobic treatment might be developed for groundwaters or industrial wastes contaminated with these or other polyhalogenated compounds.

SEQUENTIAL ANAEROBIC/AEROBIC PCB BIOREMEDIATION IN A LABORATORY MODEL

As discussed above, the actions of anaerobic bacteria produce partially dechlorinated PCBs enriched in the congeners most susceptible to aerobic degradation. The most attractive restoration for PCB contamination of sediments would be bioremediation in situ. This would require considerable stimulation of the anaerobic step, as there are still substantial quantities of the higher-chlorinated congeners remaining after many decades of natural bioremediation (21). Induction of aerobic conditions in the sediments could then promote oxidative degradation of the lower-chlorinated products.

Batch incubations and a bench-scale river model are being used to investigate the feasibility of the sequential treatment method and to examine design parameters that would be important in in situ bioremediation (4). Batch cultures were incubated anaerobically for 20 weeks in sealed serum bottles with 6.7 mg methanol per gram of Hudson River sediments. The tri-, tetra-, penta-, and hexachlorobiphenyl congeners of the added Aroclor 1242 were substantially decreased during anaerobic incubation, but not in aerobic cultures. Mono- and dichlorobiphenyl congeners increased in the anaerobic cultures but decreased slightly in the aerobic cultures. After 20 weeks of incubation, anaerobic cultures were purged with oxygen and inoculated with an aerobic bacterium isolated from the Hudson River (strain S3; biphenyl grown). During 96 hours of aerobic incubation, S3 degraded most of the monochlorobiphenyls and about one quarter of the dichlorobiphenyls; approximately 43% of the total chlorobiphenyls was degraded overall. Among the dichlorinated congeners, those with two *ortho* chlorines were most recalcitrant (ca. 18% degraded). Extension of the aerobic phase would probably have allowed further degradation of lower-chlorinated PCBs.

A river model was constructed in a large aquarium in our laboratory. Hudson River sediments spiked with 300 ppm Aroclor 1242 were added, and water was recirculated over the sediments by pumps. During the first phase, methanol (6.7 mg/g sediment/week for 3 weeks) was injected through ports spread throughout the sediments to stimulate anaerobic dechlorination. After 76 weeks of anaerobic conditions, the congener profile of sediment PCBs has shifted as in the batch cultures, with increases in mono- and dichlorobiphenyls and decreases in higher chlorinated congeners. Subsamples of the anaerobically incubated sediments have been incubated for 96 days, with hydrogen peroxide added to generate aerobic conditions. With 1714 mg of H_2O_2 added to 5 g sediment, the PCB mass decreased by 95% from the amount present at the beginning of the anaerobic phase (4a). Analyses currently in progress will reveal whether aerobic PCB degradation can be supported by injection of H_2O_2 into the sediments in the river model.

Assuming that PCB degradation is substantial during the aerobic phase of the river model experiment, consideration may be given to pilot-scale bioremediation attempts *in situ*. One scenario is to equip a barge with injection equipment to deliver substrate (e.g. methanol) and eventually oxidant (e.g. hydrogen peroxide) into the sediments. The barge would move along the contaminated section of the river, injecting substrate incrementally until monitoring revealed that dechlorination was sufficient, and then injecting oxidant. If such a process proves feasible, it may ameliorate PCB contamination at lower risk and lower cost than alternatives, such as dredging and landfilling the sediments.

SUMMARY

Several major conclusions can be drawn from the results discussed herein. One is that enzymatic cofactors are capable of catalyzing reductive dechlorination of aliphatic and aromatic compounds *in vitro*. Although this contributes to an understanding of the mechanisms involved, it is possible that dechlorination *in situ* is principally mediated by enzymes associated with intact cells. These results suggest that many species of anaerobic bacteria contribute to the dehalogenation of xenobiotic compounds. All of the substrates added to river sediments incubated anaerobically in serum bottles stimulated reductive dechlorination of PCB. The less-chlorinated congeners formed are more susceptible to complete degradation by aerobic bacteria, thus suggesting that sequential anaerobic and aerobic treatment will be required for effective bioremediation of PCB. Such a treatment has been tested successfully in microcosms and is now being tested in a bench-scale river model. Sequential anaerobic and aerobic treatment was also successful in degrading several chlorinated aliphatic compounds and hexachlorobenzene.

These findings contribute to our understanding of the processes involved in microbial dechlorination of organic compounds. It is hoped that they will also contribute to the development of practical processes for treatment and bioremediation.

Acknowledgments

Larry Montgomery was supported by NSF through the Center for Microbial Ecology, STC-8809640. Nada Assaf-Anid and Larry Nies were supported by the NIEHS Superfund Research Center (Grant #5 P42 ES04911-03). Paul Anid was supported by the Office of Research and Development, US Environmental Protection Agency under Grant R815750 to the Great Lakes and Mid-Atlantic Hazardous Substance Research Center. Partial funding of the research activities of the Center is also provided by the State of Michigan Department of Natural Resources. The content of this publication does not necessarily represent the views of any agency.

LITERATURE CITED

1. Abramowicz, D. A. 1990. Aerobic and anaerobic biodegradation of PCBs—A review. *Crit. Rev. Biotechnol.* 10:241–249.

2. Adrian, N. R., and J. M. Suflita. 1990. Reductive dehalogenation of a nitrogen heterocyclic herbicide in anoxic aquifer slurries. *Appl. Environ. Microbiol.* 56:292–294.

3. Allard, A. S., P. A. Hynning, C. Lindgren, M. Remberger, and A. H. Neilson. 1991. Dechlorination of chlorocatechols by stable enrichment cultures of anaerobic bacteria. *Appl. Environ. Microbiol.* 57:77–84.

4. Anid, P. J., L. Nies, and T. M. Vogel. 1991. Sequential anaerobic-aerobic biodegradation of PCBs in the river model. In *On-Site Bioreclamation. Processes for Xenobiotic and Hydrocarbon Treatment.* Ed. R. E., Hinchee, R. F. Olfenbuttel. Boston: Butterworth-Heinemann. 428–436.

4a. Anid, P. J., B. P. Ravest-Webster, and T. M. Vogel. Biodegradation of anaerobic PCB-contaminated sediments. *Biodegradation.* Accepted, pending modifications.

5. Assaf-Anid, N., L. Nies, and T. M. Vogel. 1992. Reductive dechlorination of a polychlorinated biphenyl congener and hexachlorobenzene by vitamin B_{12}. *Appl. Environ. Microbiol.* 58:1057–1060.

6. Assaf-Anid, N., and T. M. Vogel. Dehalogenation of chlorinated compounds by a variety of porphyrins and corrins. *Society of Environmental Toxicology and Chemistry 10th Annual Meeting.* Abstr. 404. Washington, D.C.: Soc. Environ. Toxicol. Chem. 161.

7. Assaf-Anid, N., and T. M. Vogel. 1991. Reductive dechlorination of chlorinated aliphatic compounds by metallo-organic complexes. Abstr. Q-120. *Abstr. 91st Annu. Meet. Am. Soc. Microbiol.*, Washington, D.C. 296.

8. Baek, N. H., and P. R. Jaffe. 1989. The degradation of trichloroethylene in mixed methanogenic cultures. *J. Environ. Qual.* 18:515–518.

9. Bagley, D. M., and J. M. Gossett. 1990. Tetrachloroethene transformation to trichloroethene and *cis*-1,2-dichloroethene by sulfate-reducing enrichment cultures. *Appl. Environ. Microbiol.* 56:2511–2516.

10. Battersby, N. S., and V. wilson. 1989. Survey of the anaerobic biodegradation potential of organic chemicals in digesting sludge. *Appl. Environ. Microbiol.* 55:433–439.

11. Baxter, R. M. 1989. Reductive dehalogenation of environmental contaminants: a critical review. *Water Poll. Res. J. Canada* 24:299–322.

12. Baxter, R. M. 1990. Reductive dechlorination of certain chlorinated organic compounds by reduced hematin compared with their behaviour in the environment. *Chemosphere* 21:451–458.

13. Bedard, D. L., and M. L. Haberl. 1990. Influence of chlorine substitution pattern on the degradation of polychlorinated biphenyls by 8 bacterial strains. *Microb. Ecology* 20:87–102.

14. Bedard, D. L., R. Unterman, L. H. Bopp, M. J. Brennan, M. L. Haberl, and C. Johnson. 1986. Rapid assay for screening and characterizing microorganisms for the ability to degrade polychlorinated biphenyls. *Appl. Environ. Microbiol.* 51:761–768.

15. Belay, N., and L. Daniels. 1987. Production of ethane, ethylene, and acetylene from halogenated hydrocarbons by methanogenic bacteria. *Appl. Environ. Microbiol.* 53:1604–1610.

16. Bosma, T. N. P., J. R. van der Meer, G. Schraa, M. E. Tros, and A. J. B. Zehnder. 1988. Reductive dechlorination of all trichloro-and dichlorobenzene isomers. *FEMS Microbiol. Ecol.* 53:223–229.

17. Bouwer, E. J., and P. L. McCarty. 1983. Transformations of 1-and 2-carbon halogenated aliphatic organic compounds under methanogenic conditions. *Appl. Environ. Microbiol.* 45:1286–1294.

18. Bouwer, E. J., and P. L. McCarty. 1983. Effects of 2-bromoethanesulfonic acid and 2-chloro-ethanesulfonic acid on acetate utilization in a continuous-flow methanogenic fixed-film column. *Appl. Environ. Microbiol.* 45:1408–1410.

19. Bouwer, E. J., B. E. Rittmann, and P. L. McCarty. 1981. Anaerobic degradation of halogenated 1- and 2-carbon organic compounds. *Environ. Sci. Technol.* 15:596–599.

20. Boyle, M. 1989. The environmental microbiology of chlorinated aromatic decomposition. *J. Environ. Qual.* 18:395–402.

21. Brown, J., D. L. Bedard, M. J. Brennan, J. C. Carnahan, H. Feng, and R. E. Wagner. 1987. Polychlorinated biphenyl dechlorination in aquatic sediments. *Science* 236:709–712.

22. Chaudhry, G. R., and S. Chapalamadugu. 1991. Biodegradation of halogenated organic compounds. *Microbiol. Rev.* 55:59–79.

23. Criddle, C. S., J. T. Dewitt, D. Grbic-galic, and P. L. McCarty. 1990. Transformation of carbon tetrachloride by *Pseudomonas* sp. strain KC under denitrification conditions. *Appl. Environ. Microbiol.* 56:3240–3246.

24. Criddle, C. S., J. T. Dewitt, and P. L. McCarty. 1990. Reductive dehalogenation of carbon tetrachloride by *Escherichia coli* K-12. *Appl. Environ. Microbiol.* 56:3247–3254.

25. Davis, J. W., and S. S. Madsen. 1991. The biodegradation of methylene chloride in soils. *Environ. Toxicol. Chem.* 10:463–474.

26. DeWeerd, K. A., and J. M. Suflita. 1990. Anaerobic aryl reductive dehalogenation of halobenzoates by cell extracts of "*Desulfomonile tiedjei*". *Appl. Environ. Microbiol.* 56:2999–3005.

27. Distefano, T. D., J. M. Gossett, and S. H. Zinder. 1991. Reductive dechlorination of high concentrations of tetrachloroethene to ethene by an anaerobic enrichment culture in the absence of methanogenesis. *Appl. Environ. Microbiol.* 57:2287–2292.

28. Dolfing, J., and J. M. Tiedje. 1987. Growth yield increase linked to reductive dechlorination in a defined 3-chlorobenzoate degrading methanogenic coculture. *Arch. Microbiol.* 149:102–105.

29. Dolfing, J., and J. M. Tiedje. 1991. Influence of substituents on reductive dehalogenation of 3-chlorobenzoate analogs. *Appl. Environ. Microbiol.* 57:820–824.

30. Dwyer, D. F., E. Weeg-Aerssens, D. R. Shelton, and J. M. Tiedje. 1988. Bioenergetic conditions of butyrate metabolism by a syntrophic, anaerobic bacterium in coculture with hydrogen-oxidizing methanogenic and sulfidogenic bacteria. *Appl. Environ. Microbiol.* 54:1354–1359.

31. Egli, C., R. Scholtz, A. M. Cook, and T. Leisinger. 1987. Anaerobic dechlorination of tetra-

chloromethane and 1,2-dichloroethane to biodegradable products by pure cultures of *Desulfobacterium* sp. and *Methanobacterium* sp. *FEMS Microbiol. Lett.* 43:257–261.

32. Egli, C., S. Stromeyer, A. M. Cook, and T. Leisinger. 1990. Transformation of tetrachloromethane and trichloromethane to CO_2 by anaerobic bacteria is a nonenzymic process. *FEMS Microbiol. Lett.* 68:207–212.

33. Egli, C., M. Thuer, D. Suter, A. M. Cook, and T. Leisinger. 1989. Monochloro- and dichloroacetic acids as carbon and energy sources for a stable, methanogenic mixed culture. *Arch. Microbiol.* 152:218–223.

34. Egli, C., T. Tschan, R. Scholtz, A. M. Cook, and T. Leisinger. 1988. Transformation of tetrachloromethane to dichloromethane and carbon dioxide by *Acetobacterium woodii*. *Appl. Environ. Microbiol.* 54:2819–2824.

35. Fathepure, B. Z., and S. A. Boyd. 1988. Dependence of tetrachloroethylene dechlorination on methanogenic substrate consumption by *Methanosarcina* sp. strain DCM. *Appl. Environ. Microbiol.* 54:2976–2980.

36. Fathepure, B. Z., J. P. Nengu, and S. A. Boyd. 1987. Anaerobic bacteria that dechlorinate perchloroethene. *Appl. Environ. Microbiol.* 53:2671–2674.

37. Fathepure, B. Z., J. M. Tiedje, and S. A. Boyd. 1987. Reductive dechlorination of 4-chlororesorcinol by anaerobic microorganisms. *Environ. Toxicol. Chem.* 6:929–934.

38. Fathepure, B. Z., J. M. Tiedje, and S. A. Boyd. 1988. Reductive dechlorination of hexachlorobenzene to tri- and dichlorobenzenes in anaerobic sewage sludge. *Appl. Environ. Microbiol.* 54:327–330.

39. Fathepure, B. Z., and T. M. Vogel. 1991. Complete degradation of polychlorinated hydrocarbons by a 2-stage biofilm reactor. *Appl. Environ. Microbiol.* 57:3418–3422.

40. Freedman, D. L., and J. M. Gossett. 1989. Biological reductive dechlorination of tetrachloroethylene and trichloroethylene to ethylene under methanogenic conditions. *Appl. Environ. Microbiol.* 55:2144–2151.

41. Freedman, D. L., and J. M. Gossett. 1991. Biodegradation of dichloromethane and its utilization as a growth substrate under methanogenic conditions. *Appl. Environ. Microbiol.* 57:2847–2857.

42. Furukawa, K. 1986. Modification of PCBs by bacteria and other microorganisms. In *PCBs and the Environment*, Vol. 2, Ed. J. S. Waid. Boca Raton, FL: CRC Press. 89–100.

43. Galli, R., and P. L. McCarty. 1989. Biotransformation of 1,1,1-trichloroethane, trichloromethane and tetrachloromethane by a *Clostridium* sp. *Appl. Environ. Microbiol.* 55:837–844.

44. Gantzer, C. J., and L. P. Wackett. 1991. Reductive dechlorination catalyzed by bacterial transition-metal coenzymes. *Environ. Sci. Technol.* 25:715–722.

45. Gibson, S. A., and J. M. Suflita. 1986. Extrapolation of biodegradation results to groundwater aquifers: reductive dehalogenation of aromatic compounds. *Appl. Environ. Microbiol.* 52:681–688.

46. Gorlatov, S. N. 1989. Degradation of chlorophenols by soil microflora under conditions of an Eh gradient. *Microbiology* 58:402–406.

47. Haggblom, M. M., and L. Y. Young. 1990. Chlorophenol degradation coupled to sulfate reduction. *Appl. Environ. Microbiol.* 56:3255–3260.

48. Hale, D. D., J. E. Rogers, and J. Wiegel. 1990. Reductive dechlorination of dichlorophenols by nonadapted and adapted microbial communities in pond sediments. *Microbial Ecol.* 20:185–196.

49. Klecka, G. M., and S. J. Gonsior. 1984. Reductive dechlorination of chlorinated methanes and ethanes by reduced iron (II) porphyrins. *Chemosphere* 13:391–402.

50. Kohring, G. W., J. E. Rogers, and J. Wiegel. 1989. Anaerobic biodegradation of 2,4-dichlorophenol in freshwater lake sediments at different temperatures. *Appl. Environ. Microbiol.* 55:348–353.

51. Kohring, G. W., X. M. Zhang, and J. Wiegel. 1989. Anaerobic dechlorination of 2,4-di-

chlorophenol in freshwater sediments in the presence of sulfate. *Appl. Environ. Microbiol.* 55:2735–2737.

52. Krone, U. E., R. K. Thauer, and H. P. C. Hogenkamp. 1989. Reductive dehalogenation of chlorinated C1-hydrocarbons mediated by corrinoids. *Biochemistry* 28:4908–4914.

53. Krumme, M. L., and S. A. Boyd. 1988. Reductive dechlorination of chlorinated phenols in anaerobic upflow bioreactors. *Water Res.* 22:171–177.

54. Kuhn, E. P., and J. M. Suflita. 1989. Sequential reductive dehalogenation of chloroanilines by microorganisms from a methanogenic aquifer. *Environ. Sci. Technol.* 23:848–852.

55. Kuhn, E. P., and J. M. Suflita. 1989. Dehalogenation of pesticides by anaerobic microorganisms in soils and groundwater—a review. In *Reactions and Movement of Organic Chemicals in Soils*. Madison, WI: Soil Science Society of America and American Society. Special Publication No. 22. 111–180.

56. Kuhn, E. P., G. T. Townsend, and J. M. Suflita. 1990. Effect of sulfate and organic carbon supplements on reductive dehalogenation of chloroanilines in anaerobic aquifer slurries. *Appl. Environ. Microbiol.* 56:2630–2637.

57. Linkfield, T. G., and J. M. Tiedje. 1990. Characterization of the requirements and substrates for reductive dehalogenation by strain DCB-1. *J. Ind. Microbiol.* 5:9–16.

58. McInerney, M. J., and M. P. Bryant. 1980. Syntrophic associations of H_2-utilizing methanogenic bacteria and H_2-producing alcohol and fatty acid-degrading bacteria in anaerobic degradation of organic matter. In *Anaerobes and Anaerobic Infections*. Ed. G. Gottschalk. New York: Gustav Fischer Verlag. 117–126.

59. Mikesell, M. D., and S. A. Boyd. 1986. Complete reductive dechlorination and mineralization of pentachlorophenol by anaerobic microorganisms. *Appl. Environ. Microbiol.* 52:861–865.

60. Mikesell, M. D., and S. A. Boyd. 1990. Dechlorination of chloroform by *Methanosarcina* strains. *Appl. Environ. Microbiol.* 56:1198–1201.

61. Mohn, W. W., and J. M. Tiedje. 1990. Strain DCB-1 conserves energy for growth from reductive dechlorination coupled to formate oxidation. *Arch. Microbiol.* 153:267–271.

62. Neilson, A. H. 1990. The biodegradation of halogenated organic compounds—a review. *J. Appl. Bacteriol.* 69:445–470.

63. Neilson, A. H., A. Allard, C. Lindgren, and M. Remberger. 1987. Transformations of chloroguaiacols, chloroveratroles, and chlorocatechols by stable consortia of anaerobic bacteria. *Appl. Environ. Microbiol.* 53:2511–2519.

64. Neilson, A. H., A. Allard, P. Hynning, and M. Remberger. 1988. Transformations of halogenated aromatic aldehydes by metabolically stable anaerobic enrichment cultures. *Appl. Environ. Microbiol.* 54:2226–2236.

65. Nies, L., and T. M. Vogel. 1990. Effects of organic substrates on dechlorination of Aroclor-1242 in anaerobic sediments. *Appl. Environ. Microbiol.* 56:2612–2617.

66. Nies, L., and T. M. Vogel. 1991. Identification of the proton source for the microbial reductive dechlorination of 2,3,4,5,6-pentachlorobiphenyl. *Appl. Environ. Microbiol.* 57:2771–2774.

67. Nies, L., and T. M. Vogel. Effect of substrate concentration on the rate of microbial reductive dechlorination of PCBs, Abstr. Q-40. *91st Annu. Meet. Am. Soc. Microbiol.* 283.

68. O'Connor, O. A., and L. Y. Young. 1989. Toxicity and anaerobic biodegradability of substituted phenols under methanogenic conditions. *Environ. Toxicol. Chem.* 8:853–862.

69. Quensen, J. F., S. A. Boyd, and J. M. Tiedje. 1990. Dechlorination of 4 commercial polychlorinated biphenyl mixtures (Aroclors) by anaerobic microorganisms from sediments. *Appl. Environ. Microbiol.* 56:2360–2369.

70. Quensen, J. F., J. M. Tiedje, and S. A. Boyd. 1988. Reductive dechlorination of polychlorinated biphenyls by anaerobic microorganisms from sediments. *Science* 242:752–754.

71. Scholz-Muramatsu, H., R. Szewzyk, U. Szewzyk, and S. Gaiser. 1990. Tetrachloroethylene

as electron-acceptor for the anaerobic degradation of benzoate. *FEMS Microbiol. Lett.* 66:81–85.

72. Schwarzenbach, R. P., and P. M. Gschwend. 1990. Chemical transformations of organic pollutants in the aquatic environment. In *Aquatic Chemical Kinetics.* Ed. W. Stumm. New York: Wiley-Interscience. 199–233.

73. Sharak-Genthner, B. R., W. A. Price, and P. H. Pritchard. 1989. Anaerobic degradation of chloroaromatic compounds in aquatic sediments under a variety of enrichment conditions. *Appl. Environ. Microbiol.* 55:1466–1471.

74. Sharak-Genthner, B. R., W. A. Price, and P. H. Pritchard. 1989. Characterization of anaerobic dechlorinating consortia derived from aquatic sediments. *Appl. Environ. Microbiol.* 55:1472–1476.

75. Stevens, T. O., T. G. Linkfield, and J. M. Tiedje. 1988. Physiological characterization of strain DCB-1, a unique dehalogenating sulfidogenic bacterium. *Appl. Environ. Microbiol.* 54:2938–2943.

76. Suflita, J. M., S. A. Gibson, and R. E. Beeman. 1988. Anaerobic biotransformations of pollutant chemicals in aquifers. *J. Ind. Microbiol.* 3:179–194.

77. Suflita, J. M., A. Horowitz, D. R. Shelton, and J. M. Tiedje. 1982. Dehalogenation: a novel pathway for the anaerobic biodegradation of haloaromatic compounds. *Science* 218:1115–1117.

78. Suflita, J. M., J. A. Robinson, and J. M. Tiedje. 1983. Kinetics of microbial dehalogenation of haloaromatic substrates in methanogenic environments. *Appl. Environ. Microbiol.* 45:1466–1473.

79. Traunecker, J., A. Preuß, and G. Diekert. 1991. Isolation and characterization of a methyl chloride utilizing, strictly anaerobic bacterium. *Arch. Microbiol.* 156:416–421.

80. Van Dort, H. M., and D. L. Bedard. 1991. Reductive ortho and meta dechlorination of a polychlorinated biphenyl congener by anaerobic microorganisms. *Appl. Environ. Microbiol.* 57:1576–1578.

81. Vogel, T. M., C. S. Criddle, and P. L. McCarty. 1987. Transformations of halogenated aliphatic compounds. *Environ. Sci. Technol.* 21:722–736.

82. Vogel, T. M., and P. L. McCarty. 1987. Abiotic and biotic transformations of 1,1,1-trichloroethane under methanogenic conditions. *Environ. Sci. Technol.* 21:1208–1213.

83. Wilson, B. H., G. B. Smith, and J. F. Rees. 1986. Biotransformations of selected alkylbenzenes and halogenated aliphatic hydrocarbons in methanogenic aquifer material: a microcosm study. *Environ. Sci. Technol.* 20:997–1002.

84. Zhang, X. M., and J. Wiegel. 1990. Sequential anaerobic degradation of 2,4-dichlorophenol in fresh-water sediments. *Appl. Environ. Microbiol.* 56:1119–1127.

CHAPTER 13

Biodegradation of Low-Molecular-Weight Halogenated Organic Compounds by Aerobic Bacteria

RICHARD S. HANSON and GREGORY A. BRUSSEAU

Abstract. Environmental contamination with low-molecular-weight halogenated organic compounds has been increasingly the subject of public concern and scientific inquiry. This outcry has resulted in research efforts that may lead to effective remediation. One possible solution includes the use of aerobic bacteria. Aerobic bacteria are able to grow on a variety of halogenated alkanes, alkenes, and alkanoic acids. A review of several organisms provides insight into the range of compounds degraded, enzymes responsible for activity, intermediates that may accumulate, and efforts at optimizing microbial potentials. Compounds that do not support growth of microorganisms pose greater challenges. A short historical survey of efforts to understand the degradation of the ubiquitous chlorinated hydrocarbon trichloroethylene (TCE) by cometabolism demonstrates the wide range of interest and progress. Work with methanotrophic bacteria, organisms currently believed to oxidize TCE at the highest rates, utilizes a variety of techniques, including restriction fragment length polymorphism analysis and ribosomal RNA sequencing, to study both pure cultures and environmental samples. Research presented here shows the advantages of using molecular biological techniques to understand the factors that control the distribution and growth of methanotrophs in the environment, and ultimately the potential for their exploitation in the bioremediation of low-molecular-weight halogenated hydrocarbons.

INTRODUCTION

Low-molecular-weight halogenated organic compounds are commonly detected groundwater contaminants and comprise the largest group of priority pollutants listed by the U.S. Environmental Protection Agency (71, 120). Within this group, chlorinated one-, two- and three-carbon compounds are of greatest concern because of their widespread occurrence, toxicity, carcinogenicity, and environmental persistence (5, 77, 113, 114). Chlorinated ethanes and ethenes are used as solvents for dry-cleaning, cleaning

Dr. Hanson and Mr. Brusseau are at Gray Freshwater Biological Institute, University of Minnesota, Navarre, MN 55392, U.S.A.

agents in semiconductor manufacturing, extractants of caffeine from coffee, general anaesthetics in medicine and dentistry, reagents in the production of plastics, and industrial degreasers. Dichloromethane has also been produced in large amounts for use as a propellant, paint stripper, reagent in acetate film manufacturing, solvent in pharmaceutical production, and metal cleaner. Other chloromethanes are products of reactions that occur during water chlorination (116). Some halogenated compounds, including 1,2-dibromoethane and 1,2-dibromo-3-chloropropane, are used as pesticides.

Many of these compounds, which are exclusively products of human activities, are toxic and/or carcinogenic and persist for long periods in groundwaters. For example, trichloroethylene (TCE) was observed to exhibit a half life of 300 days in one aquifer (91). Although TCE, the most frequently reported contaminant at hazardous waste sites, has not been shown to elicit a carcinogenic response in small doses in humans, it is listed as a suspected carcinogen (78). Moreover, metabolites from the degradation of the initial pollutant may also be health hazards. Tetrachloroethylene (PCE), TCE, and dichloroethylene (DCE) are converted to vinylchloride by bacteria under anaerobic conditions that exist in landfill sites and some groundwaters (115, 116). Vinylchloride is known to be a carcinogen in mammals (56, 75).

An essential step in the degradation of these compounds is dehalogenation. There are three mechanisms of dehalogenation: oxidative, reductive, and hydrolytic (72). Highly chlorinated aliphatic compounds are better electron acceptors for reductive dechlorination, and less chlorinated compounds are substrates for oxidative or hydrolytic dehalogenation (116). Anaerobic degradation of chlorinated organic compounds is described in another chapter in this book (78a).

This chapter reviews information describing the microbial biotransformations and metabolism of low-molecular-weight halogenated organic compounds by aerobic microbes. Some compounds can be metabolized to intermediates that enter the central pathways of metabolism and serve as sources of carbon and energy for the growth of microbes. Halogenated aldehydes, alcohols, and alkanoic acids are metabolic intermediates from these processes. An understanding of the metabolism of these compounds provides insight into mechanisms of dehalogenation and new applications for bioremediation. Other chemicals, including some chlorinated alkenes and chloroform, cannot be utilized as carbon and energy sources but can be biodegraded aerobically by reactions known as *fortuitous metabolism* or *cometabolism* (19, 39). In cometabolism the microbe depends on a primary substrate as a carbon and energy source and modifies the cosubstrate without benefit to the organism (17, 54). The biodegradation of halogenated organic compounds has also been reviewed recently by Chaudry and Chapalamadugu (13), Neilson (81), and Wilson (125).

GROWTH OF AEROBIC BACTERIA ON HALOGENATED ALKANES, ALKENES, AND ALKANOIC ACIDS

Halomethanes

Halomethanes are the simplest of the low-molecular-weight halogenated hydrocarbons, but, because they are one-carbon compounds, the synthesis of cell material from these substrates requires pathways not normally found in heterotrophic bacteria. Carbon tetrachloride and chloroform do not serve as carbon and energy sources for the growth of aerobic bacteria. Several bacteria have been isolated that grow on dichloromethane (DCM) as a sole source of carbon and energy (33, 106). One mole of DCM was converted to one mole of formaldehyde and two moles of hydrochloric acid in the initial reaction catalyzed by DCM dehalogenases. The bacteria that grew on DCM were facultative methylotrophic bacteria that also grow on methanol, methylamines, and other organic compounds (8, 33, 34, 68, 95, 105). These bacteria are all gram-negative and utilize formaldehyde produced from reactions catalyzed by DCM, methanol, and methylamine dehydrogenases for the synthesis of biosynthetic intermediates. The assimilation of formaldehyde proceeds via the serine pathway. Formaldehyde is also oxidized to formate and finally to CO_2 for the production of reducing power and adenosine triphosphate, which are required to sustain biosynthetic reactions. Two physiological groups of methylotrophic bacteria assimilate formaldehyde. These groups are distinguished by the pathways employed for the synthesis of cell material. The serine pathway is present in one group of methylotrophic bacteria that utilize reduced one-carbon compounds (3).

Two strains named *Pseudomonas* DM1 and *Methylobacterium* DM4 (68) are phylogenetically related to other facultative methylotrophs classified as strains or species of *Methylobacterium* (69, 112). All bacteria that are classified as *Methylobacterium* utilize the serine pathway for formaldehyde assimilation, have DNA with a moles percent guanosine plus cytosine of 62–64%, and grow on multicarbon as well as reduced one-carbon compounds. The ability to grow on DCM is not a characteristic often found in other methylotrophic bacteria (67). Two other isolates that grew on DCM were identified as *Hyphomicrobium* species (32, 68, 105). *Hyphomicrobia* are appendaged bacteria that also grow with methanol as a sole carbon and energy source, and use the serine pathway for formaldehyde assimilation (42). These bacteria contained DCM dehalogenases that were similar to each other in substrate specificity, molecular weight, N-terminal amino acid sequences, specific activities, and pH optima.

The DCM dehalogenases from *Methylobacterium* and *Hyphomicrobium* species (group A enzymes) are inducible, glutathione-dependent enzymes with a subunit molecular weight of 35,000. They belong to the glutathione S-transferase supergene family (35, 69). When the bacteria were grown on DCM, the DCM dehalogenases comprised 15–20% of the soluble protein of the cells.

Another DCM-utilizing bacterium isolated from groundwater called strain DM11 grew more rapidly with DCM than the bacteria described above. The DCM dehalogenase of this bacterium was also glutathione-dependent and had a subunit molecular weight similar to the group A enzymes; however, this enzyme had a specific activity of 97 mkat/kg of protein, which was 5.6-fold higher than that of the group A enzymes. It also differed in serological crossreactivity and N-terminal amino acid sequence from group A enzymes (95).

Strain DM11 also grew with dibromomethane or methanol as a sole carbon and energy source. It is the first bacterium known to utilize dibromomethane as a carbon and energy source. This bacterium also grows with acetate and succinate as carbon and energy sources and utilizes the ribulose monophosphate (RuMP) pathway for formaldehyde assimilation during growth on one-carbon compounds (B. J. Bratina and R. S. Hanson, unpublished results). It is phylogenetically related to other RuMP methylotrophs within the beta/gamma subgroups of the Proteobacteria (112).

The bacteria that utilize DCM are unable to grow on methylchloride. Methylchloride is a naturally occurring compound that is also manufactured in large quantities. It has been used as a methylating agent for silicones and lead (25). A strain of *Hyphomicrobium* (MC1) that utilizes this compound as a sole carbon and energy source for growth was isolated from sewage by Hartmans et al. (47). Strain MC1 grew with methanol, formate, and ethanol but not ethylchloride, methylamine, or DCM. A methylchloride monooxygenase that converted methylchloride to formaldehyde was proposed as the catalyst for the first reaction in methylchloride metabolism.

Haloalkanes

Compounds with carbon–carbon bonds have the potential for use by heterotrophic bacteria, provided that the products of dehalogenation reactions can serve as carbon and energy sources. Dehalogenases found in bacteria utilizing halogenated hydrocarbons with two or more carbons are different from halomethane dehalogenases. 1,2-Dichloroethane (DCA) is produced in greater quantities than any other industrial halogenated chemical (62), and significant amounts are released to the environment (73, 106). Janssen et al. (58) described a bacterium named *Xanthobacter autotrophicus* that utilized several halogenated hydrocarbons as sole carbon and energy sources. *Xanthobacter* are aerobic, nitrogen-fixing bacteria that grow autotrophically with hydrogen as an energy source (58). Haloalkane utilization is not a common property among members of this genus. Conversion of DCA to chloroethanol and chloride was catalyzed by a haloalkane dehalogenase, an enzyme that has been well characterized (92). The enzyme is a monomer with a molecular weight of 36,000, and shows activity toward C_1–C_4 1-halogenated n-alkanes. Chloroethanol was oxidized by a quinoprotein alcohol dehydrogenase similar to methanol dehydrogenases of methylotrophic bacteria (57–59). The product, chloroacetaldehyde, was further oxidized to chloroacetate by chloroacetaldehyde dehydrogenase. Chloroacetate undergoes dehalogenation catalyzed by a halocarboxylic acid dehalogenase to produce glycolic acid, which is a growth substrate for several bacteria. It appears that the genes for these four enzymes have been recruited by some bacteria that are normally chemoautotrophs (62).

Both gram-positive and gram-negative bacteria, isolated by enrichment on longer chain haloalkanes, grow on many different halogenated compounds (47, 93, 128). Four coryneform bacteria were isolated from enrichments with 1-chlorobutane, 1-chloropentane, or 1-chlorohexane as carbon and energy sources (94). One bacterium, *Arthrobacter* strain HA1 isolated with chlorohexane as a carbon and energy source, was shown to utilize at least 18 1-chloro-, 1 bromo-, and 1-iodoalkanes. These strains, unlike most *Arthrobacter* and related bacteria, do not utilize unsubstituted alkanes but do utilize n-hexanol. They apparently lack monooxygenases that catalyze the first reaction in hydrocarbon degradation. The ability to dehalogenate haloalkanes was inducible, and n-alcohols and halide ions were released quantitatively from the substrate when crude cell extracts were employed as sources of enzyme. The dehalogenase, which was given the name chlorohexane halidohydrolase, was a monomeric catalyst with a molecular weight of 37,000. The enzyme(s) in crude extracts dehalogenated a wider spectrum of substrates than supported growth. All 1-bromoalkanes C_2 to C_8, 1-chloroalkanes C_4 to C_8, and 1-iodoalkanes C_2 to C_7 were utilized as carbon and energy sources for growth. Over 50 compounds, including C_1 to C_7 1-iodoalkanes, C_1 to C_9 1-bromoalkanes, and C_3 to C_{10} 1-chloroalkanes, but no fluoroalkanes, were substrates for the enzyme (94). The enzyme was not stereospecific; both stereoisomers of 2-bromobutane were substrates. 2- and 3-Chloroheptane, chlorocyclohexane, and 1-bromo-2-methylpropane were all substrates, thus indicating a lack of specificity for terminally substituted alkanes. Aromatic halohydrocarbons were subject to ring dehalogenation (bromobenzene) or side-chain dehalogenation (phenylethylbromide, phenylmethylbromide, and phenylpropylbromide). Some bromoalcohols were also substrates, but haloacetates and 1-chloroacetone were not.

The highest specific activity for chlorohexane halidohydrolase was observed for 1,2-dibromopropane; however, several C1 to C10 α,ω-dihaloalkanes were also substrates. 1,2-Dibromoethane, 1,4-dichlorobutane, and 1,6-dichlorohexane were converted to monohaloalcohols, which in turn were slowly dehalogenated to yield dihydroxy alcohols. This enzyme differed greatly from dichloroethane dehalogenase in substrate specificity. For example, DCM, 1,1- and 1,2-dichloroethane, and 1,2-dichloropropane were not dechlorinated by the purified chlorohexane halidohydrolase.

2-Chlorobutane, 3-chloropentane and α,ω-dichlorohydrocarbons were dehalogenated by chlorohexane halidohydrolase but were not substrates for growth, presumably because the products were not susceptible to further metabolism by the bacterium (93). Some enzyme substrates failed to induce dehalogenase activity and may not support growth for this reason.

Another bacterium that grows on haloalkanes was isolated from activated sludge by Janssen et al. (59). This organism was originally identified as a strain of *Acinetobacter* and later found to be a gram-positive actinomycete-like organism (61). This strain, designated GJ70, was capable of growth on 1-chlorobutane, 1-chloropentane, and 1,6-dichlorohexane. The enzyme, which catalyzed the hydrolytic conversion of halogenated aliphatic hydrocarbons to the corresponding alcohols, was purified and characterized. Like the chlorohexane halidohydrolase from *Arthrobacter* strain HA1, this enzyme is a monomer with a molecular weight of 28,000. It showed activity with a wide range of halogenated hydrocarbons, secondary alkyhalides, halogenated alcohols, and chlorinated ethers. The highest specific activity was observed with dibromoethane as a substrate. Methylbromide, methyliodide, dibromomethane, 2-bromoethanol, and 3-bromopropanol were rapidly hydrolyzed, whereas the corresponding chloroalkanes were not substrates.

The three n-alkane halidohydrolases from strain GJ70, *Xanthobacter autotrophicus,* and *Arthrobacter* strain HA1 all differ from DCM dehalogenases by their lack of a requirement for glutathione. In addition, the three enzymes from the non-methylotrophic bacteria are monomeric enzymes (47, 94), whereas DCM dehalogenase has a hexameric structure (95). The enzymes from strain GJ70 and *Arthrobacter* strain HA1 do not dehalogenate chloroethanes, 3-chloropropane, and chloromethanes like the DCM dehalogenases.

Genetic studies are important for defining catabolic pathways and enhancing future applications in bioremediation of toxic wastes. The gene encoding the haloalkane dehalogenase of *Xanthobacter autotrophicus* (*dhlA*) has been cloned into a cosmid vector pLAFR1 (62). After subcloning, the gene was expressed efficiently from its own promoter in *E. coli, Pseudomonas* sp., and *Xanthobacter* sp. at levels up to 30% of the total soluble protein of host cells. The sequence of the gene was also determined. Genes encoding other enzymes involved in haloalkane metabolism, including methanol dehydrogenase, chloroacetaldehyde dehydrogenase, and halocarboxylic acid dehalogenase, were also cloned from *X. autotrophicus* GJ10 (62). Therefore, the pathway for the degradation of several haloalkanes is well established.

Unlike the preceding organisms, *Pseudomonas fluorescens* strain PFL2, a bacterium isolated from a landfill site contaminated with DCA and 1,2-dichloropropane, was reported to utilize DCA, 1,2-dichloropropane, 2,2-dichloropropane, 1,1,2-trichloroethane, and TCE as carbon sources (114). It is unclear from the report whether all of these compounds supported growth of the bacterium. To our knowledge, there have been no other reports of bacteria capable of growth on TCE as a sole source of carbon. This bacterium deserves consideration because the dehalogenase activity was not induced by benzoate, *ortho*-cresol, or *meta*-cresol, which serve as carbon and energy sources for the growth of *P. fluorescens* PFL2, but activity was present in cells grown with succinate, ethanol, acetate, or glucose. In other pseudomonads, which contain nonspecific monooxygenases and dioxygenases, the ability to degrade these compounds is induced by aromatic compounds that serve as growth substrates (82, 126). Halidohydrolases, which dehalogenate haloalkanes, were not reported to catalyze dehalogenation of TCE. The enzyme responsible for utilization of halocarbons in *P. fluorescens* PFL2 remains to be characterized.

It is clear from this body of work that all monohaloalkanes from C_1 to C_8 are biodegradable by one or more bacteria (94). Different bacteria utilize substrates of different chain lengths. It is not surprising that bacteria that utilize mono-and dihalomethanes as carbon and energy sources for growth are methylotrophs that assimilate formaldehyde, the product of the dehalogenation reactions, via pathways unique to these bacteria. Some bacteria, like *Arthrobacter* strain HA1, appear to be uniquely adapted to growth on halohydrocarbons because they apparently lack monooxygenases that are required to metabolize the corresponding n-alkanes. Aerobic organisms that grow on alkanes possess monooxygenases to initiate degradation. The bacteria that lack monooxygenases, however, can grow on halocarbons but not alkanes. Dehalogenases that degrade a wide range of C_1 to C_8 halogenated hydrocarbons exist in bacteria that range from gram-positive actinomycetes and *Arthrobacter* to gram-negative facultative autotrophs, methylotrophs, and heterotrophs.

Haloalkenes

Bacteria that grow on haloalkanes are not known to utilize haloalkenes as carbon and energy sources. Reports of successful enrichments for bacteria that degrade haloalkenes are rare. The simplest haloalkene, vinylchloride, is a commodity chemical. Seventeen million tons were produced in 1982. Large amounts have been lost to the environment during the production of polyvinylchloride. Elimination of vinylchloride in gaseous waste streams usually involves the expensive process of absorption onto activated charcoal, yet vinylchloride is degraded by aerobic bacteria present in soils and waters (46). It has been shown to be susceptible to slow degradation to ethylene under conditions that favor growth of anaerobic methanogens (6, 31, 116), and the subsequent rapid mineralization of ethylene is caused by aerobic methanotrophic bacteria (76, 83, 84, 110). The aerobic methanotrophs do not use vinylchloride as a carbon source, however. Indigenous bacteria present in unamended shallow aquifer samples have been shown to convert ^{14}C-vinylchloride slowly to $^{14}CO_2$ (23). Hartmans et al. (46) have isolated a strain of *Mycobacterium* that grew on vinylchloride as a source of carbon and energy for growth from soil that had been contaminated with vinylchloride for several years. The bacterium resembled alkane-utilizing bacteria and was shown to be capable of growth on acetate and ethylene but not propane. Cells that were grown on ethylene were also capable of degrading vinylchloride whereas acetate grown cells were not. The K_M of ethylene-grown cells for vinylchloride was low (1.7 μM), and rates of degradation were high (46). The use of these microbes may provide an attractive alternative to the absorption of vinylchloride on charcoal (46).

Haloalkanoic Acids

Alkanoic acids are intermediates in the oxidation of all haloalkanes, and some are used as herbicides and pesticides. Several gram-negative bacteria of genera *Pseudomonas* and *Alkaligenes* that grew on halogenated alkanoic acids have been isolated (43, 79). Bacteria isolated by Hardman and Slater (44) had different dehalogenating activities toward monochloroacetic acid, dichloroacetic acid, 2-chloropropionic acid, and 2,2-dichloropropionic acid. Five dehalogenases differing in relative mobilities and activity profiles were identified by discontinuous polyacrylamide gel electrophoresis. The different isolates contained different combinations of the five dehalogenases (44). Three to four different isoenzymes were detected in two isolates, and 13 isolates had two isoenzymes in several different combinations. The synthesis of the enzymes was induced by monochloracetic acid or 2-monochloropropionic acid (2-MCPA). Dehalogenase activity was established as the growth-rate limiting reaction of metabolism in these bacteria (119).

Five large plasmids with molecular weights from 99,000 to 190,000 were identified in four *Pseudomonas* species and two *Alkaligenes* species able to grow on 2-monochloroacetic acid and 2-MCPA (45). Strains cured of plasmids by growth in the presence of ethidium bromide could not utilize 2-MCPA. The ability of the isolates to grow on halogenated alkanoic acids and the presence of dehalogenases were lost during growth on media with succinate as a carbon and energy source. The evidence suggests that the dehalogenase genes were located on unstable catabolic plasmids. The authors suggested that multiple dehalogenases encoded by clusters of different dehalogenase genes provided higher dehalogenase activities, which in turn provided a competitive advantage in environments containing these compounds. The use of different isoenzymes rather than increased production of a single enzyme is an intriguing strategy to increase activity. Multiple isoenzymes may broaden the range of substrates utilized rapidly, but it is difficult to relate this hypothesis to the availability of their substrates in nature. The distribution of dehalogenase isoenzymes among several isolates suggested horizontal transfer by highly mobile genetic elements as a means of recruiting combinations of dehalogenase genes. Attempts to demonstrate self-transfer of the plasmids by conjugation or by mobilization with the broad host range plasmid R68-45 were unsuccessful (45).

Fluoroacetate is not dehalogenated by most alkanoic acid dehalogenases. This substrate is toxic to most organisms. Goldman et al. (36, 37) isolated a pseudomonad that was capable of growth in a medium containing fluoroacetate as a carbon and energy source. The pseudomonad produced a carbon–fluorine bond-cleaving enzyme constitutively. It was purified and studied (36). The enzyme preferentially utilized fluoroacetate as a substrate. The activity levels with monochloroacetate and monobromoacetate were 20% and 15% of that with fluoroacetate. Other halogenated compounds were not substrates.

Two dehalogenases were found in cells of a *Moraxella* species that used fluoroacetate as a carbon and energy source (66). One isoenzyme utilized fluoroacetate; the other resembled the chloroacetate dehalogenases (described below). Motosugi and Soda (79) have reviewed work on the degradation of haloalkanoic acids.

Enzymes that dehalogenate 2-haloalkanoic acids have been classified into haloacetate dehalogenases and 2-haloacid dehalogenases (65). These enzymes do not cleave carbon fluorine bonds, and haloalkanes are not substrates. The haloacetate dehalogenases, other than the one described by Goldman that cleaves carbon fluorine bonds, exclusively dehalogenate chloro-, bromo- and iodoacetates. In contrast, 2-haloalkanoic acid dehalogenases catalyze the dehalogenation of several 2-haloalkanoic acids, including all 2-monohaloacetates except fluoroacetate (79).

Three L-2-haloacid dehalogenases, described by Motosugi and Soda, varied greatly in terms of molecular weights (15,000, 34,000, and 68,000), pH optima, and substrate specificity (79). One of these enzymes from a strain of *Pseudomonas putida* specifically hydrolyzed L-2-haloalkanoic acid stereoisomers. Another enzyme from *Pseudomonas* sp. strain 113 dehalogenated both D and L isomers of 2-haloacids. Clearly, a variety of enzymes are able to degrade haloalkaonic acids.

COMETABOLISM

Leadbetter and Foster (70) studied oxidations of hydrocarbons by whole cells of methane-oxidizing bacteria that grew only with methane. They observed that the oxidation of the substrates, yielding methylketones and aldehydes, was incomplete. The oxidation of substrates by bacteria that do not serve as carbon or energy sources has been called *co-oxidation* (17). Jensen (63) suggested the more general term *cometabolism* to define transformations of substrates without benefit to an organism by enzymes that had another function in primary metabolism. The term *fortuitous metabolism* has been suggested by Stirling and Dalton (101) to describe this behavior.

Several studies described the oxidation of large numbers of substrates by methane-oxidizing bacteria, their extracts, and purified methane monooxygenase (11, 15, 16, 18–20, 22, 26, 30, 38, 39, 51–53, 55, 85–87, 102, 104, 110). Methane monooxygenase (MMO) catalyzes the oxidation of methane to methanol. The purified enzyme has been shown to oxidize a large number of other substrates to an equally large variety of products (22, 30). The lists of substrates for MMO included many alkenes, alkanes, aromatic and heterocyclic compounds, and halomethanes (38). The dehalogenation of halomethanes by methanotrophs was first described by Colby and Dalton (14).

Wilson and Wilson (124) demonstrated that microbes in an unsaturated soil column converted TCE to carbon dioxide under aerobic conditions. Columns that were not exposed to natural gas did not degrade TCE. It was suggested that methanotrophs or other bacteria utilized the low-molecular-weight alkanes present in natural gas for growth and energy. Trichloroethylene was then attacked because of the lack of specificity in the primary enzymes. Kampbell et al. (64) described the mineralization of TCE in a bioreactor filled with soils exposed to light hydrocarbons composed primarily of propane, n-butane and isobutane. Hou et al. (55) demonstrated that propane-utilizing bacteria caused epoxidation of short chain alkenes. As a result of this work, Wilson and Wilson (124) suggested that the enzyme system of these bacteria might be responsible for the degradation of TCE. Since the observations of Wilson and Wilson, several investigators examined the ability of methanotrophs and other bacteria that contained nonspecific monooxygenases and dioxygenases for their ability to transform a range of halogenated aliphatic compounds (13, 74, 82, 84, 103, 104, 110, 118).

These concepts were furthered by Fogel et al. (27), Henson et al. (50), and Janssen et al. (60), who examined mixed cultures enriched by growth on methane for their abilities to degrade several chlorinated derivatives of methane, ethane, and ethylene. Their studies demonstrated that the highly chlorinated hydrocarbons, such as tetrachloroethane, PCE, and carbon tetrachloride, were not transformed by consortia of methanotrophs and other bacteria. Of the compounds that were degraded, vinylchloride was oxidized faster than 1,1-dichloroethylene and TCE. Dichloromethane was transformed faster than chloroform and the rates of transformation of chlorinated ethanes decreased in the following order: 1,2-dichloroethane, 1,1-dichloroethane, 1,1,2-trichloroethane, and 1,1,1-trichloroethane (27, 50, 76).

BIODEGRADATION OF HALOGENATED HYDROCARBONS
BY PURE CULTURES OF BACTERIA THAT CONTAIN
NONSPECIFIC OXYGENASES

Nelson et al. (82) examined the potential of water samples from an industrial waste control facility to degrade TCE. Only one of 43 samples caused a decrease in the TCE concentration. Subcultures from this sample degraded TCE only when filter-sterilized or autoclaved water from the sample site was added to the culture medium. A gram-negative nonmotile rod, designated strain G4, was isolated in pure culture on agar media containing glucose and yeast extract as carbon and energy sources. TCE degradation occurred only when a component from the site water was added to the growth medium. Strain G4 did not grow with methane or methanol, and these compounds did not stimulate TCE degradation when added to the growth medium. The site-water component for induction of TCE degradation was later identified as phenol (82), although toluene, o-cresol or m-cresol could replace toluene as inducers of TCE degradation.

Subsequently, Nelson et al. (82) tested six bacterial strains capable of utilizing naphthalene, biphenyl, toluene, and phenol for their ability to degrade TCE. Two strains, *Pseudomonas putida* F1 and *P. putida* B5, caused TCE disappearance. These bacteria contain an enzyme called toluene dioxygenase, which oxidizes toluene by inserting two oxygen atoms on the aromatic ring, thus forming a *cis*-dihydrodiol. Strains that contained dioxygenases that convert naphthalene and biphenyl to *cis*-dihydrodiols, and *P. putida* that oxidizes toluene to benzoate prior to ring dioxygenation, did not degrade TCE.

Mutants of strain *Pseudomonas putida* F1 defective in toluene dioxygenase did not oxidize TCE (82, 117). Therefore, toluene dioxygenase was implicated in TCE degradation by *P. putida* F1.

Winter et al. (126) screened several bacteria that utilized aromatic compounds for their ability to degrade TCE. They discovered a toluene-oxidizing strain of *Pseudomonas mendocina* that degraded TCE. This strain utilizes a toluene monooxygenase system for the initial oxygenation of toluene to p-cresol (90, 126). A DNA fragment encoding toluene monooxygenase was cloned into *Escherichia coli* and conferred the ability to oxidize toluene and TCE upon *E. coli*. Mutants of *P. mendocina* deficient in toluene monooxygenase failed to degrade TCE.

Bacterium G4, the first toluene-utilizing isolate that degraded TCE, has been shown to degrade toluene via successive monooxygenations at the *ortho* and *meta* positions to produce 3-methylcatechol, a previously unknown pathway (28, 97). Therefore, there are known to be three oxygenases that catalyze insertion of oxygen atoms into aromatic rings that also catalyze the initial reactions in TCE degradation.

Nitrosomonas europea is an obligate, chemolithotrophic, nitrifying bacterium that derives all of its energy for growth from the oxidation of ammonia to nitrite (127). This bacterium employs an ammonia monooxygenase (AMO) for the oxidation of ammonia to hydroxylamine. This reaction catalyzed by AMO resembles the soluble methane monooxygenase (sMMO) of methanotrophs in that it reduces one atom of molecular oxygen to water while the other is incorporated into the substrate molecule (19, 127). These reactions are illustrated below:

$$\text{AMO: } NH_3 + O_2 + XH_2 \rightarrow NH_2OH + X + H_2O \qquad 1.$$

$$\text{MMO: } CH_4 + O_2 + NADH + H^+ \rightarrow CH_3OH + NAD^+ + H_2O \qquad 2.$$

Like methanotrophs, cells of *Nitrosomonas europea* are capable of cooxidizing a broad range of hydrocarbon substrates, including alkanes and alkenes. The enzymes are remarkably similar in terms of the substrates attacked. Methane is oxidized by nitrifying bacteria (7) and ammonia is known to be oxidized by several methanotrophs (7, 18, 85, 108). Cells of *N. europea* have been shown to degrade several halocarbons (89). The rate of oxidation and V_{max} values of haloethanes increased with decreasing molecular weight from iodoethane to chloroethane. Fluoroethane was also dehalogenated. Monohalogenated ethanes were oxidized to acetaldehyde. Chloroethanes were oxidized more rapidly than chloropropane and chlorobutane. Trichloroethylene was also shown to be oxidized by *N. europea* (4). It is probable that a nonspecific AMO similar to MMO is responsible, although AMO has not been purified and characterized.

Little et al. (74) described the oxidation of TCE by a pure culture of methane-utilizing bacterium and suggested that this compound was degraded by formation of TCE epoxide, similar to the conversion of propylene to propylene epoxide by MMO (20).

Wackett et al. (118) surveyed whole cells of several bacteria that produce mono- or dioxygenases, which initiate metabolism of nitropropane, cyclohexanone, 4-methoxybenzoate, hexane, and propane, for their ability to biodegrade TCE. Of the bacteria surveyed, only bacteria that possessed propane monooxygenase degraded TCE at significant rates. *Mycobacterium vaccae* JOB5 cells grown on propane displayed the highest rates of TCE oxidation, and these cells also oxidized 1,1- and 1,2-dichloroethylene and vinylchloride. Vinylchloride was oxidized more rapidly than other chlorinated ethylenes (118). Subsequently, Phelps et al. (88) isolated a gram-positive, branching bacterium that they indicated was a member of the order *Actinomycetales*. It utilized propane as a carbon and energy source. This bacterium also degraded toluene, benzene, vinylidene chloride, and vinylchloride. *Mycobacterium* is also classified in the order *Actinomycetales*.

Oldenhuis et al. (83, 84) and we (9, 110) have observed rates of biodegradation of TCE by cells of the methanotroph *Methylosinus trichosporium* OB3b that were more rapid than previously known for other bacteria. A comparison of the rates of TCE oxidation by whole bacterial cells of different species is presented in Table 1.

Some methanotrophic bacteria produce two forms of methane monooxygenase (21, 22, 96): a soluble enzyme that remains in solution after sedimentation of cell extracts at $100,000 \times$ gravity for one hour, and a particulate or membrane-bound form that sediments under the same conditions. The soluble enzyme has been purified from three methanotrophs (15, 16, 29, 87).

Cells of *Methylosinus trichosporium* OB3b oxidized TCE rapidly only when the soluble form of MMO was synthesized (83, 110). The production of sMMO has been detected by the presence of proteins that cross reacted with antisera produced in rabbits against purified protein components of the enzyme (110). The oxidation of cyclohexanol (83), chloroform (107), and naphthalene to 1- and 2-naphthols also indicate sMMO activity (9). When sMMO was present, *M. trichosporium* cells oxidized chloroform, methylenechloride, 1,1- and 1,2-dichloroethane and *trans*-1,2-dichloroethylene, 1,2-dichloropropane, and vinylchloride (9, 83, 110). The soluble form of MMO was shown to be produced when some methanotrophic bacteria were grown in media with low concentrations of copper (11, 21, 83, 96, 100, 110).

Methane-utilizing bacteria have been separated into three groups: Group I, Group II, and Group X (40, 41, 109, 121–123). Methanotrophs in Group I utilize the ribulose monophosphate pathway for formaldehyde assimilation and contain disk-shaped bundles of intracytoplasmic membranes. The predominant phospholipid fatty acids of

Table 1. Relative rates of trichloroethylene oxidation.

	Rate of oxidation[a] (nmol/min/mg)	References
Enzyme systems		
Methane monooxygenase		
Myethylosinus trichosporium OB3b	682	29
Myethylosinus capsulatus Bath	<80	38
Cytochrome P-450 rat liver microsomes	20–500	77
Live organisms		
Myethylosinus trichosporium OB3b	220	9
(Methane monooxygenase)	40–150	83, 110
Methanotrophic consortium in bioreactor	27	1
Pseudomonas cepacia strain G-4 (toluene 2-monooxygenase)	7.9	28
Pseudomonas mendocina (toluene 4-monooxygenase)	2	126
Pseudomonas putida F1 (toluene dioxygenase)	2	117
Nitrosomonas europaea (ammonia monooxygenase)	~1.0	4
Mycobacterium sp. (propane monooxygenase)	~0.5	118
Methanotroph strain 46-1 (methane monooxygenase)	<0.1	74

[a] For enzymatic oxidations, rates are listed relative to the oxygenase component; for organism oxidations, rates are listed relative to whole cell protein.

this group of bacteria contain 16 carbon atoms, and their DNA has a moles percent G + C content of 50 to 54. All but one species contain an incomplete tricarboxylic acid cycle and lack α-ketoglutarate dehydrogenase. Bacteria in Group I are not known to synthesize sMMO. Two genera of bacteria, *Methylomonas* and *Methylobacter*, have been proposed in this group.

Methanotrophs in Group X resemble the bacteria in Group I in that they have a similar fine structure of intracytoplasmic membranes, contain predominantly 16-carbon phospholipid fatty acids, and lack a complete tricarboxylic acid cycle. The one recognized species in this group, *Methylococcus capsulatus*, contains ribulosediphosphate carboxylase and is capable of autotrophic CO_2 fixation. The moles percent G + C content of DNA from *M. capsulatus* is 62.5. It is the only recognized species capable of growth above 45°C.

The bacteria in Group II employ the serine pathway for formaldehyde assimilation. Several enzymes of this pathway, like two enzymes of the RuMP pathway, are uniquely present in methylotrophic bacteria. The two pathways do not have assimilatory reactions in common. *Methanotrophic* bacteria in Group II contain intra-cytoplasmic membranes arranged in pairs aligned parallel to the cytoplasmic membrane. The predominant phospholipid fatty acids contain 18 carbon atoms, and these bacteria have a complete tricarboxylic acid cycle. Their DNA has a moles percent G + C of 62.5%. Two genera, *Methylocystis* and *Methylosinus*, have been proposed to include bacteria in this group.

Ribosomal RNAs, both 5S and 16S, from several methanotrophs have been sequenced (10, 42, 112; B. J. Bratina, G. A. Brusseau and R. S. Hanson, submitted for publication). Group I and Group X methanotrophic bacteria with completed 16S rRNAs sequences are related to bacteria within the beta/gamma subgroups of proteobacteria, whereas Group II methanotrophs with completed 16S rRNA sequences are tightly clustered within the alpha subgroup of proteobacteria (112).

We (111) have synthesized oligonucleotide signature probes complementary to 16S ribosomal RNAs of Group I and Group II methanotrophs. Probe 9-α hybridized to total RNA from Group I methanotrophs except RNA from *Methylomonas methanica*. Probe 10-γ hybridized to total RNA from all serine pathway methylotrophs tested. Neither probe hybridized to RNAs extracted from nonmethylotrophic bacteria. We have recently selected a probe with a sequence that is uniquely complementary to methane-utilizing serine-pathway methylotrophs and is not complementary to sequences of nonmethane-utilizing methylotrophs examined to date (R. S. Hanson et al., unpublished data).

Signature probes labeled with fluorescent dyes have been employed by Stahl et al. (98), Delong et al. (24), and Amman et al. (2) to detect intact cells of bacteria. We have shown that fluorescent derivatives of probes 9-α and 10-γ can detect cells of Group I or Group II methylotrophs (111).

We have surveyed several methylotrophs for the abilities to produce sMMO and to oxidize TCE (Tables 1, 2). Soluble MMOs from *Methylococcus capsulatus* Bath (a Group X methanotroph) and *Methylobacterium* sp. strain CRL-26 and *Methylosinus trichosporium* OB3b (two Group II methanotrophs) have been purified and extensively

Table 2. The relationship between the presence of soluble methane monooxygenase and TCE oxidation in methanotrophic bacteria.

	Physiological group	Soluble MMO[a]		TCE degradation	PFGE[b]-Southern hybridization with B gene probe
		Hydroxylase	Component B		
Methylomonas methanica 81Z	II	+	+	+	+
Methylosinus sporium	II	+	+	+	+
Methylosinus spp. B	II	+	+	+	+
Methylosinus trichosporium OB3b	II	+	+	+	+
Methylocytis parvus OBBP	II	−	−	−	−
Methylocytis pyriformis #14	II	−	−	−	ND
Methylococcus capsulatus Bath	X	+	−	+	−
Methylomonas albus BG8	I	−	−	−	−
Methylomonas methanica	I	−	−	−	−
Methylomonas rubra	I	−	−	−	−
Bioreactor sample	II	+	+	+	+
Bioreactor pure culture isolate	II	+	+	+[c]	+
Isolate from *L. minor*	ND	+	+	+[c]	+

[a] Detected with antibodies prepared against purified sMMO components on Western blots.
[b] PFGE: Pulse field gel electrophoresis.
[c] Determined by colorimetric assay.

characterized (16, 29, 86). Only some Group II methanotrophs and *M. capsulatus* Bath have produced this enzyme, as detected by proteins that cross react with antisera prepared against the purified *M. trichosporium* OB3b enzyme and naphthalene oxidation (Table 2). The genes encoding the five protein components of sMMOs from *M. capsulatus* Bath and *M. trichosporium* OB3b have been cloned into *Escherichia coli* and sequenced (12, 80, 99).

We have cloned a gene encoding one protein component (protein B) of the sMMO from *Methylosinus trichosporium* OB3b and have used it to detect complementary DNA sequences in DNA prepared from a variety of methanotrophs (Table 2). Some of these strains, like the strain from a bioreactor sample, are known to oxidize TCE but are otherwise not well characterized. The data in Table 2 illustrate that genes complementary to the *M. trichosporium* OB3b gene are confined to some Group II methanotrophs. The sMMO B gene of *Methylococcus capsulatus* Bath was not detected with the *M. trichosporium* gene probe. However, the *M. capsulatus* Bath sMMO was detected by using antibodies prepared against the *M. trichosporium* OB3b sMMO proteins (Table 2).

We have examined bacteria present in a consortium in a bioreactor optimized for TCE degradation that was previously described by Alvarez-Cohen and McCarty (1). Restriction fragment length polymorphism analysis and sequencing of the ribosomal RNA from the bacterium isolate from the bioreactor described by Alvarez-Cohen and McCarty (1) have shown that these Group II methanotrophs are different from other isolates in our culture collection (L. Alvarez-Cohen, H. C. Tsien, and R. S. Hanson, unpublished data). A bacterium shown to be identical to the dominant bacterium in the bioreactor was isolated in pure culture. This bacterium oxidizes TCE, hybridizes to the Group II signature probe, possesses DNA complementary to the sMMO B gene probe, and produces sMMO. The isolate had the same restriction fragment length polymorphism pattern of DNA fragments complementary to the sMMO B gene and a methanol dehydrogenase gene, and the same morphology as the dominant reactor bacterium.

We have also isolated a methanotrophic bacterium from *Lemna minor* (little duckweed). Suspensions of these floating aquatic plants were shown to oxidize methane and TCE. This bacterium also produced sMMO and contained DNA sequences complementary to the cloned sMMO B gene (Table 2).

There is good evidence that Group I methanotrophs do not synthesize a sMMO detectable with antibodies prepared against the sMMO of *Methylosinus trichosporium* OB3b or contain DNA sequences complementary to the sMMO B gene. In addition, Group I methanotrophs cannot transform naphthalene to naphthols at rates comparable to Group II methanotrophs, but can slowly oxidize TCE. The rates of TCE oxidation were slower than those reported for Group II methanotrophs grown under copper limitation (Table 1), and the oxidation of TCE was incomplete.

Little et al. (74) described two methanotrophs, strains 46-1 and 68-1. Thin sections of strain 46-1 and 68-1 clearly revealed stacked membrane arrangements typical of Group I methanotrophs and both bacteria possessed hexulose-6-phosphate synthase, an enzyme for the RuMP pathway for formaldehyde fixation. We (H. C. Tsien, R. S. Hanson, and G. S. Saylor) have failed to detect evidence for the production of sMMO by these strains 46-1 and 68-1.

Henry and Grbić-Galić (49) also isolated a Group I methanotroph, *Methylomonas* sp., that oxidized TCE in pure culture and proposed that the MMO responsible for TCE oxidation was associated with the particulate fraction of cell extracts. The K_s for TCE in the *Methylomonas* species described by them (48) is 4 μM vs. 145 μM for *Methylosinus trichosporium* OB3b (84). Therefore, type I methanotrophs containing particulate MMO may oxidize TCE more rapidly at low concentration. In addition, these cells may be less susceptible to inhibition of TCE transformation by methane and CO than type II methanotrophs.

The rates of TCE oxidation by these Group I methanotrophs were orders of magnitude less than those observed for Group II methanotrophs (Table 1). From this information, we propose as a hypothesis that only some Group II methanotrophs and *Methylococcus capsulatus* are capable of producing sMMO. We believe this enzyme oxidizes halomethanes, halogenated alkenes, and several other substrates other than methane much more rapidly than the particulate MMO and other known oxygenases.

With this as a working hypothesis, we have begun collaborations with other groups to survey microbial populations in soils and groundwaters and on plants for Group I, Group II, and Group X methanotrophs. We are employing signature probes, antibodies prepared against whole cells, gene probes (sMMO and methanol dehydrogenase genes), and the methods for detecting the activity of sMMO described above. We hope to determine environmental conditions that favor the growth of TCE-degrading methanotrophs.

We have also undertaken studies of the growth of different methanotrophs in mixed cultures where growth is restricted by limitation of one or more of several essential nutrients (copper, nitrogen, oxygen, methane, phosphorous, and sulfur). The effects of temperature, pH, etc. will also be examined. These studies have been pursued by D. Graham and R. Arnold, Arizona State University, and by us for the purpose of defining conditions that favor the growth of Group II methanotrophs.

Clearly, more work remains to be done in the investigation of the biodegradation of low molecular weight halogenated compounds. Given the potential of aerobic bacteria to remove pollutants from the environment, however, the promise of these organisms must be pursued.

Acknowledgments

Research performed at the Gray Freshwater Biological Institute that is reported in this chapter was supported in part by grants to R. S. H. (BSR 8903833 from the National Science Foundation and DE-FGO2-88ER 13862 from the US Department of Energy). We are grateful to Louise Mohn for her skillful assistance in preparation of this manuscript.

LITERATURE CITED

1. Alvarez-Cohen, L., and P. L. McCarty. 1991. Effects of toxicity, aeration and reductant supply on trichloroethylene transformation by a mixed methanotrophic culture. *Appl. Environ. Microbiol.* 57:228–235.

2. Amann, R. I., L. Krumholz, and D. A. Stahl. 1990. Fluorescent oligonucleotide probing of whole cells for determinative phylogenetic and environmental studies in microbiology. *J. Bacteriol.* 172:762–770.

3. Anthony, C. 1982. *The Biochemistry of Methylotrophs.* London: Academic Press.

4. Arciero, D., T. Vannelli, M. Logan, and A. B. Hooper. 1989. Degradation of trichloroethylene by the ammonia-oxidizing bacterium *Nitrosomonas europaea. Biochem. Biophys. Res. Commun.* 159:640–643.

5. Barbash, J., and P. V. Roberts. 1986. Volatile organic chemical contamination of groundwater resources in the U.S. *J. Water Poll. Control Fed.* 58:343–348.

6. Barrio-Lage, G., F. G. Parsons, R. S. Nassar, and P. A. Lorenzo. 1986. Sequential dehalogenation of chlorinated ethenes. *Environ. Sci. Technol.* 20:96–99.

7. Bedard, C., and R. Knowles. 1989. Physiology, biochemistry, and specific inhibitors of CH_4, $NH4^+$, and CO oxidation by methanotrophs and nitrifiers. *Microbiol. Rev.* 53:68–84.

8. Brunner, W., D. Staub, and T. Leisinger. 1980. Bacterial degradation of dichloromethane. *Appl. Environ. Microbiol.* 40:950–958.

9. Brusseau, G. A., H.-C. Tsien, R. S. Hanson, and L. P. Wackett. 1990. Optimization of trichloroethylene oxidation by methanotrophs and the use of a colorimetric assay to detect soluble methane monooxygenase activity. *Biodegradation* 1:19–29.

10. Bulygina, E. S., V. F. Galchenko, N. I. Govorukhina, A. I. Netrusov, D. I. Nikitin, Y. A. Trotsenko, and K. M. Chumakov. 1990. Taxonomic studies on methylotrophic bacteria by 5S ribosomal RNA sequencing. *J. Gen. Microbiol.* 136:441–446.

11. Burrows, K. J., A. Cornish, D. Scott, and I. J. Higgins. 1984. Substrate specificities of the soluble and particulate methane monooxygenases of *Methylosinus trichosporium* OB3b. *J. Gen. Microbiol.* 130:3327–3333.

12. Cardy, D. L. N., V. Laidler, G. P. C. Salmond, and J. C. Murrell. 1991. Molecular analysis of the methane monooxygenase gene cluster of *Methylosinus trichosporium* OB3b. *Molecular Microbiol.* 5:335–342.

13. Chaudry, G. R., and S. Chapalamadugu. 1991. Biodegradation of halogenated organic compounds. *Microbiol. Rev.* 55:59–79.

14. Colby, J., and H. Dalton. 1976. Some properties of a soluble methane monooxygenase from *Methylococcus capsulatus* strain Bath. *Biochem. J.* 157:495–497.

15. Colby, J., D. I. Stirling, and H. Dalton. 1977. The soluble methane monooxygenase of *Methylococcus capsulatus* (Bath). *Biochem. J.* 165:395–402.

16. Colby, J., and H. Dalton. 1978. Resolution of the methane monooxygenase of *Methylococcus capsulatus* (Bath) into three components: purification and properties of component C, a flavoprotein. *Biochem. J.* 171:461–468.

17. Dagley, S. 1978. Determinants of biodegradability. *Quarterly Rev. of Biophys.* 11:577–602.

18. Dalton, H. 1977. Ammonia oxidation by the methane oxidizing bacterium *Methylococcus capsulatus* strain Bath. *Arch. Microbiol.* 114:273–279.

19. Dalton, H. 1980. Oxidation of hydrocarbons by methane monooxygenases from a variety of microbes. *Adv. Appl. Mircrobiol.* 26:71–87.

20. Dalton, H., and D. I. Stirling. 1982. Co-metabolism. *Philos. Trans. R. Soc. London. Ser. B* 297:481–491.

21. Dalton, H., S. D. Prior, D. J. Leak, and S. H. Stanley. 1984. Regulation and control of methane monooxygenase. In *Microbial Growth on C_1 Compounds*. Ed. R. L. Crawford, R. S. Hanson. Washington, D.C.: American Society for Microbiology. 75–82.

22. Dalton, H., and I. J. Higgins. 1987. Physiology and biochemistry of methylotrophic bacteria. In *Microbial Growth on C_1 Compounds*. Ed. H. W. Van Verseveld, J. A. Duine. Dordrecht, the Netherlands: Martinus Nijhoff Publishers. 89–94.

23. Davis, J. W., and C. L. Carpenter. 1990. Aerobic biodegradation of vinyl chloride in groundwater samples. *Appl. Environ. Microbiol.* 56:3878–3880.

24. Delong, E. F., G. S. Wickham, and N. R. Pace. 1989. Phylogenetic stains. Ribosomal RNA-based probes for the identification of single cells. Science 243:1360–1363.

25. Edwards, R. R., J. Campbell and G. S. Milne. 1982. The impact of chloromethanes on the environment. Part 2. Methyl chloride and methylene chloride. *Chem. Indust.* 619:622–627.

26. Ferenci, T., T. Strom, and J. R. Quayle. 1975. Oxidation of carbon monoxide and methane by *Pseudomonas methanica*. *J. Gen. Microbiol.* 91:79–91.

27. Fogel, M. M., A. R. Tadeo, and S. Fogel. 1986. Biodegradation of chlorinated ethenes by a methane-utilizing mixed culture. *Appl. Environ. Microbiol.* 51:720–724.

28. Folsom, B. K., P. J. Chapman, and P. H. Pritchard. 1990. Phenol and trichloroethylene degradation by *Pseudomonas cepacia* G4: Kinetics and interactions between substrates. *Appl. Environ. Microbiol.* 56:1279–1295.

29. Fox, B. G., W. A. Froland, J. Dege, and J. D. Lipscomb. 1989. Methane monooxygenase from *Methylosinus trichosporium* OB3b. *J. Biol. Chem.* 264:10023–10033.

30. Fox, B. G., J. G. Borneman, L. P. Wackett, and J. D. Lipscomb. 1990. Haloalkene oxidation by the soluble methane monooxygenase from *Methylosinus trichosporium* OB3b: mechanistic and environmental applications. *Biochemistry* 29:6419–6427.

31. Freedman, D. L., and J. M. Gosset. 1989. Biological reductive dechlorination of tetrachloroethylene and trichloroethylene under methanogenic conditions. *Appl. Environ. Microbiol.* 55:2214–2151.

32. Gälli, R., G. Stucki, and T. Leisinger. 1982. Mechanism of dehalogenation of dichloromethane by cell extracts of *Hyphomicrobium* DM2. *Experientia* 38:1378.

33. Gälli, R., and T. Leisinger. 1985. Specialized bacterial strains for the removal of dichloromethane from industrial waste. *Conserv. Recycling* 8:91–100.

34. Gälli, R. 1987. Biodegradation of dichloromethane in wastewater using a fluidized bed bioreactor. *Appl. Microbiol. Biotechnol.* 27:206–213.

35. Gälli, R., and T. Leisinger. 1988. Plasmid analysis and cloning of the dichlormethane utilization genes of *Methylobacterium* sp. DM4. *J. Gen. Microbiol.* 134:943–952.

36. Goldman, P. 1965. The enzymatic cleavage of the carbon-fluorine bond in fluoracetate. *J. Biol. Chem.* 240:3434–3438.

37. Goldman, P., G. W. A. Milne, and M. T. Pignataro. 1967. Fluorine containing metabolites formed from 1-fluoroacetate by *Pseudomonas* species. *Arch. Biochem. Biophys.* 118:178–184.

38. Green, J., and H. Dalton. 1989. Substrate specificity of soluble methane monooxygenase: mechanistic implications. *J. Biol. Chem.* 264:17698–17703.

39. Haber, C. L., L. N. Allen, and R. S. Hanson. 1983. Methylotrophic bacteria: biochemical diversity and genetics. *Science* 221:1147–1151.

40. Hanson, R. S. 1980. Ecology and diversity of methylotrophic organisms. *Adv. Appl. Microbiol.* 26:3–39.

41. Hanson, R. S., A. I. Netrusov, and K. Tsuji. 1991. The obligate methanotrophic bacteria: *Methylococcus, Methylomonas* and *Methylosinus*. In *The Procaryotes*. Ed. A. Balows, H. G. Truper, M. Dworkin, W. Harder, K. H. Schleifers. New York: Springer-Verlag. 661–684.

42. Harder, W., and M. M. Attwood. 1978. Biology, physiology and biochemistry of *Hyphomicrobia*. *Adv. Microbiol. Physiol.* 17:303–359.

43. Hardman, D. J., and J. H. Slater. 1981. Dehalogenases in soil bacteria. *J. Gen. Microbiol.* 123:117–128.

44. Hardman, D. J., and J. H. Slater. 1981. The dehalogenase complement of a soil pseudomonad grown in closed and open cultures on alkanoic acids. *J. Gen. Microbiol.* 127–399–405.

45. Hardman, D. J., P. C. Gowland, and J. H. Slater. 1986. Large plasmids from soil bacteria enriched on alkanoic acids. *Appl. Environ. Microbiol.* 51:44–51.

46. Hartmans, S., J. A. M. deBont, J. Tramper, and K. Luyben. 1985. Bacterial degradation of vinyl chloride. *Biotech. Lett.* 7:320–325.

47. Hartmans, S., A. Schmuckle, A. M. Cook, and T. Leisinger. 1986. Methyl chloride: naturally occurring toxicant and C–1 growth substrate. *J. Gen. Microbiol.* 132:1139–1142.

48. Henry, S. M., and D. Grbić-Galić. 1990. Effect of mineral media on trichloroethylene oxidation by aquifer methanotrophs. *Microb. Ecol.* 20:15–169.

49. Henry, S. M., and D. Grbić-Galić. 1991. Influence of endogenous and exogenous electron

donors and trichloroethylene oxidation toxicity on trichloroethylene oxidation by methanotrophic cultures from a groundwater aquifer. *Appl. Environ. Microbiol.* 57:236–244.

50. Henson, J. M., M. V. Yates, and J. W. Cochran. 1989. Metabolism of chlorinated methanes, ethanes, and ehtylenes by a mixed bacterial culture growing on methane. *J. Indust. Microbiol.* 4:29–35.

51. Higgins, I. J., D. J. Best, and R. C. Hammond. 1980. New findings in methane-utilizing bacteria highlight their importance in the biosphere and their commercial potential. *Nature* 286:561–564.

52. Higgins, I. J., D. J. Best, R. C. Hammond, and D. Scott. 1981. Methane oxidizing microorganisms. *Microbiol. Rev.* 45:556–590.

53. Higgins, I. J., D. J. Best, and D. Scott. 1982. Generation of products by methanotrophs. *Basic Life Science* 19:383–402.

54. Horvath, R. S. 1972. Microbial cometabolism and the degradation of organic compounds in nature. *Bacteriol. Rev.* 36:146–155.

55. Hou, C. T., P. Patel, A. Laskin, and N. Barnabe. 1979. Microbial oxidation of gaseous hydrocarbons: epoxidation of C_2 to C_4 n-alkenes by methylotrophic bacteria. *Appl. Environ. Microbiol.* 38:127–134.

56. Infante, P. F., and Tsongas, T. A. 1987. Mutagenic and oncogenic effects of chloromethanes, chloromethanes and halogenated analogues of vinyl chloride. *Environ. Sci. Res.* 25:301–327.

57. Janssen, D. B., A. Scheper, and B. Witholt. 1984. Biodegradation of 2-chloroethanol and 1,2-dichloroethane by pure bacterial cultures. *Progr. Indust. Microbiol.* 20:169–178.

58. Janssen, D. B., A. Scheper, L. Dijkhuizen, and B. Witholt. 1985. Degradation of halogenated aliphatic compounds by *Xanthobacter autotrophicus* GJ10. *Appl. Environ. Microbiol.* 49:673–677.

59. Janssen, D. B., D. Jager, and B. Witholt. 1987. Degradation of n-haloalkanes and α-ω-dihaloalkanes by wild-type and mutants of *Acinetobacter* sp. strain GJ70. *Appl. Environ. Microbiol.* 53:561–566.

60. Janssen, D. B., G. Grobben, and B. Witholt. 1987. Toxicity of chlorinated aliphatic hydrocarbons and degradation by methanotrophic consortia. In *Proceedings of 4th European Congress on Biotechnology* 1987, 3:515–518. Ed. O. M. Neijssel, R. R. van der Meer, K. C. Luyben. Amsterdam, Netherlands: Elsevier Science Publishers B.V.

61. Janssen, D. B., J. Gerritse, J. Brackman, C. Kalk, D. Jager, and B. Witholt. 1988. Purification and characterization of a bacterial dehalogenase with activity toward halogenated alkanes, alcohols, and ethers. *Eur. J. Biochem.* 171:67–72.

62. Janssen, D. B., J. Frens Pries, J. van der Ploeg, B. Kazemier, P. Terpstra, and B. Whitholt. 1989. Cloning of 1,2-dichloroethane degradation genes of *Xanthobacter autotrophicus* GJ10 and expression and sequencing of the dhl A gene. *J. Bacteriol.* 171:6791–6799.

63. Jensen, H. L. 1963. Carbon nutrition of some microorganisms decomposing halogen-substituted aliphatic acid. *Acta. Agric. Scan.* 13:402–412.

64. Kampbell, D. H., J. T. Wilson, H. W. Reed, and T. T. Stocksdale. 1987. Removal of volatile aliphatic hydrocarbons in a soil bioreactor. *Control Technology.* 37:236–239.

65. Karlson, P., and the Nomenclature Committee of the International Union of Biochemistry. 1978. *Enzyme Nomenclature.* New York: Academic Press.

66. Kawasaki, H., N. Tone, and K. Tonomura. 1981. Plasmid determined dehalogenation of haloacetate in *Moraxella* species. *Agric. Biol. Chem.* 45:29–34.

67. Kohler-Staub, D., and T. Leisinger. 1985. Dichloromethane dehalogenase of *Hyphomicrobium* sp. strain DM1. *J. Bacteriol.* 162:676–681.

68. Kohler-Staub, D., S. Hartmans, R. Gäli, F. Suter, and T. Leisinger. 1986. Evidence for identical dichloromethane dehalogenases in different methylotrophic bacteria. *J. Gen. Microbiol.* 132:2837–2843.

69. LaRoche, S. D., and T. Leisinger. 1990. Sequence analysis and expression of the bacterial dichloromethane dehalogenase structural gene, a member of the glutathione S-transferase supergene family. *J. Bacteriol.* 172:164–171.

70. Leadbetter, E. R., and J. W. Foster. 1959. Oxidation products formed from gaseous alkanes by the bacterium *Pseudomonas methanica. Arch. Biochem. Biophys.* 82:491–492.

71. Leisinger, T. 1983. Microorganisms and xenobiotic compounds. *Experientia* 39:1183–1191.

72. Leisinger T., and W. Brunner. 1986. Poorly degradable substances. pp. 475–513. In *Biotechnology,* Vol. 8. Ed. H-J. Rehm, G. Reed. Weinheim, Federal Republic of Germany: VCH Verlagsgesellshaft mbtt.

73. Leisinger, T. 1988. Microbial degradation of problematic air components. *Biotechnology* 2:125–133.

74. Little, C. D., A. V. Palumbo, S. E. Herbes, M. E. Lidstrom, R. L. Tyndall, and P. J. Gilmer. 1988. Trichloroethylene biodegradation by a methane-oxidizing bacterium. *Appl. Environ. Microbiol.* 54:951–956.

75. Maltoni, C., and G. Lefemine. 1974. Carcinogenicity bioassays of vinylchloride. I. Research plan and early results. *Environ. Res.* 7:387–396.

76. McCarty, P. L. 1988. Bioengineering issues related to *in-situ* remediation of contaminated soils and groundwater. pp. 143–162. In *Environmental Biotechnology: Reducing Risks from Environmental Chemicals Through Biotechnology.* Ed. G. S. Omenn. New York: Plenum Press.

77. Miller, R. E., and F. P. Guengerich. 1982. Oxidation of trichloroethylene by liver microsomal cytochrome P-450: evidence for chlorine migration in a transition state not involving trichloroethylene oxide. *Biochemistry* 21:1090–1097.

78. Miller, R. E., and F. P. Guengerich. 1983. Metabolism of trichloroethylene in isolated hepatocytes, microsomes, and reconstituted enzyme systems containing cytochrome P-450. *Cancer Res.* 43:1145–1152.

78a. Montgomery, L., Asaf-Anid, N., Nies, L., Anid, P. J., Vogel, T. M. 1993. Aneorobic biodegradation of chlorinated organic compounds. In *Biological Degradation and Bioremedation of Toxic Chemicals.* Ed. G. R. Chaudhry. Portland, OR: Disocorids Press.

79. Motosugi, K., and K. Soda. 1983. Microbial degradation of organochlorine compounds. *Experentia* 39:1214–1220.

80. Mullens, I. A., and H. Dalton. 1987. Cloning of the gamma-subunit methane monooxygenase from *Methylococcus capsulatus. Biotechnology* 5:490–493.

81. Neilson, A. H. 1990. The biodegradation of halogenated organic compounds. *J. Appl. Bacteriol.* 69:445–470.

82. Nelson, M. J., S. O. Montgomery, W. R. Mahaffey, and P. H. Pritchard. 1987. Biodegradation of trichloroethylene and involvement of an aromatic biodegradative pathway. *Appl. Environ. Microbiol.* 53:949–954.

83. Oldenhuis, R., R. L. J. M. Vink, D. B. Janssen, and B. Witholt. 1989. Degradation of chlorinated aliphatic hydrocarbons by *Methylosinus trichosporium* OB3b expressing soluble methane monooxygenase. *Appl. Environ. Microbiol.* 55:2819–2926.

84. Oldenhuis, R., J. Y. Oedzes, J. J. van der Waarde, and D. B. Janssen. 1991. Kinetics of chlorinated hydrocarbon degradation by *Methylosinus trichosporium* OB3b and toxicity of trichloroethylene. *Appl. Environ. Microbiol.* 57:7–14.

85. O'Neill, J. G., and J. F. Wilkinson. 1977. Oxidation of ammonia by methane-oxidizing bacteria and the effects of ammonia on methane oxidation. *J. Gen. Microbiol.* 100:407–412.

86. Patel, R. N., C. T. Hou, A. I. Laskin, A. Felix, and P. Derelanko. 1979. Microbial oxidation of gaseous hydrocarbons: hydroxylation of n-alkanes and epoxidation of n-alkenes by cell-free particulate fractions of methane-utilizing bacteria. *J. Bacteriol.* 139:675–679.

87. Patel, R. N., C. T. Hou, A. I. Laskin, and A. Felix. 1982. Microbial oxidation of hydrocarbons: properties of a soluble methane monooxygenase from a facultative methane-utilizing organism, *Methylobacterium* sp. strain CRL-26. *Appl. Environ. Microbiol.* 44:1130–1137.

88. Phelps, T. J., K. Malachowsky, R. M. Schram, and D. C. White. 1991. Aerobic numeralization of vinylchloride by a bacterium of the order *Actinomycetales*. *Appl. Environ. Microbiol.* 57:1252–1254.

89. Rasche, M. E., R. E. Hicks, M. R. Hyman, and D. J. Arp. 1990. Oxidation of monohalogenated ethanes and n-chlorinated alkanes by whole cells of *Nitrosomonas europea*. *J. Bacteriol.* 172:5368–5373.

90. Richardson, K. L., and D. T. Gibson. 1984. A novel pathway for toluene oxidation in *Pseudomonas mendocina*. *Abst. Am. Soc. for Microbiol.* 84:K54.

91. Roberts, P. V., J. E. Schreinger, and G. C. Hopkins. 1982. Field-study of organic-water quality changes during groundwater recharge in the Palo-Alto Baylands. *Water Res.*, 1025–1035.

92. Rozenboom, H. J., J. Kingma, D. B. Janssen, and B. Dijkstra. 1988. Crystallization of haloalkane dehalogenase from *Xanthobacter autotrophicus* GJ10. *J. Mol. Biol.* 200:611–612.

93. Scholtz, R., T. Leisinger, F. Suter, and A. M. Cook. 1987a. Characterization of 1-chlorohexane halidohydrolase, a dehalogenase of wide substrate range from an *Arthrobacteri* sp. *J. Bacteriol.* 169:5016–5021.

94. Scholtz, R., A. Schmuckle, A. M. Cook, and T. Leisinger. 1987b. Degradation of eighteen 1-monohaloalkanes by *Arthrobacter* sp. strain HA1. *J. Gen. Microbiol.* 133:267–274.

95. Scholtz, R., L. Wackett, C. Egli, A. Cook, and T. Leisinger. 1988. Dichloromethane dehalogenase with improved catalytic activity isolated from a fast-growing dichloromethane-utilizing bacterium. *J. Bacteriol.* 170:5698–5704.

96. Scott, D., J. Brannan, and I. J. Higgins. 1981. The effect of growth conditions on intracytoplasmic membranes and methane monooxygenase activities in *Methylosinus trichosporium* OB3b. *J. Gen. Microbiol.* 125:63–72.

97. Shields, M. S., S. O. Montgomery, P. J. Chapman, S. M. Cuskey, and P. H. Pritchard. 1989. Novel pathway of toluene catabolism in the trichloreothylene-degrading bacterium G4. *Appl. Environ. Microbiol.* 55:1624–1629.

98. Stahl, D. A., B. Flesher, H. R. Mansfield, and L. Montgomery. 1988. Use of phylogenetically based hybridization probes for studies of ruminal microbial ecology. *Appl. Environ. Microbiol.* 54:1079–1084.

99. Stainthorpe, A. C., V. Lees, G. P. C. Salmond, H. Dalton, and J. C. Murrell. 1989. The methane monooxygenase gene cluster in *Methylococcus capsulatus* Bath. *Gene* 91:27–34.

100. Stanley, S. H., S. D. Prior, D. J. Leak, and H. Dalton. 1983. Copper stress underlies the fundamental change in intracellular location of methane monooxygenase in methane-utilizing mechanisms: studies in batch and continuous cultures. *Biotechnol. Lett.* 5:487–492.

101. Stirling, D. I., and H. Dalton. 1977. Effect of metal-binding agents and other compounds on methane oxidation by two strains of *Methylococcus capsulatus*. *Arch. Microbiol.* 114:71–76.

102. Stirling, D. I., and H. Dalton. 1979. The fortuitous oxidation and cometabolism of various carbon compounds by whole-cell suspensions of *Methylococcus capsulatus* (Bath). *FEMS Microbiol. Letters* 5:315–318.

103. Strand, S. E., and L. Shippert. 1986. Oxidation of chloroform in an aerobic soil exposed to natural gas. *Appl. Environ. Microbiol.* 52:203–205.

104. Strand, S. E., M. D. Bjelland, and H. D. Stensel. 1990. Kinetics of chlorinated hydrocarbon degradation of suspended cultures of methane-oxidizing bacteria. *J. Res. J. Water Pollut. Control. Fed.* 62:124–129.

105. Stucki, G., R. Gälli, H.-R. Ebersold, and T. Leisinger. 1981. Dehalogenation of dichloromethane by cell extracts of *Hyphomicrobium DM2*. *Arch. Microbiol.* 130:366–371.

106. Stucki, G., U. Krebser, and T. Leisinger. 1983. Bacterial growth on 1,2-dichloroethane. *Experimentia* 39:1271–1273.

107. Taylor, R. T., M. L. Hanna, S. Park, and M. W. Droege. 1990. Chloroform oxidation by *Methylosinus trichosporium* OB3b—A specific catalytic activity of the soluble form of methane monooxygenase. *Abstr. 90th Annu. Meet. Am. Soc. Microbiol.*, Anaheim, California. Abst. K-10, p. 221.

108. Topp, E., and R. S. Hanson. 1991. Metabolism of radiatively important trace gases by methane oxidizing bacteria. In *Microbial Degradation of Greenhouse Gases*. Ed. R. Whitman, E. Rogers. Washington, D.C.: Am. Soc. for Microbiol.

109. Trotsenko, Yu. A. 1983. Metabolic features of methane and methanol-utilizing bacteria. *Acta Biotechnol.* 3:269–277.

110. Tsien, H.-C., G. A. Brusseau, R. S. Hanson, and L. P. Wackett. 1989. Biodegradation of trichloroethylene by *Methylosinus trichosporium* OB3b. *Appl. Environ. Microbiol.* 55:3155–3161.

111. Tsien, H.-C., B. J. Bratina, K. Tsuji, and R. S. Hanson. 1990. Use of oligodeoxynucleotide signature probes for identification of physiological groups of methylotrophic bacteria. *Appl. Environ. Microbiol.* 56:2858–2865.

112. Tsuji, K., H.-C. Tsien, R. S. Hanson, S. R. DePalma, R. Scholtz, and S. LaRoche. 1990. 16S ribosomal RNA sequence analysis for determination of phylogenetic relationship among methylotrophs. *J. Gen. Microbiol.* 136:1–10.

113. U.S. Environmental Protection Agency. 1980. Ambient water quality criteria for vinyl chloride. Publication 440/5–80–078. National Technical Information Service, Springfield, Virginia.

114. Vandenbergh, P. A., and B. S. Kunka. 1988. Metabolism of chlorinated aliphatic hydrocarbons by *Pseudomonas fluorescens*. *Appl. Environ. Microbiol.* 54:2578–2579.

115. Vogel, T. M., and P. L. McCarty. 1985. Biotransformation of tetrachloroethylene to trichloroethylene, dichloroethylene, vinyl chloride and carbon dioxide under methanogenic conditions. *Appl. Environ. Microbiol.* 49:1080–1083.

116. Vogel, T. M., C. S. Criddle, and P. L. McCarty. 1987. Transformations of halogenated aliphatic compounds. *Environ. Sci. Technol.* 21:722–736.

117. Wackett, L. P., and D. T. Gibson. 1988. Degradation of trichloroethylene by toluene dioxygenase in whole-cell studies with *Pseudomonas putida* F1. *Appl. Environ. Microbiol.* 54:1703–1708.

118. Wackett, L. P., G. A. Brusseau, S. R. Householder, and R. S. Hanson. 1989. Survey of microbial oxygenases: Trichloroethylene degradation by propane-oxidizing bacteria. *Appl. Environ. Microbiol.* 55:2960–2964.

119. Weightman, A. J., and J. H. Slater. 1980. Selection of *Pseudomonas putida* strains with elevated dehalogenase activities by continuous culture on chlorinated alkanoic acids. *J. Gen. Microbiol.* 121:187–193.

120. Westerick, J, J., J. W. Mello, and R. F. Thomas. 1984. The groundwater supply survey. *J. Amer. Water Works Assoc.* 5:52–59.

121. Whittenbury, R., and H. Dalton. 1981. The methylotrophic bacteria. In *The Procaryotes*. Ed. M. P. Starr, H. Stolp, H. G. Truper, A. Balows, H. G. Schlegel. Berlin: Springer-Verlag KG. 894–902.

122. Whittenbury, R., K. C. Phillips, and J. F. Wilkinson. 1970. Enrichment, isolation and some properties of methane-utilizing bacteria. *J. Gen. Microbiol.* 61:205–218.

123. Whittenbury, R., J. Colby, H. Dalton, and H. L. Reed. 1986. Biology and ecology of methane oxidizers. In *Microbial Production and Utilization of Gases*. Ed. H. G. Schlegel, G. Gottschalk, N. Pfennig. Gottingen, Federal Republic of Germany: E. Goltze KG. 281–292.

124. Wilson, J. T., and B. H. Wilson. 1985. Biotransformation of trichloroethylene in soil. *Appl. Environ. Microbiol.* 49:242–243.

125. Wilson, J. T. 1988. Degradation of halogenated hydrocarbons. *Biotechnology* 2:75–77.

126. Winter, R. B., K.-M. Yen, and B. D. Ensley. 1989. Efficient biodegradation of trichloroethylene by a recombinant *Escherichia coli*. *Biotechnology* 7:282–285.

127. Wood, P. M. 1986. Nitrification as a bacterial energy source. In *Nitrification*. Ed. J. I. Prosser. Oxford: Society for General Microbiology, IRL Press. 39–62.

128. Yokota, T., H. Fuse, T. Omori, and Y. Minoda. 1986. Microbial dehalogenation of haloalkanes by oxygenase or halidohydrolase. *Agr. Biol. Chem.* 50:453–460.

CHAPTER 14

Anaerobic Utilization of Aromatic Carboxylates by Bacteria

JANE GIBSON and CAROLINE S. HARWOOD

INTRODUCTION

Very large quantities of compounds containing aromatic nuclei are produced annually from natural and industrial sources. A substantial portion of these materials accumulates in anaerobic environments, and since some of these are known or potential carcinogens, there has been growing interest in understanding how aromatics are degraded in the absence of molecular oxygen, an essential substrate in the aerobic catabolism of benzene rings. The microbiology and biochemistry involved in the degradation of natural products, mostly those derived from lignin, is considered in this chapter, with special emphasis on the role and contribution of studies with phototrophic bacteria to current understanding of these processes. Halogenated compounds are considered in detail in other chapters (those by Reineke; Hale, Jones, and Rogers; Hale, Reineke, and Wiegel), and thus are only mentioned here.

The early stages in the degradation of lignin, an extremely complex polymer that is highly resistant to microbial attack, are still not fully understood. Crawford (10) has pointed out that the extent to which dearomatization occurs within the polymer, before release of smaller substituents, may have been underestimated in studies of aerobic degradations. Although demonstrating extensive attack on natural lignin under strictly anoxic conditions has been difficult, anaerobic degradation of oligomeric forms of lignin clearly occurs to at least some extent (see 9 for a review). The analytical difficulties of detecting and quantitating small changes in bonding in an insoluble polymer have encouraged the development of an alternative approach, particularly by Young and her colleagues (64), in which smaller model compounds, whose structures correspond to major components in the natural material, are employed as substrates for selection and isolation of bacteria able to attack these molecules. The insolubility of intact lignin implies that extracellular or surface-associated enzymes must be involved in the initial attack, yielding smaller, soluble units (oligolignols) that are presumably rapidly taken up and metabolized by the depolymerizing organisms themselves, or by closely associated microorganisms that are also capable of metabolizing these compounds. Rapid utilization of lignols and lignin monomers by indigenous microflora

Dr. Gibson is in the Section of Biochemistry, Molecular and Cell Biology, Division of Biological Sciences, Cornell University, Ithaca, NY 14853, U.S.A. Dr. Harwood is at the Department of Microbiology, University of Iowa, Iowa City, IA 52242, U.S.A.

would explain why significant quantities of these probable intermediates are rarely found in natural samples. The genetic information needed for attack on these materials appears to be quite widely disseminated, because stable cultures of individual bacteria or complex consortia have been established from a number of different anaerobic environments in media containing probable lignin-derived model compounds as sole (or at least major) carbon substrates. The degradation of monomeric aromatic compounds, particularly those containing methoxyl residues and short, unsaturated side chains (for example, cinnamate, coumarate, and vanillate), has been widely investigated in recent years (reviewed in 4, 16, 51, 63, 64) in the belief that the knowledge gained will be relevant for understanding how lignin itself is degraded under anaerobic conditions.

PHYSIOLOGY OF ANAEROBIC AROMATIC-DEGRADING BACTERIA

The complete degradation of monomeric aromatic compounds can be carried out by bacteria belonging to three broad physiological classes. A large number of fermentative microorganisms can attack various molecules containing aromatic rings, but extensive degradation of these compounds is commonly dependent on interspecies hydrogen transfer, usually to methanogens, so that the bulk of the organic carbon is ultimately converted into its most oxidized and reduced forms: CO_2 and CH_4. Aromatic compounds are also attacked by bacteria that obtain energy by anaerobic respirations in which nitrate (15, 59), sulfate (57, 58), or, as recognized more recently, ferric iron (39, 40) serve as electron sinks. This mode of metabolism results in conversion of a large part of the substrate carbon into CO_2. Finally, a limited number of phototrophic bacteria are able to catabolize aromatic compounds. Light serves as the external energy source for growth of these bacteria, which are therefore able to assimilate all, or most, of their growth-substrate carbon into cell material. The metabolic activities of phototrophic bacteria may therefore go undetected in studies in which degradation is monitored by conversion of ^{14}C-labelled substrates to gaseous products.

The degradation of lignin monomers can conveniently be considered as occurring in two phases: first, modification of the ring substituents, and second, the dearomatization and opening of the aromatic ring structure. The broad distribution of methoxyl residues in lignin and its possible degradation products has prompted many investigations into their fate, which in aggregate suggest that a widely distributed strategy involves cleavage of the methoxyl ether linkages to yield products containing hydroxyl residues at the same position. Most demethylating bacteria make use of these groups as the methyl donor for generating acetate; specific examples are *Acetobacter woodii* (3, 3a); a gram-negative organism termed TH-001 (19), and *Clostridium thermoaceticum* (28, 41, 61). Reductive demethylation of aromatic compounds also serves as an electron sink for fermentative bacteria such as *Syntrophococcus sucromutans* (36). The reactions carried out by these bacteria appear to involve removal of the methyl group only; this has been demonstrated unequivocally for several bacteria that metabolized 3-[18]OCH_3 benzoate to yield stoichiometric quantities of ^{18}O-containing 3-hydroxybenzoate (12).

Other substituents of aromatics, particularly those with saturated or unsaturated carboxylated side chains, such as phenylpropionate, coumarate, or cinnamate, may be partially or fully removed by the activities of fermentative, acetogenic, or phototrophic bacteria. Examples include the recently described *Acetivibrio multivorans* (52), which grows with cinnamate and forms phenylpropionate, benzoate, and acetate, and the utilization of each of these compounds by *Rhodopseudomonas palustris* (27), which is discussed further below. Aldehydic and carboxylic substituents of methoxylated aromatic acids can also serve as electron acceptors and CO_2 donors in acetogenesis (28, 41).

STRATEGIES FOR INITIATING RING CLEAVAGE

Multiple Hydroxylations

There appear to be two distinct strategies by which hydroxylated aromatic carboxylates are further metabolized to result in ring cleavage. The first is exemplified by the metabolism of gallate (3,4,5-trihydroxybenzoate) by *Eubacterium oxidoreducens*. Here, the initial step involves a reductive decarboxylation, with either formate or hydrogen as a required cosubstrate. The product, pyrogallol, is isomerized during a complex reaction catalyzed by an oxygen-sensitive enzyme to give phloroglucinol (37), and the mechanism of the conversion has been further clarified by studies with *Pelobacter acidigallici* (8). The placement of the hydroxyl groups in this compound permits reduction of the aromatic nucleus by a dehydrogenase requiring an electron donor no more electronegative than NADPH (25). A comparable initial decarboxylation strategy for hydroxylated aromatics appears to be involved in anaerobic metabolism of resorcylic acids (2,4- or 2,6-dihydroxyresorcylic acids) by a *Clostridium* coculture (33, 55), but the decarboxylation product, resorcinol, is directly reduced *in vitro* by an oxygen-sensitive enzyme only in the presence of a much lower-potential electron donor, such as reduced methylviologen.

Thioesterification

The second major route leading to opening and further metabolism of aromatic rings, particularly those containing few substituents, requires activation of the molecule by thioesterification of a carboxyl group with coenzyme A (CoA) as a prelude to ring saturation and further modification by a β-oxidation-like sequence of reactions (see below). Compounds lacking carboxyl groups may be modified either by oxidation of a methyl substituent or direct carboxylation prior to thioesterification. A denitrifying bacterium has been shown to oxidize the methyl group of 4-cresol to carboxylate by means of a methylhydroxylase (7), although the aromatic nucleus was not further metabolized (6). Other cresol isomers were carboxylated directly to form methylbenzoates in a methanogenic consortium (46).

Phenol, with its single hydroxyl substituent, is far less easily reduced directly than are aromatics with multiple hydroxyls, and several recent investigations indicate that its degradation may be initiated by carboxylation rather than insertion of further hydroxyl groups. Transient formation of benzoate during phenol degradation has been detected in a methanogenic consortium of bacteria from a sludge reactor (34), and studies with chlorophenol-degrading bacterial consortia from freshwater sediments have shown that ^{13}C-phenol was converted to ^{13}C-benzoate (65). The conversions undergone by fluorophenols in similar consortia indicate that carboxylation occurs *para* to the hydroxyl group, and is followed by reductive dehydroxylation (50). The role of 4-hydroxybenzoate in the reactions carried out by many of these mixed cultures remains unclear. This compound was degraded by the sludge consortium but did not appear to be an intermediate in the carboxylation reaction (35). In contrast, extracts of several phenol-utilizing denitrifying bacteria were able to carry out a rapid exchange between $^{14}CO_2$ and the carboxyl group of 4-hydroxybenzoate (56), although carboxylation of phenol itself could not be detected. Detailed studies with cell extracts of a versatile denitrifying *Pseudomonas* have recently suggested that phenyl phosphate is the physiological CO_2 acceptor in the carboxylation of phenol (37a). Cell-free extracts of phenol-grown cells formed 4-hydroxybenzoyl-CoA from the free acid, and this compound was converted to benzoyl-CoA by a novel oxygen-labile reductase that required a low-potential reductant such as reduced benzylviologen (23). Studies correlating a number of enzyme activities in extracts of this bacterium grown on different aromatic substrates suggest that formation of 4-hydroxybenzoate followed by thioesterification and reductive dehydroxylation to benzoyl-CoA may also occur during degradation of phenylacetate and 4-hydroxyphenylacetate (11). Aniline degradation carried out by *Desulfobacterium anilini* also involves initial carboxylation and thioesterification, followed by reductive deamination (48). Some of the apparent differences among consortia established from different source materials may result from interspecies transfer of varied intermediates, as well as from the limited number of detailed enzymatic studies carried out with such mixtures. It is increasingly clear, however, that aromatics with simple ring substituents are frequently carboxylated prior to further metabolism, but whether this is an obligatory step in all systems remains uncertain, and the enzymology of the process has not yet been fully elucidated.

PROSPECTS FOR BIOREMEDIATION

Studies employing a number of approaches, including specific enrichment and isolation of bacteria that can grow in monoculture or in defined co-culture at the expense of substituted aromatics in the complete absence of oxygen, have thus shown that natural environments contain microorganisms able to degrade an extremely broad spectrum of aromatic compounds. Among these are many possible degradation products of plant materials containing aromatic components, and also, of course, some unnatural products currently being manufactured and released. Aromatic compounds in fact undergo an almost bewildering array of modifications in anoxic environments, but it must also be stressed that some of these reactions occur only very slowly. The specialized growth conditions required by many of the microorganisms that probably play very significant roles in natural environments may make growing large quantities of cells in the laboratory difficult, and enzymological and genetic studies laborious. Nonetheless, current interest in the potential uses of plant materials as energy sources and concern about contamination of groundwater supplies with toxic aromatics provide strong impetus for attempts to elucidate not only overall pathways but also the enzymology involved. If desirable changes, leading for example to increased rates of breakdown or to broadening the spectrum of aromatic compounds metabolized, are to be introduced into a chosen species, understanding the detailed chemistry of the enzymatic reactions catalyzed, as well as the genetics and regulation of these systems, will be necessary. Overall similarities will probably be found in the biochemistry of pathways used by different physiological types of bacteria that degrade aromatic compounds, so that some general principles will emerge from studies carried out with a limited number of species. Obviously, enzymatic studies are facilitated if pure cultures rather than consortia are used, and regulation can only be understood in such circumstances. The choice of the organism to be used for defining mechanisms of aromatic degradation will also be influenced by factors such as growth rate and metabolic flexibility. Such considerations suggest that denitrifying pseudomonads or photoorganotrophic bacteria are appropriate candidates. Our current understanding of the degradative metabolism of model aromatic acids in anaerobic environments is based in large measure on studies with phototrophic bacteria, and the remainder of this review summarizes past and current work with species that are able to supply all, or almost all, of their carbon needs from aromatic nuclei in the complete absence of oxygen by using light as energy source.

PHOTOTROPHIC BACTERIA THAT USE AROMATIC CARBOXYLATES

Nonsulfur purple bacteria are able to grow in a number of different physiological modes, ranging from photoautotrophic to aerobic chemoheterotrophic (42, 54). The most rapid growth is, however, commonly observed when the cultures are supplied with simple organic acids as the carbon source and incubated anaerobically in light. Since diazotrophy is common, including a fixed nitrogen source in the medium may be unnecessary (43, 44). Utilization of benzoate by some species of these bacteria has been known for many years (54) but is restricted to a small number, notably *Rhodopseudomonas palustris* and some species of *Rhodocyclus*. A strain of *Rhodopseudomonas acidophila* (62) and some strains of *Rhodomicrobium vannielii* have also been shown to use some aromatic compounds (60). These strains were also able to utilize benzyl alcohol for growth, although other phototrophs appear able to utilize only carboxylates. A characteristic feature of the phototrophic metabolism of these bacteria is that the growth substrate is usually very extensively assimilated into cell material; this follows from their use of light as energy source, so eliminating the need for oxidative or fermentative manipulation of a portion of the carbon substrate. Most phototrophic bacteria possess the key enzymes of the reductive pentose phosphate cycle, and are therefore able to use CO_2 as an electron sink when the growth substrate is more reduced than the average cell composition. Several recent investigations have emphasized that the range of aromatic acids metabolized by one member of the purple nonsulfur bacteria, *R. palustris,* may be broader than had previously been appreciated. Anaerobic, illuminated enrichments from a number of sources provided with *trans*-cinnamate rapidly gave rise to cultures of this bacterium (44), while several strains of *R. palustris* were found to utilize a number of different phenolic, dihydroxylated, or methoxylated derivatives, as well as aromatic acids (including cinnamate, coumarate, and phenylpropionate) that have side chains (14a, 27). One isolate also utilized lignin monomers such as caffeate, hydrocaffeate, and ferulate to completion, although other strains gave low growth yields with multiply substituted compounds and evidently were able to attack side chains only, leaving the aromatic nucleus intact. The range of substrates used, and the nature of the products that accumulated in the medium during growth of a mutant strain unable to catabolize benzoate or 4-hydroxybenzoate, suggested that these two compounds may be starting points for two distinct pathways of aromatic ring fission (Fig. 1). A single pathway initiated by reductive dehydroxylation of 4-hydroxybenzoate to benzoate, as suggested by Evans and Fuchs (16) and for a denitrifying bacterium (11, 23), cannot be ruled out at this point. We have been unable to detect benzoate in the medium of cultures growing on 4-hydroxybenzoate, however, and other considerations discussed below also suggest that 4-hydroxybenzoate is metabolized independently of benzoate. A possibly still broader role for strains of *R. palustris* in the degradation of aromatic pollutants has recently been suggested by the demonstration that a new isolate was able to degrade and incorporate most of the carbon of 3-chlorobenzoate into cell material when growing in the presence of benzoate (30). Whether dechlorination preceded further metabolism of the aromatic nucleus was not investigated, but this would be consistent with reductive dechlorinations observed in other anaerobic systems.

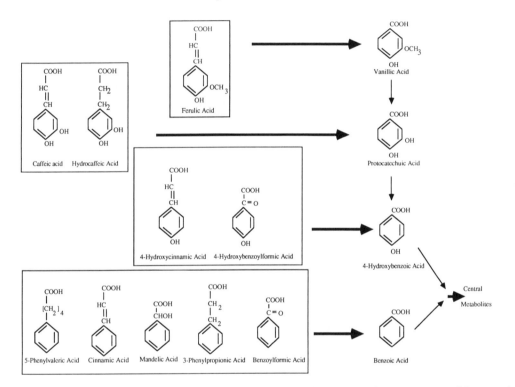

Figure 1. Aromatic compounds utilized by *Rhodopseudomonas palustris* as anaerobic growth substrates. Proposed major routes of aromatic breakdown are indicated by the *arrows*. Reprinted with permission from (27).

DEAROMATIZATION

Classical studies with benzoate-metabolizing cultures of *Rhodopseudomonas palustris* carried out by Dutton and Evans (13, 14) suggested that the aromatic ring was reductively saturated as a prelude to ring opening. Their experiments, which involved isotope trapping by potential intermediates added to cultures photometabolizing ^{14}C-benzoate, clearly showed isotope incorporation into cyclohexenecarboxylate derivatives and into pimelate, thus indicating that the carboxylic group was retained up to ring breakage. The reductive pathway proposed on the basis of these studies was supported by the growth-substrate utilization patterns of a series of mutants of *R. palustris*

(23). Further work in Evans' and others' laboratories suggested that a reductive pathway might function also in aromatic-utilizing bacteria carrying out anaerobic respiration of aromatic compounds (see 4, 15, 51). Additional work suggested that coenzyme A thioesters, rather than free acids, were the true intermediates of the β-oxidation sequence between the cyclohexene stage and ring cleavage, and extracts of benzoate-grown cells were shown to be able to form benzoyl-CoA in an appropriate reaction mix (29). Rapid and primary formation of benzoyl-CoA has also been demonstrated in intact cells (26), thus suggesting that all the enzymatic steps in the pathway probably involve coenzyme A thioesters rather than free acids. A schematic incorporating these and further additions, discussed below, to the original scheme of Dutton and Evans is shown in Figure 2. Thioesterification thus serves a function analogous to the multiple hydroxylations in aerobic or anaerobic systems in destabilizing the aromatic ring. An interesting consequence is that the CoA thioester intermediates in the pathway are unlikely to cross the cytoplasmic membrane, and so be lost to the cell—or, incidentally, be detected by the investigator—except under special conditions. As noted above, carrier trapping was required in the original experiments that led to proposal of the reductive pathway (13). Characterization of intermediates released into the medium

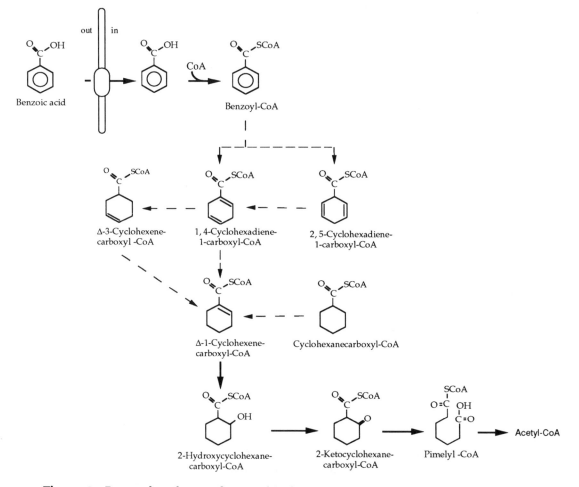

Figure 2. Proposed pathway of anaerobic benzoate metabolism in *Rhodopseudomonas palustris. Solid* arrows indicate established reaction sequences; *dashed* arrows indicate reactions whose sequence remains uncertain.

by mutants of pseudomonads with restricted ability to oxidize aromatic compounds has yielded an enormous amount of information about aerobic degradative pathways, but this approach has so far not been helpful with R. palustris; the culture supernatants of many newly isolated mutants with restricted substrate utilization ranges have been examined by high-performance liquid chromatography (HPLC), but none have contained partial degradation products (J. Gibson, unpublished experiments), except in cultures of mutants that metabolize only the side chains of aromatic compounds, discussed above. The removal of saturated or unsaturated side chains is also likely to involve thioester intermediates, as would be predicted by analogy with other β-oxidations (14a). Thus, it is necessary to postulate that the cells also contain a thioesterase to liberate free benzoate or 4-hydroxybenzoate.

Until recently, it was assumed that the ability of cell extracts to form CoA thioesters from an aromatic acid was itself a strong indication that a strictly anaerobic, reductive saturation of the ring would follow, and this has indeed been assumed for denitrifying bacteria metabolizing 2-fluorobenzoate (47) or 2-aminobenzoate (66). Closer examination of the latter reactions in extracts of a 2-amino benzoate-utilizing denitrifier, however, has revealed an entirely novel, but essentially aerobic, pathway that is also initiated by CoA derivatization. This is followed by a reaction catalyzed by an unusual flavoprotein mixed-function oxygenase/reductase, in which the aromatic ring is both hydroxylated and reduced to yield 2-amino,5-hydroxycyclohex-1-enecarboxyl-CoA (38). This compound is probably further metabolized either aerobically or anaerobically by a β-oxidation sequence similar to the one shown in Figure 2, in which molecular oxygen does not participate. Coenzyme A thioesters have also recently been shown to be intermediates in an aerobic dechlorination sequence converting 4-chlorobenzoate to 4-hydroxybenzoate (49). Participation of CoA thioesters in metabolism of aromatic carboxylates thus clearly extends into aerobic systems.

ENZYMOLOGY

Formation of benzoyl-CoA by crude extracts of Rhodopseudomonas palustris was first reported by Hutber and Ribbons (29). Their experiments, as well as our own, suggest strongly that this involves de novo formation rather than transesterification from another CoA thioester, since neither acetyl- nor succinyl-CoA could replace CoASH in the reaction mixture. It remains possible, however, that a thioester that is itself generated in the proposed pathway, such as pimelyl-CoA (which has not been tested), might serve as donor. A CoA ligase with very restricted substrate specificity was purified from benzoate-grown cells of R. palustris (21); apart from benzoate, only 2-fluorobenzoate, which does not support growth of this bacterium, was thioesterified at high rates and at low concentrations. Neither 4-hydroxybenzoate nor the cyclohexane or cyclohexene derivatives, all good growth substrates, were used. The pure enzyme has a remarkably high affinity for benzoate (K_m about 1 μM), leading to the suggestion that it plays a role not only in activating this substrate prior to reduction but also in maintaining uptake from the extracellular compartment. In short-term experiments, the apparent K_s for benzoate uptake is below 1 μM (26), thus enabling complete uptake of available substrate. This high affinity is consistent with observations that spent culture medium contains substantially less than micromolar benzoate (J. Gibson, unpublished). 4-Hydroxybenzoate is also taken up with comparably high affinity by cell suspensions, and formation of the CoA thioester in whole cells and crude extracts has been demonstrated (45). A CoA ligase that activates 4-hydroxybenzoate has been purified from extracts of cells grown on this substrate (17; J. Gibson et al., manuscript in preparation), and salient characteristics of the two ligases are given in Table 1. Extracts from 4-hydroxybenzoate-grown cells contain both enzymes, which are antigenically unrelated (J. Gibson and G. Fogg, unpublished), and the second CoA ligase activates both 4-hydroxybenzoate and benzoate. It has a much lower affinity for either substrate than the benzoate-specific enzyme, however. Although comparisons between enzyme characteristics in vivo and in vitro must of course be made only cautiously, the high uptake affinity of whole cells for 4-hydroxybenzoate (45) would not appear to be assisted by thioester formation involving this enzyme, and so our inability to detect an elevated intracellular pool of free acid during uptake of 4-hydroxybenzoate remains unexplained.

Table 1. Properties of aromatic coenzyme A ligases from *Rhodopseudomonas palustris*.

	Benzoate:CoA ligase	Aromatic acid:CoA ligase 2
Molecular properties	Monomeric; 61 kDa N-terminal blocked Stable in air	Dimeric; 60 kDa subunits N-terminal blocked Stable in air
	No antigenic crossreactivity	
Catalytic properties	Substrates and apparent K_m Benzoate (1 μM) 2-Fluorobenzoate	Benzoate (120 μM) 4-Hydroxybenzoate (130 μM)
	ATP (2 μM) CoASH (100 μM)	ATP (90 μM) CoASH (400 μM)
Induced by growth on	All aromatic carboxylates	4-Hydroxybenzoate only

Thioesterification of 4-hydroxybenzoate could not be detected in cell-free extracts of 4-hydroxybenzoate-grown *R. palustris* by Hutber and Evans (see 16), and this led them to suggest that 4-hydroxybenzoate was converted to benzoate prior to ring cleavage. The discrepancy between their experiments and ours may possibly be ascribed to inhibitory factors in crude extracts, removed during dialysis or ammonium sulfate precipitation (J. Gibson and S. J. Osmeloski, unpublished observations); a similar phenomenon has been noted with extracts of denitrifying bacteria (67). Neither of the isolated coenzyme A ligase enzymes has any significant activity with the alicyclic compounds in the pathway, which are also thought to react as thioesters rather than free acids and must therefore be activated as an initial reaction when used as carbon source. The bacteria must therefore contain further, as yet uncharacterized, CoA ligases. A comparable multiplicity of CoA ligases active with aromatic acids has been noted in a denitrifying pseudomonad (2, 67).

The least understood portion of the reductive pathway lies between the aromatic acid thioesters and the alicyclic counterparts. Dutton and Evans were unable to observe reproducible dearomatization of benzoyl-CoA in cell-free extracts (13), and our short-term uptake experiments also suggested extreme oxygen sensitivity of enzyme reactions that follow thioester formation (26). Recently, an oxygen-sensitive reduction of benzoyl-CoA has been measured in cell extracts of a denitrifying *Pseudomonas* species (35a). Small quantities of two hexadiene carboxylates that are potential reductive intermediates have been detected in the thioester pools of *Rhodospseudomonas palustris* growing exponentially on benzoate (22), thus suggesting that reduction proceeds by sequential two-equivalent additions to the ring. These findings are incorporated into the scheme shown in Figure 2. In contrast to the findings during benzoate uptake, a large number of radioactive products were formed very rapidly from ^{14}C-4-hydroxybenzoate under both anaerobic and aerobic conditions, although formation of only a single radioactive product, 4-hydroxybenzoyl-CoA, was detected in experiments with extracts (45). It seems probable that more than one pathway for metabolism of 4-hydroxybenzoyl-CoA may exist in these cells, one of which may include an oxygenase/reductase similar to that used in 2-aminobenzoate metabolism (66, 67). The final series of reactions in the pathway, the β-oxidation sequence leading to ring opening to yield pimelate, has so far been studied only in crude extracts (29).

REGULATION

Phototrophic growth on aromatic substrates appears to be necessary for the induction of most, but not all, of the enzymes involved in anaerobic ring activation and opening. Activation of benzoate, and activities associated with the β-oxidation sequence, were found in extracts of benzoate- but not succinate-grown cells, although the latter were able to oxidize octanoate (29). More detailed testing with antisera specific for benzoyl-CoA ligase (32) and the enzyme that also activates 4-hydroxybenzoate (J. Gibson, unpublished) indicated that the former enzyme was present in cells grown with benzoate, 4-hydroxybenzoate, or cyclohexene carboxylates as a carbon source but that the second enzyme was induced only during growth on 4-hydroxybenzoate.

GENETIC AND MOLECULAR APPROACHES

Guyer and Hegeman (24) noted that the chemically induced benzoate-utilization mutants of *Rhodopseudomonas palustris* had significant reversion rates. By using the classical approaches of nitrosoguanidine- or uv-induced mutagenesis followed by ampicillin enrichment, we also have isolated a collection of mutants blocked in growth on benzoate and/or 4-hydroxybenzoate, and have found that our mutants are also very unstable; in fact, the reversion rates are so high as to preclude the use of many of these mutants in biochemical or complementation studies. To circumvent this problem, we have explored possible systems of genetic exchange and the use of transposons in *R. palustris*. Cloning vectors derived from plasmids of the RP4 incompatibility group P-1, for example, pLAFR1 (20) and pRK415 (31), can be transferred and maintained in *R. palustris*. Attempts to mutagenize *R. palustris* directly, however, with the transposons Tn3, Tn5, Tn7, and Tn10 in vectors with P-1 mobilization determinants as delivery systems have met with very limited success. An indirect approach has proved much more effective. This involved mutagenizing an *R. palustris* genomic clone bank in pLAFR1 while in *Escherichia coli* with λ::Tn5, conjugating the plasmids into wild-type *R. palustris*, and then screening kanamycin-resistant colonies for defective growth on aromatic acids (53). The mutants obtained by this procedure are extremely stable, and we initially presumed that this stability derives from introduction of transposon-interrupted DNA fragments into the wild type *R. palustris* chromosome by homologous recombination. We have been unable to detect intact Tn5 insertions in many of these mutants, however, a finding that suggests that additional, or alternative, mutagenic events had occurred.

The mutants obtained by the indirect transposon mutagenesis procedure fell into several classes based on growth on benzoate, 4-hydroxybenzoate, and intermediates of the benzoate pathway. All, however, had the very interesting property (shared by the chemically induced mutants) of being leaky with respect to growth on benzoate and the free acids of proposed alicyclic degradation intermediates of benzoate metabolism. For example, some of the mutants grew anaerobically on benzoate with reproducible doubling times in excess of 100 hours, as opposed to about 15-hour doubling times for wild-type cells. One explanation for this phenotype is that one or more of the reactions involved in benzoate degradation can be catalyzed by more than one enzyme. We know that this is the case for benzoate activation, which, as noted above, can be carried out by at least two different CoA ligases with overlapping substrate specificities.

An alternative explanation for the observed phenotypes is that they reflect defects in regulatory genes required for expression of aromatic degradation enzymes. We have confirmed that this is, in fact, the case by cloning and characterizing a R. palustris gene termed aadR, for anaerobic aromatic degradation regulator (12a). The aadR gene was originally identified by its ability to complement many of our aromatic acid degradation mutants. DNA sequence analysis subsequently showed that aadR has a deduced amino acid sequence that is very similar to members of a family of transcriptional regulators that includes the Fnr and Crp proteins from Escherichia coli, and the FixK and FixK-like proteins from species of the family Rhizobiaceae. We constructed a mutant deleted in aadR, and found that it was unable to grow on 4-hydroxybenzoate under anaerobic conditions, and that it grew slowly with benzoate. In addition, immunoblot analyses revealed that the aadR mutant expressed only trace amounts of aromatic acid-CoA ligase 2. It was also defective in 4-hydroxybenzoate-induced expression of benzoate-CoA ligase, although benzoate-grown cells expressed normal levels of this enzyme. These results indicate that AadR functions as a transcriptional activator of anaerobic aromatic acid degradation. The AadR protein appears to regulate the synthesis of benzoate-CoA ligase and aromatic acid-CoA ligase 2, enzymes that catalyze the initial reactions in the degradation of benzoate and 4-hydroxybenzoate, respectively. This protein probably regulates the expression of additional benzoate degradation genes, as well. The aadR mutant did not seem to be affected in aspects of anaerobic growth other than aromatic acid degradation. The mutant grew normally on nonaromatic carbon sources and also under dinitrogen-fixing conditions.

The molecular tools that were developed for use in R. palustris as a consequence of work with the aadR gene should be very useful in future work aimed at identifying structural genes involved in anaerobic aromatic carboxylic acid degradation.

Information necessary for anaerobic benzoate degradation in Alcaligenes xylosoxidans subsp. denitrificans, a denitrifying organism previously identified as Pseudomonas PN1, was found to be carried on a 17.4 kbp plasmid (5). The genes for a CoA ligase and the oxygenase/reductase involved in the novel aerobic 2-aminobenzoate utilization pathway mentioned above are carried on a small (9.1 kbp) plasmid in another denitrifying bacterium (1). Plasmid-borne genes are also extensively involved in many aerobic pathways of aromatic degradation (18). We have screened four independently isolated strains of Rhodopseudomonas palustris for the presence of indigenous plasmids but to date can see no correlation between plasmids and aromatic utilization; indeed, although some of the strains harbor a plasmid of approximately 10 kbp, the strain that is most versatile metabolically lacks any detectable plasmid DNA. In addition, the plasmid profiles of aromatic acid utilization mutants do not appear to be altered from those of the wild-type parent.

WHAT LIMITS THE RATE OF DEGRADATION
OF AROMATIC CARBOXYLATES?

Greater understanding of the enzymes involved in the degradation of aromatic acids, and of their regulation *in vivo*, is of course needed before rational approaches for introducing desirable modifications in the range or rates of degradation of aromatics can be devised. As has been shown in many studies with heterotrophic bacteria, rate-limiting steps in substrate utilization may in practice occur at several stages, including initial entry into cells, activation, availability of inducer, or at any point in metabolism subject to product inhibition. Understanding of aromatic acid metabolism is still too fragmentary to do more than speculate about which stages may be most important.

There is no clear evidence for the involvement of specific transport systems for aromatic acids; undissociated benzoic and 4-hydroxybenzoic acids are assumed to be quite freely permeable, and have indeed been quite extensively used as permeant acids for determining intracellular pH in heterotrophic bacteria. Actual permeabilities have not, however, been measured, and it remains possible that entry into the cells may impose some rate limitation, particularly in the somewhat alkaline environments in which many phototrophs flourish and in which most of the aromatic acids would be ionized. We have been unable to detect any pH dependence of uptake affinities, so it remains possible that *R. palustris* may possess a system for facilitating aromatic anion movements; isolation of mutants would provide the clearest evidence for this possibility. However, experiments in which cells have been rapidly separated from their medium either by filtration or centrifugation through silicone oil have indicated that total intracellular concentrations of the unmodified acids were no greater than that found in the extracellular environment (26).

Although activation to yield CoA thioesters appears to be the first step following entry, very different pictures were observed with the two substrates most often used in our experiments. With benzoate, benzoyl-CoA was found to be the dominant product in small-scale experiments (26). In whole cultures, acidified rapidly and without other perturbation during exponential growth, benzoate accounted for more than 90% of the aromatic or alicyclic acid products recovered from CoA thioester pools by alkaline hydrolysis (22), and the cyclohexadiene reduction products recovered were present at less than 1% of benzoate. These experiments suggest that the reductive phase of benzoate metabolism is rate-limiting, but attempts to supplement intracellular reductant by additions to the suspending medium have not been successful. During ^{14}C-4-hydroxybenzoate uptake, on the other hand, it proved difficult to demonstrate more than trace and very transient formation of the corresponding CoA thioester. Even after a few seconds, other, as yet unidentified, compounds with mobilities characteristic of CoA thioesters in thin layer chromatograms were present in much larger quantities (45). These experiments again point to major differences between the metabolism of these two aromatic acids in our strains of *R. palustris*. The rate of metabolism of the aromatic carboxylates may also be restricted by the synthesis or pool size of coenzyme A, which could obviously limit intracellular concentrations not only of the thioester of the substrate supplied, but also of the degradative intermediates.

THE FUTURE

Much work obviously lies ahead, but there seems very good reason to expect that studies with phototrophic bacteria will continue to provide useful models for understanding both the route(s) and regulation of degradation of aromatic products, whether natural or man-made, and will suggest how bacterial systems may be modified for specific purposes.

Acknowledgments

Unpublished work from our laboratories was supported in part by Grant DE-FG02-86ER13495 from the Department of Energy, Division of Energy Biosciences to J. G. and DAALO3-89-K-0121 from the US Army Research Office (co-funded by the Department of Energy, Division of Energy Biosciences) to C.S.H.

LITERATURE CITED

1. Altenschmidt, U., C. Eckerskorn, and G. Fuchs. 1990. Evidence that enzymes of a novel aerobic 2-aminobenzoate metabolism in a denitrifying *Pseudomonas* are coded on a small plasmid. *Eur. J. Biochem.* 194:647–653.

2. Altenschmidt, U., B. Oswald, and G. Fuchs. 1991. Purification and characterization of benzoate-coenzyme A ligase and 2-aminobenzoate coenzyme A ligases from a denitrifying *Pseudomonas* sp. *J. Bacteriol.* 173:5494–5501.

3. Bache, R., and N. Pfennig. 1981. Selective isolation of *Acetobacterium woodii* on methoxylated aromatic acids and determination of growth yields. *Arch. Microbiol.* 130:255–261.

3a. Berman, M., and A. C. Frazer. 1992. Importance of tetrahydrofolate and ATP in the anaerobic O-demethylation reaction for phenylmethylethers. *Appl. Environ. Microbiol.* 58:925–931.

4. Berry, D. F., A. J. Francis, and J.-M. Bollag. 1987. Microbial metabolism of homocyclic and heterocyclic compounds under anaerobic conditions. *Microbiol. Rev.* 51:43–59.

5. Blake, C. K., and G. D. Hegeman. 1987. Plasmid pCB1 carries genes for anaerobic benzoate catabolism in *Alcaligenes xylosoxidans* subsp. *denitrificans* PN-1. *J. Bacteriol.* 169:4878-4883.

6. Bossert, I. D., and L. Y. Young. 1986. Anaerobic oxidation of p-cresol by a denitrifying bacterium. *Appl. Environ. Microbiol.* 52:1117–1122.

7. Bossert, I. D., G. Whited, D. T. Gibson, and L. Y. Young. 1989. Anaerobic oxidation of p-cresol mediated by a partially purified methylhydroxylase from a denitrifying bacterium. *J. Bacteriol.* 171:2956–2962.

8. Brune, A., and B. Schink. 1990 Pyrogallol-to-phloroglucinol conversion and other hydroxyl-transfer reactions catalyzed by cell extracts of *Pelobacter acidigallici*. *J. Bacteriol.* 172:1070-1076.

9. Colberg, P. J. 1988. Anaerobic microbial degradation of cellulose, lignin, oligolignols and monoaromatic lignin derivatives. In *Biology of Anaerobic Microorganisms*. Ed. A. J. B. Zehnder. New York: John Wiley and Sons. 333–372.

10. Crawford, R. L. 1981. *Lignin Biodegradation and Transformation*. New York: Wiley-Interscience.

11. Dangel, W., R. Brackmann, A. Lack, M. Mohamed, J. Koch, B. Oswald, B. Seyfried, A. Tschech, and G. Fuchs. 1991. Differential expression of enzyme activities initiating anoxic metabolism of various aromatic compounds via benzoyl-CoA. *Arch. Microbiol.* 155:256–262.

12. DeWeerd, K. A., A. Saxena, D. P. Nagle, and J. M. Suflita. 1988. Metabolism of the ^{18}O-methoxy substituent of 3-methoxybenzoic acid and other unlabelled methoxybenzoic acids by anaerobic bacteria. *Appl. Environ. Microbiol.* 54:1237–1242.

12a. Dispensa, M., C. T. Thomas, M.-K. Kim, J. A. Perrotta, J. Gibson, and C. S. Harwood. 1992. Anaerobic growth of *Rhodopseudomonas palustris* is dependent on AadR, a member of the cyclic AMP receptor protein family of transcriptional regulators *J. Bacteriol.* 174:5803–5813.

13. Dutton, P. L., and W. C. Evans. 1969. The metabolism of aromatic compounds by *Rhodopseudomonas palustris*. *Biochem. J.* 113:525–535.

14. Dutton, P. L., and W. C. Evans. 1978. Metabolism of aromatic compounds by Rhodospirillaceae. In *The Photosynthetic Bacteria*. Ed. R. K. Clayton, W. R. Sistrom. New York: Plenum Press. 719–726.

14a. Elder, D. J. E., P. Morgan, and D. J. Kelly. 1992. Anaerobic degradation of *trans*-cinnamate and ω-phenylalkane carboxylic acids by the photosynthetic bacterium *Rhodopseudomonas palustris*: evidence for a β-oxidation mechanism. *Arch. Microbiol.* 157:148–154.

15. Evans, W. C. 1977. Biochemistry of the bacterial catabolism of aromatic compounds in anaerobic environments. *Nature* 270:17–22.

16. Evans, W. C., and G. Fuchs. 1988. Anaerobic degradation or aromatic compounds. *Annu. Rev. Microbiol.* 42:289–317.

17. Fogg, G. C., and J. Gibson. 1990. 4-Hydroxybenzoate-coenzyme A ligase from *Rhodopseudomonas palustris*. *Am. Soc. Microbiol. 90th Annual Meeting.* K-137 (Abstract).

18. Franz, B., and A. M. Chakrabarty. 1986. Degradative plasmids in *Pseudomonas*. In The Bacteria, Vol. 10: *The Biology of Pseudomonas*. Ed. I. C. Gunsalus, J. R. Sokatch, L. N. Ornston. Orlando, FL: Academic Press. 295–323.

19. Frazer, A. C., and L. Y. Young. 1985. A gram negative anaerobic bacterium that utilizes O-methyl substituents of aromatic acids. *Appl. Environ. Microbiol.* 49:1345–1347.

20. Friedman, A. M., S. R. Long, S. E. Brown, W. J. Buikema, and F. M. Ausubel. 1982 Construction of a broad host range cosmid cloning vector and its use in the genetic analysis of *Rhizobium* mutants. *Gene* 18:289–296.

21. Geissler, J. F., C. S. Harwood, and J. Gibson. 1988. Purification and properties of benzoate-coenzymeA ligase, a *Rhodopseudomonas palustris* enzyme involved in the degradation of benzoate. *J. Bacteriol.* 170:1709–1714.

22. Gibson, K. J., and J. Gibson. 1992. Potential early intermediates in anaerobic benzoate degradation by *Rhodopseudomonas palustris*. *Appl. Environ. Microbiol.* 58:696–698.

23. Glöckler, R., A. Tschech, and G. Fuchs. 1989. Reductive dehydroxylation of 4-hydroxybenzoyl-CoA to benzoyl-CoA in a denitrifying, phenol-degrading *Pseudomonas* species. *FEBS Lett.* 251:237–240.

24. Guyer, M., and G. Hegeman. 1969. Evidence for a reductive pathway for the anaerobic metabolism of benzoate. *J. Bacteriol.* 99:906–907.

25. Haddock, J. D., and J. G. Ferry. 1989. Purification and properties of phloroglucinol reductase from *Eubacterium oxidoreducens*. *J. Biol. Chem.* 264:4423–4427.

26. Harwood, C. S., and J. Gibson. 1986. Uptake of benzoate by *Rhodopseudomonas palustris* grown anaerobically in light. *J. Bacteriol.* 165:504–509.

27. Harwood, C. S., and J. Gibson. 1988. Anaerobic and aerobic metabolism of diverse aromatic compounds by the photosynthetic bacterium *Rhodopseudomonas palustris*. *Appl. Environ. Microbiol.* 54:712–717.

28. Hsu, T., S. L. Daniel, M. F. Lux, and H. L. Drake. 1990. Biotransformations of carboxylated aromatic compounds by the acetogen *Clostridium thermoaceticum*: generation of growth-supportive CO_2 equivalents under CO_2-limited conditions. *J. Bacteriol.* 172:212–217.

29. Hutber, G. N., and D. W. Ribbons. 1983. Involvement of coenzyme A esters in the metabolism of benzoate and cyclohexanecarboxylate by *Rhodopseudomonas palustris*. *J. Gen. Microbiol.* 129:2413–2420.

30. Kamal, V. S., and R. C. Wyndham. 1990. Anaerobic phototrophic metabolism of 3-chlorobenzoate by *Rhodopseudomonas palustris* WS17. *Appl. Environ. Microbiol.* 56:3871–3873.

31. Keen, N. T., S. Tamaki, D. Kobayashi, and D. Trollinger. 1988. Improved broad-host-range plasmids for DNA cloning in gram-negative bacteria. *Gene* 70:191–197.

32. Kim, M.-K., and C. S. Harwood. 1991. Regulation of benzoate-coenzyme A ligase in *Rhodopseudomonas palustris*. *FEMS Microbiol. Lett.* 83:199–204

33. Kluge, C., A. Tschech, and G. Fuchs. 1990. Anaerobic metabolism of resorcyclic acids (m-dihydroxybenzoic acids) and resorcinol (1,3-benzenediol) in a fermenting and in a denitrifying bacterium. *Arch. Microbiol.* 155:68–74.

34. Knoll, G., and J. Winter. 1987. Anaerobic degradation of phenol in sewage sludge. Benzoate formation from phenol and CO_2 in the presence of hydrogen. *Appl. Microbiol. Biotechnol.* 25:384-391.

35. Knoll, G., and J. Winter. 1989. Degradation of phenol via carboxylation to benzoate by a defined, obligate syntrophic consortium of anaerobic bacteria. *Appl. Microbiol. Biotechnol.* 30:318–324.

35a. Koch, J., and G. Fuchs. 1992. Enzymatic reduction of benzoyl-CoA to alicyclic compounds, a key reaction in anaerobic aromatic metabolism. *Eur. J. Biochem.* 205:1915–202.

36. Krumholtz, L. R., and M. P. Bryant. 1986. *Syntrophococcus sucromutans* sp. nov. gen. nov. uses carbohydrates as electron donors and formate, methoxymonobenzenoids or *Methanobrevibacter* as electron acceptor systems. *Arch. Microbiol.* 143:313–318.

37. Krumholz, L. R., R. L. Crawford, M. E. Hemling, and M. P. Bryant. 1987. Metabolism of gallate and phloroglucinol in *Eubacterium oxidoreducens* via 3-hydroxy-5-oxohexanoate. *J. Bacteriol.* 169:1886–1890.

37a. Lack, A., and G. Fuchs. 1992. Carboxylation of phenyl phosphate by phenol carboxylase, an enzyme system of anaerobic phenol metabolism. *J. Bacteriol.* 174:3629–3636.

38. Langkau, B., S. Ghisla, R. Buder, K. Ziegler, and G. Fuchs. 1990. 2-Aminobenzoyl-CoA monooxygenase/reductase, a novel type of flavoenzyme; Identification of the reaction products. *Eur. J. Biochem.* 191:365–371.

39. Lovley, D. R., M. J. Baedeker, D. R. Lonergan, I. M. Cozzarelli, E. J. P. Phillips, and D. I. Siegel. 1989. Oxidation of aromatic contaminants coupled to microbial iron oxidation. *Nature* 399:297–300.

40. Lovley, D. R., and D. J. Lonergan. 1990. Anaerobic oxidation of toluene, phenol and p-cresol by the dissimilatory iron-reducing organism, GS-15. *Appl. Environ. Microbiol.* 56:1858-1864.

41. Lux, M. F., E. Keith, T. Hsu, and H. L. Drake. 1990. Biotransformation of aromatic aldehydes by acetogenic bacteria. *FEMS Microbiol. Lett.* 67:73–78.

42. Madigan, M. T. 1988. Microbiology, physiology and ecology of phototrophic bacteria. In *Biology of Anaerobic Microorganisms.* Ed. A. J. B. Zehnder. New York: John Wiley and Sons. 39–111.

43. Madigan, M. T., S. S. Cox, and R. A. Stegeman. 1984. Nitrogen fixation and nitrogenase activities in members of the family Rhodospirillaceae. *J. Bacteriol.* 157:73–78.

44. Madigan, M. T., and H. Gest. 1988. Selective enrichment and isolation of *Rhodopseudomonas palustris* using *trans*-cinnamate as sole carbon source. *FEMS Microbiol. Ecol.* 53:53–38.

45. Merkel, S. M., A. E. Eberhard, J. Gibson, and C. S. Harwood. 1989. Involvement of coenzyme A thioesters in anaerobic metabolism of 4-hydroxybenzoate by *Rhodopseudomonas palustris*. *J. Bacteriol.* 171:1–7.

46. Roberts, D. J., P. M. Fedorak, and S. E. Hrudey. 1990. CO_2 incorporation and 4-hydroxy-2-methylbenzoic acid formation during anaerobic metabolism of m-cresol by a methanogenic consortium. *Appl. Environ. Microbiol.* 56:472–478.

47. Schennen, U., K. Braun, and H. J. Knackmuss. 1985. Anaerobic degradation of 2-fluorobenzoate by benzoate-degrading bacteria. *J. Bacteriol.* 161:321–325.

48. Schnell, S., and B. Schink. 1991. Anaerobic aniline degradation via reductive deamination

of 4-aminobenzoyl-CoA in *Desulfobacterium anilini*. *Arch. Microbiol.* 155:183–190.

49. Scholten, J. D., K.-H. Chang, P. C. Babbitt, H. Charest, M. Sylvestre, and D. Dunaway-Mariano. 1991. Novel enzymatic hydrolytic dehydrogenation of a chlorinated aromatic. *Science* 253:182–185.

50. Sharak Genthner, B. R., G. T. Townsend, and P. J. Chapman. 1989. Anaerobic transformations of phenol to benzoate via para-carboxylation: use of fluorinated analogues to elucidate the mechanism of transformation. *Biochem. Biophys. Res. Commun.* 162:945–951.

51. Sleat, R., and J. P. Robinson. 1984. The bacteriology of anaerobic degradation of aromatic compounds. *J. Appl. Bacteriol.* 57:381–394.

52. Tanaka, K., K. Nakamura, and E. Mikami. 1991. Fermentation of cinnamate by a mesophilic strict anaerobe, *Acetivibrio multivorans* sp. nov. *Arch. Microbiol.* 155:120–124.

53. Thomas, C., M. Dispensa, C. S. Harwood, and J. Gibson. 1990. Molecular analysis of anaerobic aromatic degradation by *Rhodopseudomonas palustris*. *Am. Soc. Microbiol. 90th Annual Meeting.* K-136 (Abstract).

54. Trüper H. G., and N. Pfennig. 1981. Characterization and identification of the anoxygenic phototrophic bacteria. In *The Prokaryotes,* Vol 1. Ed. M. P. Starr, H. Stolp, H. G. Trüper, A. Balows, H. G. Schlegel. Berlin: Springer-Verlag. 299–312.

55. Tschech, A., and B. Schink. 1985. Fermentative degradation of resorcinol and resorcyclic acids. *Arch. Microbiol.* 143:52–59.

56. Tschech, A., and G. Fuchs. 1989. Anaerobic degradation of phenol via carboxylation to 4-hydroxybenzoate: *in vitro* study of isotope exchange between $^{14}CO_2$ and 4-hydroxybenzoate. *Arch. Microbiol.* 152:594–599.

57. Widdel, F. 1988. Microbiology and ecology of sulfate- and sulfur-reducing bacteria. In *Biology of Anaerobic Microorganisms.* Ed. A. J. B. Zehnder. New York: John Wiley and Sons. 469–585.

58. Widdel, F., and N. Pfennig. 1984. Dissimilatory sulfate-or sulfur-reducing bacteria. In *Bergey's Manual of Systematic Bacteriology.* Vol 1. Ed. N. R. Krieg, J. G. Holt. Baltimore, MD: Williams and Wilkins. 663–678.

59. Williams, R. J., and W. C. Evans. 1975. The metabolism of benzoate by *Moraxella* species through anaerobic nitrate respiration. *Biochem. J.* 148:1–10.

60. Wright, G. E., and M. T. Madigan. 1991. Photocatabolism of aromatic compounds by the phototrophic purple bacterium *Rhodomicrobium vannielii*. *Appl. Environ. Microbiol.* 57:2069-2073.

61. Wu, Z., S. L. Daniel, T. Hsu, and H. L. Drake. 1988. Characterization of a CO-dependent O-demethylation system from the acetogen *Clostridium thermoaceticum*. *J. Bacteriol.* 170:5747-5750.

62. Yamanaka, K., M. Moriyama, R. Minoshima, and Y. Tsuyuki. 1983. Isolation and characterization of a methanol-utilizing phototrophic bacterium *Rhodopseudomonas acidophila* M402 and its growth on vanillin derivatives. *Agric. Biol. Chem.* 47:1257–1267

63. Young, L. Y. 1984. Anaerobic degradation of aromatic compounds. In *Microbial Degradation of Organic Compounds.* Ed. D. T. Gibson. New York: Dekker. 487–523.

64. Young, L. Y., and A. C. Frazer. 1987. The fate of lignin and lignin-derived compounds in anaerobic environments. *Geomicrobiol. J.* 5:261–293.

65. Zhang, X., T. V. Morgan, and J. Wiegel. 1990. Conversion of ^{13}C-1-phenol to ^{13}C-4-benzoate, an intermediate step in the anaerobic degradation of chlorophenols. *FEMS Microbiol. Lett.* 67:63–66.

66. Ziegler, K., R. Buder, J. Winter, and G. Fuchs. 1989. Activation of aromatic acids and aerobic 2-aminobenzoate metabolism in a denitrifying *Pseudomonas* strain. *Arch. Microbiol.* 151:171–176.

67. Ziegler, K., K. Braun, A. Böckler, and G. Fuchs. 1987. Studies on the anaerobic degradation of benzoic acid and 4-aminobenzoic acid by a denitrifying *Pseudomonas* strain. *Arch. Microbiol.* 149:62–69.

CHAPTER 15

Biodegradation of Trichloroethylene in Methanotrophic Systems and Implications for Process Applications

SUSAN M. HENRY and DUNJA GRBIĆ-GALIĆ

Abstract. The laboratory research published to date has demonstrated conclusively that methane-oxidizing bacteria (methanotrophs) are capable of transforming trichloroethylene (TCE) and other chlorinated aliphatics. A field study has demonstrated that *in situ* biotransformation of chlorinated aliphatics can be achieved upon subsurface enrichment with methane and oxygen. A significant conclusion of all of these investigations is that TCE transformation rates vary considerably, depending not only upon the innate capability of the methanotrophic culture but, more importantly, on the growth conditions and those conditions imposed during, and as a consequence of, TCE transformation. In this work, mixed and pure methanotrophic cultures were enriched and isolated from groundwater and soils from an aquifer underlying the Moffett Field Naval Air Base, Mountain View, California. These cultures were evaluated for their TCE transformation capabilities by radiolabeling experiments and by gas chromatography, and have been examined by electron microscopy. Rate studies have been conducted, and kinetic coefficients for TCE, methane, and carbon monoxide oxidation have been determined. One of the objectives of this work has been to characterize the effect of factors that may influence TCE transformation in process applications. This chapter discusses the influence of methane and oxygen concentration, reductant availability, mineral nutrient bioavailability, TCE oxidation toxicity, and the TCE transformation intermediate carbon monoxide on TCE tranformation by methanotrophs.

INTRODUCTION

Methanotrophs are aerobic organisms capable of growing on methane as a sole carbon and energy source (7, 8, 67, 68, 70). The combining form "-troph" is derived from the Greek "-trophia" meaning "to nourish" (111). Methanotrophs are "nourished" by methane. Methanotrophic bacteria are currently in the spotlight because they are capable of transforming a class of fairly recalcitrant compounds, chlorinated aliphatics, including trichloroethylene (TCE). Chlorinated aliphatics are manufactured for use as industrial solvents for degreasing metals and electronic components, for dry

---

Dr. Henry is at Levine-Fricke, 1900 Powell Street, 12th floor, Emeryville, CA 94608, U.S.A.

cleaning, and in paint strippers. They are present in the majority of the contaminated aquifers throughout the nation (172). These compounds are toxic and some are known or suspected carcinogens and mutagens (17, 62, 74, 99, 133, 158–161, 163). Under the anaerobic conditions that prevail in many contaminated aquifers (18, 92), the transformation of chlorinated ethylenes can result in the production of vinyl chloride (VC) (20, 119, 166, 167), a proven mammalian carcinogen and mutagen (99, 126, 161). Under aerobic conditions, however, the reductive dechlorination of chlorinated aliphatics does not occur. Aerobic biotreatment of petroleum hydrocarbon contamination has proven to be a successful remediation method for contaminated aquifers and soils (12, 75, 92, 93, 129, 130). There is currently a strong interest in developing similar approaches to cleaning up chlorinated aliphatic-contaminated soils, groundwater, and industrial wastes (92, 125).

For many years, the prevailing dogma maintained that chlorinated aliphatics were refractory in aerobic environments and could only be degraded by reductive mechanisms under anaerobic conditions. In a 1985 publication, Wilson and Wilson first described the aerobic degradation of TCE in soil enriched with natural gas (176). These important findings launched a flurry of investigation. Many academic, government, and industrial laboratories throughout the world are now studying the aerobic biotransformation of chlorinated aliphatics in earnest. The microorganisms that have the demonstrated capability to degrade chlorinated ethanes and ethylenes include not only methanotrophs (2–5, 22, 23, 35, 41, 43, 57–61, 65, 66, 76, 77, 80, 86, 87, 96, 100, 102, 107, 112, 113, 115, 116, 118, 121, 122, 131, 135–137, 139, 148–150, 156, 157, 176), but also ammonia oxidizers (9, 127, 128, 164), ethylene oxidizers (54, 58), propane oxidizers (40, 121, 122, 168), isoprene-utilizers (37), heterotrophs that grow on aromatic compounds (42, 52, 108–110, 138, 162, 169, 171), and recombinant *Escherichia coli* (177, 181). These microorganisms are able to transform chlorinated aliphatics because they possess special oxidative enzymes, catabolic oxygenases, that incorporate oxygen into organic substrates (178). The transformations are cometabolic, in that the active microorganism is unable to grow on the chlorinated solvent. [Exceptions to this are reports of a *Mycobacterium* sp. growing on vinyl chloride (13), and different strains of soil bacteria growing on 1,2-dichloroethane (74, 151)]. In pure cultures transformation intermediates generally accumulate (58, 59, 61, 76, 96, 116), whereas in mixed cultures and soil systems that contain the active microorganisms and common soil heterotrophs, the compounds are completely mineralized (2, 40, 58, 59, 65, 76, 86, 87, 96, 100).

Most research to date concerning aerobic degradation of chlorinated solvents has focused on TCE transformation by methanotrophs. Methanotrophic transformation of TCE has been demonstrated in purified enzyme systems (43, 50), pure cultures (23, 35, 57, 59, 60, 61, 76, 80, 96, 115, 116, 118, 156), suspended mixed cultures (2, 3, 22, 41, 57–60, 65, 76, 77, 96, 118, 148, 157), fixed-film reactors (113, 118, 121, 122, 149, 150), soil systems containing aquifer material (66, 86, 87, 100, 112, 176), and *in situ* aquifer restoration field studies (102, 131, 135–137). The oxygenase in methanotrophs that transforms TCE is the methane monooxygenase enzyme system (MMO), the same enzyme that enables methanotrophs to grow on methane. Like all monooxygenases, the MMO requires for its oxidative function both molecular oxygen and a source of reductant (electrons) (8, 44, 70, 178). The MMO transforms compounds by oxygenating them—inserting an atom of molecular oxygen into the compound—while the other atom of oxygen is reduced to water (Fig. 1) (8, 44, 70). The MMO has relaxed specificity, having demonstrated activity toward a broad range of compounds that are not used as growth substrates by the methanotrophs. These compounds include alkanes, alkenes, cyclic and aromatic compounds, carbon monoxide (CO), and ammonia (13, 25, 38, 39, 44, 49, 69, 70, 120, 145, 147, 155). Recent research has demonstrated that methanotrophs can also transform recalcitrant *ortho*-chlorinated mono- and di-chlorobiphenyls (1) and linear alkylbenzene sulfonates (72). The MMO is known to exist in two forms: a membrane-associated (particulate) form, and a cytoplasm-associated (soluble) form (8, 24, 30, 155). All methanotrophs express a particulate MMO (8, 24, 30, 155). A few, most notably *Methylosinus trichosporium* OB3b and *Methylococcus capsulatus* (Bath), express the soluble form under conditions of copper limitation (24, 30, 124, 142). The soluble form of the MMO in these methanotrophs exhibits a broader specificity than their particulate form (8, 24, 25). Because only the soluble form in these two methanotrophs can oxidize TCE (23, 116, 156), it has been suggested that the ability to transform TCE in all methanotrophs is restricted to the soluble MMO; however, methanotrophs that express only a particulate form of the MMO have been shown to transform TCE at rates comparable to those expressing the soluble form (35, 59, 60, 80).

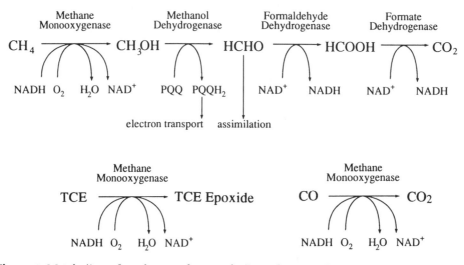

Figure 1. Metabolism of methane and cometabolism of TCE and CO by methanotrophs. Oxidation of compounds by the MMO requires molecular oxygen and reducing power (Modified from a figure provided by courtesy of B.G. Fox [44]).Metabolism of methane regenerates reducing power but cometabolism of TCE and CO does not.

A field study conducted by Roberts and colleagues of Stanford University at a test site located on the Moffett Field Naval Air Base, Mountain View, California confirmed that native methanotrophic bacterial populations present in the aquifer were capable of degrading chlorinated ethylenes, including TCE (131, 135–137). In conjunction with this study, TCE-transforming methanotrophic mixed and pure cultures have been enriched and isolated from the Moffett Field groundwater aquifer (57–61). These cultures have been evaluated by radiolabeling experiments and by gas chromatography for their capability to transform TCE, and have been examined by electron microscopy. Rate studies have been conducted, and kinetic coefficients for TCE oxidation, methane oxidation, and carbon monoxide oxidation have been determined. One of the objectives of this work has been to characterize the effect of factors that may be considered operational parameters in process applications. This review discusses the influence of mineral medium formulation, methane concentration, reductant availability, oxygen concentration, TCE oxidation toxicity, and transformation intermediates on TCE transformation by mixed and pure methanotrophic cultures.

METHANOTROPHIC CULTURES FROM
A GROUNDWATER AQUIFER

Description of Cultures

Three mixed cultures were obtained by enrichment with methane from the Moffett Field groundwater aquifer (58, 59): mixed culture MM1, enriched from aquifer solids; mixed culture MM2, from the effluent of a methane-enriched, TCE-transforming soil column (86, 87); and mixed culture MM3, from groundwater. The cultures were different, based on colony morphology and microscopic appearance. Mixed culture MM1, a stable consortium consisting of one methanotroph and three or four heterotrophs, contained predominantly Gram-negative pleomorphic coccobacilli and prosthecates, as well as some Gram-negative bacilli and cocci (Fig. 2a). The methanotroph was a pleomorphic coccobacillus that contained the internal membrane structures characteristic of type II methanotrophs [paired membranes inside the periphery of the cell (30, 65)]. Extracts of mixed culture MM1 were tested with an antibody specific to the soluble MMO of *Methylosinus trichosporium* OB3b and they cross-reacted with it (60). A pleomorphic type II methane–oxidizing bacterium, isolate CSC-1, was isolated from mixed culture MM1 (Figs. 2b and 3a). Mixed culture MM3 contained primarily Gram-negative pleomorphic coccobacilli and bacilli. This mixed

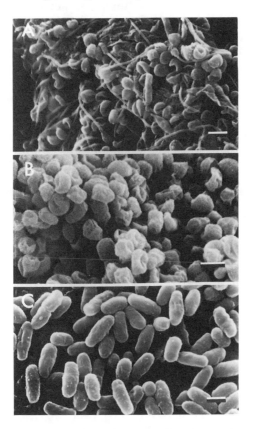

Figure 2. Scanning electron micrographs of TCE-transforming methanotrophic cultures from a groundwater aquifer. (A) Mixed culture MM1; (B) Isolate CSC-1; (C) *Methylomonas* sp. strain MM2. Bars, 1 μm.

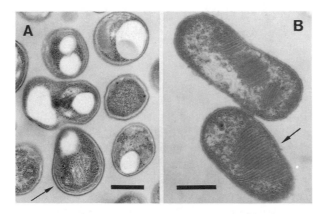

Figure 3. Transmission electron micrographs of TCE-transforming methanotrophic cultures from a groundwater aquifer. (A) Isolate CSC-1. Note type II intracytoplasmic membranes. Lipid storage granules are light-colored inclusions inside cells. (B) *Methylomonas* sp. strain MM2. Note type I intracytoplasmic membranes. Bars, 0.5 μm.

culture also contained pleomorphic type II methanotrophs. Mixed culture MM2 contained predominantly Gram-negative pleomorphic coccobacilli and motile bacilli, as well as some yeasts. Two different methanotrophs were present in mixed culture MM2, a pleomorphic type II methanotroph, and a rod-shaped methanotroph that contained the stacked internal membrane structures characteristic of type I methanotrophs (32, 68). A pure culture, *Methylomonas* sp. strain MM2, was isolated from mixed culture MM2 (Fig. 2c and 3b). The cultures were grown in continuously stirred reactors under a continuous stream of 30 to 35 percent methane (vol/vol) in air for the TCE transformation experiments (58–61). Under these growth conditions, isolate CSC-1 and methanotrophs in mixed cultures MM1 and MM3 possessed lipid inclusions (Fig. 3a), whereas *Methylomonas* sp. strain MM2 and the methanotrophs in mixed culture MM2 did not.

Methylomonas sp. strain MM2 is a pink-pigmented Gram-negative motile bacillus, approximately 0.5×1.5 μm (Fig. 2c), that contains type I internal membranes under all growth conditions (Fig. 3b). *Methylomonas* sp. strain MM2 grows well on both methane and methanol and has expressed no requirement for carbon sources other than the growth substrate. The bacterium incorporates cell carbon by way of the ribulose monophosphate pathway of carbon assimilation (59). The methane oxidation rate and growth yield of *Methylomonas* sp. strain MM2 grown on Whittenbury mineral medium (59, 173) was 1.3 mg methane mg cells (dry weight)$^{-1}$ day^{-1}, and 0.70 mg cells (dry weight) mg methane^{-1}, respectively. Under all growth conditions applied in our studies, *Methylomonas* sp. strain MM2 expressed only a membrane-associated (particulate) MMO (59, 60). Extracts of *Methylomonas* sp. strain MM2 did not cross-react with antibody specific to the soluble MMO of *Methylosinus trichosporium* OB3b (60), a finding that is characteristic of *Methylomonas* sp. (7, 179). The enzyme that is responsible for TCE transformation in *Methylomonas* sp. strain MM2 is a particulate MMO.

Summary of TCE Transformation Capabilities

All of the methanotrophic cultures enriched from the Moffett aquifer transformed TCE. Studies with radiolabeled (^{14}C) TCE confirmed that the TCE was biologically transformed and mineralized to carbon dioxide. In the mixed cultures, the TCE was almost completely degraded, whereas in the pure culture *Methylomonas* sp. strain

MM2, most of the TCE-carbon remained in aqueous intermediates (Fig. 4). Unless limited by conditions such as starvation, depletion of reductant, or toxic effects (58–61), all of the TCE was oxidized. TCE oxidation occurred in the presence and absence of methane (stationary phase transformation). Acetylene inhibited TCE oxidation and methane oxidation (58, 59, 61), a result that indicates the involvement of the MMO in the transformation (7). TCE oxidation by *Methylomonas* sp. strain MM2 grown on methanol and unexposed to methane was not detected. The TCE transformation rates were variable, depending upon growth conditions and conditions applied during the TCE transformation assays. For *Methylomonas* sp. strain MM2, the greatest rate of transformation observed was 2.3 liter mg cells (dry weight)$^{-1}$ day^{-1} at 21°C [grown in Whittenbury mineral medium (59, 173), no methane present, 2 mM formate added]. The factors that had a significant influence on TCE transformation by these aquifer cultures included mineral medium formulation, methane availability, reductant availability, concentration of competitive inhibitors (both methane and CO), oxygen concentration, and TCE oxidation toxicity.

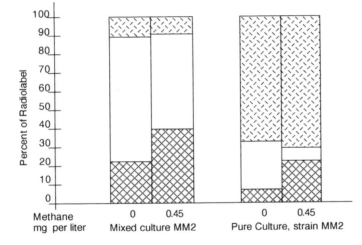

Figure 4. Percent of radiolabel from [^{14}C]-TCE incorporated into nonvolatile aqueous intermediates (◌◌◌◌) carbon dioxide (▯), and cells (▨▨▨) during TCE transformation by mixed culture MM2 and pure culture *Methylomonas* sp. strain MM2.

TCE TRANSFORMATION:
MECHANISM, INTERMEDIATES, TOXICITY

Mechanism of Transformation

The transformation of TCE involves both biotic and abiotic processes. A proposed mechanism of methanotrophic TCE transformation is illustrated in Figure 5. The MMO initiates the transformation by oxidizing the TCE to the corresponding epoxide, which is subsequently extruded from the cell. In the aqueous environment outside the cell, the TCE epoxide breaks down to CO, formate, glyoxylate, and dichloracetate. These break-down products are then subject to further metabolism by both methanotrophs and heterotrophs. In a methanotrophic consortium, the TCE is thus completely

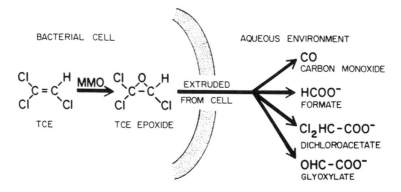

Figure 5. Proposed mechanism of TCE transformation catalyzed by the MMO in methanotrophs. TCE is epoxidated and extruded from the bacterial cell. In the aqueous environment outside the cell the TCE epoxide breaks down to CO, formate, glyoxylate, and dichloroacetate.

mineralized to carbon dioxide and cell biomass. This mechanism was originally proposed based on the findings that in mammalian oxidative systems, TCE was oxidized to the epoxide (6, 17, 62, 63, 97, 106), and that methanotrophs oxidized ethylene to ethylene epoxide, which accumulated extracellularly (69, 71, 147). Under aqueous conditions, TCE epoxide had been shown to rearrange rapidly to formate, CO, dichloracetic acid, and glyoxylic acid (64, 105). The formate and CO are produced in equimolar amounts as a result of C–C fission (64). The ratio of products formed was dependent upon the pH of the system, with CO and formate predominating at neutral and basic pH (9, 105).

Recent work by Fox and colleagues with purified soluble MMO from *Methylosinus trichosporium* OB3b has confirmed that the MMO does oxidize TCE to the epoxide, and that CO, formate, dichloracetate, and glyoxylate are transformation intermediates (43). During degradation studies with 1,2-dichloroethylene, 1,2-dichloroethylene epoxide, which has a half life in aqueous systems of 31 hr (76), has been found to accumulate extracellularly in whole-cell systems (76, 107, 136). TCE epoxide, a highly reactive species with a half life of only 12 sec (105), has not been detected in whole-cell systems, owing, at least in part, to its extremely short half life. During TCE transformation by *Methylomonas* sp. strain MM2, approximately 9 mol percent of the TCE was transformed to CO (Fig. 6) (61). Comparatively, 88 mol percent of the TCE was

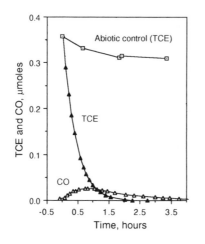

Figure 6. Production of CO during TCE transformation, and subsequent oxidation of the CO by *Methylomonas* sp. strain MM2. Total μmoles of TCE and CO are plotted against time in hours. No CO was produced in the abiotic control, which contained sterile mineral medium and TCE.

transformed to CO in the work with the purified soluble MMO (43). Water-soluble acids with retention times comparable to glyoxylate and dichloroacetate were detected in one study with a pure culture (96) but were not positively identified. Acids produced from the oxidation of TCE by *Methylomonas* sp. strain MM2 grown in Fogel mineral medium (41, 59) were detected in the culture supernatant by using ion chromatography (58). Although it was apparent that either formate or glyoxylate or both were present, however, there was no mono-, di-, or tri-chloroacetate in the culture supernatant. Instead, three unidentified acids were observed that did not coelute with mono-, di-, or tri-chloroacetate (58).

Many other products may be formed from the TCE epoxide, depending upon a number of factors, including pH and the composition of the medium. In the studies of abiotic hydrolysis of TCE epoxide, TCE epoxide was shown to form labile esters in TRIS buffers, and the observation that the level of CO was reduced in phosphate buffer prompted the authors to suggest that formyl phosphate may have formed (105). Formation of products was also dependent upon pH (64, 105). TCE epoxide breakdown also forms other reactive intermediates, such as dichloroacetyl chloride and formyl chloride (105). TCE epoxide or its reactive breakdown products could react with constituents of the medium, forming other compounds not predicted by the proposed mechanism. Furthermore, TCE epoxide is not an obligate intermediate in the oxidation of TCE by the oxidase of mammalian systems (cytochrome P-450) (105, 106). A mechanism was proposed in which CO, formate, glyoxylate, and dichloroacetate could form from a pre-epoxide transition state (105). Under *in vivo* conditions in mammalian systems, chloral, trichloroethanol, and trichloracetic acid are major products of TCE metabolism, and chloral is also highly reactive (6, 17, 97, 105). Although the formation of an epoxide as an intermediate in the methanotrophic transformation of TCE has been confirmed in one study with purified soluble MMO from *M. trichosporium* OB3b (43), it may be that, as in mammalian systems, TCE epoxide is not an obligate intermediate in methanotrophic systems. Additionally, the overall mechanisms of transformation in different methanotrophic systems may differ, depending on the nature of MMO and degree of hydrophobicity of intracellular environment. Intermediates produced as a result of oxidation by a particulate MMO located in intracytoplasmic membranes could well differ from those produced as a result of oxidation by a soluble MMO in the cytoplasm. Soluble MMO and particulate MMO differ not only with respect to intracellular location, but also with respect to the nature of their components (8). Significant differences in the levels of intermediates were observed with different forms (isozymes) of cytochrome P-450 (33, 55, 106). A similar situation may exist with the different forms of MMO.

Fate and Effect of Intermediates

The intermediates that persist in studies with purified MMO and pure cultures do not persist in mixed cultures and soil systems. Studies with ^{14}C-labeled TCE have demonstrated that in pure cultures much of the TCE carbon remains in nonvolatile aqueous intermediates, whereas in mixed cultures that contain heterotrophic bacteria as well as methanotrophs, most of the TCE is transformed to carbon dioxide and cell biomass (2, 40, 41, 58, 59, 96). For example, in *Methylomonas* sp. strain MM2, 70 to 80 percent of the radiolabel remained in the nonvolatile aqueous fraction (Fig. 4), whereas in mixed cultures MM1 and MM2, 90 to 95 percent of the radiolabel was partitioned into carbon dioxide and cell biomass (58, 59). There are very few bacteria that do not metabolize formate (88). The chlorinated acids produced during the transformation of TCE resemble naturally occurring acids (140), and many common soil bacteria have the demonstrated ability to grow on chlorinated aliphatic acids (15, 46, 51, 78, 141, 151, 152). Bacteria that oxidize and utilize CO are also common in nature (104, 180), and have been isolated from a variety of sources, including soils, waste-waters, and sewage sludges (27, 83, 104, 180).

Methanotrophs are also capable of metabolizing some of the intermediates. All methanotrophs oxidize formate to carbon dioxide, deriving energy and regenerating reductant in the process (Fig. 1). Methanotrophs also oxidize CO to carbon dioxide (13, 38, 39, 56, 61, 73, 83, 144, 146, 147). The CO that was produced during TCE oxidation by *Methylomonas* sp. strain MM2 was subsequently oxidized (Fig. 6), and *Methylomonas* sp. strain MM2 also oxidized the trace CO present in the air (61). In the studies with radiolabeled TCE, *Methylomonas* sp. strain MM2 transformed 5 to 15 per-cent of the TCE to carbon dioxide (Fig. 4). This carbon dioxide probably came from the oxidation of the CO and formate. *Methylomonas* sp. strain MM2 also incorporated radiolabel into cell biomass (58, 59), and the amount of ^{14}C incorporated into cell biomass increased, at the expense of the ^{14}C in carbon dioxide, when methane was provided (Fig. 4). The amount of radiolabel associated with the cells did not represent sorbed TCE, inasmuch as less than 3 percent of the radiolabel was found associated with inhibited or non-TCE-transforming-cells (2, 58, 59, 96). Radiolabel may be asso-ciated with the cells as a result of binding of reactive intermediates with cell molecules (see below). The amount of carbon from TCE increased significantly when the cells were metabolizing methane (Fig. 4), however, a finding that suggests that the incor-poration of carbon from TCE was associated with metabolism.

The ^{14}C from the radiolabeled TCE incorporated into the cells could have arisen from two mechanisms: incorporation of ^{14}C-labeled carbon dioxide, and incorporation of carbon from some of the nonvolatile acid intermediates. Many methanotrophs, including several strains of *Methylomonas* sp., incorporate carbon from carbon dioxide (132). As much as 30 percent of the cell carbon has been shown to come from carbon dioxide in some strains (132). In a study with ^{14}C-labeled carbon dioxide, *Methylomonas* sp. strain MM2 incorporated carbon from carbon dioxide (57). Less than 0.03 percent of the total radioactivity was associated with autoclaved controls and cells incubated without methane, whereas 3.4 percent was associated with cells incubated with methane. In another study, previously unexposed subcultures of *Methylomonas* sp. strain MM2 were incubated with supernatant containing radio-labeled nonvolatile aqueous intermediates that had been generated in an earlier experi-ment during the transformation of ^{14}C-TCE (57). Whereas only 0.06 percent of the total radioactivity was associated with autoclaved controls, 0.63 percent of the radiolabel was associated with the cells when no methane was present, and 1.57 percent when methane was present.

Methylomonas sp. strain MM2 did not grow on formate, glyoxylate, or dichloro-acetate on solid or in liquid media (58), and no measurable increase in biomass has ever been observed when *Methylomonas* sp. strain MM2 oxidized TCE. Yet, cell carbon was derived from TCE transformation intermediates, particularly during growth on methane. This raises interesting conjecture regarding the definition of TCE transforma-tion as a cometabolic process. If methanotrophs incorporate cell carbon, derive energy, and regenerate reducing power from the oxidation of TCE transformation inter-mediates, is TCE transformation a strictly cometabolic process? If "cometabolism" is defined as "the transformation of a nongrowth substrate" and "nongrowth substrates" are defined as "compounds that are unable to support cell replication" (31), then TCE transformation can be classified as a cometabolic process. That methanotrophs may derive benefit from the metabolism of nongrowth substrates, however, and that this may have significant evolutionary implications with respect to the broad specificity of the MMO, should not be overlooked.

The aqueous intermediates were not toxic to *Methylomonas* sp. strain MM2. The putative intermediates formate, gyloxylate, and dichloroacetate at 10 mM did not inhibit growth on methane in liquid or on solid media. Methane consumption was not inhibited by incubation in supernatant containing the nonvolatile aqueous fraction remaining from the transformation of 67 μM TCE (60). CO is a known toxic agent, inhibiting growth in the majority of aerobic microorganisms by binding to the terminal oxidase (29). CO at 66 μM was not toxic to *Methylomonas* sp. strain MM2 (61); however CO inhibited both TCE and methane oxidation (Fig. 7) (61) by competing for the enzyme [competitive inhibition (46)], and by exerting a demand for the reductant (see below) (61). The inhibition coefficient, K_i, for CO inhibition of TCE oxidation by *Methylomonas* sp. strain MM2 was 4.2 μM, whereas the K_i for inhibition of TCE oxidation by methane was 116 μM (61). The extent to which the CO evolved during TCE oxidation will influence TCE transformation, and growth on methane will depend on a number of factors, including the relative affinities of the enzyme for CO, TCE, and methane, the availability of the reductant, the amount of CO produced, and the removal of CO by other bacteria. If high concentrations of CO are evolved, and if the methanotrophs have a much greater affinity for CO than for methane and TCE, then CO could inhibit not only TCE oxidation, but also the growth of the culture.

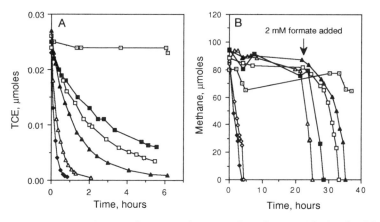

Figure 7. Competitive inhibition of TCE oxidation and methane oxidation by CO during transformation of these compounds by *Methylomonas* sp. strain MM2. (A) Total μmoles of TCE are plotted against time in hours. Bottles were amended with 2 mM formate at time 0. As the CO concentration increases, the rate of TCE oxidation decreases. Symbols: ($\blacklozenge$) no CO, rate = 0.86 liter mg cells (dry weight)$^{-1}$ day^{-1}; ($\triangle$) 6.7 μM CO, rate = 0.43; ($\blacktriangle$) 25.6 μM CO, rate = 0.11; ($\blacksquare$) 53.5 μM CO, rate = 0.04; ($\square$) 57.3 μM CO, rate = 0.05; ($\boxdot$) control (sterile mineral medium blank). The inhibition coefficient, K_i, for CO was 4.2 ± 1.7 μM. (B) Total μmoles of methane are plotted against time in hours. As the CO concentration increases, methane utilization decreases. Significant oxidation of methane did not occur until 2 mM formate was added to the bottles. Symbols: ($\blacklozenge$) no CO #1; ($\lozenge$) no CO #2; ($\triangle$) 10 μM CO; ($\blacksquare$) 18 μM; ($\square$) 25 μM; ($\blacktriangle$) 29 μM; ($\boxdot$) control (sterile mineral medium blank).

TCE Oxidation Toxicity

Although the acids and CO produced during TCE transformation are not toxic to methanotrophs, apparently transient intermediates are produced that are damaging to the cell, so that TCE oxidation itself is toxic. Exposure to high concentrations of TCE has not been observed to be toxic to methanotrophs when the compounds were not being transformed (76, 77). Exposure to tetrachloroethylene (PCE), a chemically similar solvent that is not transformed, is also apparently not toxic (60, 76, 77). During TCE oxidation, however, toxicity has been observed (2, 22, 23, 60, 77, 115, 148, 156). TCE oxidation toxicity was evaluated in mixed culture MM1 and pure culture *Methylomonas* sp. strain MM2 by incubating the cultures with relatively low (6 mg liter⁻¹) TCE concentrations, then removing all remaining TCE and providing methane (60). The subcultures that had previously oxidized TCE took three times longer to utilize methane, as illustrated for mixed culture MM1 in Figure 8. In this study, TCE oxidation by *Methylomonas* sp. strain MM2 reduced viable cells, as determined by enumeration of colony-forming units, by more than an order of magnitude (60). The extent of inactivation imposed by TCE oxidation was found to be a function of the ratio of amount of TCE to cells; less inactivation occurred when cell concentrations were greater (2, 23, 76, 115). An inactivation constant of 0.48 mg cells per μM TCE was determined for pure culture *M. trichosporium* OB3b (115).

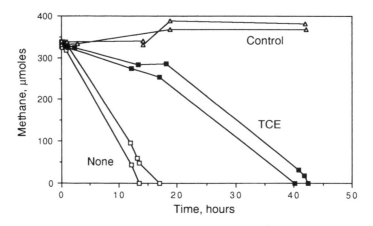

Figure 8. Effect of TCE oxidation on subsequent methane utilization by mixed culture MM1. Total μmoles of methane are plotted against time in hours. Replicate bottles containing the culture were incubated for 2 hours,after which the remaining TCE was removed before subsequent amendment with methane, so that no TCE was present during methane utilization. "None" denotes bottles incubated without TCE; "TCE" denotes bottles incubated with 6 mg per liter TCE. (Mass of TCE oxidized per cell dry weight was 0.016.) "Control" denotes sterile mineral medium blanks.

In studies with mammalian cells, TCE oxidation produced reactive metabolites that bound irreversibly to protein, DNA, and RNA and inactivated the cell (14, 16, 106). Janssen and colleagues, observing that methane-grown cells were inactivated by TCE but methanol-grown cells were not (the methanol-grown cells did not transform TCE because the MMO had not been induced), were the first to propose that toxicity might result from a product of TCE oxidation (76, 77). Wackett and colleagues observed that TCE had to be metabolically activated to produce toxic effects and suggested that reactive products of TCE oxidation modified intracellular molecules (169, 170). Fox and colleagues, studying TCE oxidation by purified MMO, observed a turnover-dependent inactivation and covalent modification of all MMO components (43). It has been suggested that the highly reactive TCE epoxide is the modifying species (2, 60, 92, 170). Based on findings of the studies of mammalian systems (106) and the purified MMO (43, 50), however, it appears that another reactive intermediate, possibly a hydrolysis product of the TCE epoxide, is the modifying species.

Regardless of the identity of the modifying species, the net effect is the same: the cells transforming TCE are inactivated. The methanotrophic biomass will therefore have a finite capacity for TCE transformation. TCE transformation capacities have been reported ranging from 0.2 μM to 2 μM TCE per mg cells (2, 115). TCE oxidation toxicity has significant implications for process applications. The decay rate of the active biomass will be a function of the amount of inactivating agent produced on a per cell per unit time basis. This is, in turn, a function of the TCE concentration and the rate of TCE oxidation (which is also a function of the TCE concentration). The TCE transformation capability of the culture will decrease as TCE is oxidized. The amount of methane required to maintain the biomass concentration at steady-state will increase as the TCE concentration and the rate of TCE oxidation increase. The oxygen demand will also increase to meet not only the needs of the methanotrophs, but particularly to meet the needs of heterotrophs catabolizing dead methanotrophic biomass. If the growth rate of the methanotrophs is insufficient to meet the decay rate during periods of peak TCE oxidation rates, shock loads of TCE could cause washout of cells from a reactor.

It may be possible to design process applications in such a way that the impact of TCE oxidation toxicity is minimized. Protecting the culture would require reducing the amount of inactivating agent and the rate at which it is produced within the cell. This would be accomplished by reducing the amount of TCE transformed per cell per unit time by reducing the amount of TCE to which individual cells were exposed. The amount of TCE relative to each cell could be reduced by increasing the cell density. It was observed that increasing the cell density increased the TCE transformation capacity of mixed and pure cultures (2, 23, 115). Reducing the concentration of TCE in the liquid phase would also reduce the amount of TCE to which each cell was exposed. Process wastes could be diluted, or TCE could be removed from solution through the use of sorptive materials. The addition of activated carbon reduced the concentration in solution and increased the TCE transformation capacity in a suspended culture of M. trichosporium OB3b (115). In soil systems, TCE oxidation toxicity may be reduced as a result of sorption of TCE to the soil. Fixed-film systems may offer protection from high TCE concentrations in the bulk solution. In a fixed-film packed-bed reactor, 20 mg liter^{-1} TCE was treated for several days with no apparent negative effect on the system (150). Protection could also be accomplished without reducing the concentration of TCE in solution by limiting the rate of TCE transformation through competitive inhibition by the growth substrate methane (see below). If solution concentrations of TCE are expected to be high, it may be desirable to develop a process design that facilitates rather than avoids competitive inhibition by methane. In this respect, plug flow systems, in which methane can be added where TCE concentrations are the highest, may be more effective than completely mixed systems.

INFLUENCE OF GROWTH CONDITIONS ON
TCE TRANSFORMATION

Methane: Growth Substrate and Competitive Inhibitor

Methane is required for the development and maintenance of the active methanotrophic biomass, as both a growth substrate and an inducer of the MMO. Since methane and TCE are oxidized by the same enzyme, they are consequently competitive inhibitors of each other's transformation. The effect of methane as a competitive inhibitor of TCE oxidation by *Methylomonas* sp. strain MM2 is illustrated in Figure 9. As the methane concentration increased, the TCE transformation rate decreased. Inhibition of TCE transformation by methane was observed in several studies that evaluated the influence of methane on TCE transformation (23, 58, 61, 86, 87, 115, 118, 137), but was not observed in others (148, 150). In some cases, high concentrations of methane were inhibitory, whereas low concentrations stimulated TCE transformation (23, 58). The extent of competitive inhibition is a function of the relative concentrations of the substrates: the greater the methane concentration, the greater the competitive inhibition. It is also a function of the affinity of the MMO for substrates, as described by the kinetic coefficient K_s (or K_m), which is the half-saturation coefficient or affinity coefficient (48). The smaller the K_s, the greater is the enzyme affinity for the compound (48). The K_i can be thought of as equivalent to the K_s (48). There is a broad variability in K_s values for methane and TCE among methanotrophs. The K_s values for methane reported for whole cells, cell extracts, and purified MMO range from 1 to 160 μM (53, 81, 82, 85, 117, 148, 155, 174). For TCE, the K_s values reported range from 4 μM to 145 μM (2, 23, 59, 115). Competitive inhibition of TCE oxidation will be less pronounced in methanotrophs that have a low K_s (high affinity) for TCE. *Methylomonas* sp. strain MM2 exhibited a K_s for TCE of 4 μM [cells grown in Whittenbury mineral medium (59, 173), no methane or formate added during transformation study] (59), and a K_i for methane of 116 μM (61). Given these relative affinities, competitive inhibition would be far less pronounced than in a methanotroph such as *M. trichosporium* OB3b that expressed a K_s for TCE of 145 μM and a K_s for methane of 92 μM (115).

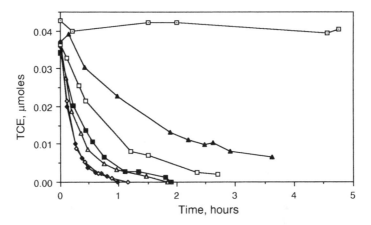

Figure 9. Competitive inhibition of TCE oxidation by methane during TCE transformation by *Methylomonas* sp. strain MM2. Total μmoles of TCE in the bottles is plotted against time in hours. Bottles were amended with 2 mM formate at time 0. As the methane concentration increases, the rate of TCE oxidation decreases. Symbols: (◆) no methane #1, rate = 1.24 liter mg cells (dry weight)$^{-1}$ day^{-1}; (◇) no methane #2, rate = 1.24; (△) 111 μM methane, rate = 0.60; (■) 225 μM, rate = 0.44; (□) 479 μM, rate = 0.26; (▲) 998 μM, rate = 0.12; (⊡) control (sterile mineral medium blank). The inhibition coefficient, K_i, for methane was 116 ± 13 μM.

In process applications, competitive inhibition could be reduced through the selection of organisms suitable to the application. If allochthonous cultures are being added to the treatment process, methanotrophs that express a high K_s (low affinity) for methane and a low K_s (high affinity) for TCE could be chosen. Adding a mixture of different methanotrophs may be valuable when seeding a reactor or augmenting a process application with additional culture: add methanotrophs with a high K_s for methane to facilitate TCE removal when methane concentrations are high, and add methanotrophs with a low K_s for methane that can grow on low concentrations and scavenge residual methane. If selecting the organisms used in the treatment process is not possible, or if the process design involves the use of indigenous organisms, enrichment techniques may help to select for methanotrophs with the desired characteristics. It may be possible to enrich for methanotrophs with low affinities for methane by enriching in the presence of high methane concentrations. This may be difficult, however, particularly with respect to *in situ* treatment of contaminated groundwater and soils, because of the low solubility of methane.

Process designs to remedy the problem of competitive inhibition of TCE oxidation by methane could include the maintainance of methane concentrations below inhibitory levels, or the spatial and/or temporal separation of the growth substrate and cometabolic substrate. Dual-reactor systems in which methanotrophs are grown in one reactor and degrade TCE in another have been proposed (5, 28, 148). Pulsed methane additions have been shown to enhance the efficiency of TCE degradation, as long as the starvation periods between pulses were not too long (122, 137). As discussed in the previous section on TCE oxidation toxicity, competitive inhibition is generally perceived to be problematic, but it could potentially be put to use in "protecting" biomass from high TCE concentrations.

Starvation and Reductant Availability

Many methanotrophic cultures exhibit the capability to transform TCE when no methane is present (2, 3, 23, 35, 58–61, 80, 115, 116, 122, 148, 150, 156), a phenomenon termed *resting-cell transformation* or *stationary-phase transformation*. There are obvious problems associated with eliminating methane, however, inasmuch as it is the growth substrate. During prolonged absence of the growth substrate (starvation), the active biomass can be expected to decline, accompanied by a decline in the TCE transformation capability. In mixed and pure cultures, declines in TCE transformation rates during starvation were observed (60, 122, 148, 150). During the *in situ* biodegradation study conducted at Moffett Field, removal of chlorinated ethylenes decreased markedly after methane ceased to be added, and Semprini et al. concluded that utilization of methane was required for biotransformation (136, 137). TCE transformation during starvation was evaluated in *Methylomonas* sp. strain MM2, mixed culture MM1, and isolate CSC-1, and an exponential decline in the rate of TCE transformation was observed (Fig. 10) (60). This decline can be attributed to a number of factors, including decay of biomass due to the lack of cell carbon and energy, inactivation of the MMO resulting from autolysis and lack of induction with methane, predation in mixed cultures, and limited availability of reductant.

The oxidation of a compound by an oxygenase is coupled to the reduction of an atom of oxygen to water, which requires electrons (178). This reducing power is provided by the electron carriers in the cell, such as NAD(P)H. In this respect, the whole oxidation process can be viewed as dependent upon the availability of reducing power within a cell. When a growth substrate such as methane is catabolized, the reductant is regenerated (Fig. 1). When the MMO oxidizes a cometabolite such as TCE or CO (Fig. 1), the reductant is not regenerated, and the cell can "run out" of reducing power. The oxidation of any one of the metabolites of methane (methanol, formaldehyde, and formate) to carbon dioxide regenerates reductant (Fig. 1). Formate has been provided as an exogenous source of electrons in many studies, and has enhanced TCE transformation during stationary-phase transformation (2, 3, 23, 60,

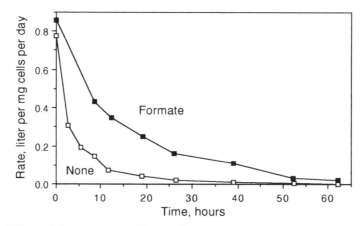

Figure 10. Effect of formate on TCE transformation rate during starvation. The ability of *Methylomonas* sp. strain MM2 to transform TCE was determined at different times during the starvation period by evaluating the TCE transformation rate [liter mg cells (dry weight)$^{-1}$ day^{-1}] of subcultures previously unexposed to TCE. "None" denotes subcultures incubated without formate addition. "Formate" denotes subcultures amended with 2 mM formate at the time of TCE addition. Formate amendment significantly increased TCE transformation rates. The decline in TCE oxidation rate followed exponential decay. The decay parameter for TCE oxidation [-ln (rate$_t$/rate$_0$) versus time] was 2.3 day^{-1} for subcultures that were not amended with formate, and 1.4 day^{-1} for those that received formate (60).

116, 156). In *Methylomonas* sp. strain MM2, addition of formate increased the TCE transformation rate during starvation (Fig. 10) (60). It also increased the CO oxidation rate from 4.6 to 10.2 liter mg cells (dry weight)$^{-1}$ day^{-1} (61). During the *in situ* biodegradation study conducted at Moffett Field, addition of formate enhanced removal of TCE, but the effect did not last beyond a few days (137). Addition of methanol was also evaluated as a source of reductant during this study, and although methanol did enhance rates, it was not as effective as formate (137). Methanol stimulated TCE oxidation by *Methylosinus trichosporium* OB3b at concentrations ranging from 0.1 to 1 mM but was inhibitory at concentrations greater than 1 mM (23). When methanol was used as a cosubstrate during TCE transformation by suspended cultures in other studies, TCE transformation was inhibited (41, 116, 157). Providing methanotrophs with methanol, formaldehyde, or formate will not obviate the need for methane. Few methanotrophs grow well on methanol, none grow well on formaldehyde, and formate is not a growth substrate (7). Additionally, in a consortium in which non-MMO–containing microorganisms—methylotrophs and heterotrophs alike—compete more successfully for such substrates, non-MMO–containing organisms would come to dominate the consortium. Such organisms would compete with methanotrophs not only for the substrate, but also for oxygen and nutrients.

Because of the effects of competitive inhibition, even though methane may be provided, the cell could still be depleted of reducing power. When CO competitively inhibited methane oxidation by *Methylomonas* sp. strain MM2, no methane oxidation was observed until formate was added as an exogenous source of reductant (Fig. 7). During this study, CO oxidation was also arrested until formate was provided (61), because CO oxidation had "used up" the available reducing power within the cell and no reductant could be regenerated as long as methane oxidation was blocked by a competitive inhibitor. CO is not the only cometabolic substrate that could have this effect. Any cometabolic substrate that is oxidized by the MMO could potentially block the regeneration of reductant by blocking methane oxidation. TCE and other halogenated aliphatics have been shown to competitively inhibit methane utilization (22).

Although formate enhanced TCE transformation by *Methylomonas* sp. strain MM2, it had no effect on transformation by mixed culture MM1 or isolate CSC-1 (60). These cultures contained lipid storage granules that may have served as an endogenous source of reductant, so that formate was not used. In another study evaluating TCE transformation during starvation, mixed cultures MM1 and MM3, which contained methanotrophs that possessed lipid inclusions, remained active toward TCE, whereas mixed culture MM2, which lacked lipid inclusions, did not (58, 60). Methane-grown methanotrophs are known to accumulate lipid and polyglucose inclusions (10, 95). Lipid storage granules have been found in numerous subsurface bacteria (45). The presence of endogenous reserves may have great significance in an environment such as the subsurface, where carbon and energy sources are scarce. Microorganisms that accumulate storage granules may be of value in treatment applications because of the endogenous reserves of reducing power and energy.

Oxygen

Methanotrophs are aerobic: they require molecular oxygen as the terminal electron acceptor for respiration. Methanotrophs also require molecular oxygen for the oxidative function of the MMO (Fig. 1) (8, 31, 70, 178). A significant problem encountered in the treatment of contaminated soils, or during *in situ* aquifer restoration, is maintaining aerobic conditions, and oxygen availability is the most significant limiting factor (18, 92, 93, 75, 130). Oxygen concentrations in aquifers are generally low, and contaminated aquifers frequently become anaerobic as a result of bacterial activity induced by the contaminants (11, 18, 47, 92, 175). The dissolved oxygen content of the groundwater at the uncontaminated Moffet Field test site before stimulation of the test zone was below 0.2 mg liter^{-1} (131). At a gasoline-contaminated site, the oxygen concentration was 0.5 mg liter^{-1} in the contaminated plume, 0.4 to 5.5 mg liter^{-1} in the zone in which biodegradation of the gasoline components was occurring, and 3.5 to 9.6 mg liter^{-1} in the uncontaminated zone (175).

Methantrophs can be expected to perform well under low oxygen concentrations. They are microaerophilic (4), and the MMO has a high affinity for oxygen, with reported K_s values ranging from 0.14 to 16.8 μM (49, 81, 91). Mixed cultures MM1 and MM2 and *Methylomonas* sp. strain MM2 were tested for the effects of oxygen availability on TCE transformation. The cultures transformed TCE under near-anoxic conditions (1 percent vol/vol oxygen in the headspace), and were not inhibited until the system ran out of oxygen (58). Other authors evaluating the influence of oxygen availability on TCE transformation have reported similiar findings (41, 116, 157). To overcome the problems associated with oxygen limitation in process applications, there may be a temptation to use pure oxygen or hydrogen peroxide. Hydrogen peroxide has been used successfully as a source of oxygen in the treatment of petroleum hydrocarbon contamination (92, 130). Inasmuch as methanotrophs are microaerophilic, however, high oxygen concentrations resulting from the use of pure oxygen or hydrogen peroxide may be inhibitory to growth and TCE transformation. In mixed cultures MM1 and MM2 and *Methylomonas* sp. strain MM2, oxygen at 50 percent vol/vol in the headspace reduced TCE transformation 20 percent relative to that observed at ambient oxygen concentrations (20 percent vol/vol in the headspace) (58). Hydrogen peroxide as an oxygen source in methanotrophic soil columns decreased methane utilization but did not decrease oxygen consumption or bacterial numbers as determined by acridine orange direct counts (100). This led the authors to suggest that the application of hydrogen peroxide was inhibitory to the methanotrophs but not the heterotrophs. No additional stimulation of TCE degradation or methane utilization was observed when hydrogen peroxide was applied instead of oxygen during the *in situ* biodegradation field study at Moffett Field (136).

Mineral Medium Formulation

Bioavailability of inorganic nutrients may affect the removal of chlorinated solvents in methanotrophic systems. Inorganic nutrients, including trace metals, that are required for metabolism may be present in limiting concentrations. The availability of nitrogen and phosphorus frequently limits biodegradation of hydrocarbons in water, sediment, soil, and groundwater (90). Amendment of soil (34, 79, 154) and groundwater (75) with nitrogen and phosphorus has been shown to increase bacterial numbers and/or to enhance biodegradation of xenobiotic substrates in some studies (34, 75, 79, 154), but not in others (94, 154). Variability in responses to amendment have been attributed to the different nutritional requirements of different microbial communities (93, 154) as well as variability in the chemical composition of soils (19, 154). No nitrogen or phosphorus or other inorganic nutrients were added to the groundwater during the field study that demonstrated *in situ* methanotrophic biodegradation of chlorinated ethylenes (136). Nitrogen and phosphorus amendment did not conclusively enhance TCE degradation in methane-enriched soil columns that contained aquifer material from the Moffett Field site (87).

The bioavailability of inorganic compounds that are inhibitory may also affect the ability of methanotrophs to degrade chlorinated solvents. In *Methylosinus trichosporium* OB3b, even small quantities of copper in the growth medium prevent TCE oxidation by inhibiting the expression of the soluble MMO (23, 116, 156). In a study with mixed groundwater cultures, high concentrations of copper were not inhibitory to TCE transformation (118). In a related study, high manganese and phosphorus concentrations were neither stimulatory nor inhibitory, whereas high ammonia concentrations were inhibitory (118). Doubling mineral concentrations in a methanotrophic continuous-recycle expanded-bed reactor resulted in 20 percent less TCE removal (122).

Mineral medium formulation, as well as copper concentration and concentration of EDTA, a metal-complexing agent, profoundly affected TCE transformation rates in both *Methylomonas* sp. strain MM2 and mixed culture MM1 (59). Rates of TCE oxidation were significantly greater in a growth medium termed "Whittenbury" (59, 173) that contained EDTA and no added copper than in a formulation termed "Fogel" (41, 59) that contained 0.06 μM copper and no EDTA. The presence of EDTA in the Whittenbury formulation significantly enhanced rates of TCE oxidation by *Methylomonas* sp. strain MM2. The kinetic coefficients for TCE oxidation by this culture were evaluated in the standard Whittenbury formulation that contained 0.67 μM EDTA and in the formulation lacking EDTA (Table 1). Both the maximal rate coefficient k and the half-saturation coefficient K_s were affected by the presence of EDTA. The rate of TCE oxidation when EDTA was present was more than two orders of magnitude greater. Omission of EDTA from the Whittenbury formulation also reduced the methane oxidation rate and growth yield (59). Addition of the same amount of EDTA to the Fogel growth medium had no significant effect on TCE oxidation rates. Addition of the same concentration of copper to the Whittenbury growth medium as that in the Fogel growth medium (0.06 μM copper) had no effect on TCE oxidation. Addition of 1.6 μM copper, however, reduced the TCE oxidation rates almost three orders of magnitude without affecting the methane oxidation rate or growth yield, and this effect was partially reversed by addition of 84 μM EDTA. EDTA is an organic ligand that com-

Table 1. Kinetic coefficients for TCE oxidation by *Methylomonas* sp. strain MM2 grown on different mineral media.

Mineral medium formulation	k, day^{-1}[a]	K_s, mg liter^{-1}[b]	k', liter mg^{-1} day^{-1}[c]
Whittenbury	0.29	0.51	0.57
Whittenbury no EDTA	0.046	1.35	0.033

[a] Maximum rate coefficient (V_{max}), mg TCE mg cells (dry weight)$^{-1}$ day^{-1}.
[b] Half-saturation coefficient, mg TCE liter^{-1}.
[c] Pseudo-first-order rate coefficient (k/K_s), liter mg cells (dry weight)$^{-1}$ day^{-1}.

plexes cations, thus increasing the bioavailability of some metals and sequestering others (21, 36, 153). EDTA may have influenced the TCE oxidaton by sequestering copper and rendering it unavailable, thus preventing inhibitory effects, although the difference in rates of TCE oxidation in the different media were probably not due entirely to copper availability (59).

Growth conditions must be taken into account when developing a methanotrophic bioremediation application for treatment of chlorinated solvents, and the biodegradability of target compounds should be evaluated in water from the process application. The bioavailability of metals could have a significant impact on the success of a process application. In the treatment of groundwater and soils, site geochemistry will influence methanotrophic activity toward cometabolites like TCE. Copper clearly has an inhibitory effect on TCE transformation. In the laboratory, rigorously limiting copper availability is possible, but this may not be possible during the treatment of soil, groundwater, and industrial wastes. Affecting mineral nutrient availability may be possible by amending process waters with mineral nutrients and chelators, including specific metal chelators. Predicting metal availability is extremely complex and is dependent upon a number of interacting factors (153), however, and addition of metals and chelators could lead to an unpredicted distribution of all mineral nutrients. Furthermore, metal chelators such as EDTA are themselves pollutants. Some complexing agents are biodegradable, depending upon their chemical speciation (89, 98), although EDTA is generally not biodegradable (98, 103).

CONCLUSIONS

The laboratory research published to date has demonstrated conclusively that methanotrophs are capable of transforming TCE and other chlorinated aliphatics. This research with cultures enriched and isolated from the Moffett Field aquifer material and groundwater has demonstrated that there are methanotrophs present in the subsurface that are capable of transforming TCE. The field study conducted at the Moffett Field test site has demonstrated that, upon subsurface enrichment with methane and oxygen, *in situ* biotransformation of chlorinated aliphatics can be achieved.

A significant conclusion of all of these investigations is that the rates of TCE transformation vary considerably, depending not only on the innate capability of the methanotrophic culture but, more importantly, on the growth conditions and the conditions imposed during, and as a consequence of, TCE transformation. The factors discussed in this chapter that can be considered operational parameters in process applications—methane concentration, oxygen concentration, reductant availability, and mineral nutrient bioavailability—influence the capability of methanotrophs to transform TCE. The very process of TCE oxidation profoundly influences the ability of methanotrophs to transform TCE because of the toxicity imposed by the reactive intermediates that are produced. Transformation intermediates, such as CO, may also influence the ability of methanotrophs to transform TCE in a process application. Other factors that have been shown to influence the degradation of xenobiotics, including temperature, pH, presence of other contaminants, sorption, and low bacterial numbers in situ (12, 47, 90, 92, 101, 114, 134, 165), will also influence TCE transformation by methanotrophs. All of these factors must be taken into consideration during the design and implementation of process applications.

Table 2 lists the optimal rates reported by several authors for different methanotrophic cultures under various growth conditions. The maximal rate coefficient k (V_{max}) is given, and so is the pseudo-first-order rate coefficient k' (k/K_s) when the authors have reported this value or the half-saturation coefficient K_s. Methylosinus trichosporium OB3b grown under optimum conditions (23, 115) has demonstrated the most rapid TCE transformation rate to date. The maximum rates demonstrated in suspended cultures in the laboratory studies cannot be taken to reflect potential rates of removal in soils, however. In soils, the desorption rate is often the rate-limiting step controlling the ultimate rate of removal (101, 114, 123, 134, 143), so high biotransformation rates may not effect enhanced removal.

Table 2. TCE tranformation rates by suspended mixed and pure methanotrophic cultures.

Culture	TCE transformation rates		Conditions[c]	Reference
	k, day $^{-1}$[a]	k', liter mg^{-1}.day^{-1}[b]		
Methylosinus trichsporium OB3b[d]	27.4	1.4	20 mM formate, copper-limited	115
M. trichsporium OB3b[d]	20.8	1.1[e]	25 mM formate, copper-limited	23
M. trichsporium OB3b[d]	1.9	0.1[e]	no formate, copper-limited	156
Methylomonas sp. strain MM2[f]	1.2[g]	2.3	2 mM formate	59
	0.3	0.6	no formate	
Mixed culture[f]	5.1	0.7[e]	20 mM formate	2
	0.6	0.08[e]	no formate	
Mixed culture[h]	n.d.[i]	0.009	no formate, closed-system reactor[j]	148

[a] Maximum rate coefficient k (V_{max}), mg TCE mg cells (dry weight)$^{-1}$ day^{-1}.
[b] Pseudo-first-order rate coefficient k' (k/K_s), liter mg cells (dry weight)$^{-1}$ day^{-1}.
[c] Resting-cell conditions (no methane) unless specified otherwise.
[d] Experiments conducted at 30°C; rate data presented here normalized to 20°C (84).
[e] Pseudo-first order rate coefficient k' calculated by dividing the given k by the given K_s.
[f] Experiments conducted at 21°C.
[g] Calculated by multiplying the given k' by the given K_s.
[h] Experiments conducted at 20°C.
[i] Not determined.
[j] Methane present; dissolved methane concentration less than 1.0 mg liter^{-1}.

The maximal rate coefficient reflects how quickly the organism can transform TCE when TCE is present in solution in saturating concentrations. It is not, however, the only important coefficient in evaluating the effectiveness of biological removal. The half-saturation coefficient must also be taken into consideration when evaluating efficiency of removal at less than saturating TCE concentrations. Reports that TCE cannot be removed below a given threshold of 0.2 mg liter^{-1} in a continuous-recycle, expanded-bed reactor (122) or below a threshold of 0.05 to 0 0.1 mg liter^{-1} in a fixed-film packed-bed reactor (150) may reflect a low affinity (high K_s) for TCE. For *Methylosinus trichosporium* OB3b grown under the conditions that optimized the TCE transformaton rate, the K_s for TCE was 138 μM (18 mg liter^{-1}) (23). A mixed culture that transformed TCE rapidly under optimal conditions exhibited a K_s of 56 μM (7.3 mg liter^{-1}) (2). *Methylomonas* sp. strain MM2, when grown in Whittenbury mineral medium (59, 173), exhibited a K_s of 3.9 μM (0.51 mg liter^{-1}) during resting-cell transformation of TCE with no formate added (Table 1) (59). The maximum allowed contaminant level for TCE in drinking water is 0.005 mg liter^{-1} (166). Under some conditions, particularly when TCE is present in μg liter^{-1} concentrations, systems that contain methanotrophs that exhibit a high affinity for TCE (low K_s) may be more effective at TCE removal.

Perhaps more important than a low K_s and a high k in determining the effectiveness of a process application is the robustness and resiliency of the methanotrophic population. The high rates presented in Table 1 are for optimum growth conditions in laboratory experiments. In "real life" applications, conditions such as copper limitation, pH, temperature, substrate availability, and biomass cannot be so rigorously controlled. The rates and removal efficiencies reported for bench-scale reactors (113, 118, 121, 122, 148, 149, 150) and soil columns (66, 86, 87, 100, 112) may more accurately reflect the performance that can be anticipated in an acutal process application.

The studies with *Methylosinus trichosporium* OB3b were conducted at 30°C (23, 115, 116, 156) [the values in Table 2 were normalized to 20°C (80)], and the temperature optimum was determined to be 28–35°C [6]. Treatment of industrial wastes and the surface treatment of soil and groundwater would realistically be performed at ambient temperatures (20–24°C), and *in situ* treatment would be conducted at lower temperatures (the temperature of the groundwater during *in situ* biodegradation study conducted at the Moffett Field test site was 18°C [123]). In scaling up from the laboratory to the field, the dependence of rates of substrate utilization and other biotransformations on temperature (Arrhenius equation) and an approximate halving of rates with each 10°C decrease in temperature (84) must be taken into account. Potential inhibition at high temperatures must also be considered. *M. trichosporium* OB3b could not sustain TCE transformation at 40°C (23). *Methylomonas* sp. strain MM2, enriched and isolated from an 18°C aquifer and cultured at 21°C, does not grow at or above 28°C.

The pH has been shown to affect the TCE transformation capability of methanotrophic cultures. The pH optimum for TCE transformation by *Methylosinus trichosporium* OB3b was pH 6.0–7.5, with a steep decline above pH 8.0 (23). A shift from pH 7.2 to 7.5 in a continuous-recycle, expanded-bed reactor decreased TCE transformation by 85 percent, but the bioreactor recovered within one day of readjustment to pH 7.2 (122). The pH of the groundwater in the biostimulated zone at the Moffett Field test site was pH 6.5 (131). The pH will affect not only mineral nutrient availability (153) but may also affect the formation of transformation intermediates, as discussed above, and in that way may have some influence on TCE oxidation toxicity.

Copper clearly has an impact on the ability of methanotrophs to transform TCE. Some methanotrophs, such as *Methylosinus trichosporium* OB3b, can only transform TCE under copper-limited conditions when expression of the soluble MMO is derepressed (23, 115, 116, 156). *M. trichosporium* OB3b exhibited the high rates listed in Table 2 only when grown to very high cell densities under conditions in which considerable precaution was taken to remove all copper from the growth medium (23, 115,

116, 156). The effect of copper on other methanotrophs that do not express the soluble MMO is apparently less pronounced (35, 59, 80, 118). The aquifer methanotrophs enriched and isolated from the Moffett Field groundwater and aquifer material could transform TCE at appreciable rates even when some copper was added to the growth medium and the cultures were grown to moderate cell densities (59–61). In actual process applications, it would be very diffcult to eliminate copper from the culture medium and grow the methanotrophs to the very high cell densities that are attainable in the laboratory.

Chlorinated ethylene oxidation toxicity is perhaps the most important consideration in the biotreatment of soil, groundwater, and wastewater contaminated with high concentrations of TCE and certain other chlorinated ethylenes. Reactivity, as reflected in mutagencity, was shown to depend upon the symmetry and hence the stability of the epoxide (62). The epoxides of PCE and *cis*- and *trans*-1,2-dichloroethylene (DCE) were symmetric, stable, and not mutagenic, whereas the epoxides of TCE, 1,1-DCE, and VC were asymetric, unstable, and mutagenic (62). Therefore, it can be anticipated that oxidation toxicity may also be encountered during the methanotrophic biotransformation of of 1,1-DCE and VC. The toxic effect resulting from the oxidation of TCE will be particularly pronounced when concentrations are in the mg liter^{-1} range, there is a large relative ratio of the amount of TCE to cell biomass, and the rate of transformation is maximized. Considerable decay of biomass would occur under such conditions, and the growth rate would have to keep pace with the decay rate. This would result in an increased demand for methane and oxygen, as discussed above. Perhaps in this respect, methanotrophs with a low k for TCE would be at an advantage: with a slower oxidation rate, less reactive intermediate would be produced per cell per unit of time. An important criterion with respect to suitability for treatment of TCE when concentrations are in the mg liter^{-1} range is resistance to inactivation. Perhaps obtaining inoculum from a historically contaminated site and enriching in the presence of TCE may result in methanotrophs more resistant to TCE oxidation toxicity.

Some research with *Methylosinus trichosporium* OB3b is focused on maximizing the expression of the soluble MMO, stabilizing the enzyme system's activity toward TCE for as long as possible, and using the culture as a reagent that is applied and "used up" during the bioremediation of TCE-contaminated aquifers (139). This application will obviate the problem of copper inhibition and the issue of biomass decay due to TCE oxidation toxicity. There may be difficulties, however, with the feasibility and economy of scaling up culturing methods that require high-grade copper-free chemicals, super-pure water, EDTA-rinsed culturing vessels, and the other precautions necessary to ensure a copper-free environment for growth (156). Also, the decayed methanotrophic biomass will exert a biological oxygen demand that could potentially exacerbate the already most frequently limiting factor in the aerobic treatment of contaminated soils: oxygen availability.

Many other factors will influence the success of methanotrophic treatment of chlorinated solvents in process applications. Unlike the conditions of the laboratory studies, methanotrophs used in the treatment of contaminated soils and groundwater will be confronted with mixtures of many different xenobiotics, both chlorinated solvents and other organics. Contaminated groundwater and soils characteristically contain a broad variety of organic and inorganic contaminants (92). Ammonia is the preferred nitrogen source for nutrient amendment during bioremediation (75, 92, 130). Ammonia is a substrate for the MMO (7, 13, 39, 117), however, and could competitively inhibit oxidation of methane and the chlorinated solvents.

One significant difference between the bioremediation of petroleum hydrocarbons and the methanotrophic treatment of chlorinated solvents is that the petroleum hydrocarbons serve as growth substrate for the active biomass (75, 90, 92, 93, 129, 130). To enrich for methanotrophs the growth substrate must be supplied. Since methane is a gas and has low water solubility, generating high cell densities is difficult, particularly *in situ* (136, 137). Also, conditions that promote high TCE transformation rates may not be optimal conditions for growth. For example, although copper limitation has been shown to enhance TCE transformation rates in several studies (23, 35, 59, 116, 156), it has also been shown to decrease the growth yields in methanotrophs (26, 91).

The success of a methanotrophic process application in remediating chlorinated solvent contamination will depend on how adequately the design and implementation of the process can address issues such as those discussed in this chapter. The capabability of methanotrophs to transform TCE is clearly dependent on the growth environment and the conditions existant during the TCE transformation. Since every contaminated site or wastewater has its own unique characteristics, every process will probably have to be adapted specifically to the application. Despite the constraints that growth conditions and oxidation toxicity may have on methanotrophic TCE transformation, it appears to be a promising approach for the biological treatment of chlorinated solvent contamination.

Acknowledgments

This chapter is dedicated to the memory of Dunja Grbić-Galić, b. April 21, 1950, d. August 20, 1993. As Associate Professor of Environmental Microbiology in the Department of Civil Engineering, Stanford University, Dunja was dissertation advisor of the research that formed the basis for this chapter. The kind and gentle manner with which she guided her students will be fondly remembered.

This work has been funded by EPA grant EPA CR 812220 awarded to Paul V. Roberts (D. Grbić-Galić was a Co-Principal Investigator on that project), by EPA grant EPA R815738-01-OCS awarded to Dunja Grbić-Galić; and by a grant from the Switzer foundation awarded to S. M. Henry.

Much appreciation is expressed to Fran Thomas for instruction and assistance with electron microscopy, and Craig Criddle for thoughtful discussions and guidance with modeling and TCE analysis.

LITERATURE CITED

1. Adriaens, P., and D. Grbić-Galić. 1991. Aerobic methanotrophic transformation of mono- and dichlorobiphenyls. Q-106, p. 294. *Abstr. 91st Annu. Meet. Am. Soc. Microbiol. 1991.*

2. Alvarez-Cohen, L., and P. L. McCarty. 1991. Effects of toxicity, aeration, and reductant supply on trichloroethylene transformation by a mixed methanotrophic culture. *Appl. Environ. Microbiol.* 57:228–235.

3. Alvarez-Cohen, L., and P. L. McCarty. 1991. Product toxicity and cometabolic competitive inhibition modeling of chloroform and trichloroethylene transformation by methanotrophic resting cells. *Appl. Environ. Microbiol.* 57:1031–1037.

4. Alvarez-Cohen, L., and P. L. McCarty. 1991. A cometabolic biotransformation model for halogenated aliphatic compounds exhibiting product toxicity. *Environ. Sci. Technol.* 25:1381–1387.

5. Alvarez-Cohen, L., and P. L. McCarty. 1991. Two-stage dispersed-growth treatment of halogenated aliphatic compounds by cometabolism. *Environ. Sci. Technol.* 25:1387–1393.

6. Anders, M. W. 1982. Aliphatic halogenated hydrocarbons. In *Metabolic Basis of Detoxification: Metabolism of Functional Groups.* Ed. W. B. Jakoby, J. R. Bend, J. Caldwell. New York: Academic Press, Inc. 29–49.

7. Anthony, C. 1982. *The Biochemistry of Methylotrophs.* London: Academic Press, Inc.

8. Anthony, C. 1986. Bacterial oxidation of methane and methanol. *Adv. in Microb. Physiol.* 27:113–209.

9. Arciero, D., T. Vannelli, M. Logan, and A. B. Hooper. 1989. Degradation of trichloroethylene by the ammonia-oxidizing bacterium *Nitrosomonas europaea. Biochem. Biophys. Res. Commun.* 159:640–643.

10. Asenjo, J. A., and J. S. Suk. 1986. Microbial conversion of methane into poly-B-hydroxybutyrate (PHB): growth and intracellular product accumulation in a type II methanotroph. *J. Ferment. Technol.* 64:271–278.

11. Baedecker, M. J., and S. Lindsay. 1984. Distribution of unstable constituents in ground water near a creosote works, Pensacola, Florida. In *Movement and Fate of Cresote Waste in Groundwater, Pensacola, Florida: U.S. Geological Survey Toxic Waste—Groundwater Contamination Program.* Ed. H. C. Mattraw, Jr., B. J. Franks. U.S. Geological Survey Open-file Report 84–466, Tallahassee, Florida. 13–22.

12. Bartha, R. 1986. Biotechnology of petroleum pollutant biodegradation. *Microb. Ecol.* 12:155–172.

13. Bedard, C., and R. Knowles. 1989. Physiology, biochemistry, and specific inhibitors of CH_4, NH_4^+, and CO oxidation by methanotrophs and nitrifiers. *Microbiol. Rev.* 53:68–84.

14. Bergman, K. 1983. Interactions of trichloroethylene with DNA *in vitro* and with RNA and DNA of various mouse tissues *in vivo. Arch. Toxicol.* 54:181–193.

15. Berry, E. K. M., N. Allison, A. J. Skinner, and R. A. Cooper. 1979. Degradation of the selective herbicide 2,2-dichloropropionate (dalapon) by a soil bacterium. *J. Gen. Microbiol.* 110: 39–45.

16. Bolt, H. M., and J. G. Filser. 1977. Irreversible binding of chlorinated ethylenes to macromolecules. *Environ. Health Perspect.* 21:107–112.

17. Bonse, G., and D. Henschler. 1976. Chemical reactivity, biotransformation, and toxicity of polychlorinated aliphatic compounds. *CRC Crit. Rev. Toxicol.* 4: 395–409.

18. Borden, R. C., P. B. Bedient, M. D. Lee, C. H. Ward, and J. T. Wilson. 1986. Transport of dissolved hydrocarbons influenced by oxygen-limited biodegradation. 2. Field application. *Water Resources Res.* 22:1983–1990.

19. Bossert, I., and R. Bartha. 1984. The fate of petroleum in soil ecosystems. In *Petroleum Microbiology.* Ed. R. M. Atlas. New York: Macmillan Publishing Co. 434–476.

20. Bouwer, E. J., and P. L. McCarty. 1983. Transformation of 1- and 2-carbon halogenated aliphatic compounds under methanogenic conditions. *Appl. Environ. Microbiol.* 45:1286–1294.

21. Bridson, E. Y., and A. Brecker. 1970. Design and formulation of microbial culture media. In *Methods in Microbiology,* vol. 3A. Ed. J. R. Norris, and D. W. Ribbons. London: Academic Press, Inc. 229–295.

22. Broholm, K., B. K. Jensen, T. H. Christensen, and L. Olsen. 1990. Toxicity of 1,1,1-trichloroethane and trichloroethylene on a mixed culture of methane-oxidizing bacteria. *Appl. Environ. Microbiol.* 56:2488–2493.

23. Brusseau, G. A., H. C. Tsien, R. S. Hanson, and L. P. Wackett. 1990. Optimization of trichloroethylene oxidation by methanotrophs and the use of a colorimetric assay to detect soluble methane monooxygenase activity. *Biodegradation* 1:19–29.

24. Burrows K. J., A. Cornish, D. Scott, and I. G. Higgins. 1984. Substrate specificities of the soluble and particulate methane monooxygenases of *Methylosinus trichosporium* OB3b. *J. Gen. Microbiol.* 130:3327–3333.

25. Colby, J., D. I. Stirling, and H. Dalton. 1977. The soluble methane monooxygenase of *Methylococcus capsulatus* (Bath), its ability to oxygenate n-alkanes, n-alkenes, ethers, and alicyclic, aromatic, and heterocyclic compounds. *Biochem. J.* 165:395–402.

26. Collins, M. L. P., L. A. Buchholz, and C. C. Remsen. Effect of copper on *Methylomonas albus* BG8. *Appl. Environ. Microbiol.* 57:1261–1264.

27. Conrad, R., and W. Seiler. 1982. Utilization of traces of carbon monoxide by aerobic oligotrophic microorganisms in ocean, lake, and soil. *Arch. Microbiol.* 132:41–46.

28. Criddle, C. S. 1989. *Anaerobic Transformation of Carbon Tetrachloride as Sole Substrate and in Mixtures with Hexachloroethane and/or Nitrate, by Subsurface Microorganisms.* Ph.D. thesis, Stanford University, Stanford, CA.

29. Cypionka, H., and O. Meyer. 1982. Influence of carbon monoxide on growth and respiration of carboxydobacteria and other aerobic organisms. *FEMS Microbiol. Lett.* 15:209–214.

30. Dalton H., S. D. Prior, D. J. Leak, and S. H. Stanley. 1984. Regulation and control of methane monooxygenase. 75–82. In *Microbial Growth on C-1 Compounds.* Ed. R. L. Crawford, R. S. Hanson Washington, D.C.: Amer. Soc. Microbiol.

31. Dalton H., and D. I. Stirling. 1982. Co-metabolism. *Philos. Trans. R. Soc. London Ser. B* 297:481–496.

32. Davies, S. L., and R. Whittenbury. 1970. Fine structure of methane and other hydrocarbon-utilizing bacteria. *J. Gen. Microbiol.* 61:227–232.

33. Dekant, W., Metzler, M., and D. Henschler. 1984. Novel metabolites of trichloroethylene through dechlorination reactions in rats, mice, and humans. *Biochem. Pharmacol.* 33:2021–2027.

34. Dibble, J. T., and R. Bartha. 1979. Effect of environmental parameters on biodegradation of oil sludge. *Appl. Environ. Microbiol.* 37:729–739.

35. DiSpirito, A. A., J. Gulledge, A. K. Shiemke, J. C. Murrell, and C. Krema. 1991. Trichloroethylene oxidation by the membrane-associated methane monooxygenase in type I, type II, and type X methanotrophs. *Abstr. 91st Annu. Meet. Am. Soc. Microbiol. 1991.* K-29, p. 216.

36. Evans, C. G. T., D. Herbert, and D. W. Tempest. 1970. The continuous cultivation of microorgansms. In *Methods in Microbiology,* vol. 2. Ed. J. R. Norris, D. W. Ribbons. London: Academic Press, Inc. 277–327.

37. Ewers, J., D. Freier-Schroder, and H. J. Knackmuss. 1990. Selection of trichloroethylene (TCE) degrading bacteria that resist inactivation by TCE. *Arch. Microbiol.* 154:410–413.

38. Ferenci, T. 1974. Carbon monoxide-stimulated respiration in methane-utilizing bacteria. *FEBS Lett.* 41:94–98.

39. Ferenci, T., T. Strom, and J. R. Quayle. 1975. Oxidation of carbon monoxide and methane by *Pseudomonas methanica. J. Gen. Microbiol.* 91:79–91.

40. Fliermans, C. B., T. J. Phelps, D. Ringelberg, A. T. Mikell, and D. C. White. 1988. Mineralization of trichloroethylene by heterotrophic enrichment cultures. *Appl. Environ. Microbiol.* 54:1709–1714.

41. Fogel, M. M., A. R. Taddeo, and S. Fogel. 1986. Biodegradation of chlorinated ethenes by a methane-utilizing mixed culture. *Appl. Environ. Microbiol.* 51:720–724.

42. Folsom, B. R., P. J. Chapman, and P. H. Pritchard. 1990. Phenol and trichloroethylene degradation by *Pseudomonas cepacia* G4: kinetics and interactions between substrates. *Appl. Environ. Microbiol.* 56:1279–1285.

43. Fox, B. G., J. G. Bourneman, L. P. Wackett, and J. D. Lipscomb. 1990. Haloalkene oxidation by the soluble methane monooxygenase from *Methylosinus trichosporium* OB3b: mechanistic and environmental implications. *Biochemistry* 29:6419–6427.

44. Fox, B. G., and J. D. Lipscomb. 1990. Methane monooxygenase: a novel biological catalyst for hydrocarbon oxidation. In *Biological Oxidation Systems,* vol. 1. Ed. G. Hamilton, C. Reddy, K. M. Madyastha. Orlando, FL: Academic Press, Inc. 367–388.

45. Ghiorse, W. C., and D. L. Balkwill. 1983. Enumeration and morphological characterization of bacteria indigenous to subsurface environments. Dev. Ind. Microbiol. 24:213–224.

46. Goldman, P. G., W. A. Milne, and D. B. Keister. 1968. Carbon-halogen bond cleavage. III. Studies on halidohydrolases. J. Biol. Chem. 243:428–434.

47. Godsy, E. M., and G. G. Ehrlich. 1978. Reconnaissance for microbial activity in the Magothy aquifer, Bay Park, New York, four years after artificial recharge. J. Res. U.S. Geol. Survey 6:829–836.

48. Grady, C. P. L., Jr., and H. C. Lim. 1980. Biological Wastewater Treatment. New York: Marcel Dekker, Inc. 306–317.

49. Green, J., and H. Dalton. 1986. Steady-state kinetic analysis of soluble methane-monooxy-genase from Methylococcus capsulatus (Bath). Biochem. J. 236:155–162.

50. Green, J., and H. Dalton. 1989. Substrate specificity of soluble methane monooxygenase, mechanistic implications. J. Biol. Chem. 264:17698–17703.

51. Hardman, D. J., and J. H. Slater. 1981. Dehalogenases in soil bacteria. J. Gen. Microbiol. 123:117–128.

52. Harker, A. R., and Y. Kim. 1990. Trichloroethylene degradation by two independent aromatic-degrading pathways in Alcaligenes eutrophus JMP134. Appl. Environ. Microbiol. 56:1179–1181.

53. Harrison, D. E. F. 1973. Studies on the affinity of methanol-and methane-utilizing bacteria for their carbon substrates. J. Appl. Bacterol. 36:308.

54. Hartsman, S., J. A. M. deBont, J. Tramper, and K. Ch.A. M. Luyben. 1985. Bacterial degradation of vinyl chloride. Biotechnol. Lett. 7:383–388.

55. Hathway, D. E. 1980. Consideration of the evidence for mechanisms of 1,1,2-trichloro-ethylene metabolism, including new identification of its dichloroacetic acid and trichloro-acetic acid metabolites in mice. Cancer Lett. 8:263–269.

56. Hegeman, G. 1980. Oxidation of carbon monoxide by bacteria. Trends Biochem. Sci. 5:256–259.

57. Henry, S. M. 1991. Trichloroethylene Transformation by Methanotrophs from a Ground-water Aquifer. Ph.D. Thesis, Stanford University, Stanford, CA.

58. Henry, S. M., and D. Grbić-Galić. 1989. TCE transformation by mixed and pure ground-water cultures. In In-situ aquifer restoration of chlorinated aliphatics by methanotrophic bacteria. Ed. P. V. Roberts, L. Semprini, G. D. Hopkins, D. Grbić-Galić, P. L. McCarty, M. Reinhard. EPA Technical Report EPA/600/2-89/033. R. S. Kerr Environmental Research Laboratory, U.S. EPA, Ada, Okla. 109–125.

59. Henry, S. M., and D. Grbić-Galić. 1990. Effect of mineral media on trichloroethylene oxidation by aquifer methanotrophs. Microb. Ecol. 20:151–169.

60. Henry, S. M., and D. Grbić-Galić. 1991. Influence of endogenous and exogenous electron donors and trichloroethylene oxidation toxicity on trichloroethylene oxidation by methanotrophic cultures from a groundwater aquifer. Appl. Environ. Microbiol. 57:236–244.

61. Henry, S. M., and D. Grbić-Galić. 1991. Inhibition of trichloroethylene oxidation by the transformation intermediate carbon monoxide. Appl. Environ. Microbiol. 57:1770–1776.

62. Henschler, D. 1977. Metabolism and mutagenicity of halogented olefins—a comparison of structure and activity. Environ. Health Perspect. 21:61–64.

63. Henschler, D. 1985. Halogenated alkenes and alkynes. In Bioactivation of Foreign Compounds. Ed. M. W. Anders. Orlando, FL: Academic Press, Inc. 317–347.

64. Henschler, D., W. R. Hoos, H. Fetz, E. Dallmeier, and M. Metzler. 1979. Reactions of tri-chloroethylene epoxide in aqueous systems. Biochem. Pharmacol. 28:543–548.

65. Henson, J. M., M. V. Yates, and J. W. Cochran. 1989. Metabolism of chlorinated methanes, ethanes and ethylenes by a mixed bacteria culture growing on methane. J. Industr. Microbiol. 4:29–35.

66. Henson, J. M., M. V. Yates, J. W. Cochran, and D. L. Shackleford. 1988. Microbial removal of halogenated methanes, ethanes, and ethylenes in aerobic soil exposed to methane. *FEMS Microbiol. Ecol.* 53:193–201.

67. Higgins, I. J. 1979. Methanotrophy. In *International Review of Biochemistry*, Vol. 21, *Microbial Biochemistry*. Ed. J. R. Quayle. Baltimore, MD: University Park Press. 300–353.

68. Higgins, I. J., D. J. Best, R. C. Hammond, and D. Scott. 1981. Methane-oxidizing microorganisms. *Microbiol. Rev.* 45:556–590.

69. Higgins, I. J., R. C. Hammond, F. S. Sariaslani, D. Best, M. M. Davies, S. E. Tryhorn, and F. Taylor. 1979. Biotransformation of hydrocarbons and related compounds by whole organism suspensions of methane-grown *Methylosinus trichosporium* OB3b. *Biochem. and Biophys. Res. Comm.* 89:671–677.

70. Hou, C. T. 1984. Microbiology and biochemistry of methylotrophic bacteria. In *Methylotrophs: Microbiology, Biochemistry, and Genetics*. Ed. C. T. Hou. Boca Raton, FL: CRC Press, Inc. 1–53.

71. Hou, C. T., R. N. Patel, A. I. Laskin, and N. Barnabe. 1979. Microbial oxidation of gaseous hydrocarbons: epoxidation of C_2 to C_4 n-alkenes by methylotrophic bacteria. *Appl. Environ. Microbiol.* 38:127–134.

73. Hubley, J. H., J. R. Mitton, and J. F. Wilkinson. 1974. The oxidation of carbon monoxide by methane-oxidizing bacteria. *Arch. Microbiol.* 95:365–368.

74. Infante, P. F., and T. A. Tsongas. 1982. Mutagenic and oncogenic effects of chloromethanes, chloroethanes, and halogenated analogues of vinyl chloride. *Environ. Sci. Res.* 25:301–327.

75. Jamison, V. W., R. L. Raymond, and J. O. Hudson, Jr. 1975. Biodegradation of high-octane gasoline in groundwater. *Dev. Ind. Microbiol.* 16:305–311.

76. Janssen, D. B., G. Grobben, R. Hoekstra, R. Oldenhuis, and B. Witholt. 1988. Degradation of *trans*-1,2-dichloroethylene by mixed and pure cultures of methanotrophic bacteria. *Appl. Microbiol. Biotechnol.* 29:392–399.

77. Janssen, D. B., G. Grobben, and B. Witholt. 1987. Toxicity of chlorinated aliphatic hydrocarbons and degradation by methanotrophic consortia. *Proc. 4th Eur. Congr. on Biotechnol.* 3:515–518.

78. Janssen, D. B., A. Scheper, L. Dijkhuizen, and B. Witholt. 1985. Degradation of halogenated aliphatic compounds by *Xanthobacter autotrophicus* GJ10. *Appl. Environ. Microbiol.* 49:673–677.

79. Jobson, A., F. D. Cook, and D. W. S. Westlake. 1972. Microbial utilization of crude oil. *Appl. Microbiol.* 23:1082–1089.

80. Joergenson, C., J. Aamand, T. Madsen, and P. Westermann. 1990. Effect of copper limitation on degradation of trichloroethylene by pure cultures of methanotrophs. *Abstr. EERO-GBF Int. Symp. Environmental Biotechnology, Braunschweig, Germany. 1990.*

81. Joergensen, L. 1985. The methane mono-oxygenase reaction system studied in vivo by membrane-inlet mass spectrometry. *Biochem. J.* 225:441–448.

82. Joergensen, L., and H. Degn. 1983. Mass spectrometric measurements of methane and oxygen utilization by methanotrophic bacteria. *FEMS Microbiol. Lett.* 20:331–335.

83. Kim, Y. M., and G. D. Hegeman. 1983. Oxidation of carbon monoxide by bacteria. *Int. Rev. Cytol.* 81:1–32.

84. Lamanna, C., and M. F. Mallete. 1965. *Basic Bacteriology*, 3rd ed. Baltimore, MD: Williams and Wilkins Co.

85. Lamb, S. C., and J. C. Garver. 1980. Batch- and continuous-culture studies of a methane-utilizing mixed culture. *Biotechnol. Bioeng.* 22:2097–2118.

86. Lanzarone, N. A., K. P. Mayer, M. E. Dolan, D. Grbić-Galić, and P. L. McCarty. 1989. Batch exchange soil column studies of biotransformation by methanotrophic bacteria. In *In-situ* aquifer restoration of chlorinated aliphatics by methanotrophic bacteria. Ed. P. V.

Roberts, L. Semprini, G. D. Hopkins, D. Grbić-Galić, P. L. McCarty, M. Reinhard. EPA Technical Report EPA/600/2-89/033. R. S. Kerr Environmental Research Laboratory, U.S. EPA, Ada, OK: 126–146.

87. Lanzarone, N. A., and P. L. McCarty. 1990. Column studies on methanotrophic degradation of trichloroethylene and 1,2 dichloroethane. *Ground Water* 28:910–919.

88. Large, P. J. 1983. *Methylotrophy and Methanogenesis.* Washington, D.C.: American Society for Microbiology.

89. Lauff, J. J., D. B. Steele, L. A. Coogan, and J. M. Breitfeller. 1990. Degradation of the ferric chelate of EDTA by a pure culture of *Agrobacterium* sp. *Appl. Environ. Microbiol.* 56:3346–3353.

90. Leahy, J. G., and R. R. Colwell. 1990. Microbial degradation of hydrocarbons in the environment. *Microbiol. Rev.* 54:305–315.

91. Leak, D. J., and H. Dalton. 1986. Growth yields of methanotrophs: 1. effect of copper on the energetics of methane oxidation. *Appl. Microbiol. Biotechnol.* 23:470–476.

92. Lee, M. D., J. M. Thomas, R. C. Borden, P. B. Bedient, C. H. Ward, and J. T. Wilson. 1988. Biorestoration of aquifers contaminated with organic compounds. *CRC Crit. Rev. Environ. Control* 18:29–89.

93. Lee, M. D., and C. H. Ward. 1985. Biological methods for the restoration of contaminated aquifers. *Environ. Toxicol. Chem.* 4:743–750.

94. Lehtomaki, M., and S. Niemela. 1975. Improving microbial degradation of oil in soil. *Ambio* 4:126–129.

95. Linton, J. D., and R. E. Cripps. 1978. The occurrence and identification of intracellular polyglucose storage granules in *Methylococcus* NCIB 11083 grown in chemostat on methane. *Arch. Microbiol.* 117:41–48.

96. Little, C. D., A. V. Palumbo, S. E. Herbes, M. E. Lidstrom, R. L. Tyndall, and P. J. Gilmer. 1988. Trichloroethylene biodegradation by a methane-oxidizing bacterium. *Appl. Environ. Microbiol.* 54:951–956.

97. Macdonald,T. L. 1983. Chemical mechanisms of halocarbon metabolism. *CRC Crit. Rev. Toxicol.* 11:85–120.

98. Madsen, E. L., and M. Alexander. 1985. Effects of chemical speciation on the mineralization of organic compounds by microorganisms. *Appl. Environ. Microbiol.* 50:342–349.

99. Maltoni, C., and G. Lefemine. 1974. Carcinogenicity bioassays of vinylchloride. I. Research plan and early results. *Environ. Res.* 7:387–396.

100. Mayer, K. P., D. Grbić-Galić, L. Semprini, and P. L. McCarty. 1988. Degradation of trichloroethylene by methanotrophic bacteria in a laboratory column of saturated aquifer material. *Wat. Sci. Technol.* 20:175–178.

101. McCarty, P. L. 1988. Bioengineering issues related to *in situ* remediation of contaminated soils and groundwater. In *Environmental Biotechnology: Reducing Risks from Environmental Chemicals through Biotechnology.* Ed. G. S. Omenn. New York: Plenum Press. 143–162.

102. McCarty, P. L., L. Semprini, M. E. Dolan, T. C. Harmon, C. Tiedeman, and S. M. Gorelick. 1991. *In situ* methanotrophic bioremediation for contaminated groundwater at St. Joseph, Michigan. *Proc. Int. Symp. In Situ and On Site Bioreclamation, San Diego, CA. 1991.*

103. Means, J. L., T. Kucak, and D. A. Crerar. 1980. Relative degradation rates of NTA, EDTA, and DTPA and environmental implications. *Environ. Pollut. Ser. B: Chem. Phys.* 1:45–60.

104. Meyer, O., and H. G. Schlegel. 1983. Biology of aerobic carbon monoxide-oxidizing bacteria. *Annu. Rev. Microbiol.* 37:277–310.

105. Miller, R. E., and F. P. Guengerich. 1982. Oxidation of trichloroethylene by liver microsomal cytochrome P-450: evidence for chlorine migration in a transition state not involving trichloroethylene oxide. *Biochemistry* 21:1090–1097.

106. Miller, R. E., and F. P. Guengerich. 1983. Metabolism of trichloroethylene in isolated hepatocytes, microsomes, and reconstituted enzyme systems containing cytochrome P-450. *Cancer Res.* 43:1145–1152.

107. Moore, A. T., A. Vira, and S. Fogel. 1989. Biodegradation of *trans*-1,2-dichloroethylene by methane-utilizing bacteria in an aquifer simulator. *Environ. Sci. Technol.* 23:403–407.

108. Nelson, M. J. K., S. O. Montgomery, W. R. Mahaffey, and P. H. Pritchard. 1987. Biodegradation of trichloroethylene and involvement of an aromatic biodegradative pathway. *Appl. Environ. Microbiol.* 53:949–954.

109. Nelson, M. J. K., S. O. Montgomery, E. J. O'Neill, and P. H. Pritchard. 1986. Aerobic metabolism of trichloroethylene by a bacterial isolate. *Appl. Environ. Microbiol.* 52:383–384.

110. Nelson, M. J. K., S. O. Montgomery, and P. H. Pritchard. 1988. Trichloroethylene metabolism by microorganisms that degrade aromatic compounds. *Appl. Environ. Microbiol.* 54:604–606.

111. Neufeldt, V., and D. B. Guralnik, eds. 1988. *Webster's New World Dictionary: Third College Edition.* New York: Simon & Schuster, Inc.

112. Nichols, P. D., G. A. Smith, C. P. Antworth, J. Parsons, J. T. Wilson, and D. C. White. 1987. Detection of a microbial consortium including type II methanotrophs, by use of phospholipid fatty acids in an aerobic halogenated hydrocarbon-degrading soil column enriched with natural gas. *Environ. Toxicol. Chem.* 6:89–97.

113. Niedzielski, J. J., R. M. Schram, T. J. Phelps, S. E. Herbes, and D. C. White. 1989. A total-recycle expanded-bed bioreactor design which allows direct headspace sampling of volatile chlorinated aliphatic compounds. *J. Microbiol. Methods* 10:215–223.

114. Ogram, A. V., R. E. Jessup, L. T. Ou, and P. S. C. Rao. 1985. Effects of sorption on biological degradation rates of (2,4-dichlorophenoxy)acetic acid in soils. *Appl. Environ. Microbiol.* 49:582–587.

115. Oldenhuis, R., J. Y. Oedzes, J. J. van der Waarde, and D. B. Janssen. 1991. Kinetics of chlorinated hydrocarbon degradation by *Methylosinus trichosporium* OB3b and toxicity of trichloroethylene. *Appl. Environ. Microbiol.* 57:7–14.

116. Oldenhuis, R., R. L. J. M. Vink, D. B. Janssen, and B. Witholt. 1989. Degradation of aliphatic hydrocarbons by *Methylosinus trichosporium* OB3b expressing soluble methane monooxygenase. *Appl. Environ. Microbiol.* 55:2819–2826.

117. O'Neill, J. G., and J. F. Wilkinson. 1977. Oxidation of ammonia by methane-oxidizing bacteria and the effects of ammonia on methane oxidation. *J. Gen. Microbiol.* 100:407–412.

118. Palumbo, A. V., W. Eng, and G. W. Strandberg. 1991. The effects of groundwater chemistry on cometabolism of chlorinated solvents by methanotrophic bacteria. In *Organic Substances and Sediments in Water,* volume 3, *Biological.* Ed. R. A. Baker. Chelsea, MI: Lewis Publishers, Inc. 225–238.

119. Parsons, F., P. R. Wood, and J. DeMarco. 1984. Transformation of tetrachloroethene and trichloroethene in microcosms and groundwater. *Am. Water Works Assoc.* 76:56–59.

120. Patel R. N., C. T. Hou, A. I. Laskin, and A. Felix. 1982. Microbial oxidation of hydrocarbons: properties of a soluble methane monooxygenase from a facultative methane-utilizing organism, *Methylobacterium* sp. strain CRL-26. *Appl. Environ. Microbiol.* 44:1130–1137.

121. Phelps, T. J., J. J. Niedzielski, K. J. Malachowsky, R. M. Schram, S. E. Herbes, and D. C. White. 1991. Biodegradation of mixed organic wastes in continuous-recycle expanded-bed reactors. *Environ. Sci. Technol.* 25:1461–1465.

122. Phelps, T. J., J. J. Niedzielski, R. M. Schram, S. E. Herbes, and D. C. White. 1990. Biodegradation of trichloroethylene in continuous-recycle expanded-bed reactors. *Appl. Environ. Microbiol.* 54:604–606.

123. Pignatello, J. J. 1986. Ethylene dibromide mineralization in soils under aerobic conditions. *Appl. Environ. Microbiol.* 51:588–592.

124. Prior, S. D., and H. Dalton. 1985. The effect of copper ions on membrane content and methane monooxygenase activity in methanol-grown cells of *Methylococcus capsulatus*, Bath. *J. Gen. Microbiol.* 131:155–163.

125. Ram, N. M., R. F. Christman, and K. P. Cantor, eds. 1990. *Significance and Treatment of Volatile Organic Compounds in Water Supplies.* Chelsea, MI: Lewis Publishers, Inc.

126. Rannug, U. A., A. Johansson, C. Ramel, and C. A. Wachtmeister. 1974. The mutagenicity of vinyl chloride after metabolic activation. *Ambio* 3:194–197.

127. Rasche, M. E., R. E. Hicks, M. R. Hyman, and D. J. Arp. 1990. Oxidation of monohalogenated ethanes and n-chlorinated alkanes by whole cells of *Nitrosomonas europaea*. *J. Bacteriol.* 172:5368–5373.

128. Rasche, M. E., M. R. Hyman, and D. J. Arp. 1990. Factors limiting aliphatic chlorocarbon degradation by *Nitrosomonas europaea*cometabolic inactivation by ammonia monooxygenase and substrate specificity. *Appl. Environ. Microbiol.* 57:2986–2994.

129. Raymond, R. L., J. O. Hudson, and V. W. Jamison. 1975. Oil degradation in soil. *Appl. Environ. Microbiol.* 31:522–535.

130. Raymond, R. L., V. W. Jamison, and J. O. Hudson, Jr. 1976. Beneficial stimulation of bacterial activity in groundwaters containing petroleum products. *AIChE Symp. Ser. 75,* 166:390–404.

131. Roberts, P. V., G. D. Hopkins, D. M. Mackay, and P. L. McCarty. 1990. A field evaluation of *in situ* biodegradation of chlorinated ethenes: part 1. methodology and field site characterization. *Ground Water* 28:591–604.

132. Romanovskaya, V. A., E. S. Lyudvichenko, T. P. Kryshtab, V. G. Zhukov, I. G. Sokolov, and Yu. R. Malashenko. 1980. Role of exogenous fixed carbon dioxide in metabolism of methane-oxidizing bacteria. *Mikrobiologiya* 49:687–694.

133. Salmon, A. G., R. B. Jones, and W. C. Mackrodt. 1981. Microsomal dechlorination of chloroethanes: structure reactivity relationships. *Xenobiotica* 11:723–734.

134. Scow, K. M., S. Simkins, and M. Alexander. 1986. Kinetics of mineralization of organic compounds at low concentrations in soil. *Appl. Environ. Microbiol.* 51:1028–1035.

135. Semprini, L., and P. L. McCarty. 1990. Comparison between model simulations and field results for *in situ* biorestoration of chlorinated aliphatics: part 1. biostimulation of methanotrophic bacteria. *Ground Water* 29:365–374.

136. Semprini, L., P. V. Roberts, G. D. Hopkins, and P. L. McCarty. 1990. A field evaluation of *in situ* biodegradation of chlorinated ethenes: part 2. results of biostimulation and biotransformation experiments. *Ground Water* 28:715–727.

137. Semprini, L., P. V. Roberts, G. D. Hopkins, D. Grbić-Galić, and P. L. McCarty. 1991. A field evaluation of *in situ* biodegradation of chlorinated ethenes: part 3. studies of competitive inhibition. *Ground Water* 29:239–250.

138. Shields, M. S., S. O. Montgomery, P. J. Chapman, S. M. Cuskey, and P. H. Pritchard. 1989. Novel pathway of toluene catabolism in the trichloroethylene-degrading bacterium G4. *Appl. Environ. Microbiol.* 55:1624–1629.

139. Shonnard, D. R., R. B. Knapp, and R. T. Taylor. 1991. Parameters affecting the motility of *Methylosinus trichosporium* OB3b in aqueous solution. *Abstr. 91st Annu. Meet. Am. Soc. Microbiol. 1991.* N-3, p. 246.

140. Siuda, J. F., and J. F. DeBernardis. 1973. Naturally occurring halogenated compounds. *Lloydia* 36:107–143.

141. Slater, H. J., D. Lovatt, A. J. Weightman, E. Senior, and A. T. Bull. 1979. The growth of *Pseudomonas putida* on chlorinated aliphatic acids and its dehalogenase activity. *J. Gen. Microbiol.* 114:125–136.

142. Stanley, S. H., S. D. Prior, D. J. Leak, and H. Dalton. 1983. Copper stress underlies the fundamental change in intracellular location of methane monooxygenase in methane-oxidizing organisms: studies in batch and continuous cultures. *Biotechnol. Lett.* 5:487–492.

143. Steinberg, S. M., J. J. Pignatello, and B. L. Sawney. 1987. Persistence of 1,2-dibromoethane in soils: entrapment in intraparticle micropores. *Environ. Sci. Technol.* 21:1201–1208.

144. Stirling, D. I., and H. Dalton. 1977. Effect of metal-binding agents and other compounds on methane oxidation by two strains of *Methylococcus capsulatus. Arch. Microbiol.* 114:71–76.

145. Stirling, D. I., J. Colby, and H. Dalton. 1979. A comparison of the substrate and electron-donor specificities of the methane monooxygenase from three strains of methane-oxidizing bacteria. *Biochem. J.* 177:361–364.

146. Stirling, D. I., and H. Dalton. 1979. Properties of the methane mono-oxygenase from extracts of *Methylosinus trichosporium* OB3b and evidence for its similiarity to the enzyme from *Methylococcus capsulatus. Eur. J. Biochem.* 96:205–212.

147. Stirling, D. I., and H. Dalton. 1979. The fortuitous oxidation and cometabolism of various carbon compounds by whole-cell suspensions of *Methylococcus capsulatus* (Bath). *FEMS Microbiol. Lett.* 5:315–318.

148. Strand, S. E., M. D. Bjelland, and H. D. Stensel. 1990. Kinetics of chlorinated hydrocarbon degradation by suspended cultures of methane-oxidizing bacteria. *Res. J. Water Pollut. Control Fed.* 62:124–129.

149. Strand, S. E., J. V. Woodrich, and H. D. Stensel. 1991. Biodegradation of chlorinated solvents in a sparged, methanotrophic biofilm reactor. *Res. J. Water Pollut. Control Fed.* 63:859–867.

150. Strandberg, G. W., T. L. Donaldson, and L. L. Farr. 1989. Degradation of trichloroethylene and trans-1,2-dichloroethylene by a methanotrophic consortium in a fixed-film, packed-bed reactor. *Environ. Sci. Technol.* 23:1422–1425.

151. Stucki, G., U. Krebser, and T. Leisinger. 1983. Bacterial growth on 1,2-dichloroethane. *Experentia* 39:1271–1273.

152. Stucki, G., and T. Leisinger. 1983. Bacterial degradation of 2-chloroethanol proceeds via 2-chloroacetic acid. *FEMS Microbiol. Lett.* 16:123–126.

153. Stumm, W., and J. J. Morgan. 1981. *Aquatic Chemistry. An Introduction Emphasizing Chemical Equilibria in Natural Waters.* 2nd ed. New York: John Wiley and Sons.

154. Swindoll, C. M., C. M. Aelion, and F. K. Pfaender. 1988. Influence of inorganic and organic nutrients on aerobic biodegradation and on adaptation response of subsurface microbial communities. *Appl. Environ. Microbiol.* 54:212–217.

155. Tonge, G. M., D. E. F. Harrison, and I. J. Higgins. 1977. Purification and properties of the methane mono-oxygenase enzyme system from *Methylosinus trichosporium* OB3b. *Biochem. J.* 161:333–344.

156. Tsien, H. C., G. A. Brusseau, R. S. Hanson, and L. P. Wackett. 1989. Biodegradation of trichloroethylene by *Methylosinus trichosporium* OB3b. *Appl. Environ. Microbiol.* 55:3155–3161.

157. Uchiyama, H., T. Nakajima, O. Yagi, and T. Tabuchi. 1989. Aerobic degradation of trichloroethylene at high concentration by a methane-utilizing mixed culture. *Agric. Biol. Chem.* 53:1019–1024.

158. U.S. Environmental Protection Agency. 1980. Ambient water quality criteria for dichloroethylenes. Publication 440/5–80–041. National Technical Information Service, Springfield, Va.

159. U.S. Environmental Protection Agency. 1980. Ambient water quality criteria for tetrachloroethylene. Publication 440/5–80–073. National Technical Information Service, Springfield, Va.

160. U.S. Environmental Protection Agency. 1980. Ambient water quality criteria for trichloroethylene. Publication 440/5–80–077. National Technical Information Service, Springfield, Va.

161. U.S. Environmental Protection Agency. 1980. Ambient water quality criteria for vinyl

chloride. Publication 440/5–80–078. National Technical Information Service, Springfield, Va.

162. Vandenbergh, P. A., and B. S. Kunka. 1988. Metabolism of volatile chlorinated aliphatic hydrocarbons by *Pseudomonas fluorescens*. *Appl. Environ. Microbiol.* 54:2578–2579.

163. Van Dyke, R. A., and C. G. Wineman. 1971. Dechlorination of chloroethanes and propanes in vitro. *Biochem. Pharmacol.* 20:463–470.

164. Vannelli, T., M. Logan, D. M. Arciero, and A. B. Hooper. 1990. Degradation of halogenated aliphatic compounds by the ammonia-oxidizing bacterium *Nitrosomonas europaea*. *Appl. Environ. Microbiol.* 56:1169–1171.

165. Ventullo, R. M., and R. J. Larson. 1985. Metabolic diversity and activity of heterotrophic bacteria in groundwater. *Environ. Toxicol. Chem.* 4:759–771.

166. Vogel, T. M., C. S. Criddle, and P. L. McCarty. 1987. Transformations of halogenated aliphatic compounds. *Environ. Sci. Technol.* 21:722–736.

167. Vogel, T. M., and P. L. McCarty. 1985. Biotranformation of tetrachloroethylene to trichloroethylene, dichloroethylene, vinyl chloride, and carbon dioxide under methanogenic conditions. *Appl. Environ. Microbiol.* 49: 1080–1083.

168. Wackett, L. P., G. A. Brusseau, S. R. Householder, and R. S. Hanson. 1989. Survey of microbial oxygenases: trichloroethylene degradation by propane-oxidizing bacteria. *Appl. Environ. Microbiol.* 55: 2960–2964.

169. Wackett, L. P., and D. T. Gibson. 1988. Degradation of trichloroethylene by toluene dioxygenase in whole-cell studies with *Pseudomonas putida* F1. *Appl. Environ. Microbiol.* 54:1703–1708.

170. Wackett, L. P., and S. R. Householder. 1989. Toxicity of trichloroethylene to *Pseudomonas putida* F1 is mediated by toluene dioxygenase. *Appl. Environ. Microbiol.* 55:2723–2725.

171. Watwood, M. E., C. S. White, and C. N. Dahm. 1991. Methodological modifications for accurate and efficient determination of contaminant biodegradation in unsaturated calcerous soils. *Appl. Environ. Microbiol.* 57:717–720.

172. Westrick, J. J., J. W. Mello, and R. G. Thomas. 1984. The ground-water supply survey. *J. Amer. Water Works Assoc.* 76:52–59.

173. Whittenbury, R., K. C. Phillips, and J. F. Wilkinson. 1970. Enrichment, isolation, and some properties of methane-utilizing bacteria. *J. Gen. Microbiol.* 61:205–218.

174. Wilkinson, T. J., and D. E. F. Harrison. 1973. The affinity for methane and methanol of mixed cultures grown on methane in continuous culture. *J. Appl. Bacterol.* 36:309–313.

175. Wilson, B. H., B. Bledsoe, and D. Kampbell. 1987. Biological processes occurring at an aviation gasoline spill site. In *Chemical Quality of Water and the Hydrologic Cycle*. Ed. R. C. Averett, D. M. McKnight. Chelsea, MI: Lewis Publishers, Inc. 125–137.

176. Wilson, J. T., and B. H. Wilson. 1985. Biotransformation of trichloroethylene in soil. *Appl. Environ. Microbiol.* 29:242–243.

177. Winter, R. B., K. M. Yen, and B. D. Ensley. 1989. Efficient degradation of trichloroethylene by a recombinant *Escherichia coli*. *Bio/Technol.* 7:282–285.

178. Wiseman, A., and D. J. King. 1982. Microbial oxygenases—and their potential application. In *Topics in Enzyme and Fermentation Biotechnology*, Vol. 6. Ed. A. Wiseman. London: Ellis Horwood Ltd. 151–206.

179. Woodland, M. P., and H. Dalton. 1984. Purification and characterization of component A of the methane monooxygenase from *Methylococcus capsulatus* (Bath). *J. Biol. Chem.* 259:53–59.

180. Zavarzin, G. A., and A. N. Nozhevnikova. 1977. Aerobic carboxydobacteria. *Microb. Ecol.* 3:305–326.

181. Zylstra, G. J., L. P. Wackett, and D. T. Gibson. 1989. Trichloroethylene degradation by *Escherichia coli* containing the cloned *Pseudomonas putida* F1 toluene dioxygenase genes. *Appl. Environ. Microbiol.* 55:3162–3166.

CHAPTER 16

Properties of Lignin-degrading Peroxidases and Their Use in Bioremediation

ANN B. ORTH, ELIZABETH A. PEASE, and MING TIEN

Abstract. In this chapter, we describe the general properties of lignin peroxidase of the white-rot fungi, especially *Phanerochaete chrysosporium,* and discuss how its characteristics make this enzyme potentially important in bioremediation. The ability of *P. chrysosporium* to degrade a large number of pollutants can be explained by a number of its unique properties; for example, the free radical catalytic mechanism of reaction for both the lignin peroxidases and the manganese peroxidases, the enzyme's stability to temperature and pH, and their high redox potential. For the potential applications of these enzymes to be realized, however, the enzymes must be produced in quantities greater than are generated by these fungi. Therefore, we also discuss lignin-degrading peroxidase production and its optimization, mutant strains of *P. chrysosporium* that may be useful in enhancing production, and scale-up production to yield high levels of enzyme. Finally, we summarize the current understanding of the molecular biology of lignin-degrading peroxidases. The ability to overexpress these enzymes in homologous or heterologous systems may greatly enhance production.

INTRODUCTION

The discovery of ligninases (lignin-degrading peroxidases) by Tien and Kirk (86) and Glenn et al. (34) revealed the mechanism by which lignin is degraded by microorganisms. Lignin is a nonlinear phenylpropanoid polymer consisting predominantly of β-O-4 type linkages, in addition to over 10 other types of bonds. Because of its complex structure, lignin is resistant to most forms of microbial degradation. Only a few groups of filamentous fungi, particularly white-rot fungi, can oxidize lignin to CO_2 in a complex mechanism that has been under investigation for 35 years. This capability also allows white-rot fungi to oxidize a broad range of toxic organic compounds to nontoxic metabolites and CO_2 (17). These compounds include, among others, various petrochemicals, such as polyhydrocarbons and polychlorinated biphenyls, and

Dr. Orth is in Discovery Research at Dow Elanco, Indianapolis, IN 46268-1053, U.S.A. Drs. Pease and Tien are at the Department of Molecular and Cell Biology, Pennsylvania State University, University Park, PA 16803, U.S.A.

agrochemicals, such as fungicides and insecticides. A partial listing of compounds that are degraded by one white-rot fungus, *Phanerochaete chrysosporium,* is given below:

Polycyclic aromatics
  Benzo[a]pyrene
  Biphenyl
  2-methylnaphthalene
  Phenanthrene
  Benzo[a]anthracene
  Pyrene
  Anthracene
  Perylene
  Dibenzo[p]dioxin

Triphenylmethane dyes
  Crystal violet
  Pararosaniline
  Cresol red
  Bromophenyl blue
  Ethyl violet
  Malachite green
  Brilliant green

Biopolymers
  Lignin
  Cellulose
  Kraft Lignin
  3-chloroaniline-lignin conjugates
  3,4-dichloroaniline-lignin conjugates

Chlorinated aromatics
  4-Chlorobenzoic acid
  Dichlorobenzoic acid
  2,4,6-trichlorobenzoic acid
  4,5-Dichloroguaiacol
  6-Chlorovanillin
  4,5,6-Trichloroguaiacol
  Tetrachloroguaiacol
  Pentachloroguaiacol
  3-chloroaniline
  3,4-dichloroaniline
  2,4,5-Trichlorophenoxyacetic acid
  Pentachlorophenol (PCP)

Polycyclic chlorinated aromatics
  DDT (1,1,1-trichloro-2,2-bis(4-chloro-phenyl)ethane
  2,3,7,8-tetrachlorodibenzo-p-dioxin
  3,4,3',4'-tetrachlorobiphenyl
  2,4,5,2',4,5'-hexachlorobiphenyl
  Aroclor 1254
  Aroclor 1242
  2-Chlorodibenzo[p]dioxin
  Dicofol (2,2,2-trichloro-1,1-bis (4-chloropheyl)ethanol

Chlorinated alkyhalides
  Lindane
  Chlordane

Bioremediation of toxic compounds by ligninase and the practical applications of this system are also discussed in another chapter in this volume (27a).

Two catalytically distinct groups of lignin-degrading enzymes, the lignin peroxidases and the Mn peroxidases (Mn-dependent peroxidases), are thought to mediate the degradation of lignin and other compounds. Many detailed reviews on ligninases and their role in lignin biodegradation have been published (52, 84, 87). In this chapter, we discuss the enzymatic properties of lignin-degrading peroxidases that impart them with the ability to degrade these recalcitrant compounds. For bioremediation of these toxic compounds to be possible on a practical level, however, ligninases must be available in large quantities. Therefore, we also address the problem of producing the enzyme in amounts sufficient to make practical applications possible by manipulation of fungal strains and culture conditions. Finally, we discuss the molecular aspects of ligninases and the possibilities for using such knowledge to enhance biodegradation.

LIGNIN-DEGRADING PEROXIDASES OF
PHANEROCHAETE CHRYSOSPORIUM

The lignin-degrading system is apparently involved in the degradation of most compounds mineralized by P. chrysosporium. Therefore, much of what is known about the enzymology and regulation of lignin degradation is also true for bioremediation. Although many organisms have been screened for their ability to degrade lignin, the white-rot fungus P. chrysosporium Burds. was found to be most promising because of its high rate of lignin degradation, rapid sporulation, and rapid growth in culture (50). Optimal growth conditions for production of ligninases were then established (49) and refined. These conditions are discussed in a later section.

P. chrysosporium produces a family of extracellular glycosylated, heme-containing lignin peroxidases that can be separated on a fast protein liquid chromatography (FPLC) mono-Q column. These peroxidases were designated according to their order of elution from this column: H1 (pI = 4.7), H2 (pI = 4.4), H6 (pI = 3.7), H7 (pI = 3.6), H8 (pI = 3.5), and H10 (pI = 3.3) are the lignin peroxidases that have veratryl alcohol-oxidizing capability (Farrell et al. 26); H3 (pI = 4.9), H4 (pI = 4.5), H5 (pI = 4.2), and H9 are the Mn-dependent peroxidases that do not have veratryl alcohol-oxidizing capability (33, 38). They oxidize primarily phenolic compounds, as well as various dyes and, amines (32, 56). These enzymes are highly homologous, with molecular weights ranging from 38,000 to 46,000 (26). Immunological cross-reactivity, DNA sequencing, and peptide mapping indicate very similar primary and tertiary structures (70). Many of the physical properties of these extracellular peroxidases are given in Table 1.

Table 1. Physical properties of ligninase isozymes.

Isozyme	Molecular weight	Carbohydrate	Extinction coefficient[a]	pI	Enzyme type[b]
H1	38,000	+	169	4.7	LP
H2	38,000	+	165	4.4	LP
H3	45,000	+	125	4.9	MnP
H4	45,000	+	127	4.5	MnP
H5	45,000	+		4.2	MnP
H6	43,000	+	162	3.7	LP
H7	42,000	+	177	3.6	LP
H8	42,000	+	168	3.5	LP
H10	43,000	+	182	3.3	LP

[a] Extinction coefficients given as mM^{-1}.
[b] LP = lignin peroxidase; MnP = manganese-dependent peroxidase.

LIGNIN-DEGRADING PEROXIDASES OF OTHER FUNGI

Although *Phanerochaete chrysosporium* is the best-characterized of the lignin-degrading fungi, many other lignin-degrading fungi are now being characterized. Another fairly well-known white-rot fungus, *Phlebia tremellosus*, selectively degrades lignin in aspen and birch wood as determined by transmission electron microscopy

(11). An efficient method for production of ligninases with carrier-bound mycelium was developed in *Phlebia radiata* on a bioreactor scale (45). Recently, a Mn peroxidase was purified from this organism as well (46). *Aspergillus flavus* can use p-coumaryl, one of the major constituents of lignin, as a sole carbon source (39). It can also grow on the insoluble lignin model compound 3,4-dimethoxy-ω-(2-methoxyphenoxy)acetophenone (10). Aspen wood blocks are selectively delignified by the white-rot fungi *Ischnoderma resinosum*, *Poria medulla-panis*, and *Xylobolus frustulatus* (68). Extracellular aromatic oxidases that may oxidize veratryl alcohol by a catalytic route different from the lignin peroxidases of *P. chrysosporium* were found in *Pleurotus sajor-caju*. Veratryl alcohol was also found to be degraded by *Penicillium simplicissimum*, a strain isolated from paper mill wastes (44). Unfortunately, very little has been done with these organisms to assess their bioremediation capabilities. They might eventually prove to be quite useful in bioremediation, since they will have growth requirements different from *P. chrysosporium*, thus giving a broader range of conditions under which lignin-degrading organisms may be utilized.

LIGNIN PEROXIDASES—MECHANISM OF REACTION

The lignin peroxidases are highly potent oxidizing agents. The role of the various isozymes in lignin biodegradation or bioremediation has not been clearly established. The initial reports of ligninase-catalyzed lignin degradation demonstrated depolymerizarion of synthetic (34) and methylated lignin by lignin peroxidase (86), but the first report of depolymerization of native lignin *in vitro* did not appear until 1991. Hammel and Moen (36) have shown that crude *Phanerochaete chrysosporium* lignin peroxidase, in the presence of H_2O_2 and veratryl alcohol, will catalyze the partial degradation of a $^{14}C_\beta$-labeled synthetic hardwood lignin to fragments with molecular weights as low as 170. The greatest progress in understanding how these enzymes work, however, has emerged came from studies of lignin model compounds (for a review, see 84). Forney et al. (28) found a temporal relationship between ligninolytic activity and the appearance of H_2O_2, which allowed the discovery of these peroxidases. Incubation of lignin model compounds with extracellular fluid from ligninolytic cultures from *P. chrysosporium* and H_2O_2 results in a C_α-C_β cleavage (34, 86).

According to kinetic and spectroscopic analyses of this family of enzymes, the catalytic mechanism is quite similar to that of other peroxidases (3). The resting enzyme is oxidized by a two-electron transfer from H_2O_2 to form compound I. Compound I oxidizes a substrate molecule by one electron, forming compound II and a free radical product. Compound II can react with another substrate molecule, forming a resting enzyme and a free radical product. These free radicals then undergo nonenzymatic reactions to form the final products.

$$\text{ferric peroxidase} + H_2O_2 \rightarrow \text{compound I} + H_2O$$

$$\text{compound I} + A^- \rightarrow \text{compound II} + A\cdot$$

$$\text{compound II} + A^- \rightarrow \text{ferric peroxidase} + A\cdot$$

The work of Ortiz de Montellano (67) has shown that the oxidation of substrates by peroxidases occurs at the heme periphery. This was demonstrated by the use of the suicide inhibitor phenylhydrazine, which covalently modified the peroxidase only at the heme periphery, in contrast to cytochrome P450, which is modified at the heme iron or pyrrole nitrogen (67). Similar results, indicating that the site of substrate oxidation is an exposed heme with a nonspecific binding site, have been obtained with lignin peroxidase (21). The lack of specificity of this site is demonstrated by the seemingly unrelated substrates oxidized by the ligninases. Besides catalyzing the oxidation of veratryl alcohol and cleavage of these model compounds, ligninases also catalyze (a) the oxidation of benzylic alcohols to the corresponding aldehydes; (b) hydroxylation of some benzylic carbons; (c) phenol dimerization; and (d) hydroxylation of C_α–C_β double bonds to form the corresponding diol (Fig. 1; 52, 84).

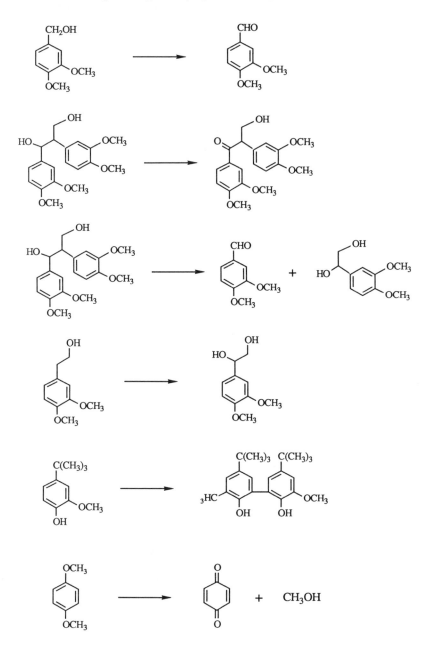

Figure 1. Reactions catalyzed by lignin peroxidase.

The elucidation of a free radical mechanism was made possible by the work of Kersten et al. (48). By electron spin resonance (ESR), these workers showed cation radicals of 1,4-dimethoxybenzene and 1,2,3,4- and 1,2,4,5-tetramethoxybenzene upon lignin peroxidase-catalyzed oxidation of methoxybenzene cogeners. The formation of a free radical was confirmed for lignin model compounds by Hammel et al. (37). ESR spectroscopy also shows formation of free radicals from the ligninase-catalyzed C_α-C_β cleavage in the alkyl side chain of the β-1 model dimethoxyhydrobenzoin (DMHB) (37). The ESR signal indicating carbon-centered radicals was only observed in anaerobic incubations of lignin peroxidase, H_2O_2, DMHB, and the spin trapping agent 5,5-dimethyl-1-pyrroline-N-oxide (DMPO). An ESR signal corresponding to peroxy radicals was observed under aerobic conditions, indicating that this radical process can be propagated. Free radicals of lignin-like compounds have been shown to be relatively stable by ESR spectroscopy and can diffuse away from the active site. Thus, they are under the influence of their environment, including pH, other radicals, and oxygen concentration. The breakdown of many other lignin model compounds has been studied (84). In summary, experimental evidence, including product profiles, stoichiometries, and kinetics, indicates cation radical chemistry for many of these substrates.

A free radical process of catalysis would explain the ability of *Phanerochaete chrysosporium* to degrade a large number of pollutants. Whether a compound is a substrate is determined by its redox potential rather than by whether it fits into an active site. This mechanism explains, in part, the nonspecificity of lignin peroxidase-catalyzed reactions.

Mn PEROXIDASES—MECHANISM OF REACTION

The Mn peroxidases have a reaction mechanism similar to the lignin peroxidases, except that they use Mn (II) as a mediator. Once Mn (II) is oxidized by the enzyme, Mn (III) can then oxidize organic substrate molecules. Mn-peroxidase compounds I and II can oxidize Mn (II) to Mn (III), but compound I can also oxidize some phenolic substrates (95, 96):

$$\text{ferric enzyme} + H_2O_2 \rightarrow \text{compound I} + H_2O$$

$$\text{compound I} + \text{Mn (II)} \rightarrow \text{compound II} + \text{Mn (III)}$$

$$\text{compound I} + AH_2 \rightarrow \text{compound II} + AH\cdot$$

$$\text{compound II} + \text{Mn (II)} \rightarrow \text{ferric enzyme} + \text{Mn (III)}$$

$$\text{Mn (III)} + AH_2 \rightarrow \text{Mn (II)} + AH\cdot$$

Suitably chelated Mn (III) is capable of oxidizing lignin model compounds and lignin. Thus, stabilization of Mn (III) seems to be a key factor in its catalytic activity (29). Many of the observed reactions of manganese peroxidase may be due to a competition between enzymatic Mn (III) formation and nonenzymatic decomposition of Mn (III) by hydrogen peroxide (1).

SIMILARITIES BETWEEN THE CATALYTIC MECHANISMS OF Mn PEROXIDASES AND LIGNIN PEROXIDASES

Mn peroxidases and lignin peroxidases share a unique feature in that they can oxidize substrates that are not readily oxidized by other peroxidases due to their high redox potentials (84). Compared to most other peroxidases, the redox potential (E_{m7} values) of lignin and Mn peroxidases are more positive. The redox potentials of the ferric/ferrous couple is about −140 mV for lignin peroxidases and −90 mV for Mn peroxidases, values that are considerably higher than the values for horseradish peroxidase (−270 mV) or cytochrome c peroxidase (−195 mV). The higher E_{m7} values for the ferric/ferrous couple suggest that the heme active site of these fungal peroxidases being somewhat more electron-deficient (62), in turn suggesting that the Mn peroxidase or lignin peroxidase compound I or II intermediates are more electron-deficient and hence have higher oxidation-reduction potential. The higher oxidation-reduction pottential would allow them to oxidize substrates of higher redox potential. These properties would prove quite useful in bioremediation because they would allow for degradation of compounds that are recalcitrant due to their high redox potentials.

FACTORS AFFECTING THE REDOX POTENTIAL OF THE HEME

The structural basis for the relatively high redox potential of the lignin peroxidases has been investigated by ^{1}H-NMR spectroscopy. ^{1}H-NMR spectroscopy is a powerful tool for characterizing the active sites of heme proteins and has shed light on structure-function relationships for several heme proteins. La Mar and coworkers (57) have shown that this technique is well suited for many heme proteins. The NMR studies show that the active site structure of lignin peroxidase is very similar to those of horseradish peroxidase and cytochrome c peroxidase. The data show evidence for the presence of a distal and axial histidine, in agreement with resonance Raman results (54) and with the amino acid sequence, deduced from the cDNA sequence, which revealed conservation of active site residues (89).

Despite the similarities between the lignin peroxidase and horseradish peroxidase spectra, some important differences are observed. Most notable are the shifted values for the protons of the proximal histidine. Considerable attention has been focused on the role of H-bonding of the proximal histidine in controlling heme electron density and reactivity. The strength of the imizadole H-bonding to neighboring amino acids has a significant influence on the imidazolate character of the axial histidine (Fig. 2). In cytochrome c peroxidase, the proximal histidine is strongly H-bonded to an aspartate residue (Asp 235), whereas in myoglobin, it is weakly H-bonded to the carbonyl of the peptide backbone.

Figure 2. Hydrogen-bonding network of imidazole of proximal histidine.

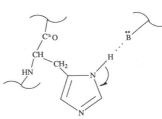

Studies of model compounds have suggested that the extent of the up-field shift of the He1 proton of the proximal histidine is related to the imidazolate character of the histidine ring (79). Indeed, the most relevant differences in the spectra of cyanide (CN) adducts of lignin peroxidase, horseradish peroxidase, and cytochrome c peroxidase are the shift values of the signal of the proximal histidine. The Hε1 protein in lignin peroxidase is less up-field-shifted than in horseradish peroxidase-CN (83) and cytochrome c peroxidase-CN (57). The shift of the Hε1 in the cyanide adducts is inversely proportional to the redox potential $(E_0 \text{ Fe}^{3+}/\text{Fe}^{2+})$ of the native proteins in the series metmyoglobin, lignin peroxidase, cytochrome c peroxidase, and horseradish peroxidase (Fig. 3). The relationship between the chemical shifts of the Hε1 proton and the redox potential indicates that lignin peroxidase has less imidazolate character than horseradish peroxidase and cytochrome c peroxidase.

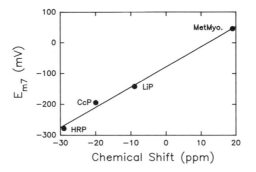

Figure 3. Relationship of the redox potential E_{my} (Fe $^{3+}/\text{Fe}^{2+}$) to chemical shift of Hε1 of the proximal histidine (His) for horseradish peroxidase, cytochrome c peroxidase, lignin peroxidase, and metmyoglobin (9).

The NMR studies are consistent with our hypothesis and the present understanding of how the protein affects the electron density of the heme. The data suggest that the proximal histidine of lignin peroxidase is more strongly H-bonded than myoglobin but less strongly bonded than horseradish peroxidase and cytochrome c peroxidase, thus imparting less imidazolate character on the proximal histidine. Consequently, the higher oxidation state of Fe^{4+} in compound I and II of lignin peroxidase would not be as stabilized as it is in horseradish peroxidase or cytochrome c peroxidase and therefore would be more electron-deficient. A more destablized Fe^{4+} would generate a more reactive higher oxidation state capable of oxidizing recalcitrant compounds.

STABILITY OF LIGNIN-DEGRADING PEROXIDASES

The study of lignin peroxidase H2 has shown that it is sufficiently stable under variable conditions of temperature and pH to be useful in chemical waste treatment systems and in biopulping (92). Without this property of stability, the practical industrial applications of any enzyme would be quite limited. Tuisel et al. (92) have shown that at pH 4.5, the enzyme had 60% of its initial activity after 2 hours; at pH 5.5, the enzyme had 95% of its initial activity after 2 hours. Since the pH optimum is 2.5, this increase in stability was obtained at the cost of lowered enzyme activity; however, as temperature was increased, the pH of optimum activity was also increased.

Lignin peroxidase H2 also showed high stability to temperature. It could be stored without great loss in activity for 48 hours at temperatures of up to 50°C. The enzyme retained approximately 80% of its activity at 40°C for 21 hours in the pH range of 4.0–6.5. Above pH 7.5 and below 4.0, all activity was lost within 5 hours.

REGULATION OF LIGNIN-DEGRADING PEROXIDASE PRODUCTION

In addition to being stable, lignin-degrading peroxidases must be readily produced by microorganisms in large quantities for them to be useful in practical applications such as bioremediation. Many studies have been performed to determine the optimal conditions for expression of the lignin peroxidases. The results are central to any strategy to use whole fungi on remediation sites. Most of the work on regulation was done with the wild type (WT, ME446, or BKM-F-1767) of *Phanerochaete chrysosporium,* the best-characterized of the white-rot fungi. In general, these enzymes are produced only under nutrient limitation during secondary metabolism. Nitrogen (49), carbon, or sulfur (43) limitation have been shown to induce ligninolytic activity. The addition of L-glutamate, glutamine, histidine, or $NH4^{4+}$ to ligninolytic cultures suppressed ligninase activity (27). In addition, excess nutrients and cycloheximide suppress ligninase activity, whereas lignin, lignin model compounds, and exposure to 100% O_2 increase activity (23). Tonon et al. (91) have studied the regulation of nitrogen metabolism pathways as they relate to the repression of the ligninolytic system by nitrogen or carbon. Their results suggest that a regulatory gene mediates nitrogen catabolite repression. When glycerol was used as a carbon source, the amount of lignin peroxidase that was produced varied indirectly with the growth of the fungal strain on this relatively poor carbon source (77).

Although veratryl alcohol is produced by the fungus during secondary metabolism (81), supplemental veratryl alcohol in the culture medium enhances ligninase activity (24, 90). This has been proposed to be due to induction of enzyme production (24), or protection of the enzyme from inactivation (90). In *Phlebia radiata,* veratric acid has a stimulatory effect on the production of extracellular enzymes (61). Veratryl alcohol and the nonphenolic β-O-4 dimer also stimulated lignin peroxidase and Mn-dependent peroxidase activities in this organism (65). Kirk et al. (51) also found an increase in ligninase activity from the addition of veratryl alcohol to shallow, stationary cultures.

Certain trace metals, including Cu^{2+} and Mn^{2+}, are known to enhance ligninolytic activity (51). This knowledge has important implications in treatment of soil that has a varying trace metal content. Brown et al. (15) saw enhanced lignin peroxidase activity upon addition of Mn^{2+} and suggested that manganese is involved in transcriptional regulation of the manganese peroxidase gene. In contrast, Kern (47) found an increase in lignin peroxidases, but not manganese peroxidases or glyoxal oxidase, when manganese(IV) oxide was added to the cultures. He postulated that manganese(IV) oxide may protect ligninase from inactivation by H_2O_2. This difference in results may be due to the different oxidation states of the manganese, since it is normally added as $MnSO_4$. Bonnarme and Jeffries (13) also found an increase in manganese peroxidases, but not lignin peroxidases, with the addition of manganese. Perez and Jeffries (71) confirmed these results, additionally showing that synthetic lignin mineralization was almost seven-fold higher at low Mn(II) concentration when compared to high Mn(II) concentrations in *Phlebia brevispora* and *Phanerochaete chrysosporium.* Both of these studies were done in shaking culture, however, not stationary culture as was used in previous studies. The effects of manganese are quite variable depending on the culture conditions and the form of manganese added. The implication for practical application is that optimal enzyme production will depend on how the organism to be used is cultured.

Temperature changes during growth of P. chrysosporium have been shown to affect lignin peroxidase production. Although maximum production is generally thought to occur at 37–40°C, there may be actually two temperature optima for lignin peroxidase synthesis (7). In P. chrysosporium strain INA-12, 37°C was optimal for mycelial growth, and 30°C was optimal for lignin peroxidase production.

Altering the culture buffer may influence the ligninolytic activity of P. chrysosporium. Substituting sodium acetate for the usual buffer, sodium 2,2-dimethyl succinate (DMS), increased ligninase production up to three-fold (20). In these cultures, however, H2 and H6 were the predominant lignin peroxidases, whereas in DMS cultures, H8 was the major protein.

Several components of ligninolytic cultures have been found to inhibit lignin peroxidase activity. During purification of the crude enzyme from Phanerochaete chrysosporium, an anionic polysaccharide-containing fraction was found that inhibited lignin peroxidase activity at low pH (53). There are also two different types of extracellular proteases present in WT P. chrysosporium cultures with maxima on day 2 or on day 8 (22). Time courses of protease and ligninase activity were negatively correlated in this study, indicating that protease activities promote the decline of ligninase activity in these types of cultures.

Other components of culture media have been found to affect ligninase production. Tween 80 (sorbitan polyoxyethylene monooleate) and oleic acid alone or oleic acid emulsified with Tween 80 have been shown to enhance lignin peroxidase activity in P. chrysosporium strain INA-12 (8). An increase in mycelial dry weight and extracellular enzyme production due to Tween 80 was also found in the white-rot fungus Pycnopous cinnabarinus (35). Lestan et al. (58) found increased ligninase production when the culture medium was supplemented with oleic acid, Tween 80, or CHAPS 3-[(cholamidopropyl)-dimethylammonio–1-propanesulphonate, which resulted in oleic acid enrichment of whole-cell and polar lipids. Although the mechanism by which detergents enhance extracellular enzyme activity in filamentous fungi is unknown, some suggest that lipases may hydrolyze ester groups, releasing free fatty acids that may affect enzyme production by changing lipid metabolism (8). It is possible, however, that Tween 80 may modify the plasma membrane, altering transport of compounds in and out of the cell, and thus in turn affect enzyme activity. Certain phospholipids have also been shown to increase lignin peroxidase secretion (19). Capdila et al. (19) postulated that a change in the physiology of the organism had occurred by altering energy metabolism. Jager et al. (40) reported that addition of detergents such as Tween 20 or Tween 80 is mandatory for ligninase production in agitated cultures, and suggested that this requirement probably has a physiological basis. Venkatadri and Irvine (93), however, found evidence to support a protective mechanism of the detergents against mechanical inactivation of the ligninolytic enzymes. These advances with shaking culture should make production of ligninase in stirred-tank fermentors possible, thus allowing for scale-up production of these enzymes.

Most of the research on regulation of the ligninolytic system and enzyme production has been focused on the application of lignin degradation. Unfortunately, very little work has been performed on regulation in response to bioremediation. Before these fungi can be utilized maximally for bioremediation, factors affecting their activity will have to be more extensively characterized.

PRODUCTION OF LIGNIN-DEGRADING PEROXIDASES

Use of the isolated ligninase rather than the organism itself for bioremediation has many advantages, including more rapid and reliable degradation under a variety of conditions. This method would eliminate the problem of having to have conditions specific to the lignin-degrading organism in the area to be treated. Thus, large-scale production of the enzyme would be quite advantageous for practical applications. A number of different culture methods have been used to produce the lignin-degrading peroxidases. These are summarized by Janshekar and Feichter (42), who provide insights into approaches for scale-up production of lignin-degrading peroxidase under stirred systems, and demonstrate such methods are possible but need further study. In a stirred tank reactor scaled up to 300 liters, they emphasize the need for low inoculum levels and polypropylene glycol, polyethylene glycol, or hexadecane to activate the peroxidase-producing ability.

Some of the difficulties in reproducibly attaining high levels of enzyme activity from agitated cultures of Phanerochaete chrysosporium wild type can be overcome by using a solid support to immobilize the pellets (41). This was successfully carried out by either entrapment and germination of spores in agarose beads or by the attachment of the mycelium on reticulated polyurethane or nylon-web cubes (97). Very high levels of lignin peroxidase activity (660 U/liter in a total of 8 liters) were achieved with this method (60).

MUTANT STRAINS OF PHANEROCHAETE CHRYSOSPORIUM

Another approach to understanding the role of ligninases in lignin degradation and in optimizing enzyme production is the isolation of mutant strains. Several mutants have been isolated that are deregulated in that they produce lignin peroxidases under nonlimiting nutrient conditions. Buswell et al. (18) reported an isolate (INA-12) that produces ligninase when in the presence of high ammonia nitrogen but not glutamate. Kuwahara and Asada (55) isolated a strain that could also produce ligninase under high nitrogen conditions. Boominathan et al. (14) isolated two nitrogen-deregulated mutants, der8-2 and der8-5. They found manganese peroxidase, lignin peroxidase and glucose oxidase activity in nitrogen-containing medium, but the activity was lower than in the wild-type under limiting conditions. A strain that was isolated by Galliano et al. (31) had enhanced cellulase activity and also showed higher xylanase, protease, and ligninase activity.

More useful would be mutants that are deregulated and produce higher amounts of the enzyme than the wild type. A strategy for obtaining such a mutant was developed by Tien et al. (85). The strategy involves covalently bonding amino acids to lignin model compounds in such a way that ligninase-catalyzed cleavage of the models releases the amino acids for growth nitrogen. A mutant isolated by this procedure, PSBL-1, is fully ligninolytic under high nutrient conditions and produces much higher ligninase activity than the wild type (BKM-F-1767) (88). This mutant was shown to have up to ten-fold higher lignin peroxidase, manganese peroxidase, and glyoxal oxidase activity than the wild type (66).

MOLECULAR BIOLOGY OF LIGNIN BIODEGRADATION

Phanerochaete chrysosporium

The study of the genes encoding lignin peroxidase would be extremely useful, both in the overexpression of the enzymes and in further characterization of their catalytic mechanism. Overexpression in a heterologous system would greatly facilitate isolation of large amounts of enzyme that could be used in bioremediation. Also, cloning the ligninase gene into various strains of bacteria may allow production of the enzyme in diverse environments because of the different optimal growth requirements of each strain. Some of these strains may even grow in extreme conditions. This approach could also allow constant production of the enzyme, since the enzyme would be applied as part of a biological system rather than as isolated enzyme.

The first sequencing and definitive identification of a cDNA encoding lignin peroxidase isozyme H8 (λML-1) was done by Tien and Tu (89). This was followed by Boer et al. (12), who sequenced the gene encoding isozymes H2 (CLG4) and H10 (pCLG5). Andrawis et al. (4) have published on the sequence for isozyme H6 (λML-5) and an allelic variant of isozyme H8 (λML-4). The cDNA clones show a high degree of similarity in the nucleotide sequences; H8, H6, and H10 are the most similar. Manganese peroxidase cDNA clones have also been isolated and reported. The cDNA sequence for the gene encoding isozyme H4 (λMP-1) (69) and an isozyme with a pI of 4.9 (possibly H3) (MnP-1) (73) were recently published. These two peroxidases were 75% similar to each other at the nucleotide level and were 60% similar to the lignin peroxidases at the nucleotide level.

Comparison of the amino acid sequence shows an even higher level of similarity due to wobble in the genetic code and substitution of similar amino acids (59). The highest degree of similarity is 95% between H6 and H8; the lowest is 63% between H4 and H6. At the active site, amino acid similarity is nearly 100% for some clones. In particular, the proximal and distal histidines that flank the active site residues, thought to be necessary for enzyme function, are conserved among many peroxidases (72).

The cDNAs cloned thus far range in size from 1285 to 1312 bp. Lignin peroxidases have a 27–28 amino acid leader sequence with a conserved dibasic Lys-Arg cleavage site, whereas Mn peroxidases have a 21 or 24 amino acid leader sequence with a cleavage site that is not conserved. Potential glycosylation sites identified by the conserved eukaryotic consensus sequence of Asn-X-Thr/Ser can be found in all of these cDNA sequences.

The genomic sequence of isozyme H8 was reported (82) that was highly homologous to the H8 cDNA (89). Other genomic sequences homologous to H8 but not identical were reported that may represent allelic variants (4, 6, 16, 94). Two clones that were 96% similar to the H8 genomic clone were found not to be allelic variants by differential hybridizations and sequence analysis (80). Most recently, the nucleotide sequence of the lignin peroxidase gene GLG3 was elucidated, which showed >96% homology to the gene encoding isozyme H8 (64). Differences in restriction length polymorphisms between GLG3 and two other genes coding for isozyme H8 indicate that GLG3 is nonallelic to previously sequenced genes. It is interesting that so many highly homologous sequences exist; these may occur because the clones were isolated from two strains of P. chrysosporium, ME446 and BKM-F-1767, and so the clones could represent allelic variants or could simply be isozyme variations between the two strains of P. chrysosporium.

The characterization of a new lignin peroxidase gene, GLG6, was reported that has

72%, 88%, and 82% homology to the lignin peroxidase isozymes H2, H3, and H10, respectively (63). N-terminal sequence comparisons to H2, H8, and H10 showed that the *GLG6*-encoded lignin peroxidase is more closely related to H8 and H10 than to H2 proteins.

All of the genes that have been sequenced thus far indicate the presence of introns of a consistent size. Isozyme H8 shows 8 intervening sequences of approximately 50 bp each and 9 coding regions that range in size from 9–424 bp. The genes contain the commonly accepted eukaryotic transcriptional upstream regulatory sequences, including a TATA box and a CAAT box at positions −75 and −107, respectively. Experimental evidence for a role of these sequences in the transcription process, or on the regulatory elements involved, remains to be found.

Other Fungi

In addition to cloning of cDNAs or genes from *Phanerochaete chrysosporium*, the nucleotide sequences from other organisms have been determined for genes encoding ligninases. From the white-rot fungus *Phlebia radiata*, the nucleotide sequence of a cDNA clone for a lignin peroxidase (*Lgp3*) was isolated whose protein product is the LIII isozyme of this organism (78). The nucleotide sequence and the deduced amino acid sequence of these foreign genes are about 60% homologous to ligninase sequences from *P. chrysosporium*. When this gene (*lgp3*) was expressed in *Trichoderma reesei*, *Lpg3* mRNA was produced but the protein could not be detected. A manganese peroxidase purified from *Lentinula edodes* showed sequence homology with *P. chrysosporium* Mn peroxidase according to its N-terminal sequence (30).

HOMOLOGOUS AND HETEROLOGOUS EXPRESSION

For the advances in molecular biology described above to assist in the practical use of lignin-degrading peroxidases, a system must be developed whereby the enzymes are expressed in high amounts either in *Phanerochaete chrysosporium* or in another organism. High-quantity production would also be useful for detailed biochemical study of the enzymes. Although many laboratories have been working toward expression of these enzymes, the task has proven difficult. Farrell et al. have reported on the expression of ligninase in *Escherichia coli* and reconstitution of active enzyme (25). In our laboratory, we have expressed ligninase as inclusion bodies in *E. coli* but have been unable to reconstitute active enzyme (5). Work in this area is ongoing in several laboratories.

Overexpression in *P. chrysosporium* may be more promising, in light of the difficulty encountered in other systems, but a transformation system and expression vector would be required. If a ligninase gene could be placed behind a strong promoter in *P. chrysosporium,* this overproducing strain could prove quite useful because the fungus is already adapted to growing in adverse environments. A transformation system would also be required for study of the regulation of ligninase genes and for detailed study of the mechanism of ligninase catalysis through site-directed mutagenesis. Two transformation systems have been reported in the literature. Alic et al. (2) used an adenine biosynthetic gene from *Schizophyllum commune* to convert an adenine auxotrophic strain of *P. chrysosporium* to a prototroph by gene integration. Randall et al. (74) used the antibiotic G418 as a selectable marker and relied on integration of a plasmid containing the resistance gene to confer antibiotic resistance. This group more recently described a stable extrachromosomally maintained transformation vector (pG12–1) for *P. chrysosporium* that could transform *P. chrysosporium* to G418 resistance (75, 76). Once these or other systems currently under development are optimized, they should prove quite interesting and useful.

CONCLUDING REMARKS

We have described the enzymatic mechanism of catalysis for lignin and manganese peroxidases, the optimization of production of these enzymes, and their known molecular biology. The work of the numerous groups involved in research on lignin biodegradation has allowed great strides toward understanding this process and developing practical applications. Ligninolytic enzymes show promise for bioremediation technology, including decolorizing chlorinated aromatic-containing E1 effluent from pulping mills, dehalogenating components of the effluent, degrading chlorinated lignin-derived by-products of the Kraft pulping process, and metabolizing PCBs and its many congeners and many other recalcitrant compounds (Table 1). Continued basic research efforts to elucidate lignin peroxidase catalytic properties and optimize enzyme production provide the key to realizing these applications.

LITERATURE CITED

1. Aitken, M. D., and R. L. Irvine. 1990. Characterization of reactions catalyzed by manganese peroxidase from *Phanerochaete chrysosporium. Arch. Biochem. Biophys.* 276:405–414.

2. Alic, M., J. R. Kornegay, D. and Pribnow, M. H. Gold. 1989. Transformation by complementation of an adenine auxotroph of the lignin-degrading basidiomycete *Phanerochaete chrysosporium. Appl. Environ. Microbiol.* 55:406–411.

3. Andrawis, A., K. A. Johnson, and M. Tien. 1988. Studies on Compound I formation of the lignin peroxidase from *Phanerochaete chrysosporium. J. Biol. Chem.* 263:1195–1198.

4. Andrawis, A., E. A. Pease, I. Kuan, E. L. Holzbaur, and M. Tien. 1989. Characterization of two lignin peroxidase clones from *Phanerochaete chrysosporium. Biochem. Biophys. Res. Commun.* 162:673–680.

5. Andrawis, A., E. A. Pease, and Tien, M. 1990. Extracellular peroxidases of *Phanerochaete chrysosporium:* cDNA cloning and expression. In *Biotechnology in Pulp and Paper Manufacture.* Ed. T. K. Kirk, H.-M. Chang. Boston, MA: Buttersworth-Heinemann. 601–613.

6. Asada, Y., Y. Kimura, M. Kuwahara, A. Tsukamoto, K. Koide, A. Oka, and M. Takanami. 1988. Cloning and sequencing of a ligninase gene from a lignin-degrading basidiomycete, *Phanerochaete chrysosporium. Appl. Microbiol. Biotechnol.* 29:469–473.

7. Asther, M., C. Capdevila, and G. Corrieu. 1988. Control of lignin peroxidase production by *Phanerochaete chrysosporium* INA-12 by temperature shifting. *Appl. Environ. Microb.* 54:3194–3196.

8. Asther M., G. Corrieu, R. Drapron, and E. Odier. 1987. Effect of Tween 80 and oleic acid on ligninase production by *Phanerochaete chrysosporium* INA-12. *Enzyme Microb. Technol.* 9:245–249.

9. Banci, L., I. Berini, P. Turano, M. Tien, and T. K. Kirk. 1991. A proton NMR investigation into the basis for the relatively high redox potential of lignin peroxidase. *Proc.Natl. Acad. Sci. USA* 88:6956–6960.

10. Betts, W. B., M. C. Ball, and R. K. Dart. 1987. Growth of *Aspergillus flavus* on an insoluble lignin model compound. *Trans. Br. Mycol. Soc.* 89:235–278.

11. Blanchette, R. A., and I. D. Reid. 1986. Ultrastructural aspects of wood delignification by *Phlebia (Merulius) tremellosus. Appl. Environ. Microbiol.* 52:239–245.

12. Boer, H. A. de, Y. Z. Zhang, C. Collins, and C. A. Reddy. 1987. Analysis of nucleotide sequences of two ligninase cDNAs from a white-rot filamentous fungus, *Phanerochaete chrysosporium. Gene* 60:93–102.

13. Bonnarme, P., and T. W. Jeffries. 1990. Mn(II) regulation of lignin peroxidases from lignin-degrading white rot fungi. *Appl. Environ. Microbiol.* 56:210–217.

14. Boominathan, K., S. B. Dass, T. A. Randall, and C. A. Reddy. 1990. Nitrogen-deregulated mutants of *Phanerochaete chrysosporium*—a lignin-degrading basidiomycete. *Arch. Microbiol.* 153:521–527.

15. Brown, J. A., J. K. Glenn, and M. H. Gold. 1990. Manganese regulates expression of manganese peroxidase by *Phanerochaete chrysosporium. J. Bacteriol.* 172:3125–3130.

16. Brown, A., P. A. G. Sims, U. Raeder, and P. Broda. 1988. Multiple ligninase-related genes from *Phanerochaete chrysosporium Gene* 73:77–85.

17. Bumpus, J. A., M. Tien, D. Wright, and S. D. Aust. 1985. Oxidation of persistent environmental pollutants by a white rot fungus. *Science* 228:1434–1436.

18. Buswell, J. A., B. Mollet, and E. Odier. 1984. Ligninolytic enzyme production by *Phanerochaete chrysosporium* under conditions of nitrogen sufficiency. *FEMS Microbiol. Lett.* 25:295–299.

19. Capdevila, C., S. Moukha, M. Ghyczy, J. Theilleux, B. Gelie, M. Delattre, G. Corrieu, and M. Asther. 1990. Characterization of peroxidase secretion and subcellular organization of *Phanerochaete chrysosporium* INA-12 in the presence of various soybean phospholipid fractions. *Appl. Environ. Microbiol.* 56:3811–3816.

20. Dass, S. B., and C. A. Reddy. 1990. Characterization of extracellular peroxidases produced by acetate-buffered cultures of the lignin-degrading basidiomycete *Phanerochaete chrysosporium. FEMS Microbiol. Lett.* 69:221–224.

21. DePillis, G. D., H. Warriishi, M. H. Gold, and P. R. Ortiz de Montellano. 1990. Inactivation of lignin peroxidase by phenylhydrazine and sodium azide. *Arch. Biochem. Biophys.* 280:217–223.

22. Dosoretz, C. G., H.-C. Chen, and H. E. Grethlein. 1990. Effect of environmental conditions on extracellular protease activity in ligninolytic cultures of *Phanerochaete chrysosporium. Appl. Environ. Microbiol.* 56:395–400.

23. Faison, B. D., and T. K. Kirk. 1985. Factors involved in the regulation of a ligninase activity in *Phanerochaete chrysosporium. Appl. Environ. Microbiol.* 49:299–304.

24. Faison, B. D., T. K. Kirk, and R. L. Farrell. 1986. Role of veratryl alcohol in regulating ligninase activity in *Phanerochaete chrysosporium. Appl. Environ. Microbiol.* 52:251–254.

25. Farrell, R. L., P. Gelep, A. Anilionis, K. Javaherian, T. E. Maione, J. R. Rusche, B. A. Sadownick, and Jackson. 1987. European Patent Application No. 87810516.2.

26. Farrell, R. L., K. E. Murtagh, M. Tien, M. D. Mozuch, and T. K. Kirk. 1989. Physical and enzymatic properties of lignin peroxidase isoenzymes from *Phanerochaete chrysosporium. Enzyme Microb. Technol.* 11:322–328.

27. Fenn, P., and T. K. Kirk. 1981. Relationship of nitrogen to the onset and suppression of ligninolytic activity and secondary metabolism in *Phanerochaete chrysosporium. Arch. Microbiol.* 130:59–65.

27a. Fernando, T., and S. D. Aust. 1993. Biodegradation of toxic chemicals by white rot fungi. In *Biological Degradation and Bioremediation of Toxic Chemicals.* Ed. G. R. Chaudhry. Portland, OR: Dioscorides Press.

28. Forney, L. J., C. A. Reddy, M. Tien, and S. D. Aust. 1982. The involvement of hydroxyl radical derived from hydrogen peroxide in lignin degradation by white-rot fungus *Phanerochaete chrysosporium J. Biol. Chem.* 257:11455–11462.

29. Forrester, I. T., A. C. Grabski, R. R. Burgess, and G. F. Leatham. 1988. Manganese, Mn-dependent peroxidases, and the biodegradation of lignin. *Biochem. Biophys. Res. Commun.* 157:992–999.

30. Forrester, I. T., A. C. Grabski, C. Mishra, B. D. Kelley, W. N. Strickland, G. F. Leatham, and R. R. Burgess. 1990. Characteristics and N-terminal amino acid sequence of a manganese peroxidase purified from *Lentinula edodes* cultures grown on a commercial wood substrate. *Appl. Microbiol. Biotechnol.* 33:359–365.

31. Galliano, H., G. Gas, and H. Durand. 1988. Ligninocellulase biodegradation and ligninase excretion by mutant strains of *Phanerochaete chrysosporium* hyperproducing cellulases. *Biotechnology Lett.* 10:655–660.

32. Glenn, J. K., L. Akileswarasn, and M. Gold. 1986. Mn(II) oxidation is the principle function of the extracellular Mn-Peroxidase from *Phanerochaete chrysosporium. Archives of Biochem. Biophys.* 251: 688–696.

33. Glenn, J. K., and M. Gold. 1985. Purification of an extracellular Mn(II)-dependent peroxidase from the lignin-degrading basidiomycete, *Phanerochaete chrysosporium. Arch. Biochem. Biophys.* 242:329–341.

34. Glenn, J. K., M. A. Morgan, M. B. Mayfield, M. Kuwahara, and M. H. Gold. 1983. An extracellular H_2O_2-requiring enzyme preparation involved in lignin biodegradation by the white rot basidiomycete *Phanerochaete chrysosporium. Biochem. Biophys. Res. Commun.* 114:1077–1083.

35. Gomez-Alarcon, G., Jimenez, C. Saiz-and R. Lahoz. 1989. Influence of Tween 80 on the secretion of some enzymes in stationary cultures of the white-rot fungus *Pycnoporus cinnabarinus. Microbios* 60:183–192.

36. Hammel, K. E., and M. A. Moen, 1991. Depolymerization of a synthetic lignin *in vitro* by lignin peroxidase. *Enzyme Microb. Technol.* 13:15–18.

37. Hammel, K. E., M. Tien, B. Kalyanaraman, and T. K. Kirk. 1985. Mechanism of oxidative C_α–C_β cleavage of a lignin model dimer by *Phanerochaete chrysosporium* ligninase. *J. Biol. Chem.* 260:8348–8353.

38. Huynh, V-B., and R. L. Crawford. 1985. Novel extracellular enzymes (ligninases) of *Phanerochaete chrysosporium. FEMS Microbiology Lett.* 28:119–123.

39. Iyayi, C. B., and R. K. Dart. 1982. The degradation of p-coumarly alcohol by *Aspergillus flavus. J. Gen. Microbiol.* 128:1473–1482.

40. Jager, A., S. Croan, and T. K. Kirk. 1985. Production of ligninases and degradation of lignin in agitated submerged cultures of *Phanerochaete chrysosporium. Appl. Environ. Microbiol.* 50:1274–1278.

41. Jager, A. G., and C. Wandrey. 1989. Immobilization of the basidiomycete *Phanerochaete chrysosporium* on sintered glass: Production of lignin peroxidases. In *Physiology of Immobilized Cells*. Proc. Inte. Symp. Wageningen, The Netherlands. 433–438.

42. Janshekar, H., and A. Feichter. 1988. Cultivation of *Phanerochaete chrysosporium* and production of lignin peroxidases on submerged stirred tank reactors. *J. Biotechnol.* 8:97–112.

43. Jeffries, T. W., S. Choi, and T. K. Kirk. 1981. Nutritional regulation of lignin degradation by *Phanerochaete chrysosporium. Appl. Environ. Microbiol.* 42:290–296.

44. Jong, E. de, E. E. Beuling, R. P. van der Zwan, and de J. A. M. Bout. 1990. Degradation of veratryl alcohol by *Penicillium simplicissmum. Appl. Microbiol. Biotechnol.* 34:420–425.

45. Kantelinen, A., A. Hatakka, and L. Viikari. 1989. Production of lignin peroxidase and laccase by *Phlebia radiata. Appl. Microbiol. Biotechnol.* 31:234–239.

46. Karhunen, E., A. Kantelinen, and M-L. Niko-Paavola. 1990. Mn-dependent peroxidase from the lignin-degrading white rot fungus *Phlebia radiata. Arch. Biochem. Biophys.* 279:25–31.

47. Kern, H. W. 1989. Improvement in the production of extracellular lignin peroxidases by *Phanerochaete chrysosporium*: Effect of solid manganese(IV)oxide. *Appl. Microbiol. Biotechnol.* 32:223–234.

48. Kersten, P. J., M. Tien, B. Kalyanaraman, and T. K. Kirk. 1985. The ligninase of *Phanerochaete chrysosporium* generates cation radicals from methoxybenzenes. *J. Biol. Chem.* 260:2609–2612.

49. Keyser, P., T. K. Kirk, and J. G. Zeikus. 1978. Ligninolytic enzyme system of *Phanerochaete chrysosporium*: synthesized in the absence of lignin in response to nitrogen starvation. *J. Bacteriol.* 135:790–797.

50. Kirk, T. K. 1981. Toward elucidating the mechanism of action of the ligninolytic system in basidiomycetes. In *Trends in the Biolgy of Fermentations for Fuels and Chemicals.* Ed. A. Hollaender New York: Plenum Press. 131.

51. Kirk, T. K., S. Croan, M. Tien, Murtagh, and R. L. Farrell. 1986. Production of multiple ligninases by *Phanerochaete chrysosporium*: effect of selected growth conditions and use of a mutant strain. *Enzyme Microb. Technol.* 8:27–32.

52. Kirk, T. K., and R. L. Farrell. 1987. Enzymatic "combustion": the microbial degradation of lignin. *Ann. Rev. Microbiol.* 41:465–505.

53. Kirkpatrick, N., and J. M. Palmer. 1989. A natural inhibitor of lignin peroxidase activity from *Phanerochaete chrysosporium*, active at low pH and inactivated by divalent metal ions. *Appl. Microbiol. Biotechnol.* 30:305–311.

54. Kuila, D., M. Tien, J. A., Fee, and M. R. Ondrias. 1985. Resonance Raman spectra of extracellular ligninase: evidence for a heme active site similar to those of peroxidases. *Biochemistry* 26:2258–2263.

55. Kuwahara, M., and Y. Asada. 1987. Production of ligninases, peroxidases and alcohol oxidases by mutants of *Phanerochaete chrysosporium*. In *Lignin enzymic and microbial degradation.* Ed. E. Odier. Versailles, France: INRA Publications. 171–176.

56. Kuwahara, M., J. K. Glenn, M. A. Morgan, and M. A. Gold. 1984. Separation and characterization of two extracellular H_2O_2-dependent oxidases from ligninolytic cultures of *Phanerochaete chrysosporium*. *FEBS* 169:247–250.

57. La Mar, G. N., J. S. de Ropp, K. M. Smith, and K. C. Langry. 1980. Proton nuclear magnetic resonance study of the electronic and molecular structure of the heme crevice in horseradish peroxidase. *J. Biol. Chem.* 255:6646–6652.

58. Lestan, D., A. Strancar, and A. Perdih. 1990. Influence of some oils and surfactants on ligninolytic activity, growth and lipid fatty acids of *Phanerochaete chrysosporium*. *Appl. Microbiol. Biotechnol.* 34:426–428.

59. Lipman, D. J., and W. R. Pearson. 1985. Rapid and sensitive protein similarity searches. *Science* 227:1435–1439.

60. Linko, S. 1988. Production of lignin peroxidase by immobilized *Phanerochaete chrysosporium* in an agitated bioreactor. *Ann. NY Acad. Sci.* 542:195–203.

61. Lundell, T., A. Leonowicz, J. Rogalski, and A. Hatakka. 1990. Formation and action of lignin-modifying enzymes in cultures of *Phlebia radiata* supplemented with veratric acid. *Appl. Environ. Microbiol.* 56:2623–2629.

62. Millis, C. D., D. Cai, M. T. Stankovich, and M. Tien. 1989. Oxidation-reduction potentials and ionization states of extracellular peroxidases from the lignin-degrading fungus *Phanerochaete chrysosporium*. *Biochemistry* 28:8484–8489.

63. Naidu, P. S., Y. Z., Zhang, and C. A. Reddy. 1990. Characterization of a new lignin peroxidase gene (*GLG6*) from *Phanerochaete chrysosporium*. *Biochem. Biophys. Res. Commun.* 173:994–1000.

64. Naidu, P. S., and C. A. Reddy. 1990. Nucleotide sequence of a new lignin peroxidase gene *GLG3* from the white-rot fungus, *Phanerochaete chrysosporium*. *Nuc. Acids Res.* 18:7173.

65. Niku-Paavola, M-L., E. Karhunen, A. Kantelinen, L. Viikari, T. Lundell, and A. Hatakka. 1990. The effect of culture conditions on the production of lignin modifying enzymes by the

white-rot fungus *Phlebia radiata*. *J. Biotechnol.* 13:211–221.

66. Orth, A. B., M. Denny, and M. Tien. 1991. Overproduction of lignin degrading enzymes by an isolate of *Phanerochaete chrysosporium*. *Appl. Environ. Microbiol.* In press.

67. Ortiz de Montellano, P. R. 1987. Control of the catalytic activity of prosthetic heme by the structure of hemoproteins. *Acc. Chem. Res.* 20:289–294.

68. Otjen, L., and R. A. Blanchette. 1985. Selective delignification of aspen wood blocks *in vitro* by three white rot basidiomycetes. *Appl. Environ. Microbiol.* 50:568–572.

69. Pease, E. A., A. Andrawis, and M. Tien. 1989. Manganese-dependent peroxidase from *Phanerochaete chrysosporium*: primary structure deduced from cDNA sequence. *J. Biol. Chem.* 264:13531–13535.

70. Pease, E. A., and M. Tien. 1991. Lignin-degrading enzymes from the filamentous fungus *Phanerochaete chrysosporium*. In *Biocatalysts for Industry*. Ed. J. S. Dodrdick. New York: Plenum Press. 115–135.

71. Perez, J., and T. W. Jeffries. 1990. Mineralization of ^{14}C-ring-labeled synthetic lignin correlates with the production of lignin peroxidase, not of manganese peroxidase or laccase. *Appl. Environ. Microbiol.* 56:1806–1812.

72. Poulos, T. L., and Kraut, J. 1980. The steriochemistry of peroxidase catalysis. *J. Biol. Chem.* 255:8199–8205.

73. Pribnow, D., M. B. Mayfield, V. J. Nipper, J. A. Brown, and M. H. Gold. 1989. Characterization of a cDNA encoding a manganese peroxidase, from the lignin-degrading basidiomycete *Phanerochaete chrysosporium*. *J. Biol. Chem.* 264:5036–5040.

74. Randall, T., T. R. Rao, and C. A. Reddy. 1989. Use of a shuttle vector for the transformation of the white rot basidiomycete, *Phanerochaete chrysosporium*. *Biochem. Biophys. Res. Commun.* 161:720–725.

75. Randall, T., and C. A.. Reddy. 1991. An improved transformation vector for the lignin-degrading white rot basidiomycete *Phanerochaete chrysosporium*. *Gene.* 103:125–130.

76. Randall, T., C. A. Reddy, and K. Boominathan. 1991. A novel extrachromasomally maintained transformation vector for the lignin-degrading basidiomycete *Phanerochaete chrysosporium*. *J. Bacteriol.* 173:776–782.

77. Roch, P., J. A. Buswell, R. B. Cain, and E. Odier. 1989. Lignin peroxidase production by strains of *Phanerochaete chrysosporium* grown on glycerol. *Appl. Microbiol. Biotechnol.* 31:587–591.

78. Saloheimo, M., V. Barajas, M.-L. Niko-Paavola, and J. K. C. Knowles. 1989. A lignin peroxidase-encoding cDNA from the white-rot fungus *Phlebia radiata*: Characterization and expression in *Trichoderma reesei*. *Gene* 85:343–351.

79. Satterlee, J. D., J. E. Erman, G. N. La Mar, K. M. Smith, and K. C. Langry. 1983. Assignment of hyperfine shifted resonances in high-spin forms of cytochrome c peroxidase by reconstitutions with deuterated hemins. *Biochim. Biophys. Acta.* 743:246–255.

80. Schalch, H., J. Gaskell, T. L. Smith, and D. Cullen. 1989. Molecular cloning and sequences of lignin peroxidase genes of *Phanerochaete chrysosporium*. *Mol. Cell. Biol.* 9:2743–2747.

81. Shimada, M., F. Nakatsubo, T. K. Kirk, and T. Higuchi. 1981. Biosynthesis of the secondary metabolite veratryl alcohol in relation to lignin degradation in *Phanerochaete chrysosporium*. *Arch. Microbiol.* 129:321–324.

82. Smith, T. L., H. Schlach, J. Gaskell, S. Covert, and D. Cullen. 1988. Nucleotide sequence of a ligninase gene from *Phanerochaete chrysosporium*. *Nuc. Acids Res.* 16:1219.

83. Thanabal, V., J. S. de Ropp, and G. N. La Mar. 1987. ^{1}H NMR study of the electronic and molecular structure of the heme cavity in horseradish peroxidase. Complete heme resonance assignments based on saturation transfer and nuclearoverhauser effects. *J. Amer. Chem. Soc.* 109:265–272.

84. Tien, M. 1987. Properties of ligninase from *Phanerochaete chrysosporium* and their possible applications. *CRC Crit. Rev. J. Microbiol.* 15:141–168.

85. Tien, M., P. J. Kersten, and T. K. Kirk. 1987. Selection and improvement of lignin-degrading microorganisms: potential strategy based on lignin model-amino acid adducts. *Appl. Environ. Microbiol.* 53:242–245.

86. Tien, M., and T. K. Kirk. 1983. Lignin-degrading enzyme from the hymenomycete *Phanerochaete chrysosporium* Burds. *Science* 221:661–663.

87. Tien, M., and T. K. Kirk. 1988. Lignin peroxidase of *Phanerochaete chrysosporium*. *Meth. Enzymol.* 161:238–249.

88. Tien, M., and S. B. Myer. 1990. Selection and characterization of mutants of *Phanerochaete chrysosporium* exhibiting ligninolytic activity under nutrient-rich conditions. *Appl. Environ. Microbiol.* 56:2540–2544.

89. Tien, M., and C.-P. D. Tu. 1987. Cloning and sequencing of a cDNA for a ligninase from *Phanerochaete chrysosporium*. *Nature* 326:520–523.

90. Tonon, F., and E. Odier. 1988. Influence of veratryl alcohol and hydrogen peroxide on ligninase activity and ligninase production by *Phanerochaete chrysosporium*. *Appl. Environ. Microbiol.* 54:466–472.

91. Tonon, F., C. P. de Castro, and E. Odier. 1990. Nitrogen and carbon catabolite regulation of lignin peroxidase and enzymes of nitrogen metabolism in *Phanerochaete chrysosporium*. *Exp. Mycol.* 14:243–254.

92. Tuisel, H., R. Sinclair, J. A. Bumpus, W. Ashbaugh, B. J. Brock, and S. D. Aust. 1990. Lignin peroxidase H_2 from *Phanerochaete chrysosporium*: Purification, characterization and stability to temperature and pH. *Arch. Biochem. Biophys.* 279:158–166.

93. Venkatadri, R., and R. Irvine. 1990. Effect of agitation on ligninase activity and ligninase production by *Phanerochaete chrysosporium*. *Appl. Environ. Microbiol.* 56:2684–2691.

94. Walther, I., M. Kälin, J. Reiser, F. Suter, B. Fritsche, M. Saloheimo, M. Leisola, T. Teeri, J. K. C. Knowles, and A. Feichter. 1988. Molecular analysis of a *Phanerochaete chrysosporium* lignin peroxidase gene. *Gene* 70:127–137.

95. Wariishi, H. W., L. Akileswaran, and M. H. Gold. 1988. Manganese peroxidase from the Basidiomycete *Phanerochaete chrysosporium*: spectral characterization of the oxidized states and the catalytic cycle. *Biochemistry* 27:5365–5370.

96. Wariishi, H., H. B. Dunford, I. D. MacDonald, and M. H. Gold. 1989. Manganese peroxidase from the lignin-degrading basidiomycete *Phanerochaete chrysosporium*. *J. Biol. Chem.* 264:3335–3340.

97. Zhong, L. C., S. Linko, N. Lindholm, and Y. Y. Linko. 1988. Lignin peroxidase production by immobilized *Phanerochaete chrysosporium*. *Ann. NY Acad. Sci.* 542:153–157.

CHAPTER 17

Biological Disposal of Lignocellulosic Wastes and Alleviation of Their Toxic Effluents

D. S. CHAHAL

Abstract. Billions of tons of lignocelluloses in the form of residues from crop and timber harvesting, wood-processing, and pulp manufacturing are produced as wastes every year. Disposal of this huge quantity of lignocellulosic wastes and their toxic effluents is an enormous environmental challenge. In some countries, crop residues are burned in the fields to clear them for subsequent crops. The burning of crop residues and of wood for energy releases a number of toxic gasses that cause environmental pollution. This chapter discusses the biodegradation and bioconversion of lignocelluloses into useful products and hence the biological alleviation of pollution from wastes from lignocelluloses. The cellulose from lignocelluloses is used to produce paper, rayon, and various cellulose derivatives through chemical processes and to produce fuel ethanol, solvents, single-cell protein (SCP), pharmaceuticals, and many more useful compounds through biological processes. Utilization of one ton of lignocelluloses for these purposes releases approximately 250 kg of hemicelluloses and 250 kg of lignin; some toxic effluents are also produced. The disposal or proper utilization of these by-products and alleviation of the toxic effluents present a challenge to biotechnologists.

INTRODUCTION

Lignocelluloses gained attention during the 1970s and 1980s as a possible source of SCPs to satisfy an increasing world-wide demand for protein and new foods (7). Lignocelluloses were also explored as an alternative source of fuel ethanol and methanol because of the dwindling supply of hydrocarbons (7) and because of the need to control pollution from hydrocarbon fuel emissions (58). Applications currently under study are the biotransformation of lignin into valuable compounds (16, 38, 47).

"Lignocelluloses" are generally defined as any of the several closely related compounds comprising plant (wood) cell walls in which cellulose is intimately associated with lignin. Lignocelluloses also contain other polysaccharides commonly called "hemicelluloses." Photosynthesis on earth results in approximately 155 billion tons dry

Dr. Chahal is at the Centre de recherche en microbiologie appliquée, Université du Québec, Institut Armand-Frappier, 531 Boulevard des Prairies, Laval, Québec H7N 4Z3, Canada.

weight of primary productivity (lignocelluloses) per year (3). Of this, about two thirds is on land, and one third in the oceans. Of the 65.5% of total productivity on land, by far the greater amount is in forests and woodlands. Some of these forests are utilized as a source of cellulose in the form of lumber and wood pulp for paper. The 2.7% of cultivated land that accounts for 5.9% of the primary productivity (mostly for food and fiber) is going to be needed entirely for agriculture (3). This land would be the major source of lignocellulosic by-products (crop residues) for production of useful products.

Ishaque and Chahal (40) have estimated that about 2.246 billion tons of cereal straw is produced world-wide annually. In countries with intensive agriculture, the crop residues are often burned to clear fields for the next crop. Burning of crop residues in the field not only causes pollution of the atmosphere (58) but also destroys the useful microflora of the topsoil. Therefore, burning of crop residues also reduces the fertility of the soil.

Rawat and Nautiyal (64) have estimated that 680 million cubic meters of visible surplus biomass are now generated every year. If only half of this could actually be made available, then over 300 million cubic meters of surplus could be used for bioconversion processes. Even considering that the total pulpwood production in the world during 1983 was 370 million cubic meters, it is apparent that at the global level, in the short run, there will be adequate availability of biomass for bioconversion processes. Although within individual geographical zones the quantity of visible surplus may be larger or smaller, depending on the volume of round wood production, there is certain positive surplus available in every part of the world. If figures for sustained surplus availability are considered (64), at the global level more than 2000 million cubic meters of forest biomass can be potentially available every year over the long run. At the regional level, however, there are long-term deficiencies in many African and Asian countries. Such shortfalls can be reduced by bringing more land under plantation and by improving productivity. Moreover, millions of tons of wood-processing residues such as bark and wastes from pulp manufacturing mills are produced every year. This chapter discusses a number of processes for the bioconversion of such lignocellulosic wastes into useful products and for the biological alleviation of the toxic effluents from the processes that produce them.

COMPOSITION AND STRUCTURE OF LIGNOCELLULOSES

Lignocelluloses are generally composed of 30 to 56% cellulose, 10 to 27% or more hemicelluloses, 3 to 30% lignin, and 3.6 to 7.2% protein. Some crop residues, like rice straw and bagasse, also contain a large quantity of silica.

All lignocelluloses are composed of plant (wood) cells with a thin primary wall that surrounds the relatively thick secondary wall (Fig. 1). The microfibrils of cellulose are variously arranged in the different layers of cell wall. They are loosely and randomly organized in the primary wall. In the secondary wall, they are arranged in crisscross fashion in the S1 layer, oriented almost parallel to the lumen axis in the S2 layer, and form a flat helix in the S3 layer. The central empty portion, the lumen, is formed after disintegration of the protoplasm. The debris of the protoplasm is seen as wartlike structures on the inner-most layer of the cell wall.

Cellulose is composed of long slender bundles of long chains of β-D-glucopyranose residues (cellulose molecules) linked by $1 \rightarrow 4$ glucosidic bonds (Fig. 2). These bundles are called *elementary fibrils*. Within each elementary fibril the cellulose molecules are

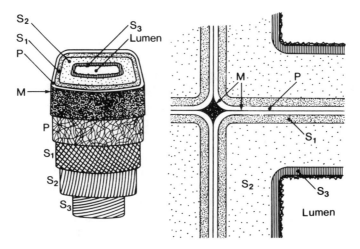

Figure 1. Diagramatic representation of various layers of plant (wood) cell wall. M, middle lamella; P, primary wall; S_1, the outer secondary layer; S_2, the middle secondary layer; S_3, the inner secondary layer. [Reproduced with permission from Chahal (7)].

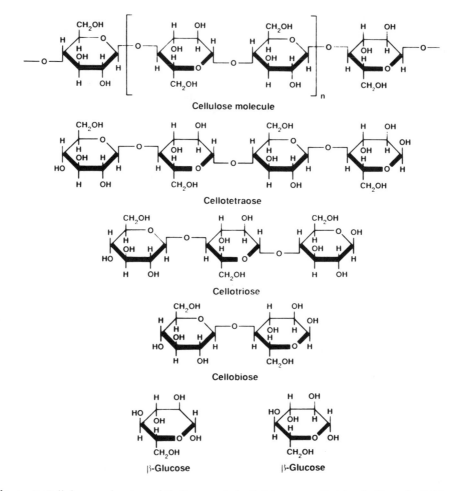

Figure 2. Cellulose molecule and its fragments (cellotetraose, cellotriose), dimer (cellobiose) and monomer (β-glucose).

bound laterally. The adjacent molecules run in opposite directions but in parallel, with various degrees of orientation. The region composed of highly oriented molecules is called *crystalline cellulose;* the region comprising less oriented molecules is called *amorphous cellulose.* Many elementary fibrils together form a microfibril; several microfibrils joined together form a macrofibril.

Hemicelluloses are composed of various sugars (the most common sugars are D-xylose, D-mannose, D-galactose, D-glucose, L-arabinose), 4-O-methyl-D-glucuronic acid, D-galacturonic acid, and D-glucuronic acid. The rare sugars of hemicelluloses are L-rhamnose, L-fucose, and various methylated neutral sugars. Xylan and mannan are the most prominent polysaccharides of hemicelluloses of hard and soft wood, respectively (79).

Xylan (4-O-acetyl-D-methylglucuronoxylan) is a polymer of β-D-xylopyranose residues linked in 1 → 4 glucosidic bonds. A 4-O-methyl-α-D-glucuronic acid residue is attached on some xylose units at C-2. Some xylose residues contain an O-acetyl group at C-2 or more frequently at C-3 (Fig. 3) (79). Mannan (galactoglucomannan), commonly found in soft wood, consists of 1 → 4 β-D-glucopyranose and β-D-mannopyranose residues distributed at random. Some hexose units carry a terminal residue of a D-galactopyranose at C-6. The acetyl groups are attached to mannose residues (Fig. 4) (79).

Lignin is formed by dehydrogenative polymerization of p-hydroxycinnamyl alcohols. The guaiacyl lignin of conifers is mainly a dehydrogenation polymer of coniferyl alcohol. Guaiacyl-syringyl lignin, which occurs in angiosperms, is composed of a mixed dehydrogenation of polymer of coniferyl and sinapyl alcohols. Grasses contain guaiacyl-syringyl-p-hydroxyphenyl lignin, which is composed of a mixed dehydrogenation polymer of coniferyl, sinapyl, and p-coumaryl alcohols (Fig. 5) (37).

The microfibrils are covered with hemicelluloses and are further encrusted with lignin. The lignin is physically and chemically bonded with hemicelluloses (25, 36). The empty spaces between the fibrils are again filled with hemicelluloses and lignin. Cellulose in nature is thus well protected against biodegradation (Fig. 6). However, physico-chemical pretreatments can be used to liberate the cellulose fibers and to reduce the crystallinity of cellulose so that the fibers can be used for the production of various useful products. For detailed information, refer to Tarkow and Feist (77), Millett et al. (59), and Chahal (8).

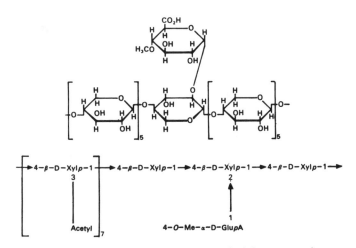

Figure 3. *Top:* Partial structure of a 4-O-methylglucuronoxylan.
Bottom: Structural formula of O-acetyl-4-O-methylglucuronoxylan (79)

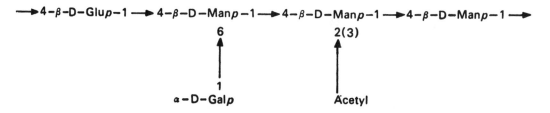

Figure 4. Structural form of O-acetylgalactoglucomannan (79)

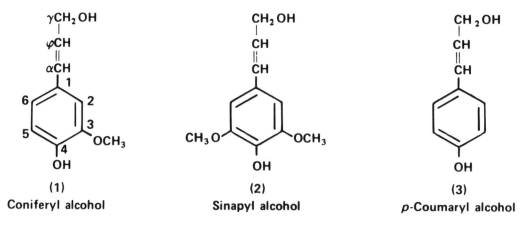

Figure 5. Lignin precursors.

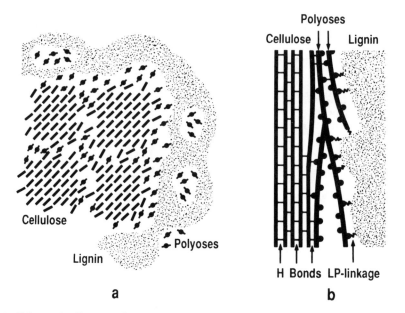

Figure 6. Schematic diagram for structural arrangements of cellulose, hemicelluloses (polyoses), and lignin in plant (wood) cell wall. [Reproduced with permission from Fengel and Wegener (25)].

BIODEGRADATION OF LIGNOCELLULOSES

An array of enzymes (ligninases, hemicellulases, and cellulases) is required for biodegradation of lignocelluloses into their simple compounds. In nature, only microorganisms that can produce such an array of enzymes and have the capability to penetrate the substrate deeply can degrade such complex organic matter (15). The majority of such microorganisms belong to the class Basidiomycetes—the "white rot" fungi. The best example is *Phanerochaete chrysosporium*, which has been studied extensively (50). This fungus can produce almost all the enzymes (ligninases, hemicellulases, and cellulases) required for biodegradation of lignocelluloses. Many miroorganisms produce hemicellulases exclusively, and a few produce hemicellulases and cellulases but not ligninases (15). Therefore, for these microorganisms, lignin must either be removed or broken down into smaller fragments physically or chemically to facilitate the direct contact of hemicellulases and cellulases with hemicelluloses and cellulose, respectively, to cause degradation.

Lignocelluloses are first cut into small chips or ground to small particles before subjecting them to various physico-chemical pretreatments to loosen their compact structure and to depolymerize or solubilize the lignin, the strongest barrier to degrading the hemicelluloses and cellulose (Fig. 6). The first step in releasing the cellulose fiber is to break the lignin-hemicellulose bonds (Fig. 7). The hypothetical enzyme "X" postulated by Cowling (15) for this reaction may be involved, but because this postulated enzyme has not yet been described, hemicellulases, especially xylanases, have been used recently to delignify the lignocelluloses. Only limited delignification has been reported by this method, however (61). Viikari et al. (81) suggested that the enzyme mixtures used may lack an essential enzyme(s), e.g. an enzyme(s) that acts on the lignin-hemicellulose linkage (Fig. 7). Further study is required to characterize the hypothetical "X" enzyme, postulated to break the bonds between lignin and hemicelluloses and to achieve complete delignification of lignocelluloses.

Individual cells are released from small particles of lignocelluloses by the combined action of pectinases and ligninases on the middle lamella. As soon as the individual cells are released, the cell wall is exposed to further biodegradation (Fig. 8). The biodegradation of the components of the cell wall—lignin, hemicelluloses, and cellulose—is discussed below.

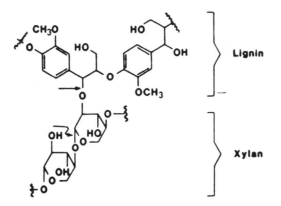

Figure 7. Possible linkage of lignin and hemicellulose (xylan) (*arrow*). Xylanase can attack at any point on the molecule of xylan as shown with the *curved arrow* but not on the linkage of lignin and xylan (*arrow*). [Reproduced with permission from Paice et al. (61)].

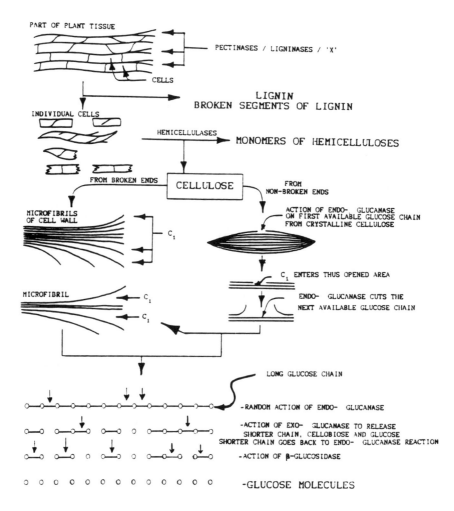

PART OF PLANT TISSUE

PECTINASES / LIGNINASES / 'X'

CELLS

LIGNIN
BROKEN SEGMENTS OF LIGNIN

INDIVIDUAL CELLS

HEMICELLULASES
MONOMERS OF HEMICELLULOSES

FROM BROKEN ENDS   CELLULOSE   FROM
NON-BROKEN ENDS

MICROFIBRILS
OF CELL WALL

C_1

ACTION OF ENDO- GLUCANASE
ON FIRST AVAILABLE GLUCOSE CHAIN
FROM CRYSTALLINE CELLULOSE

C_1 ENTERS THUS OPENED AREA

MICROFIBRIL C_1

C_1

ENDO- GLUCANASE CUTS THE
NEXT AVAILABLE GLUCOSE CHAIN

LONG GLUCOSE CHAIN

-RANDOM ACTION OF ENDO- GLUCANASE

-ACTION OF EXO- GLUCANASE TO RELEASE
SHORTER CHAIN, CELLOBIOSE AND GLUCOSE
SHORTER CHAIN GOES BACK TO ENDO- GLUCANASE REACTION

-ACTION OF β-GLUCOSIDASE

-GLUCOSE MOLECULES

Figure 8. Schematic degradation of lignocellulose. [Reproduced with permission from Chahal (7)].

Lignin

Research on lignin biodegradation increased dramatically from the late 1970s to the early 1990s. Major goals of this research have been (a) to use ligninases for pretreatment or modification of lignocelluloses for use in various processes; (b) to produce chemicals that are usually produced from petroleum; and (c) to biotransform oligolignols into high-value chemicals that are different from those produced in (b). I refer the reader a number of good books and reviews: Kirk et al. (50); Kirk and Chang (47); Kirk and Farrell (49); Crawford (16); Crawford and Crawford (17); Jansheker and Fiechter (43); and Leisola and Fiechter (53).

Kirk and Eriksson (48) point out the following developments in biodegradation of lignin:

1. description of the key features of the chemistry of biodegradation by white-rot fungi and development of an experimental system with the efficient lignin-degrading fungus *Phanerochaete chrysosporium*;

2. discovery of the first lignin-degrading enzyme (a powerful peroxidase) in *P. chrysosporium;*

3. demonstration that the *P. chrysosporium* enzyme oxidizes aromatic nuclei in lignin by one electron to cation radicals that undergo non-enzymatic degradative reactions of both a radical and ionic nature; these reactions include aromatic ring cleavage and various other cleavages that depolymerize the polymer;

4. development of improved methods of production of lignin peroxidase for fundamental and applied research,

5. discovery of a second kind of peroxidase in *P. chrysosperium* that oxidizes Mn^{2+} to Mn^{3+} and can oxidize phenolic units in lignin, thus leading to some degradation;

6. description of enzymes, also in the *P. chrysosporium* system, that generate extracellular H_2O_2 (needed by the peroxidases), including a new extracellular copper oxidase that oxidizes glyoxal and related compounds with reduction of O_2 to H_2O_2.

Despite this progress, an enzyme mixture that actually depolymerizes or solubilizes lignin has yet to be described (48), though current research is directed toward that end. Recently, however, Hammel and Moen (32) have reported that crude lignin peroxidase from *P. chrysosporium* readily catalyzes the partial depolymerization of a synthetic syringyl guaiacyl lignin *in vitro* under the following conditions: (a) the lignin is employed as a dilute dispersion in H_2O/ dimethylformamide, (b) the H_2O_2 concentration is maintained at a low level throughout the experiment, and (c) veratryl alcohol is included as a co-substrate in the reaction.

Hemicelluloses

Enzymes that attack the hemicelluloses are hydrolytic in nature and are called *hemicellulases* or *hemicellulolytic enzymes* (glycan hydrolases, EC 3.2.1) (20). Typical hemicellulases are β-D-xylanases, β-D-mannanases and β-D-galactanases. These, like most other polysaccharide-degrading enzymes, attack their substrates in two ways: exo- and endohydrolytic attack. Exo-enzymes degrade the polysaccharides (long chains of hemicelluloses) by successive removal of terminal sugar or oligosaccharide units and proceed in a stepwise manner from the reducing end of the polysaccharide chain. Endo-enzymes attack polysaccharides in a random manner, causing multiple scissions that decrease the degree of polymerization of the substrate. The polymer is progressively degraded into shorter fragments until mono- or disaccharides are formed.

The disaccharides are degraded into monosaccharides by their respective glycosidases such as β-D-xylosidases, β-D-mannosidases, α- and β-D-galactosidases, and α-L-arabinosidases. These enzymes are capable of hydrolyzing not only low-molecular-weight glycosides but also the short-chain or monosaccharide appendages from the main hemicellulosic backbone chain. The action of glycosidases is necessary to achieve total hydrolysis of the hemicelluloses (20).

Cellulose

For the complete hydrolysis of cellulose to glucose, cellulase systems must contain the following enzymes: (a) endoglucanase (1,4-β-glucan glucanohydrolase, EC 3.2.1.4)

(46); (b) exoglucanase (1,4-β-glucan cellobiohydrolase, EC 3.2.1.91) (85); and (c) β-glucosidase (β-D-glucoside glucohydrolase or cellobiase, EC 3.2.1.21) (74, 88). The synergy of these enzymes hydrolyzes cellulose to glucose (86, 87).

In addition to the activity of these enzymes, C_1 activity (cotton-hydrolyzing activity) is necessary for splitting off the elementry fiber or microfibril from the crystalline cellulose (65, 66; also see discussion on C_1 in Ref. 10). C_1 activity has been ignored in most studies on the hydrolysis of cellulose since the discovery of exoglucanase (1,4-β-glucan cellobiohydrolase) (9), because it was considered that exoglucanase is responsible for initiating hydrolysis of crystalline cellulose, the role originally assigned to C_1. However, the high C_1 activity in the cellulase system produced in solid state fermentation with Trichoderma reesei QMY-1 may be responsible for its high hydrolytic potential (9).

The step-by-step degradation of cellulose to glucose is as follows (Fig. 8): The actions of pectinases, ligninases, enzyme "X," and hemicellulases expose the cellulose fibrils. The C_1 activity enters the loosened cellulose fibrils of the plant cell wall through the broken ends and cracks formed previously during physico-chemical pretreatments. Due to its swelling and H-bond-breaking effects, C_1 activity releases microfibrils as well as macrofibrils from the cell wall. The C_1 activity continues to work from the open ends of the microfibrils as well as on macrofibrils by penetrating deep inside the cellulose structure to release elementary fibrils and single cellulose molecules. To gain entry into the crystalline cellulose portion, however, C_1 activity must be augmented by endoglucanase to break the 1,4-β linkage in the glucose chain (cellulose molecule). As soon as the two broken ends of the glucose chain are lifted, the C_1 activity enters beneath the glucose chain to release the cellulose molecule from the rest of the crystalline cellulose by its swelling and H-bond-breaking effects (65). The newly released long glucose chain (cellulose molecule) is then broken down by endoglucanase into short chains (oligosaccharides, cellotetraose, cellotriose) (Fig. 2). The action of exoglucanase releases cellobiose from the nonreducing end of the glucose chain and also from oligomers produced by endoglucanase. β-Glucosidase or cellobiase breaks down cellobiose into glucose. The β-glucosidase can also cut off glucose units from oligosaccharides (74, 88). All the above reactions occur simultaneously and synergistically on lignocelluloses to biodegrade them to their various monomers. The lignin is also degraded into oligolignols of various molecular weights by the actions of ligninases (11).

A number of reviews have been written about the production of cellulases and hydrolysis of cellulose. Readers may refer to Ryu and Mandels (70), Mandels (57), Enari (24), and Chahal et al. (10).

BIOCONVERSION OF LIGNOCELLULOSES INTO
USEFUL PRODUCTS

Pulp

Lignocelluloses, especially wood, are still the primary substances for making pulp, the raw material from which paper is produced. Nonwoody lignocelluloses, especially bagasse and cereal straws, are also used in cardboard and paper. The fiber is separated and arranged alone or with other materials to make paper products.

Mechanical pulp (ground wood) is produced by pressing debarked logs against a grinding wheel in the presence of water. The yield of pulp is over 90%. It is primarily used for newsprint, which undergoes little bleaching (31). Chemical pulp is obtained by vigorous action that removes the interfiber lignin. This produces a high strength paper but sacrifices pulp yield (about 50%) (31). This pulp requires extensive bleaching. The effluents—the spent sulfite liquor (SSL) and waste solids—from chemical and mechanical pulp processes present a serious problem of water pollution. The cellulose (pulp) thus obtained is used to make newsprints, fine quality paper, rayons, various cellulose derivatives, and many chemicals.

The effluents produced during the processing of pulp from lignocelluloses have the potential to cause considerable damage to the receiving waters, which may, in turn, be used for domestic, industrial, fishing, and recreational purposes. In some countries, the pulp and paper industry contributes as much as 75% of suspended solids and 90% of the biological oxygen demand (BOD) of the total effluent load from all industries (63). Lignin and its derivatives impart a dark brown color to the effluents that is aesthetically objectionable and may reduce the primary productivity of the receiving waters by reducing the transmission of light (62).

The chemical and physical properties of the effluents vary according to the raw materials, process schemes, and effluent treatment methods used. Still, a number of aspects are common to many effluents: suspended solids, oxygen demand, toxicity, and color. The principal toxicants and their sources are as follows (63): (a) resin acids derived from the raw materials; (b) chlorinated lignins formed in the chlorination steps of bleaching; (c) chlorinated resin acids, phenolics, and other acidic groups in the caustic extract of the bleach plant; (d) unsaturated fatty acids from woodroom and groundwood pulping; (e) diterpene alcohols from woodroom and groundwood pulping; (f) juvabiones from groundwood pulping; and (g) lignin degradation products from chemical pulping.

The effluents from pulp mills can be biologically treated to reduce their BOD, solid wastes, and toxicity before discharging into rivers or they can be fermented into useful products.

Biological Treatments.   Biological treatments will remove some of the toxicants, but mill-monitoring studies and pilot-scale tests have given inconsistent results (67, 83). Bacterial aerobic and anaerobic treatment systems can reduce the BOD and burden of suspended solids in the effluents but cannot remove the dark color of kraft bleachery, caustic extraction-stage effluent (E1 effluent) or degrade the high-molecular-weight chlorolignins they contain.

Chemical and Biological Removal of Color.   A chemical method to remove color from the effluent was developed by Croom (18) that treats the effluent with lime slurries [Ca(OH)$_2$ and CaCO$_3$]. Recently, a system based on biological decolorization and detoxification of C-and E-stage effluents by Phanerochaete chrysosporium in a rotating biological contactor has been patented by Chang et al. (14).

The biological components and conditions required for complete lignin or chlorolignin degradation by a cell-free system are still unknown. In work carried out at Paprican, Montreal, Canada and elsewhere, however, Coriolus versicolor has been shown to produce rapid decolorization of E1 effluent when applied as mycelium, mycellial pellets, or mycelial fragments immobilized in alginated beads (1). Archibald et al. (1) have reported that decolorization of E1 effluent by C. versicolor requires any of numerous mono- or disaccharides, O$_2$, a pH of 3.9–4.8, and a temperature of 29°C. Inexpensive sugar refinery or brewery wastes are excellent growth and decolorizing substrates. The inability of exogenous catalase to reduce E1 effluent decolorization and the absence of detectable exogenous peroxidase and H_2O_2 during decolorization suggest that secreted (in the medium) peroxidases are unimportant in E1 effluent decolorization; however, cell surface peroxidatic activity cannot be ruled out (1).

Cellulose

Cellulose can be hydrolyzed into glucose, as discussed above. The glucose thus produced from cellulose can be further fermented into fuel ethanol, pharmaceuticals, and many other valuable compounds. A detailed discusson of the fermentation of glucose into the above compounds is beyond the scope of this chapter, however. Cellulose can also be converted chemically into rayon and cellulose derivatives.

Single-cell Protein Production from Pulp Mill Effluents

Czechoslovakian Process.   The North Moravian Pulp Mill Paskov in Czechoslovakia uses SSL for production of fodder yeast, marketed under the tradename VITEX®, and yeast for chemical processing, marketed as VITAL® (21). SSL contains about 12% dry matter, of which about 2% comprises microbiologically utilizable substances for yeast production. The mill produces 225,000 tons/year of unbleached pulp and 27,000 tons/year of yeast. VITEX® is used as a protein-rich additive for cattle and poultry feed mixtures in Czechoslovakia and VITAL® is used for further chemical processing in the pharmaceutical industry.

The Pekilo Process.   On an industrial scale, the Pekilo process (69) is used for the production of microbial cell mass for animal feed by utilizing a strain of *Paecelomyces varioti*. The process was originally developed to use carbohydrate derivatives of SSL, but later studies have shown that other substrates, such as molasses and hydrolyzate from the prehydrolysis kraft process and hydrolyzates of wood, straw, corn stalks, corn cobs, and peat, can also be used. The SSL used in this process contains, on a dry weight basis, lignosulfonic acids 43%, hemilignin compounds 12%, incomplete hydrolyzed hemicellulose compounds and uronic acids 7%, monosaccharides 22%, acetic acid 6%, and aldonic acids and substances not investigated 10% (69). By fermenting spent sulfite liquor containing 300–400 kg of fermentable compounds per ton of pulp, 150–200 kg of Pekilo protein can be produced (27).

The crude protein content of Pekilo protein is 52 to 57%. Feeding trials have indicated that skim milk powder, fish meal, and yeast can be partly or totally replaced by Pekilo protein in the feed of growing pigs. No harmful effects on the carcass quality of lean meat, and no pathological effects have been found from use of Pekilo protein (27).

The Institut Armand-Frappier (IAF) Process.   Bioconversion of SSL into single-cell protein (SCP) and degradation of lignosulfonate in SSL is an extension of the IAF process, and is described below. Chahal et al. (13) reported that almost all the reducing sugars available in SSL supplied by the Daishowa Chemicals, Quebec (Quebec), Canada, were utilized by the two fungi tested: *Chaetomium cellulolyticum* asporogenous mutant (IAF-102) and *Pleurotus sajor-caju* (IAF-503). The final SCP product contained 43 to 50% protein.

The fermentation time was extended to 144 hr and 264 hr for of *C. cellulolyticum* asporogenous mutant and *P. sajor-caju,* respectively, to determine whether the lignosulfonate was also degraded into compounds of low molecular weight. No changes in the spectra of lignin were noticed at any time during fermentation when examined by infrared spectrophotometry or by ultraviolet spectrophotometer in the range from 250 to 330 nm. These results indicate that sodium lignosulfonate was more resistant to degradation under the cultural conditions tested than the other lignins (13).

Single-cell Protein Production from Lignocelluloses

The bioconversion of lignocelluloses into SCP has stimulated a great deal of interest since the 1970s. In fact, the first attempt to convert lignocelluloses with *Aspergillus fumigatus* into animal feed was done by Pringsheim and Lichtenstein as early as 1920 (54). The work on SCP production from lignocelluloses has been reviewed by Chahal (6), Chahal and Moo-Young (12), and Rolz (68). Chahal (6) groups these processes according to whether they convert lignocelluloses indirectly or directly.

Indirect Conversion of Lignocelluloses. Lignocelluloses are pretreated by various physico-chemical methods to obtain cellulose, to be used in the following ways.

CHEMICAL HYDROLYSIS OF CELLULOSE TO GLUCOSE AND GROWTH OF MICROORGANISMS ON THE HYDROLYZATE. Wood cellulose is hydrolyzed with acid into glucose by the Bergius or Scholler process, and *Candida utilis* is grown to produce SCP. The Scholler process was used during World Wars I and II, when production reached 15,000 tons per year (52). The recent processes based on acid hydrolysis of lignocelluloses for SCP production have been described by Callihan and Clemmer (4) and Han et al. (34).

ENZYMATIC HYDROLYSIS OF CELLULOSE TO GLUCOSE AND GROWTH OF MICROORGANISMS ON GLUCOSE. The following three processes are based on this system:

1. Indian Institute of Technology Process: The cellulosic material is pulverized to fine particles that are then hydrolyzed with cellulases. The glucose syrup is converted into SCP with yeasts (19).

2. University of California Process: Hammered mill newspaper is hydrolyzed with cellulases. Because the concentration of glucose is very low in the hydrolysate, it is concentrated before fermentation into SCP or ethanol. In this process, 59.4 tons/day of torula yeast and 8.1 tons/day of ethanol from 885 tons of newpaper were produced by Wilke et al. (84).

3. Kyoto University Process: The lignocelluloses are hydrolyzed with cellulases and xylanases into glucose, cellobiose, xylose, and xylobiose. The glucose is fermented with *Saccharomyces* species in the first tank, and yeast is removed by membrane filtration. The residual hydrolysate is fermented with *Candida* species, where xylose is consumed for SCP production. The residual disaccharides (cellobiose and xylobiose) are either recycled after hydrolyzing them into monosaccharides (glucose and xylose) with immobilized cellobiase and xylobiase, or are used in the medium for induction of enzymes (cellulases and xylanses) (78). This is a very complicated process as it involves different enzymes and microorganisms; hence synchronizing all the reactions to get the desired results would be very difficult.

Direct Conversion of Lignocelluloses. Lignocelluloses are pretreated with various physico-chemical methods, and hydrolysis of the substrate into utilizable compounds and their conversion into SCP is performed by the same organism.

LOUISIANA STATE UNIVERSITY PROCESS. The Louisiana State University process is based on the findings of Srinivasan and Han (75), Dunlap and Callihan (22), and Han and Callihan (33) and involves two bacteria. *Cellulomonas* sp. which hydrolyzes the cellulose into glucose; the glucose is consumed by the same organism to produce SCP. *Alcaligenes faecalis* is used in this fermentation to consume cellobiose, an intermediate product, to check the inhibition of cellulose hydrolysis with cellulases produced by *Cellulomonas* sp.

All of the processes described above require separation of cellulose from lignocelluloses and hydrolysis of cellulose with enzymes or acid into glucose for its conversion into SCP, except the process of Louisiana State University. In all these processes, however, the hemicelluloses, the second largest fraction of polysaccharides of lignocelluloses, and lignin, the third largest, are not utilized. Thus, these fractions become the biggest source of pollution.

In response to the above problems, I developed "An Integrated Process for Production of Food, Feed and Fuel (Ethanol) from Biomass (Lignocelluloses)" at the Institut Armand-Frappier, Laval, Quebec, Canada after many years of continuous research. This process is called the Institut Armand-Frappier (IAF) process.

THE INSTITUT ARMAND-FRAPPIER (IAF) PROCESS. The integrated IAF process was developed to utilize lignocelluloses, especially wood and crop residues. Any of the following three pretreatments can be used to lignocellulose for fractionation into its three components, cellulose, hemicelluloses, and lignin:

1. Alkali treatment: Lignocelluloses are treated with 5–20% sodium hydroxide (wt/wt of substrate) at 80–121°C for a certain time depending on the nature of the substrate (the method developed at IAF).

2. Steam treatment: The moist lignocelluloses are treated with steam at 200–230°C for a certain time (the method developed by Stake Technology, Oakville, Ontario).

3. Thermomechanical pretreatment of aqueous suspension: A 12–14% finely ground substrate in an aqueous suspension is treated at 150–230°C for certain time (the method developed by the University of Sherbrooke, Sherbrooke, Quebec).

For detailed discussion of these pretreatments, refer to Chahal (8).

The microorganisms used in this process are as follows: *Chaetomium cellulolyticum* asporogenous mutant IAF-102 *Chaetomium cellulolyticum* IAF-101, NRRL-18756; *Pleurotus sajor-caju* IAF-503, NRRL-18757; *Aspergillus* sp. IAF-201, NRRL-18758; *Penicillium* sp. IAF-603, NRRL-18759; and *Trichoderma reesei* QMY-1, IAF-702, NRRL-18760. All of these fungi except *T. reesei* are used for SCP production; the *T. reesei* is used for production of cellulases for hydrolysis of cellulose into glucose.

Figure 9 illustrates the IAF process. The lignocelluloses (**1**) are pretreated (**2**) with any one of the pretreatments described above to fractionate them into cellulose (**3**), lignin (**4**), and hemicelluloses (**5**). The pretreated lignocelluloses (**2**) are fermented with any of the above SCP-producing fungi (**7**) into protein-rich animal feed (**8**). At step (**2**), some of the hemicelluloses and lignin are solubilized. These solubilized fractions are not removed, as is done in most of the other processes explained above, but are retained in the medium where hemicelluloses are consumed simultaneously with the insoluble cellulose for production of SCP (**8**). The lignin is not fermented and is recovered from the effluents by precipitation before the effluents are discharged in water.

The fractionated cellulose at step (**3**) can be used for production of paper, rayon, cellulose derivatives, or chemicals, or it can be fermented (**12**) into high-quality protein-rich feed, SCP (**13**), which can also be used for human consumption. Part of this cellulose (**3**) fraction is used for the production of cellulases (**14**) with *T. reesei* for

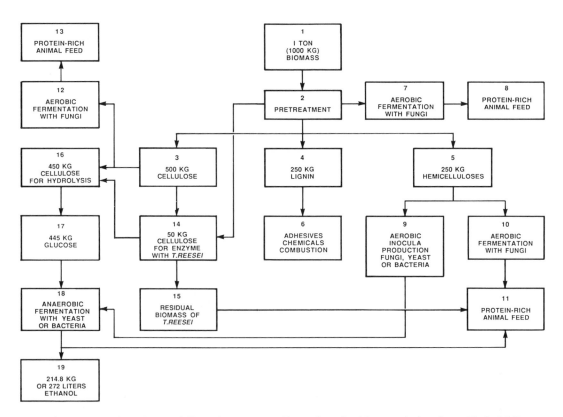

Figure 9. Institut Armand-Frappier process. [Reproduced with permission from Chahal (7)].

hydrolysis of remaining cellulose at step (**16**) into glucose (**17**). The residual biomass of
T. reesei (**15**) goes to the pool of SCP (**11**).

The fractionated hemicelluloses (**5**) are utilized for the production of inocula (**9**) of
various fungi and yeasts required for SCP production at steps (**7, 10, 12**), for cellulase
production at step (**14**), and for anaerobic fermentation (with yeasts) of glucose (**18**) into
ethanol (**19**). The remaining portion of hemicelluloses is fermented with suitable fungus
at step (**10**) for production of protein-rich feed, SCP (**11**). The residual biomass of yeasts
at step (**18**) goes to the pool of protein-rich animal feed at step (**11**). During fermentation with most of the fungi, the lignin remains unaltered and can be recovered by precipitation from the effluent. When *P. sajor-caju* is used for production of SCP from this
fraction or from unfractionated pretreated lignocelluloses, however, the lignin is
broken down into oligolignols of low molecular weights (11). These oligolignols can be
recovered from the effluents by precipitation. The lignin recovered from the effluents
or isolated at step (**4**) can be used for production of adhesives, and various chemicals
originally synthesized from petroleum. The oligolignols obtained from the fermentation effluents of *P. sajor-caju* have the potential to be further biodegraded or biotransformed into useful compounds.

Butanediol Production from Lignocelluloses

The production of 2,3-butanediol from lignocelluloses has recently been found to be
a viable approach to using it as a fuel and a potential feedstock for production of
various chemicals (26, 41, 42). It can be readily dehydrated to methyl ethyl ketone for

use as an industrial solvent and to butadiene for manufacturing of synthetic rubber and production of polyesters and polyurethane resins (56). *Klebsiella pneumoniae* (formerly known as *Aerobacter aerogenese*), a bacterium, seems to be the best candidate for production of butanediol from lignocelluloses. It can utilize and ferment all the major sugars (D-glucose, D-xylose, D-mannose, D-galactose, L-arabinose, etc.) present in lignocelluloses. It can produce butanediol at near theoretical efficiencies from both the pentose and hexose sugars.

A combined hydrolysis and fermentation (CHF) approach was studied by Yu et al. (89) in which the hydrolysis products, produced by the enzymes of *Trichoderma harzianum,* are continuously removed by the fermentative organism, *K. pneumoniae,* and are converted to fermentation products. They also reported that the product yields obtained from the unextracted steam-exploded lignocelluloses surpassed the combined yields of products obtained from the water-soluble hemicelluloses and water-insoluble cellulose.

Bioconversion of Hemicelluloses

About 250 kg of hemicelluloses are released from every ton of lignocelluloses. As discussed above, hemicelluloses can be bioconverted into SCP, and these can also be used for the production of inocula of various microorganisms required in the IAF process. Hemicelluloses are composed mainly of pentoses.

Bioconversion of Pentoses by Bacteria.   The metabolism of pentoses by bacteria has been reviewed by Horecker (39). The initial reaction may proceed in one of three ways: (a) direct isomerization of the aldopentose to its respective ketose (most species), e.g. D-xylulose is formed from D-xylose, and L-arabinose is converted to L-ribulose; (b) oxidation-reduction to the corresponding sugar alcohol, e.g. D-xylose to xylitol, or (c) direct phosphorylation of the pentose, e.g. D-ribose to D-ribose-5-phosphate. The summary of these initial routes of pentose metabolism and key enzymes involved are presented in Figure 10. The phosphorylated pentoses are subsequently converted into pyruvate through a combination of pentose phosphate and Embden-Meyerhof pathways.

Several species of the genus *Clostridium* are able to convert a mixture of cellulose and hemicelluloses into ethanol plus lactic and acetic acids. *C. thermosaccharolyticum* alone and in mixed culture with *C. thermocellum* is used for this conversion. *C. thermocellum* helps to hydrolyze xylan into xylose by way of the xylanase activity present in its cellulase system. The xylose thus produced is fermented along with glucose, produced from cellulose, into ethanol and acids by *C. thermosaccharolyticum* (26).

Acetone, butanol, and ethanol are also produced by various species of *Clostridium.* Volesky and Szczesny (82) provide a detailed review.

Bioconversion of Pentoses by Yeasts.   The bioconversion of xylose into ethanol has been reviewed by Jeffries (45). Many yeasts are capable of metabolizing pentose sugars via an oxidation-reduction pathway. Of the more than 400 species tested by Barnett (2), none could consume these carbohydrates anaerobically. The general routes of pentose metabolism by yeasts are shown in Figure 11. Although several yeasts have been shown to possess the enzyme xylose isomerase (28), the oxidation-reduction pathway appears to be obligatory in the yeasts for the breakdown of pentoses.

Schizosaccharomyces pombe is capable of producing ethanol under anaerobiosis (29). Ethanol production from pentoses by *Candida tropicalis* and *Candida* sp. XF217 has been discussed by Jeffries (44) and Gong et al. (30), respectively.

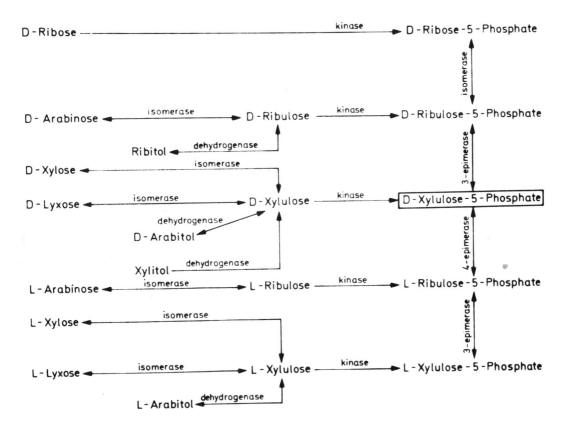

Figure 10. Pathway of pentose and pentitol metabolism in *Aerobacter aerogenes* (*Klebsiella pneumoniae*). [Reproduced with permission from Mortlock and Wood (60)].

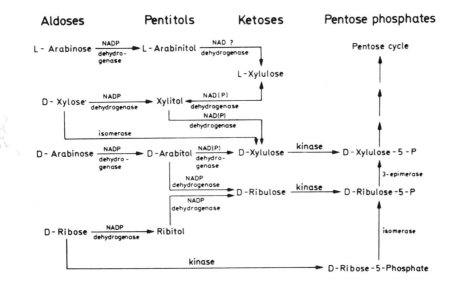

Figure 11. Pathways of pentose and pentitol metabolism in yeast. [Reproduced with permission from Barnett (2)].

Pachysolen tannophilus is another important yeast that converts hemicelluloses into ethanol. A quantitative screening procedure for xylose fermentation was conducted on 56 yeast isolates by du Preez and Prior (23). The outstanding isolate found was a strain of *Pichia stipitis* CSIR-Y633, which had an ethanol yield coefficient of 0.45 from xylose and produced no detectable amounts of xylitol. Reviews on ethanol production by yeasts have been written by Schneider (72) and Schneider et al. (73). A review of the biological and physiological properties of *P. tannophilus* has been given by Kurtzman (51).

Bioconversion of Pentoses by Fungi. Ethanol and acetic acid are produced by *Fusarium oxysporum* from xylose through D-xylulose in equimolar quantities as explained in below.

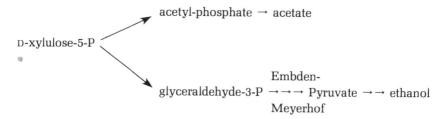

The species of *Rhizopus* can also convert pentoses into ethanol and acetic acid whereas the species of *Mucor* produce only ethanol. Ueng and Gong (80) tested several species of *Fusarium* and *Mucor* for ethanol production from the hydrolysate of hemicelluloses of sugar cane bagasse. Best yields were obtained with *Mucor* sp., because the species of *Fusarium* were subject to inhibition by an unidentified component of the hydrolysate.

Potential Products from Lignin

An enormous quantity of lignin will start accumulating as soon as the industries based on utilization of carbohydrates from lignocelluloses for various products are put in place, of about 30 millions tons of lignin already produced annually by pulp mills in the world, only 6% is recovered and marketed for industrial applications achieving a value to the pulp producer in excess of fuel value. The rest of lignin is burned or otherwise wasted (55). Some of the uses of lignin sulfonates and Kraft lignins are as emulsifiers, stabilizers, grinding aids, dispersants, and binders (17, 55).

Although research on lignin biodegradation is still at a basic level, development of lignin bioconversion processes has considerable potential for the future. Such bioconversions might produce a variety of chemicals, many of which are now derived from petroleum (17, 47). One of the uses of lignin with great potential value would be in its conversion to low-molecular-weight chemicals (17, 47); for example, vanillin, dimethylsulphide, dimethylsulphoxide, methylmercaptan, and a variety of phenols.

Chahal and Hachey (11) reported that the lignin of corn stover was degraded progressively into oligolignols of low molecular weight when the pretreated corn stover was fermented with *Pleurotus sajor-caju*. During this fermentation, the polysaccharides of corn stover were converted into a mycelial biomass rich in protein, useful as food or feed, and the biodegraded lignin became part of the effluent. This biodegraded lignin, oligolignols of low molecular weight, can be recuperated by acidifying the effluents. These oligolignols may possibly be further biologically or chemically modified into high value products. Demethylation, hydroxylation, side chain shortening and ring cleavage of lignin could alter polymeric lignin for further chemical modifications (37).

The possible involvement of lignin and lignin structures in anti-influenza virus activity (35); antitumor, antiviral, and immunopotentiating activities (71); and anticholesteremic and antihypertensive activities (76) have been reported. New uses of lignin in phrarmacology may thus emerge in the near future.

CONCLUSIONS

The ever-increasing demand for proteinaceous unconventional foods/feeds (SCP) combined with concerns over global warming, air pollution, and waste management have necessitated the search for alternatives to conventional high-calorie-foods/feeds and fossil fuels, respectively. The answer to both problems may lie in the bioconversion of lignocellulosic wastes into SCP and fuel ethanol. Ethanol is considered a cleaner fuel than all the others. A large tonnage of lignocellulosic wastes is already available as crop residues and forestry wastes. If there is a regular demand, lignocelluloses can also be raised by systematic farming and silviculture.

Many efficient processes for the bioconversion of various components of lignocelluloses into useful products are now available. Some require further improvements to make them commercially viable, however. Similarly, some biological processes are available to alleviate the pollution from effluents released by the pulp mills, the biggest consumers of lignocelluloses (wood). Research is under way on the chemical and biological conversion of lignin into additional useful chemicals.

From lignocelluloses, we may be able economically to produce single-cell protein, clean fuel (ethanol) and other chemicals formerly produced from petroleum, and new pharmaceuticals in the near future.

LITERATURE CITED

1. Archibald, F., M. G. Paice, and L. Jurasek. 1990. Decolorization of kraft bleached effluent lignins by *Coriolus versicolor*. In *Biotechnolgy in Pulp and Paper Manufacture*. Ed. T. K. Kirk and H.-M. Chang. Toronto: Butterworth-Heinemann. 253–262.

2. Barnett, J. A. 1976. The utilization of sugars by yeasts. *Adv. Carbohydrate Chem. Biochem.* 32:125–234.

3. Bassham, J. A. 1975. General consideration (Cellulose utilization: an overview). *Biotechnol. Bioeng. Symp.* No. 5:9–19.

4. Callihan, C. D., and J. E. Clemmer. 1979. Biomass from cellulosic materials. In *Microbial Biomass*. Ed. A. H. Rose. *Economic Microbiol.* New York: Academic Press. 4:271–288.

5. Chahal, D. S. 1982. Growth characteristics of microorganisms in solid state fermentation for up grading of protein values of lignocelluloses and cellulase production. *Am. Chem. Soc. Symp.* 207:421–442.

6. Chahal, D. S. 1991. Lignocellulosic wastes: biological conversion. In *Bioconversion of Waste Materials to Industrial Products*. Ed. A. M. Martin. Barking, Essex, England: Elsevier Science Publishers Ltd. 373–400.

7. Chahal, D. S., ed. 1991. *Food, Feed and Fuel from Biomass.* New Delhi: Oxford & IBH Publishing Co.

8. Chahal, D. S. 1991. Pretreatments of lignocelluloses. In *Food, Feed and Fuel from Biomass.* Ed. D. S. Chahal. New Delhi: Oxford & IBH Publishing Co. 59–83.

9. Chahal, D. S. 1991. Production of *Trichoderma reesei* cellulase system with high hydrolytic potential by solid state fermentation. *Am. Chem. Soc. Symp.* 460:111–122.

10. Chahal, D. S., P. S. Chahal, and G. Andre. 1991. Cellulase production and hydrolysis of lignocelluloses. In *Food, Feed and Fuel from Biomass.* Ed. D. S. Chahal. New Delhi: Oxford & IBH Publishing Co. 281–312.

11. Chahal, D. S., and J. M. Hachey. 1990. Use of hemicelluloses and cellulose and degradation of lignin by *Pleurotus sajor-caju* grown on corn stalks. *Am. Chem. Soc. Symp.* 433:304–310.

12. Chahal, D. S., and M. Moo-Young. 1981. Bioconversion of lignocellulosics into animal feed. *Devel. Ind. Microbiol.* 22:143–159.

13. Chahal, D. S., A. Ouellet, and J. M. Hachey. 1990. Fermentation of hemicellulose sugars from spent sulfite liquor into fungal biomass with various fungi and characterization of lignin isolated from fermentation broth. In *Biotechnology in Pulp and Paper Manufacture: Applications and Fundamental Investigations.* Ed. T. K. Kent, H.-M. Chang. Boston: Butterworth-Heinemann. 303–310.

14. Chang, H.-M., T. W. Joyce, T. K. Kirk, and V. B. Huynh. 1985. US Patent No. 4,554,075.

15. Cowling, E. B. 1958. A review of literature on enzymatic degradation of cellulose and wood. *Forest Product Laboratory Report* No. 2116, Madison, Wisconsin. 26.

16. Crawford, R. L., ed. 1981. *Lignin Biodegradation and Transformation.* New York: John Wiley.

17. Crawford, D. L., and R. L. Crawford. 1980. Microbial degradation of lignin. *Enzyme Microbiol. Technol.* 2:11–22.

18. Croom, H. C. 1973. US Patent No. 3,736,254, May 29.

19. Das, K., and T. K. Ghose. 1973. Economic evaluation of enzymic utilization of waste cellulosic materials. *J. Appl. Chem. Biotechnol.* 23:829–831.

20. Dekker, R. F. H. 1985. Biodegradation of hemicelluloses. In *Biosynthesis and Biodegradation of Wood Components.* Ed. T. Higuchi. New York: Academic Press, Inc. 505–533.

21. Dostal, J., and B. Tomis. 1990. Fodder yeast production at a Czechoslovakian pulp mill. In *Biotechnology in Pulp and Paper Manufacture.* Ed. T. K. Kirk, H.-M. Chang. Boston: Butterworth-Heinemann. 291–301.

22. Dunlap, C. E., and C. D. Callihan. 1973. Single-cell protein from waste cellulose. *Final Report on Grant EP00328-4 to the Federal Solid Waste Management Program.* Washington, DC: US Environmental Protection Agency.

23. du Preez, J. C., and B. A. Prior. 1985. A quantitative screening of some xylose-fermenting yeast isolates. *Biotechnol. Lett.* 7:241–246.

24. Enari, T. M. 1983. Microbial cellulases. In *Microbial Enzymes and Biotechnology.* Ed. W. M. Forgarty. London: Applied Science Publishers. 183–223.

25. Fengel, G., and D. Wagener. 1984. *Wood, Chemistry, Ultra structure, Reaction.* Berlin: De-Gruyter.

26. Flickinger, M. C. 1980. Current biological research in conversion of cellulosic carbohydrates into liquid fuels: How far we have come? *Biotechnol. Bioeng.* 22 (Suppl. 1):27–48.

27. Forss, K., and K. Jokinen. 1991. The Pekilo Process. In *Food, Feed and Fuel from Biomass.* Ed. D. S. Chahal. New Delhi: Oxford & IBH Publishing Company Limited. 165–173.

28. Gong, C. S., L. F. Chen, G. T. Tsao, M. C. Flickinger. 1981. Conversion of hemicellulose carbohydrate. *Adv. Biochem. Eng. Biotechnol.* 20:93–118.

29. Gong, C. S., T. A. Claypool, L. D. McCracken, C. M. Maun, P. Ueng, G. T. Tsao. 1983. Conversion of pentoses by yeasts. *Biotechnol. Bioeng.* 25:85–102.

30. Gong, C. S., L. D. McCracken, and G. T. Tsao. 1981. Direct fermentation of D-xylose to ethanol by a xylose-fermenting yeast mutant, Candida sp. XF217. Biotechnol. Lett. 3:245–250.

31. Halpern, M. G. 1975. Pulp Mill Processes: Pulping, Bleaching, Recycling. Park Ridge, New Jersey: Noyes Data Corporation.

32. Hammel, K. E., and M. A. Moen. 1991. Depolymerization of synthetic lignin in vitro by lignin peroxidase. Enzyme Microb. Technol. 13:15–18.

33. Han, Y. W., and C. D. Callihan. 1974. Cellulose fermentation: effect of substrate treatment on microbial growth. Appl. Microbiol. 27:159–165.

34. Han, Y. W., P. R. Cheeke, A. W. Anderson, and C. Lekprayoon. 1976. Growth of Aurobasidium pullulans on straw hydrolysate. Appl. Environ. Microbiol. 32:799–802.

35. Harda, H., H. Sakagami, K. Nagata, T. Oh-hara, Y. Kawazoe, A. Ishihama, N. Hata, Y. Misawa, H. Terada, and K. Konno. 1991. Possible involvement of lignin structure in anti-influenza virus activity. Antiviral Res. 15:41–50.

36. Higuchi, T. 1971. Fermentation and biological degradation of lignin. Adv. Enzymol. 34:207–277.

37. Higuchi, T. 1980. Lignin structure and morphological distribution in plant cell walls. In Lignin Biodegradation: Microbiology, Chemistry and Potential Applications. Ed. T. K. Kirk, T. Higuchi, H. M. Chang. Boca Raton, Florida: CRC Press Inc. 2:1–19.

38. Higuchi, T. 1982. Biodegradation of lignin: biochemistry and potential applications. Experientia 38:159–166.

39. Horecker, B. L. 1962. Pentose Metabolism in Bacteria. New York: John Wiley & Sons Inc.

40. Ishaque, M., and D. S. Chahal. 1991. Crop residues. In Food, Feed and Fuel from Biomass. Ed. D. S. Chahal. New Delhi: Oxford & IBH Publishing Company Limited. 19–26.

41. Jansen, M. B., M. C. Flickinger, and G. T. Tsao. 1984. Production of 2,3-butanediol from D-xylose by Klebsiella oxytoca ATCC 8724. Biotechnol. Bioeng. 26:362–369.

42. Jansen, N., and G. Tsao. 1983. Bioconversion of pentoses to 2,3 butanediol by Klebsiella pneumoniae. Adv. Biochem. Eng./Biotechnol.27:85–99

43. Janshekar, H., and A. Fiechter. 1983. Lignin: biosynthesis, application, and biodegradation. Adv. Biochem. Eng./Biotechnol. 27:119–178.

44. Jeffries, J. W. 1981. Conversion of xylose to ethanol under aerobic conditions by Candida tropicalis. Biotechnol. Lett. 3:213–218.

45. Jeffries, T. W. 1983. Utilization of xylose by bacteria, yeasts and fungi. Adv. Biochem. Eng./Biotechnol. 27:1–32.

46. King, K. W., and M. A. Vessal. 1969. Enzymes of cellulase complex. Adv. Chem. Ser. 95:7–25.

47. Kirk T. K., and H.-M. Chang. 1981. Potential applications of biolignolytic systems. Enzyme Microb. Technol. 3:189–196.

48. Kirk, T. K., and K.-E. Eriksson. 1989. Roles for biotechnology in manufacture. In World Pulp & Paper Technology—1990. Ed. F. Robert. London: The Sterling Publishing Group PLC. 23–28.

49. Kirk, T. K., and R. L. Farrell. 1987. Enzymatic "combustion": the microbial degradation of lignin. Ann. Rev. Microbiol. 41:465–505.

50. Kirk, T. K., T. Higuchi, and H.-M. Chang. 1980. Lignin Biodegradation: Microbiology, Chemistry and Potential Applications, Vols. 1 and 2. Boca Raton, Florida: CRC Press Inc.

51. Kurtzman, C. P. 1983. Biology and physiology of the D-xylose fermenting yeast Pachysolen tannophilus. Adv. Biochem. Eng./Biotechnol. 27:73–83.

52. Laskin, A. I. 1977. Single-cell protein. Ann. Rep. Fermentation Processes 1:151–180.

53. Leisola, S. A., and A. Fiechter. 1985. New trends in lignin biodegradation. Adv. Biotechnol. Processes 5:59–89.

54. Litchfield, J. 1968. The production of fungi. In *Single-cell Protein*. Ed. R. I. Mateles, S. R. Tannenbaum. Cambridge, Mass: MIT Press. 304–329.

55. Little, B. F. P. 1990. Lignin—a nuisance or an oppotunity. In *Cellulose Source and Exploitation*. Ed. J. F. Kennedy, G. O. Phillips, P. A. Williams. London: Elliswood. 473–482.

56. Long, S. K., and R. Patrick. 1963. The present status of 2,3-butylene glycol fermentation. *Adv. Appl. Microbiol.* 5:135–155.

57. Mandels, M. 1981. Cellulases. *Ann. Rep. Ferment. Processes.* 5:35–78.

58. McIntyre, T. C. 1987. An overview of the environmental impacts anticipated from large scale biomass/energy systems. In *Biomass Conversion Technology: Principles and Practice*. Ed. M. Moo-Young. Toronto: Pergamon Press. 45–52.

59. Millett, M. A., A. J. Baker, and L. D. Satter. 1970. Physical and chemical pretreatments for enhancing cellulose saccharification. *Biotechnol. Bioeng. Symp.* 6:125–153.

60. Mortlock, R. P., and W. A. Wood. 1964. Metabolism of pentoses and pentitols by *Aerobacter aerogenes. J. Bacteriol.* 88:838–844.

61. Paice, M. G., R. Bernier, and L. Jurasek. 1988. Viscosity-enhancing bleaching of hardwood Kraft pulp with xylanase from clone gene. *Biotechnol. Bioeng.* 32:235–239.

62. Parker, R. R., and J. Sibert. 1972. Effects of pulp effluent on the dissolved oxygen supply in Alberni Inlet, British Columbia. *Fish Res. Board Canada Tech. Rep.* 316.

63. Poole, N. J., D. J. Wildish, and D. D. Kristmanson. 1978. The effects of the pulp and paper industries on the aquatic environment. *CRC Critical Rev. Environ. Control* 8:153–195.

64. Rawat, J. K., and J. C. Nautiyal. 1991. Availability of forest biomass. In *Food, Feed and Fuel from Biomass*. Ed. D. S. Chahal. New Delhi: Oxford & IBH Publishing Company Limited. 27–36.

65. Reese, E. T. 1977. Degradation of polymeric carbohydrates by microbial enzymes. *Recent Adv. Phytochem.* 1:311–367.

66. Reese, E. T., S. G. H. Siu, and H. S. Levinson. 1950. The biological degradation of soluble cellulose derivatives and its relationship to the mechanism of cellulose hydrolsis. *J. Bacteriol.* 59:485–489.

67. Rogers, I. H., J. C. Davis, G. M. Kruzinski, J. W. Mahood, J. A. Servizi, and R. W. Gordon. 1975. Fish toxicants in Kraft effluents. *Tappi* 58(7):136.

68. Rolz, C. 1984. Microbial biomass from renewables: a second review of alternatives. *Ann. Rep. Ferment. Processes* 7:213–365.

69. Romantschuck, H. 1975. The Pekilo process: protein from spent sulfite liquor. In *Single-cell Protein II*. Ed. S. R. Tannenbaum, D. I. C. Wang. Cambridge, Mass: The MIT Press. 344–356.

70. Ryu, D., and M. Mandels. 1980. Cellulase: biosynthesis and applications. *Enzyme Microb. Technol.* 2:91–102.

71. Sakagami, H., Y. Kawazoe, N. Komatsu, A. Simpson, M. Nonoyama, K. Kono, T. Yoshida, Y. Kuroiwa, and S. Tanuma. 1991. Antitumor, antiviral and immunopotentiating activities of pine cone. *Anticancer Res.* 11:881–888.

72. Schneider, H. 1991. Conversion of hemicelluloses into ethanol. In *Food, Feed and Fuel from Biomass*. Ed. D. S. Chahal. New Delhi: Oxford & IBH Publishing Company Limited. 427–439.

73. Schneider, H., R. Maleszka, L. Neirinck, I. Veliky, P. Wang, and Y. K. Chan. 1983. Ethanol production from xylose and several other carbohydrates by *Pachysolen tannophilus* and other yeasts. *Adv. Biochem. Eng./Biotechnol.* 27:57–71.

74. Shewale, J. G. 1982. β-glucosidase: its role in cellulase synthesis and hydrolysis of cellulose. *Int. J. Biochem.* 14:435–443.

75. Srinivasan, V. R., and Y. W. Han. 1969. Utilization of bagasse. *Adv. Chem. Ser.* 95:447–460.

76. Takamori, T., T. Tsurumi, M. Takagi, and T. Kamiwaki. 1991. A substance having suppressing function for diseases relating to increase in cholesterol, and foods and drinks in which it is used. European Patent Appl. EP0431650 A1, 12 June 1991. 27 pp.

77. Tarkow, H., and W. D. Feist. 1969. A mechanism for improving the digestibility of lignocellulosic materials with dilute alkali and liquid ammonia. *Adv. Chem. Ser.* 95:197–218.

78. Tanaka, M., and R. Matsuno. 1985. Conversion of lignocellulosic materials to single-cell protein (SCP): review developments and problems. *Enzyme Microb. Technol.* 7:197–207.

79. Timell, T. E. 1967. Recent progress in the chemistry of hemicelluoses. *Wood Sci. Technol.* 1:45–70.

80. Ueng, P., and C. S. Gong. 1982. Ethanol production from pentoses and sugar-cane bagasse hemicellulose hydrolysate by *Mucor* and *Fuarium* species. *Enzyme Microb. Technol.* 4:169–171.

81. Viikari, L., A. Kantelinen, K. Poutanen, and M. Ranua. 1990. Characterization of pulps treated with hemicellulolytic enzymes prior to bleaching. In *Biotechnology in Pulp and Paper Manufacture: Application and Fundamental Investigations*. Ed. T. K. Kirk, H.-M. Chang. Boston: Butterworth-Heinemann. 145–151

82. Volesky, B., and T. Szczesny. 1983. Bacterial conversion of pentose sugars to acetone and butanol. *Adv. Biochem. Eng./Biotechnol.* 27:101–118.

83. Walden, C. C., and T. E. Howard. 1977. Toxicity of pulp and paper mill effluents. A review of regulations and research. *Tappi* 60 (1):122.

84. Wilke, C. R., G. R. Cysewski, R. D. Yang, and U. von Syocker. 1976. Utilization of cellulosic materials through enzymic hydrolysis. I. Preliminary essessment of an integrated processing scheme. *Biotechnol. Bioeng.* 18:1315–1323.

85. Wood, T. M. 1972. The C_1 component of cellulase complex. In *Fermentation Technology*. Ed. G. Terui. Osaka: Japan. Society of Fermentation Technology. 711–718.

86. Wood, T. M., and S. T. McRae. 1978. The cellulase of *Trichderma koningii*. Purification and properties of some endoglucanase components with special reference to their action on cellulose when acting alone and in synergism with the cellobiohydrolase. *Biochem. J.* 171:61–72.

87. Wood, T. M., and S. T. McCrae. 1979. Synergism between enzymes involved in the solubilization of native cellulose. *Adv. Chem. Ser.* 18:181–209.

88. Woodward, J., and A. Wiseman. 1983. Fungal β-glucosidase: their properties and applications. *Enzyme Microb. Technol.* 4:73–79.

89. Yu, E. K. C., L. Deschatelets, and J. N. Saddler. 1984. Combined enzymatic hydrolysis and fermentation approach to butanediol production from cellulose and hemicellulose carbohydrates of wood and agricultural residues. *Biotechnol. Bioeng. Symp.* 14:341–352.

CHAPTER 18

Biodegradation of Toxic Chemicals by White Rot Fungi

TUDOR FERNANDO and STEVEN D. AUST

INTRODUCTION

Since the beginning of the industrial revolution, an enormous number of organic compounds have been chemically synthesized. Direct application of these synthetic chemicals over a long period and indirect application through generation of chemical waste from every major economic sector, including agriculture, the oil industry, the textile and paper industries, and the defense and aerospace industries, have resulted in environmental contamination. Unlike the naturally occurring organic compounds that are readily degraded upon introduction into the environment, some of these synthetic chemicals are extremely resistant to biodegradation by native flora (62). Therefore, hazardous wastes and chemicals have become one of the major problems of modern society world-wide. Approximately 80 billion pounds of hazardous wastes are produced annually in the United States alone. According to the Environmental Protection Agency (EPA), only 10% of these wastes are disposed of safely (20). Moreover, with increased sophistication of chemical, technological, and engineering science, many new substances have been developed that are capable of causing unpredictable secondary effects on the environment and living organisms. Hazardous wastes and toxic chemicals pose complex environmental problems by directly affecting the air, water, soil, and sediment while indirectly and unpredictably affecting living organisms that use these resources. For example, many of these chemicals and waste byproducts are lipophilic and insoluble in water, and thus tend to bioconcentrate in tissues of organisms at different tropic levels of the food chain (56). Attempts have been made since the 1960s to address the magnitude of environmental contamination. Measures have been introduced to reduce contamination and to help clean the environment.

There are several types of treatment technologies. *Physical treatment* alters the hazardous material to a more convenient form for further processing or disposal. *Chemical treatment* uses chemical reactions to alter the hazardous chemicals to less

Dr. Aust is at the Biotechnology Center, Utah State University, Logan, UT 84322-4705, U.S.A. Dr. Fernando may be reached at the Metabolism, Residue, and Environmental Fate Division, Battelle, 505 King Avenue, Columbus, OH 43201-2693, U.S.A.

hazardous forms. *Thermal treatment* utilizes high temperatures to burn the hazardous wastes. The byproducts formed from these treatments, however, may be more recalcitrant, toxic, or carcinogenic than the parent chemical. Although landfills provide the cheapest method of waste disposal, they are the most dangerous form of disposal. When raw wastes or drums of wastes are buried, rainwater percolates through the landfill, leaching out toxic chemicals and contaminating soil and groundwater.

The inherent ability of white rot fungi to degrade lignin in wood suggested that such organisms may be useful for biological decontamination of toxic and hazardous chemicals. Lignin is an amorphous heteropolymer consisting of three phenyl propane subunits (coniferyl, sinapyl, and p-coumaryl alcohols) linked by a variety of carbon-carbon and ether bonds (65). Although lignin is one of the most resistant natural compounds to biological degradation, biodegradation of this complex lignin polymer in lignocellulosic plant material is accomplished by white rot fungi. The degradation of lignin is considered the rate-limiting step in bioconversion of the closely associated polysaccharides in the lignocellulosic materials, cellulose and hemicellulose (60).

The lignin degrading-system of white rot fungi cleaves the carbon–carbon and carbon–oxygen bonds that constitute the lignin molecule. During the delignification process, the bond fission or cleavage occurs regardless of the conformation of the chiral carbons of lignin. Therefore, the lignin-degrading enzyme system of these fungi must be nonstereoselective and nonspecific. This characteristic appears to be at least partly due to the free radical mechanism of degradation used by white rot fungi. Free radical species may also serve as secondary oxidant moieties that carry out the lignin depolymerization or oxidation of other compounds at sites distant from the active site of the enzyme molecule. Such a highly reactive free radical depolymerization mechanism would be ideal for the biodegradation of organic pollutants in the environment.

Typically, a good degradative enzyme must possess a low K_m value or a high affinity for a particular compound in order to continue biodegradation until the level of chemical (compound) reaches an acceptable lower limit. Therefore, an enzymatic system that involves a free radical mechanism would theoretically allow almost complete conversion of a substrate to an oxidized product. Any organism able to depolymerize a complex water-insoluble heteropolymer, i.e. lignin, should possess an enzyme system capable of catalyzing the oxidative depolymerization of lignin extracellularly. This trait is essential for use of an organism in bioremediation of recalcitrant chemicals because many environmental pollutants are either sparingly or not at all soluble in water and are usually quite tenaciously bound to the organic lignocellulosic materials found abundantly in soil, thus hindering their bioavailability. The solubilization of lignin by white rot fungi occurs extracellularly. Also, the lignin-degrading enzyme system is synthesized under nutrient-limiting conditions (carbohydrate, nitrogen, and sulfur) (45, 47). This characteristic is probably an evolutionary adaptation of the white rot fungi to the low nutrient levels in the wood. For example, the amount of nitrogen in woody tissues is usually 0.03–0.10% compared to 1.0–5.0% in nonlignified tissues. Therefore, the carbon/nitrogen ratio of most woody tissues is high, in the order of 350–500:1, and may exceed 1000:1 in some heartwood. For most fungi, a growth medium with such a high carbon/nitrogen ratio would be nitrogen-deficient and growth-limiting (33). During evolution, white rot fungi apparently adapted to this by developing efficient mechanisms of assimilating, utilizing, and conserving the meager supply of nitrogen. If the enzymes involved in the degradation of lignin are also involved in the degradation of xenobiotic molecules, prior exposure of the microorganisms, i.e. adaptation, would not be required, nor would enzyme synthesis be expected to be repressed when levels of the xenobiotics reached lower concentrations. By analogy, the enzyme system of the white rot fungi may have the ability to catalyze the initial oxidation of less complex xenobiotic molecules and have all of the steps in its degradatory pathway to produce carbon dioxide, the ultimate microbial degradation product in aerobic systems.

All of these assumptions led to the proposal of using white rot fungi for the degradation of recalcitrant toxic chemicals in the environment (13). The first experiments evaluated the ability of the white rot fungus, *Phanerochaete chrysosporium*, to degrade several dissimilar toxic recalcitrant chemicals (Fig. 1) and evaluated whether the lignin-degrading system was involved (Fig. 2). In these experiments, mineralization of uniformly [^{14}C]-ring labeled organic pollutants was selected as a method of evaluating ring cleavage and the total decomposition of ring cleavage products to $^{14}CO_2$. Mineralization is the final step in biodegradation of any chemical, however, and it is always preceded by several intermediary steps, defined as *metabolite formation*. Biodegradation, as measured by the disappearance of the parent chemical, would be expected to be greater than the amount mineralized. The results of these experiments confirmed the ability of white rot fungus, *P. chrysosporium*, to degrade recalcitrant toxic chemicals—i.e. DDT [1,1-bis (4-chlorophenyl)-2,2,2-trichloroethane], TCB (3,4,3',4'-tetrachloro-biphenyl), TCDD (2,3,7,8-tetrachlorodibenzo-p-dioxin), lindane (1,2,3,4,5,6-hexachlorocyclohexane), and benzo(a)pyrene—and also confirmed the involvement of lignin-degrading enzymes in the biodegradation of xenobiotics. The white rot fungus takes advantage of an enzyme system that evolved to degrade lignin in wood. Below we discuss some major environmental pollutants that are generally regarded as recalcitrant to biodegradation but nonetheless are biodegradable by white rot fungus.

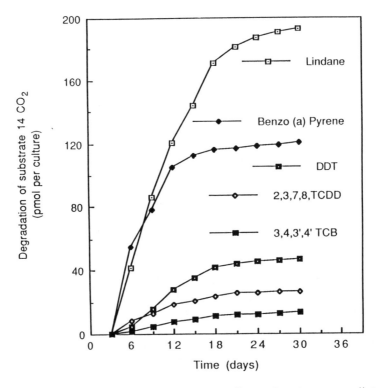

Figure 1. Rate of oxidation of five environmentally persistent organo-pollutants to CO_2 by *P. chrysosporium* (13).

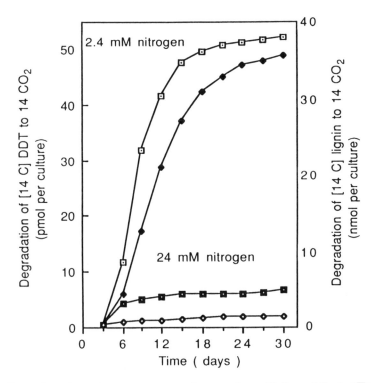

Figure 2. Effect of nutrient nitrogen concentrations on the oxidation of lignin ($\square$) and DDT ($\lozenge$) to CO_2 (13).

BIODEGRADATION OF ENVIRONMENTAL POLLUTANTS

Polychlorinated Biphenyls

The combined unique chemical and biological characteristics of polychlorinated biphenyls (PCBs), i.e. extreme insolubility in water, chemically inert/nonreactiveness, heat stability, lubricating capacity, nonpolarity and lipid solubility, and electrical resistance (34), accounts for the widespread use of PCBs in variety of products. Unfortunately, this widespread use has also resulted in their accumulation and consequently persistence in the environment. They have been implicated in adverse and toxic effects in many organisms, including humans.

PCBs disappear from the environment primarily by a combination of anaerobic and aerobic microbial activities. For example, according to Brown et al. (8, 9), PCB residues found in samples of anaerobic sediment from the upper Hudson River showed changes in PCB congener distribution, thus suggesting biologically mediated reductive dechlorination.

Although aerobic and anaerobic biodegradation of PCBs has been documented, PCBs with higher chlorination tend to be more resistant to microbial degradation. The aerobic microbial degradation of PCBs is most often limited to congeners with five or fewer chlorines and two adjacent unsubstituted carbon atoms (3, 26). Also, the aerobic degradation of most constituents of Aroclor 1242 and some of Aroclor 1254 requires an additional growth substrate (e.g. biphenyl) (4, 48).

Reductive dechlorination of PCBs by bacteria have been observed under anaerobic conditions (59). The first observation of the ability of white rot fungi to degrade PCBs was demonstrated with [U–^{14}C] Aroclor 1254, (a mixture of tetra-, penta-, and hexachlorobiphenyls) (12). Aroclor mineralization by *Phanerochaete chrysosporium* started on day 3 and continued until day 31, with 10% mineralization. Eaton (19) found that PCB mineralization by *P. chrysosporium* was proportional to concentration from 0.025 to 1000 ppm. He also observed the relationship between lignin degradation and PCB degradation and suggested that enzymes involved in lignin degradation may also be involved in PCB degradation. Bumpus et al. (12) observed the substantial mineralization of [U–^{14}C] Aroclor 1242 (a mixture of mono-, di-, tri-, and tetra-chlorobiphenyls) in nutrient nitrogen–deficient cultures of *P. chrysosporium*. Mineralization of the Aroclor mixtures proceeded more rapidly than mineralization of pure [U–^{14}C]-2,4,5,2',4',5'-hexachlorobiphenyl (12). This highly chlorinated biphenyl congener may be simply more resistant to degradation than other congeners in these complex PCB mixtures, or perhaps certain individual components of complex mixtures, such as Aroclor, exert synergistic effects on the oxidative biodegradation of other components in the mixture. There is some evidence for such interactions. For example, addition of veratryl alcohol has been shown to increase the rate and extent of degradation of benzo(a)pyrene and 4-methoxymandelic acid (27). A similar phenomenon has been observed by J. A. Bumpus and S. D. Aust (unpublished data), who found that ^{14}C-labeled naphthalene was mineralized slowly (2%) over 30 days in nutrient nitrogen–deficient cultures of *P. chrysosporium*. When ^{14}C-naphthalene was added to cultures containing coal tar contaminated soil, Tween 20, and veratryl alcohol, however, 33% of the ^{14}C-naphthalene was mineralized during the same incubation period.

Chlorinated Phenols

Chlorinated phenols and their derivatives are produced on a scale of thousands of tons annually (61). The large-scale application of chlorophenols in agriculture and as byproducts generated from industrial plants, e.g., effluents from paper pulp bleaching plants (35, 72), has led to the contamination of terrestrial and aquatic ecosystems and the listing of chlorophenols as priority pollutants by the EPA (41). The degradation of PCP by *Phanerochaete chrysosporium* showed that degradation took place via the lignin-degrading system of the fungus, because the temporal onset, time interval, and eventual decrease in the rate of mineralization were identical to the [^{14}C] lignin degradation in nutrient nitrogen–deficient cultures of the fungus (13). Furthermore, like lignin degradation, mineralization of PCP was enhanced in nutrient nitrogen–deficient cultures of the fungus, and nutrient nitrogen–sufficient cultures of the fungus suppressed PCP mineralization. *P. chrysosporium* could not be grown by inoculating spores into the culture medium in which PCP concentration was greater than 4 mg/liter; however, this toxic effect of PCP could be overcome by allowing the fungus to grow before PCP was added. This allowed the fungus to alleviate the toxicity and degrade PCP at levels up to 500 mg/liter (57).

The ligninolytic enzymes secreted by this fungus may be involved in the first oxidative dechlorination step in the degradation of several chlorinated phenols (31, 57, 71). According to Valli and Gold (71), the first step in the oxidation of 2,4-dichlorophenol by *Phanerochaete chrysosporium* is the formation of 2-chloro-1,4-benzoquinone by enzymes secreted by the fungus. 2-Chloro-1,4-benzoquinone is then intracellularly reduced to 2-chloro-1,4-hydroquinone, followed by methylation of 2-chloro-1,4-hydroquinone to form 2-chloro-1,4-dimethoxybenzene. These two compounds are suggested to be the substrates for either lignin peroxidases (LiP) or manganese-dependent peroxidases (MnP), yielding the oxidized product 2,5-dimethoxy-1,4-benzoquinone. According to Valli and Gold (71), *P. chrysosporium* can remove both chlorine atoms from 2,4-dichlorophenol before aromatic ring opening takes place. This oxidative dechlorination followed by reduction of the quinone can result in the introduction of phenolic groups that would facilitate aromatic ring opening. This scenario is in contrast with the proposed bacterial pathway in which phenolic groups are introduced by aromatic ring hydroxylation.

Polyaromatic Hydrocarbons

Polyaromatic hydrocarbons (PAHs) are well-known hazardous environmental pollutants with toxic, mutagenic, and carcinogenic properties (49, 77). They are widespread in the environment, including air (17, 18), soil (7), surface and groundwater (1), and marine environments (1). Because of their health risks to animals, including humans, PAHs are listed as priority pollutants by the EPA (41). Although the ability of microorganisms to degrade various PAHs has been studied, chemical oxidation, photolysis, and volatilization of PAHs has also been detected in nature (15). Although microorganisms have been isolated that are able to utilize PAHs as their sole carbon source, few of them are able to degrade PAHs containing four or more rings (10), probably due to solubility limits.

The ability of *Phanerochaete chrysosporium* to degrade a number of structurally diverse PAHs depends, at least in part, on the lignin-degrading system of this fungus, which is expressed under nutrient-deficient conditions. Haemmerli et al. (27) showed that purified ligninase enzyme is capable of catalyzing the initial oxidation of benzo(a)pyrene to the corresponding 1,6- 3,6- and 6,12-benzo(a)pyrene quinones. Also, Hammel, et al. (29) have reported that pyrene is oxidized to pyrene-1,6-dione and pyrene-1,8-dione by a purified ligninase enzyme. These results support the involvement of lignin degrading isozymes in the biodegradation of PAHs by *P. chrysosporium*. Also, it was observed that lignin-degrading enzymes of *P. chrysosporium* are able to oxidize PAHs with ionization potentials greater than 8.0 eV; in this respect, they are different, i.e. more electropositive or better oxidants, than other classical peroxidases, e.g. horse-radish peroxidase (30). This difference suggests that the ability of *P. chrysosporium* to degrade such a variety of structurally diverse chemical compounds stems from its ability to catalyze the rate-limiting step or initial (often in most chemicals) oxidation of these compounds.

Polychlorinated Hydrocarbons

Alkyl halide insecticides—mainly aldrin, dieldrin, kelthane, heptachlor, chlordane, lindane, methoxychlor, and mirex—are used extensively for controlling mites and other insects that serve as vectors of diseases of humans and domestic animals and for controlling agricultural insect pests. Although their use has been limited or even terminated in technologically advanced countries, they are still used in developing countries, where the disadvantages from their extensive use is considered to be outweighed by their benefits.

Alkyl halides have a multiring alicyclic structure containing olefinic bonds, e.g.

aldrin, dieldrin, heptachlor, and chlordane. Some fungi in the environment, e.g. *Trichoderma* sp., *Fusarium* sp., and *Penicillium* sp., can epoxidize one double bond of aldrin to generate dieldrin, which is then cometabolized by other organisms such as *Bacillus* sp., *Micrococcus* sp., and yeasts (67). The ability of an actinomycete *Nocardiopsis* sp. to degrade both cis- and trans-isomers of chlordane, a wood preservative used to protect wood from termites, was studied by Beeman and Matsumura (5), who found eight metabolites, e.g. dichlorochlordane, oxychlordane, heptachlor-endoperoxide, chlordane chlorohydrin, and 3-hydroxy-trans-chlordane. Under anaerobic conditions, lindane was observed to undergo reductive dehydrodechlorination more rapidly (58, 59). In the degradation of lindane by *Pseudomonas putida*, ring opening occurred only in dechlorination products with two chlorines or less (52).

The ability of *Phanerochaete chrysosporium* to mineralize [^{14}C] chlordane and lindane in silt loam soil amended with corncobs was 14.9 and 22.8%, respectively, and in liquid cultures was 9.4 and 23.4%, respectively, over a 30-day period (43). The degradation of [^{14}C]-labeled aldrin, dieldrin, and mirex was more difficult, however. This difficulty was attributed to the structure of the hexachlorocyclopentadiene rings of aldrin and dieldrin and to the cagelike perchloromethanocyclobutanpentalene structure of mirex. Chlordane was easier to degrade than heptachlor, perhaps because heptachlor has a double bond between C-2 and C-3 and is not chlorinated at C-2, whereas chlordane has a single bond between C-2 and C-3 and is chlorinated at C-2. Since this is the only difference between the two molecules, presumably the initial oxidation of chlordane by fungal enzymes occurs at either C-2 or C-3 in such a manner that the stable hexachlorocyclopentadiene portion of the molecule is subject to enzymatic attack and further degradation to CO_2. *P. chrysosporium*, however, was unable to mineralize mirex significantly ($\leq$ 2% in 30 days), and solvent extracts of cultures contaminated with mirex showed a possible mirex metabolite representing 3.6% of the total radioactivity added in liquid cultures and 9.6% of the total radioactivity in soils (43).

Chlorinated Aromatic Compounds

Chlorinated aromatic compounds have been used extensively as herbicides, e.g. 2,4-D (2,4-dichlorophenoxy acetic acid), 2,4,5-T (2,4,5-trichlorophenoxy acetic acid); insecticides, e.g. DDT [1,1,1-trichloro-2,2-bis(4-chlorophenyl)ethane]; and solvents, fumigants, and intermediates in the production of dyes, e.g. chlorobenzenes. They are toxic. Although numerous reports have shown their biodegradation, some chlorinated aromatic compounds are degraded only slowly by soil and aquatic microorganisms. For example, DDT is one of the most persistent environmental pollutants in the environment. Because of its slow degradation in the environment. DDT tends to accumulate in tissues of animals at various tropic levels of food chains (37, 76).

Degradation of DDT by *Phanerochaete chrysosporium* seems completely different from the bacterial metabolic pathway described by other workers (69). According to Subba-Rao and Alexander (69), the recalcitrance of DDT to microbial degradation can be attributed to the trichloromethyl group of the molecule. The first metabolite isolated from *P. chrysosporium*–mediated DDT degradation was DDD [1,1-dichloro-2,2-bis(4-chlorophenyl)ethane]. DDD appeared after three days of incubation of DDT with *P. chrysosporium* but disappeared from the culture medium upon continued incubation. In addition, however, the fungal enzymes, i.e. ligninases or other enzymes, also catalyzed the oxidation of the benzylic carbon of DDT by hydroxylation to form dicofol, a tertiary alcohol. As a result of this hydroxylation, the dicofol molecule has the ability to undergo subsequent oxidative degradation, because the introduction of a hydroxyl group at C-1 of DDT makes the trichloromethyl group more liable to bond cleavage and/or subsequent microbial metabolism.

Compared to primary and secondary alcohols, tertiary alcohols, in this case dicofol, are difficult to oxidize because of steric hindrance. However, substitution of a benzylic hydrogen atom of DDT by an OH group (an OH group is roughly equivalent to a methyl group in approximate size and polarizability) to form dicofol results in the formation of a tertiary alcohol in which C-2 carbon is bonded to four bulky and electrophilic groups, a situation that would be expected to favor subsequent C-1–C-2 bond cleavage to releave steric hindrance of the molecule. This profound steric hindrance of the dicofol molecule was confirmed by its rapid C-1–C-2 bond cleavage in the presence of higher pH and zinc (75). According to Walsh and Hites (75), incubation of dicofol at pH 8.2 for 24 hr at 20°–30°C has been shown to cause 100% conversion to DBP. McKinney and Fishbein (55) observed that relatively mild temperatures (125°–140°C) in the presence of zinc also cause conversion of dicofol to DBP. A product of aliphatic reductive dechlorination of dicofol, FW-152, was also found to be a major metabolite, thus suggesting that the pathway between dicofol and dibenzophenone (DBP) in *P. chrysosporium* might proceed in a manner similar to that observed in bacterial systems.

In bacterial systems, the trichloromethyl carbon undergoes successive reductive dechlorination followed by oxidation of the carboxylic acid, which then undergoes decarboxylation to form DBP. Bumpus and Aust (11) did not detect the carboxylic acid metabolite [2-hydroxy-2,2-bis(p-chlorophenyl) acetic acid] in *Phanerochaete chrysosporium* cultures, however. Similarly, they (11) did not find aromatic ring cleavage products in cultures of *P. chrysosporium* incubated with [^{14}C] DDT. They assumed, therefore, that these products were rapidly metabolized as soon as they formed, thus leaving no product accumulation (Fig. 3).

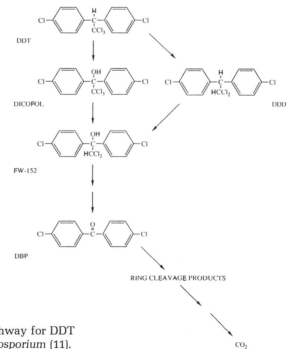

Figure 3. Proposed pathway for DDT degradation in *P. chrysosporium* (11).

We determined the effect of pH, carbon source, and xenobiotic concentration on the ability of *Phanerochaete chrysosporium* to degrade DDT in liquid cultures (22). Based on these results, we developed an amended soil system that contained ground corn cobs to support growth and [^{14}C]-DDT degradation. DDT degradation by *P. chrysosporium,* like lignin degradation, requires the presence of another carbon source to serve as a growth substrate. In practical waste treatment systems (composts or amended soils), simple carbohydrates such as glucose are unlikely to be the principal carbon source. We found that two very common glucose polymers (cellulose and starch) were able to support fungal growth and to sustain the mineralization of [^{14}C] DDT over a longer period than glucose (Fig. 4). When cellulose or starch was the carbon source, mineralization of [^{14}C]-DDT continued to occur throughout a 90 day incubation period. In contrast, when the presumably more easily metabolized simple carbohydrates—glucose, fructose, and mannose—were used, [^{14}C]-DDT mineralization virtually ceased after 20–25 days of incubation (22).

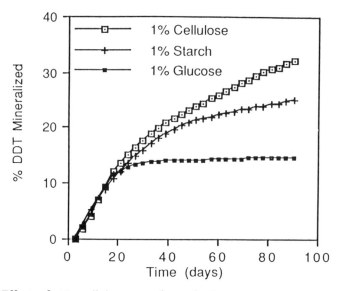

Figure 4. Effect of 1% cellulose, starch, and glucose on C^{14}-DDT mineralization by *P. chrysosporium* (22).

Ryan and Bumpus (69) studied the ability of *Phanerochaete chrysosporium* to degrade 2,4,5-trichloroacetic acid (2,4,5-T), a chemical widely used for the selective control of broad-leaved weeds and as a defoliant (51), in liquid and soil cultures. They observed extensive, vigorous degradation of [^{14}C] 2,4,5-T by the fungus *P. chrysosporium* in liquid cultures and soil amended with corn cobs, 62% mineralization in aqueous cultures over a 30-day incubation period, and polar and nonpolar metabolites in dichloromethane extracts of the cultures. The metabolites did not appear to be 2,4,5-trichlorophenol or the anisole. The time course of mineralization of 2,4,5-T paralleled that of lignin degradation. Mineralization under nitrogen–sufficient conditions was reduced to 13%, thus indicating involvement of lignin-degrading enzymes.

Munition Waste

One of the major environmental problems associated with the U.S. military is the contamination of soil, sediment, and water by toxic explosive residues at and around many military facilities. Contamination has occurred over the years by improper disposal of waste and wastewater generated during the production of munitions, burning or detonation of off-specification material, and demilitarization of out-of-date munitions. TNT (2,4,6-trinitrotoluene) and RDX (hexahydro-1,3,5-trinitro-1,3,5-triazine) are the predominant conventional explosives used by military forces (36). A homolog, octahydro-1,3,5,7-tetranitro-1,3,5,7-tetrazocine (HMX), is a byproduct of the synthesis of RDX and is also used in RDX formulations (53).

The toxic and mutagenic effects of these compounds in various organisms, including humans, have been well documented (18, 38). Exposure to TNT is known to cause pancytopenia, a disorder of the blood-forming tissues characterized by a pronounced decrease in the number of leukocytes, erythrocytes, and reticulocytes in humans and other mammals (32). Toxic effects, including liver damage and anemia, have been reported in workers engaged in large-scale manufacturing and handling practices (28, 73). Furthermore, TNT is a mutagen as assayed by the Ames test (78). RDX has been found to be a potent convulsant in the rat, with an intraperitoneal minimum lethal dose of 25 mg/kg (74). Its acute toxicity has been investigated in mice, cats, pigs, and dogs (66, 74). RDX-related toxic effects to humans have been reported world-wide. Occupationally related cases of human RDX intoxication have been reported in many countries, including the United States (38). As in animal studies, the main symptoms of RDX intoxication in humans involve the central nervous system (2, 38).

Some bacteria biotransform TNT into hydroxy-amino compounds by stepwise reduction of nitrogroups and biotransform TNT into azoxy compounds by conjugation with amino compounds; however, further biodegradation of these biotransformation products have not been observed (16, 39). According to Carpenter et al. (16), the TNT biotransformation products react with microbial lipids, fatty acids, and protein constituents to form polyamide-type macromolecular structures. The recalcitrance of these polymers to further microbial degradation parallels the documented resistance of some alkyl and aryl polyamides that contain similar linkages (63). Most previous investigations regarding the biodegradation of TNT did not indicate the cleavage and ultimate degradation of the TNT aromatic nucleus (16, 39, 54). Instead, most studies demonstrated that the TNT had been degraded by a reduction mechanism. The NO_2 groups were reduced to form hydroxyl amines, which formed dimers such as 2,2',6,6'-tetranitro-4,4'-azoxytoluene, 4,4',6,6'-tetranitro-2,2'-azoxytoluene, and 2',4,6,6'-tetranitro-2,4'-azoxytoluene (39, 56, 78). Thus, this reduction mechanism represents only a superficial modification of the molecule, not decomposition of the aromatic nucleus.

TNT reduction products may also pose environmental hazards, since they, too, are toxic and mutagenic (18, 40). McCormick, et al. (53), however, observed the biodegradation of RDX, including ring cleavage, under anaerobic conditions, which yielded hexahydro-1-nitroso-3,5-dinitro-1,3,5-triazine, hexahydro-1,3-dinitroso-5-nitro-1,3,5-triazine, hexahydro-1,3,5-trinitroso-1,3,5-triazine, etc. As was observed with TNT biotransformation, the biodegradation of RDX proceeded via successive reduction of the nitro groups; in contrast, in RDX the subsequent destabilization and fragmentation of the ring occurred. Although the RDX ring was cleaved and noncyclic degradation products were formed, McCormick et al. (53) found that several of the products that were formed were mutagenic or carcinogenic.

In contrast to other microbial degradation systems in which degradation occurs by a stepwise reduction of NO_2 groups, *Phanerochaete chrysosporium* appears to oxidize TNT and RDX. At an initial concentration of 1.25 nmol/10 ml cultures of *P. chrysosporium*, 50.8 $\pm$ 3.2% of the [^{14}C] TNT was degraded to $^{14}CO_2$ in 30 days and only about 2.8% of the initial TNT could be recovered (23). In RDX-contaminated liquid cultures (1.25 nmol/10 ml cultures), 66.6 $\pm$ 4.1% of the [14C] RDX was converted to $^{14}CO_2$ over a 30 day period and only 4% of the initial RDX was recovered. In soil, slow but sustained degradation, 6.3 $\pm$ 0.6% of the [14C] TNT initially present, i.e. 57.9 nmoles/10 g soil, was mineralized over a 30-day incubation. In contrast, 76.0 $\pm$ 3.9% of the [14C] RDX added (1.25 nmoles/10 g soil) was degraded to $^{14}CO_2$ in 30 days. When the concentration of TNT in cultures (both liquid and soil) was adjusted to high contamination levels in the environment, i.e. 10,000 mg/kg in soil and 100 mg/liter in water, 18.4 $\pm$ 2.9% and 19.6 $\pm$ 3.5% of the initial TNT was converted to $^{14}CO_2$ in 90-days in soil and liquid cultures, respectively. In both systems, about 85% of the initial TNT disappeared over a 90 day incubation period (23, 24). The enzymes involved in these oxidations appear to be LiP and MnP, which are produced under nutrient-limiting (carbon, nitrogen, and sulfur) conditions (T. Fernando and S. D. Aust, 1990, unpublished data). Most importantly, in other microbial degradations of TNT, the aromatic nucleus of the benzene ring was not cleaved (16, 39, 54, 78). As a result of the degradation of TNT, RDX or HMX by an oxidative mechanism, the possibility of conjugation of metabolites for the formation of azoxycompounds, i.e. dimers as observed in bacterial degradation of TNT, should be eliminated in *P. chrysosporium*–mediated systems. Therefore, as some other workers reported (18, 40, 53), the formation of toxic and/or mutagenic metabolites through formation of reduced nitro compounds and conjugated azoxy compounds may also be eliminated in a *P. chrysosporium*–mediated remediation process.

DEGRADATIVE ENZYMES OF *PHANEROCHAETE CHRYSOSPORIUM*

LiP and MnP are the two major peroxidases produced by *Phanerochaete chrysosporium* under nutrient nitrogen–deficient conditions. These glycosylated heme protein isozymes are identified as H1 through H10, in reference to their order of elution from a chromatographic Mono Q column. Isozymes H1, H2, H6, H7, H8, and H10 are defined as lignin peroxidases, whereas H3, H4, H5, and H9 are identified as manganese-dependent peroxidases (46). Although these LiP and MnP isozymes are usually produced in liquid cultures, the total activity and the ratios of these isozymes vary considerably depending on culture conditions. For example, in stationary 10 ml cultures, total LiP activity was lower (80–90 U/l) than in shake cultures (400–500 U/l), and LiP isozyme H8 was predominant in stationary cultures in contrast to the predominance of H2 in shake cultures.

Several studies suggest that the ability of the white rot fungus to degrade xenobiotics stems from the lignin-degrading system of this fungus. The temporal onset, time course, and eventual decrease in the rate of mineralization of these xenobiotics were similar to those of [14C]lignin in nutrient nitrogen–limited cultures. That purified lignin peroxidases are capable of partially degrading a number of xenobiotics provides more direct evidence for the involvement of the lignin-degrading system (27, 30, 31, 57, 71). Although the enzymes involved were identified, studies that considered factors that affected the metabolism of lignin by *Phanerochaete chrysosporium* (45, 47) showed that the lignin-degrading system of *P. chrysosporium* functions only in the presence of a hydrogen peroxide generating system, as hydrogen peroxide is required as a cosubstrate for lignin peroxidase. In fungal cultures, the synthesis of H_2O_2 coincides with the appearance of LiP activity at the onset of idiophasic metabolism (21, 25). The production of H_2O_2 is affected by inhibitors, O_2 concentration, and nutritional parameters such as nitrogen and carbon starvation in a manner analogous to their effect on ligninase activity. In the Ping-Pong mechanism that has been described for LiP, H_2O_2 is the first substrate to react with the enzyme (70), and its production has been described as a rate-limiting factor in the enzyme reaction mechanism (47). Glucose oxidase has been identified as a principal source of intracellular H_2O_2 in ligninolytic cultures of P. chrysosporium (42). Furthermore, both H_2O_2 production and ligninolytic activity increased when P. chrysosporium cultures were incubated under 100% oxygen. Lignin degradation was inhibited by the enzyme catalase; as a result, catalase metabolized H_2O_2 to yield O_2 and H_2O (21). The temporal correlation of the appearance of glyoxal oxidase, LiP, and oxidase substrates, i.e. glyoxal and methyl glyoxal, in P. chrysosporium cultures suggested a close physiological connection between these components, thus strengthening the extracellular H_2O_2 production hypothesis (44). In addition to ligninases and glucose oxidase, Phanerochaete chrysosporium produces several other enzymes, including cellulases, hemicellulases, and proteases (60). So far, however, LiP and MnP have been investigated more extensively than the others regarding their role in the biodegradation of xenobiotics, lignin model compounds, and lignin in wood.

CONCLUSION

The basidiomycete Phanerochaete chrysosporium Burds. (anomorph: Sporotrichum pulverulentum Novobranova) (14) is the most widely studied fungus in both basic and applied research on lignin and xenobiotic biodegradation. Compared to other potential bioremediation systems, the extracellular, nonspecific, nonstereoselective lignin-degrading system produced by P. chrysosporium has the advantage of being applicable to a variety of recalcitrant and carcinogenic chemicals. The lignin-degrading enzymes are produced extracellularly and therefore chemicals need not be absorbed or diffused into the organism. This kind of enzyme system evolved to degrade ligno-cellulosic materials in the plant kingdom. As lignin is a heteropolymer that is insoluble in water, the production of extracellular enzymes to degrade such polymers is ideal from an evolutionary standpoint. Furthermore, the ligninoytic enzyme system of white rot fungi is induced during secondary metabolism when nitrogen becomes limiting. Lignin itself is not required for induction of enzyme synthesis (60). Therefore, in bioremediation, exposure to a xenobiotic is not a criterion for the synthesis of ligninolytic enzymes.

Our toxicity studies have focused on the ability of Phanerochaeete chrysosporium to survive in the presence of different xenobiotics (57). These studies, which measured toxicity by the ability of the toxic chemical to inhibit respiration (i.e. conversion of ^{14}C-glucose to $^{14}CO_2$), demonstrated that this organism survived in the presence of a number of hazardous chemicals that were acute toxicants for other microorganisms. Taken together, these results suggest that conditions may be developed that will allow this fungus to grow and degrade hazardous pollutants in a number of potentially toxic environments. Furthermore, this organism is able to grow under diverse conditions, including a wide range of pH (3 to 6), temperature (from room temperature to 40°C), moisture (20 to 90% w/w), and oxygen (air to 100% oxygen) (22). Most importantly, P. chrysosporium has the ability to utilize different kinds of lignocellulosic materials, the most abundant material in soil that cannot be utilized as a carbon source for most organisms, i.e., ground corn cobs, wheat straw, used newspapers, wood chips, peanut shells, etc.). Soil usually contains an abundance of lignocellulosic materials on which very few soil organisms are able to thrive. The addition of lignocellulosic material into soil as a carbon source for the fungus is not impractical or expensive when the abundance of agricultural waste residues is considered, i.e. straw, corn cobs, sawdust, etc. All of these substrates have been utilized by *P. chrysosporium* in the biodegradation of hazardous waste (22).

All of our xenobiotic degradation experiments with soil were conducted under nonsterile conditions, thus demonstrating that the fungus is a good competitor under normal conditions. Many experiments were arbitrarily terminated; hence, it is possible that the extent of xenobiotic degradation could be increased simply by extending the incubation time. Furthermore, the ability of this organism to degrade several different hazardous chemicals, e.g., PAHs, creosote, PCP, HMX, TNT, from actual field samples has been investigated and the results are encouraging (50, 68; T. Fernando and D. Aust, unpublished data). Taken together, these findings suggest that *Phanerochaete chrysosporium* may be a useful and versatile organism in biotreatment systems designed to decontaminate a number of potentially toxic environments. When compared to costly, tedious physical decontamination processes, *P. chrysosporium* may provide a more economical biological treatment system that will operate *in situ* when the conditions are favorable for the growth of the fungus. The suitability of using this microorganism in a given situation will, of course, require individualized study of the site in question.

Acknowledgments

Research cited that was performed in Dr. Aust's laboratory was supported by NIH Grant ES04922. We thank Terri Maughan and Karen Ritchie for their expert secretarial assistance during the preparation of this manuscript.

LITERATURE CITED

1. Andelman, J. B., and J. E. Snodgrass, 1974. Incidence and significance of polynuclear aromatic hydrocarbons in the water environment. *Crit. Rev. Environ. Control* 5:69–83.

2. Barsotti, M., and G. Crotti. 1949. Epileptic attacks as manifestations of industrial intoxication caused by trimethylene trinitromine (T$_4$). *Medical Lavoro* 40:107–112.

3. Bedard, D. L., M. L. Haberl, R. J. May, and M. J. Brennan. 1987. Evidence for novel mechanisms of polychlorinated biphenyl metabolism in *Alcaligenes eutrophus* H 850. *Appl. Environ. Microbiol.* 53:1103–1112.

4. Bedard, D. L., R. E. Wagner, M. J. Brennan, M. L. Haberl, and J. F. Brown, Jr. 1987. Extensive degradation of Aroclors and environmentally transformed polychlorinated biphenyls by *Alcaligenes eutrophus* H 850. *Appl. Environ. Microbiol.* 53:1094–1102.

5. Beeman, R. W., and F. Matsumura. 1981. Metabolism of *cis*- and *trans*-chlordane by a soil microorganism. *J. Agric. Food Chem.* 29:84–88.

6. Beland, F. A., S. O. Farwell, A. E. Robocker, and R. D. Geer. 1976. Electrochemical reduction and anaerobic degradation of Lindane. *J. Agric. Food Chem.* 24:753–756.

7. Bossert, I., and R. Bartha. 1984. The fate of petroleum in soil ecosystems. In *Petroleum Microbiology*. Ed. R. M. Atlas. New York: MacMillan. 435–473.

8. Brown, J. F. Jr., D. L. Bedard, M. J. Brennan, J. C. Carnahan, H. Feng, and R. E. Wagner. 1987. Polychlorinated biphenyl dechlorination in aquatic sediments. *Science* 236:709–712.

9. Brown, J. F. Jr., R. E. Wagner, H. Feng, D. L. Bedard, M. J. Brennan, J. C. Carnahan, and R. J. May. 1987. Environmental dechlorination of PCBs. *Environ. Toxicol. Chem.* 6:579–593.

10. Bumpus, J. A. 1989. Biodegradation of polycyclic aromatic hydrocarbons by *Phanerochaete chrysosporium*. *Appl. Environ. Microbiol.* 55:154–158.

11. Bumpus, J. A., and S. D. Aust. 1987. Biodegradation of DDT [1,1,1-trichloro-2,2-bis (4-chlorophenyl)ethane] by the white rot fungus *Phanerochaete chrysosporium*. *Appl. Environ. Microbiol.* 53:2001–2008.

12. Bumpus, J. A., T. Fernando, G. J. Mileski, and S. D. Aust. 1987. Biodegradation of organopollutants by *Phanerochate chrysosporium*: practical considerations. In *Proceedings of the 13th Annual Research Symposium on Treatment of Hazardous Waste*. Cincinatti, Ohio: US Environmental Protection Agency.

13. Bumpus, J. A., M. Tien, D. Wright, and S. D. Aust. 1985. Oxidation of persistent environmental pollutants by a white rot fungus. *Science* 228:1434–1436.

14. Burdsall, H. H. Jr., and W. E. Eslyn. 1974. A new *Phanerochate* with a *chrysosporium* imperfect state. *Mycotaxon* 1:123–133.

15. Callahan, M. A., M. W. Slimak, N. W. Gabel, I. F. May, C. F. Fowler, J. R. Freed, P. Jennings, R. C. Durfee, F. C. Whitmore, B. Maistri, W. R. Mabey, B. R. Holt, and C. Gould. 1979. Water related environmental fate of 129 priority pollutants. Halogenated aliphatic hydrocarbons, halogenated ethers, monocyclic aromatics, phthalate esters, polycyclic aromatic hydrocarbons, nitrosoamines and miscellaneous compounds. 2:84–92, Publication 440/4–79–029 b. Cincinnati, Ohio: US Environmental Protection Agency.

16. Carpenter, D. F., N. G. McCormick, J. H. Cornell, and A. M. Kaplan. 1978. Microbial transformation of ^{14}C-labeled 2,4,6-trinitrotoluene in an activated-sludge system. *Appl. Environ. Microbiol.* 35:949–954.

17. Daisey, J. M., M. A. Leyko, and T. J. Kneip. 1979. Source identification and allocation of polynuclear aromatic hydrocarbon compounds in the New York City aerosol: methods and Applications. In *Polynuclear Aromatic Hydrocarbons*. Ed. P. W. Jones, P. Leber. Ann Arbor, Michigan: Ann Arbor Science Publishers.

18. Dilley, J. V., C. A. Tyson, and G. W. Newell. 1979. Mammalian toxicological evaluation of TNT waste waters. In *Acute and Subacute Mammalian Toxicity of Condensate Water*, Vol 8. Menlo Park, California: SRI International.

19. Eaton, D. C. 1985. Mineralization of polychlorinated biphenyls by *Phanerochaete chrysosporium*: a lignolytic fungus. *Enzyme Microbiol. Technol.* 7:194–196.

20. Epstein, S. S., L. O. Brown, and C. Pope. 1982. Hazardous wastes in America. *Sierra* (Oct.):573–594.

21. Faison, B. D., and T. K. Kirk. 1983. Relationship between lignin degradation and production of reduced oxygen species by *Phanerochaete chrysosporium*. *Appl. Environ. Microbiol.* 46:1140–1145.

22. Fernando, T., S. D. Aust, and J. A. Bumpus. 1989. Effects of culture parameters on DDT [1,1,1-trichloro-2,2-bis (4-chlorophenyl)ethane] biodegradation by *Phanerochaete chrysosporium*. *Chemosphere* 19:1387–1398.

23. Fernando, T., and S. D. Aust. 1990. Biodegradation of munition waste, TNT (2,4,6-trinitrotoluene) and RDX (hexahydro-1,3,5-trinitro-1,3,5-triazine) by *Phanerochaete chrysosporium*. In *Emerging Technologies in Hazardous Waste Mangement II*. Ed. D. W. Tedder, F. G. Pohland. *ACS Symp. Ser.* 468. Washington, D.C.: American Chemical Society. 214–232.

24. Fernando, T., J. A. Bumpus, and S. D. Aust. 1990. Biodegradation of TNT (2,4,6-trinitrotoluene) by *Phanerochaete chrysosporium*. *Appl. Environ. Microbiol.* 56:1666–1671.

25. Forney, L. J., C. A. Reddy, and H. S. Pankratz. 1982. Ultrastructural localization of hydrogen peroxide production in ligninolytic *Phanerochaete chrysosporium* cells. *Appl. Environ. Microbiol.* 44:732–736.

26. Furukawa, K., N. Tomizuka, and A. Kamibayashi. 1979. Effect of chlorine substitution on the bacterial metabolism of various polychlorinated biphenyls. *Appl. Environ. Microbiol.* 38:301–310.

27. Haemmerli, S. D., M. S. A. Leisola, D. Sanglard, and A. Fiechter. 1986. Oxidation of benzo(a)pyrene by extracellular ligninases of *Phanerochaete chrysosporium:* veratryl alcohol and stability of ligninase. *J. Biol. Chem.* 261:6900–6903.

28. Hamilton, A. 1921. Trinitrotoluene as an industrial poison. *J. Indust. Hyg.* 3:102–116.

29. Hammel, K. E., B. Kalyanaraman, and T. K. Kirk. 1986. Substrate free radicals are intermediates in ligninase catalysis. *Proc. Natl. Acad. Sci. U.S.A.* 83:3708–3712.

30. Hammel, K. E., B. Kalyanaraman, and T. K. Kirk. 1986. Oxidation of polycyclic aromatic hydrocarbons and dibenzo(p)dioxins by *Phanerochaete chrysosporium* ligninase. *J. Biol. Chem.* 261:16948–16952.

31. Hammel, K. E., and P. J. Tardone. 1988. The oxidative 4-dechlorination of polychlorinated phenols is catalyzed by extracellular fungal lignin peroxidase. *Biochemistry* 27:6563–6568.

32. Harris, J. W., and R. W. Killermeyer. 1970. *The Red Cell Production, Metabolism, Destruction: Normal and Abnormal.* Cambridge, Massachusetts: Harvard University Press.

33. Hudson, H. J., ed. 1972. Resistance of wood to decay. In *Fungal Soprophytism.* London: Arnold Publishers Limited. 10–12.

34. Hutzinger, O., S. Safe, and V. Zitko. 1974. The Chemistry of PCBs. In *Critical Reviews in Toxicology.* Boca Raton, Florida: CRC Press. 319–375.

35. Huynh, V. B., H. M. Chang, T. W. Joyce, and T. K. Kirk. 1985. Dechlorination of chloroorganics by a white rot fungus. *Tech. Assoc. Pulp Paper Indust.* 68:98–102.

36. Jenkins, T. F., and M. E. Walsh. 1987. Development of an analytical method for explosive residues in soil. Report 87-7, Cold Regions Research and Engineering Laboratory.

37. Johnson, B. T., and J. O. Kennedy. 1973. Biomagnification of p,p'-DDT and methoxychlor by bacteria. *Appl. Microbiol.* 26:66–71.

38. Kaplan, A. S., C. F. Berghout, and A. Peczenik. 1965. Human intoxication from RDX. *Arch. Environ. Health* 10:877–883.

39. Kaplan, D. L., and A. M. Kaplan. 1982. Thermophilic biotransformations of 2,4,6-trinitrotoluene under simulated composting conditions. *Appl. Environ. Microbiol.* 44:757–760.

40. Kaplan, D. L., and A. M. Kaplan. 1982. Mutagenicity of 2,4,6-trinitrotoluene-surfactant complexes. *Bull. Environ. Contam. Toxicol.* 28:33–38.

41. Keith, L. H., and W. A. Telliard. 1979. Priority pollutants I—a perspective view. *Environ. Sci. Technol.* 13:416–423.

42. Kelley, R. L., and C. A. Reddy. 1986. Identification of glucose oxidase activity as the primary source of hydrogen peroxide production in ligninolytic cultures of *Phanerochaete chrysosporium. Arch. Microbiol.* 144:248–253.

43. Kennedy, D. W., S. D. Aust, and J. A. Bumpus. 1990. Comparative biodegradation of alkyl halide insecticides by the white rot fungus, *Phanerochaete chrysosporium* (BKM-F-1767). *Appl. Environ. Microbiol.* 56:2347–2353.

44. Kersten, P. J., and T. K. Kirk. 1987. Involvement of a new enzyme, glyoxal oxidase, in extracellular H_2O_2 production by *Phanerochaete chrysosporium. J. Bacteriol.* 169:2195–2201.

45. Keyser, P., T. K. Kirk, and J. G. Zeikus. 1978. Ligninolytic enzyme system of *Phanerochaete chrysosporium:* synthesized in the absence of lignin in response to nitrogen starvation. *J. Bacteriol.* 135:790–797.

46. Kirk, T. K., S. Croan, M. Tien, K. E. Murtagh, and R. L. Farrell. 1986. Production of multiple ligninases by *Phanerochaete chrysosporium:* effect of selected growth conditions and use of a mutant strain. *Enzyme Microbiol. Technol.* 8:27–32.

47. Kirk, T. K., E. Schultz, W. J. Connors, L. F. Lorenz, and J. G. Zeikus. 1978. Influence of culture parameters on lignin metabolism by *Phanerochaete chrysosporium*. *Arch. Microbiol.* 117:277–285.

48. Kohler, H. P. E., D. Kohler-Staub, and D. D. Focht. 1988. Cometabolism of polychlorinated biphenyls: enhanced transformation of Aroclor 1254 by growing bacterial cells. *Appl. Environ. Microbiol.* 54:1940–1945.

49. Laflamme, R. E., and R. A. Hite. 1978. The global distribution of polycyclic aromatic hydrocarbons in recent sediments. *Geochim. Cosmochim. Acta* 42:289–303.

50. Lamar, R. T., and D. M. Dietrich. 1990. In situ depletion of Pentachlorophenol (PCP) by *Phanerochaete spp.* from contaminated soil. *Appl. Environ. Microbiol.* 56:3093–3100.

51. Loos, M. A. 1975. Phenoxyalkanoic acids. In *Herbicides: Chemistry, Degradation and Mode of Action,* Vol. 1. Ed. P. C. Kearney, D. D. Kaufman. New York: Dekker. 2nd edition.

52. Matsumura, F., H. J. Benezet, and K. C. Patil. 1976. Factors affecting microbial metabolism of γ-BHC. *J. Pest. Sci.* 1:3–8.

53. McCormick, N. G., J. H. Cornell, and A. M. Kaplan. 1981. Biodegradation of hexahydro-1,3,5-trinitro-1,3,5-triazine. *Appl. Environ. Microbiol.* 42:817–823.

54. McCormick, N. G., F. E. Feeherry, and H. S. Levinson. 1976. Microbial transformation of 2,4,6-trinitrotoluene and other nitroaromatic compounds. *Appl. Environ. Microbiol.* 31:949–958.

55. McKinney, J. D., and L. Fishbein. 1972. DDE formation: dehydrochlorination or dehypochlorination. *Chemosphere* 2:67–70.

56. Menzer, R. E., and J. O. Nelson. 1980. In *The basic science of poisons: Casarett and Doull's Toxicology,* 2nd ed. Ed. J. Doull, C. D. Klaassen, M. O. Amdur. New York: MacMillan Publishing Co., Inc. 632–658.

57. Mileski, G. J., J. A. Bumpus, M. A. Jurek, and S. D. Aust. 1988. Biodegradation of pentachlorophenol by the white rot fungus *Phanerochaete chrysosporium. Appl. Environ. Microbiol.* 54:2885–2889.

58. Ohisa, N., and M. Yamaguchi. 1978. Gamma-BHC degradation accompanied by growth of *Clostridium rectum* isolated from paddy field soil. *Agric. Biol. Chem.* 42:1819–1823.

59. Quensen, J. F. III, J. M. Tiedje, and S. A. Boyd. 1988. Reductive dechlorination of polychlorinated biphenyls by anaerobic microorganisms from sediments. *Science* 242:752–754.

60. Reddy, C. A. 1984. In *Physiology and biochemistry of lignin degradation.* Ed. M. J. King, C. A. Reddy. Washington, D.C.: American Society for Microbiology. 23–25.

61. Reineke, W. 1984. *Microbial Degradation of Aromatic Compounds.* Ed. D. T. Gibson. New York: Marcel Dekker, Inc.

62. Rochkind-Dubinsky, M. L., G. S. Sayler, and J. W. Blackburn. 1987. *Microbiological Decomposition of Chlorinated Aromatic Compounds.* New York: Marcel Dekker, Inc. 1–58.

63. Rogers, M. R., and A. M. Kaplan. 1971. Effects of *Penicillium janthinellus* on parachute nylon—Is there microbial deterioration. *Int. Biodeter. Bull.* 7:15–24.

64. Ryan, T. P., and J. A. Bumpus. 1989. Biodegradation of 2,4,5-trichlorophenoxyacetic acid in liquid culture and in soil by the white rot fungus *Phanerochaete chrysosporium. Appl. Microbiol. Biotechnol.* 31:302–307.

65. Sarkanen, K. U., and C. H. Ludwig, eds. 1971. Lignins: Occurence, Formation, Structure and Reactions. New York: John Wiley and Sons, Inc.

66. Schneider, N. R., S. L. Bradley, and M. E. Andersen. 1977. Toxicology of cyclotrimethylenetrinitramine: Distribution and metabolism in the rat and the miniature swine. *Toxicol. Appl. Pharmacol.* 39:531–541.

67. Singh, G. J. P. 1981. Studies on the role of microorganisms in the metabolism of dieldrin in the epicuticular wax layer of blowflies *Calliphora erythrocephala. Pest. Biochem. Physiol.* 16:256–266.

68. Stroo, H. F., M. A. Jurek, J. A. Bumpus, M. F. Torpy, and S. D. Aust. 1989. Bioremediation of wood preserving wastes using the white rot fungus *Phanerochaete chrysosporium*. *Am. Wood-Pres. Assoc.* 1:1–7.

69. Subba-Rao, R. V., and M. Alexander. 1985. Bacterial and fungal cometabolism of 1,1,1-tri-chloro-2,2-bis(4-chlorophenyl)ethane (DDT) and its breakdown products. *Appl. Environ. Microbiol.* 49:509–516.

70. Tien, M., T. K. Kirk, C. Bull, and J. A. Fee. 1986. Steady-state and transient-state kinetic studies on the oxidation of 3,4-dimethoxybenzyl alcohol catalyzed by the ligninase of *Phanerochaete chrysosporium* burds. *J. Biol. Chem.* 261:1687–1693.

71. Valli, K., and M. H. Gold. 1991. Degradation of 2,4-dichlorophenol by the lignin-degrading fungus *Phanerochaete chrysosporium*. *J. Bacteriol.* 173:345–352.

72. Valo, R., V. Kitunen, M. Salkinoja-Salonen, and S. Raisanen. 1984. Chlorinated phenols as contaminants of soil and water in the vicinity of two Finnish sawmills. *Chemosphere* 13:835–844.

73. Voegtlin, C., C. W. Hooper, and J. M. Johnson. 1919. Trinitrotoluene poisoning. *U.S. Public Health Res.* 34:1307–1313.

74. Von Oettingen, W. F., D. D. Donahue, H. Yagoda, A. R. Monaco, and M. R. Harris. 1949. Toxicity and potential dangers of cyclotrimethylene trinitramine (RDX). *J. Indust. Hyg. Toxicol.* 31:21–31.

75. Walsh, P. R., and R. A. Hites. 1979. Dicofol solubility and hydrolysis in water. *Bull. Environ. Contam. Toxicol.* 22:305–311.

76. Ware, G. W., and C. C. Roan. 1970. Interaction of pesticides with aquatic microorganisms and plankton. *Residue Rev.* 33:15–45.

77. White, K. L. 1986. An overview of immunotoxicology and carcinogenic polycyclic aromatic hydrocarbons. *Environ. Carcin. Rev.* C4:163–202.

78. Won, W. D., L. H. DiSalvo, and J. Ng. 1976. Toxicity and mutagenicity of 2,4,6-trinitrotoluene and its microbial metabolites. *Appl. Environ. Microbiol.* 31:576–580.

CHAPTER 19

Bacterial Detoxification of Toxic Chromate

HISAO OHTAKE and SIMON SILVER

Abstract. Hexavalent chromium (Cr^{6+}) is toxic for most organisms and is mutagenic. Some bacterial strains are able to reduce toxic Cr^{6+} to trivalent chromium (Cr^{3+}). Both aerobic and anaerobic reduction systems are known to occur with different bacteria. Aerobic reduction of Cr^{6+} is generally associated with soluble proteins, whereas anaerobic reduction is found to occur with membrane preparations. Cr^{3+} is much less toxic than Cr^{6+}. Cr^{3+} also forms insoluble hydroxides at a neutral pH and precipitates, thus making it less available to biological systems. The bacterial ability to reduce Cr^{6+} may be useful for cleaning up hazardous Cr^{6+} wastes. Such activities are also expected to help detoxification and immobilization to toxic Cr^{6+} in soil and water systems. Bacterial reduction of Cr^{6+} as a means of bioremediation has several potential advantages: (1) reduction occurs at a neutral pH; (2) the process requires neither chemical additives nor aeration; (3) anaerobic reduction minimizes excess sludge production in aqueous systems; (4) no toxic byproduct is formed; and (5) the activity is reproducible and reusable. This article summarizes what is known of bacterial reduction of Cr^{6+} and describes current efforts to develop biotreatment processes of toxic Cr^{6+}.

INTRODUCTION

Microbial transformations are known for many metallic minerals (49). These transformations include redox conversions of inorganic forms and conversions from inorganic to organic form and vice versa. Bacterial reduction has been found for metallic minerals such as manganese (27, 51), iron (6, 39), mercury (38, 57), selenite (13, 28), and tellurite (9, 50, 52). Some microbial transformations enable the bacteria to increase their tolerance toward toxic heavy metals (9, 57).

Hexavalent chromium is very soluble in water, and forms divalent oxyanions: chromate (CrO_4^{2-}) and dichromate ($C_2O_7^{2-}$). Experimental evidence indicates that Cr^{6+} is carcinogenic in animals (42) and mutagenic in a number of bacterial systems (41, 53). In humans, Cr^{6+} causes irritation and corrosion of the skin and respiratory tract, and is believed to be responsible for lung carcinoma (2). Mutagenic assays done in the presence of sodium sulfite, a reducing agent, resulted in dichromate losing its

Dr. Ohtake is at the Department of Fermentation Technology, Hiroshima University, Higashi-Hiroshima, Hiroshima 724, Japan. Dr. Silver is at the Department of Microbiology and Immunology, College of Medicine, University of Illinois, Chicago, IL 60612, U.S.A.

mutagenic activity, thus suggesting that the oxidized state of chromium is required for mutagenicity (32). Further studies showed that Cr^{3+} compounds were neither toxic nor mutagenic for bacterial tester strains (41). The toxicity and mutagenicity of Cr^{6+} are also likely to be related to membrane permeability. In mammalian cells, Cr^{6+} can diffuse through a faclitated transport system (presumably the nonspecific anion carrier) and across the plasma membrane, whereas Cr^{3+}, is impermeable to biological membranes (1), Cr^{6+}, which is transported by a membrane transport system, is reduced to Cr^{3+} in the cytoplasm; and Cr^{3+} generated inside the cell stably binds to proteins and interacts with nucleic acids.

Waste waters containing Cr^{6+} are generated in many industrial processes, including chrome leather tanning, chromium plating, metal cleaning and processing, wood preservation, and alloy preparation (12). To minimize environmental contamination, a large effort has been made to treat Cr^{6+} in wastewaters. Because of its broad industrial applications, however, Cr^{6+} has often been introduced into natural environments (26). Since Cr^{6+} is very soluble in water, it is easily transported by water movement, and consequently the contamination problems take place in a wide range of aqueous environments. Conventional methods for treatment of contaminated Cr^{6+} include chemical reduction followed by precipitation, ion exchange, and absorption on coal, activated carbon, alum, kaolinite, and flyash (12). Most of these methods require either high energy or large quantities of chemicals, however, and therefore more practical, cost-effective methods are being explored (14, 17, 44, 47).

Bacterial strains that are able to reduce Cr^{6+} to Cr^{3+} have been found by several workers (5, 16, 54). When reduction takes place at the cell surface, Cr^{3+} forms extracellular insoluble chromium hydroxides that subsequently precipitate in the culture medium (54, 55). Through this process, Cr^{6+} is detoxified and then immobilized. This unique ability of bacteria to reduce Cr^{6+} may be used as a promising means of bioremediation. In this chapter, we summarize what is known of bacterial reduction of Cr^{6+} and describe current efforts to develop biotreatment processes of toxic Cr^{6+}.

CHROMATE-REDUCING BACTERIA

Bioremediation is the use of biological agents to reclaim soils and waters polluted by hazardous substances. The biological agents used for bioremediation are most frequently the intact microorganisms themselves. Isolating microbes that have a desired function is the first step in developing a bioremediation process.

Chromate-reducing bacteria are also found to be resistant to Cr^{6+} (5, 45, 54). Apparently the rates of Cr^{6+} reduction are much slower than those of Cr^{6+} uptake in bacterial systems. Therefore, bacteria must be able to tolerate the toxicity of Cr^{6+} for detoxification to occur. Bacterial resistance to Cr^{6+} has been found in Pseudomonas ambigua (15), P. fluorescens (4), P. aeruginosa (7, 48), Alcaligenes eutrophus (31), Streptococcus lactis (10), and Enterobacter cloacae (54). Except for the P. ambigua and E. cloacae strains, the chromate resistance was found to be plasmid-determined. Toxic ions are generally thought to enter bacterial cells by the same transport systems as used for structurally related nutrient ions (46). In bacterial systems, CrO_4^{2-} transported mainly via the SO_4^{2-} active transport systems, and bacterial resistance toward Cr^{6+} found to be related to the lowered accumulation in cells (33). Whether this results from a direct block on uptake via the SO_4^{2-} transport system or from accelerated efflux following uptake is not yet clear. Two Cr^{6+} resistance determinants were recently cloned and sequenced from plasmids of Pseudomonas (8), and Alcaligenes (29, 30) strains. Though the DNA sequences were sufficiently different that a relationship could not be detected by Southern DNA/DNA hybridization, the gene organization and predicted protein products were closely related. Nucleotide sequence analysis revealed that a single open reading frame was sufficient to determine the resistance in *P. aeruginosa*. This open reading frame encodes a highly hydrophobic polypeptide ChrA, of 416 amino acid residues, with nine regions suitable as potential membrane spans. ChrA is presumably the inner-membrane protein responsible for the translocation of CrO_4^{2-} anions. Any additional requirements for the resistance mechanism would need to be provided by host cell genes. The *Alcaligenes eutrophus* system consists of two genes, *chrA* and *chrB*, that are needed for inducible chromate resistance plus a third partial open reading frame. Though the third open reading frame was shared by the *Alcaligenes* and *Pseudomonas* systems, it does not seem to be necessary in laboratory measurements of chromate resistance. The upstream *chrB* gene of the *Alcaligenes* system is likely to be the regulatory gene, since it is present in the cloned *Alcaligenes* system (inducible) (30) but absent in the cloned *Pseudomonas* system (constitutive) (8). A *chrA-lacZ* gene fusion has been shown to respond specifically to chromate induction but not to other tested oxyanions (D. Nies, personal communcation); however, this plasmid-determined resistance is not related to detoxification of Cr^{6+}.

There is evidence for both aerobic and anaerobic Cr^{6+} reduction systems with different microbes (16, 43, 54). Anaerobic chromate reduction occurs with membrane preparations, whereas aerobic reduction is generally associated with soluble proteins. *Pseudomonas* (43) and *Aeromonas* (24) strains have been reported to reduce Cr^{6+} anaerobically to Cr^{3+}. Lebedeva and Lyalikova (35) isolated chromate-reducing *Pseudomonas* strains from industrial sewage, and observed that the pseudomonads appeared to form aggregates and precipitates as Cr^{6+} reduction proceeded. Though these bacteria are facultative anaerobes, Cr^{6+} reduction occurred only under anaerobic conditions. Physiological studies of the anaerobic reduction of Cr^{6+} were not done with these isolates, and the basis mechanism is still unknown.

Pseudomonas strain K21 (45) and *P. ambigua* G1 (16) reduced Cr^{6+} under aerobic conditions. Horitsu et al. (16) found chromate-reducing activity in a cell-free extract of *P. ambigua* strain G1. This activity required NADH as a hydrogen donor. *P. fluorescens* strain LB300 (which provided the first chromate-resistance plasmid) also had the capacity to reduce Cr^{6+} aerobically (4). A derivative of strain LB300 cured of the plasmid retained full Cr^{6+} reductase activity, but had lost Cr^{6+} resistance. The potential for use of microbes to remove toxic Cr^{6+} from waste waters was patented (3). Ishibashi et al. (18) provided the description of a soluble chromate reductase activity that functions aerobically. The activity was highly specific for Cr^{6+}. The rate of Cr^{6+} reduction was not influenced by sulfate and nitrate, although mercury and silver cations noncompetitively inhibited the chromate reductase activity. This system may be identical to that reported with *P. ambigua* strain G1 (16). There is no evidence to support the hypothesis that aerobic chromate reduction is involved with the chromate resistance. Reduction may be a secondary activity of a soluble reductase enzyme with a quite different physiological role.

Recent progress toward development of a useful chromate reductase activity has been achieved by Ohtake and coworkers with a newly isolated *Enterobacter cloacae* strain HO1 (54). This chromate-resistant *E. cloacae* strain was isolated from activated sludge. Although it was resistant to Cr^{6+} under both aerobic and anaerobic conditions, Cr^{6+} reduction occurred only anaerobically (Fig. 1). Transformation from Cr^{6+} to Cr^{3+} was detected by electron paramagnetic resonance (EPR) spectroscopy (55). After anaerobic incubation, the culture originated a signal that could be attributed to the paramagnetic species Cr^{3+}. The intensity of the signal, calculated from the height of

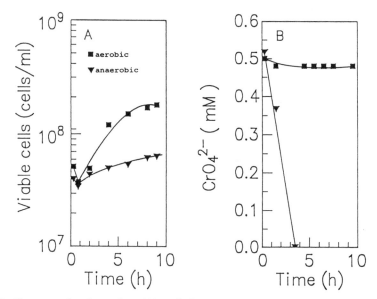

Figure 1. Changes of cell number (A) and chromate concentration (B) in aerobic (*square*) and anaerobic (*triangle*) cultures of *Enterobacter cloacae* HO1. Cells were grown overnight in KSC medium with 0.5 mM CrO_4^{2-}, harvested by centrifugation, and resuspended in the same medium as that used for preculture. Incubation was carried out at 30°C. The composition of KSC medium was described in Ref. (54).

the maximum peak of absorption, increased as the incubation proceeded, and the EPR signal was observed only when cells were grown anaerobically with Cr^{6+}. Apparently Cr^{6+} reduction required the presence of physiologically active *E. cloacae* cells. Since Cr^{6+} compounds are oxidizing agents, one may think Cr^{6+} reduction readily takes place through some chemical redox reactions without any bacterial activities; yet, no Cr^{6+} reduction was observed in the absence of bacterial cells. Neither fresh broth nor the cell-free filtrate of anaerobic cultures showed reduction of Cr^{6+}. Inhibition of cell growth by the addition of penicillin, cycloserine, or chloramphenicol resulted in the loss of Cr^{6+} reduction activity. The inhibitory effect of chloramphenicol on chromate-reducing activity suggested that this activity may require *de novo* protein synthesis. Reduction presumably takes place on the cell surface (55). The reduced chromium is found mainly in the external medium, forming insoluble chromium hydroxides. Transmission electron microscopic examination of the centrifugal pellets revealed that electron-scattering particles were deposited at the outside of bacterial cells, and energy-dispersive X-ray analysis showed that these precipitates contained exclusively chromium.

E. cloacae HO1 utilizes Cr^{6+} anaerobically as a terminal electron acceptor (54). This may be a form of bacterial respiration. In fact, the anaerobic growth of *E. cloacae* accompanied Cr^{6+} reduction. The chromate reductase activity is preferentially associated with the membrane fraction of *E. cloacae* cells prepared as right-side-out membrane vesicles (Table 1). Ascorbate-phenazine methosulfate functions as an effective electron donor with such membrane preparations. Reductase activity was lost when the membrane vesicles were heated at 100° for 1 min. Absorption spectra were examined with the membrane vesicles via potentiometric titration and low-temperature spectroscopic techniques (56). Six peaks resolved in the Cr^{6+} reduced-minus-oxidized A-B spectra were attributed to cytochromes c_{548}, c_{549}, c_{550}, b_{555}, b_{556}, and b_{558}, respectively. Among these cytochromes, c_{548} was found to be specific for the electron transfer to Cr^{6+}.

Table 1. Chromate reductase activity in various fractions prepared from *Enterobacter cloacae* and *Escherichia coli* cells.

Strain	Electron donor added	Specific activity (U/mg protein)[b]		
		Periplasmic components and outer membrane	Cytoplasm components	Right-side-out membrane visicles
E. cloacae				
HO1	None	0.00	0.00	1.64
HO1	+ Ascorbate-PMS[c]	1.40	1.22	5.47
IAM1615	None	0.00	0.00	0.17
IAM1616	+ Ascorbate-PMS	0.52	1.105	0.82
E. coli				
HB101	None	0.00	0.00	0.00
HB101	+ Ascorbate-PMS	0.68	0.98	0.53

The details are described in Ref. (55).
[a] Right-side-out membrane vesicles were prepared essentially as described by Kaback (19).
[b] One unit is defined as 1 μg of CrO_4^{2-} reduced per min. Protein was estimated by using Coomassie brilliant blue R-250 solution.
[c] Electron donor was 100 μM ascorbate plus 5 μM phenazine methosulfate (PMS).

Cr^{6+} reduction on the cell surface and formation of insoluble chromium hydroxides protect cells from the toxicity of Cr^{6+}. The Cr^{6+} reductase activity of *E. cloacae* presumably increases its tolerance toward the toxicity of Cr^{6+}, a notion that is supported by the finding that resistance and the ability to reduce Cr^{6+} were simultaneously lost when this bacterium was anaerobically grown on nitrate (37). Decreased uptake of $^{51}CrO_4^{2-}$ was also observed, however, with this organism under aerobic conditions (37), whereas chromate reduction only occurred anaerobically. Like *Pseudomonas* (33) and *Alcaligenes* (31) strains, decreased uptake of chromate is a possible basis for chromate resistance in *E. cloacae* under aerobic conditions. Whether the decreased uptake also occurs under anaerobic conditions is not yet known. There seems no doubt, however, that *E. cloacae* cells are protected from Cr^{6+} toxicity even under anaerobic conditions, because they do not lose their viability as Cr^{6+} detoxification starts.

FACTORS AFFECTING CHROMATE REDUCTION

The rate and extent of Cr^{6+} reduction by aerobic systems reported so far are too poor to assure acceptable efficiency of Cr^{6+} treatment, even though the potential for bioremediation was patented (3). Current research is directed toward using anaerobic reduction systems as a possible means of bioremediation. Therefore, the following discussion is limited to the anaerobic *Enterobacter* system, with brief or no mention of several other systems that have been reported (3, 16, 17, 45).

Factors affecting chromate reduction in the laboratory were intensively studied with anaerobic cultures of *Enterobacter cloacae* (21). Such data are essential prerequisites to develop a biological process for Cr^{6+} treatment. Fig 2 shows a typical time course of Cr^{6+} reduction in a flask culture of *E. cloacae*. Cr^{6+} reduction occurred immediately after addition of Cr^{6+}. As the concentration of Cr^{6+} decreased, the medium color changed from yellow to white and the turbidity increased. The turbidity continued to increase until Cr^{6+} reduction was complete. The rapid increase in turbidity was not accompanied by cell growth, thus indicating that insoluble chromium hydroxides were formed in the external medium. The rate of Cr^{6+} reduction was dependent on cell density (34). For example, cultures having a concentration of 3×10^9 cells per ml completely reduced 5 mM CrO_4^{2-} (260 ppm as chromium) within 3 h. The specific rate of Cr^{6+} reduction is estimated as about 1.4×10^{-6}, nmol of CrO_4^{2-} per h per cell.

The rate of Cr^{6+} reduction is dependent on the amount of added Cr^{6+}. Increasing the concentration of Cr^{6+} decreased the reduction rate and increased the time required for complete reduction (Fig. 3). Nevertheless, E. cloacae cells were able to reduce completely concentrations as high as 10 mM CrO_4^{2-} (520 ppm as chromium) within 11 h. Such a high rate of Cr^{6+} reduction has not been previously reported. The inhibition due to high Cr^{6+} itself followed typical substrate inhibition kinetics (34). It seems possible that increasing Cr^{6+} resistance in E. cloacae would enhance its ability to reduce Cr^{6+}. Chromate reduction occurs at temperatures of 10 to 50°C, with the optimum temperature around 37°C. The rate of reduction decreased by approximately 80% at 20°C, however, compared to the maximum rate of 37°C. At 10°C, only 5% of the activity

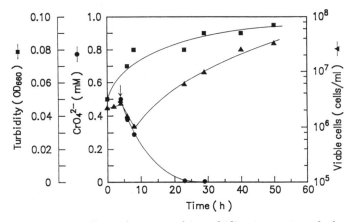

Figure 2. Changes in cell number (*triangle*), turbidity (*square*), and chromate (*circle*) of an anaerobic flask culture of *Enterobacter cloacae* strain HO1. CrO_4^{2-} was added to the culture 5 h after the start of incubation, as the *vertical arrow* indicates.

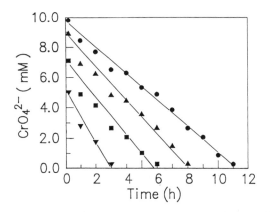

Figure 3. Chromate reduction in anaerobic flask cultures of *Enterobacter cloacae* HO1. The initial cell density 3×10^9 cells per ml. The initial concentrations of CrO_4^{2-} were: *upside-down triangle*, 5 mM; *square*, 7mM; *right-side-up triangle*, 9 mM; and *circle*, 10 mM.

remained. From a practical point of view, the most important environmental factor may be temperature, because it cannot be controlled in field situations. The effects of temperature on the reduction rate followed the Arrhenius equation over the temperature range of 10–37°C (33). At temperatures higher than 40°C, the rates of Cr^{6+} reduction decreased with increasing temperature. Chromate reduction occurs over the pH range 6.5–8.5, with the optimum at pH 7.0.

Chromate reduction by *E. cloacae* strain HO1 is sensitive to oxygen tension (20). Aeration eliminated the reductase activity of the culture grown anaerobically. The redox potential, which changed from −60 to −220 mV, did not influence the rate of Cr^{6+} reduction, and strictly anaerobic conditions were not required for Cr^{6+} reductase activity. When aeration stopped, Cr^{6+} reduction was soon restored. Flushing the culture with purified nitrogen was not required for the start of Cr^{6+} reduction. With cell densities as high as 10^7 cells per ml, anaerobic conditions could be easily established by cell respiration. This finding is particularly significant for the bioremediation process, since the use of purified nitrogen is neither practical nor economic for such processes. As aeration time increased, rates of subsequent Cr^{6+} reduction were observed to become slower. Long-time aeration presumably resulted in the replacement of Cr^{6+} reductase activity with other reductase activities that function aerobically.

A variety of organic substances have been examined for their effects on Cr^{6+} reduction (35). Amino acids, including alanine, glutamine, glycine, methionine, and threonine, were effectively utilized as electron donors for Cr^{6+} reduction. Aspartate, isocitrate, malate, oxalate and tartrate also stimulated Cr^{6+} reduction by growing cultures. Glucose, citrate, and lactate depressed Cr^{6+} reduction, though they supported bacterial growth under anaerobic conditions. Metabolic poisons, including carbonylcyanide-m-chlorophenyl hydrazone, 2,4-dinitrophenol, sodium cyanide, and formaldehde, inhibited Cr^{6+} reduction. No inhibition was observed with antimycin, sodium azide, and 2-heptylhydroxyquinolone-N-oxide.

BIOREACTORS

Through diligent use of the bacterial activities to reduce Cr^{6+} may be able to develop a bioremediation system to treat hazardous Cr^{6+} wastes. Such a system might be developed not only for Cr^{6+} waste treatment, but also for reclamation of polluted environments. In some cases, contaminated soils may be treated in on-site bioreactors. The high solubility of Cr^{6+} in water would help bacterial attack in the slurry phase of on-site bioreactors. Direct introduction of microbes to a polluted site may not be necessary to decontaminate Cr^{6+}. In other cases, polluted groundwaters might be pumped up from aquifers and passed through biological treatment facilities. In any case, chromate-reducing bacteria may be utilized in a contained reactor system that allows bacteria to perform the desired function efficiently and sustain themselves. In reactor systems, microbial activities could be maintained by optimized environmental conditions that cannot be realized in field situations. Although several workers suggested the potential of chromate-reducing bacteria to detoxify Cr^{6+} (5, 57), no work has been done on the engineering aspects until recently. Engineering studies would demonstrate how bacteria can be efficiently applied for the purpose of Cr^{6+} treatment. Such studies are essential for assessing the possibility of bioremediation of Cr^{6+} pollution. Feasibility studies were performed exclusively with pure cultures of *E. cloacae* in the laboratory (11, 22, 23). In addition, though wastewaters are variable in terms of their chemical composition, the feasibility studies were performed in a model waste solution containing Cr^{6+}. Obviously, the efficiency of biological treatment varies, depending upon the characteristics or the wastes and microbes; however, the approach used in the studies can basically be applied to biological treatment of varied types of Cr^{6+} wastes.

First consideration was given to the recovery of reduced chromium from a waste solution. From a viewpoint of pollution control, removing total chromium from wastes is desirable (though this may not be an economic benefit). Two types of bioreactors were tested: dialysis-bag (22) and anion-exchange membrane reactors (23). In the dialysis-bag reactor, E. cloacae cells were put in a semipermeable membrane bag, and the bag was submerged in Cr^{6+}-containing solution. The system has the advantage that precipitated reduced chromium can be collected inside the bag. Cr^{6+}, is allowed to diffuse into the culture, which reduces Cr^{6+} to Cr^{3+} inside the bag. Since reduced chromium readily forms insoluble chromium hydroxides, the reduced chromium is left immobilized inside the bag. Experiments with a bench-scale reactor showed that the chromium removal process was limited by Cr^{6+} diffusion through the membrane. Performance of the dialysis-bag reactor was strongly dependent on the size of the surface area of the membrane and the cell density in the dialysis bag. The dialysis-bag reactor showed about 90% removal of total chromium. About 10% or reduced chromium diffused out from the dialysis bag before precipitating.

A second type of reactor used was the dialysis culture unit with an anion-exchange membrane (23). The anion-exchange membrane allows the transfer of chromate (a divalent anion, CrO_4^{2-}), whereas reduced chromium (a trivalent cation, Cr^{3+}) is prevented from diffusing back across the membrane. The unit consisted of two chambers, between which an anion-exchange membrane (Selemion AMV; Asahi Glass Co., Tokyo, Japan) was clamped. One chamber was filled with Cr^{6+}-contaminated solution; the other chamber contained an anaerobic culture of E. cloacae. Potassium chloride was used as the counter anion for CrO_4^{2-}, diffusing across the membrane, and no electric potential was applied between the chambers. Successful removal of total chromium was observed in this type of bioreactor (Fig. 4). About 94% of the initial Cr^{6+} was removed from waste solution within 50 h. This performance could be improved by increasing the surface area of the anion-exchange membrane.

Another consideration was to minimize the Cr^{6+} toxicity toward bacterial cultures. Fed-batch culture techniques were found to be appropriate for this purpose (11). In this procedure, toxic Cr^{6+} was continuously added in small doses to minimize the toxic

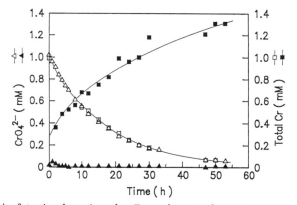

Figure 4. Removal of toxic chromium by *Enterobacter cloacae* HO1 in the anion-exchange membrane bioreactor. The initial concentration of CrO_4^{2-} in waste solution was 1.0 mM. The graph shows concentrations of CrO_4^{2-} (*open triangle*) and total chromium (*open square*) in the waste solution; and concentrations of CrO_4^{2-} (*solid triangle*) total chromium (*solid square*) in the culture of *E. cloacae* HO1. The details are described in Ref. (23).

effect. The small dosage is important, because if Cr^{6+} is overloaded, toxicity prevents the bacteria from multiplying. This type of cultivation system is not a new technology but is a common practice in fermentation processes. Performance of the fed-batch process is strongly influenced by the feeding rate of carbon and energy sources, as well as Cr^{6+}. The fed-batch process has several potential advantages: (1) it is possible to keep Cr^{6+} concentration below the toxic level; (2) the feed of Cr^{6+} is controlled by monitoring the medium color, which changes from yellow to white on CrO_4^{2-} reduction and precipitation as $Cr(OH)_3$; and (3) adequate population of Cr^{6+}-reducing cells can be maintained under Cr^{6+} stress.

CHROMATE REDUCTION IN AN INDUSTRIAL EFFLUENT

As described above, bench-scale reactor experiments established the potential for bioremediation of toxic Cr^{6+}. Although such studies with a model waste solution are valuable for examining the feasibility of biological treatment of Cr^{6+}, they cannot always be used in predicting the treatment efficiency in more complex situations. An example of the application of bacterial detoxification of Cr^{6+}-contaminated industrial effluent has recently been reported (36). The effluent, taken from a manufacturing process of copper-chromium catalysts, contained 7.0 mM CrO_4^{2-} (364 ppm as chromium), 60 mM SO_4^{2-}, and also lower concentrations of several metal cations including Cu^{2+} Mn^{2+}, Zn^{2+} and Mg^{2+}. Since the industrial effluent did not contain any organic compounds that were available to bacterial cells, a small supplement of broth was needed to provide carbon and energy sources. It was also necessary to dilute the waste solution prior to exposure to the bacteria. The original pH of the waste solution was 6.45, and changing the pH value did not allow the reduction of Cr^{6+}. Since E. gloacae is able to reduce as high as 10 mM (520 ppm) Cr^{6+} in broth, some components (probably metal cations) seemed to inhibit the ability to reduce Cr^{6+} in the waste solution. Indeed, when the waste solution was dialyzed through an anion-exchange membrane, across which cations could not diffuse, substantial reduction of Cr^{6+} occurred. A high concentration of SO_4^{2-} (60 mM in the original solution) was also inhibitory for Cr^{6+} reduction.

The rates of Cr^{6+} redeuction were dependent on the amount of carbon and energy sources available for the bacteria. The rates of Cr^{6+} reduction were also strongly dependent on the cell density (Fig. 5). No Cr^{6+} reduction was observed on the absence of bacterial cells. Increasing the cell density greatly increased the rates of C^{6+} reduction. For example, 3.5 mM (182 ppm) of Cr^{6+} was completely reduced within 3.5 h at the density of 7×10^8 cells per ml. In this case, the specific rate of Cr^{6+} reduction was estimated as 1.5×10^{-6} nmol of CrO_4^{2-} per h per cell, which was close to that obtained with broth.

After Cr^{6+} reduction was complete, an attempt was made to remove the reduced precipitated chromium by centrifugation. About 40% of the reduced chromium was removed by centrifugation. The removal of the reduced chromium was not significantly increased at increased rotation speeds (range 5000 to 20,000 rpm), and increasing the centrifugation time did not improve the removal of reduced chromium from the treated water. Increasing the precipitation of the reduced chromium seems to require use of a coagulant after the reduction is complete. Though not yet tested, dialysis culture systems with an anion-exchange membrane may be also applicable to total chromium removal from industrial effluents.

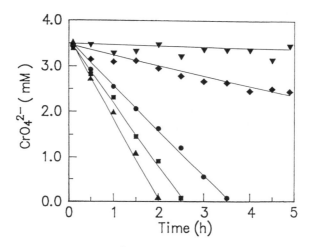

Figure 5. Bacterial reduction of CrO_4^{2-} in an industrial effluent. Initial cell densities were $\times 10^8$ cells per ml: *upside-down triangle*, 2.0; *diamond*, 4.0; *circle*, 7.0; *square*, 8.0; and *right-side up triangle*, 10.0.

CONCLUSIONS

Bacterial reduction of Cr^{6+} is of interest from a microbiological viewpoint, and it also can be a promising bioremediation process. Bacteria can reduce Cr^{6+} under mild conditions. This process may be the best mechanism to assure Cr^{6+} treatment. Methods for removal of toxic Cr^{6+} in common use today involve chemical reduction followed by precipitation, ion exchange, and absorption on coal, activated carbon, alum, kaolinite, or fly ash. Most of these methods are energy-intensive and are also costly because they require chemical additives and energy expenditure for physical treatment. Basically, waste treatment needs to be a low-cost process. Reactors must be designed to be as simple as possible. Low cost and easily available materials such as calcinated coke and sawdust have been explored as Cr^{6+} absorbents (17, 47). Absorption does not necessarily mean detoxification of Cr^{6+}, however. Further treatment is still required for absorbed Cr^{6+}. Commonly used chemical reductants are $FeSO_4$, Na_2SO_3, $NaHSO_3$, and $Na_2S_2O_5$. Since chemical reduction generally occurs only over a narrow range of pH, these methods require the use of chemical additives not only for Cr^{6+} reduction but also for pH adjustment (12). Furthermore, production of toxic gaseous byproducts can also be a problem. There have been some attempts to use microorganisms as a bioaccumulator of toxic heavy metals, including Cr^{6+} (44). In such cases, soluble toxic metals are precipitated within the boundaries of the microbial cells, and can then be removed by filtration or centrifugation. The process thus far appears successful, however, only for cleaning up dilute contaminated water. Another attempt has been made to precipitate a variety of toxic metals by using microbes that release H_2S (40). This process may be a promising means of cleaning up dilute toxic wastes, but disadvantages include slow rates of growth of obligate anaerobes and the production of unpleasant odors. Attention must also be paid to avoid the toxic effects of H_2S. Bacterial reduction of Cr^{6+} as a means of bioremediation has several potential advantages: (1) reduction occurs under mild conditions; (2) it requires neither chemical additives nor aeration; (3) anaerobic reduction minimizes excess sludge production in aqueous systems; (4) no toxic byproduct is formed; and (5) the activity is reproducible and reusable. Reduced chromium readily forms insoluble chromium hydroxides at a neutral pH. These chromium compounds are very stable, and they are unavailable to living organisms. No microbial activity has been found so far that is able to oxidize reduced chromium. Bacterial reduction of Cr^{6+} may also be a natural biological process that prevents or minimizes problems with Cr^{6+} pollution. This process may detoxify Cr^{6+} and leave it immobilized at the site, thus preventing its runoff into groundwater. Cr^{6+}-reducing bacteria may also be used in on-site bioreactors to treat contaminated soils and polluted waters. In some cases, contaminated groundwater may be pumped up from aquifers and passed through the bioreactor *in situ*. In other cases, contaminated soils may be treated in the form of slurry in a reactor, taking advantage of the high solubility of Cr^{6+}. The feasibility of the bioremediation process has been examined in the laboratory and study is now proceeding to the developmental stage. Tnough further efforts are still necessary for establishing the full-cale process, the findings reviewed here might have wide implications in environmental biotechnology, which postulates that the use of microbes creates the right methods for environmental pollution control.

LITERATURE CITED

1. Arslan, P., M. Beltrame, and A. Tomasi. 1987. Intracellular chromium reduction. *Biochim. Biophys. Acta* 931:10–15.

2. Bidstrup, P. L., and R. A. M. Case. 1956. Carcinoma of the lung in normal workmen in the bichromates-producing industry in Great Britain. *Br. J. Ind. Med.* 13:260–264.

3. Bopp, L. H. 1984. Microbial removal of chromate from contaminated waste water. U.S. patent #4,468,461. August 28, 1984.

4. Bopp, L. H., A. M. Chakrabarty, and H. L. Ehrlich. 1983. Chromate resistance plasmid in *Pseudomonas fluorescens*. *J. Bacteriol.* 155:1105–1109.

5. Bopp, L. H., and H. L. Ehrlich. 1988. Chromate resistance and reduction in *Pseudomonas fluorescens* strain LB300. *Arch. Microbiol.* 150:426–431.

6. Brock, T. D., and J. Gustafson. 1976. Ferric iron reduction by sulfur- and iron-oxidizing bacteria. *Appl. Environ. Microbiol.* 23:567–571.

7. Cervantes, C., and H. Ohtake. 1988. Plasmid-determined resistance to chromate in *Pseudomonas aeruginosa*. *FEMS Microbiol. Lett.* 56:173–176.

8. Cervantes, C., H. Ohtake, L. Chu, T. K. Misra, and S. Silver. 1990. Cloning, nucleotide sequence, and expression of the chromate resistance determinant of *Pseudomonas aeruginosa* plasmid pUM505. *J. Bacteriol.* 172:287–291.

9. Chiong, M., E. Gonzaliz, R. Barra, and C. Vasquez. 1988. Purification and biochemical characterization of tellurite-reducing activities from *Thermus thermophilus* BH8. *J. Bacteriol.* 170:3269–3273.

10. Efstathiou, J. D., and L. L. McKay. 1977. Inorganic salts resistance associated with a lactose-fermenting plasmid in *Streptococcus lactis*. *J. Bacteriol.* 130:257–265.

11. Fujii, E. K. Toda, and H. Ohtake. 1990. Bacterial reduction of toxic hexavalent chromium using a fed-batch culture of *Enterobacter cloacae* HO1. *J. Ferment. Bioeng.* 69:365–367.

12. Germain, J. E., and K. E. Patterson. 1974. Plating and cyanide wastes. *J. Water Poll. Control Fed.* 46:1301–1315.

13. Gerrard, T. L., J. N. Telford, and H. H. Williams. 1974. Detection of selenium deposits in *Escherichia coli* by electron microscopy. *J. Bacteriol.* 119:1057–1060.

14. Gupta, G. S., G. Prasad, and V. N. Singh. 1988. Removal of chrome dye from carpet effluents using coal. *Environ. Technol. Lett.* 9:153–161.

15. Horitsu, H. S. Futo, K. Ozawa, and K. Kawai. 1983. Comparison of characteristics of hexavalent chromium-tolerant bacterium, *Pseudomonas ambigua* G-1, and its hexavalent chromium-sensitive mutant. *Agric. Biol. Chem.* 47:2907–2908.

16. Horitsu, H. S. Futo, Y. Miyazawa, S. Ogai, and K. Kawai. 1987. Enzymatic reduction of hexavalent chromium by hexavalent chromium tolerant *Pseudomons ambigua* G-1. *Agric. Biol. Chem.* 51:2417–2420.

17. Huang, C., and M. Wu. 1975. Chromium removal by carbon absorption. *J. Water Poll. Control Fed.* 47:2437–2446.

18. Ishibashi, Y., C. Cervantes, and S. Silver. 1990. Chromate reduction in *Pseudomonas putida*. *Appl. Environ. Microbiol.* 56:2268–2270.

19. Kaback, H. R. 1971. Bacterial membranes. *Methods Enzymol.* 22:99–120.

20. Komori, K., K. Toda, and H. Ohtake. 1990. Effects of oxygen stress on chromate reduction in *Enterobacter cloacae* HO1. *J. Ferment. Bioeng.* 69:67–69.

21. Komori, K., P.-C. Wang, K. Toda, and H. Ohtake. 1989. Factors affecting chromate reduction in *Enterobacter cloacae* strain HO1. *Appl. Microbiol. Biotechnol.* 31:567–570.

22. Komori, K., A. Rivas, K. Toda, and H. Ohtake. 1990. A method for removal of toxic chromium using dialysis-sac cultures of a chromate-reducing strain on *Enterobacter cloacae* HO1. *Appl. Microbiol. Biotechnol.* 33:117–119.

23. Komori, K., A. Rivas, K. Toda, and H. Ohtake. 1990. Biological removal of toxic chromium using an *Enterobacter cloacae* HO1 that reduces chromate under anaerobic conditions. *Biotetchnol. Bioeng.* 35:951–954.

24. Kvasnikov, E. I., V. V. Stepanyuk, T. P. Klyushnikowa, N. S. Serpokrylov, G. A. Simonova, T. P. Kasatkina, and L. P. Panchenko. 1985. A new chromium-reducing, gram-variable bacterium with mixed type of flagellation. *Mikrobiologiya* 54:83–88.

25. Lebedeva, E. V., and N. N. Lyalikova. 1979. Reduction of crocoite by *Pseudomonas chromatophilia* sp. nov. *Mikrobiologiya* 48:517–522.

26. Leland, H. V., S. N. Luoma, J. F. Elder, and D. J. Wilkes. 1978. Heavy metals and related trace elments. *J. Water Poll. Control. Fed.* 50:1469–1514.

27. Myers, C. R., and K. H. Nealson. 1988. Bacterial manganese reduction. Growth with manganese oxide as the sole electron acceptor. *Science* 240:1319–1321.

28. Nickerson, W. J., and G. Falcone. 1963. Enzymatic reduction of selenite. *J. Bacteriol.* 85:763–771.

29. Nies, A., D. H. Nies, and S. Silver. 1989. Cloning and expression of plasmid genes encoding resistance to chromate and cobalt in *Alcaligenes eutrophus*. *J. Bacteriol.* 171:5065–5070.

30. Nies, A., D. H. Nies, and S. Silver. 1990. Nucleotide sequence and expression of a plasmid-encoded chromate resistance determinant from *Alcaligenes eutrophus*. *J. Biol. Chem.* 265:5648–5653.

31. Nies, D. H., and S. Silver. 1989. Plasmid-determined inducible efflux is responsible for resistance to cadmium, zinc, and cobalt in *Alcaligenes eutrophus*. *J. Bacteriol.* 171:896–900.

32. Nishioka, H. 1975. Mutagenic activities of metal compounds in bacteria. *Mutat. Res.* 31:185–189.

33. Ohtake, H., C. Cervantes, and S. Silver. 1987. Decreased chromate uptake in *Pseudomonas fluroescens* carrying a chromate resistance plasmid. *J. Bacteriol.* 169:3853–3856.

34. Ohtake, H., E. Fujii, and K. Toda. 1990. Bacterial reduction of hexavalent chromium: kinetic aspects of chromate reduction by *E. cloacae* HO1. *Biocatalysis* 4:227–235.

35. Ohtake, H., E. Fujii, and K. Toda. 1990. A survey of effective electron donors for reduction of toxic hexavalent chromium by *Enterobacter cloacae* (strain HO1). *J. Gen. Appl. Microbiol.* 36:203–208.

36. Ohtake, H., E. Fujii, and K. Toda. 1990. Reduction of toxic chromate in an industrial effluent by use of a chromate-reducing strain of *Enterobacter cloacae* HO1. *Environ. Technol. Lett.* 11:663–668.

37. Ohtake, H., K. Komori, C. Cervantes, and K. Toda. 1990. Chromate-resistance in a chromate-reducing strain of *Enterobacter cloacae* HO1. *FEMS Microbiol. Lett.* 67:85–88.

38. Olson, B., T. Barkay, and R. R. Colwell. 1979. Role of plasmids in mercury transformation by bacteria isolated from the aquatic environment. *Appl. Environ. Microbiol.* 38:478–485.

39. Ottow, J. C. G., and A. V. Klopotek. 1969. Enzymatic reduction of iron oxide by fungi. *Appl. Microbiol.* 16:695–702.

40. Pan-Hou, H. S. K., and N. Imura. 1981. Role of hydrogen sulfide in mercury resistance determined by plasmid of *Clostridium cochleararium* T-2. *Arch. Microbiol.* 129:49–52.

41. Petrilli, F. L., and S. D. Flora. 1977. Toxicity and mutagenicity of hexavalent chromium on *Salmonella typhimurium*. *Appl. Environ. Microbiol.* 33:805–809.

42. Role, F. J. C., and R. L. Carter. 1969. Chromium carcinogenesis: calcium chromate as a potent carcinogen for the subcutaneous tissues of the rat. *Br. J. Cancer* 23:172–176.

43. Romanenko, V. I., and V. N. Koren'kov. 1977. A pure culture of bacteria utilizing chromate and bichromates as hydrogen acceptors in growth under anaerobic conditions. *Microbiologiya* 46:414–417.

44. Sag, Y., and T. Kutsal. 1989. The use of *Zoogloea* in waste water treatment containing Cr(VI) and Cd(II) ions. *Biotechnol. Lett.* 11:145–148.

45. Shimada, K., and K. I. Matsushima. 1983. Isolation of potassium chromate-resistant bacterium and reduction of hexavalent chromium by the bacterium. *Bull. Faculty Agric. Mie Univ.* 67:101–106.

46. Silver, S. 1978. Transport of cations and anions. In *Bacterial Transport*. Ed. B. P. Rosen. New York: Marcel Dekker, Inc.

47. Srivastava H. C. P., R. P. Mathur, and I. Mehrotra. 1986. Removal of chromium from industrial effluents by absorption on sawdust. *Environ. Technol. Lett.* 7:55–63.

48. Summers, A. O., and G. A. Jacoby. 1978. Plasmid-determined resistance to boron and chromium compounds in *Pseudomonas aerugoinosa*. *Antimicrob. Agents Chemother.* 13:637–640.

49. Summers, A. O., and S. Silver. 1978. Microbial transformations of metals. *Annu. Rev. Microbiol.* 32:637–672.

50. Terai, T., T. Kamahora, and Y. Yamamura. 1958. Tellurite reductase from *Mycobacterium avium*. *J. Bacteriol.* 75:535–539.

51. Trimble, R. B., and H. L. Ehrlich. 1968. Bacteriology of manganese nodules. *Appl. Microbiol.* 16:695–702.

52. Tucker, F. L., J. F. Walper, M. D. Appleman, and J. Donohue. 1962. Complete reduction of tellurite to pure tellurium metal by microorganisms. *J. Bacteriol.* 83:1313–1314.

53. Venitt, S., and L. S. Levy. 1974. Mutagenicity of chromate in bacteria and its relevance to chromate carcinogenesis. *Nature* 250:493–495.

54. Wang, P.-C., T. Mori, K. Komori, M. Sasatsu, K. Toda, and H. Ohtake. 1989. Isolation and characterization of an *Enterobacter cloacae* strain that reduces hexavalent chromium under anaerobic conditions. *Appl. Environ. Microbiol.* 55:1665–1669.

55. Wang, P.-C., T. Mori, K. Toda, and H. Ohtake. 1990. Membrane-associated chromate reductase activity from *Enterobacter cloacae* HO1. *J. Bacteriol.* 172:1670–1672.

56. Wang, P.-C., K. Toda, H. Ohtake, I. Kusaka, and I. Yabe. 1991. Membrane-bound respiratory system of *Enterobacter cloacae* strain HO1 grown anaerobically with chromate. *FEMS Microbiol. Lett.* 78:11–16.

57. Williams, J. W., and S. Silver. 1984. Bacterial resistance and detoxification of heavy metals. *Enzyme Microb. Technol.* 6:530–537.

Chapter 20

Degradation of Chlorinated Aromatic Compounds by Bacteria: Strain Development

WALTER REINEKE

Abstract. In microbial degradation of chloroaromatic compounds, different mechanisms are used by bacteria to eliminate chlorine substituents from organic compounds. Dechlorination occurs by hydrolytic, oxygenolytic, or reductive elimination from the aromatic ring. Alternatively, degradation proceeds through chlorocatechols as central metabolites, and nonaromatic structures are generated after ring cleavage before hydrogen chloride is eliminated.

Three strategies to isolate chloroaromatics-degrading bacteria are described. In addition to the enrichment technique, strains that totally degrade chlorinated anilines, benzenes, biphenyls, phenols, salicylates, and toluenes can be developed by constructing hybrid pathways through natural genetic processes or by introducing cloned genes into a host organism.

INTRODUCTION

For decades, enormous quantities of man-made chemicals have been released into the environment. Of these, a great number are structurally related to biosynthetic compounds and are readily degraded by microorganisms from aquatic and soil environments. A significant portion, however—mainly substances with structural features or substituents rarely found in nature, the so-called xenobiotics—tend to persist and lead to environmental pollution. Ecocatastrophes of recent years illustrate the problems caused by toxic chemicals in the biosphere. The numerous dumps containing great amounts of recalcitrant pollutants also pose a high risk. Basic problems with the available physico-chemical methods for the elimination of pollutants, such as high cost and sometimes low efficiency and/or new environmental risk, clearly show that strategies for environmental protection should be improved. Furthermore, the potential of natural microbial populations to alleviate the pollution problems should be exploited.

For the development of biotechnological processes for the degradation and detoxification of pollutants, the efficiency of microorganisms in a reactor or the conditions of the *in situ* operation procedures have to be optimized. When the compounds

---

Dr. Reineke is at Chemische Mikrobiologie, Bergische Universität Gesamthochschule Wuppertal, Fachbereich 9, Gaußstrasse 20, D-42097 Wuppertal, Germany.

are very toxic or recalcitrant, however, efficient microorganisms may be unavailable or not isolated readily. It is then necessary to develop enrichment procedures to isolate a new strain with the desired catabolic potential or, if this is not possible, to construct such strains in the laboratory. This chapter discusses both strategies for the isolation of strains and reviews the total degradation of chloroaromatic compounds—one of the most spectacular classes of compounds that create environmental problems.

DECHLORINATION MECHANISMS OF CHLOROAROMATICS

The biodegradation of a halogenated arene can be considered complete only when its carbon skeleton is converted into intermediary metabolites and its organic chlorine is returned to the mineral state. The crucial point is the removal of chlorine substituents from the organic compound. This may occur at an early stage of the degradative pathway with hydrolytic, oxygenolytic, or reductive elimination of the chlorine substituent from the aromatic ring. Alternatively, degradation may proceed through chlorocatechols as central metabolites, and nonaromatic structures that spontaneously eliminate HCl may be generated after ring cleavage.

The mechanisms of the microbial dechlorination of chloroaromatics are summarized in Figure 1. The hydrolytic, oxygenolytic, and elimination of a chlorine substituent after ring cleavage are degradative enzyme sequences that allow a single organism to use the chloroaromatic compound as a sole carbon and energy source in the presence of oxygen. The reductive removal of chlorine substituents has been described to occur mostly under anaerobic conditions by consortia.

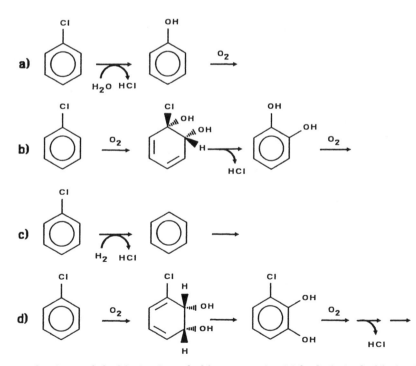

Figure 1. Mechanisms of dechlorination of chloroaromatics (a) hydrolytic dechlorination; (b) oxygenolytic dechlorination; (c) reductive dechlorination; (d) dechlorination after ring cleavage.

Replacement of Chlorine Through Hydroxyl

The mechanism of the hydrolytic dechlorination process has been clarified for 4-chlorobenzoate by labeling experiments with $^{18}O_2$ and $H_2^{18}O$ (147, 157). The data indicate that the dechlorination reaction utilizes water as the hydroxyl donor, not molecular oxygen. This mechanism has been shown for the degradation of 4-chlorobenzoate by *Micrococcus* spp., *Pseudomonas* spp., *Nocardia* sp., *Alcaligenes* sp., and *Arthrobacter* spp. to give 4-hydroxybenzoate (29, 116, 125, 126, 129, 146, 178, 197, 232, 250, 251, 253). The 4-chlorobenzoate dehalogenase from *Pseudomonas* sp. strain CBS3 has recently been shown to be a three-component enzyme complex (62). Dehalogenation is restricted to halobenzoates substituted in the *para*-position (146).

Labeling experiments seem to demonstrate that the initial dechlorination of pentachlorophenol (PCP) proceeds by a hydrolytic displacement of chlorine, rather than by an oxygenase-catalyzed mechanism, since an ^{18}O-labeled product of the dechlorination, tetrachloro-p-hydroquinone, was found only when cell extracts converted PCP in the presence of $H_2^{18}O$ (9). Schenk et al. (185), however, demonstrated that unlabeled tetrachloro-p-hydroquinone became labeled after incubation with the dehalogenase from *Arthrobacter* sp. strain ATCC 33790 in $H_2^{18}O$. Therefore, distinguishing between an oxygenolytic or a hydrolytic dehalogenation mechanism for PCP dehalogenase is impossible. In a *Flavobacterium* sp. and a coryneform-like strain, two reductive dechlorinations of tetrachloro-p-hydroquinone followed, yielding first trichlorohydroquinone and then 2,6-dichlorohydroquinone (175, 206). Degradation of tetrachloro-p-hydroquinone was found to be initiated by hydroxylation, however, yielding trichloro-1,2,4-trihydroxybenzene, which is reductively degraded to 1,2,4-trihydroxybenzene in *Rhodococcus chlorophenolicus* (10).

Oxygenolytic Chlorine-carbon Bond Cleavage

Fortuitous dechlorination by dioxygenases is another mechanism to remove chlorine from chloroaromatic compounds. The oxygen of the hydroxyl groups originates from molecular oxygen. A benzoate 1,2-dioxygenase or a phenylacetate 3,4-dioxygenase directs oxygen to the aromatic ring in such a way that after reduction, one of the *ortho*-standing hydroxyl groups is bound to the same carbon as the chlorine substituent. From the *cis*-dihydrodiol, chloride is spontaneously eliminated to yield *ortho*-diphenolic compounds such as catechol or dihydroxyphenylacetate, which are subject to further degradation by enzymes used for the mineralization of aromatic compounds. Oxygenolytic elimination has been shown for 2-chlorobenzoate and 4-chlorophenylacetate (65, 69, 70, 127, 148, 221, 252), but other haloaromatics, such as 2-fluorobenzoate, 2-bromobenzoate, and 3-fluorotoluene, have also been degraded by use of this elimination mechanism (64, 89, 106, 153, 176, 236).

Displacement of Chlorine Through Hydrogen

Aryl reductive dechlorination is the initial step in the mineralization of many chloroaromatic compounds in anoxic habitats, such as chlorobenzenes (22, 68, 228, 229), chloroanilines (132, 133, 212), chlorobenzoates (48, 51, 86, 195, 209, 213), chlorophenols (23, 24, 86, 88, 130, 151, 152, 214, 256), chlorocatechols (5, 158, 242), and chlorobiphenyls (25, 159, 164, 165). These reactions have been detected in undefined enrichments, in defined consortia, and with an isolated anaerobic bacterium as part of a methanogenic food web with hydrogen cycling (51). Little is known about the mechanism of reductive dechlorination, but the following findings have emerged from the studies:

1. The reactions are specific, since only certain congeners were used as the substrate, such as *meta*-substituted benzoates or *meta*- and *para*-substituted PCBs (49, 165, 213).

2. The more highly chlorinated congeners appear generally to be more readily dechlorinated and to yield lower substituted congeners, which are then accumulated (159, 164).

3. The dechlorination of the lower substituted congeners starts when the higher chlorinated aromatics are totally converted into the lower chlorinated ones (213).

The only detailed data on the dechlorination mechanism were obtained with the anaerobic bacterium *Desulfomonile tiedjei,* which was isolated from a methanogenic 3-chlorobenzoate-degrading consortium. The dechlorination was found to represent a novel type of anaerobic respiration (48, 195, 213). The dehalogenating activity of *Desulfomonile tiedjei* has been detected in cell extracts (47). A reductive dechlorination mechanism has also been shown to be involved in the degradation of chloroaromatic compounds by aerobic pure cultures. *Alcaligenes denitrificans* strain NTB-1 was found to degrade 2,4-dichlorobenzoate via 4-chlorobenzoate (233). Reductive dechlorination steps are also involved in the degradation of the metabolites of pentachlorophenol (see above).

Elimination of Chloride after Ring Cleavage

In the fourth mechanism by which microorganisms degrade chloroaromatic compounds—elimination of chloride after ring cleavage—a number of reactions involving chlorinated intermediates can occur before dechlorination steps are reached. Chlorocatechols formed from several chloroaromatics are the key intermediates in this type of degradation (Table 1). A common feature of these pathways (Figure 2) is the *ortho*-cleavage of the chlorocatechols by catechol 1,2-dioxygenases to produce chloromuconates (55, 56, 187). Lactonization of chloromuconates by chloromuconate cycloisomerases forms 4-carboxymethylbut-2-en-4-olides with chlorosubstitutents in the 4- or 5-position as intermediates. These intermediates spontaneously generate 4-carboxymethylenebut-2-en-4-olides by *anti*-elimination of hydrogen chloride (186, 189). The 4-carboxymethylenebut-2-en-4-olides are converted into maleylacetates by 4-carboxymethylenebut-2-en-4-olide hydrolases. Maleylacetates identified or postulated as intermediates of the chlorocatechols degradation are listed in Table 2. Maleylacetates with chlorosubstituents in the 2-position can be dechlorinated by use of a maleylacetate reductase to give 3-oxoadipates (30, 117, 117a). The modified *ortho*-cleavage pathway described above tolerates a substitution of up to three chlorine atoms at the aromatic ring. Two dechlorination steps have been described so far. Whether tetrachlorocatechol can serve as a substrate for the chlorocatechol sequence remains unknown.

Table 1. Chlorocatechols as intermediates in the degradation of chloroaromatics.

Chloroaromatic	Chlorocatechol	Reference
Chlorobenzene	3-chloro-	(172)
2-Chloroaniline		(136)
2-Chlorobenzoate		(100)
3-Chlorobenzoate		(34, 54)
3-Chlorotoluene		(24a)
3-Chloroaniline	4-chloro-	(136)
4-Chloroaniline		(136, 254, 255)
4-Chlorobenzoate		(170, 171)
4-Chlorophenoxyacetate		(67)
4-Chlorotoluene		(24a)
1,2-Dichlorobenzene	3,4-dichloro-	(95, 235)
1,3-Dichlorobenzene	3,5-dichloro-	(46)
3,5-Dichlorobenzoate		(101, 170, 171)
2,4-Dichlorophenoxyacetate		(20, 21, 66, 194, 225, 226)
3,5-Dichlorotoluene		(24a)
1,4-Dichlorobenzene	3,6-dichloro-	(162, 188, 200)
3,4-Dichloroaniline	4,5-dichloro-	(248)
1,2,4-Trichlorobenzene	3,4,6-trichloro-	(180, 234, 235)
1,2,4,5-Tetrachlorobenzene		(180)
2-Methyl-4-chloro-phenoxyacetate	3-methyl-5-chloro-	(19, 84, 85)
p-Chlorotoluene	3-chloro-6-methyl-	(96)

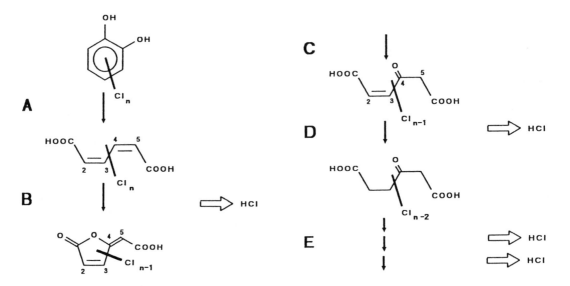

Figure 2. Degradation sequence of chlorinated catechols ($n = 1 - 4$) with ortho-cleavage and maleylacetates as intermediates. Enzymes involved: (A) catechol 1,2-dioxygenase; (B) chloromuconate cycloisomerase; (C) 4-carboxymethylenebut-2-en-4-olide hydrolase (dienelactone hydrolase); (D) maleylacetate reductase; (E) unknown sequences.

Table 2. Chloromaleylacetates as intermediates in the degradation of chlorocatechols.

Maleylacetate	Chlorocatechol	Reference
Maleylacetate	3-chloro-	(186)
	4-chloro-	(186)
2-Chloro-	3,5-dichloro-	(189)
	3,6-dichloro-	(200)
3-Chloro-	4,5-dichloro-	(248)
5-Chloro-	3,4-dichloro-	(95)
2-Methyl-	3-methyl-5-chloro-	(84)
	3-chloro-6-methyl-	(96)
2,3-Dichloro-	3,4,5-trichloro-	a
2,5-Dichloro-	3,4,6-trichloro-	(180)
2,3,5-Trichloro-	3,4,5,6-tetrachloro-	a

[a] Only postulated.

ISOLATION STRATEGIES FOR
CHLOROAROMATICS-DEGRADING BACTERIA

Chloroaromatics are mostly man-made. Since the early 1930s, they have been present in the environment at increasing concentrations. Halogenated natural products were once considered a rarity. Only about 30 organic compounds containing chlorine were characterized as natural products in 1968 (73). Suida and DeBernardis (215) later listed more than 200 naturally occurring compounds with carbon-halogen bonds, of which some 75% contained chlorine in an aromatic ring. Many new halogen-containing compounds have since been added to the list (94).

Although halogenated compounds are now known to be much more widely distributed in nature than was formerly suspected, concentrations high enough to initiate evolution of enzymes for the degradation of haloaromatics might occur only in ecological niches. That the number of pure cultures able to grow at the expense of chloroaromatics remained small for a long time supports this assumption. More recently, however, many more strains have been discovered. Does this increase result from new strategies in the isolation of bacteria to degrade chloroaromatics? Chloroaromatics-degrading strains can be obtained by (a) enrichment from nature, (b) *in vivo* genetic manipulations, and (c) *in vitro* genetic engineering.

Enrichment of Chloroaromatics-degrading Bacteria

The technique of enrichment culture is an old method, developed by Winogradsky and Beijerinck. In principle, it is a very simple one. The investigator devises a culture medium of a particular defined chemical composition, inoculates it with a mixed microbial population, such as can be found in natural waters, soils, and sewage samples, and then ascertains what kind of microorganism predominates. For each particular set of conditions, a particular kind of microorganism will predominant because of its particular ability to grow more rapidly than the other organisms in the inoculum.

To obtain organisms with degradative capacity, a chemical to be degraded is supplied as a growth-limiting source of an essential nutrient in a culture medium. Compounds with pollutant potential are usually used as the limiting carbon source. Sometimes, however, nitrogen, phosphorus, or sulfur are used as the growth-limiting source, such as for the degradation of triazines and nitroaromatics (42), methylphosphonates (40, 41), ametryne, prometryne, and naphthalenesulfonates (43, 44), respectively.

Inoculum. The choice of the inoculum seems to be important for a successful enrichment. Various papers have described the use of polluted samples of surface water, sewage, soil, and even dumps as inocula. Unfortunately, detailed data about the history of the samples, i.e. concentrations of compounds and period of exposure, are rarely given. Accordingly, studies that include controls with non-polluted samples, which might show that success depends on a population from a polluted area, have never been published. Polluted samples might be the better source to isolate degrading bacteria, however, since bacterial populations might evolve in these samples to utilize the chemicals during exposure. If the strains have not acquired the potential to use the compound as the growth substrate, they might at least survive in the presence of the compound, which may sometimes be toxic.

Crawford and Mohn (45) found that exposure has to occur for a longer period before some kind of preenrichment occurs. Stanlake and Finn (202), however, isolated pentachlorophenol-degrading bacteria from samples that apparently had not been exposed to the chemical before the enrichment. Nonetheless, any preenrichment, can be expected only as long as the concentration of the pollutant in the sample is not so high as to be toxic but only high enough to give selection pressure. Crawford and Mohn (45) were unable to isolate organisms from soil samples that were contaminated with more than 500 ppm of pentachlorophenol, concentrations that are bacteriostatic or bacteriocidal.

To "breed" microorganisms capable of utilizing 2,4,5T as the sole source of carbon and energy, Kellogg et al. (118) inoculated microorganisms from various waste-dumping sites into a chemostat (Love Canal, New York, Eglin Air Force Base, Florida, and an Arkansas dump site) along with microorganisms harboring a variety of degradative plasmids such as CAM, TOL, SAL, pAC21, and pAC25, which code the degradation of camphor, xylenes, salicylate, and chlorobenzoates. The chemostat was started with high concentrations of substrates (250 μg/ml in total) for the plasmid-harboring strains (toluate, salicylate, chlorobenzoate; isomers not specified) and a low concentration of 2,4,5T (50 μg/ml). The so-called "plasmid-assisted molecular breeding" allowed isolation of the *Pseudomonas cepacia* strain AC1100 after a period of 8 to 10 months in the chemostat and further subcultivation in batch cultures in the presence of increasing concentrations of 2,4,5T (up to 2 mg/ml). Unfortunately data concerning the operation of the chemostat, such as dilution rate and concentration of the various substrates during the enrichment, were not given. The study also failed to present data of a comparable control without the added plasmid-carrying strains; hence, the successful isolation may have been due to mutations or genetic exchanges that occurred independently from the added plasmids. The statement, that organisms could not be isolated from the dump sites without the addition of plasmid strains even when the enrichment was carried out over a prolonged incubation period, is not conclusive, since mutations leading to the ability to degrade a chemical might occur after a great variance in the time period (191).

The pathway for the degradation of 2,4,5T elucidated later (182) was quite different from pathways coded on the plasmids. In addition, when the genes coding the 2,4,5T degradation were cloned, the cloned fragments did not show homology with the plasmid DNAs, which carried degradative genes for toluene, naphthalene and 3-chlorobenzoate (182, 227). This finding indicates that the non-systematic addition of strains containing plasmids will increase the amount of genes, but whether it also increases the chance for a successful enrichment remains questionable. The addition of strains from culture collections into a enrichment population is useful, however, when the strains contain enzyme sequences that in combination create a new hybrid pathway for the enrichment substrate, as described by Hartmann et al. (101) in the enrichment of organisms degrading 3,5-dichlorobenzoate.

Batch or Continuous Culture. The technical aspect of an enrichment is another important question. Both the batch and the continuous culture techniques, have advantages and disadvantages.

Batch culture is simple; many individual experiments may be set up with little effort or expense. The disadvantages appear if the compounds are so toxic to microorganisms (172) that only low concentrations could be added. Low concentrations can be managed by adding toxic compounds like chlorobenzene via the vapor phase; however, little difference in turbidity between positive and negative controls will follow. A further drawback is the relatively low cell count present if a mutation is required to obtain outgrowing, newly arising mutant strains. The total degradation of an organochlorine compound is accompanied by the release of HCl, though growth can be shown by using an indicator to record the change in pH, which might be more sensitive than the measurement of growth. Batch cultures select for strains with high K_s value and high growth rates.

An enormous advantage of a continuous culture is the ability to maintain defined conditions, in contrast to the constantly changing conditions in batch culture. Furthermore, continuous culture allows the use of low concentrations when the chemical is toxic. The disadvantages lie in the large amounts of apparatus required per culture, the correspondingly small number of experiments that can be started, and a tendency to produce organisms that grow slowly. Selection to strains with low K_s value were obtained by this approach.

Batch cultures have been used for most successful enrichments. Enrichment by use of chemostats is reported rarely, and usually without a description of detailed data such as dilution rates. Generally, the data indicate that the great advantage of the chemostat technique—the production of great population size—has not been utilized, since in most studies only a few volume exchanges were accomplished during several months of enrichment. This might be explained by the notion that washout of the desired organism might occur at a higher dilution rate.

Analog enrichment. Some workers have used a modified enrichment technique in which not one but two carbon sources are supplied. The first substrate, which is structurally analogous to the chloroaromatic compound, is supplied as a preenrichment substrate. This substrate is stepwise replaced by increasing amounts of the chloroaromatic over a prolonged period. The addition of the "natural" substrate should maintain a steady supply of mutants. The principle of analog enrichment has been used for the isolation of bacteria for the degradation of 3-chlorobenzoate (54), chlorobenzene (172), 4-chloroaniline (254), and 2-chlorobenzoate (252) with benzoate, benzene, aniline, and salicylate, respectively, as the analogues preenrichment substrate. The degradation of chlorosubstituted aromatic compounds, however, cannot be expected to follow the known reaction pathways for the unsubstituted parent compounds, so the approach seems not to be supported by present knowledge.

Natural aromatics have been degraded via the meta-pathway or the ortho-pathway. The enzymes of the peripheral sequences involved in the degradation of aromatic compounds are normally able to transform chlorinated analogs to chlorinated catechols. Chlorinated catechols or their autoxidation products are toxic to microorganisms when accumulated at high concentrations. Accumulation of chlorinated catechols results, because "normal" ortho-pyrocatechases that function in cleaving the aromatic ring system fail to cleave the chlorocatechols at a high rate (55, 56). On the other hand, 3-chlorocatechol is converted by a meta-pyrocatechase to a highly reactive acid chloride, which irreversibly inactivates the enzyme—the so-called "suicide inactivation" (13). Other meta-pyrocatechases have been inactivated by chlorocatechols (128). The pre-enrichment substrate often selects for degradation via the meta-pathway, but degradation of chloroaromatics has to follow the ortho-pathway. As long as the preenrichment substrate remains in the medium, a degradation sequence that is only partially useful will be selected. Hence very long periods of enrichment result. The real enrichment occurs after the nonchlorinated substrate is removed. Sometimes the original meta-pathway will be lost in further selection with the chlorinated substrate (172).

The enrichment of the Pseudomonas cepacia able to grow at the expense of 2-chlorobenzoate by use of salicylate as the preenrichment substrate has to follow a quite different principle, however, because the peripheral enzyme—the salicylate hydroxylase—is unable to convert 2-chlorobenzoate. Salicylate probably functions as an inducer of an unspecific benzoate dioxygenase, as has been shown for the TOL plasmid coded toluate 1,2-dioxygenase (167).

In general, adding substrates to the enrichment medium such as glucose or acetate, seems to be more convenient than adding an analogous substrate, as these substrates do not induce a degradation pathway of aromatics, so the accumulation of chlorocatechols does not occur or occurs only at a low level. Glucose or acetate allows the maintenance of high population density over a longer period. The problem—that turbidity as a parameter of degradation cannot be used under these circumstances—might be circumvented by the addition of an indicator that shows degradation through a change in pH value.

Chloroaromatics-degrading Strains. Several pure cultures have been isolated after selective enrichment that are able to use chlorinated aromatic compounds as the carbon and energy sources; namely phenoxyacetates, benzoates, anilines, benzenes, toluenes, biphenyls, and phenols. Only the lower chlorosubstituted compounds (e.g. mono- to trisubstituted) can be used by the aerobic organisms as the growth substrates. The exception is pentachlorophenol.

CHLOROPHENOXYACETATES. A variety of microorganisms of different genera have been isolated since 1950 that are capable of degrading 2,4D and related derivatives: Achromobacter (16, 17, 203–205), Alcaligenes (52, 71, 76), Arthrobacter (11, 19–21, 58, 104, 143–145, 181, 225, 226), Corynebacterium (113, 177), Flavobacterium (113, 203, 204, 237), Pseudomonas (52, 66, 67, 76, 82–85, 90, 121, 122, 230), Streptomyces (237), and Xanthobacter (50).

CHLOROBENZOATES. The utilization of chlorinated benzoates as growth substrate has been investigated mostly with the 3- and 4-chlorosubstituted isomers (34, 54, 72, 91, 114, 244, 249) (2, 116, 125, 126, 146, 178, 197, 232, 250, 251, 253). More recently, 2-chlorobenzoate-degrading strains have been added to the list (12, 65, 69, 105, 106, 150, 199, 221, 252). Most of the isolates show striking specificity towards the isomeric chlorobenzoates. Organisms that are able to use 3-chlorobenzoate, such as

Pseudomonas sp. strain B13, fail to grow in the presence of 4-chlorobenzoate (54). Conversely, 4-chlorobenzoate-degrading organisms do not grow with 3-chlorobenzoate (125, 126, 146, 178).

CHLOROANILINES. Zeyer and Kearney (254) reported the microbial degradation of chlorinated anilines as sole carbon and nitrogen sources by a natural isolate. A pseudomonad—renamed *Moraxella* sp. strain G (255)—was enriched by use of a chemostat from an aniline-grown, mixed culture that replaced aniline stepwise by 4-chloroaniline within a few weeks. The strain G was able to use 2-chloro-, 3-chloro-, and 4-chloroaniline as sole sources of carbon and energy. Surovtseva et al. (216, 217) described a *Pseudomonas diminuta* that utilized 3-chloro-, 4-chloro-, and 3,4-dichloroaniline as the sole source of carbon, nitrogen, and energy. *Pseudomonas acidovorans* strains able to use all monochlorosubstituted anilines as the carbon, nitrogen, and energy sources have been isolated by Loidl et al. (142).

CHLOROBENZENES. Until recently, nothing was known about the pathways used by bacteria for metabolism of chlorobenzenes, because strains able to grow on such compounds in pure culture had not been isolated. The toxicity of chlorobenzenes, especially the monosubstituted one, accounts for much of the difficulty in isolating bacteria able to degrade them. Extended selected enrichment with the substrate provided at low concentrations in the vapor phase has allowed isolation of bacteria able to grow at the expense of chlorobenzenes. Reineke and Knackmuss (172) isolated a chlorobenzene-degrading strain after 9 months of chemostat selection, during which time the benzene was gradually replaced by chlorobenzene as the growth substrate. The original inoculum was obtained from a mixture of soil and sewage. Similar approaches yielded *Alcaligenes* spp. (162, 188); *Pseudomonas* spp. (162, 200) able to degrade 1,4-dichlorobenzene; an *Alcaligenes* sp. able to degrade 1,3-dichlorobenzene (46); *Pseudomonas* spp. that grow on 1,2-dichlorobenzene (95, 201); and *Pseudomonas* spp. able to grow on all the dichlorobenzenes and 1,2,4-trichlorobenzene (180, 234). Oldenhuis et al. (161) described the isolation of chloro- and 1,2-dichlorobenzene-degrading strains within days from sediment samples taken from the river Rhine by conventional batch enrichment, in contrast to the long adaptation time required for the isolation of chloro- and dichlorobenzene-degrading strains (172, 188, 234).

CHLOROTOLUENES. Strains able to grow with chlorotoluenes have only been isolated via the enrichment technique by Vandenbergh et al. (231) with inocula obtained from a landfill site in the Love Canal area.

CHLOROBIPHENYLS. Several organisms enriched for growth with biphenyl were found to be able to use monochlorosubstituted biphenyls, mostly 4-chlorobiphenyl, as the growth substrate (4, 77, 78, 80, 81, 149, 160, 163, 196, 219, 220, 223). Release of chloride was not observed, however. The organisms use the non-substituted aromatic ring as the sole source of carbon and energy; the remaining chloroaromatic was accumulated in the form of chlorobenzoates (4, 80, 149). This example illustrates that for diphenylic chloroaromatics, growth with the compound does not necessarily means that total degradation, including elimination of the chlorine substituents, has occurred.

PENTACHLOROPHENOL. Several aerobic bacteria belonging to various genera that degrade pentachlorophenol have been isolated, including *Rhodococcus* (7, 93), *Mycobacterium* (93), *Pseudomonas* (218, 238), *Flavobacterium* (179), coryneform strains, and gram-variable strains with unresolved taxonomic position (38, 39, 61, 202). Many of the pentachlorophenol-degrading strains have a wide substrate specificity for polychlorinated phenols but degrade mono- and dichlorophenols poorly (8, 93, 207). The enrichment of organisms with mono- or dichlorinated phenols as the selective substrate has not been described because the compounds are highly toxic to bacteria.

Construction of Chloroaromatics-degrading Strains by Conjugative Transfer of DNA

Principle. New degradative pathways for chloroaromatics can be developed by the judicious combination of enzymes recruited from different pathways and organisms. The *in vivo* construction of hybrid pathways via conjugative transfer of genes is illustrated in Figure 3 for the degradation of chloroaromatics. The following complementary enzyme sequences are combined to form a functioning hybrid pathway (168, 170): (a) a *peripheral degradation sequence* originating from an alkylaromatic- or an aromatic-degrading strain that is able to convert the chlorosubstituted substrate analog to the chlorocatechols; and (b) a *central chlorocatechols degradation sequence*, as found in *Pseudomonas* sp. strain B13 for the degradation of 3-chloro-, 4-chloro-, and 3,5-dichlorocatechol (55, 56, 186, 187, 189).

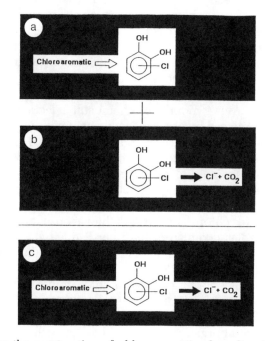

Figure 3. Concept for the construction of chloroaromatics-degrading bacteria: (a) peripheral enzyme sequences from alkylaromatics- or aromatics-degrading strains; (b) central chlorocatechols degradation sequence from strains like *Pseudomonas* sp. strain B13; (c) hybrid pathway for chloroaromatics.

Genetic and Biochemical Basis. The construction of chloroaromatics-degrading strains is feasible because several peripheral and central degradative enzyme sequences are encoded on conjugative plasmids (Table 3), or at least their genes are transferable by conjugation (174). Both sequences are combined in one organism to form a hybrid pathway for the total degradation of chloroaromatics, thus yielding chloride, carbon dioxide, and biomass. An important prerequisite for *in vivo* construction is that peripheral sequences are (a) sufficiently nonspecific to convert chloroaromatics to chlorocatechols (the size of a methylgroup and a chlorine substituent is nearly the same) and (b) are inducible by the chloroaromatics.

Figures 2 and 4 show the pathways relevant for construction in detail. The peripheral pathways, illustrated with their natural substrate, converge at the stage of

Table 3. Degradative plasmids with relevance to the construction of chloroaromatics-degrading hybrid strains.

Plasmid	Size (kb)	Conju-gative	Incompati-bility Group	Substrate	Reference
PERIPHERAL PATHWAYS					
TOL	117	+	P-9	xylenes, toluene, toluate	(15, 137-139, 240)
NAH7	83	+	P-9	naphthalene via salicylate	(15, 57, 138, 184, 246, 247)
pWW60-1	87	+	P-9	naphthalene via salicylate	(26)
pDTG1	83	+	P-9	naphthalene via salicylate	(192)
SAL1	85	+	P-9	salicylate	(15, 28, 138, 246, 247)
pKF1	82	−	ND	biphenyl via benzoate	(77)
pCITI	100	ND[a]	ND	aniline	(6)
pEB	253	ND	ND	ethylbenzene	(18)
pRE4	105	ND	ND	isopropylbenzene	(59, 60)
pWW174	200	+	ND	benzene	(241)
pHMT112	112	ND	ND	benzene	(224)
pEST1005	44	ND	ND	phenol	(124)
pVI150	mega	+	P-2	phenol, cresols, 3,4-dimethylphenol	(14, 198)
CENTRAL PATHWAYS					
pAC25	117	+	P-9	3-chlorobenzoate	(34)
pJP4	77	+	P-1	3-chlorobenzoate, 2,4D	(52)
pBR60	85	+	ND	3-chlorobenzoate	(245)
pRC10	45	ND	ND	2,4D	(37)
pP51	100	−	ND	1,2,4-trichlorobenzene	(235)

[a] Not determined.

catechol or alkylsubstituted catechols, which are mineralized via a meta-cleavage pathway. In contrast, the central degradation sequence of chlorocatechols starts with ortho-cleavage.

The in vivo construction of hybrid degradative pathways by conjugative transfer has been performed in our laboratory for various mono- and dichlorosubstituted aromatics, such as chlorinated anilines, benzenes, benzoates, biphenyls, phenols, salicylates, and toluenes (24a, 100, 102, 136, 154, 162, 173, 190). The same procedure has been followed in the laboratories of Chakrabarty, Chapman, and Ramos to isolate hybrid strains with the ability to degrade chlorinated benzenes, benzoates, and toluenes (1, 31, 32, 35, 36, 141).

Avoidance of Meta-cleavage Pathway. In addition to the desired peripheral enzyme sequence for chloroaromatics, plasmids encode meta-cleavage sequences that are unsuitable for the degradation of chloroaromatics. Therefore strategies have to be evolved by the hybrid strains to avoid misrouting the chlorocatechols into a dead-end meta-cleavage pathway, since the hybrid strains contain the entire plasmid, with desirable as well as unsuitable sequences.

The strategies used by the hybrid strains are discussed here for 3-chloro-, 4-chloro-, and 3,5-dichlorocatechol (see Figure 5). The critical element for degradation through the ortho-pathway in the hybrid strains is the presence of both ring-cleaving enzyme activities and their respective activity and affinity to the chlorocatechols. In most cases, the meta-cleaving catechol 2,3-dioxygenase (C23O) is present at a high basal

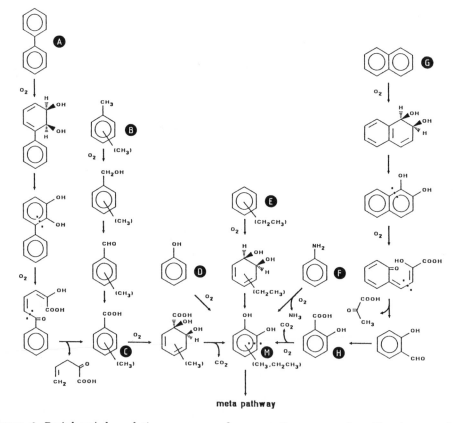

Figure 4. Peripheral degradation sequences for aromatic compounds with relevance for the construction of chloroaromatic-degrading hybrid strains: (A) biphenyl; (B) toluene or xylenes (CH_3); (C) benzoate or toluates (CH_3); (D) phenol; (E) benzene or ethylbenzene (CH_2CH_3); (F) aniline; (G) naphthalene; (H) salicylate. The central metabolites (M) catechol, methylcatechol (CH_3) or ethylcatechol (CH_2CH_3), respectively, will further be degraded via the meta-cleavage pathway. The cleavage of C–C-bonds is denoted by a line of stars.

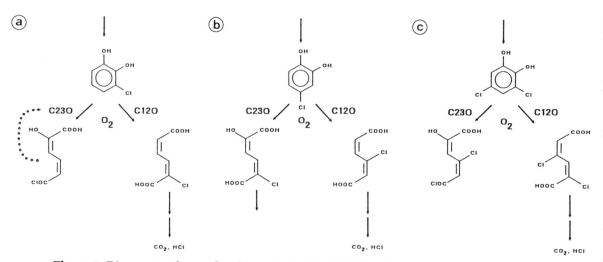

Figure 5. Divergent pathways for chlorocatechols in hybrid strains: (a) 3-chlorocatechol; (b) 4-chlorocatechol; (c) 3,5-dichlorocatechol. Left hand: meta-cleavage by catechol 2,3-dioxygenase (C23O). Right hand: ortho-cleavage by catechol 1,2-dioxygenase (C12O) and total degradation with elimination of chloride.

level and will be induced from the top of a pathway by substrates such as toluate, as has been shown for the C23O coded on the TOL plasmid of *Pseudomonas putida* strain PaW1 (243). In contrast, product induction of the ortho-cleaving enzyme, catechol 1,2-dioxygenase (C12O), has been shown. The concentration of the enzyme remains low as long as the inducing compound is absent.

Misrouting of 3-chlorocatechol into the *meta*-cleavage pathway can be avoided as follows (Figure 5a): Both enzymes are able to convert 3-chlorocatechol at a high rate; however, C23O forms an acylchloride from 3-chlorocatechol, which reacts with basic groups in the catalytic site of the enzyme, resulting in a nonreversible, so-called "suicide inactivation" of the enzyme (13). Further conversion of 3-chlorocatechol into the *meta*-cleavage pathway is prevented, and thus induction of C12O can occur.

A quite different mechanism to avoid misrouting is necessary for 4-chlorocatechol (see Figure 5b): Both enzymes show high activity and affinity with 4-chlorocatechol. Induction of C23O causes misrouting of 4-chlorocatechol into the *meta*-cleavage pathway, so no inducer of the *ortho*-cleaving enzyme can be formed. As has been shown for the degradation of 4-chlorobenzoate by the TOL plasmid carrying derivatives of *Pseudomonas* sp. strain B13, the *meta*-cleavage pathway can be prevented on the gene level, since the gene of the C23O, *xylE,* can be inactivated by an insertion of 3.2 kb or a point mutation (112; P. A. Williams, personal communication).

In analogy to the "suicide mechanism" observed for 3-chlorocatechol, the formation of an acylchloride was also expected to occur from 3,5-dichlorocatechol via C23O (see Figure 5c). The misrouting of 3,5-dichlorocatechol into the *meta*-cleavage pathway did not occur, however, because, in contrast to 3-chlorocatechol, C23O has only low activity with 3,5-dichlorocatechol and shows very low affinity to the compound (K_i value of 14 mM) (W. Reineke, unpublished results). In contrast, C12O exhibits high affinity (K_M value of 0.3 μM) and high activity (56).

In Vivo Construction, a Complex Process.   The transfer of genes into a recipient strain is often a prior condition in *in vivo* construction, which has to be followed by changes/mutations at the gene level to avoid misrouting chlorocatechols into the *meta*-cleavage pathway. As has been shown for the degradation of 3,5-dichlorobenzoate, a mutation changing the inducer specificity sometimes has to occur before TOL derivatives of strain B13 are able to use this compound as a growth substrate (see genealogy of chlorobenzoate-degrading hybrid strains in Figure 6a). For some chloroaromatics, the peripheral sequence fails to convert the chloroaromatic to the chlorocatechol stage, so two enzyme sequences from different sources have to be combined in a hybrid peripheral pathway before they give a functioning hybrid pathway for chloroaromatics with the B13 sequence. This is illustrated for chlorobiphenyl congeners in Figure 6b.

Organisms with the potential to degrade 3-chlorobiphenyl totally can be obtained by mating a biphenyl-growing strain, such as *Pseudomonas putida* strain BN10, and *Pseudomonas* sp. strain B13 (154). Both parent strains can function as the donor in the mating (although no plasmid in any of the parent strains has been shown). The resulting transconjugants fail to degrade 4-chlorobiphenyl totally, although they use the nonchlorinated ring of it as the growth substrate. The introduction of the TOL plasmid eliminates the bottleneck of the benzoate 1,2-dioxygenase for 4-chlorobenzoate by use of the toluate 1,2-dioxygenase (K. Engelberts and W. Reineke, unpublished results). Transconjugants totally degrading 4-chlorobiphenyl use the enzyme sequence that converts the compound to 4-chlorobenzoate, which is further degraded to give 4-chlorocatechol by the toluate 1,2-dioxygenase and the dihydrodihydroxytoluate dehydrogenase coded on the TOL plasmid. The benzoate dioxygenation remains the bottleneck in the mineralization of 2-chloro- and 2,4-dichlorobiphenyl in trans-conjugants such as *Pseudomonas putida* strains BN210 or KE210. Creation of 2-chloro-benzoate-degrading strains and conjugative transfer of genes coding the enzyme sequence, which brings about formation of 2-chlorobenzoate from 2-chlorobiphenyl, are the main steps involved in the development of strains degrading the *ortho*-substituted chlorobiphenyl congeners (100, 102) (see genealogy in Figure 6b).

Our data clearly indicate that the development of strains for the degradation of chloroaromatics by patchwork assembly is a complex process consisting of several steps; the steps must therefore be arranged judiciously starting with DNA transfer. The transconjugants must be able to degrade chlorocatechols so that the accumulation of these central but toxic compounds is avoided. The mating of useful strains in the presence of the new substrate often does not bring about the desired transconjugants, especially if the new compound and some metabolites are toxic to the parents as well as to the hybrid strains.

4-Chlorobenzoic acid:

1. PaW1 (4MBA$^+$) x B13 (3CBA$^+$, Sm^r) ---*---> WR211 (4MBA$^+$, 3CBA$^+$, 4CBA$^-$)

2. WR211 (4MBA$^+$, 3CBA$^+$, 4CBA$^-$) ---2*---> WR216 (4MBA$^-$, 3CBA$^+$, 4CBA$^+$)

3,5-Dichlorobenzoic acid:

WR211 (4MBA$^+$, 3CBA$^+$, 4CBA$^-$) ---3*---> WR911 (4MBA$^+$, 3CBA$^+$, 35CBA$^+$)

WR216 (4MBA$^-$, 3CBA$^+$, 4CBA$^+$) ---3*---> WR916 (4MBA$^-$, 3CBA$^+$, 35CBA$^+$)

Figure 6a. Genealogy of chlorobenzoic acids-degrading strains (112, 170, 171, 210). Abbreviations: 4MBA, 3CBA, 4CBA, 35CBA, 4-methyl-, 3-chloro-, 4-chloro-, and 3,5-dichlorobenzoic acid; Sm, streptomycin. Symbols: *, transfer of TOL plasmid; 2*, 3.2 kb insertion in *xylE*, the gene of catechol 2,3-dioxygenase; 3*, change in the inducer specificity: 3,5DCBA inducer in strains WR911 and WR916.

3-Chlorobiphenyl:

BN10 (BiP$^+$) x B13 (3CBA$^+$, Nal^r) ---*---> B131 (3CBiP$^+$)

B13 (3CBA$^+$) x BN10 (BiP$^+$, Sm^r) ---2*---> BN210 (3CBiP$^+$)

4-Chlorobiphenyl:

PaW1 (4MBA$^+$) x BN210 (3CBiP$^+$, Sm^r) ---3*---> KE210 (4CBiP$^+$)

2-Chlorobiphenyl:

1. B13 (3CBA$^+$) x WR401 (3MSA$^+$, 2MBA$^+$) ---2*---> WR612 (5CSA$^+$) ---4*---> JH230 (2CBA$^+$)

2. JHR (BiP$^+$) x JH230 (2CBA$^+$, Sm^r, Nal^r) ---*---> JHR2 (3CBiP$^+$, 4CBiP$^+$, 2CBA$^+$, 35CBA$^+$)

3. JHR2 ---5*---> JHR22 (2CBiP$^+$, 3CBiP$^+$, 4CBiP$^+$, 24CBiP$^+$, 35CBiP$^+$)

Figure 6b. Genealogy of chlorobiphenyls-degrading strains (63, 102, 154, K. Engelberts and W. Reineke, unpublished results) Abbreviations: BiP, biphenyl; 2CBiP, 3CBiP, 4CBiP, 24CBiP, 35CBiP, 2-chloro-, 3-chloro-, 4-chloro-, 2,4-dichloro-, and 3,5-dichlorobiphenyl; 2CBA, 3CBA, 4MBA, 2-chloro-, 3-chloro-, and 4-methylbenzoic acid; 3MSA, 5CSA, 3-methyl-, 5-chloro-salicylic acid; Sm, streptomycin; Nal, nalidixic acid. Symbols: *, transfer of genes coding biphenyl degradation; 2*, transfer of genes coding chlorocatechol degradation; 3*, transfer of TOL plasmid; 4*, selection for growth with 2-chlorobenzoic acid; 5*, selection for growth with 2-chlorobiphenyl.

Construction in a Chemostat-system. To overcome problems with low frequency of genetic exchange between two parents and the toxicity of new substrates to the strains, Kröckel and Focht (131) did not use the plate mating approach. Instead, they designed an experiment with a multichemostat approach for constructing chlorobenzene-degrading organisms. Two organisms, *Pseudomonas putida* R5-3 and *Pseudomonas alcaligenes* C-0, which contain the complementary enzymes for the entire chloro-benzene degradation pathway, a benzene and a chlorocatechol degradation sequence, are grown in separate chemostats. Strain R5-3 was grown on toluene, whereas strain C-0 was grown on benzoate, although strain C-0 degrades 3-chlorobenzoate. The cultures were concentrated and continuously amalgamated on a ceramic bead column. The column was subject to a continous stream of chlorobenzene vapors. Kröckel and Focht (131) point out that the reaction bed column serves several purposes: (1) the attachment of cells to the ceramic beads minimizes the probability of washout of the recombinant cells; (2) adsorption may enhance genetic exchange by immobilizing cells and concentrating them closer to one another; (3) a gradient of chlorobenzene concentrations is formed along the column. The establishment of a hybrid strain occurs when the concentration of chlorobenzene is optimal. The chlorobenzene-degrading *Pseudomonas putida* strain CB1-9 was isolated from the third chemostat following the column, which was run in the presence of chlorobenzene vapor.

The main disadvantage of this approach lies in the large number of apparatuses required. In contrast, the simple mating approach has been shown to be highly efficient in strain development even when the overall frequency of all steps is very low. Each step has to be done separately, but the strains have to establish a new function, such as a peripheral pathway, before the next selection step can be carried out.

Table 4. Cloned genes with relevance to the construction of chloroaromatics-degrading organism.[a]

Pathway	Gene product	Gene	Origin	Plasmid/characteristic	Reference
PERIPHERAL PATHWAYS					
Xyl	xylene monooxygenase	*xylA*	pWWO	pKT545: pKT231 carrying the *Xho*I G fragment of pWWO	(74)
	toluate dioxygenase	*xylD*	pWWO	pKT505: pKT231 carrying the *Sst* G fragment of pWWO	
Xyl	toluate dioxygenase	*xylD*	pWWO	pMT057: pBR322 carrying the 23.5-kb *Hind*III fragment of pWWO	(137)
	dihydrodihydroxytoluate dehydrogenase	*xylL*	pWWO	pMT057: pBR322 carrying the 23.5-kb *Hind*III fragment of pWWO	
Xyl	toluate dioxygenase	*xylD*	pTN2	pTS66: pACYC184 carrying the *Hind*III A fragment of pTN2	(108)
	positive regulator of *xylD*, E, ...	*xylS*	pTN2	pTS66: pACYC184 carrying the *Hind*III A fragment of pTN2	
Xyl	benzylalcohol dehydrogenase	*xylB*	pTN2	pTS7: pBR322 carrying the 15.4-kb *Bam*HI C fragment of pTN2	(107)

continued

Table 4, continued

Pathway	Gene product	Gene	Origin	Plasmid/characteristic	Reference
Xyl	positive regulator of *xylCAB*	*xylR*	pTN2	pTS41: pACYC184 carrying the *Bam*HI A fragment	(108, 109)
Xyl	xylene monooxygenase	*xylMA*	pWWO	pPL2: pKT231 carrying the 10-kb *Hind*III fragment of pWWO	(98)
	benzaldehyde dehydrogenase	*xylC*	pWWO	pPL2: pKT231 carrying the 10-kb *Hind*III fragment of pWWO	
	benzylalcohol dehydrogenase	*xylB*	pWWO	pGSH2819: pVL85 carrying the 11.4-kb *Bam*HI fragment of pWWO	
3,4DMP	phenol hydroxylase	*dmpA*	pVI150	EMBL3 carrying a 19-kb *Sma*I fragment of pVI150	(14)
Phe	phenol hydroxylase	*phlA*	chrom	pRO1957: pRO1727 carrying the 26-kb *Bam*HI fragment from strain PKO1	(134)
	positive regulator	*phlR*	chrom	pRO1957: pRO1727 carrying the 26-kb *Bam*HI fragment from strain PKO1	
Phe	phenol hydroxylase		pEST1005	pEST1332: pAYC32 carrying the 5.5-kb *Cla*I fragment of pEST1005	(124)
2,4D	2,4D monooxygenase	*tfdA*	pJP4	pVJH21: pVK101 carrying the 21-kb *Hind*III B fragment of pJP4	(211)
2,4D	2,4D monooxygenase	*tfdA*	pRO101	pRO2321 carrying the 3.5-kb *Bam*HI B fragment of pRO101	(99)
4CB	4CB dehalogenase		chrom	pPSA843: pPSA842 cosmid carrying a 9.5-kb *Sau*3A fragment	(183)
Tln	toluene dioxygenase	*todABC*	chrom	pDTG351: pKT230 carrying a 10.7-kb *Eco*RI fragment of strain PpF1	(257)
	dihydrodihydroxytoluene dehydrogenase	*todD*	chrom	pDTG351: pKT230 carrying a 10.7-kb *Eco*RI fragment of strain PpF1	
Tln	toluene dioxygenase		chrom	pIG: pBR322 carrying a 20-kb *Sau*3A fragment of strain NCIB11767	(208)
	dihydrodihydroxytoluene dehydrogenase		chrom	pIG: pBR322 carrying a 20-kb *Sau*3A fragment of strain NCIB11767	

Table 4, continued

Pathway	Gene product	Gene	Origin	Plasmid/characteristic	Reference
Bzn	benzene dioxygenase	bedABC1C2 pHMT112		pHMT181: pUC18 carrying the 4.5-kb PstI fragment of pHMT112	(224)
Bzn	benzene dioxygenase		chrom	pGR100: pMMB33 cosmid carrying a 10-kb EcoRI fragment	(110, 111)
1,2,4TCB	benzene dioxygenase	tcbA	pP51	pTCB60: pUC18 carrying the 12-kb HindIII fragment of pP51	(235)
	dihydrodihydroxybenzene dehydrogenase	tcbB	pP51	pTCB60: pUC18 carrying the 12-kb HindIII fragment of pP51	
Bip	biphenyl 2,3-dioxygenase	bphA	chrom	pMFB1: pKF330 carrying a 7.9-kb XhoI fragment of strain KF707	(79)
	dihydrodihydroxybiphenyl dehydrogenase	bphB	chrom	pMFB1: pKF330 carrying a 7.9-kb XhoI fragment of strain KF707	
	phenylcatechol dioxygenase	bphC	chrom	pMFB1: pKF330 carrying a 7.9-kb XhoI fragment of strain KF707	
Bip	phenylcatechol dioxygenase	bphC	chrom	pMFQ1: pKF330 cosmid carrying a 2.6-kb XhoI fragment of strain Q1	(222)
Bip	phenylcatechol dioxygenase	bphC	chrom	pOH101: pCP13 cosmid carrying a 6-kb HindIII fragment of strain OU83	(119)
Bip	phenylcatechol dioxygenase	bphC	chrom	pKH1: pKS13 cosmid carrying a 29-kb BamHI fragment of strain KKS102	(123)
	2-hydroxy-6-oxo-6-phenyl-hexa-2,4-dienoic acid hydrolase	bphD	chrom	pKH1: pKS13 cosmid carrying a 29-kb BamHI fragment of strain KKS102	
Bip	biphenyl 2,3-dioxygenase	bphA	chrom	pOH88: pCP13 cosmid carrying a 23.7-kb HindIII fragment of strain OU83	(120)
	dihydrodihydroxybiphenyl dehydrogenase	bphB	chrom	pOH88: pCP13 cosmid carrying a 23.7-kb HindIII fragment of strain OU83	
	phenylcatechol dioxygenase	bphC	chrom	pOH88: pCP13 cosmid carrying a 23.7-kb HindIII fragment of strain OU83	

continued

Table 4, continued

Pathway	Gene product	Gene	Origin	Plasmid/characteristic	Reference
	2-hydroxy-6-oxo-6-phenyl-hexa-2,4-dienoic acid hydrolase	*bphD*	chrom	pOH88: pCP13 cosmid carrying a 23.7-kb *Hind*III fragment of strain OU83	
Bip	biphenyl 2,3-dioxygenase	*bphA*	chrom	pGEM410: pMMB34 cosmid carrying a 12.4-kb *Eco*RI fragment of strain LB400	(155)
	dihydrodihydroxybiphenyl dehydrogenase	*bphB*	chrom	pGEM410: pMMB34 cosmid carrying a 12.4-kb *Eco*RI fragment of strain LB400	
	phenylcatechol dioxygenase	*bphC*	chrom	pGEM410: pMMB34 cosmid carrying a 12.4-kb *Eco*RI fragment of strain LB400	
	2-hydroxy-6-oxo-6-phenyl-hexa-2,4-dienoic acid hydrolase	*bphD*	chrom	pGEM410: pMMB34 cosmid carrying a 12.4-kb *Eco*RI fragment of strain LB400	
Bip	biphenyl 2,3-dioxygenase	*bphA*	chrom	pYH715: pHSG396 cosmid carrying a 9.4-kb *Xho*I fragment of strain KF715	(103)
	dihydrodihydroxybiphenyl dehydrogenase	*bphB*	chrom	pYH715: pHSG396 cosmid carrying a 9.4-kb *Xho*I fragment of strain KF715	
	phenylcatechol dioxygenase	*bphC*	chrom	pYH715: pHSG396 cosmid carrying a 9.4-kb *Xho*I fragment of strain KF715	
	2-hydroxy-6-oxo-6-phenyl-hexa-2,4-dienoic acid hydrolase	*bphD*	chrom	pYH715: pHSG396 cosmid carrying a 9.4-kb *Xho*I fragment of strain KF715	
Bip	biphenyl 2,3-dioxygenase	*bphA*	chrom	pDA1: pPSA842 cosmid carrying a 21.6-kb *Sau*3A fragment of strain B-356	(3)
	dihydrodihydroxybiphenyl dehydrogenase	*bphB*	chrom	pDA1: pPSA842 cosmid carrying a 21.6-kb *Sau*3A fragment of strain B-356	
	phenylcatechol dioxygenase	*bphC*	chrom	pDA1: pPSA842 cosmid carrying a 21.6-kb *Sau*3A fragment of strain B-356	
	2-hydroxy-6-oxo-6-phenyl-hexa-2,4-dienoic acid hydrolase	*bphD*	chrom	pDA1: pPSA842 cosmid carrying a 21.6-kb *Sau*3A fragment of strain B-356	
Nah	naphthalene dioxygenase	*nahA*	NAH7	pKGX505: pBR322 carrying the 30-kb *Eco*RI fragment of NAH7	(184)

Table 4, continued

Pathway	Gene product	Gene	Origin	Plasmid/characteristic	Reference
	dihydrodihydroxy-naphthalene dehydrogenase	*nahB*	NAH7	pKGX505: pBR322 carrying the 30-kb *Eco*RI fragment of NAH7	
	1,2-dihydroxynaphtha-lene dioxygenase	*nahC*	NAH7	pKGX505: pBR322 carrying the 30-kb *Eco*RI fragment of NAH7	
	2-hydroxychromene-2-carboxylate isomerase	*nahD*	NAH7	pKGX505: pBR322 carrying the 30-kb *Eco*RI fragment of NAH7	
	2-hydroxybenzalpyruvate aldolase	*nahE*	NAH7	pKGX505: pBR322 carrying the 30-kb *Eco*RI fragment of NAH7	
	salicylaldehyde dehydrogenase	*nahF*	NAH7	pKGX505: pBR322 carrying the 30-kb *Eco*RI fragment of NAH7	
	salicylate hydroxylase	*nahG*	NAH7	pKGX505: pBR322 carrying the 30-kb *Eco*RI fragment of NAH7	
Nah	naphthalene dioxygenase	*nahA*	pIG7	pAG1: pRK290 carrying the 25.2-kb *Eco*RI A fragment of pIG7	(92)
	dihydrodihydroxy-naphthalene dehydrogenase	*nahB*	pIG7	pAG1: pRK290 carrying the 25.2-kb *Eco*RI A fragment of pIG7	
	1,2-dihydroxynaphtha-lene dioxygenase	*nahC*	pIG7	pAG1: pRK290 carrying the 25.2-kb *Eco*RI A fragment of pIG7	
	2-hydroxychromene-2-carboxylate isomerase	*nahD*	pIG7	pAG1: pRK290 carrying the 25.2-kb *Eco*RI A fragment of pIG7	
	2-hydroxybenzalpyruvate aldolase	*nahE*	pIG7	pAG1: pRK290 carrying the 25.2-kb *Eco*RI A fragment of pIG7	
	salicylaldehyde dehydrogenase	*nahF*	pIG7	pAG1: pRK290 carrying the 25.2-kb *Eco*RI A fragment of pIG7	
	salicylate hydroxylase	*nahG*	pIG7	pAG1: pRK290 carrying the 25.2-kb *Eco*RI A fragment of pIG7	
	positive regulator of *nahA-G*	*nahR*	pIG7	pAG1: pRK290 carrying the 25.2-kb *Eco*RI A fragment of pIG7	
Nah	naphthalene dioxygenase	*nahA*	pWW60-1	pWW60-3002: pKT230 carrying the 6.9-kb *Hind*III E fragment of pWW60-1	(27)
	dihydrodihydroxy-naphthalene dehydrogenase	*nahB*	pWW60-1	pWW60-3002: pKT230 carrying the 6.9-kb *Hind*III E fragment of pWW60-1	
	1,2-dihydroxynaphtha-lene dioxygenase	*nahC*	pWW60-1	pWW60-3006: pKT230 carrying the 5.7-kb *Xho*I D fragment of pWW60-1	

continued

Table 4, continued

Pathway	Gene product	Gene	Origin	Plasmid/characteristic	Reference
	salicylate hydroxylase	*nahG*	pWW60-1	pWW60-3003: pKT230 carrying the 6-kb *Xho*I C fragment of pWW60-1	
Nah	naphthalene dioxygenase	*nahA*	pDTG1	pDTG113: pKT230 carrying the 15-kb *Eco*RI fragment of pDTG1	(193)
	dihydrodihydroxy-naphthalene dehydrogenase	*nahB*	pDTG1	pDTG113: pKT230 carrying the 15-kb *Eco*RI fragment of pDTG1	
	1,2-dihydroxynaphtha-lene dioxygenase	*nahC*	pDTG1	pDTG113: pKT230 carrying the 15-kb *Eco*RI fragment of pDTG1	
	2-hydroxychromene-2-carboxylate isomerase	*nahD*	pDTG1	pDTG113: pKT230 carrying the 15-kb *Eco*RI fragment of pDTG1	
	2-hydroxybenzalpyruvate aldolase	*nahE*	pDTG1	pDTG113: pKT230 carrying the 15-kb *Eco*RI fragment of pDTG1	
	salicylaldehyde dehydrogenase	*nahF*	pDTG1	pDTG113: pKT230 carrying the 15-kb *Eco*RI fragment of pDTG1	

CENTRAL PATHWAYS

Pathway	Gene product	Gene	Origin	Plasmid/characteristic	Reference
2,4D	catechol 1,2-dioxygenase	*tfdC*	pJP4	pKT231 carrying the 14-kb *Eco*RI B fragment of pJP4	(53)
	muconate cycloisomerase	*tfdD*	pJP4	pKT231 carrying the 14-kb *Eco*RI B fragment of pJP4	
	dienelactone hydrolase	*tfdE*	pJP4	pKT231 carrying the 14-kb *Eco*RI B fragment of pJP4	
2,4D	catechol 1,2-dioxygenase	*tfdC*	pRO101	pRO2334: pRO2321 carrying the 12.6-kb *Bam*HI C fragment of pRO101	(115)
	muconate cycloisomerase	*tfdD*	pRO101	pRO2334: pRO2321 carrying the 12.6-kb *Bam*HI C fragment of pRO101	
	dienelactone hydrolase	*tfdE*	pRO101	pRO2334: pRO2321 carrying the 12.6-kb *Bam*HI C fragment of pRO101	
	negative regulator of *tfdA*, *tfdCDE*	*tfdR*	pRO101	pRO1949: pRO1727 carrying the 3.9-kb *Bam*HI E fragment of pRO101	
3CB	catechol 1,2-dioxygenase	*clcA*	pAC27	pDC2: pLAFRI cosmid carrying the 27.5-kb *Eco*RI A fragment of pAC27	(33)

Table 4, continued

Pathway	Gene product	Gene	Origin	Plasmid/characteristic	Reference
	muconate cycloisomerase	*clcB*	pAC27	pDC2: pLAFRI cosmid carrying the 27.5-kb *Eco*RI A fragment of pAC27	
	dienelactone hydrolase	*clcD*	pAC27	pDC2: pLAFRI cosmid carrying the 27.5-kb *Eco*RI A fragment of pAC27	
3CB	catechol 1,2-dioxygenase	*clcA*	pAC27	pDC25: pRK290 carrying the 4.3-kb *Bgl*II E fragment of pAC27	(87)
	muconate cycloisomerase	*clcB*	pAC27	pDC25: pRK290 carrying the 4.3-kb *Bgl*II E fragment of pAC27	
	dienelactone hydrolase	*clcD*	pAC27	pDC25: pRK290 carrying the 4.3-kb *Bgl*II E fragment of pAC27	
3CB	catechol 1,2-dioxygenase	*clcA*	chrom	pMW65: pMMB33 cosmid carrying a 11.4-kb *Sau*3A fragment of strain B13	(239)
	muconate cycloisomerase	*clcB*	chrom	pMW65: pMMB33 cosmid carrying a 11.4-kb *Sau*3A fragment of strain B13	
	dienelactone hydrolase	*clcD*	chrom	pMW65: pMMB33 cosmid carrying a 11.4-kb *Sau*3A fragment of strain B13	
1,2,4TCB	catechol 1,2-dioxygenase	*tcbC*	pP51	pTCB1: pKT230 carrying the 13-kb *Sst*I fragment of pP51	(235)
	muconate cycloisomerase	*tcbD*	pP51	pTCB1: pKT230 carrying the 13-kb *Sst*I fragment of pP51	
	dienelactone hydrolase	*tcbE*	pP51	pTCB1: pKT230 carrying the 13-kb *Sst*I fragment of pP51	
	maleylacetate reductase		chrom	pRO1945: pRO1727 carrying a 6-kb *Bam*HI fragment of strain PKO1	(135)

[a]Abbreviations: Bip, biphenyl; Bzn, benzene; 3CB, 3-chlorobenzoate; 4CB, 4-chlorobenzoate; chrom, chromosome; 2,4D, 2,4-dichlorophenoxyacetate; 3,4DMP, 3,4-dimethylphenol; Nah, naphthalene; Phe, phenol; 1,2,4TCB, 1,2,4-trichlorobenzene; Tln, toluene; Xyl, xylene.

Construction of Chloroaromatics-degrading Strains by *In Vitro* Techniques

Cloned Functions with Relevance to *In Vitro* Construction. Like the strategy described above for the *in vivo* construction, it is possible to develop strains with the ability to degrade novel compounds via genetic engineering techniques. Several enzymes, enzymes sequences, and regulatory elements that function in the degradation of xylenes, phenol, phenoxyacetates, benzenes, biphenyl, naphthalene, and chlorocatechols have been cloned in recent years (Table 4). So far, however, only Lehrbach et al. (140) have described the development of strains that degrade chloroaromatics via cloning procedures.

Chlorobenzoate-degrading Strains. The expansion of the substrate range of a pathway by cloning procedures was studied with genes originating from the TOL plasmid introduced into strain B13 with the substrates 4-chloro- and 3,5-dichlorobenzoate (140). *Pseudomonas* sp. strain B13 degrades 3-chlorobenzoate, but not 4-chloro- or 3,5-dichlorobenzoate, because the first enzyme of the pathway, benzoate 1,2-dioxygenase, has a narrow substrate specificity and thus a low affinity for the later two analogues (169). The substrate specificity of the equivalent enzyme encoded by the TOL plasmid of *Pseudomonas putida* strain PaW1, toluate 1,2-dioxygenase, is broader and the enzyme is able to transform 3-chloro-, 4-chloro-, and 3,5-dichlorobenzoate (169). Franklin et al. (74, 75) have demonstrated that one SstI fragment of TOL plasmid contains the operon *xylXYZLEFG*, which encodes the enzymes of the *meta*-pathway for the degradation of toluates (97), and the operon promoter Pm. Expression of the genes requires the positive regulator encoded by the *xylS* gene, located on an adjacent DNA segment. The *xylXYZ* genes encode toluate 1,2-dioxygenase, which is able to convert 4-chloro- and 3,5-dichlorobenzoate, and the *xylL* gene encodes dihydrodihydroxytoluate dehydrogenase. In the first step, the SstI fragment D of the TOL plasmid was cloned into the SstI site of pKT231 to produce plasmid pKT505. To construct a hybrid plasmid containing *xylXYZ* and *xylL* genes and their promoter but not the *xylE* gene coding the catechol 2,3-dioxygenase, subfragments of plasmid pKT505 were generated by partial digestion with Sau3A and ligated into the BamHI site of pBR322 thus yielding the plasmid pPL401. In parallel, a DNA segment carrying *xylS* was cloned in the XhoI site of the vector pKT231 to produce the plasmid pKT570. The plasmid pPL401 was ligated through its unique HindIII site to plasmid pKT570 to give pPL403, a broad host-range hybrid plasmid that carries regulated genes encoding the enzymes required for the conversion of substituted benzoates to the respective catechols. The plasmid pPL403 was mobilized by RP1 to *Pseudomonas* sp. strain B13. Transconjugants selected directly on mineral medium containing 4-chlorobenzoate were obtained at a frequency of 1×10^{-4} per B13 cell, whereas streptomycin-resistant transconjugants occurred at a higher frequency of 2×10^{-2}. 3,5-Dichlorobenzoate-growing transconjugants were obtained at a frequency of 6×10^{-6} per recipient cell from the mating with pPL403. Analysis of the ability of derivatives of pPL403 such as pPL408 and pPL409, deleted in various segments of *xylXYZ* or *xylL*, to give rise to 3,5-dichlorobenzoate-positive derivatives showed that the acquisition of the *xylXYZ* gene alone permitted strain B13 to catabolize 4-chlorobenzoate. In contrast, the acquisition of both *xylXYZ* and *xylL* was necessary for the isolation of 3,5-dichlorobenzoate-growing derivatives of strain B13.

These results tentatively suggested that the dihydrodihydroxybenzoate dehydrogenase of the 3-chlorobenzoate-pathway of strain B13 is able to process the 4-chlorobenzoate metabolite 4-chlorodihydrodihydroxybenzoate but not the respective 3,5-dichlorobenzoate metabolite. This interpretation was found to be incorrect however.

First, the dihydrodihydroxybenzoate dehydrogenase of strain B13 is able to convert 3,5-dichlorodihydrodihydroxybenzoate to the respective catechol (A. Stolz and W. Reineke, unpublished results). Second, 4-chloro- and 3,5-dichlorobenzoate failed to function as an inducer in strain B13 (210). These results indicate that besides the presence of xylXYZ and xylS genes in strain B13, a mutation reflecting the induction of the B13 enzymes has to occur before a 4-chlorobenzoate-positive strain can appear. This additional event was not necessary when xylXYZ, xylL, and xylS were transferred to strain B13 to give 4-chlorobenzoate-growing strains.

The creation of 3,5-dichlorobenzoate-degrading strains was later interpreted in a different way. 3,5-Dichlorobenzoate failed to induce the meta-pathway in Pseudomonas putida strain PaW1 (210). This was confirmed by experiments with isolated xylS genes by Ramos et al. (167). This finding suggested that, in contrast to the data obtained by Lehrbach et al. (140), the failure of the 4-chlorobenzoate-positive derivative of B13(TOL) to grow on 3,5-dichlorobenzoate was due to a lack of synthesis of the TOL plasmid–encoded toluate 1,2-dioxygenase and dihydrodihydroxytoluate dehydrogenase. To eliminate this block, a xylS mutant whose product is activated by 3,5-dichlorobenzoate was generated by the following strategy: The construction Escherichia coli 5K (pNM185, pJLR200) was used; pNM185 is an IncP4 plasmid that carries the xylS gene, and pJLR200 is a pBR322 derivative in which the tetracycline resistance promoter has been removed and replaced by the TOL plasmid meta-pathway operon promoter Pm. Since expression of tetracyline resistance by this strain requires activation of XylS, the strain is resistant to tetracycline if a XylS protein effector is present in the culture medium. xylS mutants that are activated by the ordinarily noneffector benzoate analogue can be selected as tetracycline-resistant clones on plates containing 3,5-dichlorobenzoate. Besides clones mutated in xylS, such as pERD352, another class of clones appeared, which grew on tetracycline agar plates lacking 3,5-dichlorobenzoate. These clones contain a mutation in the Pm promoter, so that the tetracycline gene was expressed constitutively. The introduction of plasmid pERD352 carrying the mutated gene xylS352—a change from Pro to Arg at position 256 (166)—into B13(TOL) allowed the strain to grow on 3,5-dichlorobenzoate.

The trans dominance of positive-acting protein regulators of gene expression permit the experimenter to use mutant regulators directly to effect changes in the metabolic activity of a cell, without the need for prior or simultaneous elimination of the wild-type regulator gene. This example of expansion demonstrates that the alteration of catabolic pathways by replacement of narrow-specificity proteins by mutant protein is a powerful means of accelerating the evolution of new catabolic phenotypes and pathways able to mineralize a wider range of substrates.

Chlorosalicylate-degrading Strains.   A second type of expansion of the chlorocatechol pathway of strain B13 was accomplished by introducing the NAH7 plasmid gene nahG into strain B13 (140). This gene encodes a broad specificity salicylate hydroxylase (156), which is able to convert salicylate, methyl- and chlorosalicylates to their corresponding catechol derivatives. To clone the nahG gene, NAH7 plasmid DNA was cleaved with HindIII, and the fragments thereby generated were ligated with HindIII-cleaved pBR322 vector DNA. One ampicillin-resistant clone able to convert salicylate to catechol was shown to carry a pBR322 hybrid plasmid pPL299 containing a 3.1 kb HindIII fragment of NAH7 (138). For the introduction of the cloned nahG gene into Pseudomonas sp., pPL299 and the pKT231 vector were ligated at their uniqe EcoRI sites, thus resulting in plasmid pPL300 (Ap^r, Km^r, Tc^r). This hybrid plasmid was mobilized by R64-11 to Pseudomonas putida KT2442. Some salicylate-positive clones were no longer resistant to kanamycin, and their plasmids had suffered a deletion enclosing a 1.6 kb DNA fragment from the kanamycin resistance gene. One of these deletion derivatives, pPL300-1, was mobilized by RP1 to strain B13, thus producing transconjugants able to grow on 3-chloro-, 4-chloro-, and 5-chlorosalicylate.

CONCLUSION

By developing hybrid pathways in bacteria, strains can be obtained in the laboratory that will degrade a wide range of chloroaromatic compounds. Because most of the research has been done with model compounds rather than problem chemicals found in the environment, however, the following questions must be answered. How promising are the methods described for real contaminants? What are the limitations of both approaches of construction?

In principle, the in vitro method of construction, in which cloned and well-characterized genes are selectively introduced in a host organism in order to create a new pathway, has some advantages over in vivo construction: It is highly controlled, it produces predictable changes, and it enables multiple genetic changes to be effected in a single step, often without the need to select directly the final phenotype desired. The disadvantages preponderate, however. The in vitro technique does require detailed information on the genetics and biochemistry of the key elements of the pathways and enzymatic steps to be manipulated. These disadvantages prevent the application of this approach on a wide scale. In contrast, in vivo construction, whereby genes of one organism are recruited into a pathway of another organism through conjugation, can be carried out successfully without prior extensive genetic characterization if the steps in development are arranged judiciously and the direct selection of the final phenotype is avoided.

For degradation of chlorinated aromatics of environmental concern, the main limitation of both approaches appears to be enzyme availability. Both approaches are currently restricted to the degradation of dichlorocatechols. Tri- and tetrachlorocatechols can serve as substrates for the known chlorocatechol sequences, but the pathway sequences involved in the elimination of a third or fourth chlorine substituent remain speculative. Further research is needed to obtain chlorocatechol degradation sequences with broad specificity. In addition, strategies must be developed to isolate peripheral sequences that can be recruited for construction of strains for the degradation of more highly chlorinated aromatics. An enormous amount of work remains to be done on model compounds to elucidate out the ground rules of degradation of highly chlorinated aromatics and of strain development before the real problems can be addressed.

Acknowledgment

I am indepted to Dietmar Pieper for reviewing the manuscript.

Note Added in Proof

The mechanism of hydrolytic elimination of chloride from 4-chlorobenzoate to yield 4-hydroxybenzoate has recently been elucidated in several laboratories (44a, 91a, 141a, 141b, 187a): The 4-chlorobenzoate dehalogenase activity was found to be the sum of the activities of a 4-chlorobenzoate-CoA ligase, which catalyzes the formation of 4-chlorobenzoyl-CoA from 4-chlorobenzoate, coenzyme A, and ATP; a 4-chlorobenzoate-CoA dehalogenase, which catalyzes the conversion to 4-hydroxybenzoate-CoA with release of chloride; and a 4-hydroxybenzoate-CoA thioesterase.

LITERATURE CITED

1. Abril, M.-A., Michan, C., Timmis, K. N., and Ramos, J. L. 1989. Regulator and enzyme specificities of the TOL plasmid-encoded upper pathway for degradation of aromatic hydrocarbons and expansion of the substrate range of the pathway. *J. Bacteriol.* 171:6782–6790.

2. Adriaens, P., Kohler, H.-P.E, Kohler-Staub, D., and Focht, D. D. 1989. Bacterial dehalogenation of chlorobenzoates and coculture biodegradation of 4,4'-dichlorobiphenyl. *Appl. Environ. Microbiol.* 55:887–892.

3. Ahmad, D., Massé, R., and Sylvestre, M. 1990. Cloning and expression of genes involved in 4-chlorobiphenyl transformation by *Pseudomonas testosteroni*: homology to polychloro-biphenyl-degrading genes in other bacteria. *Gene* 86:53–61.

4. Ahmed, M., and Focht, D. D. 1973. Degradation of polychlorinated biphenyls by two species of *Achromobacter*. *Can. J. Microbiol.* 19:47–52.

5. Allard, A.-S., Hynning, P.-Å, Lindgren, C., Remberger, M., and Neilson, A. H. 1991. Dechlorination of chlorocatechols by stable enrichment cultures of anaerobic bacteria. *Appl. Environ. Microbiol.* 57:77–84.

6. Anson, J. G., and Mackinnon, G. 1984. Novel *Pseudomonas* plasmid involved in aniline degradation. *Appl. Environ. Microbiol.* 48:868–869.

7. Apajalahti, J. H. A., Kärpänoja, P., and Salkinoja-Salonen, M. S. 1986. *Rhodococcus chlorophenolicus* sp. nov., a chlorophenol-mineralizing actinomycete. *Int. J. Syst. Bacteriol.* 36:246–251.

8. Apajalahti, J. H. A., and Salkinoja-Salonen, M. S. 1986. Degradation of chlorinated phenols by *Rhodococcus chlorophenolicus*. *Appl. Microbiol. Biotechnol.* 25:62–67.

9. Apajalahti, J. H. A., and Salkinoja-Salonen, M. S. 1987. Dechlorination and para-hydroxy-lation of polychlorinated phenols by *Rhodococcus chlorophenolicus*. *J. Bacteriol.* 169:675–681.

10. Apajalahti, J. H. A., and Salkinoja-Salonen, M. S. 1987. Complete dechlorination of tetra-chlorohydroquinone by cell extracts of pentachlorophenol-induced *Rhodococcus chloro-phenolicus*. *J. Bacteriol.* 169:5125–5130.

11. Audus, L. J. 1950. Biological detoxication of 2,4-dichlorophenoxyacetic acid in soils: isolation of an effective organism. *Nature* 166:356.

12. Baggi, G. 1985. Richerche sulla degradazione di acidi clorobenzoici. *Ann. Microbiol.* 35:71–78.

13. Bartels, I., Knackmuss, H.-J., and Reineke, W. 1984. Suicide inactivation of catechol 2,3-dioxygenase from *Pseudomonas putida* mt-2 by 3-halocatechols. *Appl. Environ. Microbiol.* 47:500–505.

14. Bartilson, M., Nordlund, I., and Shingler, V. 1990. Location and organization of the dimethylphenol catabolic genes of *Pseudomonas* CF600. *Molec. Gen. Genet.* 220:294–300.

15. Bayley, S. A., Morris, D. W., and Broda, P. 1979. The relationship of degradative and resistance plasmids of *Pseudomonas* belonging to the same incompatibility group. *Nature* 280:338–339.

16. Bell, G. R. 1957. Some morphological and biochemical characteristics of a soil bacterium which decomposes 2,4-dichlorophenoxyacetic acid. *Can. J. Microbiol.* 3:821–840.

17. Bell, G. R. 1960. Studies on a soil *Achromobacter* which degrades 2,4-dichlorophenoxy-acetic acid. *Can. J. Microbiol.* 6:325–337.

18. Bestetti, G., and Galli, E. 1984. Plasmid-coded degradation of ethylbenzene and 1-phenylethanol in *Pseudomonas fluorescens*. *FEMS Microbiol. Lett.* 21:165–168.

19. Bollag, J.-M., Helling, C. S., and Alexander, M. 1967. Metabolism of 4-chloro-2-methylphenoxyacetic acid by soil bacteria. *Appl. Microbiol.* 15:1393–1398.

20. Bollag, J.-M., Helling, C. S., and Alexander, M. 1968. 2,4-D metabolism. Enzymatic hydroxylation of chlorinated phenols. *J. Agric. Food Chem.* 16:826–828.

21. Bollag, J.-M., Briggs, G. G., Dawson, J. E., and Alexander, M. 1968. 2,4-D metabolism: enzymatic degradation of chlorocatechols. *J. Agric. Food Chem.* 16:829–833.

22. Bosma, T. N. P., van der Meer, J. R., Schraa, G., Tros, M. E., and Zehnder, A. J. B. 1988. Reductive dechlorination of all trichloro- and dichlorobenzene isomers. *FEMS Microbiol. Ecol.* 53:223–229.

23. Boyd, S. A., and Shelton, D. R. 1984. Anaerobic biodegradation of chlorophenols in fresh and acclimated sludge. *Appl. Environ. Microbiol.* 47:272–277.

24. Boyd, S. A., Shelton, D. R., Berry, D., and Tiedje, J.-M. 1983. Anaerobic biodegradation of phenolic compounds in digested sludge. *Appl. Environ. Microbiol.* 46:50–54.

24a. Brinkmann, U., and Reineke, W. 1992. Degradation of chlorotoluenes by in vivo contructed strains: Problems of enzyme specificity, induction and prevention of meta-pathway. *FEMS Microbiol. Lett.* 96:81–88.

25. Brown, J. F. Jr., Bedard, D. L., Brennan, M. J., Carnahan, J. C., Feng, H., and Wagner, R. E. 1987. Polychlorinated biphenyl dechlorination in aquatic sediments. *Science* 236:709–712.

26. Cane, P. A., and Williams, P. A. 1982. The plasmid-coded metabolism of naphthalene and 2-methylnaphthalene in *Pseudomonas* strains: phenotypic changes correlated with structural modification of the plasmid pWW60–1. *J. Gen. Microbiol.* 128:2281–2290.

27. Cane, P. A., and Williams, P. A. 1986. A restriction map of naphthalene catabolic plasmid pWW60–1 and the location of some of its catabolic genes. *J. Gen. Microbiol.* 132:2919–2929.

28. Chakrabarty, A. M. 1972. Genetic basis of the biodegradation of salicylate in *Pseudomonas. J. Bacteriol.* 112:815–823.

29. Chapman, P. J. 1975. Bacterial metabolism of 4-chlorobenzoic acid. *Abstr. Annu. Meet. Am. Soc. Microbiol.* O2, p. 192.

30. Chapman, P. J. 1979. Degradation mechanisms. In *Microbial Degradation of Pollutants in Marine Environments.* EPA-600/9-79-012. Ed. A. W. Bourquin, P. H. Pritchard. Gulf Breeze, Fla.: US EPA. 28–66.

31. Chapman, P. J. 1988. Constructing microbial strains for degradation of halogenated aromatic hydrocarbons. In *Environmental biotechnology: Reducing risks from environmental chemicals through biotechnology.* Ed. G. S. Omenn. *Basic Life Sciences* 45:81–95. New York: Plenum Press.

32. Chatterjee, D. K., and Chakrabarty, A. M. 1982. Genetic rearrangements in plasmids specifying total degradation of chlorinated benzoic acids. *Molec. Gen. Genet.* 188:279–285.

33. Chatterjee, D. K., and Chakrabarty, A. M. 1984. Restriction mapping of a chlorobenzoate degradative plasmid and molecular cloning of the degradative genes. *Gene* 27:173–181.

34. Chatterjee, D. K., Kellogg, S. T., Hamada, S., and Chakrabarty, A. M. 1981. Plasmid specifying total degradation of 3-chlorobenzoate by a modified *ortho* pathway. *J. Bacteriol.* 146:639–646.

35. Chatterjee, D. K., Kellogg, S. T., Furukawa, K., Kilbane, J. J., and Chakrabarty, A. M. 1981. Genetic approaches to the problems of toxic chemical pollution. In *Recombinant DNA.* Ed. A. G. Walton. Amsterdam: Elsevier. 199–212.

36. Chatterjee, D. K., Kellogg, S. T., Watkins, D. R., and Chakrabarty, A. M. 1981. Plasmids in the biodegradation of chlorinated aromatic compounds. In *Molecular Biology, Pathogenicity, and Ecology of Bacterial Plasmids.* Ed. S. B. Levy, R. C. Clowes, E. L. Koenig. New York: Plenum Press. 519–528.

37. Chaudhry, G. R., and Huang, G. H. 1988. Isolation and characterization of a new plasmid from a *Flavobacterium* sp. which carries the genes for degradation of 2,4-dichlorophenoxyacetate. *J. Bacteriol.* 170:3897–3902.

38. Chu, J. P., and Kirsch, E. J. 1972. Metabolism of pentachlorophenol by an axenic bacterial culture. *Appl. Microbiol.* 23:1033–1035.

39. Chu, J. P., and Kirsch, E. J. 1973. Utilization of halophenols by a pentachlorophenol metabolizing bacterium. *Dev. Ind. Microbiol.* 14:264–273.

40. Cook, A. M., Daughton, C. G., and Alexander, M. 1978. Phosphonate utilization by bacteria. *J. Bacteriol.* 133:85–90.

41. Cook, A. M., Daughton, C. G., and Alexander, M. 1978. Phosphorus-containing pesticide breakdown products: quantitative utilization as phosphorus sources by bacteria. *Appl. Environ. Microbiol.* 36:668–672.

42. Cook, A. M., and Hütter, R. 1981. s-Triazines as nitrogen sources for bacteria. *J. Agric. Food Chem.* 29:1135–1143.

43. Cook, A. M., and Hütter, R. 1982. Ametryne and prometryne as sulfur sources for bacteria. *Appl. Environ. Microbiol.* 43:781–786.

44. Cook, A. M., Schmuckle, A., and Leisinger, T. 1986. Microbial desulfonation of multisubstituted naphthalene sulfonic acids. *Experientia* 42:95–96.

44a. Copley, S. D., and Crooks, G. P. 1992. Enzymatic dehalogenation of 4-chlorobenzoyl coenzyme A in *Acinetobacter* sp. strain 4-CB1. *Appl. Environ. Microbiol.* 58:1385–1387.

45. Crawford, R. L., and Mohn, W. W. 1985. Microbiological removal of pentachlorophenol from soil using a *Flavobacterium*. *Enzyme Microbiol. Technol.* 7:617–620.

46. De Bont, J. A. M., Vorage, M. J. A. W., Hartmans, S., and van den Tweel, W. J. J. 1986. Microbial degradation of 1,3-dichlorobenzene. *Appl. Environ. Microbiol.* 52:677–680.

47. DeWeerd, K. A., and Suflita, J. M. 1990. Anaerobic aryl reductive dehalogenation of halobenzoates by cell extracts of "*Desulfomonile tiedjei.*" *Appl. Environ. Microbiol.* 56:2999–3005.

48. DeWeerd, K. A., Mandelco, L., Tanner, R. S., Woese, C. R., and Suflita, J. M. 1990. *Desulfomonile tiedjei* gen. nov. and sp. nov., a novel anaerobic, dehalogenating, sulphate-reducing bacterium. *Arch. Microbiol.* 154:23–30.

49. DeWeerd, K. A., Suflita, J. M., Linkfield, T., Tiedje, J. M., and Pritchard, P. H. 1986. The relationship between reductive dehalogenation and other aryl substituent removal reactions catalyzed by anaerobes. *FEMS Microbiol. Lett.* 38:331–339.

50. Ditzelmüller, G., Loidl, M., and Streichsbier, F. 1989. Isolation and characterization of a 2,4-dichlorophenoxyacetic acid-degrading soil bacterium. *Appl. Microbiol. Biotechnol.* 31:93–96.

51. Dolfing, J., and Tiedje, J. M. 1986. Hydrogen cycling in a three-tiered food web growing on the methanogenic conversion of 3-chlorobenzoate. *FEMS Microbiol. Ecol.* 38:293–298.

52. Don, R. H., and Pemberton, J. M. 1981. Properties of six pesticide degradation plasmids isolated from *Alcaligenes eutrophus*. *J. Bacteriol.* 145:681–686.

53. Don, R. H., Weightman, A. J., Knackmuss, H.-J., Timmis, K. N. 1985. Transposon mutagenesis and cloning analysis of the pathways for degradation of 2,4-dichlorophenoxyacetic acid and 3-chlorobenzoate in *Alcaligenes eutrophus* JMP134(pJP4). *J. Bacteriol.* 161:85–90.

54. Dorn, E., Hellwig, M., Reineke, W., and Knackmuss, H.-J. 1974. Isolation and characterization of a 3-chlorobenzoate degrading pseudomonad. *Arch. Microbiol.* 99:61–70.

55. Dorn, E., and Knackmuss, H.-J. 1978. Chemical structure and biodegradability of halogenated aromatic compounds. Two catechol 1,2-dioxygenases from a 3-chlorobenzoate-grown pseudomonad. *Biochem. J.* 174:73–84.

56. Dorn, E., and Knackmuss, H.-J. 1978. Chemical structure and biodegradability of halogenated aromatic compounds. Substituent effects on 1,2-dioxygenation of catechol. *Biochem. J.* 174:85–94.

57. Dunn, N. W., and Gunsalus, I. C. 1973. Transmissible plasmid coding early enzymes of naphthalene oxidation in *Pseudomonas putida*. *J. Bacteriol.* 114:974–979.

58. Duxbury, J. M., Tiedje, J. M., Alexander, M., and Dawson, J. E. 1970. 2,4-D metabolism: enzymatic conversion of chloromaleylacetic acid to succinic acid. *J. Agric. Food Chem.* 18:199–201.

59. Eaton, R. W., and Timmis, K. N. 1986. Characterization of a plasmid-specified pathway for catabolism of isopropylbenzene in *Pseudomonas putida* RE204. *J. Bacteriol.* 168:123–131.

60. Eaton, R. W., and Timmis, K. N. 1986. Spontaneous deletion of a 20-kilobase DNA segment carrying genes specifying isopropylbenzene metabolism in *Pseudomonas putida* RE204. *J. Bacteriol.* 168:428–430.

61. Edgehill, R. U., and Finn, R. K. 1982. Isolation, characterization and growth kinetics of bacteria metabolizing pentachlorophenol. *Eur. J. Appl. Biotechnol.* 16:179–184.

62. Elsner, A., Löffler, F., Miyashita, K., Müller, R., and Lingens, F. 1991. Resolution of 4-chlorobenzoate dehalogenase from *Pseudomonas* sp. strain CBS3 into three components. *Appl. Environ. Microbiol.* 57:324–326.

63. Engelberts, K., Schmidt, E., and Reineke, W. 1989. Degradation of o-toluate by *Pseudomonas* sp. strain WR401. *FEMS Microbiol. Lett.* 59:35–38.

64. Engesser, K. H., Schmidt, E., and Knackmuss, H.-J. 1980. Adaptation of *Alcaligenes eutrophus* B9 and *Pseudomonas* sp. B13 to 2-fluorobenzoate as growth substrate. *Appl. Environ. Microbiol.* 39:68–73.

65. Engesser, K. H., and Schulte, P. 1989. Degradation of 2-bromo-, 2-chloro- and 2-fluorobenzoate by *Pseudomonas putida* CLB250. *FEMS Microbiol. Lett.* 60:143–148.

66. Evans, W. C., Smith, B. S. W., Fernley, H. N., and Davies, J. I. 1971. Bacterial metabolism of 2,4-dichlorophenoxyacetate. *Biochem. J.* 122:543–551.

67. Evans, W. C., Smith, B. S. W., Moss, P., and Fernley, H. N. 1971. Bacterial metabolism of 4-chlorophenoxyacetate. *Biochem. J.* 122:509–517.

68. Fathepure, B. Z., Tiedje, J. M., and Boyd, S. A. 1988. Reductive dechlorination of hexachlorobenzene to tri- and dichlorobenzenes in anaerobic sewage sludge. *Appl. Environ. Microbiol.* 54:327–330.

69. Fetzner, S., Müller, R., and Lingens, F. 1989. A novel metabolite in the microbial degradation of 2-chlorobenzoate. *Biochem. Biophys. Res. Commun.* 161:700–705.

70. Fetzner, S., Müller, R., and Lingens, F. 1989. Degradation of 2-chlorobenzoate by *Pseudomonas cepacia* 2CBS. *Biol. Chem. Hoppe-Seyler* 370:1173–1182.

71. Fisher, P. R., Appleton, J., and Pemberton, J. M. 1978. Isolation and characterization of the pesticide-degrading plasmid pJP1 from *Alcaligenes paradoxus*. *J. Bacteriol.* 135:798–804.

72. Focht, D. D., and Shelton, D. 1987. Growth kinetics of *Pseudomonas alcaligenes* C-O relative to inoculation and 3-chlorobenzoate metabolism in soil. *Appl. Environ. Microbiol.* 53:1846–1849.

73. Fowden, L. 1968. The occurrence and metabolism of carbon–halogen compounds. *Proc. R. Soc. London Ser. B* 171:5–18.

74. Franklin, F. C. H., Bagdasarian, M., Bagdasarian, M. M., and Timmis, K. N. 1981. Molecular and functional analysis of the TOL plasmid pWWO from *Pseudomonas putida* and cloning of genes for the entire regulated aromatic ring *meta* cleavage pathway. *Proc. Natl. Acad. Sci. USA.* 78:7458–7462.

75. Franklin, F. C. H., Lehrbach, P. R., Lurz, R., Rueckert, B., Bagdasarian, M., and Timmis, K. N. 1983. Localization and functional analysis of transposon mutations in regulatory genes of the TOL catabolic pathway. *J. Bacteriol.* 154:676–685.

76. Friedrich, B., Meyer, M., and Schlegel, H. G. 1983. Transfer and expression of the herbicide-degrading plasmid pJP4 in aerobic autotrophic bacteria. *Arch. Microbiol.* 134:92–97.

77. Furukawa, K., and Chakrabarty, A. M. 1982. Involvement of plasmids in total degradation of chlorinated biphenyls. *Appl. Environ. Microbiol.* 44:619–626.

78. Furukawa, K., and Matsumura, F. 1976. Microbial metabolism of polychlorinated biphenyls. Studies on the relative degradability of polychlorinated biphenyl components by *Alcaligenes* sp. *J. Agric. Food Chem.* 24:252–256.

79. Furukawa, K., and Miyazaki, T. 1986. Cloning of a gene cluster encoding biphenyl and chlorobiphenyl degradation in *Pseudomonas pseudoalcaligenes. J. Bacteriol.* 166:392–398.

80. Furukawa, K., Hayase, N., Taira, K., and Tomizuka, N. 1989. Molecular relationship of chromosomal genes encoding biphenyl/polychlorinated biphenyl catabolism: Some soil bacteria possess a highly conserved *bph* operon. *J. Bacteriol.* 171:5467–5472.

81. Furukawa, K., Matsumura, F., and Tonomura, K. 1978. *Alcaligenes* and *Acinetobacter* strains capable of degrading polychlorinated biphenyls. *Agric. Biol. Chem.* 42:543–548.

82. Gamar, Y., and Gaunt, J. K. 1971. Bacterial metabolism of 4-chloro-2-methylphenoxyacetate (MCPA). Formation of glyoxylate by side-chain cleavage. *Biochem. J.* 122:527–531.

83. Gaunt, J. K., and Evans, W. C. 1961. Metabolism of 4-chloro-2-methylphenoxyacetic acid by a soil micro-organism. *Biochem. J.* 79:25P-26P.

84. Gaunt, J. K., and Evans, W. C. 1971. Metabolism of 4-chloro-2-methylphenoxyacetate by a soil pseudomonad. Preliminary evidence for the metabolic pathway. *Biochem. J.* 122:519–526.

85. Gaunt, J. K., and Evans, W. C. 1971. Metabolism of 4-chloro-2-methylphenoxyacetate by a soil pseudomonad. Ring-fission, lactonizing and delactonizing enzymes. *Biochem. J.* 122:533–542.

86. Genthner, B. R. S., Price, W. A., and Pritchard, P. H. 1989. Anaerobic degradation of chloroaromatic compounds in aquatic sediments under a variety of enrichment conditions. *Appl. Environ. Microbiol.* 55:1466–1471.

87. Ghosal, D., You, I.-S., Chatterjee, D. K., and Chakrabarty, A. M. 1985. Genes specifying degradation of 3-chlorobenzoic acid in plasmids pAC27 and pJP4. *Proc. Natl. Acad. Sci. USA.* 82:1638–1642.

88. Gibson, S. A., and Suflita, J. M. 1986. Extrapolation of biodegradation results to groundwater aquifers: reductive dehalogenation of aromatic compounds. *Appl. Environ. Microbiol.* 52:681–688.

89. Goldman, P., Milne, G. W. A., and Pignataro, M. T. 1967. Fluorine containing metabolites formed from 2-fluorobenzoic acid by *Pseudomonas* species. *Arch. Biochem. Biophys.* 118:178–184.

90. Greer, C. W., Havari, J., and Samson, R. 1990. Influence of environmental factors on 2,4-dichlorophenoxyacetic acid degradation by *Pseudomonas cepacia* isolated from peat. *Arch. Microbiol.* 154:317–322.

91. Grishchenkov, V. G., Fedechkina, I. E., Baskunov, B. P., Anisimova, L. A., Boronin, A. M., and Golovleva, L. A. 1983. Degradation of 3-chlorobenzoic acid by a *Pseudomonas putida* strain. *Mikrobiologiya* 52:771–776. (Russian)

91a. Groenewegen, P. E. J., van den Tweel, W. J. J., and de Bont, J. A. M 1992. Anaerobic bioformation of 4-hydroxybenzoate from 4-chlorobenzoate by the coryneform bacterium NTB-1. *Appl. Microbiol. Biotechnol.* 36:541–547.

92. Grund, A. D., and Gunsalus, I. C. 1983. Cloning of genes for naphthalene metabolism in *Pseudomonas putida. J. Bacteriol.* 156:89–94.

93. Häggblom, M. M., Nohynek, L. J., and Salkinoja-Salonen, M. S. 1988. Degradation and O-methylation of polychlorinated phenolic compounds by strains of *Rhodococcus* and *Mycobacterium. Appl. Environ. Microbiol.* 54:3043–3052.

94. Hager, L. P. 1982. Mother nature likes some halogenated compounds. In *Genetic Engineering of Microorganisms for Chemicals.* Ed. A. Hollaender, R. D. De Moss, S. Kaplan, J. Konisky, D. Savage, R. S. Wolfe. New York: Plenum Press. 415–429.

95. Haigler, B. E., Nishino, S. F., and Spain, J. C. 1988. Degradation of 1,2-dichlorobenzene by a *Pseudomonas* sp. *Appl. Environ. Microbiol.* 54:294–301.

96. Haigler, B. E., and Spain, J. C. 1989. Degradation of p-chlorotoluene by a mutant of

Pseudomonas sp. strain JS6. *Appl. Environ. Microbiol.* 55:372–379.

97. Harayama, S., Lehrbach, P. R., and Timmis, K. N. 1984. Transposon mutagenesis analysis of *meta*-cleavage pathway operon genes of the TOL plasmid of *Pseudomonas putida* mt-2. *J. Bacteriol.* 160:251–255.

98. Harayama, S., Rekik, M., Wubbolts, M., Rose, K., Leppik, R. A., and Timmis, K. N. 1989. Characterization of five genes in the upper-pathway operon of TOL plasmid pWWO from *Pseudomonas putida* and identification of gene products. *J. Bacteriol.* 171:5048–5055.

99. Harker, A. R., Olsen, R. H., and Seidler, R. J. 1989. Phenoxyacetic acid degradation by the 2,4-dichlorophenoxyacetic acid (TFD) pathway of plasmid pJP4: Mapping and characterization of the TFD regulatory gene, *tfdR*. *J. Bacteriol.* 171:314–320.

100. Hartmann, J., Engelberts, K., Nordhaus, B., Schmidt, E., and Reineke, W. 1989. Degradation of 2-chlorobenzoate by in vivo constructed hybrid pseudomonads. *FEMS Microbiol. Lett.* 61:17–22.

101. Hartmann, J., Reineke, W., and Knackmuss, H.-J. 1979. Metabolism of 3-chloro-, 4-chloro-, and 3,5-dichlorobenzoate by a pseudomonad. *Appl. Environ. Microbiol.* 37:421–428.

102. Havel, J., and Reineke, W. 1991. Total degradation of various chlorobiphenyls by cocultures and in vivo constructed pseudomonads. *FEMS Microbiol. Lett.* 78:163–170.

103. Hayase, N., Taira, K., and Furukawa, K. 1990. *Pseudomonas putida* KF715 *bph ABCD* operon encoding biphenyl and polychlorinated biphenyl degradation: Cloning, analysis, and expression in soil bacteria. *J. Bacteriol.* 172:1160–1164.

104. Helling, C. S., Bollag, J.-M., and Dawson, J. E. 1968. Cleavage of the ether-bond in phenoxyacetic acid by an *Arthrobacter* sp. *J. Agric. Food Chem.* 16:538–539.

105. Hickey, W. J., and Focht, D. D. 1990. Degradation of mono-, di-, and trihalogenated benzoic acids by *Pseudomonas aeruginosa* JB2. *Appl. Environ. Microbiol.* 56:3842–3850.

106. Higson, F. K., and Focht, D. D. 1990. Degradation of 2-bromobenzoic acid by a strain of *Pseudomonas aeruginosa*. *Appl. Environ. Microbiol.* 56:1615–1619.

107. Inouye, S., Nakazawa, A., and Nakazawa, T. 1981. Molecular cloning of TOL genes *xylB* and *xylE* in *Escherichia coli*. *J. Bacteriol.* 145:1137–1143.

108. Inouye, S., Nakazawa, A., and Nakazawa, T. 1981. Molecular cloning of gene *xylS* of the TOL plasmid: evidence for positive regulation of the *xylDEGF* operon by *xylS*. *J. Bacteriol.* 148:413–418.

109. Inouye, S., Nakazawa, A., and Nakazawa, T. 1983. Molecular cloning of regulatory gene *xylR* and operator-promoter regions of the *xylABC* and *xylDEGF* operons of the TOL plasmid. *J. Bacteriol.* 155:1192–1199.

110. Irie, S., Doi, S., Yorifuji, T., Takagi, M., and Yano, K. 1987. Nucleotide sequencing and characterization of genes encoding benzene oxidation enzymes of *Pseudomonas putida*. *J. Bacteriol.* 169:5174–5179.

111. Irie, S., Shirai, K., Doi, S., and Yorifuji, T. 1987. Cloning of genes encoding oxidation of benzene in *Pseudomonas putida* and their expression in *Escherichia coli* and *Pseudomonas putida*. *Agric. Biol. Chem.* 51:1489–1493.

112. Jeenes, D. J., Reineke, W., Knackmuss, H.-J., and Williams, P. A. 1982. TOL plasmid pWWO in constructed halobenzoate-degrading *Pseudomonas* strains: Enzyme regulation and DNA structure. *J. Bacteriol.* 150:180–187.

113. Jensen, H. L., and Petersen, H. I. 1952. Detoxication of hormone herbicides by soil bacteria. *Nature* 170:39–40.

114. Johnston, H. W., Briggs, G. G., and Alexander, M. 1972. Metabolism of 3-chlorobenzoic acid by a pseudomonad. *Soil Biol. Biochem.* 4:187–190.

115. Kaphammer, B., Kukor, J. J., and Olsen, R. H. 1990. Regulation of *tfdCDEF* by *tfdR* of the 2,4-dichlorophenoxyacetic acid degradation plasmid pJP4. *J. Bacteriol.* 172:2280–2286.

116. Karasevich, Y. N., and Zaitsev, G. M. 1984. Utilization of 4-chlorobenzoic and 2,4-dichlorobenzoic acids by a mixed culture of microorganisms. *Mikrobiologiya* 53:373–380. (Russian)

117. Kaschabek, S. R. 1990. *Untersuchungen zur Dechlorierung von Chlormaleylacetaten in* Pseudomonas *sp. Stamm B13.* Diploma Thesis, University of Wuppertal.

117a. Kaschabek, S., and Reineke, W. 1992. Maleylacetate reductase of *Pseudomonas* sp. strain B13: dechlorination of chloromaleylacetates, metabolites in the degradation of chloroaromatic compounds. *Arch. Microbiol.* 159:412–417.

118. Kellogg, S. T., Chatterjee, D. K., and Chakrabarty, A. M. 1981. Plasmid-assisted molecular breeding: New technique for enhanced biodegradation of persistent toxic chemicals. *Science* 214:1133–1135.

119. Khan, A., Tewari, R., and Walia, S. 1988. Molecular cloning of 3-phenylcatechol dioxygenase involved in the catabolic pathway of chlorinated biphenyl from *Pseudomonas putida* and its expression in *Escherichia coli. Appl. Environ. Microbiol.* 54:2664–2671.

120. Khan, A., and Walia, S. 1989. Cloning of bacterial genes specifying degradation of 4-chlorobiphenyl from *Pseudomonas putida* OU83. *Appl. Environ. Microbiol.* 55:798–805.

121. Kilpi, S. 1980. Degradation of some phenoxy acid herbicides by mixed cultures of bacteria from soil treated with 2-(2-methyl-4-chloro)phenoxypropionic acid. *Microb. Ecol.* 6:261–270.

122. Kilpi, S., Backström, V., and Korhola, M. 1980. Degradation of 2-methyl-4-chlorophenoxyacetic acid (MCPA), 2,4-dichlorophenoxyacetic acid (2,4-D), benzoic acid and salicylic acid by *Pseudomonas* sp. HV3. *FEMS Microbiol. Lett.* 8:177–182.

123. Kimbara, K., Hashimoto, T., Fukuda, M., Koana, T., Takagi, M., Oishi, M., and Yano, K. 1989. Cloning and sequencing of two tandem genes involved in degradation of 2,3-dihydroxybiphenyl to benzoic acid in the polychlorinated biphenyl-degrading soil bacterium *Pseudomonas* sp. strain KKS102. *J. Bacteriol.* 171:2740–2747.

124. Kivisaar, M., Horak, R., Kasak, L., Heinaru, A., and Habicht, J. 1990. Selection of independent plasmids determining phenol degradation in *Pseudomonas putida* and the cloning and expression of genes encoding phenol monooxygenase and catechol 1,2-dioxygenase. *Plasmid* 24:25–36.

125. Klages, U., and Lingens, F. 1979. Degradation of 4-chlorobenzoic acid by *Nocardia* species. *FEMS Microbiol. Lett.* 6:201–203.

126. Klages, U., and Lingens, F. 1980. Degradation of 4-chlorobenzoic acid by a *Pseudomonas* sp. *Zbl. Bakt. Hyg. I. Abt. Orig. C* 1:215–223.

127. Klages, U., Markus, A., and Lingens, F. 1981. Degradation of 4-chlorophenylacetic acid by a *Pseudomonas* species. *J. Bacteriol.* 146:64–68.

128. Klecka, G. M., and Gibson, D. T. 1981. Inhibition of catechol 2,3-dioxygenase from *Pseudomonas putida* by 3-chlorocatechol. *Appl. Environ. Microbiol.* 41:1159–1165.

129. Köcher, H., Lingens, F., and Koch, W. 1976. Untersuchungen zum Abbau des Herbizids Chlorphenpropmethyl im Boden und durch Mikroorganismen. *Weed Res.* 16:93–100.

130. Kohring, G.-W., Rogers, J. E., and Wiegel, J. 1989. Anaerobic biodegradation of 2,4-dichlorophenol in freshwater lake sediments at different temperatures. *Appl. Environ. Microbiol.* 55:348–353.

131. Kröckel, L., and Focht, D. D. 1987. Construction of chlorobenzene-utilizing recombinants by progenitive manifestation of a rare event. *Appl. Environ. Microbiol.* 53:2470–2475.

132. Kuhn, E. P., and Suflita, J. M. 1989. Sequential reductive dehalogenation of chloroanilines by microorganisms from a methanogenic aquifer. *Environ. Sci. Technol.* 23:848–852.

133. Kuhn, E. P., Townsend, G. T., and Suflita, J. M. 1990. Effect of sulfate and organic carbon supplements on reductive dehalogenation of chloroanilines in anaerobic aquifer slurries. *Appl. Environ. Microbiol.* 56:2630–2637.

134. Kukor, J. J., and Olsen, R. H. 1990. Molecular cloning, characterization, and regulation of a *Pseudomonas picketti* PKO1 gene encoding phenol hydroxylase and expression of the gene in *Pseudomonas aeruginosa* PAO1c. *J. Bacteriol.* 172:4624–4630.

135. Kukor, J. J., Olsen, R. H., and Siak, J.-S. 1989. Recruitment of chromosomally encoded

maleylacetate reductase for degradation of 2,4-dichlorophenoxyacetic acid by plasmid pJP4. *J. Bacteriol.* 171:3385–3390.

136. Latorre, J., Reineke, W., and Knackmuss, H.-J. 1984. Microbial metabolism of chloroanilines: enhanced evolution by natural genetic exchange. *Arch. Microbiol.* 140:159–165.

137. Lehrbach, P. R., Jeenes, D. J., and Broda, P. 1983. Characterization by molecular cloning of insertion mutants in TOL catabolic functions. *Plasmid* 9:112–125.

138. Lehrbach, P. R., McGregor, I., Ward, J. M., and Broda, P. 1983. Molecular relationships between *Pseudomonas* Inc P-9 degradative plasmids TOL, NAH, and SAL. *Plasmid* 10:164–174.

139. Lehrbach, P. R., Ward, J. M., Meulien, P., and Broda, P. 1982. Physical mapping of TOL plasmids pWWO and pND2 and various R plasmid-TOL derivatives from *Pseudomonas* spp. *J. Bacteriol.* 152:1280–1283.

140. Lehrbach, P. R., Zeyer, J., Reineke, W., Knackmuss, H.-J., and Timmis, K. N. 1984. Enzyme recruitment in vitro: use of cloned genes to extend the range of haloaromatics degraded by *Pseudomonas* sp. strain B13. *J. Bacteriol.* 158:1025–1032.

141. Liu, T., and Chapman, P. J. 1983. Degradation of halogenated aromatic acids and hydrocarbons by *Pseudomonas putida*. *Abstr. Annu. Meet. Am. Soc. Microbiol.* K211, p. 212.

142a. Löffler, F., and Müller, R. 1991. Identification of a 4-chlorobenzoyl-coenzyme A as intermediate in the dehalogenation catalyzed by 4-chlorobenzoate dehalogenase from *Pseudomonas* sp. CBS3. *FEBS Lett.* 290:224–226.

141b. Löffler, F., Müller, R., and Lingens, F. 1991. Dehalogenation of 4-chlorobenzoate by 4-chlorobenzoate dehalogenase from *Pseudomonas* sp. CBS3: An ATP/coenzyme A dependent reaction. *Biochem Biophys. Res. Commun.* 176:1106–1111.

142. Loidl, M., Hinteregger, C., Ditzelmüller, G., Ferschl, A., and Streichsbier, F. 1990. Degradation of aniline and monochlorinated anilines by soil-born *Pseudomonas acidovorans* strains. *Arch. Microbiol.* 155:56–61.

143. Loos, M. A., Roberts, R. N., and Alexander, M. 1967. Phenols as intermediates in the decomposition of phenoxyacetates by an *Arthrobacter* species. *Can. J. Microbiol.* 13:679–690.

144. Loos, M. A., Roberts, R. N., and Alexander, M. 1967. Formation of 2,4-dichlorophenol and 2,4-dichloroanisole from 2,4-dichlorophenoxyacetate by *Arthrobacter* sp. *Can. J. Microbiol.* 13:691–699.

145. Loos, M. A., Bollag, J.-M., and Alexander, M. 1967. Phenoxyacetate herbicide detoxication by bacterial enzymes. *J. Agric. Food Chem.* 15:858–860.

146. Marks, T. S., Smith, A. R. W., and Quirk, A. V. 1984. Degradation of 4-chlorobenzoic acid by *Arthrobacter* sp. *Appl. Environ. Microbiol.* 48:1020–1025.

147. Marks, T. S., Wait, R., Smith, A. R. W., and Quirk, A. V. 1984. The origin of oxygen incorporated during the dehalogenation/hydroxylation of 4-chlorobenzoate by an *Arthrobacter* sp. *Biochem. Biophys. Res. Commun.* 124:669–674.

148. Markus, A., Klages, U., Krauss, S., and Lingens, F. 1984. Oxidation and dehalogenation of 4-chlorophenylacetate by a two-component enzyme system from *Pseudomonas* sp. strain CBS3. *J. Bacteriol.* 160:618–621.

149. Massé, R., Messier, F., Péloquin, L., Ayotte, C., and Sylvestre, M. 1984. Microbial degradation of 4-chlorobiphenyl, a model compound of chlorinated biphenyls. *Appl. Environ. Microbiol.* 47:947–951.

150. Miguez, C. B., Greer, C. W., and Ingram, J. M. 1990. Degradation of mono- and dichlorobenzoic acid isomers by two natural isolates of *Alcaligenes denitrificans*. *Arch. Microbiol.* 154:139–143.

151. Mikesell, M. D., and Boyd, S. A. 1985. Reductive dechlorination of the pesticides 2,4-D, 2,4,5-T, and pentachlorophenol in anaerobic sludge. *J. Environ. Qual.* 14:337–340.

152. Mikesell, M. D., and Boyd, S. A. 1986. Complete reductive dechlorination and mineralization of pentachlorophenol by anaerobic microorganisms. *Appl. Environ. Microbiol.* 52:861–865.

153. Milne, G. W. A., Goldman, P., and Holtzman, J. L. 1968. The metabolism of 2-fluorobenzoic acid. II. Studies with $^{18}O_2$. J. Biol. Chem. 243:5374–5376.

154. Mokross, H., Schmidt, E., and Reineke, W. 1990. Degradation of 3-chlorobiphenyl by in vivo constructed hybrid pseudomonads. FEMS Microbiol. Lett. 71:179–186.

155. Mondello, F. J. 1989. Cloning and expression in Escherichia coli of Pseudomonas strain LB400 genes encoding polychlorinated biphenyl degradation. J. Bacteriol. 171:1725–1732.

156. Morris, C. M., and Barnsley, E. A. 1982. The cometabolism of 1- and 2-chloronaphthalene by pseudomonads. Can. J. Microbiol. 28:73–79.

157. Müller, R., Thiele, J., Klages, U., and Lingens, F. 1984. Incorporation of [^{18}O]water into 4-hydroxybenzoic acid in the reaction of 4-chlorobenzoate dehalogenase from Pseudomonas spec. CBS3. Biochem. Biophys. Res. Commun. 124:178–182.

158. Neilson, A. H., Allard, A.-S., Hynning, P.-Å., and Remberger, M. 1988. Transformations of halogenated aromatic aldehydes by metabolically stable anaerobic enrichment cultures. Appl. Environ. Microbiol. 54:2226–2236.

159. Nies, L., and Vogel, T. M. 1990. Effects of organic substrates on dechlorination of Aroclor 1242 in anaerobic sediments. Appl. Environ. Microbiol. 56:2612–2617.

160. Ohmori, T., Ikai, T., Minoda, Y., and Yamada, K. 1973. Utilization of polyphenyl and polyphenyl-related compounds by microorganisms. Part I. Agric. Biol. Chem. 37:1599–1605.

161. Oldenhuis, R., Kuijk, L., Lammers, A., Janssen, D. B., and Witholt, B. 1989. Degradation of chlorinated and non-chlorinated aromatic solvents in soil suspensions by pure bacterial cultures. Appl. Microbiol. Biotechnol. 30:211–217.

162. Oltmanns, R. H., Rast, H. G., and Reineke, W. 1988. Degradation of 1,4-dichlorobenzene by enriched and constructed bacteria. Appl. Microbiol. Biotechnol. 28:609–616.

163. Parson, J. R., Sijm, D. T. H. M., van Laar, A., and Hutzinger, O. 1988. Biodegradation of chlorinated biphenyls and benzoic acids by a Pseudomonas strain. Appl. Microbiol. Biotechnol. 29:81–84.

164. Quensen, J. F. III, Tiedje, J. M., and Boyd, S. A. 1988. Reductive dechlorination of polychlorinated biphenyls by anaerobic microorganisms from sediments. Science 242:752–754.

165. Quensen, J. F. III, Boyd, S. A., and Tiedje, J. M. 1990. Dechlorination of four commercial polychlorinated biphenyl mixtures (Aroclors) by anaerobic microorganisms from sediments. Appl. Environ. Microbiol. 56:2360–2369.

166. Ramos, J. L., Michan, C., Rojo, F., Dwyer, D., and Timmis, K. N. 1990. Signal-regulator interactions. Genetic analysis of the effector binding site of xylS, the benzoate-activated positive regulator of Pseudomonas TOL plasmid meta-cleavage pathway operon. J. Mol. Biol. 211:373–382.

167. Ramos, J. L., Stolz, A., Reineke, W., and Timmis, K. N. 1986. New effector specificities in regulators of gene expression: TOL plasmid xylS mutants and their use to engineer expansion of the range of aromatics degraded by bacteria. Proc. Natl. Acad. Sci. USA 83:8467–8471.

168. Reineke, W. 1986. Construction of bacterial strains with novel degradative capability for chloroaromatics. J. Basic Microbiol. 26:551–567.

169. Reineke, W., and Knackmuss, H.-J. 1978. Chemical structure and biodegradability of halogenated aromatic compounds. Substituent effects on 1,2-dioxygenation of benzoic acid. Biochim. Biophys. Acta. 542:412–423.

170. Reineke, W., and Knackmuss, H.-J. 1979. Construction of haloaromatics utilising bacteria. Nature 277:385–386.

171. Reineke, W., and Knackmuss, H.-J. 1980. Hybrid pathway for chlorobenzoate metabolism in Pseudomonas sp. B13 derivatives. J. Bacteriol. 142:467–473.

172. Reineke, W., and Knackmuss, H.-J. 1984. Microbial metabolism of haloaromatics: isolation and properties of a chlorobenzene-degrading bacterium. Appl. Environ. Microbiol. 47:395–402.

173. Reineke, W., Jeenes, D. J., Williams, P. A., and Knackmuss, H.-J. 1982. TOL plasmid

pWWO in constructed halobenzoate-degrading *Pseudomonas* strains: Prevention of *meta* pathway. *J. Bacteriol.* 150:195–201.

174. Reineke, W., Wessels, S. W., Rubio, M. A., Latorre, J., Schwien, U., Schmidt, E., Schlömann, M., and Knackmuss, H.-J. 1982. Degradation of monochlorinated aromatics following transfer of genes encoding chlorocatechol catabolism. *FEMS Microbiol. Lett.* 14:291–294.

175. Reiner, E. A., Chu, J., and Kirsch, E. J. 1978. Microbial metabolism of pentachlorophenol. In *Pentachlorophenol: Chemistry, pharmacology and environmental toxicology.* Ed. K. R. Rao. New York: Plenum Press. 67–81.

176. Renganathan, V. 1989. Possible involvement of toluene-2,3-dioxygenase in defluorination of 3-fluoro-substituted benzenes by toluene-degrading *Pseudomonas* sp. strain T-12. *Appl. Environ. Microbiol.* 55:330–334.

177. Rogoff, M. H., and Reid, J. J. 1956. Bacterial decomposition of 2,4-dichlorophenoxyacetic acid. *J. Bacteriol.* 71:303–307.

178. Ruisinger, S., Klages, U., and Lingens, F. 1976. Abbau der 4-Chlorbenzoesäure durch eine *Arthrobacter*-Species. *Arch. Microbiol.* 110:253–256.

179. Saber, D. L., and Crawford, R. L. 1985. Isolation and characterization of *Flavobacterium* strains that degrade pentachlorophenol. *Appl. Environ. Microbiol.* 50:1512–1518.

180. Sander P., Wittich, R.-M., Fortnagel, P., Wilkes, H., Francke W. 1991. Degradation of 1,2,4-trichloro- and 1,2,4,5-tetrachlorobenzene by *Pseudomonas* strains. *Appl. Environ. Microbiol.* 57:1430–1440.

181. Sandmann, E. R. I. C., and Loos, M. A. 1988. Aromatic metabolism by a 2,4-D degrading *Arthrobacter* sp. *Can. J. Microbiol.* 34:125–130.

182. Sangodkar, U. M. X., Chapman, P. J., and Chakrabarty, A. M. 1988. Cloning, physical mapping and expression of chromosomal genes specifying degradation of the herbicide 2,4,5-T by *Pseudomonas cepacia* AC1100. *Gene* 71:267–277.

183. Savard, P., Péloquin, L., and Sylvestre, M. 1986. Cloning of *Pseudomonas* sp. strain CBS3 gene specifying dehalogenation of 4-chlorobenzoate. *J. Bacteriol.* 168:81–85.

184. Schell, M. A. 1983. Cloning and expression in *Escherichia coli* of the naphthalene degradation genes from plasmid NAH7. *J. Bacteriol.* 153:822–829.

185. Schenk, T., Müller, R., and Lingens, F. 1990. Mechanism of enzymatic dehalogenation of pentachlorophenol by *Arthrobacter* sp. strain ATCC 33790. *J. Bacteriol.* 172:7272–7274.

186. Schmidt, E., and Knackmuss, H.-J. 1980. Chemical structure and biodegradability of halogenated aromatic compounds. Conversion of chlorinated muconic acids into maleoylacetic acid. *Biochem. J.* 192:339–347.

187. Schmidt, E., Remberg, G., and Knackmuss, H.-J. 1980. Chemical structure and biodegradability of halogenated aromatic compounds. Halogenated muconic acids as intermediates. *Biochem. J.* 192:331–337.

187a. Scholten, J. D., Chang, K.-H., Babbitt, P. C., Charest, H., Sylvestre, M., and Dunaway-Mariano, D. 1991. Novel enzymatic hydrolytic dehalogenation of a chlorinated aromatic. *Science* 253:182–185.

188. Schraa, G., Boone, M. L., Jetten, M. S. M., van Neerven, A. R. W., Colberg, P. J., and Zehnder, A. J. B. 1986. Degradation of 1,4-dichlorobenzene by *Alcaligenes* sp. strain A175. *Appl. Environ. Microbiol.* 52:1374–1381.

189. Schwien, U., Schmidt, E., Knackmuss, H.-J., and Reineke, W. 1988. Degradation of chlorosubstituted aromatic compounds by *Pseudomonas* sp. strain B13: fate of 3,5-dichlorocatechol. *Arch. Microbiol.* 150:78–84.

190. Schwien, U., and Schmidt, E. 1982. Improved degradation of monochlorophenols by a constructed strain. *Appl. Environ. Microbiol.* 44:33–39.

191. Senior, E., Bull, A. T., and Slater, J. H. 1976. Enzyme evolution in a microbial community growing on the herbicide Dalapon. *Nature* 263:476–479.

192. Serdar, C.M, and Gibson, D. T. 1989. Isolation and characterization of altered plasmids in mutant strains of *Pseudomonas putida* NCIB 9816. *Biochem. Biophys. Res. Commun.* 164:764–771.

193. Serdar, C.M, and Gibson, D. T. 1989. Studies of nucleotide sequence homology between naphthalene-utilizing strains of bacteria. *Biochem. Biophys. Res. Commun.* 164:772–779.

194. Sharpee, K. W., Duxbury, J. M., and Alexander, M. 1973. 2,4-Dichlorophenoxyacetate metabolism by *Arthrobacter* sp.: Accumulation of a chlorobutenolide. *Appl. Microbiol.* 26:445–447.

195. Shelton, D. R., and Tiedje, J. M. 1984. Isolation and partial characterization of bacteria in an anaerobic consortium that mineralizes 3-chlorobenzoic acid. *Appl. Environ. Microbiol.* 48:840–848.

196. Shields, M. S., Hooper, S. W., Sayler, G. S. 1985. Plasmid-mediated mineralization of 4-chlorobiphenyl. *J. Bacteriol.* 163:882–889.

197. Shimao, M., Onishi, S., Mizumori, S., Kato, N., and Sakazawa, C. 1989. Degradation of 4-chlorobenzoate by facultatively alkalophilic *Arthrobacter* sp. strain SB8. *Appl. Environ. Microbiol.* 55:478–482.

198. Shingler, V., Franklin, F. C. H., Tsuda, M., Holroyd, D., and Bagdasarian, M. 1989. Molecular analysis of a plasmid-encoded phenol hydroxylase from *Pseudomonas* CF600. *J. Gen. Microbiol.* 135:1083–1092.

199. Singh, H., and Kahlon, R. S. 1989. Conjugative plasmid coding for metabolism of 2-chlorobenzoic acid by *Pseudomonas aeruginosa*. *MICERN J.* 5:255–258.

200. Spain, J. C., and Nishino, S. F. 1987. Degradation of 1,4-dichlorobenzene by a *Pseudomonas* sp. *Appl. Environ. Microbiol.* 53:1010–1019.

201. Springer, W., and Rast, H. G. 1988. Biologischer Abbau mehrfach halogenierter mono- und polyzyklischer Aromaten. *GWF Wasser Abwasser* 129:70–75.

202. Stanlake, G. J., and Finn, R. K. 1982. Isolation and characterization of a pentachlorophenol-degrading bacterium. *Appl. Environ. Microbiol.* 44:1421–1427.

203. Steenson, T. I., and Walker, N. 1956. Oberservations on the decomposition of chlorophenoxyacetic acids by bacteria. *Plant Soil* 8:17–32.

204. Steenson, T. I., and Walker, N. 1957. The pathway of breakdown of 2,4-dichloro- and 4-chloro-2-methylphenoxyacetic acid by bacteria. *J. Gen. Microbiol.* 16:146–155.

205. Steenson, T. I., and Walker, N. 1958. Adaptive patterns in the bacterial oxidation of 2,4-dichloro- and 4-chloro-2-methylphenoxyacetic acid. *J. Gen. Microbiol.* 18:692–697.

206. Steiert, J. G., and Crawford, R. L. 1986. Catabolism of pentachlorophenol by a *Flavobacterium* sp. *Biochem. Biophys. Res. Commun.* 141:825–830.

207. Steiert, J. G., Pignatello, J. J., and Crawford, R. L. 1987. Degradation of chlorinated phenols by a pentachlorophenol-degrading bacterium. *Appl. Environ. Microbiol.* 53:907–910.

208. Stephens, G. M., Sidebotham, J. M., Mann, N. H., and Dalton, H. 1989. Cloning and expression in *Escherichia coli* of the toluene dioxygenase gene from *Pseudomonas putida* NCIB11767. *FEMS Microbiol. Lett.* 57:295–300.

209. Stevens, T. O., Linkfield, T. G., and Tiedje, J. M. 1988. Physiological characterization of strain DCB-1, a unique dehalogenating sulfidogenic bacterium. *Appl. Environ. Microbiol.* 54:2938–2943.

210. Stolz, A. 1984. *Abbau von 3,5-Dichlorbenzoat durch Hybridstämme*. Diploma Thesis, University of Göttingen.

211. Streber, W. R., Timmis, K. N., and Zenk, M. H. 1987. Analysis, cloning, and high-level expression of 2,4-dichlorophenoxyacetate monooxygenase gene *tfdA* of *Alcaligenes eutrophus* JMP134. *J. Bacteriol.* 169:2950–2955.

212. Struijs, J., and Rogers, J. E. 1989. Reductive dehalogenation of dichloroanilines by

anaerobic microorganisms in fresh and dichlorophenol-acclimated pond sediment. *Appl. Environ. Microbiol.* 55:2527–2531.

213. Suflita, J. M., Horowitz, A., Shelton, D. R., and Tiedje, J. M. 1982. Dehalogenation: A novel pathway for the anaerobic biodegradation of haloaromatic compounds. *Science* 218:1115–1117.

214. Suflita, J. M., Stout, J., and Tiedje, J. M. 1984. Dechlorination of (2,4,5-trichlorophenoxy)acetic acid by anaerobic microorganisms. *J. Agric. Food Chem.* 32:218–221.

215. Suida, J. F., and De Bernardis, J. F. 1973. Naturally occurring halogenated organic compounds. *Lloydia* 36:107–143.

216. Surovtseva, E. G., Ivoilov, V. S., Karasevich, Y. N., and Vacil'eva, G. K. 1985. Chlorinated anilines, a source of carbon, nitrogen, and energy for *Pseudomonas diminuta*. *Mikrobiologiya* 54:948–952. (Russian)

217. Surovtseva, E. G., Ivoilov, V. S., and Karasevich, Y. N. 1986. Metabolism of chlorinated anilines by *Pseudomonas diminuta*. *Mikrobiologiya* 55:591–595. (Russian)

218. Suzuki, T. 1977. Metabolism of pentachlorophenol by a soil microbe. *J. Environ. Sci. Health* B12:113–127.

219. Sylvestre, M. 1980. Isolation method for bacterial isolates capable of growth on p-chlorobiphenyl. *Appl. Environ. Microbiol.* 39:1223–1224.

220. Sylvestre, M., and Fauteux, J. 1982. A new facultative anaerobe capable of growth on chlorobiphenyls. *J. Gen. Appl. Microbiol.* 28:61–72.

221. Sylvestre, M., Mailhiot, K., Ahmad, D., and Massé, R. 1989. Isolation and preliminary characterization of a 2-chlorobenzoate degrading *Pseudomonas*. *Can. J. Microbiol.* 35:439–443.

222. Taira, K., Hayase, N., Arimura, N., Yamashita, S., Miyazaki, T., and Furukawa, K. 1988. Cloning and nucleotide sequence of 2,3-dihydroxybiphenyl dioxygenase gene from the PCB-degrading strain *Pseudomonas paucimobilis* Q1. *Biochemistry* 27:3990–3996.

223. Takase, I., Omori, T., and Minoda, Y. 1986. Microbial degradation products from biphenyl-related compounds. *Agric. Biol. Chem.* 50:681–686.

224. Tan, H.-M., and Mason, J. R. 1990. Cloning and expression of the plasmid-encoded benzene dioxygenase genes from *Pseudomonas putida* ML2. *FEMS Microbiol. Lett.* 72:259–264.

225. Tiedje, J. M., and Alexander, M. 1969. Enzymatic cleavage of the ether bond of 2,4-dichlorophenoxyacetate. *J. Agric. Food Chem.* 17:1080–1084.

226. Tiedje, J. M., Duxbury, J. M., Alexander, M., and Dawson, J. E. 1969. 2,4-D metabolism: pathway of degradation of chlorocatechols by *Arthrobacter* sp. *J. Agric. Food Chem.* 17:1021–1026.

227. Tomasek, P. H., Frantz, B., Sangodkar, U. M. X., Haugland, R. A., and Chakrabarty, A. M. 1989. Characterization and nucleotide sequence determination of a repeat element isolated from a 2,4,5-T degrading strain of *Pseudomonas cepacia*. *Gene* 76:227–238.

228. Tsuchiya, T., and Yamaha, T. 1983. Reductive dechlorination of 1,2,4-trichlorobenzene on incubation with intestinal contents of rats. *Agric. Biol. Chem.* 47:1163–1165.

229. Tsuchiya, T., and Yamaha, T. 1984. Reductive dechlorination of 1,2,4-trichlorobenzene by *Staphylococcus epidermidis* isolated from intestinal contents of rats. *Agric. Biol. Chem.* 48:1545–1550.

230. Tyler, J. E., and Finn, R. K. 1974. Growth rates of a pseudomonad on 2,4-dichlorophenoxyacetic acid and 2,4-dichlorophenol. *Appl. Microbiol.* 28:181–184.

231. Vandenbergh, P. A., Olsen, R. H., and Colaruotolo, J. F. 1981. Isolation and genetic characterization of bacteria that degrade chloroaromatic compounds. *Appl. Environ. Microbiol.* 42:737–739.

232. Van den Tweel, W. J. J., Ter Burg, N., Kok, J. B., and de Bont, J. A. M. 1986. Bioformation of 4-hydroxybenzoate from 4-chlorobenzoate by *Alcaligenes denitrificans* NTB-1. *Appl. Microbiol. Biotechnol.* 25:289–294.

233. Van den Tweel, W. J. J., Kok, J. B., and de Bont, J. A. M. 1987. Reductive dechlorination of 2,4-dichlorobenzoate to 4-chlorobenzoate and hydrolytic dehalogenation of 4-chloro-, 4-bromo-, and 4-iodobenzoate by *Alcaligenes denitrificans* NTB-1. *Appl. Environ. Microbiol.* 53:810–815.

234. Van der Meer, J. R., Roelofsen, W., Schraa, G., and Zehnder, A. J. B. 1987. Degradation of low concentrations of dichlorobenzenes and 1,2,4-trichlorobenzene by *Pseudomonas* sp. strain P51 in nonsterile soil columns. *FEMS Microbiol. Ecol.* 45:333–341.

235. Van der Meer, J. R., van Neerven, A. R. W., de Vries, E. J., de Vos, W. M., and Zehnder, A. J. B. 1991. Cloning and characterization of plasmid-encoded genes for the degradation of 1,2-dichloro-, 1,4-dichloro-, and 1,2,4-trichlorobenzene of *Pseudomonas* sp. strain P51. *J. Bacteriol.* 173:6–15.

236. Vora, K. A., Singh, C., and Modi, V. V. 1988. Degradation of 2-fluorobenzoate by a pseudomonad. *Curr. Microbiol.* 17:249–254.

237. Walker, R. L., and Newman, A. S. 1956. Microbial decomposition of 2,4-dichlorophenoxy-acetic acid. *Appl. Microbiol.* 4:201–206.

238. Watanabe, I. 1973. Isolation of pentachlorophenol-decomposing bacteria from soil. *Soil Sci. Plant Nutr.* 19:109–116.

239. Weisshaar, M.-P., Franklin, F. C. H., and Reineke, W. 1987. Molecular cloning and expression of the 3-chlorobenzoate-degrading genes from *Pseudomonas* sp. strain B13. *J. Bacteriol.* 169:394–402.

240. Williams, P. A., and Murray, K. 1974. Metabolism of benzoate and methylbenzoates by *Pseudomonas putida* (arvilla) mt-2: Evidence for the existence of a TOL plasmid. *J. Bacteriol.* 120:416–423.

241. Winstanley, C., Taylor, S. C., and Williams, P. A. 1987. pWW174: A large plasmid from *Acinetobacter calcoaceticus* encoding benzene catabolism by the β-ketoadipate pathway. *Molec. Microbiol.* 1:219–227.

242. Woods, S. L., Ferguson, J. F., and Benjamin, M. M. 1989. Characterization of chloro-phenol and chloromethoxybenzene biodegradation during anaerobic treatment. *Environ. Sci. Technol.* 23:62–68.

243. Worsey, M. J., Franklin, F. C. H., and Williams, P. A. 1978. Regulation of the degradative pathway enzymes coded for by the TOL plasmid (pWWO) from *Pseudomonas putida* mt-2. *J. Bacteriol.* 134:757–764.

244. Wyndham, R. C., and Straus, N. A. 1988. Chlorobenzoate catabolism and interactions between *Alcaligenes* and *Pseudomonas* species from Bloody Run Creek. *Arch. Microbiol.* 150:230–236.

245. Wyndham, R. C., Singh, R. K., and Straus, N. A. 1988. Catabolic instability, plasmid gene deletion and recombination in *Alcaligenes* sp. BR60. *Arch. Microbiol.* 150:237–243.

246. Yen, K.-M., and Gunsalus, I. C. 1983. Plasmid gene organization: naphthalene/salicylate oxidation. *Proc. Natl. Acad. Sci. USA* 79:874–878.

247. Yen, K.-M., Sullivan, M., and Gunsalus, I. C. 1983. Electron microscope heteroduplex mapping of naphthalene oxidation genes on the NAH7 and SAL1 plasmids. *Plasmid* 9:105–111.

248. You, I.-S., and Bartha, R. 1982. Metabolism of 3,4-dichloroaniline by *Pseudomonas putida*. *J. Agric. Food Chem.* 30:274–277.

249. Zaitsev, G. M., and Baskunov, B. P. 1985. Utilization of 3-chlorobenzoic acid by *Acinetobacter calcoaceticus*. *Mikrobiologiya* 54:203–208. (Russian)

250. Zaitsev, G. M., and Karasevich, Y. N. 1980. Utilization of 4-chlorobenzoic acid by *Arthrobacter globiformis*. *Mikrobiologiya* 50:35–40. (Russian)

251. Zaitsev, G. M., and Karasevich, Y. N. 1980. Preparative metabolism of 4-chlorobenzoic acid in *Arthrobacter globiformis*. *Mikrobiologiya* 50:423–428. (Russian)

252. Zaitsev, G. M., and Karasevich, Y. N. 1982. Utilization of 2-chlorobenzoic acid by *Pseudomonas cepacia*. *Mikrobiologiya* 53:75–80. (Russian)

253. Zaitsev, G. M., and Karasevich, Y. N. 1985. Preparative metabolism of 4-chlorobenzoic and 2,4-dichlorobenzoic acids in *Corynebacterium sepedonicum*. *Mikrobiologiya* 54:356–359. (Russian)

254. Zeyer, J., and Kearney, P. C. 1982. Microbial degradation of *para*-chloroaniline as sole carbon and nitrogen source. *Pest. Biochem. Physiol.* 17:215–223.

255. Zeyer, J., Wasserfallen, A., and Timmis, K. N. 1985. Microbial mineralization of ring-substituted anilines through an *ortho*-cleavage pathway. *Appl. Environ. Microbiol.* 50:447–453.

256. Zhang, X., and Wiegel, J. 1990. Sequential anaerobic degradation of 2,4-dichlorophenol in freshwater sediments. *Appl. Environ. Microbiol.* 56:1119–1127.

257. Zylstra, G. J., McCombie, W. R., Gibson, D. T., and Finette, B. A. 1988. Toluene degradation by *Pseudomonas putida* F1: genetic organization of the *tod* operon. *Appl. Environ. Microbiol.* 54:1498–1503.

CHAPTER 21

Hazardous Waste Cleanup and Treatment with Encapsulated or Entrapped Microorganisms

WILLIAM E. LEVINSON, KEITH E. STORMO, HONG-LEI TAO, and RONALD L. CRAWFORD

Abstract. The use of immobilized or entrapped microorganisms to degrade toxic chemicals in industrial process streams or in the environment is a rapidly evolving technology that shows great promise for the hazardous waste management industry. Here we describe the state of the art in this field, including encapsulation techniques and the application of entrapped microbial systems in bioreactors to treat chemical wastes, or *in situ* to decontaminate polluted soils and natural waters. Applications of encapsulated microorganisms for treatment of hazardous chemicals appear to be nearing the point of commercial exploitation. This is a technology whose time has arrived.

BIODEGRADATION OF ORGANIC COMPOUNDS BY IMMOBILIZED MICROORGANISMS IN BIOREACTORS

Historically, immobilized cells have been widely used in the wastewater treatment industry (19), generally through the use of undefined mixed cultures immobilized by natural flocculating tendencies or as films on solid surfaces (54). More recently, interest has centered on the use of other immobilization techniques and applications of immobilization to decontamination of a variety of more specific waste streams. Among these waste applications are nitrification and denitrification of wastewater, removal or biotransformation of specific organic compounds, and removal of dissolved heavy metals (3, 19, 54).

Degradation of phenols and chlorinated phenols by pure cultures of bacteria and yeast cells immobilized by entrapment in alginate and polyacrylamide gels or adsorbed to activated carbon has been investigated by Rehm and co-workers (9, 10, 25, 26, 45, 67, 68). They found that immobilization protected cells from high phenol concentra-

Mr. Levinson, Dr. Stormo, Mr. Tao, and Dr. Crawford are at the Center for Hazardous Waste Remediation Research, University of Idaho, Moscow, ID 83843, U.S.A. Mr. Tao may be reached at EnviroSearch, 608 Wilmington Avenue, Salt Lake City, UT 84106, U.S.A.

tions in both the gel and surface adsorption systems (9, 25). Cells immobilized in alginate also tolerated high 4-chlorophenol concentrations better than did free cells (67). In gels, this was attributed to the possible formation of a diffusional barrier set up by the cells within the beads or to some type of membrane stabilization brought about by entrapment. Protection in the activated carbon system was afforded by adsorption of the phenol onto the immobilization substrate (25), which reduced the aqueous concentration to which the organisms were exposed. As the phenol in solution was degraded, desorption occurred, allowing the organisms to metabolize the substrate.

Phenol, 4-chlorophenol, and cresols were degraded in a continuous system via polyacrylamide-entrapped *Pseudomonas putida* P8 (10). When municipal wastewater was used in the system instead of sterile buffer, the degradation rate was reduced by half. This was attributed to competition for nutrients with the native microflora, or to some type of regulatory effect brought about by the presence of other carbon sources. The same *Pseudomonas* strain was used to degrade phenol through use of cells immobilized on activated carbon in a continuous reactor (26). Phenol introduced into the reactor was initially removed from the media by a combination of degradation and adsorption. As the biomass in the reactor increased, adsorption decreased and the degradation rate increased. Morsen and Rehm (45) used a defined mixed culture (of two phenol-degrading organisms with different pH optima) adsorbed on activated carbon. Due to pH changes in a closed loop reactor during phenol degradation, the mixed culture system degraded the substrate faster than did pure cultures.

Alcaligenes sp. A 7-2 was immobilized by adsorption onto "Lecaton" (light-expanded clay aggregate) by Westmeier and Rehm (68). This biocatalyst was employed to degrade 4-chlorophenol in a continuous, packed-bed fermenter supplied with sterile wastewater. Under nonsterile conditions, the *Alcaligenes* was displaced by the naturally occurring population, which accumulated a dead end metabolic product of 4-chlorophenol that appeared to be 5-chloro-2-hydroxymuconic acid semialdehyde.

Beunink and Rehm (11) co-immobilized anaerobic and aerobic microbes to achieve degradation of the insecticide DDT [1,1,1-trichloro-2,2-bis(4-chlorophenyl)ethane]. *Enterobacter cloacae* reductively dechlorinated DDT to DDD [1,1-dichloro-2,2-bis(4-chlorophenyl)ethane] during lactose fermentation. This was subsequently metabolized to 4,4′-dichlorophenylmethane (DDM) under oxidative conditions. DDM was metabolized by an *Alcaligenes* sp. growing on diphenylmethane. All reactions took place simultaneously in a reactor vessel containing both organisms co-immobilized in alginate. Though overall degradation of DDT was not impressive, these experiments indicated that carrying out simultaneous anaerobic and aerobic processes within a single matrix is quite feasible.

Biodegradation of adsorbed and liquid-phase p-nitrophenol (PNP), 2,4-dichlorophenol, and pentachlorophenol by a mixed culture adsorbed to activated carbon has been examined (55). The kinetics of a continuous reactor were affected by the differing degradation and desorption rates of the substrates. A mixed culture consisting of three strains of *Pseudomonas* immobilized on diatomaceous earth was used to evaluate microbial degradation of PNP at high concentrations in a continuous reactor (35). The authors determined that acute toxicity of PNP to the immobilized bacteria population occurred at influent concentrations between 2.1 and 2.5 g/liter at a loading rate of 0.75 g/h. A 91 to 99% degradation of influent PNP was achieved at higher mg/hour loading rates (0.93 to 0.96 g/h) by reducing the influent PNP concentration and increasing the flow rate. When the loading rate was increased to between 1.05 and 1.21 g/h, the reactor suffered periodic failures whereby the percentage of influent substrate degraded dropped to as low as 25%. Removal rates of up to 0.98 to 1.1 mg/h per gram of biocarrier were observed.

Pseudomonas strain C12B immobilized in polyacrylamide gel beads has been used to degrade alkyl sulfate and alkyl ethoxy sulfate surfactants (60, 69). When pure substrate (sodium docecyl sulfate) was used (69), loss of activity by immobilized cells and differences in the rate and extent of degradation to CO_2 between immobilized and free cells were attributed to the diffusional limitations of the beads. This system was also used to degrade a variety of mixed surfactant wastes from a hair products factory (60).

Immobilized mycelia of a strain of the fungus *Mortierella isabellina* retained the ability to biotransform dehydroabeitic acid into nontoxic metabolites for more than 110 days. This was a longer period than that achieved by freely resuspended mycelia (40).

Degradation of pentachlorophenol (PCP) by an immobilized *Flavobacterium* was accomplished in our laboratory. The bacterial cells were immobilized in alginate beads and polyurethane foams (48, 49, 51). Degradation of PCP by the *Flavobacterium* immobilized in alginate was shown to follow inhibition kinetics modeled on the equation

$$v = v_0/[1 + (K_s/S) + (S/K_i)^n].$$

A curve fit to data derived from steady-state concentrations in a continuous reactor supplemented with glutamate (0.5 g/liter) yielded the following parameters: maximum rate in absence of inhibition (v_o), 1.2 mg/g beads per hour; saturation and inhibition constants (K_s, K_i), 13 mg/liter and 70 mg/liter; order of inhibition (n), 4. These parameters resulted in the calculation of a maximum degradation rate of 0.85 mg/g beads per hour at a reactor concentration of 37 mg/liter. The reactor degraded 100 mg PCP/h per liter over a 22-day period (48).

Flavobacterium cells immobilized in polyurethane foam were capable of degrading much higher concentrations of PCP than free cells (49). Adsorption of PCP to the foam reduced the aqueous concentration to which the cells were exposed, thus reducing the toxicity of high initial concentrations in batch reactors. Free cells exposed to PCP in the presence of foam containing no cells were similarly protected. Lag times (up to four days) prior to the onset of degradation were seen, and these increased with increasing initial PCP concentration; 700 mg/liter was the maximum concentration tested (Fig. 1; 42). Cells immobilized in foam retained degradative activity for 150 days in a semi-continuous reactor experiment; however, the degradation rate declined at a slow but steady rate over that period (49). A continuous reactor exhibited steady degradation over a period of 17 days. After this time, the concentration of supplemental glutamate in the influent media was lowered, and it is not known whether the ensuing fluctuations in degradation rate were a result of this change or occurred because of limitations in the longevity of the biocatalyst (49).

A *Pseudomonas* isolate was used by O'Reilly and Crawford (50), after its immobilization in alginate, to catalyze the degradation of p-cresol (pCR) in batch reactors. Degradation rates were strongly influenced by aeration method (shake flasks versus oxygen gas-lift). Degradation by the same strain of *Pseudomonas* cells immobilized in polyurethane foam was independent of aeration method. Degradation by both alginate-and polyurethane-immobilized bacteria followed saturation kinetics in reactors that were not oxygen-limited. The maximum degradation rate (V_{max}) for the alginate system in oxygen gas-lift reactors was 1.5 mg pCR/g beads per hour. The saturation constant (K_m) was 0.22 mM. Kinetics parameters were also derived for the degradation of p-hydroxybenzoate (pHB), a metabolic intermediate in pCR degradation, in the alginate bead system and for both substrates in the foam system. Kinetic parameters for a continuous-flow column reactor containing alginate-immobilized cells were calculated.

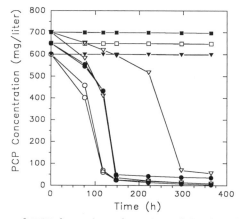

Figure 1. Degradation of PCP by polyurethane-immobilized and free *Flavobacterium* cells. Data were converted from percentage ^{14}C added trapped as $^{14}CO_2$ to mg/liter. At least 70% of the total radioactivity added was trapped from each experimental flask. Seventy-seven percent of radioactivity trapped was taken as complete mineralization to CO_2 and cell mass. Less than 0.4% was trapped from control flasks. Initial concentrations: Free cells, *solid square* = 700, *open square* = 650, *solid triangle* = 600; polyurethane-immobilized, *open triangle* = 700, *solid circle* = 650, *open circle* = 600.

TREATMENT OF CHEMICAL MIXTURES
WITH IMMOBILIZED BACTERIA

We also examined the simultaneous biodegradation of mixtures of pentachlorophenol (PCP) and p-cresol by a different system of immobilized bacteria (42). The study system consisted of two bacteria, a PCP-utilizing *Flavobacterium* and the p-cresol-degrading *Pseudomonas*. The immobilization matrices used were calcium alginate beads, agar beads, and polyurethane foam. Both batch and continuous systems were utilized. The system variables investigated included disparate metabolic rates between the two bacteria; cross-interactions among bacteria, substrates, and metabolic intermediates; and methods of ancillary carbon supply to the *Flavobacterium*.

After preliminary batch reactor experiments, the *Pseudomonas* was immobilized at one tenth of the *Flavobacterium* loading level in the matrices to compensate for metabolic rate differences between the two bacteria. The *Flavobacterium* required an ancillary carbon source (L-glutamate) to protect inoculated populations from initial PCP toxicity. Reductions in population size seen in the absence of supplemental carbon resulted in long lag times prior to the onset of PCP degradation (61). Two p-cresol metabolic intermediates were present in the reactors, p-hydroxybenzaldehyde (pBA) and p-hydroxybenzoate (pHB). These compounds were able to be utilized by the *Flavobacterium*, but did not protect populations from PCP toxicity (42).

Glutamate (0.5 g/liter) was supplied to some reactors as an ancillary carbon source. In reactors that were growth-limited by the absence of phosphate (alginate-immobilized cells), the PCP degradation rate was stable whereas PCR degradation rates declined due to a loss of viable *Pseudomonas* cells. p-Cresol degradation rates in gel bead reactors that were run under growth conditions (agar-immobilized cells) experienced an initial decline followed by recovery (Fig. 2A). Recovery of pCR degradation was accompanied by a rapid loss in PCP degradation rate. This loss in degradative ability

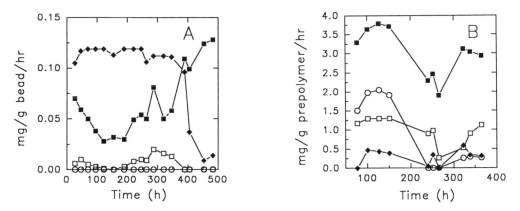

Figure 2. A. PCP and pCR degradation rates by agar-immobilized bacteria in glutamate-supplemented continuous reactor. Flow rate, 0.55 ml/min. Influent concentrations: PCP, 35 mg/liter; pCR, 40 mg/liter. B. PCP and pCR degradation rates in a glutamate-supplemented continuous reactor containing polyurethane-immobilized bacteria. Flow rate, 180 ml/hr; dilution rate, 0.45 h^{-1}. Degradation rates (mg/g beads/hr): *diamond*, PCP; *solid square*, pCR. Formation rates (mg/g beads/hr): *circle*, p-hydroxybenzaldehyde (pBA); *open square*, p-hydroxybenzoate (pHB).

was the result of competition for the glutamate supplied to the reactor between the *Flavobacterium* and the increasing *Pseudomonas* populations. In a reactor run under conditions similar to the agar reactor but containing polyurethane-immobilized bacteria, increased media flow rate allowed PCP degradation to be maintained by increasing the rate of glutamate supply (Fig. 2B). This strategy resulted in higher levels of undegraded substrate in the reactor effluent. Although each reactor run was unique, all displayed common characteristics concerning degradative ability and intermediate formation (42).

These experiments, though dealing with relatively specific reactor requirements, have implications for the design of defined multi-organism bioreactor systems. Separation of system components (organisms) into sequential reactors may allow targeting of substrates to the desired area and reduce competitive losses. Sequential reactor design could utilize combinations of high and low flow rates to maintain ancillary substrate supply while reducing target substrates to acceptable levels. Different combinations of organisms could yield noncompetitive systems or systems wherein ancillary carbon for one organism is supplied by others present within the reactor.

TREATMENT OF PCP-CONTAMINATED CLAYS WITH IMMOBILIZED BACTERIA

We recently examined the ability of an immobilized *Flavobacterium* to degrade pentachlorophenol in a clay-rich particulate material obtained as part of a soil-washing process developed by BioTrol, Inc. (Chaska, Minnesota). The material was in the form of a slurry (about 20% solids) and showed the following characteristics: pH, 7.0; total carbon, 151 g/kg; organic carbon, 93.7 g/kg; clay, 620 g/kg; sand, 113 g/kg; and pentachlorophenol, 10–13 g/kg (59).

Batch reactors (250 ml) were used; 50 ml of clay slurry was added to each reactor.

Reactors were spiked with [^{14}C]PCP. The reactors were equilibrated for 24 hours. Five grams of immobilization matrix (*Flavobacterium* immobilized in polyurethane, agar, agarose, or agarose beads secondarily entrapped in polyurethane foam) was added to provide various treatment groups. Reactors were placed on a rotary shaker at 30°C, and the evolved $^{14}CO_2$ was trapped and quantified daily. In addition, the PCP concentration in the supernatant and the PCP in the pellet were analyzed.

As shown in Figure 3, the *Flavobacterium* immobilized in agar and agarose beads mineralized high concentrations of PCP (>10 g/kg) in the clay-enriched slurry. The best mineralization was observed with agar- and agarose-immobilized bacteria: 34% and 49%, respectively, within one week. The agarose beads coated with polyurethane degraded very little PCP and no PCP degradation was detected by the *Flavobacterium* immobilized in polyurethane foam.

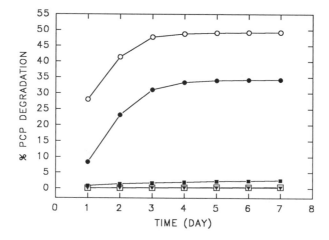

Figure 3. PCP degradation in a clay slurry with different types of immobilized *Flavobacterium* cells. *Open circle*, agarose beads; *solid circle*, agar beads; *solid triangle*, polyurethane foam; *solid square*, agarose beads entrapped in foam; *open square*, control without cells.

The time courses of PCP degradation by the *Flavobacterium* immobilized in agar and agarose were similar. Most of the PCP in the clay slurry was mineralized during the first three days. No lag phase was observed.

Previous experiments showed that PCP could not be removed by free *Flavobacterium* cells in soils when the concentration of PCP was above 0.5 g/kg (20). The data presented here show that it is possible to mineralize PCP extensively in a a clay-rich material with PCP concentrations as high as 11 g/liter (the concentration used here) by using the *Flavobacterium* immobilized in agarose or agar. In a highly contaminated soil slurry, it is very difficult for the natural microflora to become adapted to PCP mineralization, and the microbes found in the clay slurry do not significantly degrade PCP, although some cells can survive (data not shown). Therefore, the application of the immobilized *Flavobacterium* to such clay-rich soil slurries could be an effective way to remove high levels of PCP in actual waste dump soils or soil-washing residues.

PCP degradation by the *Flavobacterium* immobilized in agarose or agar beads stopped after three to four days, perhaps because it could mineralize only PCP that is found in the aqueous phase of the clay slurry or that is loosely adsorbed on the solids but not the PCP tightly bound on the solid phase. As mineralization proceeds, the adsorbed PCP is partially released to the aqueous phase, and a concentration equilibrium is established until a certain point at which the forces of adsorption become so strong that no more PCP is released to the aqueous solution. Degradation would stop at this point.

PCP degradation was not observed in the clay slurry (the solid content was 18.4%) when the added *Flavobacterium* had been immobilized in polyurethane foam. The slow rate of PCP degradation might be due to the slow release of PCP from the solid phase of the slurry to the aqueous phase, though this seems unlikely in light of the results with agarose-immobilized cells. That the polyurethane may have increased the PCP concentration near the cells to a toxic level seems more likely.

DIFFUSION LIMITATIONS OF IMMOBILIZATION MATRICES

One of the main disadvantages of using immobilized cell technology is the increased resistance to diffusion of certain immobilization matrices, especially gels (3, 19). One study (58) found that diffusion rates of substances of MW $< 2.0 \times 10^4$ into Ca-alginate beads not containing cells approached those in water. No diffusion into or out of beads was seen for larger molecules such as some proteins. The reduced diffusion of the hydrophobic compound DDT into the hydrophilic alginate matrix was postulated as a factor in reduction of transformation rates by immobilized bacteria versus free cells (11).

Oxygen supply in aerobic bioreactors causes considerable technical difficulties, and immobilization of cells can add to the problem (27). This may be turned into an advantage in the immobilization of anaerobic species, however (33). [The anaerobic transformation of DDT is discussed above (11).] In another case, a phenol-degrading methanogenic consortium was immobilized in agar (24). This system was not tested for activity under aerobic incubation conditions, but protection from high substrate concentrations was observed under anaerobic conditions.

Oxygen use by bacteria and fungi entrapped in alginate beads was investigated by Gosmann and Rehm (32). The specific rate of oxygen uptake of the immobilized microbes was dependent on the biomass concentration in the gel. Low cell concentrations allowed maximum respiration. As the concentration was increased, cells used oxygen faster than it could diffuse into the beads, until eventually the absolute oxygen uptake rate of the beads remained constant.

During the degradation of p-cresol (pCR) by alginate-immobilized *Pseudomonas* in shake flasks, p-hydroxybenzoate (pHB) was produced and not degraded until the pCR was consumed (50). This was shown to be the result of oxygen limitation, because when aeration with oxygen gas increased the pCR degradation rate, pHB formation was drastically reduced. Bacteria immobilized in polyurethane foam did not exhibit evidence of oxygen limitation. Alginate-immobilized *Serratia marcescens* producing L-arginine were also shown to be oxygen-limited (31). This limitation was relieved by supplying the biocatalyst with oxygen-enriched gas. The production of arginine was proportional to the concentration of the oxygen gas supplied.

The studies above illustrate the strategy of increasing the supply of oxygen to immobilized cells by increasing the partial pressure of oxygen in the feed gas stream. Other techniques for improving oxygen supply from external sources include increasing oxygen diffusivity by reducing the density of the matrix or increasing the surface-to-volume ratio of the particle by reducing the overall particle size (27). Oxygen carriers, such as hemoglobin or perfluorochemicals, have also been used in oxygen supply schemes (27).

In situ oxygen production has been accomplished by supplying hydrogen peroxide to organisms with high catalase activities (27). In one such system (36), initial specific productivity of dihydroxyacetone from glycerol by *Gluconobacter oxydans* increased with increasing hydrogen peroxide concentration. Overall productivity decreased, however, due to H_2O_2 toxicity. Co-immobilization with photosynthetic microbes has also been used for the in situ production of oxygen (27).

ENVIRONMENTAL APPLICATIONS OF ENCAPSULATED OR ENTRAPPED MICROBIAL CELLS

Numerous investigators have found that with rapidly metabolizing immobilized cells, active cells may be found only to a depth of 50–200 μm into a bead (12, 14–16, 52, 53, 57, 68). Thus, as much as 80% of the volume of a 2-mm bead may contain inactive or dead cells. Ignoring this large quantity of inactive cells has been a common practice; however, preparing beads containing cells that are virtually all active can offer great economic advantages as well as improved degradation rates.

In our work we are interested in promoting oxygen transfer into spherical immobilization matrices that have been loaded with aerobic pollutant-degrading bacteria. These bacteria, entrapped within polymeric beads, are to be introduced into subsurface environments where they may degrade specific toxic chemical contaminants. Both entrapment and introduction require the use of very small beads. A major limitation to current methods of entrapping bacteria within spherical immobilization matrices is that preparing beads with diameters consistently less than about 0.5 mm is difficult.

A principal method of entrapping biocatalysts within very small beads has used emulsion techniques. For example, the cells of interest may be added to molten agarose held at 45°C. The suspension is then stirred into an oil, also held at 45°C, thereby forming an emulsion of agarose beads. Cooling of the emulsion produces beads of 10–100 μm diameter (12, 37, 38, 46, 64, 65). Unfortunately, producing large quantities of beads in this manner is difficult, and washing the beads free of oil is also difficult.

We have developed a method to produce microbeads containing immobilized bacteria that overcomes many of the disadvantages of older methods (56). Large quantities of small beads of consistent diameter can be prepared in a straightforward manner, without having to separate and wash the beads after their preparation. High cell loadings (up to 50% cell volume/bead volume) of virtually 100% active cells within beads of 2–50 μm diameter are possible.

Cell suspensions were pumped through a low-pressure nozzle by using the apparatus shown in Figure 4, which introduced a fine aerosol into a stirred aqueous phase. The aqueous phase for collection of microspheres was 50 mM $CaCl_2$ for the cells in alginate suspension, or cold (4°C) buffer for the cells in agarose or prepolymer suspensions.

The low-pressure nozzle apparatus produced an aerosol of very small spheres (2–50 μm diameter). The spheres of alginate polymerized into bacteria-loaded microbeads, and the agarose spheres likewise hardened into microbeads upon contact with cold buffer. The polyurethane prepolymer polymerized into polyurethane microbeads upon contact and reaction with water. A photomicrograph of an alginate bead is shown in Figure 5. Yields of microbeads have been excellent (23.5 ml of beads/27.5 ml of suspension passed through the bead-making apparatus).

Flavobacterium cells entrapped within the three types of microbeads retained high activities with respect to biodegradation of pentachlorophenol. These microbeads have many interesting potential applications in the environmental restoration field.

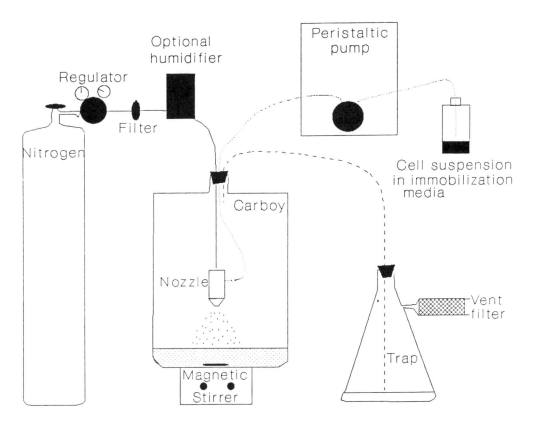

Figure 4. Apparatus for production of microbeads containing entrapped microbial cells.

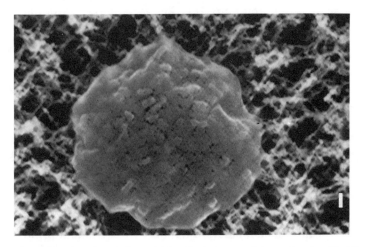

Figure 5. Scanning electron photomicrograph of a 10 μm alginate bead with *Flavobacterium* immobilized at 5% cell loading by wet weight. Bar length is 0.9 μm.

ENHANCEMENT OF MICROBIAL SURVIVAL IN SOIL
BY IMMOBILIZATION PROCEDURES

The soil, one of the earth's most complex habitats, exhibits many variables on both macroscopic and microscopic scales that cannot be controlled or even properly identified. Survival of microbes introduced into soil depends on biotic and abiotic factors, including predation by indigenous species, soil moisture content, and clay content. Here we discuss factors that influence the survival of introduced and indigenous soil microorganisms (34) and evaluate immobilization techniques for increasing the survival of microorganisms added to soil.

Much of the research in this area has concerned nitrogen-fixing bacteria or genetically engineered microorganisms, their survival, and the colonization of legume nodules (41). Experiments have been performed with free cells (4) or with cells protected by a carrier (1, 5, 8, 28, 39), and on scales ranging from small test tubes (40), pots, and microcosms (7) to large field inoculations (34, 62). Results have ranged from rapid die-off (7, 41, 44) to survival for over ten years (13) at significant levels, and from no incorporation into nodules to >50% colonization of the nodules (23). Data from numerous researchers have indicated the importance of predation (2, 17, 18, 62, 63) and the physical characteristics of the soil (6, 28, 43, 62, 71) on the survival of introduced organisms. The ability to track specific species of soil organisms (17, 21, 30, 47, 66, 70) is important, especially in determining the effect of an introduced organism (8, 28) or its fate in the soil environment (2, 13, 17, 18, 21, 43, 44).

Field- or soil-introduced organisms have a distinct advantage if they have been combined with some type of carrier material (1, 5, 8, 28, 39), such as agar (39), peat (1, 8), alginate (5), alginate-clay (29), and fluid gels (39). These carriers allow a high concentration of organisms, sometimes directly attached to the seed, to be incorporated into the soil. Carriers can also contain nutrients, serve as a moisture reserve for the cell, and isolate cells from predators. Carriers that are in the form of beads can be dehydrated to facilitate application with seeds. Lyophilization rather than air drying maintains the highest viable cell population (5), although surviving populations are not as large as those found in moist beads that have not been dehydrated.

Several abiotic factors also affect the survival of microorganisms introduced into soil, including site variability (7), timing (41), clay content (22, 62, 71), cation exchange capacity, potassium, sodium, and zinc (28). Site variability is important if the testing is done in small areas or microcosms; if the local environment varies, sampling must be adjusted. Additional clay in the soil may increase survivability because of clay's buffering ability. We immobilized a pentachlorophenol-degrading *Flavobacterium* in agarose or alginate beads of 2–50 μm diameter (Fig. 5). Many biological and abiotic factors that tend to cause losses of cell viability in introduced populations appear to have been eliminated for cells added to subsurface sand/gravel aquifer material packed into columns with groundwater flowing through the columns at *in situ* rates (0.2 to 10 cm/day). Cells within microspheres not only passed through the aquifer material (Fig. 6), they retained their abilities to degrade pentachlorophenol for many weeks in this microbial environment unfavorable (Fig. 7) to the *Flavobacterium*. Four hundred days after groundwater stopped flowing through these aquifer columns, pentachlorophenol-degrading bacteria could be enriched only in columns containing microspheres of encapsulated *Flavobacterium*. Additional research on other microbial systems is required, but thus far microencapsulation appears to have great promise as a means to protect exogenous microorganisms added to soil to foster desirable processes such as biodegradation of toxic contaminants.

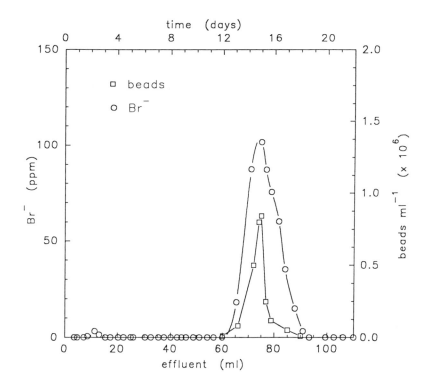

Figure 6. Tracer studies in aquifer microcosms. A suspension of *Flavobacterium* cells immobilized in agarose beads were acridine-orange-stained and a potassium bromide tracer was added. Two milliliters of this bead suspension was injected into the microcosms, and the effluent from the column was monitored for both bromide and the fluorescently stained cells inside beads.

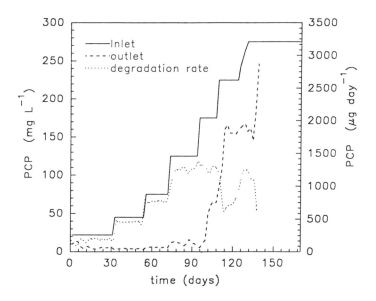

Figure 7. Pentachlorophenol degradation in aquifer microcosms. Column parameters: diameter, 2.2 cm; length, 18 cm; flow rate, 12 ml/day; residence time, 5 days; 5×10^7 *Flavobacterium* cells in 7×10^5 agarose beads/g aquifer material.

SUMMARY

Encapsulated or entrapped microorganisms will have many uses in the field of environmental cleanup and hazardous waste management. They will be used within bioreactors to treat industrial effluents or to detoxify soils and waters removed from contaminated sites. In a particularly innovative application, microspheres containing pollutant-degrading microorganisms will be used to deliver microbes to contaminated subsurface environments where traditional treatment techniques have failed. Spin-offs of this work in environmental biotechnology may have dramatic impacts on associated fields such as agriculture and mining. This review reinforces our optimistic appraisal of the future of this technology.

Acknowledgments

The authors' research was supported by grants from the U.S. Department of Energy, Subsurface Science Program; the U.S. Geological Survey; Battelle Pacific Northwest Laboratories, Richland, Washington; the U.S. Environmental Protection Agency; the U.S. Department of Agriculture (CSRS); BioTrol, Inc., Chaska, Minnesota; and the Idaho State Board of Education.

LITERATURE CITED

1. Aarons, S., and M. H. Ahmad. 1986. Examining growth and survival of cowpea rhizobia in Jamaican peat. *Lett. Appl. Microbiol.* 2:115–118.

2. Acea, M. J., C. R. Moore, and M. Alexander. 1988. Survival and growth of bacteria introduced into soil. *Soil Biol. Biochem.* 20:509–515.

3. Akin, C. 1987. Biocatalysis with immobilized cells. *Biotechnol. Genet. Eng. Rev.* 5:319–367.

4. Amner, W., A. J. McCarthy, and C. Edwards. 1988. Quantitative assessment of factors affecting the recovery of indigenous and released thermophilic bacteria from compost. *Appl. Environ. Microbiol.* 54:3107–3112.

5. Bashan, Y. 1986. Alginate beads as synthetic inoculant carriers for slow release of bacteria that affect plant growth. *Appl. Environ. Microbiol.* 51:1089–1098.

6. Bashan, Y., and H. Levanony. 1988. Adsorption the rhizosphere bacterium *Azospirillum brasilense* Cd to soil, sand and peat particles. *J. Gen. Microbiol.* 134:1811–1820.

7. Bentjen, S. A., J. K. Fredrickson, P. van Voris, and S. W. Li. 1989. Intact soil-core microcosms for evaluating the fate and ecological impact of the release of genetically engineered microorganisms. *Appl. Environ. Microbiol.* 55:198–202.

8. Berg, R. K., T. E. Loynachan, R. M. Zablotowicz, and M. T. Lieberman. 1988. Nodule occupancy by introduced *Bradyrhizobium japonicum* in Iowa soils. *Agron. J.* 80:876–881.

9. Bettman, H., and H. J. Rehm. 1984. Degradation of phenol by polymer entrapped microorganisms. *Appl. Microbiol. Biotechnol.* 20:285–290.

10. Bettman, H., and H. J. Rehm. 1985. Continuous degradation of phenol(s) by *Pseudomonas putida* P8 entrapped in polyacrylamide-hydrazide. *Appl. Microbiol. Biotechnol.* 22:389–393.

11. Beunink, J., and H. J. Rehm. 1988. Synchronous anaerobic and aerobic degradation of DDT by an immobilized mixed culture system. *Appl. Microbiol. Biotechnol.* 29:72–80.

12. Birnbaum, S., and L. Bulow. 1988. Production and release of human proinsulin by recombinant *Escherichia coli* immobilized in agarose microbeads. *Enzyme Microb. Technol.* 10:601–605.

13. Brunel, B., J. C. Cleyet-Marel, P. Normand, and R. Bardin. 1988. Stability of *Bradyrhizobium japonicum* inoculants after introduction into soil. *Appl. Environ. Microbiol.* 54:2636–2642.

14. Chang, H. N., and M.-Y. Murray. 1987. Estimation of oxygen penetration depth in immobilized cells. *Appl. Microbiol. Biotechnol.* 29:107–112.

15. Chen, K.-C., and C.-T. Huang. 1987. Effects of the growth of *Trichosporon cutaneum* in calcium alginate gel beads upon bead structure and oxygen transfer characteristics. *Enzyme Microb. Technol.* 10:284–292.

16. Chen, T.-L., and A. E. Humphrey. 1988. Estimation of critical particle diameter for optimal respiration of gel entrapped and/or pelletized microbial cells. *Biotechnol. Lett.* 10:10:699–702.

17. Compeau, G., B. J. Al-Achi, E. Platsouka, and S. B. Levy. 1988. Survival of rifampin-resistant mutants of *Pseudomonas fluorescens putida* in soil systems. *Appl. Environ. Microbiol.* 54:2432–2438.

18. Corman, A., Y. Crozat, and J. C. Cleyet-Marel. 1987. Modeling of survival kinetics of some *Bradyrhizobium japonicum* strains in soils. *Biol. Fertil. Soils* 4:79–84.

19. Coughlan, M. P., and M. P. J. Kierstan. 1988. Preparation and applications of immobilized microorganisms: a survey of recent reports. *J. Microbiol. Methods* 8:51–90.

20. Crawford, R. L., and W. W. Mohn. 1985. Microbiological removal of pentachlorophenol from soil using a *Flavobacterium*. *Enzyme Microb. Technol.* 7:617–620.

21. Crozat, Y., J. C. Cleyet-Marel, and A. Corman. 1987. Use of the fluorescent antibody technique to characterize equilibrium survival concentrations of *Bradyrhizobium japonicum* strains in soils. *Biol. Fertil. Soils* 4:85–90.

22. Dhillon, H. S., R. K. Dart, and S. P. Hitchman. 1987. Survival of soil microorganisms in storage. *Microbios* 52:73–80.

23. Dowdle, S. F., and B. B. Bohlool. 1987. Intra- and inter-specific competition in *Rhizobium fredii* and *Bradyrhizobium japonicium* as indigenous and introduced organisms. *Can. J. Microbiol.* 33:990–995.

24. Dwyer, D. F., M. L. Krumme, S. A. Boyd, and J. M. Tiedje. 1986. Kinetics of phenol biodegradation by an immobilized methanogenic consortium. *Appl. Environ. Microbiol.* 52:345–351.

25. Ehrhardt, H. M., and H. J. Rehm. 1985. Phenol degradation by microorganisms absorbed on activated carbon. *Appl. Microbiol. Biotechnol.* 21:32–36.

26. Ehrhardt, H. M., and H. J. Rehm. 1989. Semicontinuous and continuous degradation of phenol by *Pseudomonas putida* P8 adsorbed on activated carbon. *Appl. Microbiol. Biotechnol.* 30:312–317.

27. Enfors, S. O., and B. Mattiasson. 1983. Oxygenation processes involving immobilized cells. In *Immobilized Cells and Organelles,* Vol. 2. Ed. B. Mattiasson. Boca Raton, Florida: CRC Press, Inc. 41–60.

28. Fravel, D. R., and J. J. Marois. 1986. Edaphic parameters associated with establishment of the biocontrol agent *Talaromyces flavus*. *Phytopathology* 76:643–646.

29. Fravel, D. R., J. J. Marois, R. D. Lumsden, and W. J. Connick, Jr. 1985. Encapsulation of potential biocontrol agents in alginate-clay matrix. *Phytopathology* 75:774–777.

30. Fredrickson, J. K., D. F. Bezdicek, F. J. Brockman, and S. W. Li. 1988. Enumeration of Tn5 mutant bacteria in soil by using a most-probable-number-DNA hybridization procedure and antibiotic resistance. *Appl. Environ. Microbiol.* 54:446–453.

31. Fujimura, M., J. Kato, T. Tosa, and I. Chibata. 1984. Continuous production of L-arginine using immobilized growing *Serratia marcescens* cells: Effectiveness of supply of oxygen gas. *Appl. Microbiol. Biotechnol.* 19:79–84.

32. Gosmann, B., and H. J. Rehm. 1986. Oxygen uptake of microorganisms entrapped in Ca-alginate. *Appl. Microbiol. Biotechnol.* 23:163–167.

33. Haggstrom, L. 1983. Fermentations by immobilized strict anaerobes. In *Immobilized Cells and Organelles,* Vol. 2. Ed. B. Mattiasson. Boca Raton, Florida: CRC Press, Inc. 61–78.

34. Harris, J. M., J. A. Lucas, M. R. Davey, and G. Lethbridge. 1989. Establishment of *Azosprillum* inoculant in the rhizosphere of winter wheat. *Soil Biol. Biochem.* 21:59–64.

35. Heitkamp, M. A., V. Camel, T. J. Reuter, and W. J. Adams. 1990. Biodegradation of p-nitrophenol in an aqueous waste stream by immobilized bacteria. *Appl. Environ. Microbiol.* 56:2967–2973.

36. Holst, O., H. Lundback, and B. Mattiasson. 1985. Hydrogen peroxide as a oxygen source for immobilized *Gluconobacter oxydans* converting glycerol to dihydroxyacetone. *Appl. Microbiol. Biotechnol.* 22:383–388.

37. Jackson, D. A., and P. R. Cook. 1986. A cell-cycle-dependent DNA polymerase activity that replicates intact DNA in chromatin. *J. Mol. Biol.* 192:65–76.

38. Jackson, D. A., and P. R. Cook. 1986. Replication occurs at a nucleoskeleton. *EMBO J.* 5:6:1403–1410.

39. Jawson, M. D., A. J. Franzluebbers, and R. K. Berg. 1989. *Bradyrhizobium japonicum* survival in soybean inoculation with fluid gels. *Appl. Environ. Microbiol.* 55:617–622.

40. Kutney, J. P., L. S. L. Choi, G. M. Hewitt, P. J. Salisbury, and M. Singh. 1985. Biotransformation of dehydroabietic acid with resting cell suspensions and calcium alginate-immobilized cells of *Mortierella isabellina*. *Appl. Environ. Microbiol.* 49:96–100.

41. Ladha, J. K., M. Garcia, S. Miyan, A. T. Padre, and I. Watanabe. 1989. Survival of *Azorhizobium caulinodans* in the soil and rhizosphere of wetland rice under *Sesbania rostrata*-rice rotation. *Appl. Environ. Microbiol.* 55:454–460.

42. Levinson, W. E. 1991. *Simultaneous Degradation of PCP and p-Cresol by Immobilized Bacteria*. M. S. Thesis, University of Idaho, Moscow, Idaho.

43. Miller, M. S., and I. L. Pepper. 1988. Survival of a fast-growing strain of *Lupin rhizobia* in Sonoran desert soils. *Soil Biol. Biochem.* 20:323–327.

44. Morel, J. L., G. Bitton, G. R. Chaudry, and J. Awong. 1989. Fate of genetically modified microorganisms in the corn rhizosphere. *Curr. Microbiol.* 18:355–360.

45. Morsen, A., and H. J. Rehm. 1987. Degradation of phenol by a mixed culture of *Pseudomonas putida* and *Cryptococcus elinovii* adsorbed on activated carbon. *Appl. Microbiol. Biotechnol.* 26:283–288.

46. Nilsson, K., W. Scheirer, O. W. Mertin, L. Ostberg, E. Liehl, H. W. D. Katinger, and K. Mosbach. 1983. Entrapment of animal cells for production of monoclonal antibodies and other biomolecules. *Nature* 302:629–630.

47. Ogram, A. V., and G. S. Sayler. 1988. The use of gene probes in the rapid analysis of natural microbial communities. *J. Ind. Microbiol.* 3:281–292.

48. O'Reilly, K. T. 1989. *Biodegradation of Pentachlorophenol and p-cresol by Immobilized Bacteria*. Ph.D. Thesis, University of Idaho, Moscow, Idaho.

49. O'Reilly, K. T., and R. L. Crawford. 1989. Degradation of pentachlorophenol by polyurethane immobilized *Flavobacterium* cells. *Appl. Environ. Microbiol.* 55:2113–2118.

50. O'Reilly, K. T., and R. L. Crawford. 1989. Kinetics of p-cresol degradation by an immobilized *Pseudomonas* sp. *Appl. Environ. Microbiol.* 55:866–870.

51. O'Reilly, K. T., R. Kadakia, R. A. Korus, and R. L. Crawford. 1988. Utilization of immobilized-bacteria to degrade aromatic compounds common to wood-treatment wastewaters. *Water Sci. Technol.* 20:95–100.

52. Pras, N., P. G. M. Hesselink, J. ten Tusscher, and T. M. Malingr. 1989. Kinetic aspects of the bioconversion of L-tyrosine and into L-DOPA by cells of *Mucuna pruriens* L. entrapped in different matrices. *Biotechnol. Bioeng.* 34:214–222.

53. Sato, K., and K. Toda. 1983. Oxygen uptake rate of immobilized growing *Candida lipolytica*. *J. Ferment. Technol.* 61(3):239–245.

54. Scott, C. D. 1987. Immobilized cells: a review of recent literature. *Enzyme Microb. Technol.* 9:66–73.

55. Speitel, G. E. J., C.-J. Lu, M. Turakhia, and X.-J. Zhu. 1989. Biodegradation of trace concentrations of substituted phenols in granular activated carbon columns. *Environ. Sci. Technol.* 23:68–74.

56. Stormo, K. E., and R. L. Crawford. 1992. Preparation of encapsulated microbial cells for environmental application. *Appl. Environ. Microbiol.* 58(2):727–730.

57. Sun, Yan, S. Furusaki, A. Yamauchi, and K. Ichimura. 1989. Diffusivity of oxygen into carriers entrapping whole cells. *Biotechnol. Bioeng.* 34:55–58.

58. Tanaka, H., M. Matsumura, and I. A. Veliky. 1984. Diffusion characteristics of substrates in Ca-alginate gel beads. *Biotechnol. Bioeng.* 26:53–58.

59. Tao, H. 1990. *Degradation of Pentachlorophenol in Contaminated Soil Slurries by Immobilized Flavobacterium.* M. S. Thesis, University of Idaho, Moscow, Idaho.

60. Thomas, O. R. T., and G. F. White. 1991. Immobilization of the surfactant-degrading bacterium *Pseudomonas* C12B in polyacrylamide gel. III. Biodegradation specificity for raw surfactants and industrial wastes. *Enzyme Microb. Technol.* 13:338–343.

61. Topp, E., R. L. Crawford, and R. S. Hanson. 1988. Influence of readily metabolizable carbon on pentachlorophenol metabolism by a pentachlorophenol-degrading *Flavobacterium* sp. *Appl. Environ. Microbiol.* 54:2452–2459.

62. van Elsas, J. D., A. F. Dijkstra, J. M. Govaert, and J. A. van Veen. 1986. Survival of *Psuedomonas fluorescens* and *Bacillus subtilis* introduced into two soils of different texture in field microplots. *FEMS Microbiol. Ecol.* 38:151–160.

63. Vargas, R., and T. Hattori. 1986. Protozoan predation of bacterial cells in soil aggregates. *FEMS Microbiol. Ecol.* 38:233–242.

64. Vetvicka, V., and L. Fornusek. 1987. Polymer microbeads in immunology. *Biomaterials* 8:341–345.

65. Vournakis, J. N., and P. W. Runstadler. 1989. Microenvironment: The key to improved cell culture products. *Biotechnology* 7:143–145.

66. Weaver, J. C., G. B. Williams, A. Klibanov, and A. L. Demain. 1988. Gel microdroplets: rapid detection and enumeration of individual microorganisms by their metabolic activity. *Biotechnology* 6:1084–1089.

67. Westmeier, F., and H. J. Rehm. 1985. Biodegradation of 4-chlorophenol by entrapped *Alcaligenes* sp. A 7-2. *Appl. Microbiol. Biotechnol.* 22:301–305.

68. Westmeier, F., and H. J. Rehm. 1987. Degradation of 4-chlorophenol in municipal wastewater by adsorptively immobilized *Alcaligenes* sp. A 7-2. *Appl. Microbiol. Biotechnol.* 26:78–83.

69. White, G. F., and O. R. T. Thomas. 1990. Immobilization of the surfactant-degrading bacterium *Pseudomonas* C12B in polyacrylamide gel beads: I. Effect of immobilization on the primary and ultimate biodegradation of SDS, and redistribution of bacteria within beads during use. *Enzyme Microb. Technol.* 12:697–705.

70. Zeph, L. R., and G. Stotzky. 1989. Use of a biotintylated DNA probe to detect bac,teria transduced by bacteriophage P1 in soil. *Appl. Environ. Microbiol.* 55:661–665.

71. Zeph, L. R., M. A. Onaga, and G. Stotzky. 1989. Transduction of *Escherichia coli* by bacteriophage P1 in soil. *Appl. Environ. Microbiol.* 54:1731–1737.

CHAPTER 22

Recent Advances in Bioprocessing of Coal

M. V. S. MURTY, D. BHATTACHARYYA, and M. I. H. ALEEM

Abstract. Coal is a fossil fuel that comprises about three fourths of the total world energy resources. It is a heterogeneous mixture of organic and inorganic material. Although raw coal is sufficiently abundant to meet current energy demands, enhanced utilization will require upgrading of coal to meet strict environmental standards. A series of chemical processes, such as pyrolysis, gasification, and liquefaction, have been used for coal conversion under drastic reaction conditions (e.g. 300 to 400°C and 800 to 2000 psi H_2) with high capital and operating costs. With the potential of the latest developments in biotechnology to make coal an efficient source of energy, a new surge of research activity has been initiated. Because of the complex nature of solid coal, bioprocessing has to accomplish several tasks. One such task is biodesulfurization, since sulfur plays a role in the formation of acid rain. Over 90% of the pyritic sulfur in coal has been removed by the chemolithotrophic bacteria *Sulfolobus* and *Thiobacillus* spp. Similarly, biosolubilization by fungi and conversion of coal products into liquid fuels by autotrophic and anaerobic bacteria have progressed substantially. *In-situ* bioformation of ultra-fine crystal catalysts could help to produce enhanced liquefaction yields. Biohydrogenation of coal or coal liquids would also increase the fuel value of this combustible organic matter. Removal of substantial amounts of organic sulfur and nitrogen and effective leaching of trace metals from coal by microorganisms have yet to be demonstrated. Biodesulfurization, biosolubilization, and leaching of trace metals are important biodegradative and bioremediative processes to contain toxic chemicals. Design of suitable bioreactors is also essential for improving the bioprocessing of coal. The use of hyperthermophilic bacteria could also influence the rate of coal bioprocessing. Together, these aspects of bioprocessing have the potential to improve the fuel value of coal and reduce the level of toxic materials that result from its combustion.

INTRODUCTION

Bioprocessing of coal has received a resurgence of interest as one of the promising alternative technologies primarily because of its abundance (approximately 500,000 megatons). The use of microorganisms to process coal had been dormant since the 1950s [37]. Microbial action on coal was clearly suggested again in the early 1960s [95, 124]. The emphasis on current active research is mainly on three areas: desulfurization, solubilization, and conversion of coal-derived materials. The degradation of coal

Drs. Murty and Aleem are at the School of Biological Sciences, University of Kentucky, Lexington, KY 40506, U.S.A. Dr. Bhattacharyya is at the Department of Chemical Engineering, University of Kentucky. Dr. Murty is also affiliated with the Deparment of Chemical Engineering.

requires the cleavage of covalent bonds in the coal matrix, namely, the ether, methylene, and ethylene linkages. Cleavage of these linkages and hydrogenation of unsaturated bonds in coal yields water or (organic) solvent-soluble constituents. The process of cleavage could thus provide the basis for coal solubilization/liquefaction.

Recently, attention has been focused on precombustion removal of sulfur by use of microorganisms (91). One such study involves natural isolates from Yellowstone Park that were shown to remove 90% of pyritic sulfur in addition to more than 33% organic sulfur from coal (126). The rate of coal desulfurization by thermophilic bacteria, such as *Sulfolobus acidocaldarius* and *S. brierleyi* (*Acidianus brierleyi*), is higher than that by *Thiobacilli*. These thermophiles are also more resistant to the compounds present in coal (40). The structure of the coal network has been compared to that of a cross-linked polymer network, and cross-linking chains containing sulfidic bonds would be expected to be desulfurized more easily than thiophenic sulfurs in condensed ring clusters (58). Steric factors and limited accessibility of reagents and microorganisms to reaction sites may also hinder desulfurization.

Despite several studies, development in this particular biotechnology has not been outstanding. Nonetheless, a number of processes have been proposed to remove sulfur from high-sulfur coals (113) and to produce methane from low-rank coals (105). The underlying advantages and disadvantages of this process are difficult to enumerate, since bioprocessing of coal is still in its infancy. The direct advantages of the bioprocess are relatively low process temperatures, low process pressures, generation of environmentally safe effluents, and a high degree of specificity of chemical reactions. Bioprocess technology for biodepyritization has progressed in several areas, such as the search for better microbial strains, improvement of existing strains and culture conditions, and design of suitable bioreactors. It is vital to remember, however, that complete removal of organic and inorganic sulfur from coal is required to meet strict sulfur emission standards.

Various studies have shown that microorganisms have the ability to promote solubilization, depolymerization, and oxidation of coal and coal-derived materials (35, 46, 129). Most of the studies have used lower-rank coals, and some of these coals have undergone chemical pretreatments. Some genera of fungi seemed to influence the solubilization of coal (146, 148), particularly lignite. The material obtained as a result of microbial action was found to be highly polar and soluble in water up to 20 weight percent (105).

We have investigated the potential of hydrogenase-possessing bacteria to hydrogenate directly untreated coal and model compounds such as diphenyl methane (DPM) and methylene blue (MB) (5, 6). We based our study on the finding that the chemical liquefaction yield can be enhanced if microorganisms are able to transfer hydrogens into the coal. *In situ* formation of a catalyst such as FeOOH crystals on coal particles in the presence of *Acidianus brierleyi* has been demonstrated in our laboratories (24, 68). The formation of fine crystals of FeOOH would also help in transfering hydrogen during chemical liquefaction and thus appears to be responsible for the enhanced liquefaction yield. To our knowledge, the direct microbial hydrogenation of untreated coal or model compounds has not yet been demonstrated. Some of our preliminary results are discussed in this chapter (6). Indirect liquefaction with use of a mixed culture of microorganisms has been reported to convert approximately 95% of the CO and H_2 to acetate and CO_2 (53). The organisms responsible were *Peptostreptococcus productus* and *Acetobacterium woodii*. Production of more acids and alcohols from the mixed cultures isolated from wastes was also investigated. Various types of low-molecular-weight organic acids and alcohols can be produced when CO is used as the carbon source (105).

In addition to microbial desulfurization of coal, the use of microorganisms for nitrogen removal has also been considered. Aliphatic and heterocyclic nitrogen compounds were removed from shale oil by bacterial cultures (2, 3).

The ash generated during coal combustion pollutes groundwaters with toxic metals. Sulfur oxidizers are known to oxidize a variety of metal sulfides. Removal of trace elements from coal (104, 111) and reduction of trace metals in the ash content of the bioprocessed coal have also been reported (22, 24, 49).

ORIGIN OF COAL

Coal was metamorphosed from partially decomposed plant debris subjected to various physicochemical processes in the presence of high geothermic pressure and temperature over a span of several centuries. Consequently, coals contain organic and inorganic structures. The precursors of coal are organic constituents of lignin, carbohydrates, oils, fats, waxes, miscellaneous other organic materials such as resins, tannins, and alkaloides, and the mineral or inorganic constituents of plants (135).

MOLECULAR STRUCTURE OF COAL

Coal as an organic sediment can be classified in the successive ranks of peat, lignite, subbituminous, bituminous, and anthracite according to its increased aromatization and decreased oxygen content. Many models of the molecular structure of the organic material in coal have been proposed and continue to be proposed (54, 136, 149). These models should become more refined as new information is obtained (see 39 for an excellent review). The models are useful guides, but they are average structures meant to represent functional group distributions and are therefore not absolutely accurate. Coals contain a variety of elements and compounds. An analysis of some of the Kentucky coals is given in Table 1. Sulfur- and nitrogen-based compounds derived from coal influence the fuel value of coal. Detailed aspects of these compounds are discussed in the following sections.

Table 1. Analytical data of Kentucky coals.

Analysis	Coal #91182	Coal #85098
Ultimate (Wt %, dry)		
C	70.7	63.2
H	5.0	4.4
N	1.5	1.3
S	3.2	5.8
Ash	7.8	18.7
Forms of sulfur (Wt%, d)		
Pyrite	1.0	2.9
Sulfate	0.1	0.1
Organic	2.1	2.8
Volatile matter (Wt%, d)	39.1	35.5
Total iron	1.0	4.0

Sulfur Compounds

Living cells possess several organosulfur-containing amino acids, vitamins, coenzymes, penicillins of several varieties, thiostrepton with 5 sulfur rings, biotin sulfone, and chondroitin sulfate. Various amounts and selections of these compounds were transformed into different forms of sulfur in coal, depending upon the geobiology and geochemistry of sulfur cycling. The amount and form of sulfur in coals depend much more on the coal's deposition environment than on its age or rank. Several recent reviews outline the sulfur distribution in coal and the methods used to determine this distribution (9, 10, 28, 58, 63). Coals contain a mixture of inorganic and organic sulfur. There is general agreement that the inorganic form is pyrite and is present in many coals. Besides pyrite and marcasite, other mineral forms of sulfur and elemental sulfur may be present. The organically bound sulfur in coal has been suggested to include functional groups comprising benzothiophene, dibenzothiophene (DBT), thiosulfides, and disulfides. These groups have been differentiated and quantified according to their differences in reactivity. The effect of organic sulfur compounds on coal liquefaction depends somewhat upon the inherent reactivity of the structures of different organic sulfur compounds (138). Some of the heterocyclic compounds were probably formed during coal pyrolysis (58).

Several chemical and physical techniques have been employed to characterize the sulfur compounds present in coal. In wet analysis, total sulfur can be estimated via ASTM D 3177–Eshka fusion, bomb combustion, high-temperature combustion, and x-ray fluorescence. Forms of sulfur, e.g. sulfate, pyritic sulfur, and organic sulfur, are analyzed by ASTM D 2492. Analytical instrumentation techniques include x-ray and electron beam methods; energy dispersive x-ray microanalysis (XRMA), also called energy dispersive x-ray (EDAX); x-ray fluorescence spectroscopy (XRF); induced electron emission spectroscopy (IEES); and electron spectroscopy for chemical analysis (ESCA), also known as x-ray photoelectron spectroscopy (XPS), which employs an x-ray energy source. These techniques have been coupled to either transmission (TEM) or scanning (SEM) electron microscopy, or combined transmission and scanning electron microscopy (STEM) (SEM/XRMA or SEM/EDAX). When an electron beam (e.g. in an electron microscope) is used to energize a sample, it is often called electron microprobe analysis. Infrared (IR) spectroscopic analysis, Fourier transform infrared (FTIR), and different forms of FTIR such as diffuse reflectance FTIR and photoacoustic (PA) FTIR have also been used to characterize sulfur compounds present in coal. Huffman et al. (66–68) have studied the molecular structure of organic sulfur in coal by XAFS and the characterization of heteroatoms by advanced analytical techniques.

Nitrogen Compounds

Nitrogen is a minor constituent of most coals, comprising less than 2% by weight. Similar to organic sulfur, nitrogen is a structurally bound element of the coal matrix. Pyridine and pyrrol-type are the model compounds. Little is known with certainty about the forms and the amount of organic nitrogen in coal, however, because it is present in such small amounts and the necessary analyses are very difficult. Analysis of coal extracts shows a decreasing amount of basic nitrogen as the rank of the coal increases. This may be due to the conversion of basic amino (–NH$_2$) groups to nitrogen incorporated in ring structures (heterocyclic nitrogen). Knowledge of how nitrogen is incorporated in the organic matrix is necessary because nitrogen compounds appear to play a key role in coal asphaltene behavior (137), and this heteroatom probably plays a role in coal conversion chemistry (105). In addition to sulfur and nitrogen compounds, coal also contains several important metals that can be recovered from coal.

BIOCATALYSTS

A variety of microorganisms have been used to process coal. The most widely employed organisms are *Thiobacillus, Acidianus,* and the *Sulfolobus* species. The first microorganism to oxidize coal, *T. ferrooxidans,* was isolated from acid mine water in 1947 (37) and found to cause the acid pollution by oxidizing coal pyrite. Later, *T. thiooxidans* was also isolated from acid mine water. Both are associated with the oxidation of sulfur and the dissolution of iron in coal. *Thiobacillus ferrooxidans* is a sulfur- and iron-oxidizing bacterium, whereas *T. thiooxidans* is a sulfur oxidizer. Both are rod-shaped bacteria about 0.5 by 1.0 μm in size and use CO_2 as a carbon source and ammonium salt or urea as a nitrogen source. Other *Thiobacilli* have also been reported in the literature. *Thiobacillus denitrificans* grows chemolithotrophically under aerobic or anaerobic conditions at a neutral pH and on reduced sulfur compounds such as thiosulfate by using O_2 or nitrate as electron acceptor. *Thiobacillus organoparus* is an aerobic mesophile that oxidizes elemental sulfur at a pH range of 1.5–5.0 and can grow heterotrophically as well as on simple organic compounds. *Thiobacillus perometabolis* oxidizes sulfur, thiosulfate, and tetrathionate but cannot grow on inorganic compounds without the presence of organic substrate at pH 7 and at 30°C.

The second most studied organisms are the *Sulfolobus* species, which are ore-leaching bacteria found in hot spring waters. They are thermoacidophilic archaebacteria with a pH optimum of 2.5 and an optimum temperature of 60°C. They have been reported to be chemoautotrophic bacteria (29). *Sulfolobus* cells usually have irregular sphere shapes and range from 0.8 to 1.0 μm in size (31). The mesophilic and thermophilic sulfur-oxidizing bacteria were reviewed by Bos and Kuenen (26) and Kelly and Deming (82). Extremely thermophilic anaerobic archaebacteria such as *Desulfurococcus, Pyrobaculum, Pyrococcus, Pyrodictium, Staphylothermus, Thermococcus, Thermodiscus, Thermofilum,* and *Thermoproteus* were reported to metabolize sulfur, which serves as electron acceptor from H_2 under anaerobic growth conditions. The optimum temperature for growth of these bacteria ranges from 75 to 105°C. The metabolic features, habitat, and morphological and biochemical characteristics of this special group of bacteria were summarized by Kelly and Deming (82).

A *Pseudomonas* strain isolated from soil by Atlantic Research Corporation was claimed to be able to oxidize major types of organic sulfur into water soluble compounds (71). The fungi *Cunninghamella elegans, Rhizopus arrhizus,* and *Mortierella isabellina* are also known to oxidize DBT to its sulfoxides and sulfone (60). Fungi produced organic-containing liquids when grown in the presence of low-rank coals (129, 146, 148). Isolation of thermophilic microorganisms capable of biosolubilizing leonardite has also been reported (127). Besides fungi (147, 150), bacteria, particularly *Streptococcus* species, have the ability to degrade solubilized coal. In addition to studies with whole cells, studies have been carried out with cell-free enzyme extracts to promote coal solubilization and degradation of the resulting solubilized polymer (141, 147). Enzyme laccase (105) and peroxidase (141) have been reported to enhance the rate of coal solubilization.

We have used an aerobic sulfur-oxidizing thermoacidophilic archaebacterium *Acidianus brierleyi* and an anaerobic sulfate-reducing mesophilic bacterium *Desul-*

fovibrio desulfuricans. Both organisms possess the enzyme hydrogenase, which is capable of transfering reducing equivalents, e.g. electrons from H_2 to either methylene blue or unprocessed coal or some model compounds (5, 6), whereas *Peptostreptococcus productus* and *Acetobacterium woodii* have been employed to convert coal-derived materials into liquid fuels (105).

Microorganisms have been known to remove aliphatic nitriles and amines (2) and heterocyclic nitrogen compounds from shale oil (3). Mixed bacterial cultures were also reported to degrade nitrogen-containing compounds present in oils (48, 51). The sulfur-oxidizing microorganisms used in hydrometallurgical mining for metal extraction, concentration, and recovery were discussed by Monroe (106). Microorganisms (e.g. *Thiobacillus* and *Sulfolobus*) have been reported to remove trace elements from coal (104, 111).

BIODESULFURIZATION OF COAL

The biochemistry of sulfur has been discussed in detail by Huxtable (69), and desulfurization of coal has been reviewed by Kargi (73, 74). Sulfur-oxidizing micro-organisms play an important role in the sulfur cycle. Sulfidic minerals and sulfur deposits are oxidized by aerobic or anaerobic bacteria to sulfate. Sulfate is then assimilated by plants and microorganisms. Through the mineralization process, sulfidic minerals and sulfur are reproduced, and the cycle starts again.

Chemistry of Microbial Depyritization

In chemolithotrophic bacteria such as *Thiobacillus ferrooxidans, T. thiooxidans,* and *Sulfolobus* species, reduced sulfur compounds and ions such as Fe (2^+), U (4^+), Sn (2^+), Sb (3^+), and Cu (1^+) are commonly the electron sources, whereas oxygen is the main oxidant. In Figure 1, Khalid et al. (86) have outlined the different oxidation processes

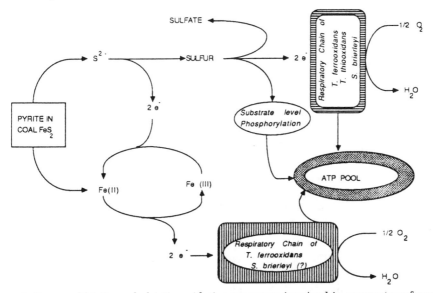

Figure 1. Proposed biotic and abiotic oxidative processes involved in energy transfer reactions during desulfurization of coal by *Thiobacillus* and *Sulfolobus* species (reprinted from Reference 86).

involved in energy transduction during desulfurization of coals by these bacterial species. The electrons generated through the oxidation–reduction reactions are passed through the respiratory chain of the bacteria and produce ATP by oxidative phosphorylation. Elemental sulfur may be produced through biological or abiological means. Thus *T. ferrooxidans* biologically oxidizes iron sulfides (pyrite), producing elemental sulfur and ferric sulfate as the main end products (56). Chemically, elemental sulfur can also be regenerated by interaction of acidic ferric (3) ions with sulfides, or polysulfides, under aerobic conditions. The elemental sulfur thus formed is also oxidized by *T. ferrooxidans, T. thiooxidans,* and *Sulfolobus* and *Acidianus* species to sulfuric acid. This reaction is critical, since it prevents the accumulation of sulfur on pyrite surfaces, which is believed to inhibit oxidation (144).

Elemental sulfur thus produced can be oxidized by these bacteria to sulfate, and ATP is generated by oxidative phosphorylation. The respiratory chain therefore is the dominant feature of ATP generation in these organisms; some aspects of this have been investigated (89). As reported by Kelly (81), the energy generation from sulfur oxidation (−389 KJ/mol) is much more favorable than from iron oxidation (−47 KJ/mol). Mechanisms for oxidation of pyrite or ferrous iron by *T. ferrooxidans* are presented in Figure 2 [modified after Ingledew (70)]. In *Sulfolobus* species, a similar mechanism is operational, except that cytochrome c is replaced by b-type cytochromes and cytochrome a_1 is replaced by the cytochrome oxidase components such as aa_3. The reduced species of iron or pyrite are oxidized at or near the surface, thus reducing rusticyanin, a blue copper protein present in *T. ferrooxidans*. The presence of rusticyanin in *Sulfolobus* has not been investigated.

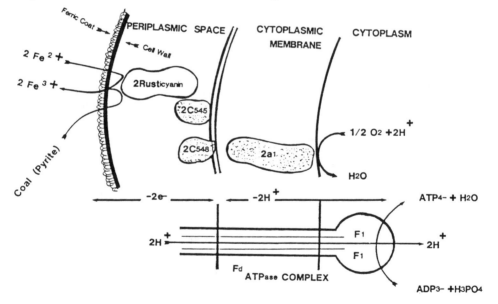

Figure 2. Proposed mechanisms of microbial ferrous iron oxidation and chemiosmotic coupling of ATP synthesis in *Thiobacillus ferrooxidans* [Modified after J. W. Ingledew (70)].

Mössbauer spectroscopy provided a quantitative measurement of the reactions of pyrite and its transformations in Western Kentucky #11 coal treated for biodesulfurization by *Acidianus brierleyi* (5, 24). The technique was essentially the spectroscopy based on the resonant absorption of nuclear gamma rays. Of the approximately 80 nuclear isotopes that exhibit the Mössbauer effect, ^{57}Fe has by far the best characteristics. Consequently, ^{57}Fe Mössbauer spectroscopy is usually the best technique available for investigating the iron-bearing phases in a complex sample. The behavior and transformation of pyrite in the bioprocessed Western Kentucky #11 coal were demonstrated by ^{57}Fe Mössbauer spectroscopy in collaboration with Huffman et al. (67) The pyrite content was significantly reduced, and the iron-bearing oxidation products were precipitated to a large extent. This techninque yielded ferric sulfate (Jarosite) as the main oxidation product but also some superparamagnetic FeOOH. Such paramagnetic phases have very fine particle sizes (50–200 Å) and exhibit enhanced liquefaction yield of coal. The factors that control the formation of the FeOOH catalyst are pH, temperature, and CO_2 concentration.

Bacterial Depyritization

The inorganic sulfur in coal is present as iron disulfide minerals such as pyrite or marcassite. Various physical, chemical, and microbiological methods have been applied to reduce the pyritic content of coal (113). Bacterial dissolution of pyrite has been studied both theoretically (8, 25, 94, 102, 103) and experimentally (25, 94, 102, 103, 114). The relative contributions of biological and chemical reactions to the overall rate of pyrite oxidation at different temperatures have been reported (25). Various patterns of microbial leaching on the surface of pyrite crystals have also been studied (15). Bioleaching of nonferrous sulfides, namely chalcocite, sphalerite, and galena, was shown to be effective with adapted thiophilic strains of Thiobacillus ferrooxidans (12). Olson (114) compared the different laboratory results on the rate of pyrite bioleaching by T. ferrooxidans. Continuous bacterial leaching of iron pyrite and related growth models of T. ferrooxidans has been studied (32). The mean rate of bioleaching of the pyrite reference material was 12.4 mg of Fe per liter per hr with a coefficient of variation (percent relative standard deviation) of 32% as determined by eight laboratories (114). The pyrite oxidation rates were derived with empirical equations obtained from the stoichiometry and kinetics of the spontaneous chemical reaction between pyrite and ferric iron at different temperatures. At the highest temperature studied (70°C), 43% (highest value) of the pyrite was oxidized chemically by ferric iron. Therefore, it is expected that only reactors operating at highest temperatures with extremely thermophilic bacteria will cause a significant deviation of the first-order overall kinetics of biological pyrite oxidation (25). The study of the thermophilic archaebacteria of the genus Sulfolobus (S. acidocaldarius, S. sulfataricus, and Acidianus brierleyi) showed that only A. brierleyi was able to oxidize and grow autotrophically on pyrite (101). In fact, the rate of pyrite removal by thermophiles was much higher than that of mesophiles, particularly when residence time was considered.

Process Technology for Biodepyritization

Dissolution of pyrite can be influenced by several parameters, such as pH, temperature, dissolved oxygen tension, pulp density, microorganisms, adaptation of microorganisms, substrate characteristics, design of bioreactor, and composition of the reaction medium. Each of these parameters must be carefully controlled. Depyritization of coal by mesophilic and thermophilic bacteria is carried out at an optimum pH value of 2.0–2.5 (40, 79, 112). Formation of complex hydroxy sulfate (jarosite) from oxidation of pyrite is hindered at pH 1.8. Therefore, many workers recommend this pH value as the most suitable pH for sulfur removal (19, 123). Bacterial cultures isolated from coal mines have been found to solubilize pyrite at higher rates than the respective axenic cultures (59, 96). Adaptation of microbial cultures to coal pyrite has been found to shorten the lag phase (59). An optimum inoculum size for depyritization has been reported to be within 10^{10}–10^{13} cells/g pyrite (7, 40). However, similar solubilization was obtained at a much lower inoculum concentration (17). Interestingly, competitive inhibition was observed in the kinetics of sulfur and pyrite oxidation with increased cell concentration of *T. ferrooxidans* (102, 103). Dugan (42) reported the highest pyrite oxidation rates from using a microbial consortium.

The elements required for the metabolic activity of microorganisms can be generated *in situ* because most of them are present in coal or associated minerals (96). Kos et al. (96) reported that a medium for desulfurization of 20% pulp density containing 1% pyrite must contain at least 0.5 mM ammonium and 0.1 mM phosphate. Low phosphate concentrations with a N/P molar ratio of 90:1 were recommended to prevent precipitation of undesirable compounds on the surface of the coal (59).

The dramatic influence of CO_2 was observed during the depyritization of bituminous coal by *Acidianus brierleyi* (24). Although the usual lag time of 2–3 days in desulfurization was delayed when 18% CO_2 was used, an enhanced microbial desulfurization rate was obtained with 7% CO_2 concentration. An initial desulfurization rate of 3.4 mg S/g/d was observed with 7% CO_2, which was an improvement over that of 2.3 mg S/g/d at 18% CO_2. Thus, the gas mass transfer of oxygen and CO_2 concentrations may be a limiting step in operating large scale reactors and leaching heaps (27). Another important parameter in assessing the extent of pyrite exposure is the particle size and surface area of the coal. The optimal pulp density of coal slurry for microbial oxidation was found to be 20–30% (17, 40, 65, 112). At higher concentrations, the reaction rates diminished, perhaps because of a lower bacteria-to-solid ratio. Mechanical shearing can also be a factor in controlling the biodepyritization (11, 61, 83).

Several types of bioreactors have been used for bioprocessing coal for depyritiza- tion: shake-flask, small-tank reactors or stir-tank reactors (59, 64, 112); air lift reactors (13, 19, 77); pachuca tank reactors (40, 61, 117); percolation columns (120); two-inch pipeline loop reactors (121); and packed column reactors (142). Scale-up studies for pyrite removal have been carried out in a multistage bioreactor of 200 liter volume run continuously for 200 days (20). Heap leaching conditions were simulated in percola- tion columns with coal particles up to 20 mm in size (18, 104, 142). Coal percolation in ditch and ring heaps has also been suggested (14). Microbial oxidation of pyrite is an exothermic reaction that could increase the temperatures and affect the heat balance, thus leading to variations in the rate of pyrite oxidation within a single system. This effect could be pronounced during heap-leaching where heat transfer is limited. Patents to remove pyrite from coal have already been reported (41, 44). In addition to simple estimates (109), detailed studies of the cost effectiveness of different configura- tions of microbial coal desulfurization processes have also been published (14, 21, 27, 40).

Removal of Organic Sulfur

Organic sulfur often accounts for 33 to 50% or more of the sulfur in most coals. Because organic sulfur is generally not removed by physical coal cleaning methods, it must be removed by either chemical or microbial processes. Attempts were made to desulfurize crude oil by using different axenic and natural bacterial isolates (51, 75, 107). These attempts were not successful. The superior performance of thermophilic organisms in desulfurizing coal has been established (79, 110). *Sulfolobus acidocaldarius* has been employed in developing an efficient desulfurization process (109). Bhattacharyya et al. (23) established that biological coal desulfurization can be made economical. In contrast to an earlier report (113), Khalid and Aleem (84, 85) succeeded in demonstrating the desulfurization potential of *Acidianus brierleyi. A. brierleyi,* when acclimatized with high sulfur coals, removed 30% of the organic sulfur from hard bituminous Kentucky coals with 5.8% total sulfur (Fig. 3) (86, 88, 89). During the acclimatization procedure, the pyrite oxidation capability (sulfur oxidase system) of these cells was repressed significantly. *Sulfolobus acidocaldarius* has also been reported to remove organic sulfur from bituminous coal and DBT (80).

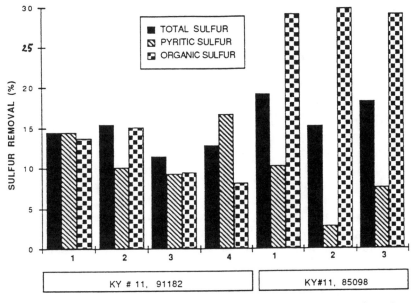

Figure 3. Desulfurization of bituminous (high-sulfur) coals by acclimatized *Acidianus brierleyi.* *Bar number* indicates the transfer number (reprinted from Referece 89).

The ability of coal-solubilizing fungi to remove sulfur from coal has also been investigated (45). Bioprocessing of bituminous coal was attempted in a two-stage system. Bituminous coal was first treated with *Poria placenta* and then desulfurized with *A. brierleyi* at 60°C. The fungal treatment facilitated desulfurization but not significantly (87).

Dibenzothiophenes (DBT) were used as model compounds for isolation and enrichment studies of bacterial cultures capable of desulfurizing crude oils (151). *Pseudomonas jianii* and *P. abikonensis* were isolated and found to degrade DBT into water-soluble compounds, which were identified by Kodama et al. (92) for the first time, who proposed a DBT degradation pathway (93). Other studies carried out independently have also provided evidence for the existence of this catabolic pathway (62, 100). So far no microorganism has been isolated that degrades DBT completely. Partial degradation of DBT was reported by Holland et al. (60) and Fortnagel et al. (52). Dibenzothiophene was shown to be oxidized to the corresponding sulfoxide and sulfone, which were metabolized further by soil- and sediment-enriched cultures (108). Nevertheless, sulfite or sulfate production has not been detected during these experiments. Sulfate accumulation during microbial attack on DBT and thioxanthane by *S. acidocaldarius* has been reported (72, 76, 78). Henderson et al. (57) found that soil isolate, *Pseudomonad* CB1, when grown on benzoate was capable of desulfurizing DBT, producing sulfate and 2,2′-dihydroxybiphenyl. They proposed a desulfurization mechanism of DBT in which 2,2′-dihydroxybiphenyl was the end product of DBT. The proposed pathway for DBT degradation by *Brevibacterium* sp. does not support this finding, however, since benzoate, instead of 2-hydroxybenzoate, was identified as an intermediate (145).

Because many organisms degrade thiophene (sulfide) to sulfoxide, to sulfone, to sulfonic acid, and finally to free-inorganic sulfate, this postulated pathway has been called the 4-S pathway (122). The 4-S pathway has attracted the attention of coal biotechnologists because of its desulfurization traits. Dugan (43) calculated the Gibbs free energy released during each step involved in this pathway. Incorporation of a single oxygen in DBT to form DBT-5-oxide (sulfoxide) releases 20 kcal per mole when oxygen is the oxidant; however, this energy release is enhanced to 50 kcal per mole if the microorganisms employ peroxide. Conversion of sulfoxide to DBT-sulfone is an endergonic reaction that requires 26 kcal per mole. In an organism that is capable of utilizing a peroxide, 3 kcal of energy per mole is released, an amount that is thought sufficient to overcome this uphill reaction slowly. Therefore, to degrade DBT through the 4-S pathway, microorganisms should substitute a peroxy enzyme for an oxygenase. Once DBT-sulfone has been formed, further conversion becomes easy, since a great amount of energy is released during the subsequent reactions. An assay has been developed at Lehigh University that permits a rapid screening of microorganisms capable of desulfurizing DBT (98). Another study revealed that selected isolates degraded DBT and accumulated DBT-5-oxide in culture fluids. One of the cultures degraded DBT-5-oxide, but none of the isolates degraded DBT-sulfone. Modification of water-soluble Illinois #6 and Ugljevik coal-derived products by DBT-degrading microorganisms has also been observed (139).

Oil-2, a new isolate (90), was found to degrade DBT through the putative Kodama's pathway (92, 110) to 3-hydroxy-2-formyl-benzothiophene (HFB). Co-metabolism of DBT was monitored by measuring the optical density (OD) of the aqueous phase at 395 nm, which indicated the formation of HFB. Two other metabolites with absorption maxima at 470 nm [trans-4-2-(3-hydroxy) benzothiophene-2-oxo-3 butonic acid] and 310 nm were also detected. When employed for desulfurizing oil-water emulsion of 4% DBT solution in mineral oil, this culture was able to remove 24% of the organic sulfur from the oil phase into the soluble aqueous phase. When crude oil was employed as the substrate, however, both oil-2 and axenic culture of *Pseudomonas putida* removed 70% of organic sulfur present in the crude oil-water emulsions (90).

A significant portion (64–66%) of the desulfurized sulfur was found to be present as sulfate when pure DBT was used as the substrate. A lesser proportion (47%) of desulfurized sulfur could, however, be recoverable as water-soluble sulfate during desulfurization of crude oil-water emulsions. Presumably, accumulation of this sulfate may be considered an indication that in addition to the ring-destructive mechanism (Kodama's pathway), another alternative metabolic pathway capable of attacking C–S bonds during desulfurization also exists in these organisms (90). This is the 4-S pathway postulated by Rhee and Campbell (122) (see above).

Some fungi are known to oxidize DBT to its sulfoxides and sulfone (60) but they do not desulfurize it. Kargi and Robinson (78) observed desulfurization of DBT by an archaebacterium *Sulfolobus acidocaldarius*. Khalid et al. (89) have demonstrated that *Acidianus brierleyi* was capable of desulfurizing the organic sulfur present in bituminous coal. Cleavage of C–S bonds leading to the removal of the sulfonic acid group during rearomatization of aryl sulfonates has been reported (30). The reaction is initiated by dioxigenase, which takes part in the naphthalene catabolism. *Escherichia coli* has been shown to degrade aromatic compounds, and it has been extensively characterized at the genetic level. Isolated successive mutants of *E. coli*, NAR 10, 20, 30, and 40, acquired the ability to use various compounds, e.g. furans and thiophenes, as the sole source of carbon and energy (1). Genes for thiophene degradation were designated as *thd* A (5–15 min), *thd* B (30–45 min), and *thd* C (85–100 min). These mutants (*thd* A, *thd* B, and *thd* C) were further improved for thiophene degradation (4, 33). Ultimately, the genes responsible for thiophene degradation could be transferred to the *E. coli* strain, thus creating a single organism that can degrade a broad spectrum of organic compounds.

LIQUEFACTION

Coal can be liquefied by a number of methods: (a) biological conversion of coal to liquid (bioconversion); (b) conversion of coal by hydrogenation to liquid fuels (direct liquefaction); (c) conversion of synthesis gas (mixture primarily of CO, H_2, and CO_2) to liquid fuels (indirect liquefaction); (d) pyrolysis and mild gasification to produce liquid fuels from coal; (e) production of liquid fuels from combined coal and petroleum feedstocks (coprocessing). Because this review is confined to the bioprocessing of coal, we discuss only the biological aspects of coal liquefaction below.

Solubilization of coal by Biological Systems

McIlwain (105) extensively reviewed the various aspects of the biological coal solubilization process. Although microbial action on coal was suggested as early as the 1960s (95, 124), only in the 1980s was it observed that fungi were capable of producing coal-derived liquid materials (35, 47). These findings were confirmed by Faison and Lewis (46), Scott et al. (129), Ward (146), and Wilson et al. (148). Organic-containing liquids were produced by numerous types of fungi grown in the presence of low-rank coals. Newly isolated thermophilic microorganisms were also capable of bio-solubilizing a lignite coal called leonardite (127). It was found that leonardite produced the highest yields of water-soluble organic products. The oxygen content of leonardite was considerably higher than that of the other coal varieties. These findings prompted investigations on the influence of oxidative pretreatment of coals in relation to microbial solubilization. The influence of oxidoreductases in coal solubilization has been investigated both *in vivo* and *in vitro* (36, 118). Interestingly, an increase of enzyme activity in the presence of organic solvent systems has also been observed (130). Coal solubilization has also been attributed to the production of alkaline metabolites (119, 140).

Aerobic microorganisms produced high molecular weight, polar, heterogeneous materials with high oxygen content when treated with low-ranked coals (129). Aerobic solubilization produced low molecular weight, oxidized coal "monomers" such as sub-stituted aromatic acids or phenols (45). Unfortunately, these oxidized products are of less fuel value; however, oxidized aromatic compounds obtained from biosolubiliza-tion of coal can be degraded anaerobically to produce methane (38). Low-rank coals have been considered to have a low fuel value that cannot be improved even by coal-cleaning technology. Hence, for bioprocessing technology to break-down these coals into low molecular weight compounds that can serve as valuable source of feed-stock for petrochemical/fermentation industry is a justifiable objective. Compounds present in coal, which are the secondary byproducts of chemical or biological reactions, may also adversely affect the process of coal biosolubilization (97, 125).

Biohydrogenation of Coal and Model Compounds

The hydrogenase activity involving H_2 uptake by methylene blue under strict anaerobic conditions was measured by gas chromatography. The enzyme activity was linear for the duration of the experiment (144 hr). The data show the relative rates of hydrogen uptake by coal (Fig. 4) in the presence and absence of bacteria incubated anaerobically at 37°C. In most cases, the microbially catalyzed H_2 uptake by methylene blue and coal was greater than that in the absence of the microbes. One of the coals treated with *Desulfovibrio desulfuricans* under hydrogen enhanced the liquefaction yield by about 5%. Coal treated with 50% benzene in the presence of bacteria increased the liquefaction yield by 7.3%; this result indicates that *D. desulfuricans* is metabolically active in an environment containing H_2 atmosphere and a concentration of 50% benzene (5, 6).

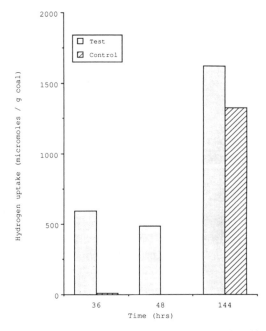

Figure 4. Hydrogen uptake in the presence of *Desulfovibrio desulfuricans* on coal #91182.

Chemical liquefaction of coal was achieved with a 50 cc stainless steel micro-autoclave reactor at 385°C for 15 min in the presence of hydrogen gas at 800 psi with tetralin as the solvent. After the reaction, different fractions such as oils, asphaltenes, and preasphaltenes were quantified based on the soluble property of these fractions in organic solvents (e.g. benzene, pyridine). The amount of the formation of the organic solvent solubles against insoluble matter is referred to as *conversion*. The liquefaction product yield of the coal desulfurized by *Acidianus brierleyi* is given in Table 2. The data show that the biotreatment of coal with *A. brierleyi* enhances the coal liquefaction yield by about 6% when coal particles are impregnated with the superparamagnetic FeOOH formed during the desulfurization process. The liquefaction yield was increased by approximately 25% with dimethyl disulfide, however, when the biotreated coal sample was subjected to chemical liquefaction (D. Bhattacharyya, M. V. S. Murty, R. I. Kermode, V. Madhu, M. I. H. Aleem, unpublished results). The coal-liquefaction product yields (wt.%) were determined at the University of Kentucky Center for Applied Energy.

Table 2. Chemical (direct) liquefaction product yield (wt%) of coal bio-desulfurized by *Acidianus brierleyi*.

	Before FeOOH precipitation	After FeOOH precipitation
Gas	5	6
Oil	41	39
Asphaltenes	29	33
Preasphaltenes	10	12
Conversion	84	90

Liquefaction of Coal-derived Products

Synthesis gas (syn-gas) may be produced from several sources, including coal, shale oil, tar sands, heavy residues, biomass, or natural gas. Indirect liquefaction of coal-derived products by microorganisms has been performed by a group at the University of Arkansas (53). The products of microbial conversion of syn-gas were found to be primarily methane and a small quantity of acetate. When *Peptostreptococcus productus* was adopted to a test gas (65% CO_2, 22% H_2, 11% CO, and 2% CH_4), approximately 90% of the CO was converted to acetate in 42 min. The sensitivity of *P. productus* to hydrogen sulfide and carbon disulfide, common catalyst poisons, did not inhibit acetate production. Mixed cultures produced small amounts of ethanol, butanol, and methanol as well as acetate. A pure isolate of *Clostridium* sp. has been found to produce only acetate and ethanol, as much as 4.3 g/l (105).

OTHER CONSIDERATIONS

Removal of Nitrogen Compounds

Biochemical investigations of the removal of organic nitrogen from N heterocyclic model compounds such as pyrrol or pyridine have been carried out. During microbial degradation, pyridine or substituted pyridine are reduced to the respective dihydroquinones, which are further oxygenated or hydrated. Nitrogen is then released as ammonium and the resulting product is metabolized through the tricarboxylic acid cycle or by the beta-oxidation pathway. Similarly, metabolic routes through which pyrrole can be degraded by microorganisms are also known. The pyrrol ring is initially oxidized and cleaved to form anthranilic acid; the amino group is then eliminated and the product is further degraded via gentisic acid (34, 131).

Shale oils are known to contain high concentrations of nitrogen, thus making them unsuited for direct refining into useful fuels (143). Two early reports described the degradation of nitrogen-containing compounds in oils by mixed bacterial cultures (48, 51). Aliphatic nitriles and amines were selectively removed from Stuart shale oil bacterial cultures without attacking the hydrocarbon or heterocyclic nitrogen fractions of the oil (2). Later studies demonstrated that microorganisms can remove heterocyclic nitrogen compounds present in shale oil, namely quinolines, methylquinolines, and isoquinolines, without attacking aromatic or aliphatic hydrocarbons (3). Heterocyclic nitrogen compounds are not recalcitrant to microbial attack; bacteria have previously been reported to biotransform pyridines, pyrroles, quinolines, indoles, and carbazoles (16, 48, 50, 51, 55, 99, 115, 116, 128, 131–134).

Bioleaching of Trace Metals

Large quantities of fly and bottom ash are produced during coal combustion in thermal power plants. Because the ash causes pollution of ground water resources by leaching of the arsenic, antimony, lead, cadmium and other heavy toxic metals contained in the waste, its disposal must be carried out under constraints imposed by the environmental protection agencies.

Sulfur oxidizers such as *Thiobacillus ferrooxidans, T. thiooxidans, Sulfolobus acidocaldarius,* and *Acidianus brierleyi* are known to oxidize a variety of metal sulfides, in addition to pyrite. Therefore, it is possible that the various associated metal sulfides may also be oxidized to soluble sulfates during microbial depyritization of coal. We noticed a significant reduction in the ash content of the bioprocessed bituminous coal that was treated with *A. brierleyi* (22, 24). Trace elements such as Be, Ga, Co, Ni, V (104), Cu, Zn, and Mo (111) from coal of various types have been removed by microbial action. Ferraiolo et al. (49) reviewed different methods of fly ash disposal and utilization and suggested that bioprocessing of coal before combustion may reduce the ash content. Reducing the ash content of coal and thereby the toxic metal content would substantially control the problem of water pollution.

SUMMARY

Current methods for bioprocessing of coal include desulfurization, solubilization/liquefaction, conversion of coal-derived materials, removal of nitrogen compounds, reduction of ash content, and removal of toxic metals. Some sulfur-oxidizing bacteria such as *Thiobacillus ferrooxidans* and *Sulfolobus acidocaldarius* can alter pyrite in coals to water-soluble sulfate ion. Their activities depend upon pH, temperature, the particle size of the coal, the concentration of the slurry, and the strain's metabolic potential. At the optimized conditions, about 90% of inorganic sulfur in coal can be removed in 1 or 2 weeks in bench-scale experiments. Organic sulfur is more difficult to remove than the inorganic form. Very few strains can substantially remove organic sulfur. *Sulfolobus acidocaldarius* can remove both inorganic and organic sulfur simultaneously. One mutant strain of *Pseudomonas* is notabale for its enhanced rate of desulfurization. Some fungi and bacteria can degrade some low-rank coals into a water-soluble black liquid. This product contains roughly 95% of the original energy and may have a decreased ash content.

Microbial solubilization results from the action of a biological catalyst, such as an enzyme released from the microorganism. Therefore, research on a cell-free reaction system has begun. The bioconversion of low-rank coals to methane gas has been proposed. *Thiobacillus ferrooxidans* and *S. acidocaldarius* can remove most inorganic sulfur in bench-scale experiments. More extensive studies will be required, however, before the reactors can be applied on a commercial scale. Removal of organic sulfur through bioprocessing is very difficult, since efficient microorganisms have not yet been found. Microbial liquefaction and gasification of coal are promising technologies for coal utilization (91). Although knowledge of the distribution of the organic sulfur in the various macerals is important for developing any plan for sulfur removal, little detailed information is currently available.

The direct involvement of microbes in coal solubilization has not yet been documented, but results obtained so far indicate that microorganisms do promote the solubilization of pretreated coals. The subsequent product may be further degraded by microbial action. The bioprocessing of coal with the aerobic thermoacidophilic archaebacterium *Acidianus brierleyi* leads to a 25% enhancement in the chemical liquefaction yield. It appears to serve as an important biocatalyst in the microbial coal conversion processes by virtue of its ability to produce superparamagnetic FeOOH fine particles that form a coat on coal particles and thus act as an important catalyst. *Desulfovibrio desulfuricans* is metabolically active even with 50% benzene in the medium; 50% benzeme actually increases the liquefaction yield from 5% to over 7%. Whether any paramagnetic catalysts are produced by the organism remains to be determined. This organism possesses hydrogenases, with iron, selenium, and nickle constituting the prosthetic group. *Acetobacterium woodii* and *Peptostreptococcus productus* have been reported to convert carbon monoxide to acetate and ethanol. High conversion yields and short contact times have been found for synthetic gas mixtures representative of syn-gas. Economic analysis indicates that microbial based processes may be comparable in cost to similar conventional chemical processes, but neither type of process is viable in today's energy market.

Though a few microorganisms have been identified that can degrade some of the aliphatic and heterocyclic nitrogen compounds present in oils, organisms are still sought that degrade other compounds such as methylisoquinolines, methylpyridines, and methylquinolines to upgrade shale oils. Biological reduction of the nitrogen content from coal has been elusive until recently because of the complex structure of coal. Nevertheless, bioleaching of toxic metals from coal should be pursued because the initial findings are promising. New and innovative technologies are needed to improve coal utilization and to make coal not only the most abundant but also the most economical fuel and energy source for the future.

Acknowledgments

This study was supported by a U.S. Department of Energy contract (DE-FC22-86PC90017). We thank all previous coworkers of our laboratories, especially Dr. A. M. Khalid, for conducting research in the bioprocessing of coal.

LITERATURE CITED

1. Abdulrahid, N., and D. P. Clark. 1987. Isolation and genetic analysis of mutations allowing the degradations of furans and thiophenes by *Escherichia coli*. *J. Bacteriol.* 169:1267–1271.

2. Aislabie, J., and R. M. Atlas. 1988. Biodegradation of nitriles in shale oil. *Appl. Environ. Microbiol.* 54:2197–2202.

3. Aislabie, J., and R. M. Atlas. 1989. Microbial removal of heterocyclic nitrogen compounds from shale oil. In *Proceedings on Bioprocessing of Fossil Fuels Workshop*. Virginia, USA.

4. Alam, K. Y., M. J. Worland, and D. P. Clark. 1990. Analysis and molecular cloning of genes involved in thiophenes and furan oxidation by *Escherichia coli*. *Appl. Biochem. Biotechnol.* 24/25:843–855.

5. Aleem, M. I. H., D. Bhattacharyya, G. P. Huffman, R. I. Kermode, M. V. S. Murty, H. Venkatachalam, and M. Ashraf. 1991. Microbial hydrogenation of coal and diphenyl methane. *ACS Div. Fuel Chem.* 36:53–57.

6. Aleem, M. I. H., D. Bhattacharyya, R. I. Kermode, M. V. S. Murty, H. Venkatachalam, and M. Ashraf. 1991. Bioprocessing of coal: Microbial hydrogenation of coal and effect on liquefaction. In *EPRI-Second International Symposium on Biological Processing of Coal*. San Diego, Calif. Palo Alto, Calif: Electric Power Research Institute (EPRI).

7. Andrews, G. F., and J. Maczuga. 1982. Bacterial coal desulfurization. *Biotechnol. Bioeng. Symp.* 12:337–348.

8. Andrews, G. 1991. A model for coal depyritization. *Resour. Conserv. Recycl.* 5:285–296.

9. Attar, A. 1978. Chemistry, thermodynamics and kinetics of reactions of sulfur in coal-gas reactions: A review. *Fuel* 57:201–212.

10. Attar, A., and G. T. Hendrickson. 1984. Functional groups and heteroatoms in coal. In *Coal Structure*. Ed. R. A. Meyers. New York: Academic Press.

11. Attia, Y. A., and M. El-zeky. 1985. Biosurface modification in the separation of pyrite from coal by froth flotation. In *Processing and Utilization of High Sulfur Coals*. Ed. Y. A. Attia. Amsterdam: Elsevier.

12. Attia, Y. A., and M. El-zeky. 1990. Bioleaching of non-ferrous sulfides with adapted thiophilic bacteria. *Chem. Eng. J.* 44:B31–B40.

13. Beier, E. 1985. Removal of pyrite from coal using bacteria. In *Processing and Utilization of High Sulfur Coals*. Ed. Y. A. Attia. Amsterdam: Elsevier.

14. Beier, E. 1988. Microbial pyrite removal from hard coal. *Resour. Conserv. Recycl.* 1:223–250.

15. Bennett, J. C., and H. J. Tributsch. 1978. Bacterial leaching patterns on pyrite crystal surfaces. *J. Bacteriol.* 134:310–317.

16. Bennett, J. L., D. M. Updegraff, W. E. Pereira, and C. E. Rostad. 1985. Isolation of four species of quinoline-degrading *Pseudomonads* from cresol-contaminated site at Pensacola, Florida. *Microbios Lett.* 29:147–154.

17. Beyer, M., H. G. Ebner, and J. Klein. 1986. Influence of pulp density and bioreactor design on microbial desulfurized coals. *Appl. Microbiol. Biotechnol.* 24:342–346.

18. Beyer, M. 1987. Microbial removal of pyrite from coal using a percolation bioreactor. *Biotechnol. Lett.* 9:19–24.

19. Beyer, M., H. G. Ebner, H. Assenmacher, and J. Frigge. 1987. Elemental sulphur in microbiologically desulfurized coals. *Fuel* 66:551–555.

20. Beyer, M., H. J. Hone, and J. Klein. 1987. A multistage bioreactor for continuous desulfurization of coals by *Thiobacillus ferrooxidans*. In *Biohydrometallurgy*. Ed. P. R.Norris, D. P. Kelly. London: Science and Technology Letters.

21. Beyer, M., J. Klein, K. Vaupel, and D. Weigand. 1988. Microbial desulfurization of coal: Calculation of costs. In *International Symposium on Biotechnology to Fuel, Chemical Production and Waste Management*. Helsingor, Denmark.

22. Bhattacharyya, D., M. Hseieh, A. M. Khalid, H. Francis, R. I. Kermode, and M. I. H. Aleem. 1988. Biological desulfurization of coals by mesophilic and thermophilic microorganisms. In *Proceedings Bioprocessing of Coals Workshop-III*. Idaho Falls, Idaho: Idaho National Engineering Laboratory Bioprocessing Center, Department of Energy.

23. Bhattacharyya, D., R. I. Kermode, M. I. H. Aleem, M. Hseieh, and A. M. Khalid. 1988. Biological desulfurization of coal shows promises. *Liquefact. Sci. Update* 4:1–4.

24. Bhattacharyya, D., M. Hseieh, H. Francis, R. I. Kermode, A. M. Khalid, and M. I. H. Aleem. 1990. Biological desulfurization of coals by mesophilic and thermophilic microorganisms. *Resour. Conserv. Recycl.* 3:81–96.

25. Boogerd, F. C., C. Van-Den-Beemed, T. Stoelwinder, P. Bos, and L. G. Kuenen. 1991. Relative contributions of biological and chemical reactions to the overall rate of pyrite oxidation at temperatures between 30°C and 70°C. *Biotechnol. Bioeng.* 38:109–155.

26. Bos, P., and J. G. Kuenen. 1983. Microbiology of sulfur oxidizing bacteria. In *Microbial Corrosion*. Ed. N. P. L. Teddington. London: The Metals Society.

27. Bos, P., T. F. Huber, C. H. Kos, C. Ras, and G. J. Kunen. 1986. A dutch feasibility study on microbial coal desulfurization. In *Fundamental and Applied Biohydrometallurgy*. Ed. R. W. Lawrence, R. M. R. Branion, and H. G. Ebner. Amsterdam: Elsevier.

28. Boudou, J. P., J. Boulegue, L. Malechjaux, M. Nip, J. W. DeLeeuw, and J. Boon. 1987. Identification of some sulfur species in a high organic sulfur coal. *Fuel* 66:1558–1569.

29. Brierley, C. L., and J. A. Brierley. 1973. A chemoautotrophic and thermophilic microorganism isolated from hot spring. *Can. J. Microbiol.* 19:183–188.

30. Brilon, C., W. Beckamnn, and H. J. Knackmuss. 1981. Catabolism of naphthalene sulfonic acid by *Pseudomonas* sp. A3 and *Pseudomonas* sp. C22. *Appl. Environ. Microbiol.* 42:44–55.

31. Brock, T. D. 1978. The genus *Sulfolobus*. In *Thermophilic Microorganisms and Life at High Temperature*. New York: Springer-Verlag.

32. Chang, Y. C., and A. S Myerson. 1982. Growth models of continuous bacterial leaching of iron pyrite by *Thiobacillus ferrooxidans*. *Biotechnol. Bioeng.* 24:889–902.

33. Clark, D., D. Alam, N. Abdulrashid, and B. Klubek. 1988. Successive mutation of *Escherichia coli* for improved thiophene degradation. *Appl. Biochem. Biotechnol.* 18:393–401.

34. Claus, G., and H. T. Kutzner. 1983. Degradation of indole by *Alcaligenes* sp. *System. Appl. Microbiol.* 4:169–180.

35. Cohen, M. S., and P. D. Gabriele. 1982. Degradation of coal by the fungi *Polyporus versicolor* and *Poria placenta*. *Appl. Environ. Microbiol.* 44:23–27.

36. Cohen, M. S., W. C. Bowers, H. Aronson, and E. T. Gray, Jr. 1987. Cell-free solubilization of coal by *Polyporus versicolor*. *Appl. Environ. Microbiol.* 3:2840–2843.

37. Colmer, A. R., and M. E. Hinkle. 1947. The role of microorganisms in acid mine drainage. *Science* 106:253–257.

38. Davidson, B. H., D. M. Nicklaus, A. Misra, S. N. Lewis, and B. D. Faison. 1990. Utilization of microbially solubilized coal. Preliminary studies on aerobic conversion. *Appl. Biochem. Biotechnol.* 24/25:447–456.

39. Davidson, R. M. 1982. Coal. In *Coal Science and Technology*. Vol. 1., Ed. M. L. Gorbaty, J. W. Larsen, I. Wender. New York: Academic Press.

40. Detz, C. M., and G. Barvinchak. 1979. Microbial desulfurization of coal. *Mining Congr. J.* 65:75–86.

41. Detz, C. M., and G. Barvinchak. 1980. Microbial desulfurization of coal. US Patent 4,206,288.

42. Dugan, P. R. 1984. Desulfurization of coal by mixed microbial cultures. *Ohio State Univ. Bioscience Colloq.* 1982. 8:3–9.

43. Dugan, P. R. 1988. Processing of coals at the INEL. In *Proceedings Bioprocessing of Coals Workshop—III*. Idaho Falls, Idaho: Idaho National Engineering Laboratory Bioprocessing Center.

44. Dugan, P. R., and W. A. Apel. 1984. Microbial desulfurization of coal. US Patent 4,456,688.

45. Faison, B. D., T. M. Clark, S. N. Lewis, C. Y. Ma, D. M. Sharkey, and C. A. Woodward. 1993. Degradation of organic sulphur compounds by a coal solubilizing fungus. *Appl. Biochem. Biotechnol.* In press.

46. Faison, B. D., and S. N. Lewis. 1990. Microbial coal solubilization in defined cultures systems. Biochemical and physiological studies. *Resour. Conserv. Recycl.* 3:59–67.

47. Fakoussa, R. M. 1981. Kohl als Substract für Microorganismem: Untersuchungen zur Mikrobiellen Unsetzung Nativer Steinkohle. (Coal as a substrate for microorganisms: Investigations of the microbial decompositions of untreated hard coal). Ph.D Thesis, University of Bonn, Germany.

48. Fedorak, P. M., and D. W. S. Westlake. 1984. Microbial degradation of alkyl carbazoles in

Norman Wells Crude Oil. *Appl. Environ. Microbiol.* 47:858–862.

49. Ferraiolo, G., M. Zilli, and A. Converti. 1990. Fly ash disposal and utilization. *J. Chem. Technol. Biotechnol.* 47:281–305.

50. Finnerty, W. R. 1982. Microbial desulfurization and denitrogenation of fossil fuels. In *Energy Technology Proceedings of the Ninth Energy Technology Conference.* Washington, D.C.: Government Institutes Inc.

51. Finnerty, W. R., K. Shocley, and H. Attaway. 1982. Microbial desulfurization and denitrogenation of hydrocarbons. In *Microbial Enhanced Oil Recovery.* Ed. J. E. Zeijic, D. C. Cooper, T. R. Jack, N. Kosaric. Oklahoma: Renn Well Books Tulsa.

52. Fortnagel, P., H. Harms, R. M. Wittich. 1989. Cleavage of dibenzofuran and dibenzodioxin ring systems by a *Pseudomonas* bacterium. *Naturwissenschaften* 76:1135–1141.

53. Gaddy, J. L. 1986. Production of methane from coal synthesis gas. In *Proceedings: Biological Processing of Coals Workshop.* Herndon, VA.

54. Given, P. H. 1960. The distribution of hydrogen in coals and its relation to coal structure. *Fuel* 39:147–153.

55. Grant, D. J. W., and Al-Najjar. 1976. Degradation of quinoline by a soil bacterium. *Microbios* 15:177–189.

56. Hazeu, W., W. H. Batenburg-van der Vegte, P. Bos, R. K. Van der Pas, and J. G. Kuenen. 1988. The production and utilization of intermediary elemental sulfur during the oxidation of reduced sulfur compounds by *Thiobacillus ferrooxidans.* *Arch. Microbiol.* 150:574–579.

57. Henderson, C. B., J. D. Isbister, and E. A. Kobylinski. 1985. Microbial desulfurization of coal. In *Proceeding of the 10th International Conference on Slurry Technology.* Washington, D.C.

58. Hippo, E. J., J. C. Crelling, S. R. Palmer, and M. A. Kruge. 1990. Organic sulfur compounds in coal. In *Proceedings of Fourteenth Annual EPRI Conference on Fuel Science.* Palo Alto, California: EPRI.

59. Hoffmann, M. R., B. G. Faust, F. A. Panda, H. H Koo, and H. M. Tsuchiya. 1981. Kinetics of the removal of iron pyrite from coal by microbial catalysis. *Appl. Environ. Microbiol.* 42:259–271.

60. Holland, H. L., S. H. Khan, D. Richards, E. Riemland. 1986. Biotransformation of polycyclic aromatic compounds by fungi. *Xenobiotica* 16:733–741.

61. Hone, H. J., M. Beyer, H. J. Ebner, J. Klein, and H. Juntgen. 1987. Microbial desulfurization of coal-development and application of a slurry reactor. *Chem. Eng. Technol.* 3:173–179.

62. Hou, C. T., and A. I. Laskin. 1976. Microbial conversion of dibenzothiophene. *Dev. Ind. Microbiol.* 17:351–362.

63. Hsien, K. C., and C. A. Wert. 1986. Direct measurement of organic sulfur in coal. *Fuel* 64:255–262.

64. Huber, T. F., N. W. F. Kossen, P. Bos, and J. C. Kuenen. 1983. Modelling design and scale-up of a reactor for microbial desulfurization of coal. In *Recent Progress in Biohydrometallurgy.* Ed. G. Rossi and A. E. Torma. Iglesias, Italy: Associazione Mineraria Sarda.

65. Huber, T. F., C. Ras, and N. W. F. Kossen. 1984. Design and scale-up of a reactor for microbial desulfurization of coal. A kinetic model for bacterial growth and pyrite oxidation. In *Proceedings of Third European Congress on Biotechnology.* Vol. 3. New York: VCH (Verlagesellschaft).

66. Huffman, G. P., F. E. Huggins, S. Mitra, N. Shah, R. J. Pugmire, B. Davis, F. W. Lytle, and R. B. Greegor. 1989. Investigations of the molecular structure of organic sulfur in coal by XAFS spectroscopy. *Energy Fuels* 3:200–205.

67. Huffman, G. P., N. Shah, and F. Huggins. 1989. Characterization of inorganic constituents and heteroatoms in coal by advanced analytical techniques. In *Proceedings: 1989 Symposium on Biological Processing of Coal and Coal-derived Substances.* Palo Alto, California: EPRI ER-6572.

68. Huffman, G. P., F. P. Huggins, H. E. Francis, H. E. Mitra, and N. Shah. 1990. Structural characterization of sulfur in bioprocessed coal. In *Third International Conference on Processing and Utilization of High-Sulfur Coals III*. Ed. R. Markuszewski, T. D. Wheelock. Amsterdam: Elsevier.

69. Huxtable, R. J. 1986. *Biochemistry of Sulfur*. New York/London: Plenum Press.

70. Ingledew, W. J. 1986. Ferrous iron oxidation by *Thiobacillus ferrooxidans*. *Biotechnol. Bioeng. Symp.* 16:23–33.

71. Isbister, J. D., and R. C. Doyle. 1985. Mutant microorganisms and its use in removing organic sulfur compounds. US Patent 4,562,156.

72. Isbister, J. D., and E. A. Kobylinski. 1985. Microbial desulfurization of coal. In *International Conference on Processing and Utilization of High Sulfur Coals*. Ed. Y. A. Attia. Amsterdam: Elsevier.

73. Kargi, F. 1982. Microbial coal desulfurization. *Enzyme Microb. Technol.* 4:13–19.

74. Kargi, F. 1984. Microbial desulfurization of coal. *Biotechnol. Progress/Processes* 3:241–272.

75. Kargi, F. 1986. Microbial methods for desulfurization of coal. *Trends Biotechnol.* 4:293–297.

76. Kargi, F. 1987. Biological oxidation of thianthrene, thioxanthane and dibenzothiophene by the thermophilic organisms *Sulfolobus acidocaldarius*. *Biotechnol. Lett.* 9:478–482.

77. Kargi, F., and Cervoni, T. D. 1983. An air-lift-recycle fermentor for microbial desulfurization of coal. *Biotechnol. Lett.* 5:33–38.

78. Kargi, F., and J. M. Robinson. 1984. Microbial oxidation of dibenzothiophene by the thermophilic organisms *Sulfolobus acidocaldarius*. *Biotechnol. Bioeng.* 25:687–690.

79. Kargi, F., and J. M. Robinson. 1985. Biological removal of pyritic sulfur from coal by the thermophilic organisms *Sulfolobus acidocaldarius*. *Biotechnol. Bioeng.* 27:41–49.

80. Kargi, F., and J. M. Robinson. 1986. Removal of organic sulfur from bituminous coal and dibenzothiophene by the thermophilic organisms *Sulfolobus acidocaldarius*. *Fuel* 65:397–399.

81. Kelly, D. P. 1990. Energetics of chemolithotrophs. In *The Bacteria: Bacterial Energetics*. Vol. 12. Ed. T. A. Krulwich. New York: Academic Press Inc.

82. Kelly, R. M., and J. D. Deming. 1988. Extremely thermophilic archaebacteria: biological and engineering considerations. *Biotechnol. Progr.* 4:47–62.

83. Kempton, A. G., N. Moneib, R. G. L. McCready, and C. E. Capes. 1980. Removal of pyrite from coal by conditioning with *Thiobacillus ferrooxidans* followed by oil agglomeration. *Hydrometallurgy* 5:115–117.

84. Khalid, A. M., and M. I. H. Aleem. 1988. Desulfurization of coal by *Sulfolobus* and *Thiobacillus* species, abstr. 51, p. 270. *Abstr. 88th Annu. Meet. Am. Soc. Microbiol.* 1988.

85. Khalid, A. M., and M. I. H. Aleem. 1988. Oxidation of sulfur compounds by the archaebacterium *Sulfolobus brierleyi*. Abstr.I-103, p. 198. *Abstr. 88th. Annu. Meet. Am. Soc. Microbiol.* 1988.

86. Khalid, A. M., D. Bhattacharyya, and M. I. H. Aleem. 1989. Sulfur metabolism and the coal desulfurization potential of *Sulfolobus brierleyi* and *Thiobacilli*. In *Proceedings: 1989 Symposium on Biological Processing of Coal and Coal-Derived Substances*. Palo Alto, California: EPRI ER-6572.

87. Khalid, A. M., D. Bhattacharyya, and M. I. H. Aleem. 1989. Biological desulfurization of coal. In *Processing and Utilization of High Sulfur Coals*. Ed. R. Markuszewski, T. D. Wheelock. Amsterdam: Elsevier.

88. Khalid, A. M., M. Hsieh, R. I. Kermode, D. Bhattacharyya, and M. I. H. Aleem. 1989. Biological desulfurization of coal. Paper No: 98h *Abstr. Annu. Meet. Am. Inst. Chem. Eng.*, San Francisco, California. 1989.

89. Khalid, A. M., D. Bhattacharyya, and M. I. H. Aleem. 1990. Coal desulfurization and

electron transport linked oxidation of sulfur compounds by *Thiobacillus* and *Sulfolobus* species. *Dev. Ind. Microbiol.* 31:115–126.

90. Khalid, A. M., M. I. H. Aleem, R. I. Kermode, and D. Bhattacharyya. 1991. Bioprocessing of coal and oil-water emulsions and microbial metabolism of dibenzothiophene (DBT). *Resour. Conserv. Recycl.* 5:167–181.

91. Kimura, K., N. Ohmura, H. Saiki, and Y. Khono. 1989. Microorganisms for coal processing. *Denryoku Chuo Kenkyusho Hokoku* 0 (U89032), 1–6, 1–32 (Japanese).

92. Kodama, K., S. Nakatani, K. Umehara, K. Shimizu, Y. Minoda, and K. Yamada. 1970. Microbial conversion of petro-sulfur compounds. *Agric. Biol. Chem.* 34:1320–1324.

93. Kodama, K., K. Umehara, K. Shimizu, S. Nakatani, S., Y. Minoda, and K. Yamada. 1973. Identification of microbial products from dibenzothiophenes and its proposed oxidative pathway. *Agric. Biol. Chem.* 37:45–50.

94. Konishi, Y., S. Asai, and H. Katoh. 1990. Bacterial dissolution of pyrite by *Thiobacillus ferrooxidans*. *Bioprocess Eng.* 5:231–237.

95. Korburg, J. A. 1964. Microbiology of coal: Growth of bacreria in plain and oxidized slurries. *Proc. West Virginia Acad. Sci.* 36:26–30.

96. Kos, C. H., W. Bijleveld, T. Grotenhuis, P. Bos, P. R. E. Poorter, and J. G. Kuenen. 1983. Composition of mineral salts medium for microbial desulfurization of coal. In *Recent Progress in Biohydrometallurgy*. Ed. G. Rossi, A. E. Torma. Iglesias, Italy: Associazione Mineraria Sarda.

97. Kosanke, R. M. 1954. A bacteriostatic substance extracted from vitrain ingredients of coal. *Science* 119:214–216.

98. Krawiec, S. 1988. Detection, isolation and initial characterization of bacteria with the ability to desulfurize dibenzothiophene to o-o'-biphenyl. In *Proceedings Bioprocessing of Coals Workshop—III*. Idaho Falls, Idaho: Department of Energy, Idaho National Engineering Laboratory Bioprocessing Center.

99. Kucher, R. V., A. Turovskii, N. V. Dzumdedzei, and A. G. Shevchenko. 1980. Microbiological transformation of quinoline by *Pseudomonas putida* bacteria. *Mikrobiol. Zh.* 42:284–287.

100. Laborde, A. L., and D. T. Gibson. 1977. Metabolism of dibenzothiophene by *Beijerinckia* species. *Appl. Environ. Microbiol.* 34:783–790.

101. Larsson, L., G. Olsson, O. Holst, and H. T. Karlosson. 1990. Pyrite oxidation by thermophilic archaebacteria. *Appl. Environ. Microbiol.* 56:697–701.

102. Lizama, H. M., and I. Suzuki. 1989. Rate equations and kinetic parameters of the reactions involved in pyrite oxidation by *Thiobacillus ferrooxidans*. *Appl. Environ. Microbiol.* 55:2918–2923.

103. Lizama, H. M., and I. Suzuki. 1991. Kinetics of sulfur and pyrite oxidation by *Thiobacillus thiooxidans*: Competitive inhibition by increasing concentrations of cells. *Can. J. Microbiol.* 37:182–187.

104. McCready, R. G. L., and M. Zentilli. 1985. A feasibility study on the reclamation of coal waste dumps by bacterial leaching. *CIM Bull.* 78:67–78.

105. McIlwain, M. E. 1989. Review of bioconversion of coal. In *Coal Liquefaction—Research Needs, Assessment Technical Background*. Vol. 2. DOE.DE-AC01-84ER30110. McLean, Virginia: DOE.

106. Monroe, D. 1985. Microbial metal mining. *American Biotechnology Lab*. Jan/Feb., 10.

107. Monticello, D. J., and W. R. Finnerty. 1985. Microbial desulfurization of fossil fuel. *Annu. Rev. Microbiol.* 39:371–389.

108. Morimile, M. R., and R. M. Atlas. 1988. Mineralization of the dibenzothiophene biodegradation products, 3-hydroxy-2-formyl-benzothiophene and dibenzothiophene sulfone. *Appl. Environ. Microbiol.* 54:3183–3184.

109. Murphy, J., E. Riestenberg, R. Mohler, D. Marek, B. Beck, and D. Skidmore. 1985. Coal

desulfurization by microbial processing. In *Proceeding and Utilization of High Sulfur Coals*. Ed. Y. A. Attia. Amsterdam: Elsevier.

110. Murr, L. E., and P. Mehta. 1982. Coal desulfurization by leaching involving acidophilic and thermophilic microorganisms. *Biotechnol. Bioeng.* 24:743–748.

111. Norton, G. A., and R. Markuszewski. 1989. Trace element removal during physical and chemical coal cleaning. *Coal Preparation* 7:55–68.

112. Olsen, T., D. Ashman, A. E. Torma, and L. E. Murr. 1980. Desulfurization of coal by *Thiobacillus ferrooxidans*. In *Biogeochemistry of Ancient and Modern Environments*. Ed. A. P. Trudinger, M. R. Walters, B. J. Ralph. Berlin: Springer.

113. Olson, G. J., and F. E. Brinckman. 1986. Bioprocessing of coal. *Fuel* 65:1638–1646.

114. Olson, G. J. 1991. Rate of pyrite bioleaching by *Thiobacillus ferrooxidans*: Results of an inter laboratory comparison. *Appl. Environ. Microbiol.* 57:642–644.

115. Periera, W. E., C. E. Rostar, D. M. Updergraff, and J. L. Bennet. 1987. Fate and movement of azaarenes and their anaerobic transformation products in an aquifer contaminated wood-treatment chemicals. *Environ. Toxicol. Chem.* 6:163–176.

116. Periera, W. E., C. E. Rostar, T. J. Leiker, D. M. Updergraff, and J. L. Bennet. 1988. Microbial hydroxylation of quinoline in contaminated ground water: evidence for incorporation of oxygen atom of water. *Appl. Environ. Microbiol.* 54:827–829.

117. Pooley, F. D., and A. S. Atkins. 1983. Desulfurization of coal using bacteria by both drum and process plant techniques. In *Recent Progress in Biohydrometallurgy*. Ed. G. Rossi, A. E. Torma. Inglesias, Italy: Associazione Mineraria Sarda.

118. Pyne, J. W., D. L. Stewart, J. Fredrickson, and B. W. Wilson. 1987. Solubilization of leonardite by an extracellular fraction from *Cordius versicolor*. *Appl. Environ. Microbiol.* 53:2844–2848.

119. Quigley, D. R., B. Ward, D. L. Crawford, H. J. Hatcher, and P. R. Dugan. 1989. Evidence that microbially produced alkaline materials are involved in coal solubilization. *Appl. Biochem. Biotechnol.* 20/21:753–763.

120. Radway, J. C., L. H. Tuttle, N. J. Fendinger, and J. C. Means. 1987. Microbially mediated leaching of low-sulfur coal in experimental coal columns. *Appl. Environ. Microbiol.* 53:1056–1063.

121. Rai, C. 1985. Microbial desulfurization of coals in a slurry pipeline reactor using *Thiobacillus ferrooxidans*. *Biotechnol. Progr.* 1:200–204.

122. Rhee, K. H., and I. M. Campbell. 1988. Overview of the DOE/PETC microbiological coal preparation R & D program: A voyage on the flood tide. In *Proceedings Bioprocessing of Coals Workshop-III*. Idaho Falls, Idaho: Idaho National Engineering Laboratory Bioprocessing Center.

123. Rinder, G., and E. Beier. 1983. Mikrobiologische entpyritisierung von kohlen ins suspension. *Erdo. Kohl Erdgas Petrochem. Brennst. Ceh.* 36:170–174.

124. Rogoff, M. H., I. Wender, and R. B. Anderson. 1962. *Microbiology of Coal*. Information circular 8075, US Department of Interior, Bureau of Mines, USA.

125. Rogoff, M. H., and I. Wender. 1963. *Materials in coal inhibitory to Growth of Microorganisms*. Report of Investigations 6279, Department of Interior, Bureau of Mines, USA.

126. Runnion, K. N., and J. D. Combie. 1990. Microbial removal of organic sulfur from coal. In *First International Symposium of the Biological Processing of Coal*. Orlando, Florida, USA.

127. Runnion, K. N., and J. D. Combie. 1990. Thermophilic microorganisms for coal biosolubilization. *Appl. Biochem. Biotechnol.* 24/25:817–829.

128. Schwartz, G., E. Senghas, A. Erben, B. Schafer, F. Lingens, and H. Hoke. 1988. Isolation and characterization of quinoline-degrading bacteria. *System. Appl. Microbiol.* 10:185–190.

129. Scott, C. D., G. W. Strandberg, and S. N. Lewis. 1986. Microbial solubilization of coal. *Biotechnol. Progr.* 2:131–139.

130. Scott, C. D., C. A. Woodward, J. E. Thompson, and S. L. Blankinship. 1990. Coal solubilization by enhanced enzyme activity in organic solvents. *Appl. Biochem. Biotechnol.* 24/25:799–815.

131. Shukla, O. P. 1986. Microbial transformation of quinoline by *Pseudomonas* sp. *Appl. Environ. Microbiol.* 51:1332–1342.

132. Shukla, O. P., and S. M. Kaul. 1986. Microbial transformation of pyridine-n-oxide and pyridine by *Nocardia* sp. *Can. J. Microbiol.* 32:330–341.

133. Sims, G. K., and L. E. Sommers. 1986. Biodegradation of pyridine derivatives in soil suspensions. *Environ. Toxicol. Chem.* 5:503–509.

134. Sims, G. K., L. E. Sommers, and A. Konopa. 1986. Degradation of pyridine by *Micrococcus luteus* isolated from soil. *Appl. Environ. Microbiol.* 51:963–968.

135. Speight, J. G. 1983. *The Chemistry and Technology of Coal.* New York: Marcel Dekker Inc.

136. Spiro C. L., and P. G. Kosky. 1982. Space-filling models for coal. 2. Extension of coals of various ranks. *Fuel* 61:1080–1084.

137. Sternberg, H. W., R. Raymond, and F. K. Schweighardt. 1975. Acid-based structure of coal-derived asphaltenes. *Science* 188:49–51.

138. Stock, L. M., J. E. Duran, C. B. Huang, B. R. Srinivas, and R. S. Willis. 1985. Aspects of donor solvent coal dissolution reactions. *Fuel* 64:754–760.

139. Stoner, D. L., J. E. Wey, K. B. Barrett, J. G. Lolley, R. B. Wright, and P. R. Dugan. 1990. Modification of water-soluble coal-derived products by dibenzothiophene-degrading microorganisms. *Appl. Environ. Microbiol.* 56:2667–2676.

140. Strandberg, C. W., and S. N. Lewis. 1987. Solubilization of coal by an extracellular product from *Streptomyces setonii* 75 v. 2. *J. Ind. Microbiol.* 1:371–375.

141. Szanto, M., L. Wondrack, and W. Wood. 1988. Depolymerization of soluble coal polymer derived from lignite and subbituminous coal by lignin peroxidase of *Phaenerochete chrysosporium.* In *Tenth Symposium on Biotechnology for Fuels and Chemicals.* Gatlinburg, Tennesee.

142. Tillet, D. M., and A. S. Meyerson. 1987. The removal of pyritic sulfur from coal employing *Thiobacillus ferrooxidans* in a packed column reactor. *Biotechnol. Bioeng.* 29:146–150.

143. Tissot, B. P., and D. H. Welte. 1984. *Petroleum Formation and Occurrence.* New York: Springer Verlag.

144. Tributsch, H., and J. C. J. Bennett. 1981. Semiconductor electrochemical aspects of bacterial leaching. II—Survey of rate controlling sulfide properties. *J. Chem. Technol. Biotechnol.* 31:627–639.

145. Van Afferden, M., S. Schacht, J. Klein, and H. G. Truper. 1990. Degradation of dibenzothiophene by *Brevibacterium* sp. DO. *Arch. Microbiol.* 153:324–328.

146. Ward, H. B. 1985. Lignite degrading fungi isolated from a weathered outcrop. *System. Appl. Microbiol.* 6:236–238.

147. Wilson, B. W., and M. S. Cohen. 1987. Biosolubilization of coal. An overview. In *Proceedings: First Annual Workshop on Biological Processing of Coal.* EPRI, Monterey, Calif. Palo Alto, Calif: EPRI.

148. Wilson, B. W., R. M. Beans, J. A. Franz, and B. L. Thomas. 1987. Microbial conversion of low rank coal: Characterization of biodegraded products. *J. Energy Fuels* 1:80–84.

149. Wiser, W. H. 1973. *Proceedings of the Electric Power Research Institute Conference on Coal Catalysis.* Santa Monica, Calif. Palo Alto, Calif.: EPRI.

150. Wyza, R. E., A. E. Desouza, and J. D. Isbister. 1987. Depolymerization of low-rank coals by a unique microbial consortium. In *Proceedings: Biological Treatment of Coals Workshop.* Vienna, Virginia.

151. Yamada, K., Y. Minoda, S. Nikatani, and T. Akasaki. 1968. Microbial conversion of petrosulfur compounds. *Agric. Biol. Chem.* 32:840–845.

CHAPTER 23

Microbial Desulfurization of Coal with Emphasis on Inorganic Sulfur

LISELOTTE LARSSON, GUNNEL OLSSON, HANS T. KARLSSON
and OLLE HOLST

Abstract. Sulfur dioxide plays an important role in the formation of acid rain. To avoid the effects of acid rain, the amount of sulfur dioxide emitted into the atmosphere must be reduced. A large source of sulfur dioxide emission is the combustion plants for coal. Flue-gas cleaning helps to reduce emissions and is becoming more common in the large power plants of industrialized countries. An alternative method to reduce sulfur dioxide emissions would be to provide combustion plants with desulfurized coal. Low-sulfur coal would then be available even for small combustion plants, for which flue-gas cleaning is too expensive. The use of micro-organisms to desulfurize coal has been the subject of research in several countries since the late 1970s. Bacteria have been isolated that are able to remove as much as 90% of the sulfur found in coal. The most commonly studied desulfurizing bacteria are of the genera *Thiobacillus,* *Sulfolobus, Acidianus,* and *Pseudomonas.* The bacteria convert the sulfur to water-soluble sulfate. A great advantage of using a microbiological cleaning process is the mild conditions required; in contrast, removal of sulfur from coal by harsh, chemical means causes a loss in the potential heating value of the coal.

INTRODUCTION

Fossil fuels such as coal, oil, and natural gas comprise the major energy sources. Because of the foreseeable exhaustion of oil resources, exploration for coal will probably further increase. The utilization of fossil fuels has a large impact on the environment. The flue gases from combustion consist of many hazardous compounds, including carbon dioxide, sulfur dioxide, nitrogen oxides, metals, and fly ash. Sulfur dioxide and nitrogen oxides cause deposition of acids in the environment. This acid rain has detrimental effects on various ecosystems. For example, changes in pH alter the solubility of metals and cause increased leaching of metal ions from minerals. Environmental legislation concerning emissions from combustion sites is becoming more stringent world-wide, thus requiring more efficient cleaning processes for fossil fuel utilization. The sulfur content of different coals varies from essentially zero to 12%. Naturally, sulfur content determines the extent of the cleaning efforts required to comply with legislation.

Drs. Larsson and Holst are at the Department of Biotechnology, Lund University, S-221 00 Lund, Sweden. Drs. Olsson and Karlsson are at the Department of Chemical Engineering II, Lund University, S-221 00 Lund, Sweden.

MICROBIAL DESULFURIZATION

Three strategies can be used to control sulfur emissions from coal-utilization plants (38), as shown in Figure 1. Sulfur can be removed from the coal *before* combustion, thus obtaining "clean coal." Injection of lime or limestone into the furnace will capture the sulfur dioxide as it is formed *during* combustion. Most commonly, sulfur is removed from the flue gases *after* combustion by various cleaning processes, including wet scrubbing, dry scrubbing, and spray-dry scrubbing.

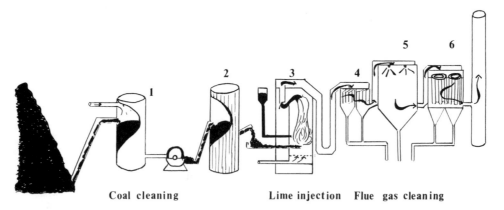

Coal cleaning   Lime injection   Flue gas cleaning

Figure 1. Options for sulfur emission control from fossil fuel combustion plants: (1) coal cleaning device; (2) coal storage; (3) furnace; (4) solids precipitator (ESP); (5) scrubber; (6) bag filter.

An alternative method for cleaning coal is to use microorganisms that utilize the sulfur compounds, mainly as an energy source, in the coal, thereby solubilizing the sulfur. To achieve this, the coal has to be pulverized and mixed with water and nutrient salts before inoculating the microorganisms. The microorganisms convert the inorganic sulfur to sulfate, which is soluble in water. After cleaning, the coal is separated from the liquid that contains the sulfate and washed with additional water. The coal can be dried and burned as pulverized clean coal or used for the preparation of coal water mixtures (CWM), which may be used as a substitute for oil. Microbial desulfurization is well-adapted to the CWM preparation process because drying the coal is unnecessary; the water content is merely adjusted and the stabilizing agents added. Figure 2 illustrates a microbial coal cleaning process.

Coal contains both organic and inorganic sulfur. By using microbial cleaning, 80 to 90% of the inorganic sulfur can be easily removed; however, removing the organic sulfur is much more difficult. The inorganic sulfur consists mainly of pyrite (FeS_2), which is distributed in the coal matrix as crystals of different sizes (20). The organic sulfur is bound to the polymeric structure of the coal as thiophenes, thiols, or sulfides. Some microorganisms are claimed to remove 20 to 40% of the organic sulfur from coal.

Physical and chemical coal cleaning processes are also available, such as flotation, magnetic separation, and chemical oxidation. In addition, microorganisms have been used to enhance froth flotation of coal by altering the surface properties (2).

The advantages of microbial coal desulfurization are the low temperature and atmospheric pressure used during operation and the low cost of the chemicals required. The harsh conditions required for chemical removal of sulfur often decrease the heating value of coal. Physical separation of the coal and pyrite is difficult, owing to the small size of the pyrite particles enclosed in the coal matrix. The potential of removing organic sulfur renders the microbial approach even more interesting. The main disadvantage of microbial desulfurization is the long residence time needed to achieve the desired reduction of sulfur.

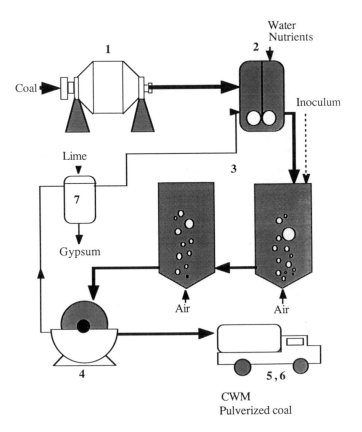

Figure 2. Process scheme of microbial desulfurization; (1) coal grinding; (2) preparation of the coal slurry; (3) reactors for coal desulfurization; (4) filtration and washing of the coal; (5,6) preparation of dry pulverized coal or coal water mixtures (CWM); (7) removal of sulfate and other byproducts from the recirculated process water.

MICROORGANISMS

The most commonly studied microorganism for coal desulfurization is *Thiobacillus ferrooxidans,* which is a mesophilic and acidophilic bacterium growing at 30°C and pH 2. It was isolated in 1947 (37) from old mining sites, where it was found to be responsible for leakage of acidic water. The bacterium was soon associated with the dissolution of metal sulfides, including pyrite (FeS_2), copper sulfide, and arsenic sulfide. Its ability to desulfurize stems from its capacity to dissolve pyrite. The bacterium has also been used for microbial leaching of low-grade ores (11). Thermophilic archaebacteria of the genera *Sulfolobus (Sulfolobus acidocaldarius)* and *Acidianus (Acidianus brierleyi)* are other interesting bacteria that have been studied more frequently. These bacteria are isolated from sulfuric hot springs in volcanic areas (12), where they utilize the elemental sulfur in the springs. Some strains are able to oxidize ferrous iron to ferric iron and to oxidize pyrite. In addition to the utilization of inorganic sulfur, *Sulfolobus acidocaldarius* has been reported to degrade pure organic sulfur compounds (31). Pure organic sulfur compounds are often used as model compounds in studies of the removal of organic sulfur from coal. The diversity among the strains of *Sulfolobus* is great, however, and some strains are not able to oxidize pyrite or remove sulfur from coal (39). In addition to its ability to oxidize pyrite, *Acidianus brierleyi* has been shown to remove some of what is reported to be organic sulfur from coal (46). *A. brierleyi* was formerly called *Sulfolobus brierleyi* (49).

Pseudomonas-type bacteria have been of great interest for their ability to remove organic sulfur. For example, the mutagenically altered bacterium *Coal Bug One (CB1)* (24) was isolated from coal-contaminated soil, using dibenzothiophene for the selection. *CB1* was reported to remove 20–40% of the organic sulfur, but no oxidation of pyrite was recorded. Other strains of *Pseudomonas (P. putida)* have been claimed to remove both organic and inorganic sulfur (48). Conflicting results were obtained in our studies: we were unable to find any evidence for significant removal of sulfur with several different strains of *Pseudomonas* species (40). The discrepancies might be explained, however, by our use of different coal types and use of microbial strains obtained from different sources.

Recently, *Leptospirillum*-like bacteria were reported to remove 85% of the pyrite from coal at the same temperature and pH as *Thiobacillus ferrooxidans* (43).

Table 1 summarizes some results of microbial desulfurization of coal. The results were obtained under various conditions with respect to microbial strain, coal type, and coal particle size.

Table 1. Summary of results from microbial desulfurization of coal.

Microorganism	Pyrite removal (%)	Total S in coal (%)	Residence time (days)	References
Thiobacillus ferrooxidans	57	3.31	10	19
T. ferrooxidans	72	6.04	10	14
T. ferrooxidans	40–60	1.66	2	44
T. ferrooxidans	96	6.93	57	42
T. ferrooxidans	80		9	47
T. ferrooxidans	43	4.50	3	50
T. ferrooxidans	66		10	21
T. ferrooxidans	69	11.00	5	29
T. ferrooxidans	80	3.30	24	3
T. ferrooxidans	83	6.64	26	15
Mixed culture from coal	95	3.15	21	36
Sulfolobus acidocaldarius	48	11.00	16	25
S. acidocaldarius	96	4.00	16	25
S. acidocaldarius	92	3.87	10	17
Acidianus brierleyi	75	0.43	10	L. Larsson et al., unpublished
A. brierleyi	85	0.85	10	46
A. brierleyi	90	3.43	10	8
A. brierleyi	90	5.80	7	32
Pseudomonas putida	69	4.00	5	48
IGTS7	74	2.85	212	34

MICROBIAL OXIDATION OF PYRITE

Chemical Reactions

Pyrite (FeS_2) is oxidized to sulfate (SO_4^{2-}) and ferric iron (Fe^{3+}) by both *Thiobacillus ferrooxidans* and the thermophilic archaebacteria. Sulfates are often soluble in water, and the sulfur can hence be removed from the coal during washing. Acidic conditions are required by the microorganisms and also to prevent precipitation of iron sulfates, especially jarosites.

Two mechanisms for microbial oxidation of pyrite have been proposed: the direct mechanism and the indirect mechanism. The direct mechanism involves direct oxidation of pyrite by the microorganisms following Reaction 1 (see below). Physical contact between the microorganisms and the pyrite is required. As can be seen from Reaction 1, sulfuric acid is produced, which tends to decrease the pH in the reaction vessel.

$$\text{Microorganisms}$$
$$4FeS_2 + 15O_2 + 2H_2O \rightarrow 2Fe_2(SO_4)_3 + 2H_2SO_4 \tag{1}$$

The indirect mechanism involves chemical oxidation of the pyrite by ferric iron (Reaction 2). The ferric iron is produced by the microorganisms by the oxidation of ferrous iron (reaction 3).

$$FeS_2 + 14Fe^{3+} + 8H_2O \rightarrow 15Fe^{2+} \ 2SO_4^{2-} + 16H^+ \tag{2}$$

$$\text{Microorganisms}$$
$$2Fe^{2+} + 2H^+ + \tfrac{1}{2}O_2 \rightarrow 2Fe^{3+} + H_2O \tag{3}$$

The oxidation of ferrous iron in the absence of microorganisms is a slow process. It is considered to be the rate-limiting step for the oxidation of pyrite with ferric iron.

Another option for the indirect mechanism is that the ferric iron oxidizes the ferrous iron in the pyrite, leaving elemental sulfur behind (Reaction 4).

$$FeS_2 + 2Fe^{3+} \rightarrow 3Fe^{2+} + 2S^0 \tag{4}$$

The elemental sulfur is then oxidized to sulfate by the microorganisms (Reaction 5). An increase of elemental sulfur has been found in coal treated with *Thiobacillus ferrooxidans* (6). Chemical oxidation of pyrite with ferric iron resulted in the leaching of more iron than sulfur according to the stoichiometry. This supports the latter option for the indirect mechanism. It is more energetically favorable, however, for the microorganisms to oxidize elemental sulfur than to oxidize ferrous iron (35). Furthermore, later results (10) showed that no elemental sulfur was produced during oxidation of pure pyrite.

$$\text{Microorganisms}$$
$$2S^0 + 3O_2 + 2H_2O \rightarrow 2H_2SO_4 \tag{5}$$

As can be seen in Figure 3, the indirect chemical reactions are important for the efficiency of the process, because the micropores of the coal are too small for the microorganisms to enter. At the elevated temperatures used for the thermophilic bacteria, the chemical reactions are faster and affect the overall pyrite oxidation rate to a larger

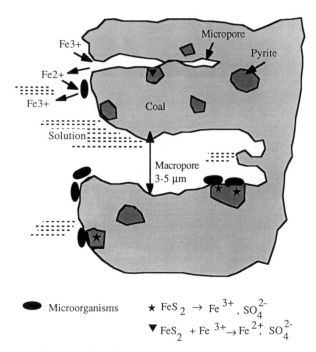

Figure 3. Suggested mechanisms of microbial pyrite oxidation (23).

extent than at temperatures used for the mesophilic bacteria. The oxidation rate of pure pyrite is nearly five times higher for *Acidianus brierleyi* than for the mesophilic *Thiobacillus ferrooxidans* (Figure 4); however, during oxidation of pyrite in coal there is no significant difference in the rate of sulfur removal between *A. brierleyi* and *T. ferrooxidans*. The rate of pyrite oxidation in coal might be mass-transport-limited rather than substrate-limited for the microorganisms. Pyrite oxidation in coal seems to be a first-order reaction in pyrite concentration, whereas pyrite oxidation with pure pyrite is more complicated and depends on the ferric iron concentration as well as the pyrite concentration.

Another important factor in considering microbial oxidation of pyrite is the formation of iron precipitates, mainly jarosites. The general chemical formula of jarosite is $MFe_3(SO_4)_2(OH)_6$, where M stands for either hydronium, potassium, sodium, or

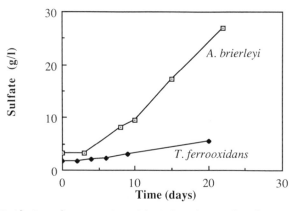

Figure 4. Oxidation of pure pyrite with *A. brierleyi* and *T. ferrooxidans* under similar conditions.

ammonium. The presence of microorganisms and elevated temperatures increases the formation of jarosites (41). The addition of alkali to control the pH in pyrite oxidation experiments severely increases the formation of jarosite; during pyrite oxidation, as much as 30% of the ferric iron may precipitate (39). The formation of these precipitates counteracts the desulfurization, since the precipitates stick to the coal even after the washing step. The concentration of soluble ferric iron also decreases, a result that has a large impact on the chemical reactions involved in the indirect mechanism.

Kinetics

Several kinetic models have been suggested to evaluate the kinetics of pyrite oxidation in coal (Table 2). These kinetic approaches can be divided into complex and simple models. The complex models consider microbial growth, adsorption of bacteria on the coal surfaces, growth on the surface and in the bulk liquid, formation of precipitates, etc. The simple models relate the pyrite oxidation rates to pyrite concentration or pyrite surface area. The kinetic models can be used in simulation and scaling up of the process.

Figure 5 compares the sulfur removal rate to initial pyrite concentration in coal slurry. All experiments were done with thermophilic bacteria of the genera *Sulfolobus* and *Acidianus,* but different types and particle sizes of coal were used. The rate of pyrite removal, calculated as released sulfur, seems to correlate quite well with the initial pyrite concentration.

Table 2. Kinetic models.

Model	Reference
COMPLEX MODELS	
Biomass/substrate-limited regions	22
Direct/indirect mechanism, monod-dependent	16
Surface adsorption	13
Surface adsorption, preferably on pyrite	27
SIMPLE MODELS	
Pyrite concentration, first-order reaction	18
Pyrite concentration, first-order reaction	1
Sulfur concentration, first-order reaction	46b
Pyrite surface area, first-order reaction	21

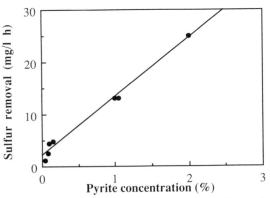

Figure 5. Sulfur removal rate versus initial pyrite concentration in the coal slurry for different experiments performed with thermophilic archaebacteria. [Results were obtained from the work of Kargi and Robinson (25), Chen and Skidmore (17), Khalid et al. (32), and Larsson et al. (39).]

INDUSTRIAL APPLICATION

Process Parameters

The efficiency of a process for microbial desulfurization on an industrial scale depends not only on the capability of the microorganisms but also on the properties of the coal. A number of parameters have a large impact on the efficiency of the process (35): for example, the particle size of the pulverized coal, the sulfur content and the distribution of sulfur forms (i.e. inorganic or organic), nutrient media composition, pH, temperature, aeration and agitation, and reactor design. The different parameters are discussed briefly below. So far, the majority of studies have been done on a laboratory scale.

Microorganisms

In addition to the type of bacteria used, the size of the inoculum affects the rate of pyrite removal. A larger inoculum tends to speed up the reaction in the beginning of a batch experiment (26), a result that is quite predictable. The desulfurization process is thought to be biomass-limited initially; as the pyrite is consumed, it becomes substrate-limited (22). Adaptation of the microorganisms to pyrite or coal prior to inoculation diminishes the lag phase and tends to give a higher rate of sulfur removal.

Thiobacillus ferrooxidans and *Acidianus brierleyi* grow under extreme conditions that are hospitable to only a small number of competing microorganisms. Bacteria used in a larger scale process must be able to compete with the naturally occurring microflora on the coal, since sterilization of the coal would be too expensive. Mixed cultures of *T. ferrooxidans* and naturally occurring microorganisms have been reported to be the most effective pyrite oxidizers (36). The elevated temperature used for *A. brierleyi* decreases the risk of contamination. Microscopic studies of the coal slurries in our experiments have shown a large number of rods, probably *Thiobacillus*-like bacteria. The rods disappear 1 to 2 days after inoculation with *A. brierleyi*, when *A. brierleyi* enters the growth phase.

Because coals from different mines possess quite variable properties, the microorganisms must be able to adapt themselves to different coal types. They must also tolerate high concentrations of metals and other compounds leached from the coal. *Thiobacillus ferrooxidans* is reported to tolerate high concentrations of metal ions (11). One strain of *Sulfolobus acidocaldarius* showed some sensitivity towards compounds leached from coal but *Acidianus brierleyi* did not (45). The growth of the microorganisms, also including *S. solfataricus,* was tested in media prepared from water in which coal had been leached.

Coal Properties

As discussed above, coal is a very inhomogeneous material with large differences in its physical and chemical properties. The distribution of the different sulfur forms affects the efficiency of desulfurization, since pyrite is much easier to remove than organically bound sulfur. The accessibility of pyrite to microorganisms depends on the size and distribution of the pyrite crystals in the coal matrix. The porosity and the distribution of pore sizes of the coal also affect accessibility, because the micropores are too small for the microorganisms to enter. The size of the microorganisms is between 1 and 5 μm, while the micropores are less than 0.1 μm in diameter. Also, in experiments with *Pseudomonas fluorescens* and *A. brierleyi*, the BET-surface of the coal decreased during biotreatment; this indicated that the pores were clogged, perhaps by precipitated jarosite (40).

pH and Media Composition

The acidophilic bacteria require a pH below 2 for optimal growth. This requirement coincides well with the acidic pH necessary to prevent precipitation of iron. Iron precipitates as iron hydroxides [FeO(OH)] at neutral pH and as iron sulfates, for example jarosite, even at pH below 3. The formation of jarosite is also affected by the media composition. High ionic strength promotes the formation of jarosite. Phosphate, which may form precipitates with iron, seems to decrease the amount of jarosite formed during desulfurization of low-sulfur coal, in which the iron concentration is low in the liquid phase. Moderate changes in media compositions do not affect the rate of pyrite oxidation. Conditions of high salt concentration or low salt concentration (no addition of nutrient salts) both result in decreasing maximum pyrite oxidation rate (17).

Autotrophic growth of the bacteria is desired, because the addition of an organic carbon source would be very expensive. *T. ferrooxidans* grows strictly autotrophically, whereas the archaebacteria can grow heterotrophically as well. Addition of small amounts of yeast extract improves the growth of *A. brierleyi*, especially on low-sulfur coals with a low pyritic content.

Particle Size and Slurry Concentration

The size of the coal particles is an important factor in the microorganisms' ability to desulfurize coal, as it largely determines the pyrite exposure to the microorganisms. A marked increase occurs in the rate of sulfur removal as the particle size of coal is reduced to 100 μm (28, 30); however, reduction of the particle size from 100 μm to 20 μm does not significantly affect the rate of sulfur removal with *A. brierleyi* on low-sulfur coal. A typical size range is 90% less than 100 μm for the purpose of production of coal water mixtures and pulverized coal.

How economical the process is depends largely on the concentration of coal in the slurry during the cleaning process. High concentration decreases the amount of slurry to be processed and the water required, but a concentration above 20% (w/V) coal is reported to inhibit the growth of the microorganisms (4). Several explanations are possible: (a) poor mass and heat transfer due to difficulties in agitating the slurry and agglomeration of coal particles; (b) extensive shear stress on the microorganisms due to attrition; and (c) buildup of compounds leached from the coal. Microorganisms, especially archaebacteria, are fragile and sensitive to shear forces because they lack a rigid cell wall.

Reactor Design

The design and construction of reactors for microbial desulfurization are somewhat problematic. A steel construction can be severely corroded in a short time, since the microorganisms oxidize iron. In addition, the required residence time is long, 5 to 10 days: therefore, the reactor must have a large volume. The contents of the reactor must be well mixed to prevent settling of the solids; however, microorganisms are sensitive to shear forces and can be damaged by too-vigorous mixing.

Various solutions to the latter problem have been suggested. Air-lift reactors have been shown to be beneficial for microbial growth. The cell yield in mechanically stirred reactors is much lower (23). In air-lift reactors, aeration and agitation are achieved simultaneously. The process gas must be cleaned before it is emitted into the environment, however, since it may contain high concentrations of sulfuric acid. This problem is more pronounced when thermophilic bacteria are used, because evaporation at the temperature used, 70°C, is quite extensive. In addition, heat is lost from evaporation of water.

Enrichment of the process gas with carbon dioxide has been discussed frequently, with conflicting reports as to whether excess carbon dioxide increases the efficiency of desulfurization. Some studies have shown that carbon dioxide does not significantly affect the rate of sulfur removal, but others claim that the carbon dioxide becomes limiting at high biomass concentrations. *Thiobacillus ferrooxidans* and *A. brierleyi* both use carbon dioxide as a source of carbon.

A type of plug-flow reactor has also been proposed in the form of a pipeline through which the coal slurry is pumped (48). Air is bubbled through the coal slurry in a vessel outside the pipeline. A plug-flow reactor would be consistent with the proposed first-order kinetics of pyrite oxidation, but there is a risk that the process will suffer from washout because the growth of the microorganisms is quite slow. To avoid the need for continuous reinoculation, the mixture must be back-mixed or the bacterial cells will have to be recycled.

A cheap but not as efficient process is heap leaching. The process water is percolated through a pile of crushed coal. This requires a longer time for processing but does not incur additional cost because the storage time for coal often is already quite long. If the microorganisms grow on the coal surface, the process water can be continuously regenerated and there is not any buildup of end products. The required residence time for percolation was estimated to be 70 days for 75% removal of pyritic sulfur (5).

Waste Products

The process water from a microbial desulfurization plant would contain large amounts of sulfate and iron and can be contaminated by heavy metals leached from the coal. The sulfate could be precipitated as gypsum, which is a common byproduct of all types of desulfurization plants, including flue-gas cleaning plants. So far the gypsum is either used as constructing material, if clean enough or deposited at specially designed disposal sites. The gypsum may also be used to produce elemental sulfur in regenerative processes. Because large amounts of process water would be used the water should be recirculated. The process gas must be cleaned before it is emitted into the environment, since it would contain variable amounts of sulfuric acid.

Economy

Cost estimates for microbial desulfurization have so far been based on rather small-scale experimental studies. An economic feasibility study of three processes was performed based on German conditions (7): *Thiobacillus ferrooxidans* was utilized for the removal of pyritic sulfur, and the plants were designed for 100,000 annual tons of coal. The cost varied between 121 and 132 DM/ton coal ($65–77/ton coal), which can be compared with 70 DM/ton coal ($40/ton coal) for flue-gas desulfurization in a large power plant. For small power plants, flue-gas desulfurization is more expensive. The cost of flue-gas desulfurization has also been found to depend very much on site-specific conditions (33). Boone et al. (9) estimated the cost for microbial desulfurization with moderately thermophilic bacteria with an additional chemical oxidation step as $38/ton coal. Only pyritic sulfur was expected to be removed, and the capacity of the plant was 100,000 tons of coal/year. Earlier cost estimates have varied between $15 and $70/ton coal depending on the process chosen.

For large power plants, flue-gas cleaning would probably be most economical, but microbial desulfurization plants could be used to provide small combustion plants with "clean coal."

CONCLUSIONS

Both thermophilic and mesophilic bacteria of the genera *Sulfolobus, Acidianus,* and *Thiobacillus* remove 80–90% of the pyritic sulfur from bituminous coals in 5–10 days. The search for interesting, new bacteria continues. Bacteria that would remove both organic and inorganic sulfur simultaneously would be most favorable. The rate of pyrite removal depends on the initial pyrite content of the coal. The extent of pyrite removal depends on the accessibility of the pyrite, i.e. the pyrite distribution in the coal particles and the porosity of the coal. Smaller particle sizes (<100 μm) allow a substantial increase in the sulfur removal rate.

It is desirable to find optimal operating conditions to minimize the required residence time and to improve the efficiency and hence economy of the process. Technical solutions need to be further investigated.

Microbial desulfurization of low-sulfur coals containing mostly pyritic sulfur would remove sufficient sulfur to satisfy the requirements of most legislation. For high sulfur coals, additional removal of the organic sulfur is necessary to reduce the sulfur content sufficiently.

Acknowledgment

Financial support is provided by the Swedish National Board for Industrial and Technical Development.

LITERATURE CITED

1. Andrews, F. G., and J. Maczuga. 1984. Bacterial removal of pyrite from coal. *Fuel* 63:297.

2. Attia, Y. A., and M. A. Elzeky. 1985. Biosurface modification in the separation of pyrite from coal by froth flotation. *Coal Sci. Technol.* 9:673.

3. Beier, E. 1985. Removal of pyrite from coal using bacteria. *Coal Sci. Technol.* 9:653.

4. Beyer, M., H. G. Ebner, and J. Klein. 1986. Influence of pulp density and bioreactor design on microbial desulfurization of coal. *Appl. Microbiol. Biotechnol.* 24:342.

5. Beyer, M. 1987. Microbial removal of pyrite from coal using a percolation bioreactor. *Biotechnol. Lett.* 9:19.

6. Beyer, M., H. G. Ebner, H., Assenmacher, and J. Frigge. 1987. Elemental sulphur in microbiologically desulphurized coals. *Fuel* 66:551.

7. Beyer, M., M. Pietsch, and T. Tebrügge. 1990. Technical and economical feasibility of microbial pyrite removal from coal. DMT—Gesellschaft für Forschung and Prüfung mbH, Essen, FRG.

8. Bhattacharyya, D., M. Hsieh, H. Francis, R. I. Kermode, A. M. Khalid, and H. M. I. Aleem. 1990. Biological desulfurization of coal by mesophilic and thermophilic microorganisms. *Resources Conserv. Recycl.* 3:81.

9. Boone, M., T. A. Meeder, and K. Ch. A. M. Luyben. 1990. High slurry densities for the microbial desulfurization of coal. *5th European Congress on Biotechnology*, Copenhagen, Denmark.

10. Bos, P., and F. C. Boogerd. 1991. The relative contributions of biological and chemical reactions to the overall rate of pyrite oxidation. *2nd Int. Symp. on Biological Processing of Coal*, San Diego, California.

11. Brierley, C. 1978. Bacterial leaching. *CRC Crit. Rev. Microbiol.* 6:207.

12. Brock, T.D. 1981. Extreme thermophiles of the genus *Thermus* and *Sulfolobus*. In *The Prokaryotes*, Vol. 1. Ed. M. P. Starr, H. Stolp, H. G. Trüper, A. Balows, H. G. Schegel. New York: Springer-Verlag. 978.

13. Chang, Y. C., and A. S. Meyerson. 1982. Growth models of the continuous bacterial leaching of iron pyrite by *Thiobacillus ferrooxidans*. *Biotechnol. Bioeng.* 24:889.

14. Chandra, D., P. Roy, A. K. Mishra, J. N. Chakrabarti, N. K. Prasad, and S. G. Chaudhuri. 1988. Removal of sulphur from coal by *Thiobacillus ferrooxidans* and mixed acidophilic bacteria present in coal. *Fuel* 59:249.

15. Chandra, D., and A. K. Mishra. 1990. In *Bioprocessing and Biotreatment of Coal*. Ed. Wise D. L. New York: Marcel Dekker Inc. 631.

16. Chen, C. 1986. *Kinetic study of Microbial Desulfurization using Thermophilic Microorganisms*. Ph.D. Thesis, Ohio State University.

17. Chen, C., and D. Skidmore. 1990. Microbial coal desulfurization with thermophilic microorganisms. In *Bioprocessing and Biotreatment of Coal*. Ed. D. L. Wise. New York: Marcel Dekker, Inc. 653.

18. Detz, C. M., and G. Barvinchak. 1979. Microbial desulfurization of coal. *Mining Congr. J.* 7:75.

19. Dogan, M. Z., G. Özbayogli, C. Hicyilmaz, M. Sarihaya, and G. Özengiz. 1985. Bacterial leaching versus bacterial conditioning and flotation in desulphurization of coal. *Congr. Int. Mineral.* 2:165

20. Greer, R. T. 1977. Coal microstructure and pyrite distribution. 1977. In *Coal Desulphurization. Chemical and Physical Methods*. Ed. T. D. Wheelock. *ACS Symp. Ser.* 64:3.

21. Hoffman, M. R., B. Faust, F. A. Panda, H. H. Koo, and H. M. Tsuchia. 1981. Kinetics of the removal of iron and pyrite from coal by microbial catalysis. *Appl. Env. Microbiol.* 42(2):259.

22. Huber, T. F., N. W. F. Kossen, P. Bos, and J. G. Kuenen. 1984. Design and scale up of a reactor for microbial desulphurization of coal: A regime analysis. In *Innovations in Biotechnology*. Ed. E. H. Houwink, R. R. van der Meer. Amsterdam: Elsevier Science Publishers B.V.

23. Höne, H-J., M. Beyer, H. G. Ebner, J. Klein, and H. Jüntgen. 1987. Microbial desulphurization of coal—Development and application of a slurry reactor. *Chem. Eng. Technol.* 10:173.

24. Isbister, J. D., and R. C. Doyle. 1985. A novel mutant microorganism and its use in removing organic sulphur from coal. European Patent Appl. 85112450.3 02.10.85 Publ. No. 0218734.

25. Kargi, F., and J. M. Robinson. 1982. Removal of sulphur compounds from coal by the thermophilic organism *Sulfolobus acidocaldarius*. *Appl. Environ. Microbiol.* 44(4):878.

26. Kargi, F. 1982. Enhancement of microbial removal of pyritic sulphur from coal. *Biotech. Bioeng.* 24:749.

27. Kargi, F., and J. G. Weissman. 1984. A dynamic mathematical model for microbial removal of pyritic sulfur from coal. *Biotechnol. Bioeng.* 26:687.

28. Kargi, F. 1984. Microbial Desulphurization of Coal. In *Advances in Biotechnology Processes*, Vol. 3. New York: Alan R. Liss, Inc.

29. Kargi, F., and J. M. Robinson. 1985. Biological removal of pyritic sulphur from coal by the thermophilic organism *Sulfolobus acidocaldarius*. *Biotech. Bioeng.* 27:41.

30. Kargi, F. 1986. Microbial methods for desulphurization of coal. *Trends Biotechn.*, Nov, 293.

31. Kargi, F. 1987. Biological oxidation of thianthrene, thioxanthene and dibenzothiophene by the thermophilic organism *Sulfolobus acidocaldarius*. *Biotech. Lett.* 9:478.

32. Kahlid, A. M., D. Bhattacharyya, and M. H. I. Aleem. 1990. Coal desulfurization and electron transport-linked oxidation of sulfur compounds by *Thiobacillus* and *Sulfolobus* species. *Dev. Ind. Microbiol.* 31:115.

33. Karlsson, H. T., O. Holst, L. Larsson, G. Olsson, B. Mattiasson, B. Nilsson, and I. Ericsson. 1987. Prospects for microbial desulfurization of coal for CWM production. *Int. Chem. Eng. Symp. Ser.* 107:25.

34. Kilbane, J. J. 1989. Desulphurization of coal: the microbial solution. *Trends Biotechnol.* 7:79.

35. Klein, J., M. Beyer, M. Afferden, W. Hodek, F. Pfeifer, H. Seewald, and Ed. Wolff-Fischer. 1988. Coal in biotechnology. In *Biotechnology* 6b:523. Ed. H-J. Rehm. Weinheim, FRG: VCH Verlagsgesellschaft mbH.

36. Kos, G. H., R. P. E. Porter, P. Bos, and J. G. Kuenen. 1981. Geochemistry of sulfides in coal and microbial leaching experiments. *Proc. Int. Conf. on Coal Science,* Dusseldorf. Essen, FRG: Verlang Glückauf Gmbh. 842.

37. Kuenen, J. G., and O. H. Touvinen. 1981. The genera *Thiobacillus* and *Thiomicrospira*. In *The Procaryotes* 81:1023.

38. Kyte, W. S. 1989. Technologies for the removal of sulphur dioxide from coal combustion. *Int. Chem. Eng. Symp. Ser.* 106:15.

39. Larsson, L., G. Olsson, O. Holst, and H. T. Karlsson. 1990. Pyrite oxidation by thermophilic bacteria. *Appl. Environ. Microbiol.* 56:697.

40. Larsson, L., O. Holst, G. Olsson, and H. T. Karlsson. 1991. Prospects on the use of unaltered *Pseudomonas* for desulfurization of coal. *2nd Int. Symp. on Bioproc. of Coal,* San Diego, California.

41. Lazaroff, N., W. Sigal, and A. Wasserman. 1982. Iron oxidation and precipitation of ferric hydroxysulphates by resting *Thiobacillus ferrooxidans* cells. *Appl. Env. Microbiol.* 43(4):924.

42. McCready, R. G. L., and M. Zentilli. 1985. Benefication of coal by bacterial leaching *Can. Metall. Q.* 24:135. .

43. Merrettig, U., P. Wlotzka, and U. Onken. 1989. The removal of pyritic sulphur from coal by *Leptospirillum*-like bacteria. *Appl. Microbiol. Biotechnol.* 31:626.

44. Myerson, A. S., and P. C. Kline. 1984. Continuous bacterial coal desulphurization employing *Thiobacillus ferrooxidans*. *Biotech. Bioeng.* 26:92.

45. Olsson, G., L. Larsson, O. Holst, and H. T. Karlsson. 1989a. Microorganisms for desulphurization of coal: the influence of leaching compounds on their growth. *Fuel* 68:1270.

46. Olsson, G., L. Larsson, O. Holst, and H. T. Karlsson. 1989b. Benefication of coal by microbial desulfurization. *Proc. 1989 Int. Conf. on Coal Science,* Japan.

46b. Olsson, G., L. Larsson, O. Holst, and H. T. Karlsson. 1993. Kinetics of coal desulphurization by *Acidianus brierleyi*. *Chem. Eng. Technol.* 16:180.

47. Rai, C. 1986. Microbial desulphurization of bituminous coals fossil fuels utilization. *ACS Symp. Ser.* 319:86.

48. Rai, C., and J. P. Reyniers. 1988. Microbial desulphurization of coals by the organisms of the genus *Pseudomonas*. *Biotechnol. Progr.* 4(4):225.

49. Segerer, A., A. Neuner, J. K. Kristjansson, and K. O. Stetter. 1986. *Acidianus infernus* gen. nov., spec. nov. and *Acidianus brierleyi* comb. nov.: Facultatively aerobic, extremely acidophilic thermophilic sulfur metabolizing archaebacteria. *Int. J. Syst. Bacteriol.* 36:559.

50. Silverman, M. P., R. H. Rogoff, and I. Wender. 1962. Removal of pyritic sulphur from coal by bacterial action. *Fuel* 42:113.

Index